T0396164

Progress in Brain Research
Volume 252

Recent Advances in Parkinson's Disease

Serial Editor

Vincent Walsh
Institute of Cognitive Neuroscience
University College London
17 Queen Square
London WC1N 3AR UK

Editorial Board

Progress in Brain Research
Volume 252

Recent Advances in Parkinson's Disease

Edited by

Anders Björklund
*Department of Experimental Medical Science, Wallenberg
Neuroscience Centre, Lund University, Lund, Sweden*

M. Angela Cenci
*Basal Ganglia Pathophysiology Unit, Department of Experimental
Medical Science, Wallenberg Neuroscience Centre,
Lund University, Lund, Sweden*

ELSEVIER

Elsevier
Radarweg 29, PO Box 211, 1000 AE Amsterdam, Netherlands
The Boulevard, Langford Lane, Kidlington, Oxford, OX5 1GB, United Kingdom
50 Hampshire Street, 5th Floor, Cambridge, MA 02139, United States

First edition 2020

Notices
Knowledge and best practice in this field are constantly changing. As new research and experience broaden
our understanding, changes in research methods, professional practices, or medical treatment may
become necessary.

Practitioners and researchers must always rely on their own experience and knowledge in evaluating and
using any information, methods, compounds, or experiments described herein. In using such information
or methods they should be mindful of their own safety and the safety of others, including parties for whom
they have a professional responsibility.

To the fullest extent of the law, neither the Publisher nor the authors, contributors, or editors, assume any
liability for any injury and/or damage to persons or property as a matter of products liability, negligence or
otherwise, or from any use or operation of any methods, products, instructions, or ideas contained in the
material herein.

ISBN: 978-0-444-64260-8
ISSN: 0079-6123

For information on all Academic Press publications
visit our website at https://www.elsevier.com/books-and-journals

Publisher: Zoe Kruze
Acquisitions Editor: Sam Mahfoudh
Editorial Project Manager: Peter Llewellyn
Production Project Manager: Abdulla Sait
Cover Designer: Greg Harris

Typeset by SPi Global, India

Working together
to grow libraries in
developing countries

www.elsevier.com • www.bookaid.org

Contributors

Dilan Athauda
Department of Clinical and Movement Neurosciences, UCL Institute of Neurology, London, United Kingdom

Daniela Berg
Department of Neurology, Christian-Albrechts-University of Kiel, Kiel, Germany

Anders Björklund
Department of Experimental Medical Science, Wallenberg Neuroscience Centre, Lund University, Lund, Sweden

Salvatore Bonvegna
Parkinson Institute, ASST Gaetano Pini-CTO, Milan, Italy

Inês Caldeira Brás
Department of Experimental Neurodegeneration, Center for Biostructural Imaging of Neurodegeneration, University Medical Center Göttingen, Göttingen, Germany

Patrik Brundin
Center for Neurodegenerative Science, Van Andel Institute, Grand Rapids, MI, United States

Maria Claudia Caiazza
Oxford Parkinson's Disease Centre, Department of Physiology, Anatomy and Genetics, University of Oxford, Oxford, United Kingdom

Anna R. Carta
Department of Biomedical Sciences, University of Cagliari, Cittadella Universitaria di Monserrato, Cagliari, Italy

M. Angela Cenci
Basal Ganglia Pathophysiology Unit; Department of Experimental Medical Science, Wallenberg Neuroscience Centre, Lund University, Lund, Sweden

Roberto Cilia
Fondazione IRCCS Istituto Neurologico Carlo Besta, Movement Disorders Unit, Milan, Italy

Mark R. Cookson
Laboratory of Neurogenetics, National Institute on Aging, National Institutes of Health, Bethesda, MD, United States

Lindsey A. Cunningham
Van Andel Institute Graduate School; Center for Neurodegenerative Science, Van Andel Institute, Grand Rapids, MI, United States

Rubens Gisbert Cury
Movement Disorders Center, Department of Neurology, School of Medicine, University of São Paulo, São Paulo, Brazil

Phillip Engen
Department of Internal Medicine, Division of Digestive Disease and Nutrition, Rush University Medical Center, Chicago, IL, United States

Tom Foltynie
Department of Clinical and Movement Neurosciences, UCL Institute of Neurology, London, United Kingdom

Romulo Fuentes
Universidad de Chile, Santiago, Chile

Patricia Gonzalez-Rodriguez
Department of Physiology, Feinberg School of Medicine, Northwestern University, Chicago, IL, United States

Ashley S. Harms
Center for Neurodegeneration and Experimental Therapeutics, Department of Neurology, The University of Alabama at Birmingham, Birmingham, AL, United States

Ali Keshavarzian
Department of Internal Medicine, Division of Digestive Disease and Nutrition, Rush University Medical Center, Chicago, IL, United States

Andrea A. Kühn
Charité—Universitätsmedizin Berlin, Berlin, Germany

Katarzyna Z. Kuter
Department of Neuropsychopharmacology, Maj Institute of Pharmacology, Polish Academy of Sciences, Krakow, Poland

Charmaine Lang
Oxford Parkinson's Disease Centre, Department of Physiology, Anatomy and Genetics, University of Oxford, Oxford, United Kingdom

Rebekah G. Langston
Laboratory of Neurogenetics, National Institute on Aging, National Institutes of Health, Bethesda, MD, United States

Sara Meoni
Movement Disorders Unit, Division of Neurology, CHU of Grenoble, Grenoble Alpes University; INSERM U1216, Grenoble Institute of Neurosciences, Grenoble, France

Darren J. Moore
Center for Neurodegenerative Science, Van Andel Institute, Grand Rapids, MI, United States

Elena Moro
Movement Disorders Unit, Division of Neurology, CHU of Grenoble, Grenoble Alpes University; INSERM U1216, Grenoble Institute of Neurosciences, Grenoble, France

Wolf-Julian Neumann
Charité—Universitätsmedizin Berlin, Berlin, Germany

Tiago Fleming Outeiro
Department of Experimental Neurodegeneration, Center for Biostructural Imaging of Neurodegeneration, University Medical Center Göttingen; Max Planck Institute for Experimental Medicine, Göttingen, Germany; Institute of Neuroscience, The Medical School, Newcastle University, Newcastle upon Tyne, United Kingdom

Per Petersson
Umeå University, Umeå; Lund University, Lund, Sweden

Ronald B. Postuma
Department of Neurology, Montreal General Hospital, Montreal, QC, Canada

Eva Schaeffer
Department of Neurology, Christian-Albrechts-University of Kiel, Kiel, Germany

Aubrey M. Schonhoff
Center for Neurodegeneration and Experimental Therapeutics, Department of Neurology, The University of Alabama at Birmingham, Birmingham, AL, United States

David G. Standaert
Center for Neurodegeneration and Experimental Therapeutics, Department of Neurology, The University of Alabama at Birmingham, Birmingham, AL, United States

Simon R.W. Stott
The Cure Parkinson's Trust, London, United Kingdom

D. James Surmeier
Department of Physiology, Feinberg School of Medicine, Northwestern University, Chicago, IL, United States

Richard Wade-Martins
Oxford Parkinson's Disease Centre, Department of Physiology, Anatomy and Genetics, University of Oxford, Oxford, United Kingdom

Zachary D. Wallen
Center for Neurodegeneration and Experimental Therapeutics, Department of Neurology, The University of Alabama at Birmingham, Birmingham, AL, United States

Gregory P. Williams
Center for Neurodegeneration and Experimental Therapeutics, Department of Neurology, The University of Alabama at Birmingham, Birmingham, AL, United States

Richard K. Wyse
The Cure Parkinson's Trust, London, United Kingdom

Mary Xylaki
Department of Experimental Neurodegeneration, Center for Biostructural Imaging of Neurodegeneration, University Medical Center Göttingen, Göttingen, Germany

Enrico Zampese
Department of Physiology, Feinberg School of Medicine, Northwestern University, Chicago, IL, United States

Contents

PART II Translational therapeutics

CHAPTER 10 Prodromal PD: A new nosological entity.................329

Eva Schaeffer, Ronald B. Postuma, and Daniela Berg

CHAPTER 11 The gut microbiome in Parkinson's disease: A culprit or a bystander?.................................357

Ali Keshavarzian, Phillip Engen, Salvatore Bonvegna, and Roberto Cilia

Preface: The evolving scenario of Parkinson's research

The two first volumes on *Recent Advances in Parkinson's Disease* were published in the *Progress in Brain Research* series 10 years ago. In the intervening decade our understanding of Parkinson's disease (PD), its causes and pathogenesis have made tremendous progress. The field has grown and expanded, and this relatively common neurological condition has now become the focus of advanced investigations in a wide range of disciplines. For these reasons we have found it timely to add a new volume to this series with the goal to summarize some of the most interesting advances that have taken place in experimental and translational PD research during the last decade. The 14 contributions to this volume of *Progress in Brain Research* capture some of the richness and complexity of PD as a subject of investigation within the fields of molecular genetics, cellular biology, systems neuroscience, clinical physiology translational therapeutics. The topics have been selected to represent lines of research that have undergone major developments during the past decade and helped to uncover new aspects of etiopathogenic factors, disease mechanisms, and therapeutic opportunities.

The chapters have been grouped into two parts. In Part I: "*Basic science*" we include nine chapters that cover a range of concepts and experimental approaches to the study of disease mechanisms, including in vitro and in vivo modeling, molecular and cellular mechanisms of neurodegeneration, and the involvement of innate and adaptive immune mechanisms. In Part II: "*Translational therapeutics*" we include five chapters that highlight some of the most interesting advances made in the pursuit of novel ideas and approaches to treatment. Collectively, they attest to the breadth, excitement and dynamism that characterize PD research today.

The PD research field has over the last decades gained in momentum thanks to wider public attention and increased involvement of patient organizations, and the rapid advancement in PD research that is now taking place has been made possible by improved funding initiatives offered not only by public institutions, but also, importantly, by non-profit organisations and private foundations. The capacity to attract research resources to PD gives hope to the patient community, and further boosts interest in this area on the part of investigators with different backgrounds. The progress made in the identification and exploration of novel therapeutic targets for intervening in key disease mechanisms is very promising. In contrast to the treatments available today, which are mostly aimed at ameliorating the symptoms, the next generation of therapies for PD will be targeting the disease itself, its causes and the mechanisms underlying both the neurodegenerative processes and the maladaptive changes underlying the onset of neurological deficits. The progress along these lines is indeed very encouraging and there is good reason to hope that, during the coming decades, we will see these efforts come to fruition.

We thank all the authors for their outstanding contributions and their willingness to share their knowledge and insights into their respective research fields. We also want to express our thanks to the Developmental Editors at Elsevier, Peter Llewellyn, Sam Mahfoudh, and most recently Hilal Johnson, for their assistance and support.

Anders Björklund
M. Angela Cenci
Department of Experimental Medical Science,
Wallenberg Neuroscience Centre, Lund University,
Lund, Sweden

Basic Science

What we can learn from iPSC-derived cellular models of Parkinson's disease

Maria Claudia Caiazza, Charmaine Lang, Richard Wade-Martins*

Oxford Parkinson's Disease Centre, Department of Physiology, Anatomy and Genetics, University of Oxford, Oxford, United Kingdom
*Corresponding author: Tel.: +44-1865-282837,
e-mail address: richard.wade-martins@dpag.ox.ac.uk

Abstract

Parkinson's disease (PD) is an age-related neurodegenerative disorder with no known cure. In order to better understand the pathological mechanisms which lead to neuronal cell death and to accelerate the process of drug discovery, a reliable *in vitro* model is required. Unfortunately, research into PD and neurodegeneration in general has long suffered from a lack of adequate *in vitro* models, mainly due to the inaccessibility of live neurons from vulnerable areas of the human brain. Recent reprogramming technologies have recently made it possible to reliably derive human induced pluripotent stem cells (iPSCs) from patients and healthy subjects to generate specific, difficult to obtain, cellular sub-types. These iPSC-derived cells can be employed to model disease to better understand pathological mechanisms and underlying cellular vulnerability. Therefore, in this chapter, we will discuss the techniques involved in the reprogramming of somatic cells into iPSCs, the evolution of iPSC differentiation methods and their application in neurodegenerative disease modeling.

Keywords

iPSC, Dopaminergic neurons, Reprogramming, Parkinson's disease, α-Synuclein, SNCA, LRRK2, GBA, PINK1, Parkin

Abbreviations

DAn	dopaminergic neuron
EB	embryoid body
ER	endoplasmic reticulum

ERAD   ER-associated degradation
ETC   electron transport chain
iPSC   induced pluripotent stem cell
PD   Parkinson's disease
ROS   reactive oxygen species
SNc   substantia nigra pars compacta
TH   tyrosine hydroxylase
UPR   unfolded protein response

1 Introduction

PD is the second most common neurodegenerative disease affecting 0.1–0.2% of the world's population. It is an age-related condition with a prevalence increasing to 1% in those over 60 (De Lau and Breteler, 2006). PD is classically associated with motor symptoms such as bradykinesia, resting tremor and rigidity, but encompasses a number of non-motor symptoms such as sleep disorders, olfactory dysfunction and gastrointestinal symptoms all of which severely effect one's quality of life (Jankovic, 2008). Histologically, PD is characterized by two hallmarks: (1) the degeneration of dopaminergic neurons (DAns) located in a part of the midbrain called the *substantia nigra pars compacta* (SNc) and (2) by the formation of protein aggregates called Lewy bodies mainly consisting of α-synuclein, thought likely to be toxic (Spillantini et al., 1997).

The etiology of PD is largely unknown, but for 5–10% of patients it is possible to identify a genetic factor (De Lau and Breteler, 2006). Several genetic risk factors and mutations have been described in PD, each with differing penetrance. Among the most significant genes identified in PD are *SNCA* and *LRRK2*. *Mutations* in both genes can cause monogenic forms of PD with dominant inheritance, whereas common *polymorphisms* at both loci can act as genetic risk factors. The *SNCA* gene encodes α-synuclein, a synaptic protein that accumulates and forms cytotoxic inclusions in PD (Bellucci et al., 2012; Spillantini et al., 1997). LRRK2 is a multidomain protein containing kinase, GTPase and scaffolding functions; it is proposed to be involved in protein trafficking and autophagy, both of which are altered in PD. Heterozygous mutations in *GBA* also act as a strong genetic risk factor for PD, whereas homozygous mutations in *GBA* cause the lysosomal storage disorder Gaucher's disease. *GBA* encodes for the enzyme β-glucocerebrosidase, a lysosomal hydrolytic enzyme mutations in which are associated with lysosome and autophagy dysfunction in PD. Recessive mutations in the genes *PINK1* and *PRKN*, encoding for the proteins PINK1 and parkin, respectively, impact the clearance of damaged mitochondria by mitophagy, leading to early-onset PD (Billingsley et al., 2018).

PD has a complex pathophysiology and there are no disease-modifying drugs to treat the condition. Therefore, accelerating the process of drug discovery requires accurate disease models. The most common disease models are divided into *in vitro* models including cell lines and primary cultures and *in vivo* models including worms, flies, fish, rodent models, and non-human primates. *In vitro* models are easily tractable, and are therefore the most suitable for the earliest phases of drug discovery as they allow large-scale phenotypic screening of compound libraries. Cell lines, on the other hand, are usually immortalized cells often with aberrant cell biology. Additionally, primary cultures, derived from animal models, may not completely share the same biology as humans.

Embryonic stem cell or iPSC-derived neurons from patients offer an excellent opportunity to recapitulate *in vitro* features of human pathology. Many protocols have been developed for the differentiation of neurons and glia bearing features of different brain regions. iPSC-derived cortical neurons can be employed to study dementia associated with PD and glial cells to study non-cell-autonomous mechanisms of degeneration, but the focus in PD research has been the development of iPSC-derived DAns from PD patients. So far, this model has identified phenotypes that characterize PD and have enabled the investigation of underlying PD-relevant molecular mechanisms in human DAns and gathering evidence for new therapeutic targets.

The aim of this chapter is to describe some crucial phenotypes identified so far in iPSC-derived neurons from patients and to highlight some future directions of this technology.

2 Establishment of iPSC technology and methods of differentiation into DAns

There are two principal methods to obtain DAns from patients. One approach involves reprogramming patient somatic cells (most commonly fibroblasts) into iPSCs and differentiating these iPSCs into DAns. Another, more recent, approach involves the direct reprogramming of fibroblasts to DAns. Each method encompasses a number of benefits and drawbacks which will be discussed here.

2.1 Establishment of induced pluripotent stem cells

The advent of iPSC technology by Yamanaka's group revolutionized the field of human *in vitro* disease modeling. Pluripotent stem cells can be derived from adult somatic cells by inducing the expression of the Yamanaka factors: Oct3/4, Sox2, Klf4, c-Myc (Takahashi and Yamanaka, 2006). Oct3/4 and Sox2 are required for pluripotency, whereas, c-Myc and Klf4 increase the efficiency of iPSC production by interfering with apoptosis and senescence. Originally, reprogramming was achieved by employing integrating viruses such as lentiviruses and retroviruses to induce the expression of the Yamanaka factors (Takahashi et al., 2007; Takahashi and Yamanaka, 2006). However, integration of the viral genome carries a risk of inducing insertional mutagenesis in the host genome. Therefore a non-integrating approach is now favored, through the use of non-integrating viruses such as Sendai and adenovirus (Fusaki et al., 2009; Stadtfeld et al., 2008; Zhou and Freed, 2009), episomal plasmids (Okita et al., 2008), piggyBac transposons (Kaji et al., 2009), proteins (Kim et al., 2009) or small molecules (Hou et al., 2013).

2.2 **Differentiation into DAns**

The earliest, and still popular, technique for the generation of DAns from iPSCs goes through the generation of embryoid bodies (EBs). EBs are 3D colonies of pluripotent stem cells which undergo spontaneous differentiation toward the three germ layers (endoderm, mesoderm and ectoderm) recapitulating the early phases of *in vivo* development. Neurons derive from the ectoderm and in order to enrich for neuroectodermal cells, it is possible to plate embryoid bodies in an adherent culture so that the culture starts forming neural tube-like rosettes which can be isolated. These neuronal precursors express the neuroectoderm marker PAX6. A key study demonstrated that exposure of these PAX6+ cells to FGF8 and SHH could produce midbrain DAns (Yan et al., 2005). In particular, early exposure to FGF8 was necessary to pattern the precursors toward a more posterior fate (caudalization) and subsequent exposure to FGF8 and SHH allowed differentiation into midbrain DAns (Yan et al., 2005). This protocol generated a rather mixed population of electrophysiologically active cells of which 32% were positive for tyrosine hydroxylase (TH), a key enzyme for the production of dopamine and therefore a marker for DAns.

In the following years there has been an increasing interest in improving the regionalization of the cells. Recent neurodevelopmental studies demonstrated that SNc DAns derive from the midbrain floor plate rather than PAX6+ precursors (Ono et al., 2007). Therefore, a number of studies aimed at deriving DAns through the *in vitro* generation of midbrain floor plate precursors. Neural conversion is a default program, but efficiency can be increased by blocking mesoderm and endoderm conversions. In 2009, the Studer laboratory developed the dual inhibition of SMAD signaling to reliably induce neural conversion. The system employs Noggin (or the agonist dorsomorphin or LDN193189) to block BMP signaling and SB431542 to block Activin and Nodal signaling (Chambers et al., 2009). The neural precursor cells generated by means of the dual SMAD inhibition default to an anterior neural fate. However, midbrain neurons derive from the floor plate and, therefore, in 2011 the Studer laboratory slightly modified this protocol to pattern the precursors toward a more posterior and lower fate (caudalization and ventralization) therefore inducing dopaminergic neuron progenitors in the first phases of the differentiation (Kriks et al., 2011). Ventralization is obtained by activating the Sonic hedgehog signaling pathway using recombinant SHH and puromorphamine. Caudalization and the final patterning of dopaminergic neural progenitors requires FGF8a and Wnt signaling activation (achieved with CHIR99021, a potent inhibitor of GSK3B able to activate Wnt signaling). The resultant dopaminergic neural progenitors express the midbrain dopaminergic neural progenitor markers FOXA2 and LMX1A. Furthermore, neuronal differentiation and maturation is achieved using DAPT to inhibit Notch signaling and push toward neuronal differentiation, db-cAMP to stimulate neurite elongation, survival and further maturation, and a cocktail of neurotrophic factors (BDNF, GDNF, ascorbic acid and TGFb3) for trophic support (Kriks et al., 2011). The resultant neuronal culture yields 75% TH+ neurons with a ventral midbrain identity (where the SNc is located). The cells are electrophysiologically active and were successfully transplanted in a murine model of PD (Kriks et al., 2011).

An alternative to differentiating DAns from iPSCs or other stem cell sources is direct reprogramming; this approach aims at generating DAns directly from other somatic cell types such as fibroblasts. This approach relies on the overexpression of lineage-specific transcription factors in order to force trans-differentiation. DAns can be obtained through the viral transduction and overexpression of *ASCL1* with *LMX1A* and *NURR1* with an efficiency of 3% TH+ neurons from human adult fibroblasts (Caiazzo et al., 2011). *ASCL1* is a neural-lineage-specific transcription factor that has been employed for direct reprogramming of fibroblasts to neurons (Vierbuchen et al., 2010); *NURR1* is a DAn-specific transcription factor that is essential for the development of these cells (Vierbuchen et al., 2010); *LMX1A* is another DAn-specific transcription factor that was selected by screening a set of DAn-specific transcription factors for their reprogramming capacity and their ability to induce TH expression. Efficiency was improved to achieve 60% TH+ neurons from fetal human fibroblasts and 8% from adult human fibroblasts by introducing mir124, a microRNA which instructs changes in chromatin remodeling complexes, a process essential for neuronal differentiation, and p53 shRNA to increase cell identity plasticity (Jiang et al., 2015). Other work employed Ascl1, Brn2, and Myt1l—all used for the direct reprogramming of fibroblasts to neurons (Vierbuchen et al., 2010)—and *ASCL1* with *LMX1A* obtaining 10% TH+ neurons from fetal human fibroblasts (Pfisterer et al., 2011). An advantage of the direct reprogramming method is that, unlike iPSC-derived neurons from somatic cells, the pluripotent stage is skipped. Eliminating the Yamanaka factors, some of which may cause tumor formation, therefore makes direct reprogramming of DAns safer for transplantation. One of the concerns about the use of iPSC-derived neurons in the study of neurodegenerative diseases is that the process of reprogramming may reset the cellular epigenetic clock, which might mask *in vitro* age-related phenotypes. Direct reprogramming may circumvent this problem as cells may retain their age during this process and therefore allow the ability to reproduce *in vitro* age-related phenotypes (Kim et al., 2018). However, direct reprogramming is still very inefficient yielding only 3% TH+ neurons and requires a consistent supply of potentially scarce primary cells.

3 iPSC-derived neurons for disease modeling

One of the major advantages of iPSC-derived neurons is that they enable researchers to study phenotypes in neurons derived directly from patients with different genetic backgrounds to highlight the shared pathogenic mechanisms in PD. Here, we will focus on the recent cellular phenotypes discovered in iPSC-derived neurons from different genetic backgrounds (Fig. 1).

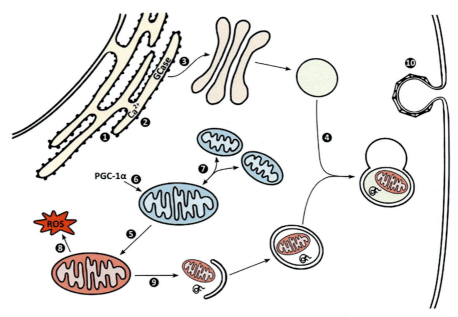

FIG. 1

Major phenotypes identified in iPSC-derived DAns from PD patients. (1) Neurodegenerative diseases are characterized by an accumulation of misfolded protein in an activation of the ER which results in the unfolded protein response (UPR), which in brief is aimed at reducing the amount of misfolded proteins and at restoring ER homeostasis. The ER stress response can be monitored by measuring the levels of mediators (such as BiP, PDI, IRE1, phospho-PERK, phospho-eIF2α, spliced XBP1, ERO1 and FKBP9), or the accumulation of immature or misfolded proteins in the ER, such as GCase. In *GBA*-PD and *SNCA*-PD iPSC-derived DAns, the ER stress response is chronically activated. (2) The ER is the major calcium store of the cell and its calcium levels are altered in *GBA-L444P* and *LRRK2-G2019S* iPSC-derived DAns. (3) *SNCA*-triplication causes an accumulation of immature lysosomal hydrolytic enzymes such as GCase in the ER. (4) In *GBA*-PD iPSC-derived DAns an increase in the lysosomal marker LAMP1 is observed, suggesting an accumulation of lysosomes; however, these lysosomes appear to be dysfunctional in both *GBA*-PD and *SNCA*-PD lines as long-lived proteins accumulate. (5) Mitochondrial deficits have a role in PD. In particular, the activity of the electron transport chain (ETC) is altered in *SNCA*-PD, *GBA*-PD, *LRRK2*-PD and *PINK*-PD neurons. (6) Mitochondrial biogenesis is compromised in *SNCA*-PD neurons as demonstrated by altered levels of the PGC-1α protein, a key regulator of mitochondrial biogenesis. (7) *SNCA*-PD neurons also display dysfunctional mitochondrial fusion and fission. (8) An increase in ROS burden has been described in *SNCA*-PD, *PINK1*-PD and *PRKN*-PD neurons. (9) Autophagy is the process by which the cell degrades dysfunctional proteins and organelles; *GBA*-PD, *PARK2*-PD, *SNCA*-PD and *LRRK2*-PD neurons display alterations in the autophagy and mitophagy (degradation of mitochondria via autophagy) processes, which causes dysfunctional proteins and organelles to build up. (10) *LRRK2*-PD neurons display impaired clathrin-mediated endocytosis.

Related gene	Cell type	Phenotype	References
MAPT	DAn	Decreased mitochondrial motility	Beevers et al. (2017)
LRRK2	Astrocytes	Compromised neuroprotective potential	Booth et al. (2019)
SNCA	DAn	Increased ROS levels	Byers et al. (2011)
SNCA	Cortical neurons	ER stress, accumulation of ERAD substrates	Chung et al. (2013)
PINK1	DAn	Altered mitochondrial structure, increased ROS levels	Chung et al. (2016)
PRKN	DAn	Altered mitochondrial structure, increased ROS levels	Chung et al. (2016)
LRRK2	DAn	Altered endocytosis	Connor-Robson et al. (2019)
LRRK2	DAn	Altered mitochondrial respiration, decreased mitochondrial motility	Cooper et al. (2012)
PINK1	DAn	Altered mitochondrial respiration	Cooper et al. (2012)
LRRK2	Astrocytes	Compromised neuroprotective potential	di Domenico et al. (2019)
GBA	DAn	ER stress increase, UPR increase, lysosomal impairment, increase α-synuclein release	Fernandes et al. (2016)
LRRK2	DAn	Altered mitophagy	Hsieh et al. (2016)
Tau	Cortical neurons	Altered mitochondrial motility	Iovino et al. (2015)
LRRK2	DAn	Altered calcium content	Korecka et al. (2019)
GBA	DAn	ER stress increase, UPR increase, HDAC4 mislocalization	Lang et al. (2019)
SNCA	Cortical neurons	Altered mitochondrial respiration	Ludtmann et al. (2018)
SNCA	DAn	Lysosomal impairment, altered protein trafficking	Mazzulli et al. (2016)
LRRK2	DAn	Increased sensitivity to oxidative insults	Nguyen et al. (2011)
SNCA	DAn	Altered mitochondrial respiration, defective mitochondrial biogenesis, increased ROS levels	Ryan et al. (2013)
GBA	DAn	Altered calcium function, altered calcium content	Schöndorf et al. (2014)

Continued

—cont'd

Related gene	Cell type	Phenotype	References
GBA	DAn	ER stress increase, UPR increase, lysosomal impairment, increased α-synuclein levels, alteration in mitochondrial structure, altered mitochondrial respiration, decreased mitochondrial dynamics	Schöndorf et al. (2018)
SNCA	DAn	Altered mitophagy	Shaltouki et al. (2018)
Tau	Cortical neurons	Altered Tau splicing	Sposito et al. (2015)
PRKN	DAn	Altered mitophagy	Suzuki et al. (2017)
SNCA	DAn	ER stress increase, UPR increase, altered mitochondrial structure, altered mitochondrial respiration, decreased mitochondrial dynamics, defective mitochondrial biogenesis, increased ROS levels	Zambon et al. (2019)

3.1 Endoplasmic reticulum stress

The endoplasmic reticulum (ER) is responsible for the proper folding and post-translational modification of membrane proteins. Misfolded proteins are trafficked to the ER to undergo proteolytic degradation in a process called ER-associated degradation (ERAD) (Smith et al., 2011). The ER is also involved in lipid biosynthesis and is the major calcium store of the cell.

In neurodegenerative diseases, misfolded proteins build up and accumulate in the ER. The ER responds to the increased misfolded protein load with a stress response called the unfolded protein response (UPR) which induces a general decrease in protein translation and an increase in the expression of particular protein chaperones and ERAD proteins in order to restore ER homeostasis (Scheper and Hoozemans, 2015; Walter and Ron, 2011). There are three main stress sensors in the membrane of the ER including RNA-activated protein kinase R (PKR)-like ER kinase (PERK), activating transcription factor 6 (ATF6) and inositol requiring enzyme 1 (IRE1). These activate downstream mediators of the UPR response such as eIF2α, ATF6 and XBP1 which are commonly used to measure UPR activation levels (Scheper and Hoozemans, 2015). ER stress levels can also be monitored looking at the levels of ERAD substrates such as nicastrin and GCase.

ER stress has been thoroughly investigated in PD human iPSC-derived neurons. iPSC-derived neurons DAns from patients bearing the *N370S* mutation in the *GBA* gene display increased levels of ER stress markers including: BiP, PDI, IRE1, phospho-PERK, phospho-eIF2α and spliced XBP1 (Fernandes et al., 2016; Schöndorf et al., 2014). Moreover, it has been demonstrated that these neurons express higher levels of the ER-resident chaperones and ER stress mediators

ERO1 and FKBP9 (Lang et al., 2019). Cortical neurons derived from patients with a triplication of the *SNCA* locus accumulated the ERAD substrates GCase and nicastrin suggesting the induction of ER stress (Chung et al., 2013). In addition, iPSC-derived DAns mutated in the *SNCA* locus (*A53T* and *Triplication*) displayed increased levels of BiP and IRE1 (Zambon et al., 2019).

3.2 Protein trafficking and degradation deficits

PD is characterized by the formation of Lewy bodies mainly comprising aggregated α-synuclein (Spillantini et al., 1997). α-Synuclein, encoded by the *SNCA* gene, is a natively unfolded protein which shows a higher tendency to aggregate and to form ordered β-sheet structures with increasing concentration and when the protein is mutated or undergoes post-translational modifications, such as hyperphosphorylation (Gallegos et al., 2015). An increasing body of evidence suggests that alterations in the degradation of α-synuclein and also wider alterations in the autophagy/lysosomal pathway, which degrades many proteins, may cause α-synuclein to build up in the cell and therefore be involved in cytotoxicity in PD. Accordingly, lysosomal impairment is observed in *SNCA-Triplication* iPSC-derived DAns and pulse-chase experiments displayed that long-lived proteolysis rates are decreased (Mazzulli et al., 2016). Additionally, it has been discovered that α-synuclein interacts with the lysosomal protein GCase forming a positive feedback loop that results in accumulation of α-synuclein. In particular, α-synuclein inhibits lysosomal GCase activity and loss of GCase function contributes to α-synuclein accumulation (Mazzulli et al., 2011).

Mutations in the *GBA* gene encoding for the lysosomal protein GCase are linked to an increased risk of developing PD (Sidransky and Lopez, 2012). DAns derived from PD *GBA* patients display perturbations in protein degradation and autophagy pathways. iPSC-derived DAns from *GBA-L444P* and *GBA-N370S* PD patients show increased levels in the lysosomal marker LAMP1 suggesting lysosomal accumulation as well as an observed increase in LC3, an autophagosomal marker (Fernandes et al., 2016; Schöndorf et al., 2014). These patients also displayed decreased activity of GCase, the lysosomal protein encoded by the *GBA* gene, and a general defect in lysosomal degradation monitored by increased levels of p62 (a ubiquitin-binding protein that targets cargo to autophagosomes for later degradation). Additionally, electron microscope observations highlighted an accumulation of electron-dense material within the lysosomes which probably represents undegraded cargo, and an enlargement of the lysosomal compartment; these two observations suggest a lysosomal impairment (Fernandes et al., 2016). A decrease in lysosomal function in these neurons was also associated with increased α-synuclein levels (Schöndorf et al., 2014).

Intracellular protein trafficking is the mechanism through which proteins are sorted to their correct localization after translation and is crucial for optimal cellular function. Alterations in protein trafficking have linked neurodegenerative diseases with the accumulation of proteinaceous inclusions (Hunn et al., 2015; Wang et al., 2014).

Intracellular trafficking is mainly involved in the transport of protein and lipids to and from cell membranes. Protein trafficking can be divided into the exocytic and the endocytic pathways. The former is involved in the shuttling of proteins from the ER to the Golgi and then to cellular membranes. The latter starts with endocytosis (a cellular process through which molecules are taken up by the cell by membrane pinch-off), therefore molecules are surrounded by the cell membrane and internalized within a vesicle. Subsequently, the vesicle fuses with the endosome, a membrane-bound compartment of the cell. Cargo are then sorted and can either be recycled or follow the degradation pathway which requires that endosomes fuse with lysosomes. It is therefore clear that intracellular protein trafficking and protein degradation are tightly linked.

Alterations in the exocytic pathway have been extensively described in PD DAns. α-Synuclein overexpression or *SNCA-Triplication* cause alterations in the trafficking of lysosomal hydrolases from the ER to the Golgi causing an accumulation of immature enzymes in the ER. In turn, this causes alterations in lysosomal activity which causes accumulation of α-synuclein in a feedforward loop (Mazzulli et al., 2016). iPSC-derived DAns bearing mutations in the *LRRK2* gene (*G2019S* and *R1441C*) express altered levels of key endocytic protein which functionally resulted in impaired clathrin-mediated endocytosis of the lipophilic dye FM1-43 (Connor-Robson et al., 2019).

3.3 Mitochondrial defects and oxidative stress

Although the etiology of PD has been elusive since the first description of the disease in 1817 (reproduced in Parkinson (2002)), an increasing body of evidence suggests a role for mitochondrial dysfunction in the progression of the disease. Mitochondrial toxins such as the insecticide rotenone, the fungicide maneb, the herbicide paraquat and MPTP (a by-product of the manufacture of the recreational drug MPPP) recapitulate many of the physiological phenotypes in murine models of PD (Ryan et al., 2015; reviewed in Dauer and Przedborski (2003)). However, their involvement in the development of human sporadic PD is still debated. Additionally, mutations in mitochondrial genes *PINK1* and *PRKN* cause early-onset PD (Kitada et al., 1998; Valente et al., 2004). Mitochondrial deficits have been investigated in iPSC-derived DAns where changes in mitochondrial structure have been observed in *GBA-L444P* (Schöndorf et al., 2014), *PINK1-Q456X, PRKN-V324A* (Chung et al., 2016) and *SNCA*-PD neurons (Zambon et al., 2019).

Research has highlighted alterations in mitochondrial respiration, assessed by measuring oxygen consumption rate in the presence of toxins inhibiting different steps of the electron transport chain (ETC). iPSC-derived DAns with mutations in the *SNCA* gene (*A53T* and *Triplication*) display decreased basal respiration, ATP production, maximal respiration and spare respiratory capacity (Ryan et al., 2013; Zambon et al., 2019). It has been shown that oligomeric α-synuclein interacts with ATP-synthase in *SNCA-Triplication* iPSC-derived cortical neurons; moreover, these cells show an increased NADH redox index, a measure of NADH resting levels which suggests an impairment of complex I of the ETC (Ludtmann et al., 2018). For these reasons, the authors speculated that the interaction between oligomeric α-synuclein and ATP-synthase might be detrimental and be one of the causative factors for mitochondrial respiration disruption in PD (Ludtmann et al., 2018). Mitochondrial respiration also appears to be compromised in *GBA*-PD (*N370S, L444P, RecNcil*) (Schöndorf et al., 2018), in *LRRK2* (*G2019S* and *R1441C*), in *PINK1* (*Q456X*) (Cooper et al., 2012) and *SNCA*-PD (Zambon et al., 2019) iPSC-derived DAns.

In addition to alterations in mitochondrial respiration, there also appear to be alterations in mitochondrial dynamics, biogenesis and degradation. Mitochondria are highly dynamic organelles that undergo fusion and fission (division) events. These processes are tightly regulated and involve a plethora of specific proteins whose levels can be used as a proxy to monitor the two phenomena. Alterations in these processes have been described in *GBA*-PD iPSC-derived DAns, where decreases in mitochondrial fission proteins (DRP1 and Fis1) and mitochondrial fusion proteins (OPA1 and Mfn1) were observed (Schöndorf et al., 2018). Decreases in DRP1-pS616 have also been observed in iPSC-derived *SCNA-A53T* and *Triplication* iPSC-derived DAns (Zambon et al., 2019).

Defective mitochondrial biogenesis might also account for altered mitochondrial function in PD. In a study conducted in *SNCA-A53T* DAns, stress induced *via* mitochondrial toxins or hydrogen peroxide (H_2O_2) produced aberrant post-translational modifications of the transcription factor MEF2C (Ryan et al., 2013). These modifications decreased its function thus causing a decrease in the expression of the genes it regulates including PGC-1α (a key factor in mitochondrial biogenesis). In the absence of stressors; however, cells bearing the same mutation appear to have increased levels of PGC-1α (Zambon et al., 2019). This might be explained by cells trying to cope with altered respiration by boosting biogenesis in a compensatory fashion, but when stress increases the compensatory mechanism fails. Damaged mitochondria are usually degraded by mitophagy which prevents the accumulation of dysfunctional mitochondria. As previously mentioned, mutations in some regulators of this process (PINK1 and parkin) cause early-onset PD. Furthermore, *PRKN*-PD (Suzuki et al., 2017), *SNCA-A53T* (Shaltouki et al., 2018) and *LRRK2*-G2019S (Hsieh et al., 2016) DAns all display decreased or slowed mitophagic activity upon mitochondrial stress.

Neurons are highly polarized cells so the transport of cargo along these cells is particularly important for their function (Hunn et al., 2015). Moreover, given the high energy demand of neurons, mitochondrial transport is absolutely crucial. A decrease in mitochondrial motility has been observed in *LRRK2-G2019S* iPSC-derived DAns (Cooper et al., 2012). This might provide mechanistic insight into the role of the Tau protein encoded by the *MAPT* locus—a known risk factor for PD identified by genome-wide association studies (International Parkinson Disease Genomics Consortium, 2011)—in PD. Tau is a microtubule-associated protein mainly involved in microtubule stability. In general, correct Tau splicing has been associated with neuronal health and Tau mutations have been demonstrated to alter Tau splicing in iPSC-derived cortical neurons (Sposito et al., 2015). More specifically, a study conducted in iPSC-derived DAns observed that Tau can also interfere with axonal transport of mitochondria as Tau knock-down results in an overall increase of mitochondrial motility (Beevers et al., 2017). Mitochondrial motility is also altered in iPSC-derived cortical neurons harboring *MAPT-P301L* mutation (Iovino et al., 2015).

Mitochondria are the main source of reactive oxygen species (ROS) and many ROS scavengers are located in the mitochondria. However, increase in the ROS burden might damage the mitochondria itself. *SNCA-A53T* and *SNCA-Triplication* DAns show increased ROS levels and upregulated genes encoding for anti-oxidant proteins such as catalase and heme oxygenase (Byers et al., 2011; Ryan et al., 2013; Zambon et al., 2019). *PINK1*-PD and *PRKN*-PD DAns show increased ROS production and *PRKN*-PD DAns are more susceptible to rotenone-induced oxidative stress (Chung et al., 2016; Suzuki et al., 2017). Moreover, *LRRK2-G2019S* DAns appear to be more sensitive to oxidative insults (Nguyen et al., 2011).

3.4 Alterations in calcium homeostasis

Calcium is an important signaling molecule involved in the regulation of many cellular functions such as gene expression, cell bioenergetics, protein catabolism and other signaling cascades (reviewed in Brini et al. (2013)). In bioenergetically active cells such as neurons, calcium has a role in synaptic release and plasticity (reviewed in Zucker (1999) and Brini et al. (2013)) and in particular, L-type calcium channels support the crucial pacemaker activity in SNc DAns (Guzman et al., 2009). The continuous calcium waves occurring in DAns, however, place them in an environment where even small alterations in calcium homeostasis might impact on cellular function and have been linked to mitochondrial dysfunction (Foehring et al., 2009; Guzman et al., 2010).

Mass spectroscopy experiments revealed that DAns derived from iPSCs bearing *GBA-L444P* mutations upregulate the calcium binding protein NECAB2 suggesting alterations in calcium function (Schöndorf et al., 2014). Moreover, calcium imaging highlighted changes in calcium content in DAns derived from patients with *GBA-L444P* (Schöndorf et al., 2014) and *LRRK2-G2019S* mutations (Korecka et al., 2019). Intriguingly, a calcium channel antagonist is able to prevent rotenone-induced death in *PRKN*-PD iPSC-derived DAns (Tabata et al., 2018).

4 iPSC-derived glial cells for disease modeling

Increasing evidence suggests a potential role for glial cells in the pathogenesis of PD. Astrocytes exert many functions aimed at the maintenance of neuron integrity. They maintain fluid homeostasis in the synaptic area as they regulate ion concentration through several ion channels and water content through aquaporins that take up water from blood vessels (Sofroniew and Vinters, 2010). Astrocytes are involved in neurotransmitter homeostasis at the synapse as they express transporters for glutamate, GABA, and glycine that are used to re-uptake neurotransmitters after synaptic transmission (Sofroniew and Vinters, 2010). Additionally, they modulate synaptic transmission through the release of signaling molecules including glutamate, GABA and purines. These gliotransmitters are released in response to neuronal activity and are able to modulate neuronal activity itself; these studies brought forward the theory of the "tripartite synapse" which suggests that astrocytes are directly involved in synaptic function (Sofroniew and Vinters, 2010). Additionally, astrocytes provide metabolic support to neurons as they provide the neurons with metabolites and trophic factors such as GDNF, which is crucial for cell survival. Moreover, astrocytes are involved in the maintenance of the blood-brain barrier, a highly selective semipermeable barrier that divides the blood from the brain and its fluid (Sofroniew and Vinters, 2010). Alterations in some of these functions can be detrimental for neurons and brain tissue. Indeed, alterations in neurotransmitter re-uptake and alterations in the blood-brain barrier have been described in PD (Booth et al., 2017). For these reasons, there is an increasing interest in the study of astrocytes derived from PD patients.

Recently, a protocol to generate midbrain-patterned astrocytes from human induced pluripotent stem cells has been described (Booth et al., 2019). The first phase of the protocol consists of the induction of midbrain floor plate progenitors according to the Studer protocol with minor modifications (Beevers et al., 2017; Kriks et al., 2011). Then the neural progenitors are converted into astroglial progenitors by culturing them in the presence of EGF and LIF, to induce proliferation. Finally astrocyte progenitors are expanded and matured in the presence of EGF, FGF2 and heparin, to maintain proliferation (Tropepe et al., 1999).

A transcriptomic study conducted on midbrain-patterned astrocytes bearing the *LRRK2-G2019S* mutation showed that the mutation might compromise their neuroprotective potential. In fact, the genes encoding for TGFB1 (involved in the inhibition of microglia in a PD model) and MMP2 (shown to be able to catabolize α-synuclein aggregates *in vitro* and *in vivo*) were downregulated (Booth et al., 2019). Indeed, co-culture of DAns and astrocytes derived from patients with the *LRRK2-G2019S* mutation, showed that astrocytes might account for a form of non-cell-autonomous neurotoxicity in PD (di Domenico et al., 2019). It has been demonstrated that PD iPSC-derived astrocytes can induce neurite alterations and α-synuclein accumulation in healthy iPSC-derived cortical neurons. On the other hand, healthy astrocytes could revert some PD neuronal phenotypes such as neurite alterations and α-synuclein accumulation (di Domenico et al., 2019).

5 Co-culture systems, 3D cultures and organoids for disease modeling

2D cultures of a single-cell type are a simplified model sufficient to recapitulate cellular dysfunction in a dish. However, as described previously, this model may be over-simplified as different cell types may interact in a complex way.

Therefore, there is an increasing interest in more complex disease models that can account for the cellular diversity in the brain and the interaction between different cell types.

Besides the degeneration of DAns of the SNc, PD is characterized by a dysfunction of the circuitry in which these cells are involved, the basal ganglia loop, involved in motor control. The striatum receives information from the cortex which is then relayed to the output nuclei. The DAns of the SNc modulate the activity of the neurons of the striatum and therefore of this network. It would be crucial to reproduce in the laboratory parts of this network to better understand PD. One way of doing this would be to reconstruct the network using a microfluidic system, which has been achieved in murine primary cultures (Lassus et al., 2018). This microfluidic device recapitulates the cortico-striatal network using separated culture chambers linked by asymmetric channel microfluidics. This system could be also implemented using iPSC-derived neurons from patients to investigate circuitry disruption in PD.

Another problem faced when dealing with 2D cultures is that they do not fully recapitulate the interactions that occur in the three-dimensional environment of a tissue. Some attempts to reproduce this 3D environment using microfluidic systems or biomaterial scaffolding have been made (Adil et al., 2017; Moreno et al., 2015). Organoids are a simplified model of an organ that can account for cell variability and 3D interactions. In fact, organoids should contain multiple organ-specific cell types and be able to self-organize in such a way that resembles the cytoarchitecture of the organ that they model. Moreover, the organoids should recapitulate at least some of the specific functions of the organ. Recently, two protocols for the differentiation of midbrain-like organoids from human stem cells have been described (Jo et al., 2016; Monzel et al., 2017). The Jo et al. protocol includes the formation of ventral midbrain precursors in the first phase of the differentiation using SB431542, Noggin and CHIR99021, recombinant SHH and FGF8. Organoid development is then obtained with db-cAMP, BDNF, GDNF and ascorbic acid. During their development, the organoids are kept on an orbital shaker to make sure that the core of the organoid is well perfused. The Monzel et al. protocol starts from neural stem cells that are then patterned using CHIR99021, puromorphamine and ascorbic acid. They are then differentiated with db-cAMP, BDNF, GDNF and ascorbic acid. In both cases, DAns, oligodendrocytes and glial cells were differentiated within the organoid. DAns in the organoids were elecrophysiologically active and displayed a pacemaker-like activity, consistent with the electrophysiological profile of DAns in the SNc. Additionally, the midbrain organoids described by Jo et al. had neuromelanin deposition and dopamine release, thus recapitulating SNc physiology. Organoids derived from *LRRK2-G2019S* iPSCs were also obtained using the Monzel et al. protocol (Smits et al., 2019). In the paper, the investigators employed high content imaging to study possible phenotypes in the organoids. After 35 days of differentiation, there was a significant reduction in the number of DAns in the patient-derived organoids compared to controls and in general the complexity of the arborization of the dendritic network appeared to be decreased.

6 iPSC-derived neurons for cell-based treatments

In the context of neurodegeneration, interest in replacing lost cells is very high. For this reason, in 1987 the first attempt to transplant healthy dopaminergic neuroblasts into PD patient brains was attempted (Lindvall et al., 1989) followed by a second improved trial in 1989 (Lindvall et al., 1990). The grafted neurons derived from ventral midbrains of 8–11-week-old human fetuses produced very promising results. Dopamine release was restored, and the patients displayed clinical improvement. Follow up of the patients 10 years after the transplant observed that the patients displayed only minor parkinsonian symptoms (Piccini et al., 1999). Subsequent trials however, revealed that the procedure produced quite variable results and that the procurement of the transplanted DAns was a very difficult procedure to standardize (Freed et al., 2001). These difficulties brought the scientific community to consider the application of iPSC-derived DAns in this process. As a result, a clinical trial by Prof. Jun Takahashi at the Kyoto University Hospital, is currently being conducted, involving 14 patients who will be transplanted with DAns derived from iPSCs from a healthy donor (ID: UMIN000033564).

7 Limitations of iPSC-based models

The advent of iPSC technology has completely revolutionized translational research. Traditional *in vitro* models rely on cell lines, primary cultures from animal models or primary cultures from patients. Cell lines are immortalized and as a result often carry aberrant chromosomes, whereas primary cultures from animal models have the caveat of being from another species. Primary cultures of relevant cell types from patients can be very difficult to handle and for neurodegenerative diseases, almost impossible to obtain. iPSC-derived DAns therefore afford the opportunity to culture in a dish the very same neurons that degenerate in the brains of PD patients. However, this technology also has some drawbacks.

It has been demonstrated that iPSCs show a high level of variability reflecting the genetic differences among donors (Rouhani et al., 2014). On the one hand, the variability due to different genetic backgrounds makes iPSC-based cell models highly physiological and may help highlight the shared pathogenic mechanisms in PD; on the other hand, this phenotypic noise may mask some specific phenotypes in these cells. This raises the question of what is the best control for iPSC-derived cellular models. Many studies use cells derived from control subjects; however, it may be necessary to include a high number of patient and control lines per experiment in order to separate the control and the disease state to study idiopathic disease. In the case of a known mutation, one can employ genome editing and generate isogenic controls. Although this poses considerable technical challenges it can reduce experimental variability and identify phenotypes that can be definitely attributed to the mutation.

Another concern is cell maturity and age. During the process of reprogramming, stem cells may be rejuvenated and signs of cellular aging such as inactivation of telomerase expression, are erased (Marión and Blasco, 2010). Therefore, the differentiation process starts with immature neurons, moves onto young neurons, then mature neurons and possibly old neurons. However, neuronal maturation is a slow process and transplantation experiments have shown that there is something like an autonomous clock that seems to be species-dependent (Isacson and Deacon, 1997). This might also be true *in vitro*, therefore making it difficult to obtain fully mature cortical neurons derived from human iPSCs. However, this seems not to be the case in iPSC-derived DAns which may be purified using the marker TH and shown to possess a mature transcriptomic profile indistinguishable to that of isolated human post-mortem DAns (Sandor et al., 2017). Utilizing iPSC-derived neurons to model PD generates added challenges as PD is an age-related disease. It is common to observe cellular phenotypes such as DNA-damage and mtDNA-damage, iron accumulation and neuromelanin accumulation in aged cells, which may be quite challenging to reproduce *in vitro*.

Overall, iPSC-derived neuronal models of Parkinson's are showing great value in highlighting the molecular and cellular defects responsible for dopamine neuron vulnerability early in the disease process and provide a platform to develop treatments to slow or prevent disease.

8 Future applications

iPSC-derived DAns have begun to provide new insights into the cellular phenotypes involved in PD that could be used for target-directed drug discovery. In combination with the advent of new unbiased, high-throughput techniques, new cytopathic mechanisms in many models may be uncovered. In particular, next generation sequencing has become widely employed to study iPSC-derived DAns. A recent study conducted on PD *GBA-N370S* iPSC-derived DAns employed single-cell RNA-sequencing to discover a potentially new druggable cytotoxic mechanism (Lang et al., 2019). In this work, single-cell RNA-sequencing was employed to create a pseudo-timeline of the changes in gene expression. This way it was possible to highlight a group of early differentially-expressed genes that were more likely to be responsible for early pathological events. These genes appeared to be enriched for targets regulated by the transcription factor HDAC4. Further validation highlighted that this protein is mislocalized to the nucleus in PD patient iPSC-derived DAns and that re-localization of HDAC4 back to the cytoplasm is able to revert some of the phenotypes described in these DAns. This study highlighted HDAC4 as a new potential therapeutic target for PD. In another recent study, the Connectivity Map resource was employed to identify drugs that influence the expression of differentially-expressed genes in iPSC-derived DAns from *LRRK2-G2019S* patients (Sandor et al., 2017). The study highlighted that clioquinol—a drug able to rescue DAn loss in a PD murine model—is also predicted to restore a normal gene expression profile in PD iPSC-derived DAns.

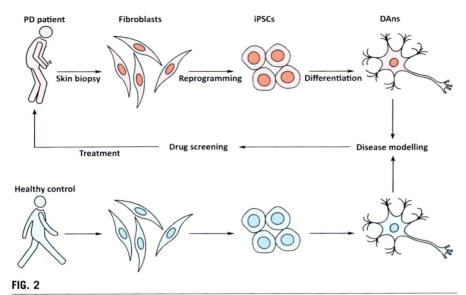

FIG. 2

Applications for iPSC-derived neurons in PD. IPSCs are derived from patient somatic cells (typically fibroblasts) by the expression of the Yamanaka factors (Oct3/4, Sox2, Klf4, c-Myc). iPSCs are then differentiated into DAns or into other disease relevant cells. These cells can be employed to study phenotypes and discover pathogenic mechanisms relevant to PD. This study can be pursued with classic cellular and molecular assays or by novel, unbiased, high-throughput techniques such as next generation sequencing or transcriptomic profiling. High-throughput preclinical drug screening can be performed on patient cells in order to discover potential therapeutic agents able to reverse the disease phenotypes.

There are currently no disease-modifying drugs available for PD. The long process of drug discovery requires reliable *in vitro* platforms for rapid screening of candidate compounds before starting preclinical and clinical testing and iPSC-derived neurons could be the answer for both target-directed drug screening and phenotypic screening (Fig. 2).

9 Conclusions

Human iPSCs derived from patients represent a highly valuable tool to help us unveil the molecular mechanisms underlying PD. Although there are still some challenges around the use of iPSCs in PD research, new co-culture and organoid methods may even increase the potential of this unique model to study neurodegenerative diseases. iPSC-derived models have the major benefit of allowing generation of the very same neurons that show vulnerability to the progression of PD, which has made them the most physiological *in vitro* model available. These models have already shown their value for the investigation of the pathogenic mechanisms active in the early phases of the disease and have revealed potential new therapeutic targets. Human iPSC-models have an enormous potential and might increase the translatability of basic research in PD and neurodegenerative diseases as a whole.

Acknowledgments

Work in the Wade-Martins laboratory is funded by the Monument Trust Discovery Award from Parkinson's UK (J-1403) and by the Michael J Fox Foundation. M.C.C. holds the Joan Pitts-Tucker/Heyman Moritz Studentship.

References

Adil, M.M., Rodrigues, G.M.C., Kulkarni, R.U., Rao, A.T., Chernavsky, N.E., Miller, E.W., et al., 2017. Efficient generation of hPSC-derived midbrain dopaminergic neurons in a fully defined, scalable, 3D biomaterial platform. Sci. Rep. 7, 40573. Nature Publishing Group.

Beevers, J.E., Lai, M.C., Collins, E., Booth, H.D.E., Zambon, F., Parkkinen, L., et al., 2017. MAPT genetic variation and neuronal maturity alter isoform expression affecting axonal transport in iPSC-derived dopamine neurons. Stem Cell Rep. 9 (2), 587–599. Elsevier.

Bellucci, A., Zaltieri, M., Navarria, L., Grigoletto, J., Missale, C., Spano, P., 2012. From α-synuclein to synaptic dysfunctions: new insights into the pathophysiology of Parkinson's disease. Brain Res. 1476, 183–202. Elsevier.

Billingsley, K.J., Bandres-Ciga, S., Saez-Atienzar, S., Singleton, A.B., 2018. Genetic risk factors in Parkinson's disease. Cell Tissue Res. 373 (1), 9–20. Springer.

Booth, H.D.E., Hirst, W.D., Wade-Martins, R., 2017. The role of astrocyte dysfunction in Parkinson's disease pathogenesis. Trends Neurosci. 40 (6), 358–370. Elsevier.

Booth, H.D.E., Wessely, F., Connor-Robson, N., Rinaldi, F., Vowles, J., Browne, C., et al., 2019. RNA sequencing reveals MMP2 and TGFB1 downregulation in LRRK2 G2019S Parkinson's iPSC-derived astrocytes. Neurobiol. Dis. 129, 56–66. Elsevier.

Brini, M., Ottolini, D., Calì, T., Carafoli, E., 2013. Calcium in health and disease. In: Sigel, A., Helmut, R.K. (Eds.), Interrelations Between Essential Metal Ions and Human Diseases, Metal Ions in Life Sciences. Springer, pp. 81–137.

Byers, B., Cord, B., Nguyen, H.N., Schüle, B., Fenno, L., Lee, P.C., et al., 2011. SNCA triplication Parkinson's patient's iPSC-derived DA neurons accumulate α-synuclein and are susceptible to oxidative stress. PLoS One 6 (11), e26159. Public Library of Science.

Caiazzo, M., Dell'Anno, M.T., Dvoretskova, E., Lazarevic, D., Taverna, S., Leo, D., et al., 2011. Direct generation of functional dopaminergic neurons from mouse and human fibroblasts. Nature 476 (7359), 224. Nature Publishing Group.

Chambers, S.M., Fasano, C.A., Papapetrou, E.P., Tomishima, M., Sadelain, M., Studer, L., 2009. Highly efficient neural conversion of human ES and iPS cells by dual inhibition of SMAD signaling. Nat. Biotechnol. 27 (3), 275. Nature Publishing Group.

Chung, C.Y., Khurana, V., Auluck, P.K., Tardiff, D.F., Mazzulli, J.R., Soldner, F., et al., 2013. Identification and rescue of α-synuclein toxicity in Parkinson patient-derived neurons. Science 342 (6161), 983–987. American Association for the Advancement of Science.

Chung, S.Y., Kishinevsky, S., Mazzulli, J.R., Graziotto, J., Mrejeru, A., Mosharov, E.V., et al., 2016. Parkin and PINK1 patient iPSC-derived midbrain dopamine neurons exhibit mitochondrial dysfunction and α-synuclein accumulation. Stem Cell Rep. 7 (4), 664–677. Elsevier.

Connor-Robson, N., Booth, H., Martin, J.G., Gao, B., Li, K., Doig, N., et al., 2019. An integrated transcriptomics and proteomics analysis reveals functional endocytic dysregulation caused by mutations in LRRK2. Neurobiol. Dis. 127, 512–526. Elsevier.

Cooper, O., Seo, H., Andrabi, S., Guardia-Laguarta, C., Graziotto, J., Sundberg, M., et al., 2012. Pharmacological rescue of mitochondrial deficits in iPSC-derived neural cells from patients with familial Parkinson's disease. Sci. Transl. Med. 4 (141), 141ra90. American Association for the Advancement of Science.

Dauer, W., Przedborski, S., 2003. Parkinson's disease: mechanisms and models. Neuron 39 (6), 889–909. Elsevier.

De Lau, L.M.L., Breteler, M.M.B., 2006. Epidemiology of Parkinson's disease. Lancet Neurol. 5 (6), 525–535. Elsevier.

di Domenico, A., Carola, G., Calatayud, C., Pons-Espinal, M., Muñoz, J.P., Richaud-Patin, Y., et al., 2019. Patient-specific iPSC-derived astrocytes contribute to non-cell-autonomous neurodegeneration in Parkinson's disease. Stem Cell Rep. 12 (2), 213–229. Elsevier.

Fernandes, H.J.R., Hartfield, E.M., Christian, H.C., Emmanoulidou, E., Zheng, Y., Booth, H., et al., 2016. ER stress and autophagic perturbations lead to elevated extracellular α-synuclein in GBA-N370S Parkinson's iPSC-derived dopamine neurons. Stem Cell Rep. 6 (3), 342–356. Elsevier.

Foehring, R.C., Zhang, X.F., Lee, J.C.F., Callaway, J.C., 2009. Endogenous calcium buffering capacity of substantia nigral dopamine neurons. J. Neurophysiol. 102 (4), 2326–2333. American Physiological Society Bethesda, MD.

Freed, C.R., Greene, P.E., Breeze, R.E., Tsai, W.-Y., DuMouchel, W., Kao, R., et al., 2001. Transplantation of embryonic dopamine neurons for severe Parkinson's disease. N. Engl. J. Med. 344 (10), 710–719. Mass Medical Soc.

Fusaki, N., Ban, H., Nishiyama, A., Saeki, K., Hasegawa, M., 2009. Efficient induction of transgene-free human pluripotent stem cells using a vector based on Sendai virus, an RNA virus that does not integrate into the host genome. Proc. Jpn. Acad. Ser. B 85 (8), 348–362. The Japan Academy.

Gallegos, S., Pacheco, C., Peters, C., Opazo, C.M., Aguayo, L.G., 2015. Features of alpha-synuclein that could explain the progression and irreversibility of Parkinson's disease. Front. Neurosci. 9, 59. Frontiers.

Guzman, J.N., Sánchez-Padilla, J., Chan, C.S., Surmeier, D.J., 2009. Robust pacemaking in substantia nigra dopaminergic neurons. J. Neurosci. 29 (35), 11011–11019. Soc Neuroscience.

Guzman, J.N., Sanchez-Padilla, J., Wokosin, D., Kondapalli, J., Ilijic, E., Schumacker, P.T., et al., 2010. Oxidant stress evoked by pacemaking in dopaminergic neurons is attenuated by DJ-1. Nature 468 (7324), 696. Nature Publishing Group.

Hou, P., Li, Y., Zhang, X., Liu, C., Guan, J., Li, H., et al., 2013. Pluripotent stem cells induced from mouse somatic cells by small-molecule compounds. Science 341 (6146), 651–654. American Association for the Advancement of Science.

Hsieh, C.-H., Shaltouki, A., Gonzalez, A.E., da Cruz, A.B., Burbulla, L.F., Lawrence, E.S., et al., 2016. Functional impairment in miro degradation and mitophagy is a shared feature in familial and sporadic Parkinson's disease. Cell Stem Cell 19 (6), 709–724. Elsevier.

Hunn, B.H.M., Cragg, S.J., Bolam, J.P., Spillantini, M.-G., Wade-Martins, R., 2015. Impaired intracellular trafficking defines early Parkinson's disease. Trends Neurosci. 38 (3), 178–188. Elsevier.

International Parkinson Disease Genomics Consortium, 2011. Imputation of sequence variants for identification of genetic risks for Parkinson's disease: a meta-analysis of genome-wide association studies. Lancet 377 (9766), 641–649. Elsevier.

Iovino, M., Agathou, S., González-Rueda, A., Del Castillo Velasco-Herrera, M., Borroni, B., Alberici, A., et al., 2015. Early maturation and distinct tau pathology in induced pluripotent stem cell-derived neurons from patients with MAPT mutations. Brain 138 (11), 3345–3359. Oxford University Press.

Isacson, O., Deacon, T., 1997. Neural transplantation studies reveal the brain's capacity for continuous reconstruction. Trends Neurosci. 20 (10), 477–482. Elsevier.

Jankovic, J., 2008. Parkinson's disease: clinical features and diagnosis. J. Neurol. Neurosurg. Psychiatry 79 (4), 368–376. BMJ Publishing Group Ltd.

Jiang, H., Xu, Z., Zhong, P., Ren, Y., Liang, G., Schilling, H.A., et al., 2015. Cell cycle and p53 gate the direct conversion of human fibroblasts to dopaminergic neurons. Nat. Commun. 6, 10100. Nature Publishing Group.

Jo, J., Xiao, Y., Sun, A.X., Cukuroglu, E., Tran, H.-D., Göke, J., et al., 2016. Midbrain-like organoids from human pluripotent stem cells contain functional dopaminergic and neuromelanin-producing neurons. Cell Stem Cell 19 (2), 248–257. Elsevier.

Kaji, K., Norrby, K., Paca, A., Mileikovsky, M., Mohseni, P., Woltjen, K., 2009. Virus-free induction of pluripotency and subsequent excision of reprogramming factors. Nature 458 (7239), 771. Nature Publishing Group.

Kim, D., Kim, C.-H., Moon, J.-I., Chung, Y.-G., Chang, M.-Y., Han, B.-S., et al., 2009. Generation of human induced pluripotent stem cells by direct delivery of reprogramming proteins. Cell Stem Cell 4 (6), 472. NIH Public Access.

Kim, Y., Zheng, X., Ansari, Z., Bunnell, M.C., Herdy, J.R., Traxler, L., et al., 2018. Mitochondrial aging defects emerge in directly reprogrammed human neurons due to their metabolic profile. Cell Rep. 23 (9), 2550–2558. Elsevier.

Kitada, T., Asakawa, S., Hattori, N., Matsumine, H., Yamamura, Y., Minoshima, S., et al., 1998. Mutations in the parkin gene cause autosomal recessive juvenile parkinsonism. Nature 392 (6676), 605. Nature Publishing Group.

Korecka, J.A., Talbot, S., Osborn, T.M., de Leeuw, S.M., Levy, S.A., Ferrari, E.J., et al., 2019. Neurite collapse and altered ER Ca2+ control in human parkinson disease patient iPSC-derived neurons with LRRK2 G2019S mutation. Stem Cell Rep. 12 (1), 29–41. Elsevier.

Kriks, S., Shim, J.-W., Piao, J., Ganat, Y.M., Wakeman, D.R., Xie, Z., et al., 2011. Dopamine neurons derived from human ES cells efficiently engraft in animal models of Parkinson's disease. Nature 480 (7378), 547. Nature Publishing Group.

Lang, C., Campbell, K.R., Ryan, B.J., Carling, P., Attar, M., Vowles, J., et al., 2019. Single-cell sequencing of iPSC-dopamine neurons reconstructs disease progression and identifies HDAC4 as a regulator of Parkinson cell phenotypes. Cell Stem Cell 24 (1), 93–106. Elsevier.

Lassus, B., Naudé, J., Faure, P., Guedin, D., Von Boxberg, Y., La Cour, C.M., et al., 2018. Glutamatergic and dopaminergic modulation of cortico-striatal circuits probed by dynamic calcium imaging of networks reconstructed in microfluidic chips. Sci. Rep. 8 (1), 17461. Nature Publishing Group.

Lindvall, O., Rehncrona, S., Brundin, P., Gustavii, B., Åstedt, B., Widner, H., et al., 1989. Human fetal dopamine neurons grafted into the striatum in two patients with severe

Parkinson's disease: a detailed account of methodology and a 6-month follow-up. Arch. Neurol. 46 (6), 615–631. American Medical Association.

Lindvall, O., Brundin, P., Widner, H., Rehncrona, S., Gustavii, B., Frackowiak, R., et al., 1990. Grafts of fetal dopamine neurons survive and improve motor function in Parkinson's disease. Science 247 (4942), 574–577. American Association for the Advancement of Science.

Ludtmann, M.H.R., Angelova, P.R., Horrocks, M.H., Choi, M.L., Rodrigues, M., Baev, A.Y., et al., 2018. α-Synuclein oligomers interact with ATP synthase and open the permeability transition pore in Parkinson's disease. Nat. Commun. 9 (1), 2293. Nature Publishing Group.

Marión, R.M., Blasco, M.A., 2010. Telomere rejuvenation during nuclear reprogramming. Curr. Opin. Genet. Dev. 20 (2), 190–196. Elsevier.

Mazzulli, J.R., Xu, Y.-H., Sun, Y., Knight, A.L., McLean, P.J., Caldwell, G.A., et al., 2011. Gaucher disease glucocerebrosidase and α-synuclein form a bidirectional pathogenic loop in synucleinopathies. Cell 146 (1), 37–52. Elsevier.

Mazzulli, J.R., Zunke, F., Isacson, O., Studer, L., Krainc, D., 2016. α-Synuclein–induced lysosomal dysfunction occurs through disruptions in protein trafficking in human midbrain synucleinopathy models. Proc. Natl. Acad. Sci. U. S. A. 113 (7), 1931–1936. National Acad Sciences.

Monzel, A.S., Smits, L.M., Hemmer, K., Hachi, S., Moreno, E.L., van Wuellen, T., et al., 2017. Derivation of human midbrain-specific organoids from neuroepithelial stem cells. Stem Cell Rep. 8 (5), 1144–1154. Elsevier.

Moreno, E.L., Hachi, S., Hemmer, K., Trietsch, S.J., Baumuratov, A.S., Hankemeier, T., et al., 2015. Differentiation of neuroepithelial stem cells into functional dopaminergic neurons in 3D microfluidic cell culture. Lab Chip 15 (11), 2419–2428. Royal Society of Chemistry.

Nguyen, H.N., Byers, B., Cord, B., Shcheglovitov, A., Byrne, J., Gujar, P., et al., 2011. LRRK2 mutant iPSC-derived DA neurons demonstrate increased susceptibility to oxidative stress. Cell Stem Cell 8 (3), 267–280. Elsevier.

Okita, K., Nakagawa, M., Hyenjong, H., Ichisaka, T., Yamanaka, S., 2008. Generation of mouse induced pluripotent stem cells without viral vectors. Science 322 (5903), 949–953. American Association for the Advancement of Science.

Ono, Y., Nakatani, T., Sakamoto, Y., Mizuhara, E., Minaki, Y., Kumai, M., et al., 2007. Differences in neurogenic potential in floor plate cells along an anteroposterior location: midbrain dopaminergic neurons originate from mesencephalic floor plate cells. Development 134 (17), 3213–3225. The Company of Biologists Ltd.

Parkinson, J., 2002. An essay on the shaking palsy. J. Neuropsychiatry Clin. Neurosci. 14 (2), 223–236. Am Neuropsych Assoc.

Pfisterer, U., Kirkeby, A., Torper, O., Wood, J., Nelander, J., Dufour, A., et al., 2011. Direct conversion of human fibroblasts to dopaminergic neurons. Proc. Natl. Acad. Sci. U. S. A. 108 (25), 10343–10348. National Acad Sciences.

Piccini, P., Brooks, D.J., Björklund, A., Gunn, R.N., Grasby, P.M., Rimoldi, O., et al., 1999. Dopamine release from nigral transplants visualized in vivo in a Parkinson's patient. Nat. Neurosci. 2 (12), 1137. Nature Publishing Group.

Rouhani, F., Kumasaka, N., de Brito, M.C., Bradley, A., Vallier, L., Gaffney, D., 2014. Genetic background drives transcriptional variation in human induced pluripotent stem cells. PLoS Genet, 10 (6), e1004432. Public Library of Science.

Ryan, S.D., Dolatabadi, N., Chan, S.F., Zhang, X., Akhtar, M.W., Parker, J., et al., 2013. Isogenic human iPSC Parkinson's model shows nitrosative stress-induced dysfunction in MEF2-PGC1α transcription. Cell 155 (6), 1351–1364. Elsevier.

Ryan, B.J., Hoek, S., Fon, E.A., Wade-Martins, R., 2015. Mitochondrial dysfunction and mitophagy in Parkinson's: from familial to sporadic disease. Trends Biochem. Sci. 40 (4), 200–210. Elsevier.

Sandor, C., Robertson, P., Lang, C., Heger, A., Booth, H., Vowles, J., et al., 2017. Transcriptomic profiling of purified patient-derived dopamine neurons identifies convergent perturbations and therapeutics for Parkinson's disease. Hum. Mol. Genet. 26 (3), 552–566. Oxford University Press.

Scheper, W., Hoozemans, J.J.M., 2015. The unfolded protein response in neurodegenerative diseases: a neuropathological perspective. Acta Neuropathol. 130 (3), 315–331. Springer.

Schöndorf, D.C., Aureli, M., McAllister, F.E., Hindley, C.J., Mayer, F., Schmid, B., et al., 2014. iPSC-derived neurons from GBA1-associated Parkinson's disease patients show autophagic defects and impaired calcium homeostasis. Nat. Commun. 5, 4028. Nature Publishing Group.

Schöndorf, D.C., Ivanyuk, D., Baden, P., Sanchez-Martinez, A., De Cicco, S., Yu, C., et al., 2018. The NAD+ precursor nicotinamide riboside rescues mitochondrial defects and neuronal loss in iPSC and fly models of Parkinson's disease. Cell Rep. 23 (10), 2976–2988. Elsevier.

Shaltouki, A., Hsieh, C.-H., Kim, M.J., Wang, X., 2018. Alpha-synuclein delays mitophagy and targeting Miro rescues neuron loss in Parkinson's models. Acta Neuropathol. 136 (4), 607–620. Springer.

Sidransky, E., Lopez, G., 2012. The link between the GBA gene and Parkinsonism. Lancet Neurol. 11 (11), 986–998. Elsevier.

Smith, M.H., Ploegh, H.L., Weissman, J.S., 2011. Road to ruin: targeting proteins for degradation in the endoplasmic reticulum. Science 334 (6059), 1086–1090. American Association for the Advancement of Science.

Smits, L.M., Reinhardt, L., Reinhardt, P., Glatza, M., Monzel, A.S., Stanslowsky, N., et al., 2019. Modeling Parkinson's disease in midbrain-like organoids. NPJ Parkinsons Dis. 5 (1), 5. Nature Publishing Group.

Sofroniew, M.V., Vinters, H.V., 2010. Astrocytes: biology and pathology. Acta Neuropathol. 119 (1), 7–35. Springer.

Spillantini, M.G., Schmidt, M.L., Lee, V.M.-Y., Trojanowski, J.Q., Jakes, R., Goedert, M., 1997. α-Synuclein in Lewy bodies. Nature 388 (6645), 839. Nature Publishing Group.

Sposito, T., Preza, E., Mahoney, C.J., Setó-Salvia, N., Ryan, N.S., Morris, H.R., et al., 2015. Developmental regulation of tau splicing is disrupted in stem cell-derived neurons from frontotemporal dementia patients with the 10 + 16 splice-site mutation in MAPT. Hum. Mol. Genet. 24 (18), 5260–5269. Oxford University Press.

Stadtfeld, M., Nagaya, M., Utikal, J., Weir, G., Hochedlinger, K., 2008. Induced pluripotent stem cells generated without viral integration. Science 322 (5903), 945–949.

Suzuki, S., Akamatsu, W., Kisa, F., Sone, T., Ishikawa, K., Kuzumaki, N., et al., 2017. Efficient induction of dopaminergic neuron differentiation from induced pluripotent stem cells reveals impaired mitophagy in PARK2 neurons. Biochem. Biophys. Res. Commun. 483 (1), 88–93. Elsevier.

Tabata, Y., Imaizumi, Y., Sugawara, M., Andoh-Noda, T., Banno, S., Chai, M., et al., 2018. T-type calcium channels determine the vulnerability of dopaminergic neurons to mitochondrial stress in familial Parkinson disease. Stem Cell Rep. 11 (5), 1171–1184. Elsevier.

Takahashi, K., Yamanaka, S., 2006. Induction of pluripotent stem cells from mouse embryonic and adult fibroblast cultures by defined factors. Cell 126 (4), 663–676.

Takahashi, K., Tanabe, K., Ohnuki, M., Narita, M., Ichisaka, T., Tomoda, K., et al., 2007. Induction of pluripotent stem cells from adult human fibroblasts by defined factors. Cell 131 (5), 861–872. Elsevier.

Tropepe, V., Sibilia, M., Ciruna, B.G., Rossant, J., Wagner, E.F., van der Kooy, D., 1999. Distinct neural stem cells proliferate in response to EGF and FGF in the developing mouse telencephalon. Dev. Biol. 208 (1), 166–188. Elsevier.

Valente, E.M., Abou-Sleiman, P.M., Caputo, V., Muqit, M.M.K., Harvey, K., Gispert, S., et al., 2004. Hereditary early-onset Parkinson's disease caused by mutations in PINK1. Science 304 (5674), 1158–1160. American Association for the Advancement of Science.

Vierbuchen, T., Ostermeier, A., Pang, Z.P., Kokubu, Y., Südhof, T.C., Wernig, M., 2010. Direct conversion of fibroblasts to functional neurons by defined factors. Nature 463 (7284), 1035. Nature Publishing Group.

Walter, P., Ron, D., 2011. The unfolded protein response: from stress pathway to homeostatic regulation. Science 334 (6059), 1081–1086. American Association for the Advancement of Science.

Wang, X., Huang, T., Bu, G., Xu, H., 2014. Dysregulation of protein trafficking in neurodegeneration. Mol. Neurodegener. 9 (1), 31. BioMed Central.

Yan, Y., Yang, D., Zarnowska, E.D., Du, Z., Werbel, B., Valliere, C., et al., 2005. Directed differentiation of dopaminergic neuronal subtypes from human embryonic stem cells. Stem Cells 23 (6), 781–790. Wiley Online Library.

Zambon, F., Cherubini, M., Fernandes, H.J.R., Lang, C., Ryan, B.J., Volpato, V., et al., 2019. Cellular α-synuclein pathology is associated with bioenergetic dysfunction in Parkinson's iPSC-derived dopamine neurons. Hum. Mol. Genet. 28 (12), 2001–2013.

Zhou, W., Freed, C.R., 2009. Adenoviral gene delivery can reprogram human fibroblasts to induced pluripotent stem cells. Stem Cells 27 (11), 2667–2674. Wiley Online Library.

Zucker, R.S., 1999. Calcium-and activity-dependent synaptic plasticity. Curr. Opin. Neurobiol. 9 (3), 305–313. Elsevier.

Animal models for preclinical Parkinson's research: An update and critical appraisal

M. Angela Cenci[*], **Anders Björklund**

Department of Experimental Medical Science, Wallenberg Neuroscience Centre, Lund University, Lund, Sweden

[*]*Corresponding author: Tel.: +46-72-5007879, e-mail address: angela.cenci_nilsson@med.lu.se*

Abstract

Animal models of Parkinson's disease (PD) are essential to investigate pathogenic pathways at the whole-organism level. Moreover, they are necessary for a preclinical investigation of potential new therapies. Different pathological features of PD can be induced in a variety of invertebrate and vertebrate species using toxins, drugs, or genetic perturbations. Each model has a particular utility and range of applicability. Invertebrate PD models are particularly useful for high throughput-screening applications, whereas mammalian models are needed to explore complex motor and non-motor features of the human disease. Here, we provide a comprehensive review and critical appraisal of the most commonly used mammalian models of PD, which are produced in rats and mice. A substantial loss of nigrostriatal dopamine neurons is necessary for the animal to exhibit a hypokinetic motor phenotype responsive to dopaminergic agents, thus resembling clinical PD. This level of dopaminergic neurodegeneration can be induced using specific neurotoxins, environmental toxicants, or proteasome inhibitors. Alternatively, nigrostriatal dopamine degeneration can be induced via overexpression of α-synuclein using viral vectors or transgenic techniques. In addition, protein aggregation pathology can be triggered by inoculating preformed fibrils of α-synuclein in the substantia nigra or the striatum. Thanks to the conceptual and technical progress made in the past few years a vast repertoire of well-characterized animal models are currently available to address different aspects of PD in the laboratory.

Keywords

Movement disorders, Basic science, Pathophysiology, Neurodegeneration, Neuroinflammation

1 Introduction

Research on animal models of neurological disease is often questioned on ethical grounds (LaFollette and Shanks, 1996), conceptual grounds (Diederich et al., 2019; Drummond and Wisniewski, 2017; Gomez-Marin and Ghazanfar, 2019), or pragmatic grounds. On a pragmatic level, it is sometimes argued that studies in animal models are unable to predict the outcome of new treatments for human disease (Pound and Ritskes-Hoitinga, 2018), although it is often pointed out that clinical trial failure does not necessarily depend on the intrinsic limitations of animal models (Bespalov et al., 2016; van der Worp et al., 2010). In addition to these general considerations, the recent availability of disease models in patient-derived cells (Caiazza et al., 2020) is stimulating a debate around the necessity of animal research, at least for certain applications.

In spite of the ongoing debate, the past few years have witnessed a remarkable progress in developing and characterizing animal models of PD. This progress has been inspired by an increased understanding of the complex etiopathogenesis and multisystem pathology of the human disease. Thanks to both conceptual and technical advances, we now have unprecedented opportunities to recreate key pathological aspects of PD in laboratory animals. In this review we have sought to provide an up-to-date overview of the main rodent PD models available today, appraising both their advantages and their limitations. The wide range of PD models now available offers new opportunities, but it is at the same time a challenge for the researcher to select the most suitable model for the questions under study. Our aim is to offer both a critical reflection and an updated resource that can inform on the use of suitable models for different research applications.

2 Models in different species

Comparing results across animal species serves as a powerful approach to promote scientific rigor and to discover biological principles of universal validity (Yartsev, 2017). The possibility to model PD in a multitude of species should thus be regarded as an asset to research, ultimately leading to a better understanding of the human disease.

Using either toxins or genetic perturbations, PD-like conditions can be induced in invertebrate organisms, the most common ones being the fruit fly *Drosophila Melanogaster* (Guo, 2010) and the nematode *Caenorhabditis Elegans* (Maulik et al., 2017). These models are particularly useful for high-throughput genetic analyses (such as, experimental mutagenesis to identify genetic modifiers of α-synuclein pathology or toxicant exposure). When more advanced behavioral or functional analyses are needed, investigators usually prefer to produce PD models in vertebrate species, the most common being either small fishes (Matsui and Takahashi, 2018) or rodents. The particular strength of small fish models (such as zebrafish) is their amenability to high-throughput in vivo drug screening studies (Flinn et al., 2008), which can be aided by new automated methods of phenotypic analysis (Palmer et al., 2017). On the other hand, rodents show a significant degree of human homology regarding the organization of cortico-basal ganglia-thalamocortical loops (Reiner et al., 1998) and their corresponding behavioral functions (Redgrave et al., 2010). Moreover, rodents can produce complex movements homologous to those in humans (Sacrey et al., 2009) and they exhibit functionally similar motor deficits after nigrostriatal dopamine (DA) lesions, as well as analogous motor responses to dopamine (DA) replacement therapy (Cenci et al., 2002). Being less expensive than non-human primates (NHPs) and ethically less problematic to use, rodents continue to provide the most widely used models in PD research, particularly for studies that require an analysis of brain functions including movement, cognition, sleep, affective behaviors. In addition, rodent models are used increasingly often in studies addressing the functionality of peripheral organs (in particular, bladder, heart, gastrointestinal tract) in the setting of experimental parkinsonism or synucleinopathy. Because of the above reasons, most of the literature review provided in this chapter is based on rodent studies.

Models in NHP, particularly those in macaque monkeys, offer the specific advantage of a striking similarity to humans regarding the phenomenology of different movement disorders (Cenci and Crossman, 2018; Johnston and Fox, 2015). This makes it possible to quantify parkinsonian and dyskinetic features in the animals using similar principles to those used in patients, streamlining the translational path from the lab to the clinic (Fox and Brotchie, 2019). Moreover, the larger brain and body size of macaque monkeys conceivably facilitates the experimental evaluation of therapeutic interventions requiring surgery, such as those needed to infuse trophic factors and implant cells or stimulation devices. The main disadvantages of NHP models are a high cost and the necessity of highly specialized housing facilities. For these reasons, NHP models are currently used only in few research centers worldwide.

3 The importance of nigrostriatal dopaminergic degeneration

Although PD is clinically and pathologically heterogeneous (Berg et al., 2014; Erro et al., 2016), a severe loss of putaminal dopaminergic innervation is a necessary prerequisite for the appearance of motor symptoms that lead to clinical diagnosis. Parkinsonian motor features become manifest when more than 50% of putaminal DA contents are lost (Fearnley and Lees, 1991), and a rapid loss of the residual putaminal DA input appears to occur during the first 5 years following clinical diagnosis (Kordower et al., 2013). Accordingly, in both rodent and macaque models of PD, motor deficits start to become manifest when striatal motor regions have lost more than 50% of their dopaminergic input (Boix et al., 2018; Decressac et al., 2012b), and a full-blown parkinsonian-like syndrome appears only after removing more than 80% of putaminal dopaminergic fibers (Francardo et al., 2011; Guigoni et al., 2005; Winkler et al., 2002). Therefore, reports of hypokinetic features in animals that exhibit only a modest degree of DA cell loss and/or mild deficits in striatal DA contents should raise suspicion of a systemic disease or pervasive neurological intoxication depending on the model at hand (in both instances, the animal would move less). To ascertain the parkinsonian character of motor features observed in the animal model, it is recommended to evaluate the effects of L-DOPA (Cenci et al., 2002; Xu et al., 2012). Indeed, treatment with L-DOPA improves gross hypokinetic deficits (Francardo et al., 2011; Lundblad et al., 2002), although it may not improve tasks requiring a high degree of motor precision (Metz and Whishaw, 2002; Winkler et al., 2002). If the denervation of striatal motor regions exceeds 90%, the majority of animals treated with therapeutic-like doses of L-DOPA will develop abnormal involuntary movements analogous to L-DOPA-induced dyskinesia (LID) (Francardo et al., 2011; Winkler et al., 2002). A similar relationship between degree of putaminal DA denervation and incidence of LID has been reported in macaque models of PD (Schneider, 1989).

The crucial importance of striatal DA depletion to the appearance of PD-relevant motor deficits explains the continuing interest in developing experimental approaches to selectively damage dopaminergic neurons. As reviewed below, mitochondrial and oxidant toxins have been in use for many years. Additional and more recent methods involve an intracerebral delivery of proteasome inhibitors. Moreover, efficient approaches have been developed to induce α-synuclein pathology using viral vectors, inoculation of α-synuclein fibrils, or transgenic technologies. A graphic summary of these different approaches is presented in Fig. 1.

4 6-Hydroxydopamine

The first toxin-based animal model of PD consisted of rats sustaining intracerebral injections of 6-hydroxydopamine (6-OHDA) (Ungerstedt, 1968). This chemical is a hydroxylated analog of DA that also occurs in the brain (Jellinger et al., 1995). 6-OHDA is a catecholamine-selective neurotoxin because it enters neurons via the dopamine or noradrenaline transporter. Once inside the neuron, 6-OHDA undergoes auto-oxidation and conversion to reactive oxygen species (ROS) (Rotman and Creveling, 1976). Neurons rapidly die because of oxidative damage to cellular constituents and mitochondrial dysfunction (Kupsch et al., 2014), and there is wide consensus that such mechanisms are relevant to the pathogenesis of the human disease (Grunewald et al., 2019). Moreover, the degeneration of DA cell bodies and axon terminals triggers proinflammatory glial reactions that contribute to the neurodegenerative process [reviewed in (Kuter et al., 2020)], and this mechanism is also relevant to the pathogenesis of PD.

6-OHDA does not cross the blood-brain barrier (BBB) and therefore necessitates a direct delivery to the nigrostriatal system, which leads to dopaminergic degeneration in all animal species. For the sake of producing PD models, the three most common injection targets are the substantia nigra, the medial forebrain bundle (MFB), and the striatum. Injection of 6-OHDA into the MFB is the preferred

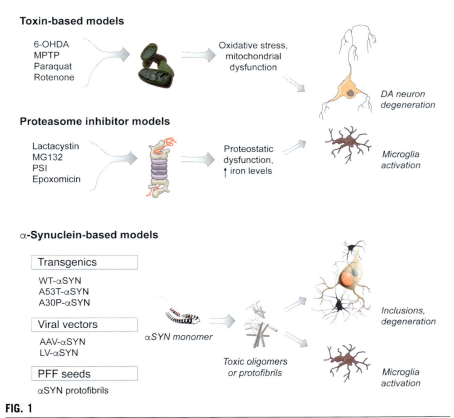

Toxin-based models

6-OHDA
MPTP
Paraquat
Rotenone

Oxidative stress,
mitochondrial
dysfunction

DA neuron
degeneration

Proteasome inhibitor models

Lactacystin
MG132
PSI
Epoxomicin

Proteostatic
dysfunction,
↑ iron levels

Microglia
activation

α-Synuclein-based models

Transgenics

WT-αSYN
A53T-αSYN
A30P-αSYN

Viral vectors

AAV-αSYN
LV-αSYN

PFF seeds

αSYN protofibrils

αSYN monomer

Toxic oligomers
or protofibrils

Inclusions,
degeneration

Microglia
activation

FIG. 1

Overview of the main methods currently used to obtain animal models of PD exhibiting degeneration of nigrostriatal DA neurons.

procedure to obtain a model of severe and reproducible dopaminergic degeneration with negligible tendency for animals to spontaneously compensate even at very long survival times [reviewed in (Francardo et al., 2017)]. Thanks to the predictability, stability, and severity of the DA lesion, this model is particularly useful for studies evaluating the effects of long-term pharmacological treatments or neural transplants. On the other hand, models based on intrastriatal 6-OHDA delivery afford a remarkable flexibility in modulating the severity and regional distribution of DA denervation by varying toxin dose and injection coordinates (Francardo et al., 2011; Winkler et al., 2002). Intrastriatal 6-OHDA models have proven particularly useful to study the effects of neuroprotective and neurorestorative treatments (Bjorklund et al., 1997; Francardo et al., 2017).

A lot has been learned by studying 6-OHDA lesion models of PD. Beside the elucidation of many potential treatment principles, including circuit restoration (Thompson and Bjorklund, 2012) and neuroprotection (Francardo et al., 2017), research carried out on these models has elucidated questions of fundamental importance, such as the relationship between nigrostriatal damage and motor dysfunction (Decressac et al., 2012b; Kirik et al., 1998; Winkler et al., 2002), the postsynaptic consequences of DA denervation (Cenci and Konradi, 2010; Kostrzewa, 1995; Simola et al., 2007), and compensatory responses to dopaminergic damage (Lee et al., 2008; Zigmond, 1997). Moreover, rats with 6-OHDA lesions still provide the best validated rodent model to study L-DOPA-induced dyskinesia in the laboratory (Cenci and Crossman, 2018). Like any other approach targeting the nigrostriatal dopaminergic pathway, 6-OHDA lesions do not mimic the multisystem pathology of PD. However, it should be noted that PD-relevant pathological features are usually found also in non-dopaminergic neuronal systems, including serotonergic and noradrenergic projections and striatal neuron dendrites [reviewed in (Cenci, 2014, Fieblinger and Cenci, 2015)]. Moreover, it is technically possible to combine the injection of 6-OHDA with other genetic or chemical lesions in the same animal. Notwithstanding these possibilities, 6-OHDA-based models do not mimic two characterizing features of nigrostriatal neurodegeneration in PD, that is, the progressive time course and the formation of intracellular α-synuclein aggregates.

5 1-Methyl-4-phenyl-1,2,3,6-tetrahydropyridine (MPTP)

MPTP was discovered as a contaminant of synthetic heroin following the report of a severe parkinsonian syndrome developing in some users of illicit drugs in California (Langston et al., 1983). MPTP is a lipophilic compound that can cross the BBB. Once in the brain, it is metabolized by monoamine oxidase B (MAO-B) to the potent dopaminergic neurotoxin 1-methyl-4-phenylpyridinium ion (MPP^+) (Chiba et al., 1984), which is a structural analog of DA and can therefore be taken up by dopaminergic neurons via the DA transporter (DAT). After entering the neuron, MPP^+ becomes highly concentrated in the mitochondria and inhibits complex 1 of the electron transport chain (Ramsay et al., 1986), causing both inhibition of mitochondrial respiration and ROS accumulation. MPTP has been used to induce dopaminergic degeneration in both invertebrate and vertebrate species, the latter including non-human primates (Johnston and Fox, 2015), minipigs (Nielsen et al., 2016), and mice (Meredith and Rademacher, 2011). Rats are, however, resistant to MPTP toxicity (Sundstrom and Samuelsson, 1997), which is partly due to their increased capacity for vesicular sequestration of this toxin (Staal et al., 2000). Several protocols have been established to induce MPTP lesions in mice, consisting of acute, subchronic, or chronic regimens of MPTP intoxication [reviewed in (Meredith and Rademacher, 2011)]. The generally large interest in MPTP models can perhaps be attributed to the fact that the toxin is technically easy to administer (at least compared to toxins requiring intracerebral delivery), and that MPTP has been found to cause parkinsonism in humans.

Collectively, MPTP lesion models have had a remarkable scientific impact because they have been widely used to test hypotheses regarding both pathogenic mechanisms and neuroprotective treatments for PD (Langston, 2017; Przedborski and Vila, 2003). Moreover, MPTP-lesioned monkeys have been essential to identify new symptomatic treatments based on circuit modulation [reviewed in (Wichmann et al., 2018)]. The discovery of MPTP and its dopaminergic neurotoxicity has also spurred a new wave of epidemiological research on the role of environmental toxicants in the etiopathogenesis of PD [reviewed in (Langston, 2017)].

Like 6-OHDA-based models, MPTP-lesioned animals do not reproduce the multisystem pathology of PD nor the formation of intracellular protein aggregates (Johnston and Fox, 2015). Although continuous systemic MPTP infusion with osmotic minipumps has been proposed as progressive PD model featuring α-synuclein inclusions in DA neurons (Fornai et al., 2005), these findings have been difficult to replicate (Alvarez-Fischer et al., 2008). In addition, MPTP models in mice may entail a high mortality, variability in behavioral and biochemical outcomes, and a potential for spontaneous compensation already within few months (Francardo, 2018; Meredith and Rademacher, 2011; Rousselet et al., 2003). This may explain why 6-OHDA is preferred to MPTP for the sake of producing mouse models to evaluate symptomatic and/or antidyskinetic treatments for PD. For these applications, the animal model must exhibit reproducible motor deficits that remain stable under a sufficiently long time.

6 Environmental toxicants

Several epidemiological studies have demonstrated an association between rural residence, pesticide exposure, and an increased risk of PD (Ascherio and Schwarzschild, 2016; Chade et al., 2006). Accordingly, some environmental toxicants present in rural environments have been tested for their capacity to induce nigrostriatal DA degeneration in animals [partly reviewed in (Jiang and Dickson, 2018)]. Among these toxicants, the herbicide paraquat and the pesticide rotenone have now become well-established research tools for both in vitro and in vivo applications.

6.1 Rotenone

Rotenone (a natural extract from plants) is a broad-spectrum insecticide and pesticide. Because of its hydrophobicity, rotenone can easily cross the blood-brain barrier, and once in DA neurons, it inhibits mitochondrial complex I and activates the production of ROS (Cannon and Greenamyre, 2010). Greenamyre and collaborators were the first to develop an animal model of PD based on the continuous administration of rotenone via osmotic minipumps (Betarbet et al., 2000). In part of the animals, rotenone administration induced loss of nigrostriatal DA neurons associated with formation of α-synuclein inclusions and development of hypokinetic-rigid features. Locus coeruleus noradrenergic neurons were mildly affected too (Betarbet et al., 2000). This seminal publication was followed by other studies reporting toxic effects of rotenone on multiple neuronal systems. In particular, Höglinger and colleagues reported that chronic infusion of rotenone causes damage to striatal serotonergic fibers, striatal projection neurons and cholinergic interneurons, pedunculopontine tegmental nucleus and locus coeruleus, concluding that rotenone intoxication is more suitable to model atypical parkinsonian syndromes than PD (Hoglinger et al., 2003). In a similar vein, using different rotenone doses and administration routes in rats, other studies reported lack of correlation between loss of striatal DA innervation and motor deficits, concluding that the motor phenotype induced by rotenone intoxication may depend on some pervasive neurological effects of the toxin (Fleming et al., 2004; Lapointe et al., 2004).

Since this earlier controversy, some successful attempts have been made to increase the reproducibility and specificity of rotenone models by carefully titrating the toxicant dose and using a more lipophilic injection vehicle (Cannon et al., 2009). Moreover, unilateral infusion of rotenone into the medial forebrain bundle was reported to produce progressive dopaminergic degeneration, accompanied by increased expression and aggregation of α-synuclein, in the absence of peripheral toxicity (Ravenstijn et al., 2008).

6.2 Paraquat

The herbicide paraquat has a structure similar to MPP^+. Like MPTP, paraquat can cross the BBB, it is taken up by the DAT, and induces dopaminergic degeneration via oxidative stress and mitochondrial dysfunction (Fei et al., 2008; Powers et al., 2017). Paraquat is usually administered orally or intraperitoneally to rats or mice, and it is most often combined with the fungicide maneb, which has been found to potentiate paraquat toxicity toward nigrostriatal DA neurons (Thiruchelvam et al., 2000). The dose and duration of the treatment have varied between studies, and so has the behavioral-histopathological phenotype of the corresponding animal models. Nevertheless, most studies have shown that paraquat can induce a dose-dependent partial degeneration of nigrostriatal DA neurons, although the effects on striatal dopaminergic fibers and DA levels have been quite variable, possibly due to compensatory mechanisms [for a recent review see (Cenci and Sgambato, 2020)]. Interestingly, low-dose chronic administration of paraquat has been reported to cause upregulation and aggregation of α-synuclein in wild-type mice (Manning-Bog et al., 2002) and to exacerbate markers of α-synuclein aggregation in the enteric nervous system in a transgenic synucleinopathy model (Naudet et al., 2017).

In summary, herbicides and pesticides are very interesting research tools because of their relevance to the environmental components of PD etiopathogenesis, although they induce only partial dopaminergic degeneration and entail a generally high risk of systemic toxicity if used at effective doses. If applied at low doses, herbicides and pesticides can provide a valuable approach to probe the vulnerability of nigrostriatal DA neurons under different conditions, such as aging (Cannon et al., 2009; Thiruchelvam et al., 2003), stress and gut dysbiosis (Dodiya et al., 2020), α-synuclein pathology and neuroinflammation (Ling et al., 2004), or any other factor that may be relevant to the etiopathogenesis of PD.

7 Proteasome inhibitors

Deficits in protein degradation are attributed a key role in the pathogenesis of PD as they are reciprocally linked with the accumulation and misfolding of α-synuclein (Xilouri et al., 2013). Since the early 2000s, there has been an increasing interest in modeling PD by administering proteasome inhibitors to a variety of species (*C. Elegans*, small fishes, mice, rats, minipigs, and non-human primates) (Bentea et al., 2017; Lillethorup et al., 2018). Proteasome inhibitors are drugs originally developed for the treatment of cancer (myeloma, in particular) based on their capacity to induce programmed cell death. Compounds of this class have been administered to rodents using systemic or intracerebral delivery methods. The most successful results have been obtained by injecting potent and irreversible proteasome inhibitors (such as lactacystin) into the substantia nigra or the MFB [reviewed in (Bentea et al., 2017)]. The corresponding rodent models exhibit a rapid, dose-dependent degeneration of nigral DA neurons (McNaught et al., 2002; Xie et al., 2010) that can be associated with L-DOPA-responsive motor deficits (Konieczny et al., 2014). The mechanisms underlying neurodegeneration in these models include apoptotic cell death, mitochondrial dysfunction, iron dysregulation, oxidative and nitrosative stress [reviewed in (Bentea et al., 2017; Le, 2014)].

The doses of lactacystin needed to obtain ≥50% loss of nigrostriatal DA neurons produce various degrees of extranigral pathology. Interestingly, in rats sustaining intranigral injections of lactacystin, substantial neurodegeneration has been detected in the ipsilateral pedunculopontine nucleus (Elson et al., 2016), causing >60% loss of cholinergic neurons and somatic hypotrophy of the remaining neurons in this region (Pienaar et al., 2015). This observation is in keeping with the reported high sensitivity of cholinergic neurons to proteasomal inhibition, as demonstrated in a dose-response study of lactacystin delivery to the basal forebrain (MacInnes et al., 2008). Furthermore, rats sustaining unilateral injections of lactacystin in the MFB have been found to exhibit a progressive pattern of brain structural changes including striatal atrophy, cortical thinning, and enlargement of the lateral ventricles, which were already significant by 3 weeks but became more severe by 5 weeks post injection (Vernon et al., 2011). The progressive extranigral pathology may depend on inflammatory mechanisms because intracerebral injections of lactacystin have been shown to produce widespread, pronounced and sustained activation of astroglia and microglia (Elson et al., 2016; Savolainen et al., 2017). These glial reactions appear to greatly exceed those reported in neurotoxin-based PD models and may stem from a direct action of proteasome inhibitors on glial cells (Ding et al., 2004).

One interesting feature of proteasome inhibitor models is the occurrence of neuronal α-synuclein accumulation within the affected regions, with the appearance of a diffusely increased cellular immunostaining for α-synuclein (Elson et al., 2016, Savolainen et al., 2017), sometimes associated with the formation of small inclusion bodies (Elson et al., 2016; MacInnes et al., 2008; McNaught et al., 2002) that may be immunopositive for Ser129-phosphorylated α-synuclein (Bentea et al., 2015). It is, however, unclear whether α-synuclein accumulation plays a causal role in the neuronal death caused by proteasome inhibitors [reviewed in (Bentea et al., 2017)].

In conclusion, although proteasome inhibitor models may be associated with nonspecific neuronal and glial cell toxicity, they provide useful tools to investigate pathways of proteostatic dysfunction and iron dyshomeostasis, and to evaluate neuroprotective treatments targeting these pathways (Bentea et al., 2017, Le, 2014). Moreover, proteasome inhibitors can be administered to animals overexpressing α-synuclein as an approach to trigger or aggravate the neurodegenerative process (Stefanova et al., 2012).

8 Alpha-synuclein models

The role of α-synuclein in PD pathogenesis goes back to the identification of a mutation in the corresponding gene (SNCA) as a cause of familiar parkinsonism (Polymeropoulos et al., 1997). Interest in the pathogenic role of this protein was further reinforced by the observation that α-synuclein is a major component of Lewy bodies and Lewy neurites (Spillantini et al., 1998), and that elevated expression of the non-mutated protein is sufficient to cause a PD-like disorder in individuals carrying SNCA duplication or triplication (Chartier-Harlin et al., 2004; Singleton et al., 2003). Attempts to replicate α-synuclein-related pathology in animals were first made using transgenic techniques, soon followed by a development of viral vectors (recombinant adeno-associated virus, AAV, or lentivirus, LV) and, most recently, intracerebral injections of α-synuclein fibrils. The tools generated in this way have added an important new dimension to the modeling of PD pathogenesis involving protein misfolding and aggregation, and they have also made it possible to study mechanisms of disease progression related to the cell-to-cell spreading of toxic α-synuclein species. The transgenic and viral vector models are complementary. The transgenic models, obtained in mice, offer opportunities to model systemic disease, but are less useful for the study of cell-type specific pathogenic processes. The viral models, on the other hand, are applicable to both mice, rats and monkeys. They offer the opportunity to study α-synuclein -related toxic processes specific to midbrain DA neurons, while also being applicable to other brain regions or neuron types.

8.1 The AAV-α-synuclein model

Like the 6-OHDA toxin, AAV mediated α-synuclein overexpression requires that the vector is injected locally in the brain using stereotactic surgery. Although this may seem a limitation, it offers distinct advantages in that the overexpression of α-synuclein (wild-type or mutated) can be selectively targeted to the midbrain region encompassing substantia nigra (SN)-ventral tegmental area (VTA) and restricted to one side of the brain, leaving the contralateral side as an internal control. A range of AAV vector serotypes have been explored for this purpose. The early studies made use of recombinant vectors of the AAV2 serotype (Kirik et al., 2002;

Yamada et al., 2004), which have later been replaced by vector serotypes with better tissue spread and transduction efficiency for midbrain DA neurons, notably AAV2/5 (Gorbatyuk et al., 2008); AAV2/6 (Decressac et al., 2012b), AAV2/7 (Van der Perren et al., 2015a), AAV2/8 (McFarland et al., 2009) and AAV2/9 (Bourdenx et al., 2015). The transduction efficiency in midbrain DA neurons achieved with LV vectors (usually not more than 50%) is clearly lower than that obtained with the more efficient AAV vectors, and the extent of DA neuron cell loss is also less pronounced (typically between 25% and 35%) (Lauwers et al., 2003, 2007; Lo Bianco et al., 2002).

AAV mediated overexpression of α-synuclein induces progressive degenerative changes in midbrain DA neurons that replicate some of the key features of the human disease, most prominently the development of α-synuclein-containing protein aggregates positive for Ser129-phosphorylated α-synuclein (p-Syn+), accompanied by prominent axonal pathology and a progressive loss of nigral DA neurons. Neuritic changes develop early and precede DA neuron cell loss. Thus, the dendritic projections of DA neurons in the SN pars reticulata are truncated with distorted morphology, and the pre-terminal axons display swollen and p-Syn+ distorted profiles. As in the human disease, these degenerative changes are associated with an early activation of microglia, an increase in pro-inflammatory cytokines, and lymphocyte infiltration preceding cell loss [for review see (Ulusoy et al., 2010, Van der Perren et al., 2015b, Volpicelli-Daley et al., 2016)].

The progressive time-course of neurodegeneration is an attractive feature of this model, making it possible to distinguish between an *early presymptomatic stage* and a later *symptomatic stage*. The *presymptomatic stage* corresponds to the first month after vector injection, and it is characterized by the development of inclusions, axonal pathology and impaired DA synthesis and release. The *symptomatic stage* develops over the subsequent months, when a significant portion (>50%) of the nigral DA neurons have degenerated and part of the still surviving neurons express p-Syn+ pathology (Decressac et al., 2012b; Lundblad et al., 2012). With this level of cell loss (50–80%) the animals show impairments in standard motor tests similar to what is typically seen following intrastriatal 6-OHDA lesions (Bourdenx et al., 2015, Decressac et al., 2012a,b, Van der Perren et al., 2015a). From a direct comparison between the two models (Decressac et al., 2012b), we have suggested that the motor impairment in the toxin-based model is well correlated with the magnitude of DA neuron loss, while the motor deficits seen in the AAV model result from the combination of DA cell death and dysfunction of the remaining nigrostriatal neurons. While both of the two models have been developed to mimic DA neuron deficiency, they differ in their temporal and neuropathological characteristics, and replicate different pathophysiological aspects of the human disease. The early developing axonal pathology and striatal DA dysfunction preceding overt nigral cell loss (Butler et al., 2015; Chung et al., 2009; Lundblad et al., 2012) is a particular feature of the AAV-α-synuclein model that mimics the disease progression seen in human PD (Burke and O'Malley, 2013; Kordower et al., 2013).

The main weakness of the AAV-α-synuclein model is the variability in the magnitude of the neurodegenerative response, which has made it difficult to obtain consistent behavioral impairments. The magnitude of DA neuron cell loss varies considerably depending on vector types and batches. As discussed in further detail in a recent review (Volpicelli-Daley et al., 2016), this variability is due to many factors: the viral vector serotype, the promoter used, the production process, and the quality and purity of the final product. A further complicating factor is that the common measure of vector titer, genome copies/µL, does not reliably predict the in vivo transduction efficiency of the AAV vector. For this reason, it is necessary to establish the optimal working titer for each individual production round before it is used in a planned experiment. For each vector batch, however, the transduction efficiency can be expected to be consistent from animal to animal, and the variability in outcome will be the same as for 6-OHDA models (i.e., due to investigator skills in stereotactic targeting).

8.2 The PFF inoculation model

This model builds on the finding that oligomeric fibrillar α-synuclein can act as a seed to recruit the monomeric form of the protein into pathogenic aggregates. Using preformed fibrils of recombinant α-synuclein (PFFs), this property has been demonstrated in cell cultures in vitro (Lu et al., 2009; Volpicelli-Daley et al., 2011, 2014), as well as after injection into the brain (Luk et al., 2012a,b; Osterberg et al., 2015). The formation of protein aggregates and the progressive development of cellular dysfunction and cell death are not caused by the fibrils themselves, but by recruitment of endogenous α-synuclein into cellular inclusions. Thus, the injected PFFs are not pathogenic when applied to cells lacking α-synuclein (Luk et al., 2012b; Volpicelli-Daley et al., 2011, 2014). Moreover, their toxicity is increased and accelerated in the presence of elevated levels of monomeric α-synuclein (Peelaerts et al., 2015; Thakur et al., 2017), and the formation of aggregates is more efficient if PFFs and monomeric α-synuclein are from the same animal species (Luk et al., 2016; Peelaerts et al., 2015).

The robust formation of inclusions resembling Lewy bodies and Lewy neurites is a characteristic feature of the PFF model not present in AAV-α-synuclein models. This feature makes this model highly useful for studying the formation and spread of aggregated α-synuclein species. The PFF-induced inclusions share many features with those found in human PD brains: they have a filamentous structure and they are insoluble, hyperphosphorylated, ubiquitinated and morphologically similar to the spheroid inclusions seen in Lewy bodies and neurites (Volpicelli-Daley et al., 2011, 2014).

The pathogenic process is quite fast in cultured neurons, developing within 1–2 weeks (Volpicelli-Daley et al., 2011), but it progresses very slowly when PFFs are injected into the brain of normal animals not overexpressing α-synuclein. Thus, it may take up to 6 months for significant neurodegenerative changes to appear in midbrain DA neurons when the PFFs are injected into the striatum or SN (Espa et al., 2019; Luk et al., 2012a; Paumier et al., 2015; Peelaerts et al., 2015). The same protracted time course of aggregate formation and toxicity has been observed following injections of PFFs into the cortex (Osterberg et al., 2015).

Fibril inoculation and viral vector models can have complementary applications due to differences in their associated pathogenic processes. The PFF model allows for studying the seeding process and the formation, spread and impact of toxic aggregates, while models with vector-mediated α-synuclein overexpression make it possible to study successive stages in the development of cellular and functional changes, including presymptomatic-predegenerative changes. The protracted time-course seen in PFF models is experimentally disadvantageous for evaluating the effects of neuroprotective treatments on behavioral impairments and neuronal cell loss. An additional potential concern is the limited spread of the PFFs within the tissue, which limits the number of DA neurons that can be targeted by an injection in the striatum. This may be less of an issue in the small mouse brain, but more of a problem when applying this approach to rats. In the most extensive study thus far performed in rats, Caryl Sortwell's group have studied the effects of unilateral intrastriatal PFF injections for up to 6 months (Patterson et al., 2019). At the highest PFF dose (16 μg) they observed a progressive downregulation of tyrosine hydroxylase (TH) expression in about 30% and 50% of nigral neurons at 4 and 6 months, respectively. Actual cell loss was evident only at 6 months, with a loss of approximately 30% nigral neurons (identified using the pan-neuronal marker NeuN). Notably, a reduction in nigral NeuN+ cell number was observed also on the contralateral side. The accumulation of p-Syn, which was observed in about one third of the nigral neurons, peaked at 2 months and was largely gone by 6 months. Consistent with the fairly modest degenerative changes, no or only minimal impairment in motor behavior was observed in the PFF-injected rats even at the longest time point. These data reinforce the impression that it is difficult to obtain consistent and significant behavioral impairments using intrastriatal PFF injections in rats [see (Volpicelli-Daley et al., 2016) for further discussion of this issue].

As already mentioned, the most interesting use of the PFF model consists in mechanistic studies of seeding, aggregation and spread of α-synuclein-related pathology in the brain. PFFs injected into the brain parenchyma are internalized by neurons and axons, and efficiently transported retrogradely in neurons whose axons terminate in the injected area. Thus, the initial spread of synuclein pathology is due to retrograde axonal transport of the PFFs, which will seed the formation of p-Syn+ aggregates in the parent neurons. Following injections of PFFs into the striatum, p-Syn+ inclusions will appear within 1–2 months not only in the striatum itself, but also in the principal regions projecting to the striatum, i.e., substantia nigra, cortex, thalamus, amygdala (Abdelmotilib et al., 2017; Luk et al., 2012a; Masuda-Suzukake et al., 2013; Paumier et al., 2015). Similarly, PFF injections into the hippocampus will result, at 3 months, in the appearance of pSyn+ pathology in some though not all neuronal populations projecting to the injected area (Nouraei et al., 2018). In the same study, some limited anterograde transport was observed in two of the major efferent projections from the hippocampus, entorhinal cortex and septum.

It remains unclear, however, whether any significant cell-to-cell transfer of synuclein pathology occurs in this model, and if so, how long it would take for this mechanism to become functionally relevant. Transfer of α-synuclein between cells, in monomeric or aggregated form, has been clearly demonstrated in cell culture systems, suggesting that this is likely to occur also in the brain [see (Tyson et al., 2016) for review]. The most careful investigation of this issue has been performed by Patrik Brundin's group using mice that received injections of human or mouse PFFs into the olfactory bulb (Rey et al., 2013, 2016, 2018). In this model, cellular immunoreactivity for p-Syn+ was found to spread from the olfactory bulb to anatomically connected olfactory and non-olfactory regions, a process that progressed gradually over 12 months. Over the first 9 months, the spread of synuclein inclusions appeared to occur via retrograde axonal transport as it was limited to first-order afferents to the injected area (Mason et al., 2016; Mezias et al., 2020). At longer time points, a reduction in the density of synuclein aggregates was observed in some regions, and no further spread was detected (Rey et al., 2018). Thus, in contrast to the continued disease progression seen in the advanced stages of human PD, the progression of PFF-induced pathology seems to taper off after about a year.

Taken together, these data indicate that cell-to-cell spread of PFF-induced pathology is a late event. The initial spread is clearly due to retrograde transport along afferent connections. At a later stage, only many months later, monomeric or fibrillar α-synuclein species released from the affected neurons may be taken up by adjacent cells or axons to act as seeds for further propagation of synuclein pathology to other interconnected brain regions. This slow and protracted progression limits the usefulness of the PFF model for evaluating therapies that target the spread of PD pathology (e.g., antibodies directed against α-synuclein oligomers). If the goal is to counteract or block cell-to-cell transfer, the use of assessment end-points shorter than 1 year will be irrelevant—at such short survival times the role of cell-to-cell transfer in the spread of p-Syn+ pathology is likely to be negligible. For this type of studies, the more efficient spread of pathology seen in PFF-injected transgenic α-synuclein overexpressing mice may provide a more attractive model. In a study performed by Luk et al. (2012b), human α-synuclein PFFs were injected unilaterally into the striatum in transgenic mice overexpressing human mutated (A53T) α-synuclein. The injection was given at a time point well before any pathology related to the transgene had appeared. With this approach, p-Syn+ pathology developed more rapidly and was much more widespread, than that obtained after similar injections in wild-type mice (Luk et al., 2012b). Motor deficits appeared after about 3 months, a time point when extensive p-Syn+ pathology had already emerged. In this transgenic model, p-Syn+ inclusions developed not only in structures directly connected with the injected area (including cortex, thalamus and substantia nigra), but notably also in more distant sites, such as deep cerebellar nuclei, remote brainstem areas, and spinal cord, indicating that an efficient cell-to-cell transfer had occurred in this model. Whether this represents actual trans-synaptic transmission of α-synuclein seeds, rather than passive diffusion within the extracellular space, is currently unclear and remains to be investigated (Luk and Lee, 2014).

8.3 Combined AAV-PFF α-synuclein models

A shortcoming of the AAV overexpression method is that the α-synuclein levels needed to induce sufficient DA neuron cell death (linked to significant motor impairments) are quite high, in the order of 4–5-fold above the endogenous levels (Decressac et al., 2012b; Faustini et al., 2018). This is well above what may be seen in human PD, raising the question whether the cellular toxicity associated with these high expression levels is predictive for the clinical condition. One way to circumvent this limitation is to combine viral vector-mediated α-synuclein overexpression with PFF inoculation, delivered to the SN either as two separate injections (Thakur et al., 2017) or mixed in a single injection (our yet unpublished data). As explained above, PFFs can act as seeds for the recruitment of monomeric α-synuclein into toxic fibrillar aggregates, the speed by which this happens is dependent on the level of monomeric α-synuclein (Volpicelli-Daley et al., 2011, 2014), and the seeding is most efficient if PFFs and monomeric α-synuclein are from the same species (Luk et al., 2016). In the combined AAV-PFF model, (Thakur et al., 2017) we expressed human WT α-synuclein at a low level (closer to that seen in patients with SNCA triplication) and Lewy-like pathology, neurodegeneration and DA neuron cell death were triggered by injection of human PFF seeds into the SN. Animals receiving both the AAV and the PFF exhibited an enhanced and accelerated development of pathology where the formation of p-Syn+ cellular and neuritic inclusions was evident already by 3 weeks after PFF injection. This pathology was accompanied by a prominent inflammatory response involving both activation of resident microglia and infiltration of CD4+ and CD8+ T-lymphocytes. The degeneration of nigral DA neurons was progressive, leading to a 50–60% cell loss by 24 weeks (Thakur et al., 2017). The progressive nature of the combined AAV-PFF model may offer the possibility to pre-screen the animals at an early time-point using sensitive behavioral tests (as currently done in 6-OHDA-lesioned rats and mice) in order to identify those animals that will become fully symptomatic at a later time point. In this way, it would become possible to evaluate potential neuroprotective treatments in well-matched groups of animals that are on the way to developing a significant disease phenotype.

This combined approach is applicable also for the induction of cortical Lewy-like pathology analogous to that seen in human Lewy body disease. In a recent study (Espa et al., 2019), we induced overexpression of human wild-type α-synuclein bilaterally in rat medial prefrontal cortex using an AAV2/6 synuclein vector, followed 3 weeks later by an injection of human PFFs. The PFF injection targeted the rostromedial striatum based on the expectation that PFFs would be efficiently transported from this region to the AAV-targeted cortical areas via retrograde axonal transport (which proved to be the case). While neither the α-synuclein overexpression nor the PFFs induced any behavioral phenotype if given alone, their combined application induced significant impairments in tests of working memory, attention and inhibitory control, accompanied by the development of prominent proteinase K-resistant, p-Syn+ inclusions, swollen and distorted cortical dendrites, and cortical neuronal loss by 24 weeks (Espa et al., 2019). These results further support the notion that inoculating fibril seeds into a brain region expressing high levels of monomeric α-synuclein leads to an accelerated and amplified development of pathology (which was found to also involve frontocortical afferent regions). Thanks to this approach, it became possible to experimentally reproduce cognitive and pathological features relevant to Lewy body disease using the rat frontocortical circuits as a model system.

In the studies by Thakur et al. (2017) and Espa et al. (2019), vector and PFFs were administered with a 3–4-week interval. In a recent follow-up study (Hoban et al., under revision) we have now proceeded to combine the two preparations in a single, mixed injection, which is experimentally more convenient. Following injection into the rat SN, the same, accelerated and enhanced p-Syn+ pathology, inflammatory response, and progressive DA neuron cell loss are obtained also in this version of the AAV-PFF model. Significant motor impairments, as assessed in tests of forelimb use (cylinder and stepping tests), are observed already by 3–4 weeks after injection, at a time when most of the affected neurons still survive but in a down-regulated state characterized by impaired striatal DA release and reduced expression of TH and vesicular monoamine transporter 2 (VMAT2), as well as downregulation of nuclear receptor related-1 (Nurr1) transcription factor (which controls the expression of DA phenotype genes). At 3–4 months, when DA neurodegeneration is complete, marked impairments in forelimb use is seen in at least half of the injected animals, those with >50% nigral cell loss.

8.4 Transgenic α-synuclein overexpressing mice

Considerable efforts have been made to generate transgenic models of PD based on overexpression of human α-synuclein in its wild-type or mutated forms. A database assembled by the Joint Programme for Neurodegenerative diseases (JPND) lists a total of 24 transgenic α-synuclein models published and characterized to date [list available through (Joint Programme for Neurodegenerative Diseases, 2019)]. Only few of them develop dopaminergic dysfunction and DA cell death of a magnitude justifying their use as models of PD [see (Jiang and Dickson, 2018, Magen and Chesselet, 2010) for review]. The first transgenic models in this category were generated by Eliezer Masliah's group and expressed wild-type human α-synuclein under either the platelet-derived growth factor beta (PDGF-beta) promoter (Masliah et al., 2000) or the Thy-1 promoter (Chesselet et al., 2012; Rockenstein et al., 2002). In these mice, expression of the α-synuclein transgene is widespread, and cytoplasmic and nuclear inclusions containing human α-synuclein develop in several brain areas, including cortex, hippocampus, olfactory bulb, and to some extent also in the SN.

In the PDGF-beta transgenic model, α-synuclein-positive inclusions at the level of the SN were limited to only a few scattered cells. Nevertheless, the mice with the highest level of transgene expression (Line D) were impaired in motor performance, as assessed at 12 months in the rotarod test. This was accompanied by a 50%

reduction in striatal TH immunostaining and TH protein levels, but no DA neuron cell loss. The extent of nigral pathology was more prominent in the Thy-1 mice, but the impact on the integrity of the nigrostriatal system remained quite modest: a 40% reduction in striatal DA and 17% reduction in striatal TH, seen at 14 months, without any measurable nigral cell loss (Chesselet et al., 2012). The changes in motor behavior were bi-phasic: an early phase of locomotor hyperactivity at 4–5 months of age was followed by a slow development of sensorimotor impairments that became evident only at 14 months. The time-dependent changes seen in these two transgenic strains seem compatible with a slowly developing axonopathy in nigrostriatal DA neurons accompanied by a reduction in TH expression in the surviving neurons, although the presence of neurodegenerative changes in several brain regions makes it difficult to attribute the observed deficits exclusively to the pathology of nigrostriatal DA neurons. Indeed, these mice also developed gut dysfunction and changes in circadian rhythm that preceded the motor impairments, suggesting an early impact of the α-synuclein transgene in areas outside the nigrostriatal system, akin to the prodromal phase of human PD.

In a similar way, a constipation-like phenotype preceding motor impairments has been observed in a bacterial artificial chromosome (BAC) transgenic mouse line expressing wild-type α-synuclein from the complete SNCA locus at disease-relevant levels (Janezic et al., 2013). These mice showed a 30% loss of nigral DA neurons at 18 months, preceded by a 30% reduction in striatal DA release, without any observable α-synuclein pathology. These results suggest that the functional impairments seen in these mice were caused by their elevated levels of monomeric α-synuclein rather than a formation of toxic aggregates.

A more interesting PD-like phenotype has been obtained in transgenic mice expressing an aggregation-prone, truncated version of human α-synuclein (1–120 α-synuclein) driven by the TH promoter (Tofaris et al., 2006; Wegrzynowicz et al., 2019). In these mice the transgene expression is confined to DA neurons in the SN and the olfactory bulb (though also occurring in locus coeruleus noradrenaline neurons), having a pronounced impact on the integrity and function of the nigrostriatal pathway. In a first version of this mouse model, the α-syn120 line (Tofaris et al., 2006), dense α-synuclein cytoplasmic inclusions were observed in atrophic TH-positive nigral neurons at 12–14 months, accompanied by a 30% reduction in striatal DA levels, while the number of TH cell bodies remained unchanged. At the longest time-point analyzed, 18 months, the mice exhibited reduced spontaneous locomotor activity.

The Spillantini lab has recently published an improved version of this TH promoter-driven 1–120 synuclein mouse, referred to as the MI2 line (Wegrzynowicz et al., 2019). These mice show more prominent, progressive aggregation of α-synuclein in nigral DA neurons and striatal DA terminals. The first protein aggregates appear at 1.5 month of age in the form of small puncta, and then develop into larger Lewy body-like aggregates at 6–12 months, some of which are ubiquitin-positive and proteinase K-resistant. In these mice there was a significant loss of nigral TH neurons starting at 9–12 months and amounting to about 50% at 20 months of age. The first signs of motor impairment (affecting gait pattern) developed at 9 months of age, at a time when there was a marked reduction in striatal DA release measured with microdialysis, but preceding the appearance of any significant nigral cell loss. A more overt behavioral phenotype was observed only later (at 20 months) when 50% of nigral TH-positive neurons were lost. This sequence of events—axon terminal dysfunction followed by cellular pathology and DA cell death—is reminiscent of the histopathological progression seen in human PD, and supports the view that α-synuclein-induced degenerative changes start at the level of the axon terminals.

A large number of transgenic mouse lines have been generated carrying mutated versions of α-synuclein (A53T, A30P or E46K). Many of these develop motor impairments linked to a widespread synucleinopathy with intraneuronal α-synuclein inclusions, but no overt damage to nigrostriatal DA neurons [see (Jiang and Dickson, 2018, Magen and Chesselet, 2010) for review]. One notable exception is a conditional transgenic model obtained using a tetracyclin-dependent inducible system to overexpress human A53T α-synuclein selectively in midbrain DA neurons (Lin et al., 2012). In these mice there is a 2–4-fold increase in α-synuclein protein and mRNA in the midbrain, with a development of granular α-synuclein deposits in nigral TH neurons and α-synuclein aggregates in striatal axons and terminals by 12–18 months of age. This is accompanied by a gradual loss of nigral DA neurons, amounting to 15% at 1 month and 40% at 12–20 months of age. In this mouse model, a motor impairment in open field and rotarod tests was reported to develop already at 1–2 months, before any major cell loss had occurred. At this early time point, Lin et al. (2012) observed a marked downregulation of both TH, DAT, VMAT2 and Nurr1 in the α-synuclein-overexpressing nigral neurons, similar to what has been found in human PD (Chu et al., 2006) and in the AAV-α-synuclein model at a similar early time point (Decressac et al., 2012a). In line with these findings, Lin et al. detected a 70–80% reduction in baseline and evoked striatal DA release at 3–4 months. The protracted time course of SN degeneration in this model is consistent with observations made in other transgenic or viral α-synuclein models (see above) and indicates that the earliest impact of elevated levels of wild-type or mutated α-synuclein occurs at the level of axons and presynaptic terminals (reflected in a downregulation of the DA synthesis machinery and a reduction in striatal DA release). These changes precede cell death and may be sufficient to cause motor deficits. Nevertheless, caution should be exerted when attributing motor deficits to nigrostriatal DA pathology if a transgenic model also exhibits non-PD-specific phenotypic features. For example, in the above-mentioned study (Lin et al., 2012), the transgenic human α-synuclein protein was expressed not only in the midbrain but also, at very high levels, in cerebellum and hippocampus. Moreover, compared to their wild-type controls, the α-synuclein transgenic mice exhibited a significant and progressive body weight reduction starting at young adult age, whose implications remain unclear.

9 Other genetic models of PD

As already mentioned, the discovery of gene mutations associated with autosomal recessive and autosomal dominant forms of PD has prompted the development of a large number of transgenic rodent lines expressing PD-causing mutations [for review see (Blesa and Przedborski, 2014, Creed and Goldberg, 2018, Konnova and Swanberg, 2018, Xu et al., 2012)]. In this section, we have chosen to focus on four well-established genetic PD models exhibiting nigrostriatal DA deficits, the *Engrailed1, Nurr1, Pitx3-Aphakia*, and *mitochondria-deficient "MitoPark"* mouse models. Engrailed1, Nurr1 and Pitx3 are transcription factors involved in the development and survival of midbrain DA neurons during development. They are, however, expressed not only during development but also in adulthood, playing a role in the maintenance and survival of mature DA neurons. The MitoPark mouse, by contrast, carries a genetic defect that will impair mitochondrial function and replication selectively within midbrain DA neurons. These models are interesting and valuable as they can replicate a PD-like progressive dopaminergic neurodegeneration; at the same time, their usefulness to mimic the disease process is limited by the lack of α-synuclein-related pathology.

In the *heterozygous Engrailed1 (En1$^{+/-}$) knock-out mouse,* ablating one of the two Engrailed1 genes induces a progressive degenerative process in nigrostriatal DA neurons that starts at the level of axon terminals at about 4 weeks of age and progresses over the subsequent months, resulting in a significant loss of DA neurons at 3–5 months of age (Nordstroma et al., 2015; Sonnier et al., 2007). The magnitude of DA neuron loss is, however, too small to induce any significant motor phenotype. In this model, the early axonopathy/axon terminal loss followed by nigral DA neuron degeneration is reminiscent of the disease process seen in human PD (Burke and O'Malley, 2013; Kordower et al., 2013). Similar to the human disease, the En1$^{+/-}$ mice exhibit decreased mitochondrial complex-I activity and signs of impaired autophagy, although they lack α-synuclein aggregate formation, a hallmark of human PD. In a recent extension of the model by the Brundin group, PFFs were injected into the striatum to induce synuclein pathology in En1$^{+/-}$ mice (Chatterjee et al., 2019). In the absence of one Engrailed allele, the induction of pathological α-synuclein aggregates was found to be more pronounced than in wild-type mice, suggesting that the mitochondrial and autophagic deficits seen in En1$^{+/-}$ mice may help to accelerate the seeding and aggregation process. This combined approach thus brings the Engrailed model one step closer to mimicking PD-like pathology.

In the *Nurr1 model*, one or both of the Nurr1 genes are ablated. Because homozygous Nurr1 knockout (KO) mice do not survive beyond birth, Nurr1 hypomorphic models are obtained by using either heterozygous mice lacking one allele through life (Jiang et al., 2005) or conditional KO mice where one or both alleles are selectively removed in mature DA neurons (Kadkhodaei et al., 2009, 2013). There is evidence to suggest that Nurr1 is involved in the pathophysiology of PD. Indeed, the expression of Nurr1 is reduced in DA neurons affected by synuclein pathology

(Chu et al., 2006; Decressac et al., 2012a) and polymorphisms in the Nurr1 gene point to reduced Nurr1 expression being a risk factor for the development of PD (Grimes et al., 2006; Xu et al., 2002; Zheng et al., 2003). In line with these findings, reduced Nurr1 expression, as seen in heterozygous Nurr1$^{+/-}$ mice, is associated with a slowly developing nigrostriatal dysfunction including DA neuron loss and reduced striatal DA levels, which becomes evident at 15–24 months of age (Jiang et al., 2005), as well as an increased sensitivity to the DA neurotoxin MPTP (Le et al., 1999). These changes are even more pronounced in conditional knock-out mice where the Nurr1 gene is ablated by administering tamoxifen treatment at 5 weeks of age. In this model, the Nurr1 target genes TH, VMAT2, and DAT are markedly down-regulated by 4 months, and tissue DA levels are reduced by over 80% in both striatum and nucleus accumbens at 11 months, accompanied by motor impairments in the open field and vertical pole tests developing gradually over time (Kadkhodaei et al., 2013). No loss of cell bodies was detected for up to 11 months after tamoxifen administration, while TH-positive axons and dendrites developed clear signs of pathology, exhibiting a swollen and fragmented morphology. This pattern of pathological changes—axon terminal degeneration and reductions in striatal DA preceding DA neuron cell loss—suggests similarity with early stage PD. However, similarly to the En1$^{+/-}$ model, Nurr1 deficient mice do not develop any signs of α-synuclein pathology.

The Pitx3-Aphakia mouse is the best characterized and most commonly used of the three transcription factor-related models. The expression of Pitx3 (paired-like homeodomain transcription factor 1) is restricted to the developing eye and midbrain DA neurons from embryonic day 11 throughout adult life (Smidt et al., 1997). Aphakia mice carry a spontaneous deletion at the *Pitx3* locus causing microphthalmia and aphakia (i.e., absence of the lens of the eye). Mice homozygous for this mutation (termed ak/ak mice) exhibit an almost complete loss of DA neurons in the pars compacta of the SN that is present already at birth, while DA neurons in the VTA are relatively spared up to about 6 weeks of age. At later time points (100 days), about 50% of the VTA neurons are lost (Hwang et al., 2003; Nunes et al., 2003; van den Munckhof et al., 2003). As a result of the marked DA cell depletion in the SN, the motor part of the striatum is severely denervated and exhibits an over 90% reduction in DA levels. This is associated with supersensitivity of DA receptor-mediated signaling in striatal neurons (Hwang et al., 2005), which enables the induction of dyskinesia by repeated L-DOPA administration (Ding et al., 2007; Suarez et al., 2018). Differently from the dorsolateral (motor) striatum, nucleus accumbens and ventral striatal areas exhibit only approximately 70% DA loss in adult mice (3 months old) (van den Munckhof et al., 2003). Ak/ak mice do not show any gross alterations in motor behavior, but display clear L-DOPA-reversible defects in sensitive measures of nigrostriatal motor function, such as longer latency and shorter steps in the beam walking test and impaired performance in the vertical pole test (Hwang et al., 2005). Moreover, these mice show impairments in striatum-dependent cognitive tests including rotarod learning, T-maze and inhibitory avoidance tasks (Ardayfio et al., 2008). The Pitx3-Aphakia mouse has been proposed as valid model for PD because of its conspicuous loss of nigral DA neurons with relative sparing of the VTA, resembling the pattern of dopaminergic degeneration in the human disease. Nevertheless, this model has limited utility for studying PD pathogenesis and treatments thereof, since the loss of SN DA neurons results from a developmental defect as opposed to an adult degenerative process. The applications benefiting the most from this mouse model are pathophysiological-behavioral studies addressing the effects of severe nigrostriatal DA depletion and the associated compensatory mechanisms. Due to its bilateral phenotype, the Pitx3-Aphakia mouse offers an interesting complement to the common, unilateral 6-OHDA lesion models for this type of studies.

The MitoPark mouse is a conditional knock-out mouse where the gene for mitochondrial transcription factor A (Tfam) is disrupted selectively in DA neurons. The TFAM protein is a regulator of mitochondrial replication and decreased levels of this protein result in a reduction of mitochondrial DNA copy number (Ekstrand et al., 2004), similar to what has been observed in nigral DA neurons in human PD (Grunewald et al., 2019). In the affected DA neurons, Tfam disruption induces a respiratory chain deficiency, which in turn causes a progressive degenerative phenotype. The dopaminergic deficiency yields a L-DOPA-responsive motor impairment that is first observed at around 12–15 weeks of age, accompanied by a gradual loss of DA neurons in both SN and VTA, which reaches about 80–90% at 10 months of age (Ekstrand et al., 2007). Prior to the onset of cell loss these mice show a presymptomatic impairment in striatal DA release (Good et al., 2011) and also changes in somatodendritic morphology in DA neurons (Lynch et al., 2018). The protracted time-course of degenerative changes, leading to a near-complete loss of midbrain DA neurons within a reasonable time span, is an attractive feature of the MitoPark mouse. Mitochondrial dysfunction is a characteristic feature of human PD and the MitoPark mouse provides a highly useful model for this aspect of the disease.

10 Concluding remarks

During the past decades, an increased understanding of pathological features, genetic and environmental factors underlying PD has prompted the development of a vast and diversified repertoire of animal models. Today we have unprecedented opportunities to recreate and study virtually all critical aspects of PD pathogenesis in laboratory animals. This is a very active research field in continuous communication with other research disciplines, in particular, molecular genetics, protein biochemistry, pharmacology, physiology, and comparative anatomy. In spite of the criticism often raised on the validity of animal models, there is no question that disease models in adult living organisms will continue to be indispensable for many fundamental applications. Although no single animal model replicates all pathogenic and clinical features of PD, the range of rodent models available today offers opportunities to reproduce specific disease features within a tightly controlled in vivo system. If carefully chosen and correctly applied, animal models are requisite for scientific progress and invaluable for the exploration and development of novel therapeutic ideas. To this end, selecting the most suitable model for the questions under study is essential, and a continuous, bidirectional dialogue between experimentalists and clinical researchers is of vital importance.

Acknowledgments

The authors' work in this area is funded by research grants from the Swedish Research Council, the Swedish Parkinson's Foundation, the Swedish Brain Foundation, the Åhlén Foundation, and the Swedish Governmental Funding for Clinical Research. We thank Bengt Mattsson and Elena Espa for excellent graphic support.

References

Abdelmotilib, H., Maltbie, T., Delic, V., Liu, Z., Hu, X., Fraser, K.B., Moehle, M.S., Stoyka, L., Anabtawi, N., Krendelchtchikova, V., Volpicelli-Daley, L.A., West, A., 2017. alpha-Synuclein fibril-induced inclusion spread in rats and mice correlates with dopaminergic neurodegeneration. Neurobiol. Dis. 105, 84–98.

Alvarez-Fischer, D., Guerreiro, S., Hunot, S., Saurini, F., Marien, M., Sokoloff, P., Hirsch, E.C., Hartmann, A., Michel, P.P., 2008. Modelling Parkinson-like neurodegeneration via osmotic minipump delivery of MPTP and probenecid. J. Neurochem. 107, 701–711.

Ardayfio, P., Moon, J., Leung, K.K., Youn-Hwang, D., Kim, K.S., 2008. Impaired learning and memory in Pitx3 deficient aphakia mice: a genetic model for striatum-dependent cognitive symptoms in Parkinson's disease. Neurobiol. Dis. 31, 406–412.

Ascherio, A., Schwarzschild, M.A., 2016. The epidemiology of Parkinson's disease: risk factors and prevention. Lancet Neurol. 15, 1257–1272.

Bentea, E., Van Der Perren, A., Van Liefferinge, J., El Arfani, A., Albertini, G., Demuyser, T., Merckx, E., Michotte, Y., Smolders, I., Baekelandt, V., Massie, A., 2015. Nigral proteasome inhibition in mice leads to motor and non-motor deficits and increased expression of Ser129 phosphorylated alpha-synuclein. Front. Behav. Neurosci. 9, 68.

Bentea, E., Verbruggen, L., Massie, A., 2017. The proteasome inhibition model of Parkinson's disease. J. Parkinsons Dis. 7, 31–63.

Berg, D., Postuma, R.B., Bloem, B., Chan, P., Dubois, B., Gasser, T., Goetz, C.G., Halliday, G.M., Hardy, J., Lang, A.E., Litvan, I., Marek, K., Obeso, J., Oertel, W., Olanow, C.W., Poewe, W., Stern, M., Deuschl, G., 2014. Time to redefine PD? Introductory statement of the MDS Task Force on the definition of Parkinson's disease. Mov. Disord. 29, 454–462.

Bespalov, A., Steckler, T., Altevogt, B., Koustova, E., Skolnick, P., Deaver, D., Millan, M.J., Bastlund, J.F., Doller, D., Witkin, J., Moser, P., O'donnell, P., Ebert, U., Geyer, M.A., Prinssen, E., Ballard, T., Macleod, M., 2016. Failed trials for central nervous system disorders do not necessarily invalidate preclinical models and drug targets. Nat. Rev. Drug Discov. 15, 516.

Betarbet, R., Sherer, T.B., Mackenzie, G., Garcia-Osuna, M., Panov, A.V., Greenamyre, J.T., 2000. Chronic systemic pesticide exposure reproduces features of Parkinson's disease. Nat. Neurosci. 3, 1301–1306.

Bjorklund, A., Rosenblad, C., Winkler, C., Kirik, D., 1997. Studies on neuroprotective and regenerative effects of GDNF in a partial lesion model of Parkinson's disease. Neurobiol. Dis. 4, 186–200.

Blesa, J., Przedborski, S., 2014. Parkinson's disease: animal models and dopaminergic cell vulnerability. Front. Neuroanat. 8, 155.

Boix, J., Von Hieber, D., Connor, B., 2018. Gait analysis for early detection of motor symptoms in the 6-OHDA rat model of Parkinson's disease. Front. Behav. Neurosci. 12, 39.

Bourdenx, M., Dovero, S., Engeln, M., Bido, S., Bastide, M.F., Dutheil, N., Vollenweider, I., Baud, L., Piron, C., Grouthier, V., Boraud, T., Porras, G., Li, Q., Baekelandt, V., Scheller, D., Michel, A., Fernagut, P.O., Georges, F., Courtine, G., Bezard, E., Dehay, B., 2015. Lack of additive role of ageing in nigrostriatal neurodegeneration triggered by alpha-synuclein overexpression. Acta Neuropathol. Commun. 3, 46.

Burke, R.E., O'malley, K., 2013. Axon degeneration in Parkinson's disease. Exp. Neurol. 246, 72–83.

Butler, B., Saha, K., Rana, T., Becker, J.P., Sambo, D., Davari, P., Goodwin, J.S., Khoshbouei, H., 2015. Dopamine transporter activity is modulated by alpha-synuclein. J. Biol. Chem. 290, 29542–29554.

Caiazza, M.C., Lang, C., Wade-Martins, R., 2020. What we can learn from iPSC-derived cellular models of Parkinson's disease. Prog. Brain Res. 252, this volume.

Cannon, J.R., Greenamyre, J.T., 2010. Neurotoxic in vivo models of Parkinson's disease recent advances. Prog. Brain Res. 184, 17–33.

Cannon, J.R., Tapias, V., Na, H.M., Honick, A.S., Drolet, R.E., Greenamyre, J.T., 2009. A highly reproducible rotenone model of Parkinson's disease. Neurobiol. Dis. 34, 279–290.

Cenci, M.A., 2014. Presynaptic mechanisms of l-DOPA-induced dyskinesia: the findings, the debate, and the therapeutic implications. Front. Neurol. 5, 242.

Cenci, M.A., Crossman, A.R., 2018. Animal models of l-dopa-induced dyskinesia in Parkinson's disease. Mov. Disord. 33, 889–899.

Cenci, M.A., Konradi, C., 2010. Maladaptive striatal plasticity in L-DOPA-induced dyskinesia. Prog. Brain Res. 183, 209–233.

Cenci, M.A., Sgambato, V., 2020. Toxin-based rodent models of Parkinson's disease. In: Perez-Lloret, S., Rascol, O. (Eds.), Clinical Trials in Parkinson's Disease. Springer Science+Business Media, New York.

Cenci, M.A., Whishaw, I.Q., Schallert, T., 2002. Animal models of neurological deficits: how relevant is the rat? Nat. Rev. Neurosci. 3, 574–579.

Chade, A.R., Kasten, M., Tanner, C.M., 2006. Nongenetic causes of Parkinson's disease. J. Neural Transm. Suppl. 70, 147–151.

Chartier-Harlin, M.C., Kachergus, J., Roumier, C., Mouroux, V., Douay, X., Lincoln, S., Levecque, C., Larvor, L., Andrieux, J., Hulihan, M., Waucquier, N., Defebvre, L., Amouyel, P., Farrer, M., Destee, A., 2004. Alpha-synuclein locus duplication as a cause of familial Parkinson's disease. Lancet 364, 1167–1169.

Chatterjee, D., Sanchez, D.S., Quansah, E., Rey, N.L., George, S., Becker, K., Madaj, Z., Steiner, J.A., Ma, J., Escobar Galvis, M.L., Kordower, J.H., Brundin, P., 2019. Loss of one Engrailed1 allele enhances induced alpha-synucleinopathy. J. Parkinsons Dis. 9, 315–326.

Chesselet, M.F., Richter, F., Zhu, C., Magen, I., Watson, M.B., Subramaniam, S.R., 2012. A progressive mouse model of Parkinson's disease: the Thy1-aSyn ("Line 61") mice. Neurotherapeutics 9, 297–314.

Chiba, K., Trevor, A., Castagnoli Jr., N., 1984. Metabolism of the neurotoxic tertiary amine, MPTP, by brain monoamine oxidase. Biochem. Biophys. Res. Commun. 120, 574–578.

Chu, Y., Le, W., Kompoliti, K., Jankovic, J., Mufson, E.J., Kordower, J.H., 2006. Nurr1 in Parkinson's disease and related disorders. J. Comp. Neurol. 494, 495–514.

Chung, C.Y., Koprich, J.B., Siddiqi, H., Isacson, O., 2009. Dynamic changes in presynaptic and axonal transport proteins combined with striatal neuroinflammation precede dopaminergic neuronal loss in a rat model of AAV alpha-synucleinopathy. J. Neurosci. 29, 3365–3373.

Creed, R.B., Goldberg, M.S., 2018. New developments in genetic rat models of Parkinson's disease. Mov. Disord. 33, 717–729.

Decressac, M., Kadkhodaei, B., Mattsson, B., Laguna, A., Perlmann, T., Bjorklund, A., 2012a. alpha-Synuclein-induced down-regulation of Nurr1 disrupts GDNF signaling in nigral dopamine neurons. Sci. Transl. Med. 4, 163ra156.

Decressac, M., Mattsson, B., Bjorklund, A., 2012b. Comparison of the behavioural and histological characteristics of the 6-OHDA and alpha-synuclein rat models of Parkinson's disease. Exp. Neurol. 235, 306–315.

Diederich, N.J., James Surmeier, D., Uchihara, T., Grillner, S., Goetz, C.G., 2019. Parkinson's disease: is it a consequence of human brain evolution? Mov. Disord. 34, 453–459.

Ding, Q., Dimayuga, E., Markesbery, W.R., Keller, J.N., 2004. Proteasome inhibition increases DNA and RNA oxidation in astrocyte and neuron cultures. J. Neurochem. 91, 1211–1218.

Ding, Y., Restrepo, J., Won, L., Hwang, D.Y., Kim, K.S., Kang, U.J., 2007. Chronic 3,4-dihydroxyphenylalanine treatment induces dyskinesia in aphakia mice, a novel genetic model of Parkinson's disease. Neurobiol. Dis. 27, 11–23.

Dodiya, H.B., Forsyth, C.B., Voigt, R.M., Engen, P.A., Patel, J., Shaikh, M., Green, S.J., Naqib, A., Roy, A., Kordower, J.H., Pahan, K., Shannon, K.M., Keshavarzian, A., 2020. Chronic stress-induced gut dysfunction exacerbates Parkinson's disease phenotype and pathology in a rotenone-induced mouse model of Parkinson's disease. Neurobiol. Dis. 135, 104352.

Drummond, E., Wisniewski, T., 2017. Alzheimer's disease: experimental models and reality. Acta Neuropathol. 133, 155–175.

Ekstrand, M.I., Falkenberg, M., Rantanen, A., Park, C.B., Gaspari, M., Hultenby, K., Rustin, P., Gustafsson, C.M., Larsson, N.G., 2004. Mitochondrial transcription factor A regulates mtDNA copy number in mammals. Hum. Mol. Genet. 13, 935–944.

Ekstrand, M.I., Terzioglu, M., Galter, D., Zhu, S., Hofstetter, C., Lindqvist, E., Thams, S., Bergstrand, A., Hansson, F.S., Trifunovic, A., Hoffer, B., Cullheim, S., Mohammed, A.H., Olson, L., Larsson, N.G., 2007. Progressive parkinsonism in mice with respiratory-chain-deficient dopamine neurons. Proc. Natl. Acad. Sci. U. S. A. 104, 1325–1330.

Elson, J.L., Yates, A., Pienaar, I.S., 2016. Pedunculopontine cell loss and protein aggregation direct microglia activation in parkinsonian rats. Brain Struct. Funct. 221, 2319–2341.

Erro, R., Picillo, M., Vitale, C., Palladino, R., Amboni, M., Moccia, M., Pellecchia, M.T., Barone, P., 2016. Clinical clusters and dopaminergic dysfunction in de-novo Parkinson disease. Parkinsonism Relat. Disord. 28, 137–140.

Espa, E., Clemensson, E.K.H., Luk, K.C., Heuer, A., Bjorklund, T., Cenci, M.A., 2019. Seeding of protein aggregation causes cognitive impairment in rat model of cortical synucleinopathy. Mov. Disord. 34, 1699–1710.

Faustini, G., Longhena, F., Varanita, T., Bubacco, L., Pizzi, M., Missale, C., Benfenati, F., Bjorklund, A., Spano, P., Bellucci, A., 2018. Synapsin III deficiency hampers alpha-synuclein aggregation, striatal synaptic damage and nigral cell loss in an AAV-based mouse model of Parkinson's disease. Acta Neuropathol. 136, 621–639.

Fearnley, J.M., Lees, A.J., 1991. Ageing and Parkinson's disease: substantia nigra regional selectivity. Brain 114 (Pt. 5), 2283–2301.

Fei, Q., McCormack, A.L., Di Monte, D.A., Ethell, D.W., 2008. Paraquat neurotoxicity is mediated by a Bak-dependent mechanism. J. Biol. Chem. 283, 3357–3364.

Fieblinger, T., Cenci, M.A., 2015. Zooming in on the small: the plasticity of striatal dendritic spines in L-DOPA-induced dyskinesia. Mov. Disord. 30, 484–493.

Fleming, S.M., Zhu, C., Fernagut, P.O., Mehta, A., Dicarlo, C.D., Seaman, R.L., Chesselet, M.F., 2004. Behavioral and immunohistochemical effects of chronic intravenous and subcutaneous infusions of varying doses of rotenone. Exp. Neurol. 187, 418–429.

Flinn, L., Bretaud, S., Lo, C., Ingham, P.W., Bandmann, O., 2008. Zebrafish as a new animal model for movement disorders. J. Neurochem. 106, 1991–1997.

Fornai, F., Schluter, O.M., Lenzi, P., Gesi, M., Ruffoli, R., Ferrucci, M., Lazzeri, G., Busceti, C.L., Pontarelli, F., Battaglia, G., Pellegrini, A., Nicoletti, F., Ruggieri, S., Paparelli, A., Sudhof, T.C., 2005. Parkinson-like syndrome induced by continuous MPTP infusion: convergent roles of the ubiquitin-proteasome system and alpha-synuclein. Proc. Natl. Acad. Sci. U. S. A. 102, 3413–3418.

Fox, S.H., Brotchie, J.M., 2019. Viewpoint: developing drugs for levodopa-induced dyskinesia in PD: lessons learnt, what does the future hold? Eur. J. Neurosci. 49, 399–409.

Francardo, V., 2018. Modeling Parkinson's disease and treatment complications in rodents: potentials and pitfalls of the current options. Behav. Brain Res. 352, 142–150.

Francardo, V., Recchia, A., Popovic, N., Andersson, D., Nissbrandt, H., Cenci, M.A., 2011. Impact of the lesion procedure on the profiles of motor impairment and molecular responsiveness to L-DOPA in the 6-hydroxydopamine mouse model of Parkinson's disease. Neurobiol. Dis. 42, 327–340.

Francardo, V., Schmitz, Y., Sulzer, D., Cenci, M.A., 2017. Neuroprotection and neurorestoration as experimental therapeutics for Parkinson's disease. Exp. Neurol. 298, 137–147.

Gomez-Marin, A., Ghazanfar, A.A., 2019. The life of behavior. Neuron 104, 25–36.

Good, C.H., Hoffman, A.F., Hoffer, B.J., Chefer, V.I., Shippenberg, T.S., Backman, C.M., Larsson, N.G., Olson, L., Gellhaar, S., Galter, D., Lupica, C.R., 2011. Impaired nigrostriatal function precedes behavioral deficits in a genetic mitochondrial model of Parkinson's disease. FASEB J. 25, 1333–1344.

Gorbatyuk, O.S., Li, S., Sullivan, L.F., Chen, W., Kondrikova, G., Manfredsson, F.P., Mandel, R.J., Muzyczka, N., 2008. The phosphorylation state of Ser-129 in human alpha-synuclein determines neurodegeneration in a rat model of Parkinson disease. Proc. Natl. Acad. Sci. U. S. A. 105, 763–768.

Grimes, D.A., Han, F., Panisset, M., Racacho, L., Xiao, F., Zou, R., Westaff, K., Bulman, D.E., 2006. Translated mutation in the Nurr1 gene as a cause for Parkinson's disease. Mov. Disord. 21, 906–909.

Grunewald, A., Kumar, K.R., Sue, C.M., 2019. New insights into the complex role of mitochondria in Parkinson's disease. Prog. Neurobiol. 177, 73–93.

Guigoni, C., Dovero, S., Aubert, I., Li, Q., Bioulac, B.H., Bloch, B., Gurevich, E.V., Gross, C.E., Bezard, E., 2005. Levodopa-induced dyskinesia in MPTP-treated macaques is not dependent on the extent and pattern of nigrostrial lesioning. Eur. J. Neurosci. 22, 283–287.

Guo, M., 2010. What have we learned from Drosophila models of Parkinson's disease? Prog. Brain Res. 184, 3–16.

Hoglinger, G.U., Feger, J., Prigent, A., Michel, P.P., Parain, K., Champy, P., Ruberg, M., Oertel, W.H., Hirsch, E.C., 2003. Chronic systemic complex I inhibition induces a hypo-kinetic multisystem degeneration in rats. J. Neurochem. 84, 491–502.

Hwang, D.Y., Ardayfio, P., Kang, U.J., Semina, E.V., Kim, K.S., 2003. Selective loss of do-paminergic neurons in the substantia nigra of Pitx3-deficient aphakia mice. Brain Res. Mol. Brain Res. 114, 123–131.

Hwang, D.Y., Fleming, S.M., Ardayfio, P., Moran-Gates, T., Kim, H., Tarazi, F.I., Chesselet, M.F., Kim, K.S., 2005. 3,4-dihydroxyphenylalanine reverses the motor deficits in Pitx3-deficient aphakia mice: behavioral characterization of a novel genetic model of Parkinson's disease. J. Neurosci. 25, 2132–2137.

Janezic, S., Threlfell, S., Dodson, P.D., Dowie, M.J., Taylor, T.N., Potgieter, D., Parkkinen, L., Senior, S.L., Anwar, S., Ryan, B., Deltheil, T., Kosillo, P., Cioroch, M., Wagner, K., Ansorge, O., Bannerman, D.M., Bolam, J.P., Magill, P.J., Cragg, S.J., Wade-Martins, R., 2013. Deficits in dopaminergic transmission precede neuron loss and dysfunction in a new Parkinson model. Proc. Natl. Acad. Sci. U. S. A. 110, E4016–E4025.

Jellinger, K., Linert, L., Kienzl, E., Herlinger, E., Youdim, M.B., 1995. Chemical evidence for 6-hydroxydopamine to be an endogenous toxic factor in the pathogenesis of Parkinson's disease. J. Neural Transm. Suppl. 46, 297–314.

Jiang, P., Dickson, D.W., 2018. Parkinson's disease: experimental models and reality. Acta Neuropathol. 135, 13–32.

Jiang, C., Wan, X., He, Y., Pan, T., Jankovic, J., Le, W., 2005. Age-dependent dopaminergic dysfunction in Nurr1 knockout mice. Exp. Neurol. 191, 154–162.

Johnston, T.M., Fox, S.H., 2015. Symptomatic models of Parkinson's disease and L-DOPA-induced dyskinesia in non-human primates. Curr. Top. Behav. Neurosci. 22, 221–235.

Joint Programme For Neurodegenerative Diseases, 2019. JPND Database of Experimental Models for Parkinson's Disease: In Vivo Mammalian Models/Alpha-Synuclein [Online]. Available:https://www.neurodegenerationresearch.eu/models-for-parkinsons-disease/in-vivo-mammalian-models/α-synuclein. Accessed 12 February 2020.

Kadkhodaei, B., Ito, T., Joodmardi, E., Mattsson, B., Rouillard, C., Carta, M., Muramatsu, S., Sumi-Ichinose, C., Nomura, T., Metzger, D., Chambon, P., Lindqvist, E., Larsson, N.G., Olson, L., Bjorklund, A., Ichinose, H., Perlmann, T., 2009. Nurr1 is required for mainte-nance of maturing and adult midbrain dopamine neurons. J. Neurosci. 29, 15923–15932.

Kadkhodaei, B., Alvarsson, A., Schintu, N., Ramskold, D., Volakakis, N., Joodmardi, E., Yoshitake, T., Kehr, J., Decressac, M., Bjorklund, A., Sandberg, R., Svenningsson, P., Perlmann, T., 2013. Transcription factor Nurr1 maintains fiber integrity and nuclear-encoded mitochondrial gene expression in dopamine neurons. Proc. Natl. Acad. Sci. U. S. A. 110, 2360–2365.

Kirik, D., Rosenblad, C., Bjorklund, A., 1998. Characterization of behavioral and neurodegen-erative changes following partial lesions of the nigrostriatal dopamine system induced by intrastriatal 6-hydroxydopamine in the rat. Exp. Neurol. 152, 259–277.

Kirik, D., Rosenblad, C., Burger, C., Lundberg, C., Johansen, T.E., Muzyczka, N., Mandel, R.J., Bjorklund, A., 2002. Parkinson-like neurodegeneration induced by targeted overexpression of alpha-synuclein in the nigrostriatal system. J. Neurosci. 22, 2780–2791.

Konieczny, J., Czarnecka, A., Lenda, T., Kaminska, K., Lorenc-Koci, E., 2014. Chronic L-DOPA treatment attenuates behavioral and biochemical deficits induced by unilateral lactacystin administration into the rat substantia nigra. Behav. Brain Res. 261, 79–88.

Konnova, E.A., Swanberg, M., 2018. Animal models of Parkinson's disease. In: Stoker, T.B., Greenland, J.C. (Eds.), Parkinson's Disease: Pathogenesis and Clinical Aspects Brisbane (AU).

Kordower, J.H., Olanow, C.W., Dodiya, H.B., Chu, Y., Beach, T.G., Adler, C.H., Halliday, G.M., Bartus, R.T., 2013. Disease duration and the integrity of the nigrostriatal system in Parkinson's disease. Brain 136, 2419–2431.

Kostrzewa, R.M., 1995. Dopamine receptor supersensitivity. Neurosci. Biobehav. Rev. 19, 1–17.

Kupsch, A., Schmidt, W., Gizatullina, Z., Debska-Vielhaber, G., Voges, J., Striggow, F., Panther, P., Schwegler, H., Heinze, H.J., Vielhaber, S., Gellerich, F.N., 2014. 6-Hydroxydopamine impairs mitochondrial function in the rat model of Parkinson's disease: respirometric, histological, and behavioral analyses. J. Neural Transm. (Vienna) 121, 1245–1257.

Kuter, K.Z., Cenci, M.A., Carta, A.R., 2020. The role of glia in Parkinson's disease: emerging concepts and therapeutic applications. Prog. Brain Res. 252, this volume.

Lafollette, H., Shanks, N., 1996. The origin of speciesism. Philosophy 71, 41–61.

Langston, J.W., 2017. The MPTP story. J. Parkinsons Dis. 7, S11–S19.

Langston, J.W., Ballard, P., Tetrud, J.W., Irwin, I., 1983. Chronic Parkinsonism in humans due to a product of meperidine-analog synthesis. Science 219, 979–980.

Lapointe, N., St-Hilaire, M., Martinoli, M.G., Blanchet, J., Gould, P., Rouillard, C., Cicchetti, F., 2004. Rotenone induces non-specific central nervous system and systemic toxicity. FASEB J. 18, 717–719.

Lauwers, E., Debyser, Z., Van Dorpe, J., De Strooper, B., Nuttin, B., Baekelandt, V., 2003. Neuropathology and neurodegeneration in rodent brain induced by lentiviral vector-mediated overexpression of alpha-synuclein. Brain Pathol. 13, 364–372.

Lauwers, E., Beque, D., Van Laere, K., Nuyts, J., Bormans, G., Mortelmans, L., Casteels, C., Vercammen, L., Bockstael, O., Nuttin, B., Debyser, Z., Baekelandt, V., 2007. Non-invasive imaging of neuropathology in a rat model of alpha-synuclein overexpression. Neurobiol. Aging 28, 248–257.

Le, W., 2014. Role of iron in UPS impairment model of Parkinson's disease. Parkinsonism Relat. Disord. 20 (Suppl. 1), S158–S161.

Le, W., Conneely, O.M., He, Y., Jankovic, J., Appel, S.H., 1999. Reduced Nurr1 expression increases the vulnerability of mesencephalic dopamine neurons to MPTP-induced injury. J. Neurochem. 73, 2218–2221.

Lee, J., Zhu, W.M., Stanic, D., Finkelstein, D.I., Horne, M.H., Henderson, J., Lawrence, A.J., O'connor, L., Tomas, D., Drago, J., Horne, M.K., 2008. Sprouting of dopamine terminals and altered dopamine release and uptake in Parkinsonian dyskinaesia. Brain 131, 1574–1587.

Lillethorup, T.P., Glud, A.N., Alstrup, A.K.O., Mikkelsen, T.W., Nielsen, E.H., Zaer, H., Doudet, D.J., Brooks, D.J., Sorensen, J.C.H., Orlowski, D., Landau, A.M., 2018. Nigrostriatal proteasome inhibition impairs dopamine neurotransmission and motor function in minipigs. Exp. Neurol. 303, 142–152.

Lin, X., Parisiadou, L., Sgobio, C., Liu, G., Yu, J., Sun, L., Shim, H., Gu, X.L., Luo, J., Long, C.X., Ding, J., Mateo, Y., Sullivan, P.H., Wu, L.G., Goldstein, D.S., Lovinger, D., Cai, H., 2012. Conditional expression of Parkinson's disease-related mutant alpha-synuclein in the midbrain dopaminergic neurons causes progressive neurodegeneration and degradation of transcription factor nuclear receptor related 1. J. Neurosci. 32, 9248–9264.

Ling, Z., Chang, Q.A., Tong, C.W., Leurgans, S.E., Lipton, J.W., Carvey, P.M., 2004. Rotenone potentiates dopamine neuron loss in animals exposed to lipopolysaccharide prenatally. Exp. Neurol. 190, 373–383.

Lo Bianco, C., Ridet, J.L., Schneider, B.L., Deglon, N., Aebischer, P., 2002. alpha - Synucleinopathy and selective dopaminergic neuron loss in a rat lentiviral-based model of Parkinson's disease. Proc. Natl. Acad. Sci. U. S. A. 99, 10813–10818.

Lu, X.H., Fleming, S.M., Meurers, B., Ackerson, L.C., Mortazavi, F., Lo, V., Hernandez, D., Sulzer, D., Jackson, G.R., Maidment, N.T., Chesselet, M.F., Yang, X.W., 2009. Bacterial artificial chromosome transgenic mice expressing a truncated mutant parkin exhibit age-dependent hypokinetic motor deficits, dopaminergic neuron degeneration, and accumulation of proteinase K-resistant alpha-synuclein. J. Neurosci. 29, 1962–1976.

Luk, K.C., Lee, V.M., 2014. Modeling Lewy pathology propagation in Parkinson's disease. Parkinsonism Relat. Disord. 20 (Suppl. 1), S85–S87.

Luk, K.C., Kehm, V., Carroll, J., Zhang, B., O'brien, P., Trojanowski, J.Q., Lee, V.M., 2012a. Pathological alpha-synuclein transmission initiates Parkinson-like neurodegeneration in nontransgenic mice. Science 338, 949–953.

Luk, K.C., Kehm, V.M., Zhang, B., O'brien, P., Trojanowski, J.Q., Lee, V.M., 2012b. Intracerebral inoculation of pathological alpha-synuclein initiates a rapidly progressive neurodegenerative alpha-synucleinopathy in mice. J. Exp. Med. 209, 975–986.

Luk, K.C., Covell, D.J., Kehm, V.M., Zhang, B., Song, I.Y., Byrne, M.D., Pitkin, R.M., Decker, S.C., Trojanowski, J.Q., Lee, V.M., 2016. Molecular and biological compatibility with host alpha-synuclein influences fibril pathogenicity. Cell Rep. 16, 3373–3387.

Lundblad, M., Andersson, M., Winkler, C., Kirik, D., Wierup, N., Cenci, M.A., 2002. Pharmacological validation of behavioural measures of akinesia and dyskinesia in a rat model of Parkinson's disease. Eur. J. Neurosci. 15, 120–132.

Lundblad, M., Decressac, M., Mattsson, B., Bjorklund, A., 2012. Impaired neurotransmission caused by overexpression of alpha-synuclein in nigral dopamine neurons. Proc. Natl. Acad. Sci. U. S. A. 109, 3213–3219.

Lynch, W.B., Tschumi, C.W., Sharpe, A.L., Branch, S.Y., Chen, C., Ge, G., Li, S., Beckstead, M.J., 2018. Progressively disrupted somatodendritic morphology in dopamine neurons in a mouse Parkinson's model. Mov. Disord. 33, 1928–1937.

MacInnes, N., Iravani, M.M., Perry, E., Piggott, M., Perry, R., Jenner, P., Ballard, C., 2008. Proteasomal abnormalities in cortical Lewy body disease and the impact of proteasomal inhibition within cortical and cholinergic systems. J. Neural Transm. (Vienna) 115, 869–878.

Magen, I., Chesselet, M.F., 2010. Genetic mouse models of Parkinson's disease The state of the art. Prog. Brain Res. 184, 53–87.

Manning-Bog, A.B., McCormack, A.L., Li, J., Uversky, V.N., Fink, A.L., Di Monte, D.A., 2002. The herbicide paraquat causes up-regulation and aggregation of alpha-synuclein in mice: paraquat and alpha-synuclein. J. Biol. Chem. 277, 1641–1644.

Masliah, E., Rockenstein, E., Veinbergs, I., Mallory, M., Hashimoto, M., Takeda, A., Sagara, Y., Sisk, A., Mucke, L., 2000. Dopaminergic loss and inclusion body formation in alpha-synuclein mice: implications for neurodegenerative disorders. Science 287, 1265–1269.

Mason, D.M., Nouraei, N., Pant, D.B., Miner, K.M., Hutchison, D.F., Luk, K.C., Stolz, J.F., Leak, R.K., 2016. Transmission of alpha-synucleinopathy from olfactory structures deep into the temporal lobe. Mol. Neurodegener. 11, 49.

Masuda-Suzukake, M., Nonaka, T., Hosokawa, M., Oikawa, T., Arai, T., Akiyama, H., Mann, D.M., Hasegawa, M., 2013. Prion-like spreading of pathological alpha-synuclein in brain. Brain 136, 1128–1138.

Matsui, H., Takahashi, R., 2018. Parkinson's disease pathogenesis from the viewpoint of small fish models. J. Neural Transm. (Vienna) 125, 25–33.

Maulik, M., Mitra, S., Bult-Ito, A., Taylor, B.E., Vayndorf, E.M., 2017. Behavioral phenotyping and pathological indicators of Parkinson's disease in C. elegans models. Front. Genet. 8, 77.

McFarland, N.R., Lee, J.S., Hyman, B.T., Mclean, P.J., 2009. Comparison of transduction efficiency of recombinant AAV serotypes 1, 2, 5, and 8 in the rat nigrostriatal system. J. Neurochem. 109, 838–845.

Mcnaught, K.S., Bjorklund, L.M., Belizaire, R., Isacson, O., Jenner, P., Olanow, C.W., 2002. Proteasome inhibition causes nigral degeneration with inclusion bodies in rats. Neuroreport 13, 1437–1441.

Meredith, G.E., Rademacher, D.J., 2011. MPTP mouse models of Parkinson's disease: an update. J. Parkinsons Dis. 1, 19–33.

Metz, G.A., Whishaw, I.Q., 2002. Drug-induced rotation intensity in unilateral dopamine-depleted rats is not correlated with end point or qualitative measures of forelimb or hindlimb motor performance. Neuroscience 111, 325–336.

Mezias, C., Rey, N., Brundin, P., Raj, A., 2020. Neural connectivity predicts spreading of alpha-synuclein pathology in fibril-injected mouse models: involvement of retrograde and anterograde axonal propagation. Neurobiol. Dis. 134, 104623.

Naudet, N., Antier, E., Gaillard, D., Morignat, E., Lakhdar, L., Baron, T., Bencsik, A., 2017. Oral exposure to paraquat triggers earlier expression of phosphorylated alpha-synuclein in the enteric nervous system of A53T mutant human alpha-synuclein transgenic mice. J. Neuropathol. Exp. Neurol. 76, 1046–1057.

Nielsen, M.S., Glud, A.N., Moller, A., Mogensen, P., Bender, D., Sorensen, J.C., Doudet, D., Bjarkam, C.R., 2016. Continuous MPTP intoxication in the Gottingen minipig results in chronic parkinsonian deficits. Acta Neurobiol. Exp. (Wars) 76, 199–211.

Nordstroma, U., Beauvais, G., Ghosh, A., Pulikkaparambil Sasidharan, B.C., Lundblad, M., Fuchs, J., Joshi, R.L., Lipton, J.W., Roholt, A., Medicetty, S., Feinstein, T.N., Steiner, J.A., Escobar Galvis, M.L., Prochiantz, A., Brundin, P., 2015. Progressive nigrostriatal terminal dysfunction and degeneration in the engrailed1 heterozygous mouse model of Parkinson's disease. Neurobiol. Dis. 73, 70–82.

Nouraei, N., Mason, D.M., Miner, K.M., Carcella, M.A., Bhatia, T.N., Dumm, B.K., Soni, D., Johnson, D.A., Luk, K.C., Leak, R.K., 2018. Critical appraisal of pathology transmission in the alpha-synuclein fibril model of Lewy body disorders. Exp. Neurol. 299, 172–196.

Nunes, I., Tovmasian, L.T., Silva, R.M., Burke, R.E., Goff, S.P., 2003. Pitx3 is required for development of substantia nigra dopaminergic neurons. Proc. Natl. Acad. Sci. U. S. A. 100, 4245–4250.

Osterberg, V.R., Spinelli, K.J., Weston, L.J., Luk, K.C., Woltjer, R.L., Unni, V.K., 2015. Progressive aggregation of alpha-synuclein and selective degeneration of lewy inclusion-bearing neurons in a mouse model of parkinsonism. Cell Rep. 10, 1252–1260.

Palmer, T., Ek, F., Enqvist, O., Olsson, R., Astrom, K., Petersson, P., 2017. Action sequencing in the spontaneous swimming behavior of zebrafish larvae—implications for drug development. Sci. Rep. 7, 3191.

Patterson, J.R., Duffy, M.F., Kemp, C.J., Howe, J.W., Collier, T.J., Stoll, A.C., Miller, K.M., Patel, P., Levine, N., Moore, D.J., Luk, K.C., Fleming, S.M., Kanaan, N.M., Paumier, K.L., El-Agnaf, O.M.A., Sortwell, C.E., 2019. Time course and magnitude of alpha-synuclein inclusion formation and nigrostriatal degeneration in the rat model of synucleinopathy triggered by intrastriatal alpha-synuclein preformed fibrils. Neurobiol. Dis. 130, 104525.

Paumier, K.L., Luk, K.C., Manfredsson, F.P., Kanaan, N.M., Lipton, J.W., Collier, T.J., Steece-Collier, K., Kemp, C.J., Celano, S., Schulz, E., Sandoval, I.M., Fleming, S., Dirr, E., Polinski, N.K., Trojanowski, J.Q., Lee, V.M., Sortwell, C.E., 2015. Intrastriatal injection of pre-formed mouse alpha-synuclein fibrils into rats triggers alpha-synuclein pathology and bilateral nigrostriatal degeneration. Neurobiol. Dis. 82, 185–199.

Peelaerts, W., Bousset, L., Van Der Perren, A., Moskalyuk, A., Pulizzi, R., Giugliano, M., Van Den Haute, C., Melki, R., Baekelandt, V., 2015. alpha-Synuclein strains cause distinct synucleinopathies after local and systemic administration. Nature 522, 340–344.

Pienaar, I.S., Harrison, I.F., Elson, J.L., Bury, A., Woll, P., Simon, A.K., Dexter, D.T., 2015. An animal model mimicking pedunculopontine nucleus cholinergic degeneration in Parkinson's disease. Brain Struct. Funct. 220, 479–500.

Polymeropoulos, M.H., Lavedan, C., Leroy, E., Ide, S.E., Dehejia, A., Dutra, A., Pike, B., Root, H., Rubenstein, J., Boyer, R., Stenroos, E.S., Chandrasekharappa, S., Athanassiadou, A., Papapetropoulos, T., Johnson, W.G., Lazzarini, A.M., Duvoisin, R.C., Di Iorio, G., Golbe, L.I., Nussbaum, R.L., 1997. Mutation in the alpha-synuclein gene identified in families with Parkinson's disease. Science 276, 2045–2047.

Pound, P., Ritskes-Hoitinga, M., 2018. Is it possible to overcome issues of external validity in preclinical animal research? Why most animal models are bound to fail. J. Transl. Med. 16, 304.

Powers, R., Lei, S., Anandhan, A., Marshall, D.D., Worley, B., Cerny, R.L., Dodds, E.D., Huang, Y., Panayiotidis, M.I., Pappa, A., Franco, R., 2017. Metabolic investigations of the molecular mechanisms associated with Parkinson's disease. Metabolites 7, 22.

Przedborski, S., Vila, M., 2003. The 1-methyl-4-phenyl-1,2,3,6-tetrahydropyridine mouse model: a tool to explore the pathogenesis of Parkinson's disease. Ann. N. Y. Acad. Sci. 991, 189–198.

Ramsay, R.R., Salach, J.I., Singer, T.P., 1986. Uptake of the neurotoxin 1-methyl-4-phenylpyridine (MPP+) by mitochondria and its relation to the inhibition of the mitochondrial oxidation of NAD+-linked substrates by MPP+. Biochem. Biophys. Res. Commun. 134, 743–748.

Ravenstijn, P.G., Merlini, M., Hameetman, M., Murray, T.K., Ward, M.A., Lewis, H., Ball, G., Mottart, C., De Ville De Goyet, C., Lemarchand, T., Van Belle, K., O'neill, M.J., Danhof, M., De Lange, E.C., 2008. The exploration of rotenone as a toxin for inducing Parkinson's disease in rats, for application in BBB transport and PK-PD experiments. J. Pharmacol. Toxicol. Methods 57, 114–130.

Redgrave, P., Rodriguez, M., Smith, Y., Rodriguez-Oroz, M.C., Lehericy, S., Bergman, H., Agid, Y., Delong, M.R., Obeso, J.A., 2010. Goal-directed and habitual control in the basal ganglia: implications for Parkinson's disease. Nat. Rev. Neurosci. 11, 760–772.

Reiner, A., Medina, L., Veenman, C.L., 1998. Structural and functional evolution of the basal ganglia in vertebrates. Brain Res. Brain Res. Rev. 28, 235–285.

Rey, N.L., Petit, G.H., Bousset, L., Melki, R., Brundin, P., 2013. Transfer of human alpha-synuclein from the olfactory bulb to interconnected brain regions in mice. Acta Neuropathol. 126, 555–573.

Rey, N.L., Steiner, J.A., Maroof, N., Luk, K.C., Madaj, Z., Trojanowski, J.Q., Lee, V.M., Brundin, P., 2016. Widespread transneuronal propagation of alpha-synucleinopathy triggered in olfactory bulb mimics prodromal Parkinson's disease. J. Exp. Med. 213, 1759–1778.

Rey, N.L., Wesson, D.W., Brundin, P., 2018. The olfactory bulb as the entry site for prion-like propagation in neurodegenerative diseases. Neurobiol. Dis. 109, 226–248.

Rockenstein, E., Mallory, M., Hashimoto, M., Song, D., Shults, C.W., Lang, I., Masliah, E., 2002. Differential neuropathological alterations in transgenic mice expressing alpha-synuclein from the platelet-derived growth factor and Thy-1 promoters. J. Neurosci. Res. 68, 568–578.

Rotman, A., Creveling, C.R., 1976. A rationale for the design of cell-specific toxic agents: the mechanism of action of 6-hydroxydopamine. FEBS Lett. 72, 227–230.

Rousselet, E., Joubert, C., Callebert, J., Parain, K., Tremblay, L., Orieux, G., Launay, J.M., Cohen-Salmon, C., Hirsch, E.C., 2003. Behavioral changes are not directly related to striatal monoamine levels, number of nigral neurons, or dose of parkinsonian toxin MPTP in mice. Neurobiol. Dis. 14, 218–228.

Sacrey, L.A., Alaverdashvili, M., Whishaw, I.Q., 2009. Similar hand shaping in reaching-for-food (skilled reaching) in rats and humans provides evidence of homology in release, collection, and manipulation movements. Behav. Brain Res. 204, 153–161.

Savolainen, M.H., Albert, K., Airavaara, M., Myohanen, T.T., 2017. Nigral injection of a proteasomal inhibitor, lactacystin, induces widespread glial cell activation and shows various phenotypes of Parkinson's disease in young and adult mouse. Exp. Brain Res. 235, 2189–2202.

Schneider, J.S., 1989. Levodopa-induced dyskinesias in parkinsonian monkeys: relationship to extent of nigrostriatal damage. Pharmacol. Biochem. Behav. 34, 193–196.

Simola, N., Morelli, M., Carta, A.R., 2007. The 6-hydroxydopamine model of Parkinson's disease. Neurotox. Res. 11, 151–167.

Singleton, A.B., Farrer, M., Johnson, J., Singleton, A., Hague, S., Kachergus, J., Hulihan, M., Peuralinna, T., Dutra, A., Nussbaum, R., Lincoln, S., Crawley, A., Hanson, M., Maraganore, D., Adler, C., Cookson, M.R., Muenter, M., Baptista, M., Miller, D., Blancato, J., Hardy, J., Gwinn-Hardy, K., 2003. alpha-Synuclein locus triplication causes Parkinson's disease. Science 302, 841.

Smidt, M.P., Van Schaick, H.S., Lanctot, C., Tremblay, J.J., Cox, J.J., Van Der Kleij, A.A., Wolterink, G., Drouin, J., Burbach, J.P., 1997. A homeodomain gene Ptx3 has highly restricted brain expression in mesencephalic dopaminergic neurons. Proc. Natl. Acad. Sci. U. S. A. 94, 13305–13310.

Sonnier, L., Le Pen, G., Hartmann, A., Bizot, J.C., Trovero, F., Krebs, M.O., Prochiantz, A., 2007. Progressive loss of dopaminergic neurons in the ventral midbrain of adult mice heterozygote for Engrailed1. J. Neurosci. 27, 1063–1071.

Spillantini, M.G., Crowther, R.A., Jakes, R., Hasegawa, M., Goedert, M., 1998. alpha-Synuclein in filamentous inclusions of Lewy bodies from Parkinson's disease and dementia with Lewy bodies. Proc. Natl. Acad. Sci. U. S. A. 95, 6469–6473.

Staal, R.G., Hogan, K.A., Liang, C.L., German, D.C., Sonsalla, P.K., 2000. In vitro studies of striatal vesicles containing the vesicular monoamine transporter (VMAT2): rat versus mouse differences in sequestration of 1-methyl-4-phenylpyridinium. J. Pharmacol. Exp. Ther. 293, 329–335.

Stefanova, N., Kaufmann, W.A., Humpel, C., Poewe, W., Wenning, G.K., 2012. Systemic proteasome inhibition triggers neurodegeneration in a transgenic mouse model expressing human alpha-synuclein under oligodendrocyte promoter: implications for multiple system atrophy. Acta Neuropathol. 124, 51–65.

Suarez, L.M., Alberquilla, S., Garcia-Montes, J.R., Moratalla, R., 2018. Differential synaptic remodeling by dopamine in direct and indirect striatal projection neurons in Pitx3(-/-) mice, a genetic model of Parkinson's disease. J. Neurosci. 38, 3619–3630.

Sundstrom, E., Samuelsson, E.B., 1997. Comparison of key steps in 1-methyl-4-phenyl-1,2,3,6-tetrahydropyridine (MPTP) neurotoxicity in rodents. Pharmacol. Toxicol. 81, 226–231.

Thakur, P., Breger, L.S., Lundblad, M., Wan, O.W., Mattsson, B., Luk, K.C., Lee, V.M.Y., Trojanowski, J.Q., Bjorklund, A., 2017. Modeling Parkinson's disease pathology by combination of fibril seeds and alpha-synuclein overexpression in the rat brain. Proc. Natl. Acad. Sci. U. S. A. 114, E8284–E8293.

Thiruchelvam, M., Brockel, B.J., Richfield, E.K., Baggs, R.B., Cory-Slechta, D.A., 2000. Potentiated and preferential effects of combined paraquat and maneb on nigrostriatal dopamine systems: environmental risk factors for Parkinson's disease? Brain Res. 873, 225–234.

Thiruchelvam, M., McCormack, A., Richfield, E.K., Baggs, R.B., Tank, A.W., Di Monte, D.A., Cory-Slechta, D.A., 2003. Age-related irreversible progressive nigrostriatal dopaminergic neurotoxicity in the paraquat and maneb model of the Parkinson's disease phenotype. Eur. J. Neurosci. 18, 589–600.

Thompson, L., Bjorklund, A., 2012. Survival, differentiation, and connectivity of ventral mesencephalic dopamine neurons following transplantation. Prog. Brain Res. 200, 61–95.

Tofaris, G.K., Garcia Reitbock, P., Humby, T., Lambourne, S.L., O'connell, M., Ghetti, B., Gossage, H., Emson, P.C., Wilkinson, L.S., Goedert, M., Spillantini, M.G., 2006. Pathological changes in dopaminergic nerve cells of the substantia nigra and olfactory bulb in mice transgenic for truncated human alpha-synuclein(1-120): implications for Lewy body disorders. J. Neurosci. 26, 3942–3950.

Tyson, T., Steiner, J.A., Brundin, P., 2016. Sorting out release, uptake and processing of alpha-synuclein during prion-like spread of pathology. J. Neurochem. 139 (Suppl. 1), 275–289.

Ulusoy, A., Decressac, M., Kirik, D., Bjorklund, A., 2010. Viral vector-mediated overexpression of alpha-synuclein as a progressive model of Parkinson's disease. Prog. Brain Res. 184, 89–111.

Ungerstedt, U., 1968. 6-Hydroxy-dopamine induced degeneration of central monoamine neurons. Eur. J. Pharmacol. 5, 107–110.

Van Den Munckhof, P., Luk, K.C., Ste-Marie, L., Montgomery, J., Blanchet, P.J., Sadikot, A.F., Drouin, J., 2003. Pitx3 is required for motor activity and for survival of a subset of midbrain dopaminergic neurons. Development 130, 2535–2542.

Van Der Perren, A., Toelen, J., Casteels, C., Macchi, F., Van Rompuy, A.S., Sarre, S., Casadei, N., Nuber, S., Himmelreich, U., Osorio Garcia, M.I., Michotte, Y., D'hooge, R., Bormans, G., Van Laere, K., Gijsbers, R., Van Den Haute, C., Debyser, Z., Baekelandt, V., 2015a. Longitudinal follow-up and characterization of a robust rat model for Parkinson's disease based on overexpression of alpha-synuclein with adeno-associated viral vectors. Neurobiol. Aging 36, 1543–1558.

Van Der Perren, A., Van Den Haute, C., Baekelandt, V., 2015b. Viral vector-based models of Parkinson's disease. Curr. Top. Behav. Neurosci. 22, 271–301.

Van Der Worp, H.B., Howells, D.W., Sena, E.S., Porritt, M.J., Rewell, S., O'collins, V., Macleod, M.R., 2010. Can animal models of disease reliably inform human studies? PLoS Med. 7, e1000245.

Vernon, A.C., Crum, W.R., Johansson, S.M., Modo, M., 2011. Evolution of extra-nigral damage predicts behavioural deficits in a rat proteasome inhibitor model of Parkinson's disease. PLoS One 6, e17269.

Volpicelli-Daley, L.A., Luk, K.C., Patel, T.P., Tanik, S.A., Riddle, D.M., Stieber, A., Meaney, D.F., Trojanowski, J.Q., Lee, V.M., 2011. Exogenous alpha-synuclein fibrils induce Lewy body pathology leading to synaptic dysfunction and neuron death. Neuron 72, 57–71.

Volpicelli-Daley, L.A., Gamble, K.L., Schultheiss, C.E., Riddle, D.M., West, A.B., Lee, V.M., 2014. Formation of alpha-synuclein Lewy neurite-like aggregates in axons impedes the transport of distinct endosomes. Mol. Biol. Cell 25, 4010–4023.

Volpicelli-Daley, L.A., Kirik, D., Stoyka, L.E., Standaert, D.G., Harms, A.S., 2016. How can rAAV-alpha-synuclein and the fibril alpha-synuclein models advance our understanding of Parkinson's disease? J. Neurochem. 139 (Suppl. 1), 131–155.

Wegrzynowicz, M., Bar-On, D., Calo, L., Anichtchik, O., Iovino, M., Xia, J., Ryazanov, S., Leonov, A., Giese, A., Dalley, J.W., Griesinger, C., Ashery, U., Spillantini, M.G., 2019. Depopulation of dense alpha-synuclein aggregates is associated with rescue of dopamine neuron dysfunction and death in a new Parkinson's disease model. Acta Neuropathol. 138, 575–595.

Wichmann, T., Bergman, H., Delong, M.R., 2018. Basal ganglia, movement disorders and deep brain stimulation: advances made through non-human primate research. J. Neural Transm. (Vienna) 125, 419–430.

Winkler, C., Kirik, D., Bjorklund, A., Cenci, M.A., 2002. L-DOPA-induced dyskinesia in the intrastriatal 6-hydroxydopamine model of parkinson's disease: relation to motor and cellular parameters of nigrostriatal function. Neurobiol. Dis. 10, 165–186.

Xie, W., Li, X., Li, C., Zhu, W., Jankovic, J., Le, W., 2010. Proteasome inhibition modeling nigral neuron degeneration in Parkinson's disease. J. Neurochem. 115, 188–199.

Xilouri, M., Brekk, O.R., Stefanis, L., 2013. alpha-Synuclein and protein degradation systems: a reciprocal relationship. Mol. Neurobiol. 47, 537–551.

Xu, P.Y., Liang, R., Jankovic, J., Hunter, C., Zeng, Y.X., Ashizawa, T., Lai, D., Le, W.D., 2002. Association of homozygous 7048G7049 variant in the intron six of Nurr1 gene with Parkinson's disease. Neurology 58, 881–884.

Xu, Q., Shenoy, S., Li, C., 2012. Mouse models for LRRK2 Parkinson's disease. Parkinsonism Relat. Disord. 18 (Suppl. 1), S186–S189.

Yamada, M., Iwatsubo, T., Mizuno, Y., Mochizuki, H., 2004. Overexpression of alpha-synuclein in rat substantia nigra results in loss of dopaminergic neurons, phosphorylation of alpha-synuclein and activation of caspase-9: resemblance to pathogenetic changes in Parkinson's disease. J. Neurochem. 91, 451–461.

Yartsev, M.M., 2017. The emperor's new wardrobe: rebalancing diversity of animal models in neuroscience research. Science 358, 466–469.

Zheng, K., Heydari, B., Simon, D.K., 2003. A common NURR1 polymorphism associated with Parkinson disease and diffuse Lewy body disease. Arch. Neurol. 60, 722–725.

Zigmond, M.J., 1997. Do compensatory processes underlie the preclinical phase of neurodegenerative disease? Insights from an animal model of parkinsonism. Neurobiol. Dis. 4, 247–253.

Selective neuronal vulnerability in Parkinson's disease

3

Patricia Gonzalez-Rodriguez, Enrico Zampese, D. James Surmeier*

Department of Physiology, Feinberg School of Medicine, Northwestern University, Chicago, IL, United States

*Corresponding author: Tel.: +1-312-503-4904, e-mail address: j-surmeier@northwestern.edu

Abstract

Parkinson's disease (PD) is the second most common neurodegenerative disease, disabling millions worldwide. Despite the imperative PD poses, at present, there is no cure or means of slowing progression. This gap is attributable to our incomplete understanding of the factors driving pathogenesis. Research over the past several decades suggests that both cell-autonomous and non-cell autonomous processes contribute to the neuronal dysfunction underlying PD symptoms. The thesis of this review is that an intersection of these processes governs the pattern of pathology in PD. Studies of substantia nigra pars compacta (SNc) dopaminergic neurons, whose loss is responsible for the core motor symptoms of PD, suggest that they have a combination of traits—a long, highly branched axon, autonomous activity, and elevated mitochondrial oxidant stress—that predispose them to non-cell autonomous drivers of pathogenesis, like misfolded forms of alpha-synuclein (α-SYN) and inflammation. The literature surrounding these issues will be briefly summarized, and the translational implications of an intersectional hypothesis of PD pathogenesis discussed.

Keywords

Parkinson's disease, Neurodegeneration, Axon, Mitochondrial dysfunction, Oxidant stress, Calcium, α-synuclein, Synapse, Lewy pathology, Propagation, Aging

Abbreviations

ATP	adenosine triphosphate
BF	basal forebrain
DMV	dorsal motor nucleus of the vagus
ER	endoplasmic reticulum
ETC	electron transfer chain
ILT	intralaminar thalamus
IMM	inner mitochondrial membrane

Progress in Brain Research, Volume 252, ISSN 0079-6123, https://doi.org/10.1016/bs.pbr.2020.02.005

LC	locus coeruleus
LP	Lewy pathology
MAMs	mitochondria-associated membranes
MCI	mitochondrial complex I
mtDNA	mitochondrial DNA
OXPHOS	oxidative phosphorylation
PD	Parkinson's disease
PFFs	pre-formed α-SYN fibrils
PPN	pedunculopontine nucleus
RN	raphe nucleus
ROS	reactive oxygen species
RYRs	ryanodine receptors
SNc	substantia nigra pars compacta
TCA	tricarboxylic acid
α-SYN	-alpha-synuclein

1 Introduction

James Parkinson described the "shaking palsy" now referred to as PD more than 200 years ago (Parkinson, 2002). PD is a progressive, aging-dependent, psychomotor disorder, afflicting at least 0.3% of the worldwide population and over 3% of those over 80 years old (Pringsheim et al., 2014). The causes of PD remain to be unequivocally determined and there is no proven strategy for slowing disease progression (Fahn and Sulzer, 2004; Hung and Schwarzschild, 2007).

There are two defining features of PD. One is the degeneration of SNc dopaminergic neurons. It was recognized some 60 years ago that this loss was responsible for the slowness of movement (bradykinesia) and stiffness (rigidity) characteristic of PD (Poewe et al., 2017). This recognition also underlies the use of the dopamine precursor levodopa to treat the motor symptoms of the disease. Although the degeneration of SNc dopaminergic neurons is an early sign of PD, neuronal loss in other regions of the brain is a common feature of PD. Neurons in the locus coeruleus (LC), basal forebrain (BF), pedunculopontine nucleus (PPN), intralaminar thalamus (ILT) and the dorsal motor nucleus of the vagus (DMV) are among those regions with well-documented cell loss late in the progression of PD (Surmeier et al., 2017a). However, the magnitude and timing of this degeneration is poorly defined because of the constraints of working with post-mortem human tissue (Giguere et al., 2018).

Another key feature of PD is the accumulation of intracellular, proteinaceous aggregates in neurons (Goedert, 2001). Named after Fritz Lewy, who discovered them early in the last century, this pathology is distributed in the brain, as well as in the peripheral autonomic and enteric nervous systems (Braak and Del Tredici, 2017). Lewy pathology (LP) is heterogeneous in composition, including misfolded α-SYN, lipids, mitochondrial fragments and a range of other proteins (Shults, 2006). Based upon a comparative analysis of LP in the brains of symptomatic PD patients and nominally healthy control subjects, Braak and colleagues have advanced the hypothesis that LP is seeded by pathology in either the gastrointestinal tract and/or the olfactory epithelium and then spreads through synaptically coupled neural networks into the brain (Braak et al., 2003, Brundin and Melki, 2017). Although the pathogen underlying the spreading was initially undefined, in recent years the attention of the research community has focused on α-SYN. A natively unfolded protein that plays a role in synaptic transmission, α-SYN can misfold and aggregate into oligomeric and fibrillar forms (Goedert et al., 2013). Because these misfolded forms of α-SYN can recruit unfolded forms of α-SYN and can spread from one cell to another, it has been proposed that α-SYN oligomers or fibrils act like prions to spread from peripheral seeding sites through the brain, assuming the role of the pathogen originally proposed by Braak (Brundin and Melki, 2017). Once established, intracellular α-SYN aggregates are hypothesized to cause neuronal dysfunction and death.

As appealing as this hypothesis is in its simplicity, there are several key gaps in the evidence supporting it. Two prominent, yet unanswered questions will be explored here. First, while it is clear that α-SYN pathology can spread through synaptically coupled networks in vivo, in humans it does so in a restricted fashion; that is, LP is evident in only a subset of neurons in synaptically coupled networks, suggesting that there are other factors governing the propagation or the persistence of pathology. Second, the relationship between LP, neuronal dysfunction, and cell death is unclear. LP can apparently exist for decades in many parts of the brain without causing obvious neuronal death or dysfunction. Moreover, neuronal degeneration occurs in PD without concomitant LP (Schneider and Alcalay, 2017; Simon et al., 2020; Surmeier et al., 2017a). In fact, in some well-characterized PD cases, particularly those linked to genetic mutations, there is no sign of LP even at late stages of the disease (Berg et al., 2014).

The heterogeneity in PD pathology has led to the hypothesis that PD has multiple etiologies (Espay et al., 2017). While this ultimately may turn out to be the case, a more parsimonious position at this stage in our pursuit of a deeper understanding of PD pathogenesis is that there are common threads that tie the disease together.

2 Cell autonomous determinants of vulnerability

SNc dopaminergic neurons have a collection of traits that appear to render them vulnerable to aging, and environmental toxins that compromise mitochondrial function, as well as to genetic mutations associated with PD. The most prominent of these is their massive axonal arbor. In a mouse, the axon of a single SNc dopaminergic neuron innervates much of the basal ganglia, having a length of over 40 cm with several hundred thousand dopamine release sites (Andén et al., 1966; Arbuthnott and Wickens, 2007; Gauthier et al., 1999; Groves et al., 1994; Matsuda et al., 2009; Prensa and Parent, 2001). In humans, the disproportionate growth of the telencephalon innervated by SNc dopaminergic neurons has pushed this arbor even more, increasing the estimated number of release sites in an individual axon to 1–2 million (Bolam and Pissadaki, 2012; Diederich et al., 2019; Hardman et al., 2002;

Pissadaki and Bolam, 2013). It is of some note that Braak and colleagues suggested early on that a long, highly branched and poorly myelinated axon was a common feature of neurons manifesting LP in PD (Braak et al., 2004; Braak and Del Tredici, 2004).

Why should a long, unmyelinated axon be a risk-factor? This question has been the subject of considerable speculation. One set of ideas is focused on the regional bioenergetic demands posed by the axon. Computational studies suggest that regenerative activity in the axon is at the heart of vulnerability (Pissadaki and Bolam, 2013). Another possibility is that the energetic costs associated with vesicular recycling, sequestration of transmitter and release underlie vulnerability (Bolam and Pissadaki, 2012; Harris et al., 2012). For SNc dopaminergic neurons, both of these bioenergetic demands are sustained, as the maintenance of dopaminergic tone in the basal ganglia is a key feature of this network (Gerfen and Surmeier, 2011; Howe and Dombeck, 2016; Surmeier et al., 2014). While these regional bioenergetic demands are undoubtedly substantial, it is not clear that in and of themselves, these demands create "stress" or increased vulnerability. There have been no direct measurements of bioenergetic demand in dopaminergic axons, although the advent of new genetically encoded optical sensors put these measurements within reach (Berg et al., 2009; Giguere et al., 2019; Graves et al., 2020; Pacelli et al., 2015).

Another consideration is the global anabolic burden created by the sheer size of dopaminergic axons. Although the turnover rates for axonal proteins and organelles are in general poorly defined, it is reasonable to assume that the proteostatic burden created by a massive axon (for production, degradation and transport of axonal proteins) is substantial (Gennerich and Vale, 2009; Millecamps and Julien, 2013; Misgeld and Schwarz, 2017; Schmieg et al., 2014). Indeed, recent work has shown that manipulating the size of the axonal arbor shapes the vulnerability of SNc dopaminergic neurons to stressors (Giguere et al., 2019; Pacelli et al., 2015).

This risk-factor is shared by many of the other neurons manifesting signs of pathology in PD. Adrenergic LC neurons have massive axonal (Aston-Jones and Waterhouse, 2016; Jones and Moore, 1977; Schwarz and Luo, 2015; Ungerstedt, 1971), as do serotonergic raphe nucleus (RN) neurons (Hornung, 2003; Jacobs and Azmitia, 1992; Maeda et al., 1989; Michelsen et al., 2007; Muzerelle et al., 2016; Vertes and Linley, 2007), cholinergic neurons in the DMV (Greene, 2014), cholinergic neurons in the PPN (Dautan et al., 2014, 2016; Lavoie and Parent, 1994; Mena-Segovia et al., 2008; Ros et al., 2010; Tubert et al., 2019) and cholinergic neurons in the BF (Wu et al., 2014).

Another unusual feature of SNc dopaminergic neurons is that they are slow, autonomous pacemakers (1–4 Hz) with broad spikes (Surmeier et al., 2017a). This regular pacemaking is interrupted by synaptically driven bursts or pauses (Fujimura and Matsuda, 1989; Grace and Bunney, 1983; Grace and Onn, 1989; Hainsworth et al., 1991; Nedergaard et al., 1993). Pacemaking is accompanied by large fluctuations in cytosolic Ca^{2+} concentration that depend upon opening of plasma membrane, L-type (Ca_v1) Ca^{2+} channels (Chan et al., 2007; Guzman et al., 2009; Mercuri et al., 1994; Nedergaard et al., 1993). Although it was originally thought that these channels were essential for pacemaking, more recent work using channel-specific concentrations of dihydropyridines and molecular approaches have shown that this is not the case and pacemaking relies upon voltage-dependent Na^+ and cation channels (e.g., NALCN, HCN channels), as in other pacemaking neurons (Guzman et al., 2009, 2018; Puopolo et al., 2007).

What role do L-type, Ca_v1 Ca^{2+} channels in SNc dopaminergic neurons play? The engagement of Ca_v1 Ca^{2+} channels undoubtedly enhances the robustness of pacemaking by providing a depolarizing, inward current late in the interspike interval (Koschak et al., 2001; Putzier et al., 2009; Scholze et al., 2001). However, Ca_v1 channels play other roles as well. One of the key roles Ca_v1 channels serve in neurons is to provide an intracellular signal that this correlated with spiking. For example, Ca_v1 channels are critical to homeostatic plasticity and regulate activity-induced gene expression (Berridge, 1998; Brini et al., 2014; Clapham, 2007; Frank, 2014). Another closely related role, which was first described in muscle (Diaz-Vegas et al., 2018; Viola et al., 2009), is to regulate mitochondrial oxidative phosphorylation (OXPHOS). A clue that this was happening came from the recognition that while the cytosolic oscillation depends upon opening of plasma membrane $Ca_v1.3$ channels, it also depends upon release of Ca^{2+} from intracellular, endoplasmic reticulum (ER) stores (Sanchez-Padilla et al., 2014). Unpublished work by our group (Zampese et al.) has demonstrated that opening of $Ca_v1.3$ channels triggers ER release of Ca^{2+} through ryanodine receptors (RYRs). The low intrinsic Ca^{2+} buffering capacity of SNc dopaminergic neurons (Foehring et al., 2009) ensures that this Ca^{2+} "wave" propagates throughout the somatodendritic region, leading to elevations in $[Ca^{2+}]$ that reach into the high nanomolar or possibly micromolar range with each action potential (Guzman et al., 2018). Although this oscillation ensures efficient activation of Ca^{2+}-activated K^+ channels that regulate repetitive spiking (Guzman et al., 2009; Iyer et al., 2017; Ping and Shepard, 1996; Shepard and Bunney, 1991), is also promotes the entry of Ca^{2+} into mitochondria at specialized junctions between the ER and mitochondria called mitochondria-associated membranes (MAMs) (Csordas et al., 2018). Because of the restricted diffusion at these junctions, Ca^{2+} concentration rises high enough to induce permeation through channels in the inner mitochondrial membrane (IMM). The IMM mitochondrial Ca^{2+} uptake complex (MCUC) is formed by a channel protein (MCU) (Baughman et al., 2011; De Stefani et al., 2011) and several regulatory proteins, that are required for the correct assembly and gating (Kamer and Mootha, 2015; Penna et al., 2018). The MCUC opens only when exposed to high Ca^{2+} concentrations that can be achieved at the MAM (Csordas et al., 2010, 2018; Giacomello et al., 2010). Ca^{2+} entry into the mitochondrial matrix through the MCUC dis-inhibits enzymes of the tricarboxylic acid (TCA) cycle, leading to the generation of reducing equivalents for the electron transfer chain (ETC) and adenosine triphosphate (ATP) production. Elevated cytosolic $[Ca^{2+}]$ also increases mitochondrial transport of substrates necessary for OXPHOS (e.g., malate/aspartate shuttle) (Denton, 2009; Gellerich et al., 2013; Griffiths and Rutter, 2009; Rossi et al., 2019). Thus, the Ca^{2+} oscillation in SNc dopaminergic neurons serves to stimulate mitochondrial ATP generation in anticipation of the demands associated with activity; that is, Ca^{2+} entry through Ca_v1 channels serves as a feed-forward (anticipatory) bioenergetic control signal.

It is likely that this feed-forward control mechanism is present in most, if not all, excitable cells. What appears to vary is the gain of the Ca^{2+} signal. In SNc dopaminergic neurons the coupling between $Ca_v1.3$ channels, ER RYRs and mitochondria is very robust, leading to a high feedforward gain. Why is the gain so high? One reason is that pacemaking creates a high basal demand because of the need to re-establish ionic gradients dissipated by regenerative activity (Harris et al., 2012; Pissadaki and Bolam, 2013; Surmeier et al., 2017a; Surmeier and Schumacker, 2013). Another, less well appreciated demand, may be the anabolic demand created by the axon. It is reasonable to assume that these demands scale with activity and the size of the axonal arbor. Viewed in this way, the massive axonal arbor of SNc dopaminergic neurons would pose an enormous bioenergetic burden per spike. Meeting this demand, without allowing ATP levels to fall or without failing to meet the anabolic needs of the axon (resulting in impaired synaptic transmission), appears to require that feedforward stimulation of mitochondrial OXPHOS have a high gain.

The downside of this feedforward control system is mitochondrial oxidant stress. It is inescapable that the flux of electrons through the ETC leads to the generation of reactive oxygen species (ROS) (Brookes et al., 2004; Murphy, 2009). Moreover, because stimulation of OXPHOS is not always accompanied by a drop in cytosolic ATP, substrate inhibition of mitochondrial complex V will create periods of both increased generation of reducing equivalents and hyperpolarized IMM, leading to "stalling" of electrons along the ETC and increase superoxide generation by complex I and III (Murphy, 2009; Nickel et al., 2014). While mitochondrial have efficient means of scavenging ROS, these mechanisms are imperfect and there will inevitably be damages to mitochondrial DNA (mtDNA) and proteins that impair mitochondrial function (Nickel et al., 2014; Scheibye-Knudsen et al., 2015). Indeed, high levels of Ca^{2+}-dependent mitochondrial oxidant stress have been observed in SNc dopaminergic neurons (Dryanovski et al., 2013; Guzman et al., 2010, 2018; Lieberman et al., 2017), as well as in LC and DMV neurons (Goldberg et al., 2012; Sanchez-Padilla et al., 2014). This oxidant stress is sufficient to damage mitochondrial proteins in vivo and to promote their trafficking into lysosomes (Guzman et al., 2018). Reducing the gain of the feedforward control system by systemic administration of dihydropyridines lowers mitochondrial oxidant stress, diminishes damage of mitochondrial proteins and elevates the abnormally low mitochondrial mass in SNc dopaminergic neurons (Guzman et al., 2018; Liang et al., 2007).

Another factor that may contribute to the vulnerability of SNc dopaminergic neurons is dopamine. Although there has been a great deal of speculation about the toxicity of auto-oxidation products of dopamine or metabolites (Sulzer and Surmeier, 2013), the absence of a clear mechanistic framework for how they might be generated in vivo has undermined enthusiasm for this idea. However, recent work has put this hypothesis on new footing. First, dopamine metabolism by monoamine oxidase, which is anchored to the outer membrane of mitochondria, leads to shuttling of electrons into the ETC and OXPHOS (Graves et al., 2020). While this helps support the bioenergetic demands associated with dopaminergic neurotransmission, it can also lead to mitochondrial oxidant stress, particularly in axons where dopamine turnover rates are high. Mitochondrial oxidant stress also promotes oxidation of dopamine, which not only increases the aggregation of α-SYN, but also impairs lysosomal function (Burbulla et al., 2017). As cytosolic dopamine levels appear to be higher in human dopaminergic neurons than in mouse dopaminergic neurons (Burbulla et al., 2017), this cascade of events could be a key factor in the selective vulnerability of human dopaminergic neurons in PD. The absence of a disease accelerating effect of systemic levodopa in symptomatic PD patients (Fahn and Parkinson Study, 2005) may stem from the fact that at this stage in disease, dopaminergic axons—where this cascade manifests itself most clearly—have already lost their phenotype or have degenerated (Kordower et al., 2013).

Although this model is based upon work in rodents, there are good reasons to think it is relevant to the human condition. First, postmortem analysis of PD brains has revealed that there is a loss of functional mitochondrial complex I (MCI) in SNc and in surviving SNc dopaminergic neurons (Cooper et al., 1992; Reeve et al., 2014; Schapira et al., 1990) and an elevation in mtDNA deletions (Bender et al., 2006; Dolle et al., 2016; Elstner et al., 2011a, 2011b; Kraytsberg et al., 2006; Muller et al., 2013; Sanders et al., 2014)—both are telltale signs of sustained mitochondrial oxidant stress. Second, recessive genetic mutations associated with PD impair either mitochondrial oxidant defenses or impair the turnover of mitochondrial proteins (either by disrupting macroautophagy or trafficking of mitochondria-derived vesicles to lysosomes) (Ashrafi et al., 2014; Krebiehl et al., 2010; McLelland et al., 2014; Narendra et al., 2008; Pickrell and Youle, 2015). Clearly, in the presence of a demand on these mechanisms, loss of function mutations are likely to have an impact on neuronal viability. It is also worth noting that mitochondrial damage can trigger innate immunity and inflammation, which also has been implicated in PD pathogenesis (Pickrell and Youle, 2015; Matheoud et al., 2016, 2019).

Many of the other neurons that are at-risk in PD appear to have characteristics that are close to those of SNc dopaminergic neurons (Surmeier et al., 2017a,b). LC neurons manifest spontaneous rhythmic spiking (0.5–5 Hz) (Alreja and Aghajanian, 1991; Graham and Aghajanian, 1971; Masuko et al., 1986; Williams et al., 1984) that are correlated with waking and arousal (Aston-Jones and Bloom, 1981; Chu and Bloom, 1973; Foote et al., 1980; Gervasoni et al., 1998). These neurons express $Ca_v1.3$ channels, have large cytosolic Ca^{2+} oscillations and elevated mitochondrial oxidant stress (Imber and Putnam, 2012; Matschke et al., 2015; Sanchez-Padilla et al., 2014; Sukiasyan et al., 2009). DMV neurons also are spontaneously active (0.5-5 Hz), have large cytosolic Ca^{2+} oscillations, $Ca_v1.3$ channels and elevated mitochondrial oxidant stress (Goldberg et al., 2012; Marks et al., 1993; Mo et al., 1992; Travagli et al., 1991, 1992). RN neurons manifest a sustained, regular spiking (0.2–5 Hz) during the waking state (McGinty and Harper, 1976; Trulson and Jacobs, 1979; Urbain et al., 2006). PPN neurons manifest spontaneous activity correlated with the waking state; particularly vulnerable PPN cholinergic neurons have a regular, sustained spiking rate (Petzold et al., 2015; Scarnati et al., 1987; Takakusaki and Kitai, 1997).

3 Non-cell autonomous determinants and α-SYN

In 1997, the linkage between mutations in *SNCA*, the gene coding for α-SYN, and familial forms of PD was first established (Polymeropoulos et al., 1997). In the same year, Spillantini et al. (1997) reported that α-SYN was a major component of LP. Together, these observations put α-SYN and LP center stage in the pursuit of pathogenic mechanisms in PD. Subsequently, the genetic evidence implicating α-SYN has grown to include other point mutations, as well as duplication and triplication variants (Ibáñez et al., 2004; Krüger et al., 1998; Pasanen et al., 2014; Proukakis et al., 2013; Singleton et al., 2003; Thakur et al., 2019; Zarranz et al., 2004).

Using the new immunocytochemical tools for studying LP in postmortem tissue, Braak et al. introduced the idea that LP emerges in a staged manner, arising initially in the olfactory bulb or the DMV and then spreading to other brain regions through synaptically coupled networks (Braak et al., 2003; Brundin and Melki, 2017). The initiation of the pathology was hypothesized to be triggered by a pathogen that was either inhaled or consumed in food. The proposition that PD was due to an infectious agent, just like a cold, had obvious conceptual appeal. It was a simple explanation for what appeared to be a complicated, mysterious disease. Moreover, the recognition that LP was not restricted to the SNc, fundamentally changed thinking about the clinical picture of PD; it was no longer simply a movement disorder, but rather a multi-faceted disease with a variety of concomitant symptoms reflecting the distributed LP in PD patients (Braak et al., 2004; Hawkes et al., 2007).

As ground-breaking as Braak's idea was, the limitations in the data upon which it was based needs to be recognized. First, there was no longitudinal data demonstrating that one stage of pathology actually led to another. This is particularly important for the assertion about where the pathology starts, as this was inferred from the analysis of brains taken from asymptomatic patients; there still is no reliable way of identifying prodromal PD patients. Second, the sample size used to draw conclusions was small. Subsequent work has found that a large percentage (∼50%) of PD patients do not conform to Braak staging, arguing that the evolution of LP is not as stereotyped as proposed (Halliday et al., 2011; Jellinger, 2009; Kalaitzakis et al., 2008). In fact, some PD patients lack any discernible LP (Berg et al., 2014). The variability also extends to the relationship between clinical state and underlying LP (Espay et al., 2017; Jellinger, 2009).

Although Braak staging has its limitations, the proposition that LP spreads in the brain has continued to gain support. One of the most influential discoveries in this regard came from postmortem analysis of fetal transplants to the striatum of PD patients. In 2008, two studies reported that dopaminergic neurons in transplants had proteinaceous inclusions that strongly resembled LP (Kordower et al., 2008; Li et al., 2008), although another study failed to find similar pathology (Mendez et al., 2008). It was inferred that LP spread from the host to the transplant. While this was possible, it was also possible that the stress created by transplantation led to de novo protein misfolding, particularly given the fact that the LP was seen in a small percentage of dopaminergic neurons in the graft and that the pathology was not evident in other cell types within the graft. Nevertheless, subsequent work has provided unequivocal support for the idea that misfolded forms of α-SYN can be released, taken up by neighboring neurons and seed the formation of Lewy-like intracellular inclusions (Desplats et al., 2009; Hansen et al., 2011). The most influential demonstration of this idea initially came from the studies Lee and colleagues showing that pre-formed α-SYN fibrils (PFFs), when injected into the brain, spread from the inoculation site (Luk et al., 2012). This landmark study has been followed by a number of confirmations (Henderson et al., 2019a; Masuda-Suzukake et al., 2013; Peelaerts et al., 2015), including the demonstration that extracts of human LP can spread in a similar way (Recasens et al., 2014). Moreover, the reliance of propagation upon endogenous α-SYN has led to the hypothesis that the spread of LP is prion-like; that is, misfolded α-SYN not only spreads from cell to cell, but it templates unfolded α-SYN to form new pathological seeds (Brundin et al., 2010; Brundin and Melki, 2017). Recent work has gone so far as to demonstrate that injection of α-SYN PFFs into the gastrointestinal tract of rodents recapitulates key features of Braak staging, including the eventual degeneration of SN dopaminergic neurons and motor disability (Kim et al., 2019).

While the evidence that misfolded forms of α-SYN can spread in the brain is unambiguous, there are key questions that remain unanswered. One of them is why there is a divergence from the pattern of LP in PD patient brains and the brain connectome (Surmeier et al., 2017a,b). Recent work by Lee's group suggests that while spreading of PFFs is constrained by synaptic connections (that is, spreading does not occur between cells that are simply nearby), there are other constraints (Henderson et al., 2019b). The expression level of α-SYN is one factor that might contribute, with higher expression levels enhancing the probability of spread (Henderson et al., 2019a; Luna et al., 2018). If the expression of α-SYN scales with axonal arbor size, then this would increase the probability that neurons with massive axonal arbors, like SNc dopaminergic neurons, would manifest LP that was seeded elsewhere. However, it is not clear that brain inoculation of PFFs accurately models naturally occurring spread of α-SYN pathology. Infection of vagal axons with a virus expressing human α-SYN induces LP-like pathology in the DMV that spreads rostrally (Ulusoy et al., 2017). But the propagation of the pathology in this case was dependent upon oligomeric forms of α-SYN (not fibrillar forms) that had been modified by processes that were dependent upon nitrosative or oxidative stress (Burai et al., 2015; Schildknecht et al., 2013). Interestingly, α-SYN pathology increase cytosolic oxidant stress in dopaminergic neurons (Dryanovski et al., 2013; Subramaniam et al., 2014), creating a potential environment for generation of spreading competent forms of α-SYN; in fact, PFF-induced cytosolic oxidant stress was diminished by *N*-acetyl-cysteine, as was spreading of α-SYN pathology (Dryanovski et al., 2013; Luna et al., 2018). C-terminal truncation of α-SYN by the Ca^{2+}-dependent proteases calpain 1 and 2 also have been implicated in α-SYN seeding (Diepenbroek et al., 2014; Mahul-Mellier et al., 2018). These results point to the importance of post-translational modifications of misfolded forms of α-SYN that enable aggregation and spreading (Mahul-Mellier et al., 2018). The engagement of these mechanisms (e.g., calpain activation) may be very cell-type specific and help explain the pattern of LP seen in humans.

Another key set of questions is the relationship between LP and neurodegeneration. In experimental models, there is compelling evidence that α-SYN PFFs can cause neuronal degeneration in a variety of brain regions (Froula et al., 2018; Henderson et al., 2019b; Kam et al., 2018; Luk et al., 2012; Luna et al., 2018; Osterberg et al., 2015; Rey et al., 2018). But precisely how this occurs in most circumstances remains largely undefined. The study by Kam et al. provides one of the most detailed examinations of this question, implicating activation of poly (adenosine 5′-diphosphate-ribose) polymerase-1 (PARP-1) and parthanatos in PFF-induced cell death (Kam et al., 2018). But this elegant and ubiquitous cellular mechanism does not explain the cell-to-cell variability in the toxicity of PFFs (Henderson et al., 2019b; Luna et al., 2018; Osterberg et al., 2015; Rey et al., 2016). This variability is not simply stochastic noise but appears to be governed by cellular phenotype (Damier et al., 1999; Dopeso-Reyes et al., 2014; German et al., 1992; Liss and Roeper, 2008; McCormack et al., 2006; Pissadaki and Bolam, 2013; Surmeier et al., 2012). In our hands, modest injections of α-SYN PFFs into the PPN results in rapid propagation of α-SYN pathology to dozens of innervating regions (as judged by the appearance of phosphorylated forms of α-SYN), but in most of these regions the pathology clears within weeks without neuronal loss (Henrich unpublished observations). Similar results have been reported by others (Osterberg et al., 2015).

Does the PFF-induced pathology accurately reflect what is happening in PD patients? In humans, loss of SNc dopaminergic neurons precedes the appearance of local LP (Dijkstra et al., 2014; Milber et al., 2012; Surmeier et al., 2017a,b). Moreover, some forms of PD lack any discernible LP, despite there being frank loss of dopaminergic neurons (Berg et al., 2014; Espay et al., 2017, 2019). In many parts of the brain, LP is apparently present for decades without their being any significant loss of neurons (e.g., DMV) (Surmeier et al., 2017a,b), although it must be acknowledged that the assessment of neuronal loss has not been rigorously distinguished from phenotypic down-regulation in the vast majority of studies (Giguere et al., 2018). Nevertheless, this mismatch and studies exploring the toxicity of various forms of misfolded α-SYN have led to the proposition that oligomeric forms of α-SYN, rather than the fibrillar forms found in LP, are in fact the toxic species of α-SYN in PD (Bengoa-Vergniory et al., 2017; Helwig et al., 2016; Winner et al., 2011). From this perspective, the neurodegeneration in experimental models produced by inoculation with fibrillar forms of α-SYN is a pharmacological artifact. But, until there is a reliable means of quantitatively measuring the abundance of oligomeric species of α-SYN in human tissue, this issue will remain unresolved.

Before concluding this section, the potential role of inflammation in the early stages of PD must be acknowledged. There is a growing body of evidence that innate immunity may contribute to pathogenesis in both familial and idiopathic forms of the disease (Garretti et al., 2019). As compelling as this evidence is, it remains to be seen how it might contribute to selective vulnerability, although there are some clues emerging (Cebrián et al., 2014).

4 Toward a consensus view of PD pathogenesis

As outlined above, there is a growing consensus that both cellular phenotype and propagation through brain circuits of a peripherally seeded α-SYN pathology contribute to PD pathogenesis. While it is possible that there are two (or more) types of PD in which one or the other of these factors is dominant, it seems more likely that there is an interaction between these two that underlies most forms of PD.

How might cellular phenotype contribute to the seeding of α-SYN pathology and spreading? Our reasoning will be based upon examination of SNc dopaminergic neurons, but should extend to most of the other neurons at-risk in PD. As outlined above, SNc dopaminergic neurons manifest a collection of traits that experimental studies suggest should promote the seeding of α-SYN aggregation (Henderson et al., 2019b). First, they have massive axonal arbors with hundreds of thousands of synaptic release sites that should demand high levels of α-SYN expression, as α-SYN is primarily a synaptic protein (Braak and Del Tredici, 2004; Braak et al., 2004). Although α-SYN expression levels in at-risk neurons has not been rigorously quantified to our knowledge, this seems to be a reasonable assumption. Second, SNc dopaminergic neurons have high basal oxidant and nitrosative stress, as well as dramatic moment to moment elevations in cytosolic Ca^{2+} concentration—all factors that are known to directly promote α-SYN aggregation (Dryanovski et al., 2013; Follett et al., 2013; Guzman et al., 2010, 2018; Lieberman et al., 2017, Nath et al., 2011; Rcom-H'cheo-Gauthier et al., 2016; Votyakova and Reynolds, 2001). Elevated mitochondrial oxidant stress promotes the formation of damaging dopamine quinones, which promote α-SYN aggregation, impair chaperone-mediated autophagy and reduce lysosomal activity (Burbulla et al., 2017; Schneider and Cuervo, 2014). In addition, the high basal rate of mitophagy in SNc dopaminergic neurons (Guzman et al., 2018) could diminish "spare" lysosomal capacity and the ability to clear misfolded α-SYN. High cytosolic Ca^{2+} concentration also can promote α-SYN aggregation indirectly by activating calpain (Diepenbroek et al., 2014; Dufty et al., 2007; Melachroinou et al., 2013) and calcineurin (Caraveo et al., 2014), as well as by diminishing lysosomal activity (Gomez-Sintes et al., 2016). Third, mitochondrial dysfunction in dopaminergic neurons, particularly with aging (Chen et al., 2019; Elstner et al., 2011a, 2011b; Mullin and Schapira, 2015; Schapira et al., 1990; Simcox et al., 2013) is likely to result in altered mitochondrial retrograde signaling that could result in post-translational modifications in α-SYN, like acetylation, that promote aggregation (Fauvet et al., 2012; Martinez-Reyes et al., 2016). Lastly, although not discussed in any detail here, genetic mutations that compromise mitochondrial, lysosomal, and synaptic function, all of which are strained in at-risk neurons, could create a tipping point for α-SYN pathology (Beilina and Cookson, 2016; de Vries and Przedborski, 2013; Duda et al., 2016; Gegg and Schapira, 2016).

How might α-SYN pathology increase the vulnerability of neurons, specifically those known to be lost in PD? Although it is evident that the over-expression of

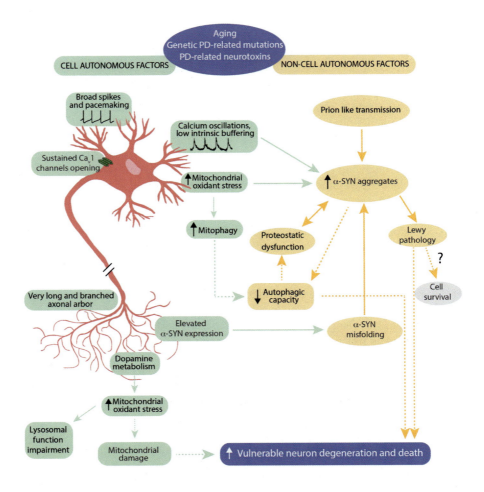

FIG. 1

Cell autonomous and non-cell autonomous factors driving degeneration in vulnerable neurons in PD. Schematic diagram summarizing how cell autonomous factors, non-cell autonomous factors and their interaction can contribute to neurodegeneration and cell death in PD. Intrinsic properties of vulnerable neurons (cell autonomous factors—green boxes) like pacemaking activity with broad spikes, very long and branched axonal arbor, elevated α-SYN expression in axons, elevated cytosolic Ca^{2+} and dendritic Ca^{2+} oscillations and dopamine metabolism increase mitochondrial oxidant stress and put these vulnerable neurons at risk. Due to this "vulnerable phenotype," aging and PD-related neurotoxins can promote mitochondrial and lysosomal dysfunction, neurodegeneration and cell death in PD. In parallel, non-cell autonomous factors (orange boxes) are reinforced by these cell autonomous factors. Mitochondrial oxidant stress and elevated cytosolic Ca^{2+} promote formation of intracellular α-SYN aggregates, leading to LP. In parallel, genetic mutations or aging can enhance α-SYN misfolding and α-SYN aggregation. Additionally, increase in mitophagy (due to mitochondrial damage) and α-SYN aggregates contribute to decrease autophagic capacity resulting also in neuronal death. Finally, an increase in α-SYN aggregation can lead proteostatic dysfunction promoted also by mitochondrial oxidant stress and elevated cytosolic Ca^{2+}. (Solid arrows describe well established links and broken arrows indicate links that need further verification.)

α-SYN and α-SYN PFFs preferentially impact SNc dopaminergic neurons (Kirik et al., 2002, 2003; Luk et al., 2012; Maingay et al., 2006), precisely why this is the case is unclear. There are some important clues, however. Several lines of study argue that misfolded forms of α-SYN disrupt mitochondrial function, either by inhibiting MCI activity (Bose and Beal, 2016; Devi et al., 2008; Schapira, 2007), by disrupting cytochrome C (Bayir et al., 2009), or by impairing protein import (Di Maio, 2016). Indeed, lowering α-SYN expression in dopaminergic neurons diminishes the toxicity of the mitochondrial toxin rotenone (Zharikov et al., 2015). In addition, the proteostatic burden created by misfolded α-SYN, particularly fibrillar forms that are degraded by macroautophagy (Cuervo et al., 2004), should compromise the ability to keep up with the high rate of mitochondrial damage in SNc dopaminergic neurons (Guzman et al., 2010, 2018), leading to an elevation in oxidant/nitrosative stress. The accumulation of damaged mitochondria with a compromised ability to generate ATP also could compromise axonal transport and dynamics, resulting in deficits in transmitter release (Berthet et al., 2014). In fact, an early feature of PD is the disruption of dopaminergic axons innervating the striatum (Burke and O'Malley, 2013; Kordower and Burke, 2018).

Taken together, the experimental literature suggests that phenotype of neurons at-risk in PD sets the stage for α-SYN pathology. Not only does this phenotype appear to promote α-SYN aggregation and propagation, it renders neurons more susceptible to the consequences of α-SYN pathology (Fig. 1). What are the translational implications of this conclusion? To date, all of the clinical trials aimed at slowing the progression of PD have failed (Sardi et al., 2018; Stern et al., 2012). This includes a recent Phase 3 clinical trial with the Ca_v1 channel inhibitor isradipine (Biglan et al., 2017; Parkinson Study Group et al., 2013). While it could be the case that these trials, most of which have targeted mechanisms implicated in the discussion here, have failed because of inadequate target engagement, it is also possible that targeting any one mechanism will not be enough. As outlined above, PD pathogenesis seems to be multifactorial, involving both cell autonomous and non-cell autonomous processes. It may be that targeting complementary mechanisms may be necessary to significantly slow progression.

References

Alreja, M., Aghajanian, G.K., 1991. Pacemaker activity of locus coeruleaus neurons: whole-cell recordings in brain slices show dependence on cAMP and protein kinase A. Brain Res. 556 (2), 339–343.

Andén, N.E., Hfuxe, K., Hamberger, B., Hökfelt, T., 1966. A quantitative study on the nigro-neostriatal dopamine neuron system in the rat. Acta Physiol. Scand. 67 (3), 306–312.

Arbuthnott, G.W., Wickens, J., 2007. Space, time and dopamine. Trends Neurosci. 30 (2), 62–69. https://doi.org/10.1016/j.tins.2006.12.003.

Ashrafi, G., Schlehe, J.S., LaVoie, M.J., Schwarz, T.L., 2014. Mitophagy of damaged mitochondria occurs locally in distal neuronal axons and requires PINK1 and Parkin. J. Cell Biol. 206 (5), 655–670. https://doi.org/10.1083/jcb.201401070.

Aston-Jones, G., Bloom, F.E., 1981. Activity of norepinephrine-containing locus coeruleus neurons in behaving rats anticipates fluctuations in the sleep-waking cycle. J. Neurosci. 1 (8), 876–886.

Aston-Jones, G., Waterhouse, B., 2016. Locus coeruleus: from global projection system to adaptive regulation of behavior. Brain Res. 1645, 75–78. https://doi.org/10.1016/j.brainres.2016.03.001.

Baughman, J.M., Perocchi, F., Girgis, H.S., Plovanich, M., Belcher-Timme, C.A., Sancak, Y., … Mootha, V.K., 2011. Integrative genomics identifies MCU as an essential component of the mitochondrial calcium uniporter. Nature 476 (7360), 341–345. https://doi.org/10.1038/nature10234.

Bayir, H., Kapralov, A.A., Jiang, J., Huang, Z., Tyurina, Y.Y., Tyurin, V.A., … Kagan, V.E., 2009. Peroxidase mechanism of lipid-dependent cross-linking of synuclein with cytochrome C: protection against apoptosis versus delayed oxidative stress in Parkinson disease. J. Biol. Chem. 284 (23), 15951–15969. https://doi.org/10.1074/jbc.M900418200.

Beilina, A., Cookson, M.R., 2016. Genes associated with Parkinson's disease: regulation of autophagy and beyond. J. Neurochem. 139 (Suppl. 1), 91–107. https://doi.org/10.1111/jnc.13266.

Bender, A., Krishnan, K.J., Morris, C.M., Taylor, G.A., Reeve, A.K., Perry, R.H., … Turnbull, D.M., 2006. High levels of mitochondrial DNA deletions in substantia nigra neurons in aging and Parkinson disease. Nat. Genet. 38 (5), 515–517. https://doi.org/10.1038/ng1769.

Bengoa-Vergniory, N., Roberts, R.F., Wade-Martins, R., Alegre-Abarrategui, J., 2017. Alpha-synuclein oligomers: a new hope. Acta Neuropathol. 134 (6), 819–838. https://doi.org/10.1007/s00401-017-1755-1.

Berg, J., Hung, Y.P., Yellen, G., 2009. A genetically encoded fluorescent reporter of ATP: ADP ratio. Nat. Methods 6 (2), 161–166. https://doi.org/10.1038/nmeth.1288.

Berg, D., Postuma, R.B., Bloem, B., Chan, P., Dubois, B., Gasser, T., … Deuschl, G., 2014. Time to redefine PD? Introductory statement of the MDS task force on the definition of Parkinson's disease. Mov. Disord. 29 (4), 454–462. https://doi.org/10.1002/mds.25844.

Berridge, M.J., 1998. Neuronal calcium signaling. Neuron 21, 13–26.

Berthet, A., Margolis, E.B., Zhang, J., Hsieh, I., Zhang, J., Hnasko, T.S., … Nakamura, K., 2014. Loss of mitochondrial fission depletes axonal mitochondria in midbrain dopamine neurons. J. Neurosci. 34 (43), 14304–14317. https://doi.org/10.1523/JNEUROSCI.0930-14.2014.

Biglan, K.M., Oakes, D., Lang, A.E., Hauser, R.A., Hodgeman, K., Greco, B., … Parkinson Study Group STEADY-PD III Investigators, 2017. A novel design of a phase III trial of isradipine in early Parkinson disease (STEADY-PD III). Ann. Clin. Transl. Neurol. 4 (6), 360–368. https://doi.org/10.1002/acn3.412.

Bolam, J.P., Pissadaki, E.K., 2012. Living on the edge with too many mouths to feed: why dopamine neurons die. Mov. Disord. 27 (12), 1478–1483. https://doi.org/10.1002/mds.25135.

Bose, A., Beal, M.F., 2016. Mitochondrial dysfunction in Parkinson's disease. J. Neurochem. 139 (Suppl. 1), 216–231. https://doi.org/10.1111/jnc.13731.

Braak, H., Del Tredici, K., 2004. Poor and protracted myelination as a contributory factor to neurodegenerative disorders. Neurobiol. Aging 25 (1), 19–23. https://doi.org/10.1016/j.neurobiolaging.2003.04.001.

Braak, H., Del Tredici, K., 2017. Neuropathological staging of Brain pathology in sporadic Parkinson's disease: separating the wheat from the chaff. J. Parkinsons Dis. 7 (s1), S71–S85. https://doi.org/10.3233/JPD-179001.

Braak, H., Del Tredici, K., Rüb, U., de Vos, R.A., Jansen Steur, E.N., Braak, E., 2003. Staging of brain pathology related to sporadic Parkinson's disease. Neurobiol. Aging 24, 197–211.

Braak, H., Ghebremedhin, E., Rub, U., Bratzke, H., Del Tredici, K., 2004. Stages in the development of Parkinson's disease-related pathology. Cell Tissue Res. 318 (1), 121–134. https://doi.org/10.1007/s00441-004-0956-9.

Brini, M., Cali, T., Ottolini, D., Carafoli, E., 2014. Neuronal calcium signaling: function and dysfunction. Cell. Mol. Life Sci. 71 (15), 2787–2814. https://doi.org/10.1007/s00018-013-1550-7.

Brookes, P.S., Yoon, Y., Robotham, J.L., Anders, M.W., Sheu, S.S., 2004. Calcium, ATP, and ROS: a mitochondrial love-hate triangle. Am. J. Physiol. Cell Physiol. 287 (4), C817–C833. https://doi.org/10.1152/ajpcell.00139.2004.

Brundin, P., Melki, R., 2017. Prying into the prion hypothesis for Parkinson's disease. J. Neurosci. 37 (41), 9808–9818. https://doi.org/10.1523/JNEUROSCI.1788-16.2017.

Brundin, P., Melki, R., Kopito, R., 2010. Prion-like transmission of protein aggregates in neurodegenerative diseases. Nat. Rev. Mol. Cell Biol. 11 (4), 301–307. https://doi.org/10.1038/nrm2873.

Burai, R., Ait-Bouziad, N., Chiki, A., Lashuel, H.A., 2015. Elucidating the role of site-specific nitration of alpha-Synuclein in the pathogenesis of Parkinson's disease via protein Semisynthesis and mutagenesis. J. Am. Chem. Soc. 137 (15), 5041–5052. https://doi.org/10.1021/ja5131726.

Burbulla, L.F., Song, P., Mazzulli, J.R., Zampese, E., Wong, Y.C., Jeon, S., Santos, D.P., Blanz, J., Obermaier, C.D., Strojny, C., Savas, J.N., Kiskinis, E., Zhuang, X., Krüger, R., Surmeier, D.J., Krainc, D., 2017. Dopamine oxidation mediates mitochondrial and lysosomal dysfunction in Parkinson's disease. Science 357, 1255–1261.

Burke, R.E., O'Malley, K., 2013. Axon degeneration in Parkinson's disease. Exp. Neurol. 246, 72–83. https://doi.org/10.1016/j.expneurol.2012.01.011.

Caraveo, G., Auluck, P.K., Whitesell, L., Chung, C.Y., Baru, V., Mosharov, E.V., … Lindquist, S., 2014. Calcineurin determines toxic versus beneficial responses to alpha-synuclein. Proc. Natl. Acad. Sci. U. S. A. 111 (34), E3544–E3552. https://doi.org/10.1073/pnas.1413201111.

Cebrián, C., Loike, J.D., Sulzer, D., 2014. Neuroinflammation in Parkinson's disease animal models: a cell stress response or a step in neurodegeneration? Curr. Top. Behav. Neurosci. 22, 237–270.

Chan, C.S., Guzman, J.N., Ilijic, E., Mercer, J.N., Rick, C., Tkatch, T., … Surmeier, D.J., 2007. Rejuvenation' protects neurons in mouse models of Parkinson's disease. Nature 447 (7148), 1081–1086. https://doi.org/10.1038/nature05865.

Chen, C., Turnbull, D.M., Reeve, A.K., 2019. Mitochondrial dysfunction in Parkinson's disease-cause or consequence? Biology (Basel) 8 (2), 38. https://doi.org/10.3390/biology8020038.

Chu, N., Bloom, F.E., 1973. Norepinephrine-containing neurons: changes in spontaneous discharge patterns during sleeping and waking. Science 179 (4076), 908–910.

Clapham, D.E., 2007. Calcium signaling. Cell 131 (6), 1047–1058. https://doi.org/10.1016/j.cell.2007.11.028.

Cooper, J.M., Mann, V.M., Krige, D., Schapira, A.H.V., 1992. Human mitochondrial complex I dysfunction. Biochim. Biophys. Acta 1101 (2), 198–203.

Csordas, G., Varnai, P., Golenar, T., Roy, S., Purkins, G., Schneider, T.G., … Hajnoczky, G., 2010. Imaging interorganelle contacts and local calcium dynamics at the ER-mitochondrial interface. Mol. Cell 39 (1), 121–132. https://doi.org/10.1016/j.molcel.2010.06.029.

Csordas, G., Weaver, D., Hajnoczky, G., 2018. Endoplasmic reticulum-mitochondrial Contactology: structure and signaling functions. Trends Cell Biol. 28 (7), 523–540. https://doi.org/10.1016/j.tcb.2018.02.009.

Cuervo, A.M., Stefanis, L., Fredenburg, R., Lansbury, P.T., Sulzer, D., 2004. Impaired degradation of mutant α-synuclein by chaperone-mediated autophagy. Science 305, 1292–1295.

Damier, P., Hirsch, E.C., Agid, Y., Graybiel, A.M., 1999. The substantia nigra of the human brain. II. Patterns of loss of dopamine-containing neurons in Parkinson's disease. Brain 122, 1437–1448.

Dautan, D., Huerta-Ocampo, I., Witten, I.B., Deisseroth, K., Bolam, J.P., Gerdjikov, T., Mena-Segovia, J., 2014. A major external source of cholinergic innervation of the striatum and nucleus accumbens originates in the brainstem. J. Neurosci. 34 (13), 4509–4518. https://doi.org/10.1523/JNEUROSCI.5071-13.2014.

Dautan, D., Hacioglu Bay, H., Bolam, J.P., Gerdjikov, T.V., Mena-Segovia, J., 2016. Extrinsic sources of cholinergic innervation of the striatal complex: a whole-Brain mapping analysis. Front. Neuroanat. 10, 1. https://doi.org/10.3389/fnana.2016.00001.

Desplats, P., Lee, H.J., Bae, E.J., Patrick, C., Rockenstein, E., Crews, L., Spencer, B., Masliah, E., Lee, SJ., 2009. Inclusion formation and neuronal cell death through neuron-to-neuron transmission of alpha-synuclein. Proc. Natl. Acad. Sci. U. S. A 106, 13010–13015.

De Stefani, D., Raffaello, A., Teardo, E., Szabo, I., Rizzuto, R., 2011. A forty-kilodalton protein of the inner membrane is the mitochondrial calcium uniporter. Nature 476 (7360), 336–340. https://doi.org/10.1038/nature10230.

de Vries, R.L., Przedborski, S., 2013. Mitophagy and Parkinson's disease: be eaten to stay healthy. Mol. Cell. Neurosci. 55, 37–43. https://doi.org/10.1016/j.mcn.2012.07.008.

Denton, R.M., 2009. Regulation of mitochondrial dehydrogenases by calcium ions. Biochim. Biophys. Acta 1787 (11), 1309–1316. https://doi.org/10.1016/j.bbabio.2009.01.005.

Devi, L., Raghavendran, V., Prabhu, B.M., Avadhani, N.G., Anandatheerthavarada, H.K., 2008. Mitochondrial import and accumulation of alpha-synuclein impair complex I in human dopaminergic neuronal cultures and Parkinson disease brain. J. Biol. Chem. 283 (14), 9089–9100. https://doi.org/10.1074/jbc.M710012200.

Diaz-Vegas, A.R., Cordova, A., Valladares, D., Llanos, P., Hidalgo, C., Gherardi, G., … Jaimovich, E., 2018. Mitochondrial calcium increase induced by RyR1 and IP3R channel activation after membrane depolarization regulates skeletal muscle metabolism. Front. Physiol. 9, 791. https://doi.org/10.3389/fphys.2018.00791.

Diederich, N.J., James Surmeier, D., Uchihara, T., Grillner, S., Goetz, C.G., 2019. Parkinson's disease: is it a consequence of human brain evolution? Mov. Disord. 34 (4), 453–459. https://doi.org/10.1002/mds.27628.

Diepenbroek, M., Casadei, N., Esmer, H., Saido, T.C., Takano, J., Kahle, P.J., … Nuber, S., 2014. Overexpression of the calpain-specific inhibitor calpastatin reduces human alpha-Synuclein processing, aggregation and synaptic impairment in [A30P]alphaSyn transgenic mice. Hum. Mol. Genet. 23 (15), 3975–3989. https://doi.org/10.1093/hmg/ddu112.

Dijkstra, A.A., Voorn, P., Berendse, H.W., Groenewegen, H.J., Netherlands Brain, B., Rozemuller, A.J., van de Berg, W.D., 2014. Stage-dependent nigral neuronal loss in incidental Lewy body and Parkinson's disease. Mov. Disord. 29 (10), 1244–1251. https://doi.org/10.1002/mds.25952.

Di Maio, R., Barrett, P.J., Hoffman, E.K., Barrett, C.W., Zharikov, A., Borah, A., Hu, X., McCoy, J., Chu, C.T., Burton, E.A., Hastings, T.G., Greenamyre, J.T., 2016. α-Synuclein binds to TOM20 and inhibits mitochondrial protein import in Parkinson's disease. Sci. Transl. Med. 8 (342), 342ra378.

Dolle, C., Flones, I., Nido, G.S., Miletic, H., Osuagwu, N., Kristoffersen, S., ... Tzoulis, C., 2016. Defective mitochondrial DNA homeostasis in the substantia nigra in Parkinson disease. Nat. Commun. 7, 13548. https://doi.org/10.1038/ncomms13548.

Dopeso-Reyes, I.G., Rico, A.J., Roda, E., Sierra, S., Pignataro, D., Lanz, M., ... Lanciego, J.L., 2014. Calbindin content and differential vulnerability of midbrain efferent dopaminergic neurons in macaques. Front. Neuroanat. 8, 146. https://doi.org/10.3389/fnana.2014.00146.

Dryanovski, D.I., Guzman, J.N., Xie, Z., Galteri, D.J., Volpicelli-Daley, L.A., Lee, V.M., ... Surmeier, D.J., 2013. Calcium entry and alpha-synuclein inclusions elevate dendritic mitochondrial oxidant stress in dopaminergic neurons. J. Neurosci. 33 (24), 10154–10164. https://doi.org/10.1523/JNEUROSCI.5311-12.2013.

Duda, J., Potschke, C., Liss, B., 2016. Converging roles of ion channels, calcium, metabolic stress, and activity pattern of substantia nigra dopaminergic neurons in health and Parkinson's disease. J. Neurochem. 139 (Suppl. 1), 156–178. https://doi.org/10.1111/jnc.13572.

Dufty, B.M., Warner, L.R., Hou, S.T., Jiang, S.X., Gomez-Isla, T., Leenhouts, K.M., ... Rohn, T.T., 2007. Calpain-cleavage of alpha-synuclein: connecting proteolytic processing to disease-linked aggregation. Am. J. Pathol. 170 (5), 1725–1738. https://doi.org/10.2353/ajpath.2007.061232.

Elstner, M., Morris, C.M., Heim, K., Bender, A., Mehta, D., Jaros, E., ... Prokisch, H., 2011a. Expression analysis of dopaminergic neurons in Parkinson's disease and aging links transcriptional dysregulation of energy metabolism to cell death. Acta Neuropathol. 122 (1), 75–86. https://doi.org/10.1007/s00401-011-0828-9.

Elstner, M., Müller, S.K., Leidolt, L., Laub, C., Krieg, L., Schlaudraff, F., ... Bender, A., 2011b. Neuromelanin, neurotransmitter status and brainstem location determine the differential vulnerability of catecholaminergic neurons to mitochondrial DNA deletions. Mol. Brain 4, 43. https://doi.org/10.1186/1756-6606-4-43.

Espay, A.J., Brundin, P., Lang, A.E., 2017. Precision medicine for disease modification in Parkinson disease. Nat. Rev. Neurol. 13 (2), 119–126. https://doi.org/10.1038/nrneurol.2016.196.

Espay, A.J., Vizcarra, J.A., Marsili, L., Lang, A.E., Simon, D.K., Merola, A., ... Leverenz, J.B., 2019. Revisiting protein aggregation as pathogenic in sporadic Parkinson and Alzheimer diseases. Neurology 92 (7), 329–337. https://doi.org/10.1212/WNL.0000000000006926.

Fahn, S., Parkinson Study, G., 2005. Does levodopa slow or hasten the rate of progression of Parkinson's disease? J. Neurol. 252 (Suppl. 4), IV37–IV42. https://doi.org/10.1007/s00415-005-4008-5.

Fahn, S., Sulzer, D., 2004. Neurodegeneration and neuroprotection in Parkinson disease. NeuroRx 1, 139–154.

Fauvet, B., Fares, M.B., Samuel, F., Dikiy, I., Tandon, A., Eliezer, D., Lashuel, H.A., 2012. Characterization of semisynthetic and naturally Nalpha-acetylated alpha-synuclein in vitro and in intact cells: implications for aggregation and cellular properties of alpha-synuclein. J. Biol. Chem. 287 (34), 28243–28262. https://doi.org/10.1074/jbc.M112.383711.

Foehring, R.C., Zhang, X.F., Lee, J.C., Callaway, J.C., 2009. Endogenous calcium buffering capacity of substantia nigral dopamine neurons. J. Neurophysiol. 102 (4), 2326–2333. https://doi.org/10.1152/jn.00038.2009.

Follett, J., Darlow, B., Wong, M.B., Goodwin, J., Pountney, D.L., 2013. Potassium depolarization and raised calcium induces alpha-synuclein aggregates. Neurotox. Res. 23 (4), 378–392. https://doi.org/10.1007/s12640-012-9366-z.

Foote, S.L., Aston-Jones, G., Bloom, F.E., 1980. Impulse activity of locus coeruleus neurons in awake rats and monkeys is a function of sensory stimulation and arousal. Proc. Natl. Acad. Sci. U. S. A. 77 (5), 3033–3037.

Frank, C.A., 2014. How voltage-gated calcium channels gate forms of homeostatic synaptic plasticity. Front. Cell. Neurosci. 8, 40. https://doi.org/10.3389/fncel.2014.00040.

Froula, J.M., Henderson, B.W., Gonzalez, J.C., Vaden, J.H., McLean, J.W., Wu, Y., … Volpicelli-Daley, L.A., 2018. Alpha-Synuclein fibril-induced paradoxical structural and functional defects in hippocampal neurons. Acta Neuropathol. Commun. 6 (1), 35. https://doi.org/10.1186/s40478-018-0537-x.

Fujimura, K., Matsuda, Y., 1989. Autogenous oscillatory potentials in neurons of the Guinea pig substantia nigra pars compacta in vitro. Neurosci. Lett. 104 (1–2), 53–57.

Garretti, F., Agalliu, D., Lindestam Arleham, C.S., Sette, A., Sulzer, D., 2019. Autoimmunity in Parkinson's disease: the role of alpha-Synuclein-specific T cells. Front. Immunol. 10, 303. https://doi.org/10.3389/fimmu.2019.00303.

Gauthier, J., Parent, M., Lévesque, M., Parent, A., 1999. The axonal arborization of single nigrostriatal neurons in rats. Brain Res. 834 (1–2), 228–232.

Gegg, M.E., Schapira, A.H., 2016. Mitochondrial dysfunction associated with glucocerebrosidase deficiency. Neurobiol. Dis. 90, 43–50. https://doi.org/10.1016/j.nbd.2015.09.006.

Gellerich, F.N., Gizatullina, Z., Gainutdinov, T., Muth, K., Seppet, E., Orynbayeva, Z., Vielhaber, S., 2013. The control of brain mitochondrial energization by cytosolic calcium: the mitochondrial gas pedal. IUBMB Life 65 (3), 180–190. https://doi.org/10.1002/iub.1131.

Gennerich, A., Vale, R.D., 2009. Walking the walk: how kinesin and dynein coordinate their steps. Curr. Opin. Cell Biol. 21 (1), 59–67. https://doi.org/10.1016/j.ceb.2008.12.002.

Gerfen, C.R., Surmeier, D.J., 2011. Modulation of striatal projection systems by dopamine. Annu. Rev. Neurosci. 34, 441–466. https://doi.org/10.1146/annurev-neuro-061010-113641.

German, D.C., Manaye, K.F., Sonsalla, P.K., Brooks, B.A., 1992. Midbrain dopaminergic cell loss in Parkinson's disease and MPTP-induced parkinsonism: sparing of Calbindin-D28k-containing cells. Ann. N. Y. Acad. Sci. 11 (648), 42–62.

Gervasoni, D., Darracq, L., Fort, P., Soulière, F., Chouvet, G., Luppi, P.-H., 1998. Electrophysiological evidence that noradrenergic neurons of the rat locus coeruleus are tonically inhibited by GABA during sleep. Eur. J. Neurosci. 10 (3), 964–970.

Giacomello, M., Drago, I., Bortolozzi, M., Scorzeto, M., Gianelle, A., Pizzo, P., Pozzan, T., 2010. Ca2+ hot spots on the mitochondrial surface are generated by Ca2+ mobilization from stores, but not by activation of store-operated Ca2+ channels. Mol. Cell 38 (2), 280–290. https://doi.org/10.1016/j.molcel.2010.04.003.

Giguere, N., Burke Nanni, S., Trudeau, L.E., 2018. On cell loss and selective vulnerability of neuronal populations in Parkinson's disease. Front. Neurol. 9, 455. https://doi.org/10.3389/fneur.2018.00455.

Giguere, N., Delignat-Lavaud, B., Herborg, F., Voisin, A., Li, Y., Jacquemet, V., … Trudeau, L.E., 2019. Increased vulnerability of nigral dopamine neurons after expansion of their axonal arborization size through D2 dopamine receptor conditional knockout. PLoS Genet. 15 (8), e1008352. https://doi.org/10.1371/journal.pgen.1008352.

Goedert, M., 2001. Alpha-synuclein and neurodegenerative diseases. Nat. Rev. Neurosci. 2 (7), 492–501.

Goedert, M., Spillantini, M.G., Del Tredici, K., Braak, H., 2013. 100 years of Lewy pathology. Nat. Rev. Neurol. 9 (1), 13–24. https://doi.org/10.1038/nrneurol.2012.242.

Goldberg, J.A., Guzman, J.N., Estep, C.M., Ilijic, E., Kondapalli, J., Sanchez-Padilla, J., Surmeier, D.J., 2012. Calcium entry induces mitochondrial oxidant stress in vagal neurons at risk in Parkinson's disease. Nat. Neurosci. 15 (10), 1414–1421. https://doi.org/10.1038/nn.3209.

Gomez-Sintes, R., Ledesma, M.D., Boya, P., 2016. Lysosomal cell death mechanisms in aging. Ageing Res. Rev. 32, 150–168. https://doi.org/10.1016/j.arr.2016.02.009.

Grace, A.A., Bunney, B.S., 1983. Intracellular and extracellular electrophysiology of nigral dopaminergic neurons—1. Identification and characterization. Neuroscience 10 (2), 301–315.

Grace, A.A., Onn, S.P., 1989. Morphology and electrophysiological properties of immunocytochemically identified rat dopamine neurons recorded in vitro. J. Neurosci. 9 (10), 3463–3481.

Graham, A.W., Aghajanian, G.K., 1971. Effects of amphetamine on single cell activity in a catecholamine nucleus, the locus coeruleus. Nature 234 (5324), 100–102.

Graves, S.M., Xie, Z., Stout, K.A., Zampese, E., Burbulla, L.F., Shih, J.C., … Surmeier, D.J., 2020. Dopamine metabolism by a monoamine oxidase mitochondrial shuttle activates the electron transport chain. Nat. Neurosci. 23 (1), 15–20. https://doi.org/10.1038/s41593-019-0556-3.

Greene, J.G., 2014. Causes and consequences of degeneration of the dorsal motor nucleus of the vagus nerve in Parkinson's disease. Antioxid. Redox Signal. 21 (4), 649–667. https://doi.org/10.1089/ars.2014.5859.

Griffiths, E.J., Rutter, G.A., 2009. Mitochondrial calcium as a key regulator of mitochondrial ATP production in mammalian cells. Biochim. Biophys. Acta 1787 (11), 1324–1333. https://doi.org/10.1016/j.bbabio.2009.01.019.

Groves, P.M., Linder, J.C., Young, S.J., 1994. 5-hydroxydopamine-labeled dopaminergic axons: three-dimensional reconstructions of axons, synapses and postsynaptic targets in rat neostriatum. Neuroscience 58 (3), 593–604.

Guzman, J.N., Sanchez-Padilla, J., Chan, C.S., Surmeier, D.J., 2009. Robust pacemaking in substantia nigra dopaminergic neurons. J. Neurosci. 29 (35), 11011–11019. https://doi.org/10.1523/JNEUROSCI.2519-09.2009.

Guzman, J.N., Sanchez-Padilla, J., Wokosin, D., Kondapalli, J., Ilijic, E., Schumacker, P.T., Surmeier, D.J., 2010. Oxidant stress evoked by pacemaking in dopaminergic neurons is attenuated by DJ-1. Nature 468 (7324), 696–700. https://doi.org/10.1038/nature09536.

Guzman, J.N., Ilijic, E., Yang, B., Sanchez-Padilla, J., Wokosin, D., Galtieri, D., … Surmeier, D.J., 2018. Systemic isradipine treatment diminishes calcium-dependent mitochondrial oxidant stress. J. Clin. Invest. 128 (6), 2266–2280. https://doi.org/10.1172/JCI95898.

Hainsworth, A.H., Röper, J., Kapoor, R., Ashcroft, F.M., 1991. Identification and electrophysiology of isolated pars-compacta neurons from Guinea-pig substantia nigra. Neuroscience 43 (1), 81–93.

Halliday, G.M., Song, Y.J., Harding, A.J., 2011. Striatal beta-amyloid in dementia with Lewy bodies but not Parkinson's disease. J. Neural Transm. (Vienna) 118 (5), 713–719. https://doi.org/10.1007/s00702-011-0641-6.

Hansen, C., et al., 2011. alpha-Synuclein propagates from mouse brain to grafted dopaminergic neurons and seeds aggregation in cultured human cells. J. Clin. Invest. 121 (2), 715–725.

Hardman, C.D., Henderson, J.M., Finkelstein, D.I., Horne, M.K., Paxinos, G., Halliday, G.M., 2002. Comparison of the basal ganglia in rats, marmosets, macaques, baboons, and humans: volume and neuronal number for the output, internal relay, and striatal modulating nuclei. J. Comp. Neurol. 445 (3), 238–255.

Harris, J.J., Jolivet, R., Attwell, D., 2012. Synaptic energy use and supply. Neuron 75 (5), 762–777. https://doi.org/10.1016/j.neuron.2012.08.019.

Hawkes, C.H., Del Tredici, K., Braak, H., 2007. Parkinson's disease: a dual-hit hypothesis. Neuropathol. Appl. Neurobiol. 33 (6), 599–614. https://doi.org/10.1111/j.1365-2990.2007.00874.x.

Helwig, M., Klinkenberg, M., Rusconi, R., Musgrove, R.E., Majbour, N.K., El-Agnaf, O.M., … Di Monte, D.A., 2016. Brain propagation of transduced alpha-synuclein involves non-fibrillar protein species and is enhanced in alpha-synuclein null mice. Brain 139 (Pt. 3), 856–870. https://doi.org/10.1093/brain/awv376.

Henderson, M.X., Cornblath, E.J., Darwich, A., Zhang, B., Brown, H., Gathagan, R.J., … Lee, V.M.Y., 2019a. Spread of alpha-synuclein pathology through the brain connectome is modulated by selective vulnerability and predicted by network analysis. Nat. Neurosci. 22 (8), 1248–1257. https://doi.org/10.1038/s41593-019-0457-5.

Henderson, M.X., Trojanowski, J.Q., Lee, V.M., 2019b. Alpha-Synuclein pathology in Parkinson's disease and related alpha-synucleinopathies. Neurosci. Lett. 709, 134316. https://doi.org/10.1016/j.neulet.2019.134316.

Hornung, J.-P., 2003. The human raphe nuclei and the serotonergic system. J. Chem. Neuroanat. 26 (4), 331–343. https://doi.org/10.1016/j.jchemneu.2003.10.002.

Howe, M.W., Dombeck, D.A., 2016. Rapid signalling in distinct dopaminergic axons during locomotion and reward. Nature 535 (7613), 505–510. https://doi.org/10.1038/nature18942.

Hung, A.Y., Schwarzschild, M.A., 2007. Clinical trials for neuroprotection in Parkinson's disease: overcoming angst and futility? Curr. Opin. Neurol. 20, 477–483.

Ibáñez, P., Bonnet, A.M., Débarges, B., Lohmann, E., Tison, F., Agid, Y., … Pollak, P., 2004. Causal relation between α-synuclein locus duplication as a cause of familial Parkinson's disease. Lancet 364 (9440), 1169–1171. https://doi.org/10.1016/s0140-6736(04)17104-3.

Imber, A.N., Putnam, R.W., 2012. Postnatal development and activation of L-type Ca2+ currents in locus ceruleus neurons: implications for a role for Ca2+ in central chemosensitivity. J. Appl. Physiol. (1985) 112 (10), 1715–1726. https://doi.org/10.1152/japplphysiol.01585.2011.

Iyer, R., Ungless, M.A., Faisal, A.A., 2017. Calcium-activated SK channels control firing regularity by modulating sodium channel availability in midbrain dopamine neurons. Sci. Rep. 7 (1), 5248. https://doi.org/10.1038/s41598-017-05578-5.

Jacobs, B.L., Azmitia, E.C., 1992. Structure and function of the brain serotonin system. Physiol. Rev. 72 (1), 165–229.

Jellinger, K.A., 2009. A critical evaluation of current staging of alpha-synuclein pathology in Lewy body disorders. Biochim. Biophys. Acta 1792 (7), 730–740. https://doi.org/10.1016/j.bbadis.2008.07.006.

Jones, B.E., Moore, R.Y., 1977. Ascending projections of the locus coeruleus in the rat. II. Autoradiographic study. Brain Res. 127 (1), 23–53. https://doi.org/10.1016/0006-8993(77)90378-x.

Kalaitzakis, M.E., Graeber, M.B., Gentleman, S.M., Pearce, R.K., 2008. The dorsal motor nucleus of the vagus is not an obligatory trigger site of Parkinson's disease: a critical analysis of alpha-synuclein staging. Neuropathol. Appl. Neurobiol. 34 (3), 284–295. https://doi.org/10.1111/j.1365-2990.2007.00923.x.

Kam, T.-I., Mao, X., Park, H., Chou, S.-C., Karuppagounder, S.S., Umanah, G.E., Yun, S.P., Brahmachari, S., Panicker, N., Chen, R., Andrabi, S.A., Qi, C., Poirier, G.G., Pletnikova, O., Troncoso, J.C., Bekris, L.M., Leverenz, J.B., Pantelyat, A., Ko, H.S., Rosenthal, L.S., Dawson, T.M., Dawson, V.L., 2018. Poly(ADP-ribose) drives pathologic a-synuclein neurodegeneration in Parkinson's disease. Science 362 (557), eaat8407. https://doi.org/10.1126/science.aat8407.

Kamer, K.J., Mootha, V.K., 2015. The molecular era of the mitochondrial calcium uniporter. Nat. Rev. Mol. Cell Biol. 16 (9), 545–553. https://doi.org/10.1038/nrm4039.

Kim, S., Kwon, S.H., Kam, T.I., Panicker, N., Karuppagounder, S.S., Lee, S., … Ko, H.S., 2019. Transneuronal propagation of pathologic alpha-Synuclein from the gut to the brain models Parkinson's disease. Neuron 103 (4), 627–641.e627. https://doi.org/10.1016/j.neuron.2019.05.035.

Kirik, D., Rosenblad, C., Burger, C., Lundberg, C., Johansen, T.E., Muzyczka, N., Mandel, R.J., Björklund, A., 2002. Parkinson-like neurodegeneration induced by targeted overexpression of alpha-synuclein in the nigrostriatal system. J. Neurosci 22 (7), 2780–2791.

Kirik, D., Annett, L.E., Burger, C., Muzyczka, N., Mandel, R.J., Björklund, A., 2003. Nigrostriatal alpha-synucleinopathy induced by viral vector-mediated overexpression of human alpha-synuclein: a new primate model of Parkinson's disease. Proc. Natl. Acad. Sci. U. S. A. 100 (5), 2884–2889.

Kordower, J.H., Burke, R.E., 2018. Disease modification for Parkinson's disease: axonal regeneration and trophic factors. Mov. Disord. 33 (5), 678–683. https://doi.org/10.1002/mds.27383.

Kordower, J.H., Chu, Y., Hauser, R.A., Freeman, T.B., Olanow, C.W., 2008. Lewy body-like pathology in long-term embryonic nigral transplants in Parkinson's disease. Nat. Med. 14 (5), 504–506. https://doi.org/10.1038/nm1747.

Kordower, J.H., Olanow, C.W., Dodiya, H.B., Chu, Y., Beach, T.G., Adler, C.H., … Bartus, R.T., 2013. Disease duration and the integrity of the nigrostriatal system in Parkinson's disease. Brain 136 (Pt. 8), 2419–2431. https://doi.org/10.1093/brain/awt192.

Koschak, A., Reimer, D., Huber, I., Grabner, M., Glossmann, H., Engel, J., Striessnig, J., 2001. Alpha 1D (Cav1.3) subunits can form l-type Ca2+ channels activating at negative voltages. J. Biol. Chem. 276 (25), 22100–22106. https://doi.org/10.1074/jbc.M101469200.

Kraytsberg, Y., Kudryavtseva, E., McKee, A.C., Geula, C., Kowall, N.W., Khrapko, K., 2006. Mitochondrial DNA deletions are abundant and cause functional impairment in aged human substantia nigra neurons. Nat. Genet. 38 (5), 518–520. https://doi.org/10.1038/ng1778.

Krebiehl, G., Ruckerbauer, S., Burbulla, L.F., Kieper, N., Maurer, B., Waak, J., … Kruger, R., 2010. Reduced basal autophagy and impaired mitochondrial dynamics due to loss of Parkinson's disease-associated protein DJ-1. PLoS One 5 (2), e9367. https://doi.org/10.1371/journal.pone.0009367.

Krüger, R., Kuhn, W., Müller, T., Woitalla, D., Graeber, M., Kösel, S., Przuntek, H., Epplen, J.T., Schöls, L., Riess, O., 1998. Ala30Pro mutation in the gene encoding a-synuclein in Parkinson's disease. Nat. Genet. 18 (2), 106–108.

Lavoie, B., Parent, A., 1994. Pedunculopontine nucleus in the squirrel monkey: projections to the basal ganglia as revealed by anterograde tract-tracing methods. J. Comp. Neurol. 344 (2), 210–231.

Li, J.Y., Englund, E., Holton, J.L., Soulet, D., Hagell, P., Lees, A.J., … Brundin, P., 2008. Lewy bodies in grafted neurons in subjects with Parkinson's disease suggest host-to-graft disease propagation. Nat. Med. 14 (5), 501–503. https://doi.org/10.1038/nm1746.

Liang, C.L., Wang, T.T., Luby-Phelps, K., German, D.C., 2007. Mitochondria mass is low in mouse substantia nigra dopamine neurons: implications for Parkinson's disease. Exp. Neurol. 203 (2), 370–380. https://doi.org/10.1016/j.expneurol.2006.08.015.

Lieberman, O.J., Choi, S.J., Kanter, E., Saverchenko, A., Frier, M.D., Fiore, G.M., … Mosharov, E.V., 2017. Alpha-Synuclein-dependent calcium entry underlies differential sensitivity of cultured SN and VTA dopaminergic neurons to a parkinsonian neurotoxin. eNeuro 4 (6). e0167-17.2017. https://doi.org/10.1523/ENEURO.0167-17.2017.

Liss, B., Roeper, J., 2008. Individual dopamine midbrain neurons: functional diversity and flexibility in health and disease. Brain Res. Rev. 58 (2), 314–321. https://doi.org/10.1016/j.brainresrev.2007.10.004.

Luk, K.C., Kehm, V., Carroll, J., Zhang, B., O'Brien, P., Trojanowski, J.Q., Lee, V.M., 2012. Pathological alpha-synuclein transmission initiates Parkinson-like neurodegeneration in nontransgenic mice. Science 338 (6109), 949–953. https://doi.org/10.1126/science.1227157.

Luna, E., Decker, S.C., Riddle, D.M., Caputo, A., Zhang, B., Cole, T., … Luk, K.C., 2018. Differential alpha-synuclein expression contributes to selective vulnerability of hippocampal neuron subpopulations to fibril-induced toxicity. Acta Neuropathol. 135 (6), 855–875. https://doi.org/10.1007/s00401-018-1829-8.

Maeda, T., Fujimiya, M., Kitahama, K., Imai, H., Kimura, H., 1989. Serotonin neurons and their physiological roles. Arch. Histol. Cytol. 52 (Suppl), 113–120.

Mahul-Mellier, A.-L., Altay, F., Burtscher, J., Maharjan, N., Ait Bouziad, N., Chiki, A., … Lashuel, H.A., 2018. The making of a Lewy body: the role of α-synuclein post-fibrillization modifications in regulating the formation and the maturation of pathological inclusions. bioRxiv. https://doi.org/10.1101/500058.

Maingay, M., Romero-Ramos, M., Carta, M., Kirik, D., 2006. Ventral tegmental area dopamine neurons are resistant to human mutant alpha-synuclein overexpression. Neurobiol. Dis. 23 (3), 522–532. https://doi.org/10.1016/j.nbd.2006.04.007.

Marks, J.D., Donnelly, D.F., Haddad, G.G., 1993. Adenosine-induced inhibition of vagal motoneuron excitability: receptor subtype and mechanisms. Am. J. Physiol. 264 (2 Pt. 1), L124–L132.

Martinez-Reyes, I., Diebold, L.P., Kong, H., Schieber, M., Huang, H., Hensley, C.T., … Chandel, N.S., 2016. TCA cycle and mitochondrial membrane potential are necessary for diverse biological functions. Mol. Cell 61 (2), 199–209. https://doi.org/10.1016/j.molcel.2015.12.002.

Masuda-Suzukake, M., Nonaka, T., Hosokawa, M., Oikawa, T., Arai, T., Akiyama, H., … Hasegawa, M., 2013. Prion-like spreading of pathological alpha-synuclein in brain. Brain 136 (Pt. 4), 1128–1138. https://doi.org/10.1093/brain/awt037.

Masuko, S., Nakajima, Y., Nakajima, S., Yamaguchi, K., 1986. Noradrenergic neurons from the locus ceruleus in dissociated cell culture: culture methods, morphology, and electrophysiology. J. Neurosci. 6 (11), 3229–3241.

Matheoud, D., Sugiura, A., Bellemare-Pelletier, A., Laplante, A., Rondeau, C., Chemali, M., … Desjardins, M., 2016. Parkinson's disease-related proteins PINK1 and Parkin repress mitochondrial antigen presentation. Cell 166 (2), 314–327. https://doi.org/10.1016/j.cell.2016.05.039.

Matheoud, D., Cannon, T., Voisin, A., Penttinen, A.M., Ramet, L., Fahmy, A.M., … Desjardins, M., 2019. Intestinal infection triggers Parkinson's disease-like symptoms in Pink1(−/−) mice. Nature 571 (7766), 565–569. https://doi.org/10.1038/s41586-019-1405-y.

Matschke, L.A., Bertoune, M., Roeper, J., Snutch, T.P., Oertel, W.H., Rinne, S., Decher, N., 2015. A concerted action of L- and T-type Ca(2+) channels regulates locus coeruleus pacemaking. Mol. Cell. Neurosci. 68, 293–302. https://doi.org/10.1016/j.mcn.2015.08.012.

Matsuda, W., Furuta, T., Nakamura, K.C., Hioki, H., Fujiyama, F., Arai, R., Kaneko, T., 2009. Single nigrostriatal dopaminergic neurons form widely spread and highly dense axonal arborizations in the neostriatum. J. Neurosci. 29 (2), 444–453. https://doi.org/10.1523/JNEUROSCI.4029-08.2009.

McCormack, A.L., Atienza, J.G., Langston, J.W., Di Monte, D.A., 2006. Decreased susceptibility to oxidative stress underlies the resistance of specific dopaminergic cell populations to paraquat-induced degeneration. Neuroscience 141 (2), 929–937. https://doi.org/10.1016/j.neuroscience.2006.03.069.

McGinty, D.J., Harper, R.M., 1976. Dorsal raphe neurons: depression of firing during sleep in cats. Brain Res. 101 (3), 569–575.

McLelland, G.L., Soubannier, V., Chen, C.X., McBride, H.M., Fon, E.A., 2014. Parkin and PINK1 function in a vesicular trafficking pathway regulating mitochondrial quality control. EMBO J. 33 (4), 282–295. https://doi.org/10.1002/embj.201385902.

Melachroinou, K., Xilouri, M., Emmanouilidou, E., Masgrau, R., Papazafiri, P., Stefanis, L., Vekrellis, K., 2013. Deregulation of calcium homeostasis mediates secreted alpha-synuclein-induced neurotoxicity. Neurobiol. Aging 34 (12), 2853–2865. https://doi.org/10.1016/j.neurobiolaging.2013.06.006.

Mena-Segovia, J., Sims, H.M., Magill, P.J., Bolam, J.P., 2008. Cholinergic brainstem neurons modulate cortical gamma activity during slow oscillations. J. Physiol. 586 (12), 2947–2960. https://doi.org/10.1113/jphysiol.2008.153874.

Mendez, I., Vinuela, A., Astradsson, A., Mukhida, K., Hallett, P., Robertson, H., … Isacson, O., 2008. Dopamine neurons implanted into people with Parkinson's disease survive without pathology for 14 years. Nat. Med. 14 (5), 507–509. https://doi.org/10.1038/nm1752.

Mercuri, N.B., Bonci, A., Calabresi, P., Stratta, F., Stefani, A., Bernardi, G., 1994. Effects of dihydropyridine calcium antagonists on rat midbrain dopaminergic neurones. Br. J. Pharmacol. 113 (3), 831–838.

Michelsen, K.A., Schmitz, C., Steinbusch, H.W., 2007. The dorsal raphe nucleus–from silver stainings to a role in depression. Brain Res. Rev. 55 (2), 329–342. https://doi.org/10.1016/j.brainresrev.2007.01.002.

Milber, J.M., Noorigian, J.V., Morley, J.F., Petrovitch, H., White, L., Ross, G.W., Duda, J.E., 2012. Lewy pathology is not the first sign of degeneration in vulnerable neurons in Parkinson disease. Neurology 79 (24), 2307–2314.

Millecamps, S., Julien, J.P., 2013. Axonal transport deficits and neurodegenerative diseases. Nat. Rev. Neurosci. 14 (3), 161–176. https://doi.org/10.1038/nrn3380.

Misgeld, T., Schwarz, T.L., 2017. Mitostasis in neurons: maintaining mitochondria in an extended cellular architecture. Neuron 96 (3), 651–666. https://doi.org/10.1016/j.neuron.2017.09.055.

Mo, Z.L., Katafuchi, T., Muratani, H., Hori, T., 1992. Effects of vasopressin and angiotensin II on neurones in the rat dorsal motor nucleus of the vagus, in vitro. J. Physiol. 458, 561–577.

Muller, S.K., Bender, A., Laub, C., Hogen, T., Schlaudraff, F., Liss, B., … Elstner, M., 2013. Lewy body pathology is associated with mitochondrial DNA damage in Parkinson's disease. Neurobiol. Aging 34 (9), 2231–2233. https://doi.org/10.1016/j.neurobiolaging.2013.03.016.

Mullin, S., Schapira, A.H., 2015. Pathogenic mechanisms of neurodegeneration in Parkinson disease. Neurol. Clin. 33 (1), 1–17. https://doi.org/10.1016/j.ncl.2014.09.010.

Murphy, M.P., 2009. How mitochondria produce reactive oxygen species. Biochem. J. 417 (1), 1–13. https://doi.org/10.1042/BJ20081386.

Muzerelle, A., Scotto-Lomassese, S., Bernard, J.F., Soiza-Reilly, M., Gaspar, P., 2016. Conditional anterograde tracing reveals distinct targeting of individual serotonin cell groups (B5-B9) to the forebrain and brainstem. Brain Struct. Funct. 221 (1), 535–561. https://doi.org/10.1007/s00429-014-0924-4.

Narendra, D., Tanaka, A., Suen, D.F., Youle, R.J., 2008. Parkin is recruited selectively to impaired mitochondria and promotes their autophagy. J. Cell Biol. 183 (5), 795–803. https://doi.org/10.1083/jcb.200809125.

Nath, S., Goodwin, J., Engelborghs, Y., Pountney, D.L., 2011. Raised calcium promotes alpha-synuclein aggregate formation. Mol. Cell. Neurosci. 46 (2), 516–526. https://doi.org/10.1016/j.mcn.2010.12.004.

Nedergaard, S., Flatman, J.A., Engberg, I., 1993. Nifedipine- and omega-conotoxin-sensitive Ca2+ conductances in Guinea-pig substantia nigra pars compacta neurones. J. Physiol. 466, 727–747.

Nickel, A., Kohlhaas, M., Maack, C., 2014. Mitochondrial reactive oxygen species production and elimination. J. Mol. Cell. Cardiol. 73, 26–33. https://doi.org/10.1016/j.yjmcc.2014.03.011.

Osterberg, V.R., Spinelli, K.J., Weston, L.J., Luk, K.C., Woltjer, R.L., Unni, V.K., 2015. Progressive aggregation of alpha-synuclein and selective degeneration of lewy inclusion-bearing neurons in a mouse model of parkinsonism. Cell Rep. 10 (8), 1252–1260. https://doi.org/10.1016/j.celrep.2015.01.060.

Pacelli, C., Giguere, N., Bourque, M.J., Levesque, M., Slack, R.S., Trudeau, L.E., 2015. Elevated mitochondrial bioenergetics and axonal Arborization size are key contributors to the vulnerability of dopamine neurons. Curr. Biol. 25 (18), 2349–2360. https://doi.org/10.1016/j.cub.2015.07.050.

Parkinson, J., 2002. An essay on the shaking palsy. Neuropsychiatry Classics 14 (2), 223–236.

Parkinson Study Group, Simuni, T., Biglan, K., Oakes, D., Bakris, G., Hauser, R.A., A, J., Lang, A., Surmeier, J.D., Gardiner, N., Deeley, C., Brandabur, M., Burke, D., Factor, S.A., Fernandez, H.H., Goudreau, J.L., Grimes, D.A., Jain, S., Kompoliti, K., Marshall, F., Miyasaki, J., Nance, M., Parkinson, S., Pfeiffer, R.F., Ross, G., Russell, D.S., Sahay, A., Saint-Hilaire, M.H., Tuite, P., Van Gerpen, J., Zadikoff, C., Becker, P., Berry, D., Blasucci, L.M., Conway, J., Deeley, C., Deppen, P., Duderstadt, K., Ede, P., Gartner, M., Ivanco, L.S., McMurray, R., Pecoraro, M.,

Pfeiffer, B., Quesada, M., Gables, C., Reys, L., Rolandelli, S., Russell, D., Sieren, J., Sommerfeld, B., Strongosky, A., Terashita, S., Thomas, C.A., Wessels, S., Siderowf, A., Langbehn, D., Sorrentino, M., 2013. Phase II safety, tolerability, and dose selection study of isradipine as a potential disease-modifying intervention in early Parkinson's disease (STEADY-PD). Mov. Disord. 28 (13), 1823–1831.

Pasanen, P., Myllykangas, L., Siitonen, M., Raunio, A., Kaakkola, S., Lyytinen, J., … Paetau, A., 2014. Novel alpha-synuclein mutation A53E associated with atypical multiple system atrophy and Parkinson's disease-type pathology. Neurobiol. Aging 35 (9), 2180. e1–2180.e5. https://doi.org/10.1016/j.neurobiolaging.2014.03.024.

Peelaerts, W., Bousset, L., Van der Perren, A., Moskalyuk, A., Pulizzi, R., Giugliano, M., … Baekelandt, V., 2015. Alpha-Synuclein strains cause distinct synucleinopathies after local and systemic administration. Nature 522 (7556), 340–344. https://doi.org/10.1038/nature14547.

Penna, E., Espino, J., De Stefani, D., Rizzuto, R., 2018. The MCU complex in cell death. Cell Calcium 69, 73–80. https://doi.org/10.1016/j.ceca.2017.08.008.

Petzold, A., Valencia, M., Pal, B., Mena-Segovia, J., 2015. Decoding brain state transitions in the pedunculopontine nucleus: cooperative phasic and tonic mechanisms. Front Neural Circuits 9, 68. https://doi.org/10.3389/fncir.2015.00068.

Pickrell, A.M., Youle, R.J., 2015. The roles of PINK1, parkin, and mitochondrial fidelity in Parkinson's disease. Neuron 85 (2), 257–273. https://doi.org/10.1016/j.neuron.2014.12.007.

Ping, H.X., Shepard, P.D., 1996. Apamin-sensitive Ca(2+)-activated K+ channels regulate pacemaker activity in nigral dopamine neurons. Neuroreport 7 (3), 809–814. https://doi.org/10.1097/00001756-199602290-00031.

Pissadaki, E.K., Bolam, J.P., 2013. The energy cost of action potential propagation in dopamine neurons: clues to susceptibility in Parkinson's disease. Front. Comput. Neurosci. 7, 13. https://doi.org/10.3389/fncom.2013.00013.

Poewe, W., Seppi, K., Tanner, C.M., Halliday, G.M., Brundin, P., Volkmann, J., … Lang, A.E., 2017. Parkinson disease. Nat. Rev. Dis. Primers. 3, 17013. https://doi.org/10.1038/nrdp.2017.13.

Polymeropoulos, M.H., Lavedan, C., Leroy, E., Ide, S.E., Dehejia, A., Dutra, A., Pike, B., Root, H., Rubenstein, J., Boyer, R., Stenroos, E.S., Chandrasekharappa, S., Athanassiadou, A., Papapetropoulos, T., Johnson, W.G., Lazzarini, A.M., Duvoisin, R.C., Di Iorio, G., Golbe, L.I., Nussbaum, R.L., 1997. Mutation in the α-synuclein gene identified in families with Parkinson's disease. Science 276, 2045–2047.

Prensa, L., Parent, A., 2001. The nigrostriatal pathway in the rat: a single-axon study of the relationship between dorsal and ventral tier nigral neurons and the striosome/matrix striatal compartments. J. Neurosci. 21 (18), 7247–7260.

Pringsheim, T., Jette, N., Frolkis, A., Steeves, T.D., 2014. The prevalence of Parkinson's disease: a systematic review and meta-analysis. Mov. Disord. 29 (13), 1583–1590. https://doi.org/10.1002/mds.25945.

Proukakis, C., Dudzik, C.G., Brier, T., MacKay, D.S., Cooper, J.M., Millhauser, G.L., Houlden, H., Schapira, A.H., 2013. A novel α-synuclein missense mutation in Parkinson disease. Neurology 80 (11), 1062–1064. https://doi.org/10.1212/WNL.0b013e31828727ba.

Puopolo, M., Raviola, E., Bean, B.P., 2007. Roles of subthreshold calcium current and sodium current in spontaneous firing of mouse midbrain dopamine neurons. J. Neurosci. 27 (3), 645–656. https://doi.org/10.1523/JNEUROSCI.4341-06.2007.

Putzier, I., Kullmann, P.H., Horn, J.P., Levitan, E.S., 2009. Cav1.3 channel voltage dependence, not Ca2+ selectivity, drives pacemaker activity and amplifies bursts in nigral dopamine neurons. J. Neurosci. 29 (49), 15414–15419. https://doi.org/10.1523/JNEUROSCI.4742-09.2009.

Rcom-H'cheo-Gauthier, A.N., Davis, A., Meedeniya, A.C.B., Pountney, D.L., 2016. Alpha-synuclein aggregates are excluded from calbindin-D28k-positive neurons in dementia with Lewy bodies and a unilateral rotenone mouse model. Mol. Cell. Neurosci. 77, 65–75. https://doi.org/10.1016/j.mcn.2016.10.003.

Recasens, A., Dehay, B., Bove, J., Carballo-Carbajal, I., Dovero, S., Perez-Villalba, A., ... Vila, M., 2014. Lewy body extracts from Parkinson disease brains trigger alpha-synuclein pathology and neurodegeneration in mice and monkeys. Ann. Neurol. 75 (3), 351–362. https://doi.org/10.1002/ana.24066.

Reeve, A., Simcox, E., Turnbull, D., 2014. Ageing and Parkinson's disease: why is advancing age the biggest risk factor? Ageing Res. Rev. 14, 19–30. https://doi.org/10.1016/j.arr.2014.01.004.

Rey, N.L., Steiner, J.A., Maroof, N., Luk, K.C., Madaj, Z., Trojanowski, J.Q., ... Brundin, P., 2016. Widespread transneuronal propagation of alpha-synucleinopathy triggered in olfactory bulb mimics prodromal Parkinson's disease. J. Exp. Med. 213 (9), 1759–1778. https://doi.org/10.1084/jem.20160368.

Rey, N.L., George, S., Steiner, J.A., Madaj, Z., Luk, K.C., Trojanowski, J.Q., ... Brundin, P., 2018. Spread of aggregates after olfactory bulb injection of alpha-synuclein fibrils is associated with early neuronal loss and is reduced long term. Acta Neuropathol. 135 (1), 65–83. https://doi.org/10.1007/s00401-017-1792-9.

Ros, H., Magill, P.J., Moss, J., Bolam, J.P., Mena-Segovia, J., 2010. Distinct types of non-cholinergic pedunculopontine neurons are differentially modulated during global brain states. Neuroscience 170 (1), 78–91. https://doi.org/10.1016/j.neuroscience.2010.06.068.

Rossi, A., Pizzo, P., Filadi, R., 2019. Calcium, mitochondria and cell metabolism: a functional triangle in bioenergetics. Biochim. Biophys. Acta, Mol. Cell Res. 1866 (7), 1068–1078. https://doi.org/10.1016/j.bbamcr.2018.10.016.

Sanchez-Padilla, J., Guzman, J.N., Ilijic, E., Kondapalli, J., Galtieri, D.J., Yang, B., ... Surmeier, D.J., 2014. Mitochondrial oxidant stress in locus coeruleus is regulated by activity and nitric oxide synthase. Nat. Neurosci. 17 (6), 832–840. https://doi.org/10.1038/nn.3717.

Sanders, L.H., McCoy, J., Hu, X., Mastroberardino, P.G., Dickinson, B.C., Chang, C.J., ... Greenamyre, J.T., 2014. Mitochondrial DNA damage: molecular marker of vulnerable nigral neurons in Parkinson's disease. Neurobiol. Dis. 70, 214–223. https://doi.org/10.1016/j.nbd.2014.06.014.

Sardi, S.P., Cedarbaum, J.M., Brundin, P., 2018. Targeted therapies for Parkinson's disease: from genetics to the clinic. Mov. Disord. 33 (5), 684–696. https://doi.org/10.1002/mds.27414.

Scarnati, E., Proia, A., Di Loreto, S., Pacitti, C., 1987. The reciprocal electrophysiological influence between the nucleus tegmenti pedunculopontinus and the substantia nigra in normal and decorticated rats. Brain Res. 423 (1–2), 116–124.

Schapira, A.H., 2007. Mitochondrial dysfunction in Parkinson's disease. Cell Death Differ. 14 (7), 1261–1266. https://doi.org/10.1038/sj.cdd.4402160.

Schapira, A.H., Cooper, J.M., Dexter, D., Clark, J.B., Jenner, P., Marsden, C.D., 1990. Mitochondrial complex I deficiency in Parkinson's disease. J. Neurochem. 54 (3), 823–827.

Scheibye-Knudsen, M., Fang, E.F., Croteau, D.L., Wilson 3rd, D.M., Bohr, V.A., 2015. Protecting the mitochondrial powerhouse. Trends Cell Biol. 25 (3), 158–170. https://doi.org/10.1016/j.tcb.2014.11.002.

Schildknecht, S., Gerding, H.R., Karreman, C., Drescher, M., Lashuel, H.A., Outeiro, T.F., … Leist, M., 2013. Oxidative and nitrative alpha-synuclein modifications and proteostatic stress: implications for disease mechanisms and interventions in synucleinopathies. J. Neurochem. 125 (4), 491–511. https://doi.org/10.1111/jnc.12226.

Schmieg, N., Menendez, G., Schiavo, G., Terenzio, M., 2014. Signalling endosomes in axonal transport: travel updates on the molecular highway. Semin. Cell Dev. Biol. 27, 32–43. https://doi.org/10.1016/j.semcdb.2013.10.004.

Schneider, S.A., Alcalay, R.N., 2017. Neuropathology of genetic synucleinopathies with parkinsonism: review of the literature. Mov. Disord. 32 (11), 1504–1523. https://doi.org/10.1002/mds.27193.

Schneider, J.L., Cuervo, A.M., 2014. Autophagy and human disease: emerging themes. Curr. Opin. Genet. Dev. 26, 16–23. https://doi.org/10.1016/j.gde.2014.04.003.

Scholze, A., Plant, T.D., Dolphin, A.C., Nürnberg, B., 2001. Functional expression and characterization of a voltage-gated CaV1.3 (alpha1D) calcium channel subunit from an insulin-secreting cell line. Mol. Endocrinol. 15 (7), 1211–1221. https://doi.org/10.1210/mend.15.7.0666.

Schwarz, L.A., Luo, L., 2015. Organization of the locus coeruleus-norepinephrine system. Curr. Biol. 25 (21), R1051–R1056. https://doi.org/10.1016/j.cub.2015.09.039.

Shepard, P.D., Bunney, B.S., 1991. Repetitive firing properties of putative dopamine-containing neurons in vitro: regulation by an apamin-sensitive Ca2+−activated K + conductance. Exp. Brain Res. 86, 141–150.

Shults, C.W., 2006. Lewy bodies. PNAS 103 (6), 1661–1668.

Simcox, E.M., Reeve, A., Turnbull, D., 2013. Monitoring mitochondrial dynamics and complex I dysfunction in neurons: implications for Parkinson's disease. Biochem. Soc. Trans. 41 (6), 1618–1624. https://doi.org/10.1042/BST20130189.

Simon, D.K., Tanner, C.M., Brundin, P., 2020. Parkinson disease epidemiology, pathology, genetics, and pathophysiology. Clin. Geriatr. Med. 36 (1), 1–12. https://doi.org/10.1016/j.cger.2019.08.002.

Singleton, A.B., Farrer, M., Johnson, J., Singleton, A., Hague, S., Kachergus, J., … Gwinn-Hardy, K., 2003. Alpha-Synuclein locus triplication causes Parkinson's disease. Science 302 (5646), 841. https://doi.org/10.1126/science.1090278.

Spillantini, M.G., Schmidt, M.L., Lee, V.M., Trojanowski, J.Q., Jakes, R., Goedert, M., 1997. a-Synuclein in Lewy bodies. Nature 388, 839–840.

Stern, M.B., Lang, A., Poewe, W., 2012. Toward a redefinition of Parkinson's disease. Mov. Disord. 27 (1), 54–60. https://doi.org/10.1002/mds.24051.

Subramaniam, M., Althof, D., Gispert, S., Schwenk, J., Auburger, G., Kulik, A., … Roeper, J., 2014. Mutant α-Synuclein enhances firing frequencies in dopamine substantia Nigra neurons by oxidative impairment of A-type potassium channels. J. Neurosci. 34 (41), 13586–13599. https://doi.org/10.1523/jneurosci.5069-13.2014.

Sukiasyan, N., Hultborn, H., Zhang, M., 2009. Distribution of calcium channel Ca(V)1.3 immunoreactivity in the rat spinal cord and brain stem. Neuroscience 159 (1), 217–235. https://doi.org/10.1016/j.neuroscience.2008.12.011.

Sulzer, D., Surmeier, D.J., 2013. Neuronal vulnerability, pathogenesis, and Parkinson's disease. Mov. Disord. 28 (1), 41–50. https://doi.org/10.1002/mds.25095.

Surmeier, D.J., Schumacker, P.T., 2013. Calcium, bioenergetics, and neuronal vulnerability in Parkinson's disease. J. Biol. Chem. 288 (15), 10736–10741. https://doi.org/10.1074/jbc.R112.410530.

Surmeier, D.J., Guzman, J.N., Sanchez, J., Schumacker, P.T., 2012. Physiological phenotype and vulnerability in Parkinson's disease. Cold Spring Harb. Perspect. Med. 2 (7), a009290.

Surmeier, D.J., Graves, S.M., Shen, W., 2014. Dopaminergic modulation of striatal networks in health and Parkinson's disease. Curr. Opin. Neurobiol. 29, 109–117. https://doi.org/10.1016/j.conb.2014.07.008.

Surmeier, D.J., Obeso, J.A., Halliday, G.M., 2017a. Selective neuronal vulnerability in Parkinson disease. Nat. Rev. Neurosci. 18 (2), 101–113.

Surmeier, D.J., Obeso, J.A., Halliday, G.M., 2017b. Parkinson's disease is not simply a prion disorder. J. Neurosci. 37 (41), 9799–9807. https://doi.org/10.1523/JNEUROSCI.1787-16.2017.

Takakusaki, K., Kitai, S.T., 1997. Ionic mechanisms involved in the spontaneous firing of tegmental pedunculopontine nucleus neurons of the rat. Neuroscience 78 (3), 771–794.

Thakur, P., Chiu, W.H., Roeper, J., Goldberg, J.A., 2019. Alpha-Synuclein 2.0—moving towards cell type specific pathophysiology. Neuroscience 412, 248–256. https://doi.org/10.1016/j.neuroscience.2019.06.005.

Travagli, R.A., Gillis, R.A., Rossiter, C.D., Vicini, S., 1991. Glutamate and GABA-mediated synaptic currents in neurons of the rat dorsal motor nucleus of the vagus. Am. J. Physiol. 260 (3 Pt. 1), G531–G536.

Travagli, R.A., Gillis, R.A., Vicini, S., 1992. Effects of thyrotropin-releasing hormone on neurons in rat dorsal motor nucleus of the vagus, in vitro. Am. J. Physiol. 263 (4 Pt. 1), G508–G517.

Trulson, M.E., Jacobs, B.L., 1979. Raphe unit activity in freely moving cats: correlation with level of behavioral arousal. Brain Res. 163 (1), 135–150.

Tubert, C., Galtieri, D., Surmeier, D.J., 2019. The pedunclopontine nucleus and Parkinson's disease. Neurobiol. Dis. 128, 3–8. https://doi.org/10.1016/j.nbd.2018.08.017.

Ulusoy, A., Phillips, R.J., Helwig, M., Klinkenberg, M., Powley, T.L., Di Monte, D.A., 2017. Brain-to-stomach transfer of alpha-synuclein via vagal preganglionic projections. Acta Neuropathol. 133 (3), 381–393. https://doi.org/10.1007/s00401-016-1661-y.

Ungerstedt, U., 1971. Stereotaxic mapping of the monoamine pathways in the rat Brain. Acta Physiol. Scand. 82 (S367), 1–48. https://doi.org/10.1111/j.1365-201X.1971.tb10998.x.

Urbain, N., Creamer, K., Debonnel, G., 2006. Electrophysiological diversity of the dorsal raphe cells across the sleep-wake cycle of the rat. J. Physiol. 573 (Pt. 3), 679–695. https://doi.org/10.1113/jphysiol.2006.108514.

Vertes, R.P., Linley, S.B., 2007. Comparison of projections of the dorsal and median raphe nuclei, with some functional considerations. Int. Congr. Ser. 1304, 98–120. https://doi.org/10.1016/j.ics.2007.07.046.

Viola, H.M., Arthur, P.G., Hool, L.C., 2009. Evidence for regulation of mitochondrial function by the L-type Ca2+ channel in ventricular myocytes. J. Mol. Cell. Cardiol. 46 (6), 1016–1026. https://doi.org/10.1016/j.yjmcc.2008.12.015.

Votyakova, T.V., Reynolds, I.J., 2001. DeltaPsi(m)-dependent and -independent production of reactive oxygen species by rat brain mitochondria. J. Neurochem 79 (2), 266–277.

Williams, J.T., North, R.A., Shefner, S.A., Nishi, S., Egan, T.M., 1984. Membrane properties of rat locus coeruleus neurones. Neuroscience 13 (1), 137–156.

Winner, B., Jappelli, R., Maji, S.K., Desplats, P.A., Boyer, L., Aigner, S., … Riek, R., 2011. In vivo demonstration that alpha-synuclein oligomers are toxic. Proc. Natl. Acad. Sci. U. S. A. 108 (10), 4194–4199. https://doi.org/10.1073/pnas.1100976108.

Wu, H., Williams, J., Nathans, J., 2014. Complete morphologies of basal forebrain cholinergic neurons in the mouse. Elife 3, e02444. https://doi.org/10.7554/eLife.02444.

Zarranz, J.J., Alegre, J., Gomez-Esteban, J.C., Lezcano, E., Ros, R., Ampuero, I., Vidal, L., Hoenicka, J., Rodriguez, O., Atarés, B., Llorens, V., Tortosa, E.G., del Ser, T., Munoz, D.G., de Yebenes, J.G., 2004. The new mutation, E46K, of α-Synuclein causes Parkinson and Lewy body dementia. Ann. Neurol. 55, 164–173.

Zharikov, A.D., Cannon, J.R., Tapias, V., Bai, Q., Horowitz, M.P., Shah, V., … Burton, E.A., 2015. shRNA targeting alpha-synuclein prevents neurodegeneration in a Parkinson's disease model. J. Clin. Invest. 125 (7), 2721–2735. https://doi.org/10.1172/JCI64502.

Mechanisms of alpha-synuclein toxicity: An update and outlook

4

Inês Caldeira Brás[a,†], Mary Xylaki[a,†], Tiago Fleming Outeiro[a,b,c,*]

[a]*Department of Experimental Neurodegeneration, Center for Biostructural Imaging of Neurodegeneration, University Medical Center Göttingen, Göttingen, Germany*
[b]*Max Planck Institute for Experimental Medicine, Göttingen, Germany*
[c]*Institute of Neuroscience, The Medical School, Newcastle University, Newcastle upon Tyne, United Kingdom*
[*]*Corresponding author: Tel.: +49-551-3913544; Fax: +49-551-3922693, e-mail address: touteir@gwdg.de*

Abstract

Alpha-synuclein (aSyn) was identified as the main component of inclusions that define synucleinopathies more than 20 years ago. Since then, aSyn has been extensively studied in an attempt to unravel its roles in both physiology and pathology. Today, studying the mechanisms of aSyn toxicity remains in the limelight, leading to the identification of novel pathways involved in pathogenesis.

In this chapter, we address the molecular mechanisms involved in synucleinopathies, from aSyn misfolding and aggregation to the various cellular effects and pathologies associated. In particular, we review our current understanding of the mechanisms involved in the spreading of aSyn between different cells, from the periphery to the brain, and back. Finally, we also review recent studies on the contribution of inflammation and the gut microbiota to pathology in synucleinopathies. Despite significant advances in our understanding of the molecular mechanisms involved, we still lack an integrated understanding of the pathways leading to neurodegeneration in PD and other synucleinopathies, compromising our ability to develop novel therapeutic strategies.

Keywords

Alpha-synuclein, Synucleinopathies, Toxicity, Spreading, Microbiota, Inflammation, Parkinson's disease, Dementia Lewy bodies, Multiple system atrophy, Pure autonomic failure

---

[†]Equal contribution.

Progress in Brain Research, Volume 252, ISSN 0079-6123, https://doi.org/10.1016/bs.pbr.2019.10.005

Abbreviations

6-OHDA	6-hydroxydopamine
AD	Alzheimer's disease
aSyn	alpha-synuclein
ATF6	activating transcription factor 6
ATP13A2	ATPase Cation Transporting 13A2
BBB	blood–brain barrier
Bip	binding immunoglobulin protein
bSyn	beta-synuclein
Ca^{2+}	calcium
CJD	Creutzfeldt-Jakob disease
CMA	chaperone mediated autophagy
CNS	central nervous system
CSPα	cysteine-string protein-α
C-terminus	carboxyl-terminus
DAMP	damage-associated molecular pattern
DAT	dopamine active transporter
DJ-1	protein deglycase 1
DLB	dementia with Lewy bodies
ENS	enteric nervous system
ER	endoplasmic reticulum
ERAD	ER-associated degradation
FBX07	F-box only protein 7
GBA	glucosylceramidase
GCIs	glial cytoplasmic inclusions
GD	Gaucher's disease
gSyn	gamma-synuclein
GWAS	genome wide association studies
IL	interleukin
INF-γ	interferon-γ
iNOS	inducible nitric oxide synthase
iPSCs	induced pluripotent stem cells
KO	knockout
LAG3	lymphocyte activation gene 3
Lamp2A	lysosome associated membrane protein 2A
LBs	Lewy bodies
LNs	Lewy neurites
LPS	lipopolysaccharide
LRP 10	LDL receptor related protein 10
LRRK2	leucine rich repeat kinase 2
MAMs	mitochondria-associated ER membranes
MAPS	misfolding-associated protein secretion pathway
MAPT	microtubule-associated protein tau
MPTP	1-methyl-4-phenyl-1,2,3,4-tetrahydropyridine
MSA	multiple system atrophy
mtDNA	mitochondrial DNA

NAC	non-amyloid-β component
NADPH	nicotinamide adenine dinucleotide phosphate
NLRs	NOD like receptors
NO	nitric oxide
NOD	nucleotide-binding oligomerization
NSF	*N*-ethylmaleimide-sensitive factor
N-terminus	amino-terminus
Omi/HtrA2	mitochondrial protease Omi/HtrA2
PAF	pure autonomic failure
PARK 10, 11, 12 and 16	Parkinson disease 10, 11, 12 and 16
PD	Parkinson's disease
PDD	Parkinson's disease dementia
PFFs	pre-formed fibrils
PINK1	PTEN-induced kinase 1
PLA2G6	phospholipase A2 Group VI
PRRs	pattern recognition receptors
pS129	phosphorylation at serine 129
PTMs	post-translational modifications
RBD	REM sleep behavior disorder
REM	rapid eye movement
ROS	reactive oxygen species
SCFAs	short-chain-fatty-acids
SN	substantia nigra
SNARE	soluble NSF attachment protein receptor
TH	tyrosine hydroxylase
TLRs	Toll-like receptors
TNF-α	tumor necrosis factor alpha
TNTs	tunneling nanotubes
TSE	transmissible spongiform encephalopathies
UCH-L1	ubiquitin C-terminal hydrolase L1
UPR	unfolded protein response
UPS	ubiquitin proteasome system
VPS35	VPS35 retromer complex component
WT	wild type

1 Introduction

Alpha-synuclein (aSyn) was initially identified as a presynaptic protein in the electric organ of *Torpedo californica* (Maroteaux et al., 1988). About 10 years later, it was found to associated with familial forms of PD (Polymeropoulos et al., 1997), and to be a major component of the characteristic abnormal deposits of proteins that define synucleinopathies, namely, Lewy bodies (LBs), Lewy neurites (LNs), and glial cytoplasmic inclusions (GCIs) (Spillantini, 1999; Spillantini et al., 1998a,b). Synucleinopathies are defined by the accumulation of misfolded aSyn, and include Parkinson's disease (PD), dementia with Lewy bodies (DLB), multiple system atrophy (MSA) and pure autonomic failure (PAF).

The function of aSyn remains poorly understood, but previous studies indicate that it is involved in neurotransmitter release due to its association to synaptic vesicles and it is located in pre-synaptic termini. Inside the cells, it is present in the cytoplasm, nucleus, mitochondria, endoplasmic reticulum (ER) and membranes (Devi et al., 2008; Goers et al., 2003; Guardia-Laguarta et al., 2014; Li et al., 2007; McLean et al., 2000a,b; Pinho et al., 2019). Although the role aSyn plays in the different organelles is not yet clear, the abnormal folding, accumulation, and aggregation of the protein disturbs the physiological functions of the of those organelles and leads to cellular pathologies thought to culminate in cell death, especially in dopaminergic in the substantia nigra, for reasons we do not fully understand.

aSyn misfolding and aggregation is currently seen as a culprit leading to synaptic dysfunction. The transition from the unfolded form to a beta-sheet-rich conformation is a necessary step in the LB formation (Miraglia et al., 2018). However, it is still unclear whether pathology is a consequence of a gain of toxic function, or a loss of the normal physiological function of aSyn—most likely, it is a consequence of both.

Several factors are thought to contribute to the aggregation of aSyn and pathology in synucleinopathies. These include aging, environmental factors, toxins, mutations and alterations in aSyn expression, the interplay with other aggregated proteins, or post-translational modifications (PTMs) (Kramer and Schulz-Schaeffer, 2007; Kuusisto et al., 2003; Lázaro et al., 2014; Saito et al., 2003; Tenreiro et al., 2014; Vicente Miranda et al., 2017).

Recently, it was demonstrated that different aggregated forms of aSyn, known as strains, display specific effects in cells and have different seeding capacity, possibly leading to distinct synucleinopathies (Bousset et al., 2013; Peelaerts et al., 2018; Peng et al., 2018; Pieri et al., 2016). These aSyn strains template the aggregation of native aSyn, leading to the spreading and progression of disease pathology. The progression of Lewy pathology in PD was initially proposed as consequence of the observation of aggregated aSyn in different tissues and brain areas that correlated with clinical features (Braak et al., 2002, 2003a; Del Tredici and Braak, 2012).

Given the increases in life expectancy we have witnessed in the last decades, the number of patients with synucleinopathies will most likely continue to increase over the coming years. Therefore, a better understanding of the factors that contribute to disease risk and progression is desperately needed in order to enable the development of novel therapeutic strategies that can alter disease progression or even prevent disease onset.

In this chapter, we review our current understanding of the molecular underpinnings of synucleinopathies, including those associated with the spreading of aSyn pathology and disease progression.

2 Parkinson's disease: Clinical features and genetic factors

The first clinical description of PD was presented by James Parkinson in an "Essay on the Shaking Palsy," over 200 years ago (Parkinson, 2002). The typical motor features of PD include slowness of movement, rigidity, resting tremor, and postural instability (Jenner et al., 2013; Postuma et al., 2015b), and arise due to the loss of dopaminergic neurons in the substantia nigra (SN), which leads to the deregulation in the activity of the basal ganglia activity (Dauer and Przedborski, 2003). However, patients often display a variety of non-motor symptoms including constipation, hyposmia, rapid eye movement (REM) sleep disorder (RBD), and cognitive impairment (Gibb and Lees, 1988; Poewe et al., 2017). Neuropathologically, PD is characterized by the presence of LBs, first described in the dorsal motor nucleus of the vagus nerve in 1912 (Lewy, 1912).

In sporadic PD, Lewy pathology seems to present in a sequential order in different brain regions (Braak et al., 2003a). In most cases, pathology is detected in the anterior olfactory bulb and/or the dorsal motor nucleus of the vagus nerve in the medulla. From the medulla, the neuropathological process progresses to the medulla oblongata and pontine tegmentum, followed by the amygdala and SN. At this point, patients exhibit the typical motor symptoms and, later, pathology reaches the temporal cortex and neocortex, possibly contributing to the cognitive impairment associated with the disease. In later stages, immunoreactive astrocytes are also detected (Braak et al., 2007).

Although most PD cases are sporadic, several genes have been associated with both familial forms of the disease. In fact, the field of PD genetics has been extremely important for our understanding of the molecular mechanisms associated (Billingsley et al., 2018). To date, six missense mutations have been identified in the *SNCA* gene, encoding for aSyn, in dominantly inherited forms of PD. These include A53T (Polymeropoulos et al., 1997), A30P (Kruger et al., 1998), E46K (Zarranz et al., 2004), H50Q (Appel-Cresswell et al., 2013), G51D (Lesage et al., 2013) and A53E (Pasanen et al., 2014). Patients with mutations G51D, A53E and A53T seem to show earlier age of disease onset but, given the low numbers of patients carrying these mutations, it is difficult to draw strong conclusions.

Duplications and triplications of the *SNCA* gene have also been found in familial forms of PD (Chartier-Harlin et al., 2004; Ibanez et al., 2004; Singleton et al., 2003). While duplications are associated with a phenotype resembling sporadic PD, triplications are associated with earlier age of disease onset and with more prominent cognitive symptoms, indicating a dose-dependent effect (Farrer et al., 2004).

More recently, genome wide association studies (GWAS) identified *SNCA* as a major risk factor for idiopathic PD (Satake et al., 2009; Simon-Sanchez et al., 2009), again confirming the central role aSyn is likely to play in PD and other synucleinopathies. Other autosomal-dominant and autosomal-recessive forms of familial PD have been identified. These include mutations in Parkin (*Parkin*), ubiquitin C-terminal hydrolase L1 (*UCH-L1*), PTEN-induced kinase 1 (*PINK1*), protein deglycase 1 (*DJ-1*), leucine rich repeat kinase 2 (*LRRK2*), ATPase Cation Transporting 13A2 (*ATP13A2*), mitochondrial protease Omi/HtrA2 (*Omi/HtrA2*), phospholipase A2 Group VI (*PLA2G6*), F-box only protein 7 (*FBX07*), VPS35 retromer complex component, or LDL receptor related protein 10 (*LRP10*) (Quadri et al., 2018). Additional genes linked with disease, namely, Parkinson disease 10, 11, 12 and 16 (*PARK 10*, *11*, *12* and *16*) exhibit an uncertain mode of inheritance (Bonifati, 2014; Houlden and Singleton, 2012; Mullin and Schapira, 2015; Wirdefeldt et al., 2011).

Additionally, GWAS studies have also identified the genes encoding for microtubule-associated protein tau (MAPT) and glucocerebrosidase (GBA) as significant risk factors for PD (Aharon-Peretz et al., 2004; Edwards et al., 2010; Gan-Or et al., 2008, 2015; Gegg et al., 2012; Sidransky et al., 2009).

Covering each of these genes and proteins in detail is beyond the scope of this manuscript, so we will only touch upon those that may relate to aSyn toxicity since this is the topic of this article.

3 Other synucleinopathies
3.1 Dementia with Lewy bodies

DLB resembles idiopathic PD in terms of Lewy pathology. However, the two diseases are clinically distinguished based on the time of onset of dementia in relation to motor symptoms (Outeiro et al., 2019). DLB is characterized by the presence of Lewy body pathology in the neocortex. In DLB patients, there is accelerated cognitive decline, parkinsonism, hallucinations and fluctuation of attention, and dementia starts within 1 year since the onset of motor symptoms (Mayo and Bordelon, 2014; McKeith et al., 2017). Patients that develop cognitive impairment at least 1 year after the diagnosis of PD are classified with PD dementia (PDD) (Kovari et al., 2009; Vasconcellos and Pereira, 2015).

3.2 Multiple system atrophy

In MSA, patients exhibit parkinsonism, motor weakness, autonomic failure and cognitive decline (Fanciulli and Wenning, 2015; Quinn, 2015) and aSyn accumulates in GCIs (Spillantini et al., 1998a,b; Tu et al., 1998). Cytoplasmic inclusions in Schwann cells are also common (Nakamura et al., 2015). However, it is still unknown how these aggregates are formed in oligodendrocytes, since they are thought to express low levels of aSyn. It is possible that these cells cannot clear aSyn, leading to its accumulation or, alternatively, that aSyn is taken up from the extracellular space. Recent studies showed that aSyn can indeed be transferred from neurons to oligodendrocytes, astrocytes, or microglia, via the extracellular space (Reyes et al., 2014; Steiner et al., 2018).

3.3 Pure autonomic failure

PAF is a rare sporadic neurodegenerative disorder of the peripheral autonomic nervous system. It is characterized by progressive failure of autonomic regulation, sympathetic and parasympathetic deregulation, and by orthostatic hypotension (Kaufmann, 1996). Pathologically, LBs and LNs are present in the sympathetic ganglia and postganglionic sympathetic axons, as well as in the central nervous system (CNS). PAF can progress to PD, DLB, or MSA (Arai et al., 2000; Hague et al., 1997; Kaufmann and Goldstein, 2010). Recently, aSyn deposits were identified in skin biopsies of PAF patients (Donadio et al., 2016).

3.4 REM sleep behavior disorder

REM sleep behavior disorder (RBD) is normally considered as a separate disease (Schenck et al., 1986), but it is frequently associated with synucleinopathies (McKenna and Peever, 2017; Vilas et al., 2016). In fact, RBD is a common clinical feature in patients with PD, DLB and MSA (Iranzo et al., 2016; Postuma et al., 2015a) and, with disease progression, RBD patients usually develop motor and cognitive symptoms similar to PD, DLB and MSA. RBD is characterized by the occurrence of vivid dreams where the patient is agitated and enacts the dreams, by muscle paralysis, and desynchronized electroencephalographic activity. RBD patients do not display cognitive or motor dysfunction but exhibit hyposmia (loss of smell), constipation, and reduced dopamine levels in the nigrostriatal pathway (Iranzo, 2018).

3.5 Genetic factors in other synucleinopathies

Over the last years, variants in *SNCA* and *MAPT* were also identified as risk factors for MSA (Al-Chalabi et al., 2009; Scholz et al., 2009; Vilarino-Guell et al., 2011). However, MSA is rarely familial, and aSyn mutations have not been associated with familial forms of the disease (Ozawa et al., 2006; Soma et al., 2006).

In DLB, *SNCA* and *GBA* mutations are also known risk factors (Chiba-Falek, 2017; Han et al., 2015). In fact, the association of *GBA* mutations with DLB is stronger than with PD, while the relationship with MSA is still unclear (Blandini et al., 2019; Goker-Alpan et al., 2004; Nalls et al., 2013).

4 Alpha-synuclein
4.1 Alpha-synuclein physiology

aSyn is a small, 14.5 kDa protein whose biological function is still not fully understood. It was initially described as a nuclear and synaptic protein (Maroteaux et al., 1988), and is mainly localized in presynaptic terminals, although it can also be found in cell bodies and axons of neuronal cells (Iwai et al., 1995). Although the presence of aSyn in the nucleus has been repeatedly described (Gonçalves and Outeiro, 2013;

McLean et al., 2000a; Pinho et al., 2019), the exact role aSyn may play in the nucleus remains unclear. However, the synaptic localization and the connection with neuro-degenerative disorders has biased most studies to focus on the study of aSyn in the context of synaptic function. In this context, aSyn has been shown to play a role in synaptic vesicle reuptake and neurotransmitter release. While aSyn does not seem to affect synaptic biogenesis, aSyn null mice display accelerated dopaminergic neuro-transmission (Abeliovich et al., 2000). In contrast, mice overexpressing aSyn display decreased neurotransmitter release as a result of the reduced reuptake of synaptic vesicles (Nemani et al., 2010). The role of aSyn in synaptic homeostasis is also evident in the involvement of soluble N-ethylmaleimide-sensitive factor (NSF) attachment protein receptor (SNARE) complex formation. Ablation of all three synucleins in mice inhibits SNARE complex formation (Burré et al., 2010), and accumulation of aSyn results in mislocalization of the SNARE proteins (Garcia-Reitböck et al., 2010), both disturbing exocytosis. The importance of aSyn in SNARE assembly is highlighted by its capability to eliminate assembly disturbances installed by ablation of the chaperon cysteine-string protein-α (CSPα) (Chandra et al., 2005). Collectively, these data suggest that a fine balance in the levels of aSyn is a prerequisite for the maintenance of SNARE complex assembly physiology and, thereby, of endocytosis-exocytosis.

aSyn displays high homology (67–83%) with proteins that bind fatty acids, suggesting a possible role in lipid transport between cytosolic and membranous cellular compartments (Sharon et al., 2001). aSyn shows a high preference in binding lipid rafts with high content in unsaturated and polyunsaturated fatty acids (Fortin, 2004; Kubo et al., 2005). Together, these observations suggest a regulatory role of aSyn in lipid metabolism. In fact, this has been verified in studies using mammalian and yeast cell models, where increased levels of aSyn lead to the accumulation of certain lipids (Outeiro and Lindquist, 2003; Perrin et al., 2001; Sharon et al., 2003; Yallapu et al., 2012).

Once embedded on membranes, aSyn forces them to bend, leading fusion of highly curved surfaces, like vesicles, with flat membranes. In addition, vesicle clustering mediated by dimeric aSyn seems to regulate the number of vesicles in the reserve pool (Auluck et al., 2010). aSyn seems to be acting as a synaptic vesicle connector regulator, clustering vesicles to regulate the readily released pool (Laugks et al., 2017). Besides these functions, aSyn was also reported to sense membrane defects and to act by smoothing their surfaces, supporting proper vesicle shapes (Ouberai et al., 2013).

4.2 Protein sequence

Several domains can be defined in the primary amino acid sequence of aSyn, and these have been correlated with both physiology and pathobiology (Fig. 1A). Amino acids 1–102 consist of a succession of 7- or 11-residue segments, containing a preserved "KTKEGV" consensus sequence that folds into amphipathic helix (Fig. 1B). This structure is similar to that found in apolipoprotein A2, possible explaining its ability to interact with certain lipids and proteins (Davidson et al., 1998; Segrest et al., 1990).

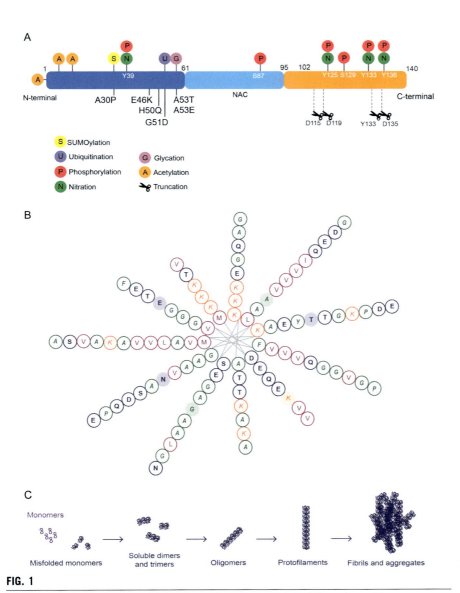

FIG. 1

aSyn features leading toxicity. (A) Schematic representation of the polypeptide sequence of aSyn. Amino acids 1–61 (blue) consist the N-terminus that has apolipoprotein function and bares the known mutations A30P, E46K, H50Q, G51D, A53E, and A53T that lead to familial disease forms. Residues 61–95 (light blue), recognized as the non-amyloid-β component (NAC), are implemented in aSyn oligomerization. The C-terminus (orange), aminoacids 95–140, is highly acidic and plastic, interacting with other proteins and metals. This part includes most of the tyrosines (Y) that are subject to phosphorylation (P) and nitration (N). Also truncations linked with enhanced aggregation are evident in this region. Acetylation (A) is located at N-terminus, and lastly, SUMOylation (S), ubiquitination (U)

(Continued)

In the amino-terminal (N-terminal) region of aSyn, a domain with KAKEGV-VAAAE repeats enables interactions with mitochondrial membranes (Zigoneanu et al., 2012). The fifth 11-residue segment, consisting of residues 61–95, was termed as non-amyloid-β component (NAC) as it was first isolated from amyloid plaques from AD brains (Uéda et al., 1993). The NAC domain is thought to be important for aSyn aggregation, as ablation of this region reduces oligomer and fibril formation (Giasson et al., 2001).

The carboxyl-terminal (C-terminal) region, consisting of amino acids 95–140, is less well understood. It is rich in acidic and charged amino acids, and is highly dynamic, possibly explaining chaperone-like functions of aSyn (Cookson, 2005). The C-terminus is natively disordered, lacking defined secondary structure, and contributes to aSyn solubility, protecting against aggregation (Eliezer et al., 2001). Even upon fibrillization, the C-terminus of aSyn seems to remain unfolded (Murray et al., 2003).

4.3 Tertiary and quaternary structure of aSyn

The actual conformation of aSyn in the cell is still a matter of controversy and, most likely, consists of an ensemble of conformations. Initially, aSyn was reported to be natively unfolded in aqueous solutions (Eliezer et al., 2001; Theillet et al., 2016). More recently, it was described to also occur as a folded and organized tetramer that was resistant to oligomerization (Bartels et al., 2011; Wang et al., 2011). Despite the high solubility, aSyn binds to membranes and vesicles, showing high preference for vesicles of small diameter and high curvature (Davidson et al., 1998; Ouberai et al., 2013). Upon binding to lipid micelles, aSyn folds into two curved antiparallel helixes connected by a linker that consists of amino acids 38–44 (Bussell and Eliezer, 2003). Although aSyn seems to naturally bind micelles, the ratio between the two components is critical, and when the threshold is exceeded oligomerization is driven by aSyn condensation on the micelle (Narayanan and Scarlata, 2001; Zhu et al., 2003).

FIG. 1—Cont'd

and glycation (G) can be found anywhere in the polypeptide sequence but usually concentrate in the N-terminus. (B) Helical wheel plot of aSyn. The residues are plotted onto a helical wheel in which 11 amino acids make up five turns. Aliphatic residues ILVM are marked in violet (regular font), hydroxyl and acidic residues DENQST are marked in blue (bold font), and the positively charged residues HKR are marked in orange (italic font). Residues with the PD familial mutations are shown in shadow. Helical wheel was created at the EMBOSS Website: http://emboss.open-bio.org/. (C) Steps toward aSyn fibrillation. Aberrantly folded aSyn species (misfolded monomers) or natively folded aSyn (monomers) under yet not clearly understood circumstances start to form soluble dimers and trimmers toward higher order oligomers. These initial steps seem to be subject to an equilibrium that is lost once aSyn structure changes to a beta-sheet (protofilament). Aggregation is then a fast marching or slow process depending on the applicable strains eventually leading to fibrils and amyloid aggregates.

During in vitro oligomerization, the structure of aSyn changes to a beta-sheet-rich conformation, forming aggregates with amyloid properties (Li et al., 2002; Uversky et al., 2001a,b). First, protofilaments are formed and the reaction proceeds to forming other species known as oligomers and protofibrils, that eventually assemble into amyloid fibrils (Fig. 1C) (Kessler et al., 2003; Vilar et al., 2008).

Although aggregated aSyn is traditionally associated with pathology, given the accumulation of protein inclusions in the brains of patients with synucleinopathies, it is still debatable whether these species are toxic or not. In fact, data from studies with post-mortem tissue suggest that soluble oligomeric species might be more toxic that fibrillar species (Sharon et al., 2003; Tofaris et al., 2003). Several in vitro studies have also supported the hypothesis that oligomeric species of aSyn are responsible for cellular damage (Cremades et al., 2012; Outeiro et al., 2008; Vekrellis et al., 2009). However, and although this hypothesis has more traction, since it also fits with the idea that different strains of oligomers may explain different disease phenotypes (Bousset et al., 2013), the field suffers from the ambiguity of the terms presently used. For example, the term "oligomer" is used to define a broad range of species that may differ significantly depending on the laboratory, on the method of detection and analysis, and on the sample. Therefore, defining the tertiary and quaternary structure of aSyn in greater detail should be a major goal of the field.

4.4 The effect of PD-associated mutations in aSyn

Interestingly, all known mutations in aSyn associated with PD are located in the N-terminal region, between residues 30 and 53. Thus far, six mutations have been identified in the gene encoding for aSyn (*SNCA*), and result in the substitutions: A30P, E46K, H50Q, G51D, A53E, and A53T (Fig. 1A). These mutations lead to early-onset (A30P, E46K, G51D, A53E, A53T) or late-onset (H50Q) disease, and are thought to disturb the protein's structure, leading to changes in oligomerization, aggregation, and/or cytotoxicity (Flagmeier et al., 2016; Lázaro et al., 2014).

The A30P aSyn mutation affects lipid and membrane binding, promotes oligomerization but slows fibril formation when compared to wild-type (WT) aSyn (Conway et al., 2000). The mutations A53E, A53T, and E46K seem to enhance polymerization (Conway et al., 1998, 2000). Finally, H50Q and G51D seem to alter the release and/or uptake of aSyn, possible increasing toxicity in the extracellular space (Fares et al., 2014; Khalaf et al., 2014; Villar-Piqué et al., 2016, 2017).

4.5 Post-translational modifications

The primary sequence of aSyn is rich in residues that can undergo different types of PTMs including, but not limited to, ubiquitination (Oueslati et al., 2010), phosphorylation (Anderson et al., 2006; Paleologou et al., 2010), SUMOylation (Plotegher and Bubacco, 2016), glycation (Vicente Miranda and Outeiro, 2010), or acetylation (de Oliveira et al., 2017) (Fig. 1A).

Polyubiquitination leads to aSyn degradation (Tofaris et al., 2011) while monoubiquitination seems to induce oligomerization and the formation of amorphous polymers (Rott et al., 2008). On the other hand, SUMOylation seems to have no direct effect on aSyn oligomerization (Krumova et al., 2011), and may block poly-ubiquitination, resulting in the inhibition of degradation and in aggregation as a result of the accumulation of the protein (Rott et al., 2017).

Glycation in aSyn is a spontaneous non-enzymatic oxidative modification that is believed to contribute to neurodegeneration (Vicente Miranda and Outeiro, 2010). Although it has not yet been extensively studied, glycation of aSyn by MGO results in increased oligomerization and reduced fibril formation (Paik et al., 2004; Vicente Miranda et al., 2017).

aSyn is natively acetylated in the N-terminus (Bartels et al., 2014; Fauvet et al., 2012), and in residues K6 and K10, a modification that contributes to the proper protein folding and membrane interactions (Bartels et al., 2014; de Oliveira et al., 2017; Fauvet et al., 2012). Upon deacetylation by Sirt2, the aggregation propensity and cytotoxicity of aSyn increase (Bu et al., 2017; de Oliveira et al., 2017).

The disordered C-terminal region of aSyn contains a high number of aspartic, glutamic, and prolines that can undergo several PTMs and modulate interactions with metals and other proteins. The PTMs observed in this region include phosphorylation, (Oueslati et al., 2010), nitration (He et al., 2018), or truncation (Li et al., 2005) (Fig. 1A).

Phosphorylation is the most studied PTM of aSyn. Phosphorylation at S129 (pS129) is identified in almost 90% of aggregated aSyn in the brains of patients, while only 4% is identified in healthy brain (Fujiwara et al., 2003). Apart from S129, other serines, threonines and tyrosines are candidate sites for phosphorylation at the C-terminus. Whether phosphorylation modulates aSyn aggregation is still unclear (Tenreiro et al., 2014).

aSyn can be nitrated in tyrosine residues (Y39, Y125, Y133, or Y136). Nitration leads to reduced membrane binding and accelerated fibril formation (Hodara et al., 2004).

The acidic nature of the C-terminus of aSyn favors the formation of complexes with di- and trivalent metal cations like aluminum (III), copper (II), iron (III), cobalt (III), and manganese (II) and results in increased fibril formation. Moreover, increased levels of metals are identified in the SN of PD patients (Uversky et al., 2001a,b, 2002).

LBs comprise full length (Spillantini et al., 1998a,b) and also truncated versions of aSyn at aspartic acid (D) residues: D119, D115, Y133, and D135 (Anderson et al., 2006). The truncated versions of aSyn increase aSyn aggregation and fibrillation (Hoyer et al., 2004). Truncation is mediated by different enzymes in distinct positions, and the effects of enzymatic cleavage may vary depending on the intrinsic state of aSyn (Mishizen-Eberz et al., 2003, 2005).

5 Alpha-synuclein pathology in the synapse

One of the many open questions in the field pertains to the actual time and space of LB formation. In post-mortem tissue from DLB patients, aSyn aggregates are evident in synapses prior to the appearance of LBs, suggesting that pathology may progress from the synapse to the cell body (Marui et al., 2002). Biochemical characterization of brain extracts from DLB and PD patients confirmed the presence of small aggregates, rather than mature LBs, degenerating synapses (Kramer and Schulz-Schaeffer, 2007; Schulz-Schaeffer, 2010). Similar observations were reported using genetic mouse models of synucleinopathies, supporting the idea that synaptic dysfunction precedes neuronal death (Schirinzi et al., 2016; Wu et al., 2019) Indeed, interactions between aSyn and key components of synaptic function, such as SNARE complex proteins and synaptic vesicles, suggest alterations in such interactions may contribute to synaptic pathology. aSyn aggregates bind synaptobrevin-2 and disrupt SNARE complex formation by obstruction of vesicle docking (Choi et al., 2013). Furthermore, the aggregates seem to bind vesicles and form clusters, thereby constraining normal physiological trafficking (Wang et al., 2014).

It is widely accepted that aSyn plays an important role on synaptic vesicle biology. Overexpression of aSyn in cell models reduces the reserve pool of vesicles and inhibits vesicle priming, resulting in inhibition of neurotransmitter release (Larsen et al., 2006; Nemani et al., 2010). Studies in transgenic animals expressing aSyn showed redistribution of synaptic vesicles in sites further away from the active zone, which appears to be larger but with reduced postsynaptic density (Janezic et al., 2013). How these abnormalities in synaptic vesicle priming, docking, and fusion occur remains elusive.

In transgenic mice expressing truncated aSyn the distribution and levels of the core SNARE proteins SNAP-25, syntaxin-1 and synaptobrevin-2 was altered (Garcia-Reitböck et al., 2010).

Increased aSyn levels disrupt dopaminergic neurotransmission by modulating the activity of tyrosine hydroxylase (TH), the rate limiting enzyme in dopamine production (Kirik et al., 2002; Perez et al., 2002). In addition, aSyn can bind the dopamine active transporter (DAT) and inhibit dopamine reuptake (Paxinou et al., 2001; Wersinger and Sidhu, 2003). Based on recent findings on the interplay between dopamine and aSyn (Choi et al., 2013; Mor et al., 2017; Outeiro et al., 2009), the accumulation of aSyn in the synaptic cleft can be detrimental. Intriguingly, oligomeric aSyn was found to interact with vesicles, perhaps introducing pores on their surface (Danzer et al., 2007; Lashuel et al., 2002; Volles et al., 2001), leading to neurotransmitter leakage. In the case of dopamine leakage, further aSyn oligomerization and oxidative stress may occur, given the high reactivity of this neurotransmitter (Mosharov, 2006).

6 aSyn pathology in mitochondria

Impairments in mitochondrial complex 1 activity and levels (Hattori et al., 1991; Parker et al., 1989; Schapira et al., 1990) or oxidative damage on the catalytic subunits (Keeney, 2006), were identified in PD, suggesting a mitochondrial component in the disease. Similar effects were reported in MSA patients (Blin et al., 1994; Gu et al., 1997).

aSyn was identified inside mitochondria in tissue from PD patients or from transgenic mouse models (Li et al., 2007; Zhang et al., 2008), suggesting it may disrupt the physiological architecture of the membranes (Cole et al., 2008; Devi and Anandatheerthavarada, 2010) and affect complex 1 activity (Chinta et al., 2010; Devi et al., 2008). The association of aSyn with mitochondrial membranes was also shown to promote abnormal mitochondrial fragmentation (Nakamura et al., 2011).

Impaired production and handling of ROS leads to cellular damage, but was also shown to induce aSyn expression, oligomerization, and aggregation via oxidative damage (Betarbet et al., 2006; Giasson, 2000). Interestingly, aSyn knock out (KO) mice show reduced ROS production and resistance to toxic byproducts of mitochondrial metabolism (Klivenyi et al., 2006). In addition, aSyn was associated with mitochondrial DNA (mtDNA) damage (Martin, 2006) and mitophagy, the selective degradation of mitochondria by autophagy (Chen et al., 2015a; Choubey et al., 2011). These impairments seem to be more prominent in synaptic mitochondria, and they also seem to occur prior to more generalized impairments of mitochondrial functions in neuronal cell bodies (Szegő et al., 2019), consistent with the idea that synaptic dysfunction may precede other cellular pathologies that culminate with cell death.

7 aSyn pathology in the ER-Golgi compartments

The ER is an important compartment enabling the folding and maturation of secretory proteins. Therefore, it plays a central role in proteostasis, and evolved two major protein quality mechanisms: the ER-associated degradation (ERAD) pathway, and the unfolded protein response (UPR). If these pathways fail, proteins accumulate in the ER and cause ER stress, which can be detected by the upregulation of various ER-stress response proteins (Görlach et al., 2006; Smith et al., 2011).

In cell models, ER stress can be elicited by PD-associated toxins such as MPTP, 6-hydroxydopamine (6-OHDA), or rotenone, inducing an UPR (Holtz and O'Malley, 2003; Ryu et al., 2018). Consistently, activation of ERAD is observed in postmortem brain tissue from PD and DLB patients, and several ER chaperones appear trapped in LBs (Conn et al., 2004). While this may suggest an attempt of the cell for clearing aberrantly folded aSyn, it may also occur as a consequence of the failure to deal with aggregated and toxic species of aSyn (Colla et al., 2018).

Oligomeric aSyn accumulates in the ER in the brains of PD patients and in the brains of animal models of synucleinopathy (Bellucci et al., 2011; Colla et al., 2012; Heman-Ackah et al., 2017). Similarly, the PD-associated aSyn mutants A53T and E46K were shown to induce ER stress in cell models (Lázaro et al., 2014; Smith et al., 2005). aSyn interacts with several chaperones and ER-related proteins mediating further toxic events and induce the UPR (Bellucci et al., 2011; Credle et al., 2015; Salganik et al., 2015).

Studies in yeast revealed, for the first time, that aSyn accumulation can induce defects in ER-Golgi trafficking, by sequestering Rab1, a central regulator of membrane trafficking (Cooper et al., 2006). aSyn was also found to interact with activating transcription factor 6 (ATF6) and prevent the vesicular transport to the Golgi. More recently, aSyn was also shown to directly affect the Golgi apparatus, by inducing fragmentation, an effect that is part of apoptotic cascade (Paiva et al., 2018).

Importantly, aSyn is enriched in the mitochondria-associated ER membranes (MAMs) (Guardia-Laguarta et al., 2015), structures that mediate several key processes such as calcium (Ca^{2+}) signaling, mitochondrial and ER stress, macroautophagy, and lipid synthesis and that may explain how aSyn may lead to cellular pathologies (Gómez-Suaga et al., 2018).

7.1 Degradation and clearance of aSyn

Excessive production of aSyn is genetically associated with synucleinopathies. Therefore, excessive accumulation, due to impaired degradation, are also of great interest for the understanding of the molecular basis of these diseases. aSyn is degraded by the two major protein degradation systems in the cell: the ubiquitin proteasome system (UPS) and the autophagy-lysosome pathways (ALP). The ALP can be further subdivided into several sub-types but, for the purpose of this chapter, we will focus on macroautophagy and chaperone mediated autophagy (CMA) (Webb et al., 2003).

Ubiquitination can regulate the fate of aSyn in the cell. Monoubiquitination of aSyn leads to proteasomal degradation but, in parallel, also seems to promote its aggregation (Haj-Yahya et al., 2013; Hejjaoui et al., 2011). Polyubiquitination of aSyn signals for proteasomal degradation, but can also be targeted by deubiquitylases.

Interestingly aSyn was found to be degraded by the proteasome even without the standard ubiquitin tag (Tofaris et al., 2001). Non-ubiquitinated species of aSyn that are targeted to the proteasome can cause proteasomal dysfunction due to the formation of soluble oligomers (Emmanouilidou et al., 2010b). Proteasomal dysfunction is evident in transgenic mice expressing A53T mutant aSyn in brain extracts from PD patients (Emmanouilidou et al., 2010b; Tofaris et al., 2003).

Although the proteasome seems to play a role in the degradation of aSyn, it was then proposed that lysosomal degradation my constitute the major pathway for aSyn clearance (Cuervo et al., 2004; Lee, 2004). Interestingly, selective inhibition of macroautophagy alone had no effect suggesting that the CMA might be the major pathway for aSyn degradation. This was further confirmed by mutation of the aSyn lysosome-recognizing motif 95VKKDQ99 that resulted in decreased internalization of aSyn into the lysosomes (Cuervo et al., 2004). Later it was demonstrated that the lysosome associated membrane protein 2A (Lamp2A) is the limiting step for aSyn degradation by the lysosome (Vogiatzi et al., 2008).

Several studies suggest that aSyn accumulation and aggregation may impair the ALP, although existing data are conflicting. Overexpression of aSyn in SKNSH cells disturbs autophagy, reducing autophagosome synthesis by sequestering Rab1 (Winslow et al., 2010). Expression of A53T aSyn but not WT aSyn in PC12 cells induces autophagosome accumulation (Stefanis et al., 2018). Treatment of HEK cells with aSyn fibrils results in the accumulation of intracellular aggregates and of autophagosomes as a result of impaired macroautophagy (Tanik et al., 2013). These findings are consistent with studies using brain tissue from DLB patients where autophagy is normally found activated despite the accumulation aSyn aggregates (Crews et al., 2010; Yu et al., 2009).

7.2 aSyn pathology and the cytoskeleton

aSyn interacts with tau, a microtubule stabilizing protein associated with AD, PD, and other neurodegenerative diseases. This interaction affects the phosphorylation state of tau, thereby regulating tubulin binding and stabilization (Jensen et al., 1999). Tubulin also occurs in LBs in post-mortem tissue from DLB patients. The interactions between aSyn and alpha and beta tubulin suggest a physiological interaction, in addition to the possible pathological interplay that modulates aSyn aggregation and microtubule polymerization (Alim et al., 2002; Chen et al., 2007). Co-aggregation of aSyn and tubulin may impair synaptic vesicle trafficking, may lead to Golgi fragmentation, and to neurite degeneration (Lee et al., 2006).

aSyn also interacts with actin, impacting on its polymerization and on the integrity of the cytoskeletal network (Sousa et al., 2009). This interaction is in part regulated by sequestration of cofilin 1, an actin-binding protein that can modulate the toxic effects of aSyn on actin filaments (Tilve et al., 2015). In contrast, aSyn may be sequestered by PrP^C, resulting in the loss of actin dynamics and excessive formation of cofilin/actin rods, a process implicated in neurodegeneration (Bras et al., 2018).

7.3 Effects of aSyn in the nucleus

Considering the size of aSyn, it might be able to passively diffuse across the nuclear pore complex, into the nucleus (Specht et al., 2005; Wang and Brattain, 2007). Indeed, aSyn can occur in the nucleus, and PD-associated mutations (A30P, G51D, and A53T) have been shown to increase nuclear localization in model systems (Fares et al., 2014; Kontopoulos et al., 2006). In addition, we have recently reported that phosphorylation on S129 (pS129) is increased when aSyn is in the nucleus (Pinho et al., 2019). As previously mentioned, the role of aSyn in the nucleus remains unclear, but it was found to interact with histones and to affect gene expression (Goers et al., 2003). Consistently, aSyn inhibits histone acetylation, leading to neuronal degeneration (Kontopoulos et al., 2006).

aSyn was also found to bind DNA affecting its fragmentation and repair (Pinho et al., 2019; Schaser et al., 2019).

Finally, aSyn has also been shown to affect gene expression by modulating various epigenetic processes. Such findings have been extensively reviewed elsewhere, and will not be covered here (Pavlou et al., 2016; Sturm and Stefanova, 2014).

8 Spreading of aSyn pathology

The Braak staging model suggests that PD pathology progresses in a somewhat ste-reotypic manner across neuroanatomically connected regions in the brain (Braak et al., 2003b). Environmental factors, such as toxins, microbes/viruses, or inflamma-tory agents, might induce LB pathology in the enteric nervous system (ENS), or in the olfactory bulb, and these may constitute starting sites for the spreading of aSyn pathology (Hawkes et al., 2009; Holmqvist et al., 2014; Sampson et al., 2016; Shannon and Vanden Berghe, 2018; Ulusoy et al., 2013) (Fig. 2).

In experimental models, aSyn pathology was shown to spread from the periphery, e.g., muscle, and stomach, to the spine and brainstem (Breid et al., 2016; Holmqvist et al., 2014; Peelaerts et al., 2015; Rey et al., 2013; Sacino et al., 2014).

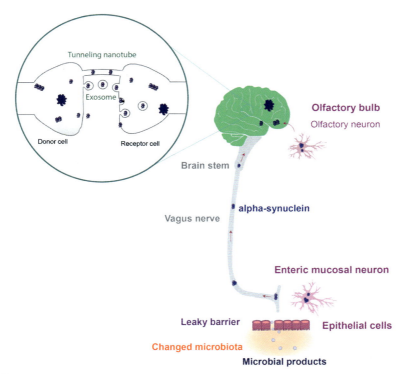

FIG. 2

Pathways involved in synucleinopathies pathology progression. Alterations in the gut microbiome might create intestinal leakage, allowing the passage of microbial products and inflammatory cytokines to the intestines. aSyn deposition can start in the olfactory bulb or in the enteric nervous system and then propagate to the central nervous system via the vagus nerve. In the brain, aSyn assemblies can be transferred between cells via tunneling nanotubes, exosomes or in free form, contributing for the pathology propagation in the brain.

In DLB, cognitive impairment can be associated with the presence of aSyn pathology in brain areas responsible cognition. However, in DLB pathology is thought pathology to start in the olfactory bulb and propagate to the primary olfactory cortex. In fact, the olfactory bulb is connected to the limbic system, contributing to the development of cognitive dysfunction before Parkinsonism (Beach et al., 2009; Mason et al., 2016; Rey et al., 2016). A fraction of DLB cases also display aSyn pathology in the ENS and in the vagus nerve (Beach et al., 2010), suggesting perhaps a dual propagation route for DLB, as happens for PD (Cersosimo, 2018).

The prion-like hypothesis was supported from the observation of Lewy pathology in grafts of fetal mesencephalic dopaminergic neurons in the striatum of PD patients (Kordower et al., 2008; Li et al., 2008). These results, together with Braak staging, suggested that misfolded aSyn can template the conversion of WT endogenous aSyn into the aggregated state, thereby perpetuating the propagation of pathology (Goedert, 2015; Recasens et al., 2014). However, it is also possible that a toxic environment may induce LB pathology in grafted young neurons over time (Surmeier et al., 2017; Walsh and Selkoe, 2016).

Studies in experimental models showed that injection of aSyn assemblies (preformed fibrils—PFFs) into animals induces the formation of intracellular inclusions at the injection site, propagating to distant brain regions (Abounit et al., 2016a; Bousset et al., 2013; Chen et al., 2015b; Kim et al., 2016b; Ma et al., 2016; Peelaerts and Baekelandt, 2016; Peng et al., 2018; Vasili et al., 2019).

The transfer of aSyn between cells may follow different pathways, but the relative contribution of each of those pathways toward transfer is still unclear. Some of these pathways involved the release of aSyn to the extracellular space followed by uptake by neighboring cells (Grozdanov and Danzer, 2018) (Fig. 2). Some studies suggest that monomeric, but not oligomeric or aggregated aSyn, can diffuse through the cell membrane and be released to the extracellular space (Ahn et al., 2006; Lee et al., 2008b). This process can occur in both directions, possibly involving a membrane translocator (Lee et al., 2005), and would not result from neuronal death or injury (Ulusoy et al., 2015).

Another possibility is that aSyn is released in exosomes, small vesicles released from cells to transport different types of cargo (Alvarez-Erviti et al., 2011; Bliederhaeuser et al., 2016; Danzer et al., 2012; Emmanouilidou et al., 2010a). The levels of aSyn in exosomes are low but, nevertheless, they hold great value as putative biomarkers for PD (Tofaris, 2017).

An alternative pathway is the misfolding-associated protein secretion pathway (MAPS), that involves the recruitment of aSyn to the ER-associated deubiquitylase USP19, which is later encapsulated into late endosomes and secreted to the extracellular space (Lee et al., 2016).

Transfer of aggregated aSyn can also occur via tunneling nanotubes (TNTs), possibly within lysosomes (Abounit et al., 2016a; Dieriks et al., 2017; Rostami et al., 2017). TNTs are thin tubular structures directly connecting the cytoplasm of two adjacent cells, enabling the exchange of their content (Abounit et al., 2016b).

In vitro, exogenously applied aSyn species (monomers, oligomers, or PFFs) can be uptaken by cells and induce the formation of protein inclusions (Danzer et al., 2009; Desplats et al., 2009; Luk et al., 2009; Masaracchia et al., 2018). The uptake of pathological forms of aSyn can be mediated by receptors, such as lymphocyte activation gene 3 (LAG3) (Mao et al., 2016), or heparan sulfate proteoglycans (Holmes et al., 2013). Furthermore, other membrane proteins are known to interact with different aSyn species at the cell surface, including the α3 subunit of the Na$^+$/K$^+$-ATPase (Shrivastava et al., 2015), the Fc gamma receptor IIb (Choi et al., 2018), or neurexin 1α (Mao et al., 2016; Shrivastava et al., 2015). Interestingly, the prion protein was also identified as a possible sensor for aSyn (Aulic et al., 2017; Bras et al., 2018; Ferreira et al., 2017).

aSyn can also be actively internalized via endocytosis, and be delivered to different organelles (Desplats et al., 2009; Hansen et al., 2011; Kim et al., 2008; Lee et al., 2008a,b; Volpicelli-Daley et al., 2011).

Different cell types in the brain, such as oligodendrocytes, astrocytes, and microglia, can uptake aSyn assemblies (Braidy et al., 2013; Hoffmann et al., 2016; Kim et al., 2013; Lindstrom et al., 2017; Reyes et al., 2014). As a consequence, transfer of aSyn species can occur not only between neurons, but also between different cell types. Whether the transfer is bidirectional or only unidirectional, as shown between neurons and astrocytes, is still unclear and will require additional investigation (Loria et al., 2017; Reyes et al., 2014).

Microglial cells appear to be more efficient than neurons or astrocytes in the uptake and degradation of extracellular aSyn, suggesting they may play a physiological role in removing toxic aSyn assemblies after aSyn is released from neurons (Zhang et al., 2005). Astrocytes can take up aSyn assemblies and target them to the lysosomal pathway for degradation (Loria et al., 2017). After uptake, aggregated aSyn can escape to the cytoplasm from endolysosomal compartments, inducing mitochondrial dysfunction, oxidative stress and inflammation (Flavin et al., 2017; Freeman et al., 2013).

As a consequence of the transfer of aSyn between cells, PD, DLB and MSA were suggested to be prion-like disorders. The existence of different types of aSyn assemblies with unique features in different synucleinopathies suggests a possible explanation for the heterogeneity observed in these disorders (Peelaerts et al., 2018). However, studies demonstrating that aSyn can be transmitted between individuals, in an infectious manner, are still lacking (Brundin and Melki, 2017; Irwin et al., 2013), and the results described in the literature are not always consistent (Steiner et al., 2018). This may be due to variability in the methodologies used, including in the amount of exogenous aSyn assemblies, due to differences in the genetic background of the animal models, or due to interference with the expression of mouse aSyn (Fares et al., 2016). Furthermore, batch-to-batch variability in the type of aSyn assemblies can also result in poor propagating and/or seeding capacities (Polinski et al., 2018). Therefore, additional studies and standardizations, as well as studies in human populations, will be essential to enable further clarifications in the field with respect to the prion-like character of synucleinopathies.

9 The interplay between aSyn and the microbiome

Recent studies have focused on the potential mechanisms by which the microbiome may contribute to aSyn pathology in the ENS and CNS (Fitzgerald et al., 2019; Perez-Pardo et al., 2017a,b; Scheperjans et al., 2018) (Fig. 2). Changes in the gastrointestinal tract composition may play an important role in the pathogenesis of PD. Neurodegenerative diseases are often associated with disturbances in the gastrointestinal tract, which can result in an imbalance of the immune system. Consequently, mucosal and systemic inflammation might reach the brain and cause, or at least contribute to neurodegeneration (Lubomski et al., 2019).

Short chain fatty acids (SCFAs) are produced by bacteria in the gut and affect the permeability of the blood-gut and blood–brain barriers (Manfredsson et al., 2018). The damage of these barriers, and consequent translocation of bacterial metabolites to the brain, has been proposed to contribute to neuroinflammation in the brain (Perez-Pardo et al., 2017a). A previous study demonstrated that PD patients have higher levels of *Enterobacteriaceae* and described a possible correlation with the severity of the postural instability and gait difficulties (Scheperjans et al., 2015). Also, analysis of fecal samples from PD patients and controls revealed decreased amounts of *Faecalibacterium* and SCFAs (Keshavarzian et al., 2015; Unger et al., 2016). More recently, aSyn was identified in the submucosa of the sigmoid colon in PD patients, and aSyn expression was observed during acute and chronic gastrointestinal infection (Stolzenberg et al., 2017; Visanji et al., 2017). PD patients with RBD exhibit more phosphorylated aSyn pathology in the colon and in the skin compared to PD patients without RBD (Leclair-Visonneau et al., 2017).

Administration of SCFAs into a mouse model overexpressing aSyn induces motor deficits and causes aSyn reactive microglia in the brain (Sampson et al., 2016). Colonic dysmotility in the gastrointestinal tract is also correlated with the presence of aSyn species in the ENS of animal models (Manfredsson et al., 2018; Paumier et al., 2015). Exposure to lipopolysaccharide (LPS) positive bacteria produced aSyn fibrillar forms in mice that received intracerebral injection of aSyn, in contrast to mice exposed to LPS-negative bacteria (Kim et al., 2016a). This indicates that contact with bacterial metabolites in the gut can induce pathology in the CNS (Svensson et al., 2015). Furthermore, some PD patients with vagotomies still develop PD suggesting there may be alternative routes, in addition to the vagus nerve, for the transport of pathological aSyn from the ENS to the CNS. Therefore, a better understanding of the gut-brain axis and the role of the intestinal microbiota in the regulation of immune responses might bring new insights into the pathobiology of PD and other neurodegenerative diseases.

10 Inflammation

Inflammation has gained significant weight in the context of the pathogenesis of synucleinopathies. Although inflammation is activated in response to pathogens or damage in order to restore physiology, chronic activation of inflammatory pathways may lead to broad cellular deregulation and damage.

The first link of synucleinopathies and inflammation resulted from the observation of activated microglia in the SN of PD patients (McGeer et al., 1988). Excessive microglial activation results in the production of inflammatory cytokines, such as tumor necrosis factor alpha (TNF-α), interleukin-1-β (IL-1β), interleukin-6 (IL-6), interferon-γ (INF-γ), and this may induce deleterious effects for dopaminergic neurons (Zhang et al., 2017). Interestingly, epidemiological studies suggest a link between increased risk for PD and polymorphisms in the genes encoding for TNF-α and IL-6 (Krüger et al., 2000). Elevated levels of these cytokines have been identified in the brains of PD patients, and to contribute for a pro-apoptotic environment (Mogi et al., 2000). Increased levels of cytokines have been also identified in bowel biopsies of PD patients, contributing to the gastrointestinal symptoms of PD, and supporting the involvement of the gut-brain axis in disease progression (Devos et al., 2013).

Innate immunity is actively involved in the regulation and progression of aSyn pathology. aSyn is recognized by microglial Toll-like receptors (TLRs) as a damage-associated molecular pattern (DAMP), activating innate immunity and possibly contributing to pathogenesis (Roodveldt et al., 2013). Another class of innate immune system receptors with a possible role in PD are the nucleotide-binding oligomerization (NOD)-like receptors (NLRs). NLRs can recognize prion-like proteins and recruit the inflammasome complex, which in turn activates interleukins (Hafner-Bratkovič et al., 2012). Finally, A2A adenosine receptors are also important for neuronal-glia interactions. PD patients exhibit an upregulation of these receptors, and a proportional increase in disease severity (Casetta et al., 2014). Collectively, pathological aSyn and microglial activation can potentiate each other, leading to an endless cycle (Zhang et al., 2017).

The adaptive immune system is also involved in PD, as peripheral leukocytes may migrate from the blood into the brain. CD8[+] and CD4[+] cells, but not B cells, invade the brain in PD and accumulate in the SN. This phenomenon was shown to be active, and not simply due BBB degradation, suggesting that T cells may target dopaminergic neurons and actively participate in neuronal death and inflammation (Brochard et al., 2008). Another connection is the crosstalk of microglia with CD4[+] T cells. Activated microglia express MHCII and can present antigens, thus activating both innate and adaptive immune responses in PD (Appel et al., 2010; Harms et al., 2011).

11 Conclusions and outlook

In recent years, considerable progress has been made in our understanding of the molecular mechanisms underlying aSyn-mediated toxicity. Excessive expression or accumulation of aSyn, due to impairments in protein quality control systems, may lead to the formation of abnormal forms of aSyn, which can become toxic and impact on cellular physiology. Inside the cell aSyn may disturb the proper function of major organelles by sequestering important proteins or by physically disturbing physiological processes due to aggregation. The course of events leading to aSyn pathology is not yet clear. A growing body of evidence suggests that pathology may start at the synapse and move toward other cellular compartments and organelles. What is clear so far is that the deleterious effects of aSyn on a particular cellular site can induce additional cellular pathologies and toxicity. The establishment of the timeline of events between dysfunction of the different cellular pathways during synucleinopathies progression would be an important step in the field.

In the quest for understanding the mechanisms leading toxicity in the various cellular sites, different studies have tried to correlate aSyn toxicity with particular species (e.g., oligomers), and with the subsequent distribution of pathology throughout the brain. However, in both physiological and pathological conditions, a multitude of aSyn species are likely to coexist inside and outside cells. While the existence of different assemblies may contribute to disease pathogenesis and explain disease heterogeneity, the identification of such assemblies is technically challenging. Nevertheless, the identification of toxic Syn species in the different synucleinopathies would constitute important steps forward. Since PTMs directly influence the folding, accumulation, aggregation, and localization of proteins, including those of aSyn, additional studies may also lead to the identification of putative targets for therapeutic intervention.

Additionally, recent findings linking aSyn pathology in the brain with events in the periphery, possibly due to retrograde transport into the brain, the microbiome and inflammation at the periphery may play important roles in the misfolding and aggregation of aSyn, creating a positive feedback loop that enhances pathology progression.

In total, additional research on the molecular mechanisms associated with aSyn pathobiology, and factors that modulate aSyn synthesis, aggregation, spreading and clearance, will be required in order to open new avenues for the development of accurate diagnostic tools and of novel therapeutic strategies.

Acknowledgments

T.F.O. is supported by DFG SFB1286 (projects B6 and B8), and by European Union's Horizon 2020 research and innovation program under grant agreement No. 721802.

References

Abeliovich, A., Schmitz, Y., Fariñas, I., et al., 2000. Mice lacking α-synuclein display functional deficits in the nigrostriatal dopamine system. Neuron 25 (1), 239–252.

Abounit, S., Bousset, L., Loria, F., et al., 2016a. Tunneling nanotubes spread fibrillar alpha-synuclein by intercellular trafficking of lysosomes. EMBO J. 35 (19), 2120–2138.

Abounit, S., Wu, J.W., Duff, K., et al., 2016b. Tunneling nanotubes: a possible highway in the spreading of tau and other prion-like proteins in neurodegenerative diseases. Prion 10 (5), 344–351.

Aharon-Peretz, J., Rosenbaum, H., Gershoni-Baruch, R., 2004. Mutations in the glucocerebrosidase gene and Parkinson's disease in Ashkenazi Jews. N. Engl. J. Med. 351 (19), 1972–1977.

Ahn, K.J., Paik, S.R., Chung, K.C., et al., 2006. Amino acid sequence motifs and mechanistic features of the membrane translocation of alpha-synuclein. J. Neurochem. 97 (1), 265–279.

Al-Chalabi, A., Durr, A., Wood, N.W., et al., 2009. Genetic variants of the alpha-synuclein gene SNCA are associated with multiple system atrophy. PLoS One 4 (9), e7114.

Alim, M.A., Hossain, M.S., Arima, K., et al., 2002. Tubulin seeds α-synuclein fibril formation. J. Biol. Chem. 277 (3), 2112–2117.

Alvarez-Erviti, L., Seow, Y., Schapira, A.H., et al., 2011. Lysosomal dysfunction increases exosome-mediated alpha-synuclein release and transmission. Neurobiol. Dis. 42 (3), 360–367.

Anderson, J.P., Walker, D.E., Goldstein, J.M., et al., 2006. Phosphorylation of Ser-129 is the dominant pathological modification of α-synuclein in familial and sporadic Lewy body disease. J. Biol. Chem. 281 (40), 29739–29752.

Appel, S.H., Beers, D.R., Henkel, J.S., 2010. T cell-microglial dialogue in Parkinson's disease and amyotrophic lateral sclerosis: are we listening? Trends Immunol. 31 (1), 7–17.

Appel-Cresswell, S., Vilarino-Guell, C., Encarnacion, M., et al., 2013. Alpha-synuclein p.H50Q, a novel pathogenic mutation for Parkinson's disease. Mov. Disord. 28 (6), 811–813.

Arai, K., Kato, N., Kashiwado, K., et al., 2000. Pure autonomic failure in association with human alpha-synucleinopathy. Neurosci. Lett. 296 (2–3), 171–173.

Aulic, S., Masperone, L., Narkiewicz, J., et al., 2017. Alpha-synuclein amyloids hijack prion protein to gain cell entry, facilitate cell-to-cell spreading and block prion replication. Sci. Rep. 7 (1), 10050.

Auluck, P.K., Caraveo, G., Lindquist, S., 2010. α-synuclein: membrane interactions and toxicity in Parkinson's disease. Annu. Rev. Cell Dev. Biol. 26, 211–233.

Bartels, T., Choi, J.G., Selkoe, D.J., 2011. α-Synuclein occurs physiologically as a helically folded tetramer that resists aggregation. Nature 477 (7362), 107–110.

Bartels, T., Kim, N.C., Luth, E.S., et al., 2014. N-alpha-acetylation of α-synuclein increases its helical folding propensity, GM1 binding specificity and resistance to aggregation. PLoS One 9 (7), e103727.

Beach, T.G., White 3rd, C.L., Hladik, C.L., et al., 2009. Olfactory bulb alpha-synucleinopathy has high specificity and sensitivity for Lewy body disorders. Acta Neuropathol. 117 (2), 169–174.

Beach, T.G., Adler, C.H., Sue, L.I., et al., 2010. Multi-organ distribution of phosphorylated alpha-synuclein histopathology in subjects with Lewy body disorders. Acta Neuropathol. 119 (6), 689–702.

Bellucci, A., Navarria, L., Zaltieri, M., et al., 2011. Induction of the unfolded protein response by α-synuclein in experimental models of Parkinson's disease. J. Neurochem. 116 (4), 588–605.

Betarbet, R., Canet-Aviles, R.M., Sherer, T.B., et al., 2006. Intersecting pathways to neurodegeneration in Parkinson's disease: effects of the pesticide rotenone on DJ-1, α-synuclein, and the ubiquitin-proteasome system. Neurobiol. Dis. 22 (2), 404–420.

Billingsley, K.J., Bandres-Ciga, S., Saez-Atienzar, S., et al., 2018. Genetic risk factors in Parkinson's disease. Cell Tissue Res. 373 (1), 9–20.

Blandini, F., Cilia, R., Cerri, S., et al., 2019. Glucocerebrosidase mutations and synucleinopathies: toward a model of precision medicine. Mov. Disord. 34 (1), 9–21.

Bliederhaeuser, C., Grozdanov, V., Speidel, A., et al., 2016. Age-dependent defects of alpha-synuclein oligomer uptake in microglia and monocytes. Acta Neuropathol. 131 (3), 379–391.

Blin, O., Desnuelle, C., Rascol, O., et al., 1994. Mitochondrial respiratory failure in skeletal muscle from patients with Parkinson's disease and multiple system atrophy. J. Neurol. Sci. 125 (1), 95–101.

Bonifati, V., 2014. Genetics of Parkinson's disease—state of the art, 2013. Parkinsonism Relat. Disord. 20 (Suppl. 1), S23–S28.

Bousset, L., Pieri, L., Ruiz-Arlandis, G., et al., 2013. Structural and functional characterization of two alpha-synuclein strains. Nat. Commun. 4, 2575.

Braak, H., Del Tredici, K., Bratzke, H., et al., 2002. Staging of the intracerebral inclusion body pathology associated with idiopathic Parkinson's disease (preclinical and clinical stages). J. Neurol. 249 (Suppl. 3), III/1–5.

Braak, H., Rub, U., Gai, W.P., et al., 2003a. Idiopathic Parkinson's disease: possible routes by which vulnerable neuronal types may be subject to neuroinvasion by an unknown pathogen. J. Neural Transm. (Vienna) 110 (5), 517–536.

Braak, H., Del Tredici, K., Rüb, U., et al., 2003b. Staging of brain pathology related to sporadic Parkinson's disease. Neurobiol. Aging 24 (2), 197–211.

Braak, H., Sastre, M., Del Tredici, K., 2007. Development of alpha-synuclein immunoreactive astrocytes in the forebrain parallels stages of intraneuronal pathology in sporadic Parkinson's disease. Acta Neuropathol. 114 (3), 231–241.

Braidy, N., Gai, W.P., Xu, Y.H., et al., 2013. Uptake and mitochondrial dysfunction of alpha-synuclein in human astrocytes, cortical neurons and fibroblasts. Transl. Neurodegener. 2 (1), 20.

Bras, I.C., Lopes, L.V., Outeiro, T.F., 2018. Sensing alpha-synuclein from the outside via the prion protein: implications for neurodegeneration. Mov. Disord. 33 (11), 1675–1684.

Breid, S., Bernis, M.E., Babila, J.T., et al., 2016. Neuroinvasion of alpha-synuclein prionoids after intraperitoneal and intraglossal inoculation. J. Virol. 90 (20), 9182–9193.

Brochard, V., Combadière, B., Prigent, A., et al., 2008. Infiltration of CD4+ lymphocytes into the brain contributes to neurodegeneration in a mouse model of Parkinson disease. J. Clin. Investig. 119 (1), 182–192.

Brundin, P., Melki, R., 2017. Prying into the prion hypothesis for Parkinson's disease. J. Neurosci. 37 (41), 9808–9818.

Bu, B., Tong, X., Li, D., et al., 2017. N-terminal acetylation preserves α-synuclein from oligomerization by blocking intermolecular hydrogen bonds. ACS Chem. Nerosci. 8 (10), 2145–2151.

Burré, J., Sharma, M., Tsetsenis, T., et al., 2010. Alpha-synuclein promotes SNARE-complex assembly in vivo and in vitro. Science 329 (5999), 1663–1667.

Bussell, R., Eliezer, D., 2003. A structural and functional role for 11-mer repeats in alpha-synuclein and other exchangeable lipid binding proteins. J. Mol. Biol. 329 (4), 763–778.

Casetta, I., Vincenzi, F., Bencivelli, D., et al., 2014. A2A adenosine receptors and Parkinson's disease severity. Acta Neurol. Scand. 129 (4), 276–281.

Cersosimo, M.G., 2018. Propagation of alpha-synuclein pathology from the olfactory bulb: possible role in the pathogenesis of dementia with Lewy bodies. Cell Tissue Res. 373 (1), 233–243.

Chandra, S., Gallardo, G., Fernández-Chacón, R., et al., 2005. α-Synuclein cooperates with CSPα in preventing neurodegeneration. Cell 123 (3), 383–396.

Chartier-Harlin, M.C., Kachergus, J., Roumier, C., et al., 2004. Alpha-synuclein locus duplication as a cause of familial Parkinson's disease. Lancet 364 (9440), 1167–1169.

Chen, L., Jin, J., Davis, J., et al., 2007. Oligomeric α-synuclein inhibits tubulin polymerization. Biochem. Biophys. Res. Commun. 356 (3), 548–553.

Chen, L., Xie, Z., Turkson, S., et al., 2015a. A53T human α-synuclein overexpression in transgenic mice induces pervasive mitochondria macroautophagy defects preceding dopamine neuron degeneration. J. Neurosci. 35 (3), 890–905.

Chen, S.W., Drakulic, S., Deas, E., et al., 2015b. Structural characterization of toxic oligomers that are kinetically trapped during alpha-synuclein fibril formation. Proc. Natl. Acad. Sci. U. S. A. 112 (16), E1994–E2003.

Chiba-Falek, O., 2017. Structural variants in SNCA gene and the implication to synucleinopathies. Curr. Opin. Genet. Dev. 44, 110–116.

Chinta, S.J., Mallajosyula, J.K., Rane, A., et al., 2010. Mitochondrial alpha-synuclein accumulation impairs complex I function in dopaminergic neurons and results in increased mitophagy in vivo. Neurosci. Lett. 486 (3), 235–239.

Choi, B., Choi, M., Kim, J., et al., 2013. Large α-synuclein oligomers inhibit neuronal SNARE-mediated vesicle docking. Proc. Natl. Acad. Sci. U. S. A. 110 (10), 4087–4092.

Choi, Y.R., Cha, S.H., Kang, S.J., et al., 2018. Prion-like propagation of α-synuclein is regulated by the FcγRIIB-SHP-1/2 signaling pathway in neurons. Cell Rep. 22 (1), 136–148.

Choubey, V., Safiulina, D., Vaarmann, A., et al., 2011. Mutant A53T α-Synuclein induces neuronal death by increasing mitochondrial autophagy. J. Biol. Chem. 286 (12), 10814–10824.

Cole, N.B., DiEuliis, D., Leo, P., et al., 2008. Mitochondrial translocation of α-synuclein is promoted by intracellular acidification. Exp. Cell Res. 314 (10), 2076–2089.

Colla, E., Jensen, P.H., Pletnikova, O., et al., 2012. Accumulation of toxic α-synuclein oligomer within endoplasmic reticulum occurs in α-synucleinopathy in vivo. J. Neurosci. 32 (10), 3301–3305.

Colla, E., Panattoni, G., Ricci, A., et al., 2018. Toxic properties of microsome-associated alpha-synuclein species in mouse primary neurons. Neurobiol. Dis. 111, 36–47.

Conn, K.J., Gao, W., McKee, A., et al., 2004. Identification of the protein disulfide isomerase family member PDIp in experimental Parkinson's disease and Lewy body pathology. Brain Res. 1022 (1–2), 164–172.

Conway, K.A., Harper, J.D., Lansbury, P.T., 1998. Accelerated in vitro fibril formation by a mutant alpha-synuclein linked to early-onset Parkinson disease. Nat. Med. 4 (11), 1318–1320.

Conway, K.A., Lee, S.J., Rochet, J.C., et al., 2000. Acceleration of oligomerization, not fibrillization, is a shared property of both alpha-synuclein mutations linked to early-onset Parkinson's disease: implications for pathogenesis and therapy. Proc. Natl. Acad. Sci. U. S. A. 97 (2), 571–576.

Cookson, M.R., 2005. The biochemistry of Parkinson's disease. Annu. Rev. Biochem. 74, 29–52.

Cooper, A.A., Gitler, A.D., Cashikar, A., et al., 2006. Alpha-synuclein blocks ER-Golgi traffic and Rab1 rescues neuron loss in Parkinson's models. Science (New York, N.Y.) 313 (5785), 324–328.

Credle, J.J., Forcelli, P.A., Delannoy, M., et al., 2015. α-Synuclein-mediated inhibition of ATF6 processing into COPII vesicles disrupts UPR signaling in Parkinson's disease. Neurobiol. Dis. 76, 112–125.

Cremades, N., Cohen, S.I.A., Deas, E., et al., 2012. Direct observation of the interconversion of normal and toxic forms of α-synuclein. Cell 149 (5), 1048–1059.

Crews, L., Spencer, B., Desplats, P., et al., 2010. Selective molecular alterations in the autophagy pathway in patients with Lewy body disease and in models of α-synucleinopathy. PLoS One 5 (2), e9313.

Cuervo, A.M., Stafanis, L., Fredenburg, R., et al., 2004. Impaired degradation of mutant α-synuclein by chaperone-mediated autophagy. Science 305 (5688), 1292–1295.

Danzer, K.M., Haasen, D., Karow, A.R., et al., 2007. Different species of alpha-synuclein oligomers induce calcium influx and seeding. J. Neurosci. 27 (34), 9220–9232.

Danzer, K.M., Krebs, S.K., Wolff, M., et al., 2009. Seeding induced by alpha-synuclein oligomers provides evidence for spreading of alpha-synuclein pathology. J. Neurochem. 111 (1), 192–203.

Danzer, K.M., Kranich, L.R., Ruf, W.P., et al., 2012. Exosomal cell-to-cell transmission of alpha synuclein oligomers. Mol. Neurodegener. 7, 42.

Dauer, W., Przedborski, S., 2003. Parkinson's disease: mechanisms and models. Neuron 39 (6), 889–909.

Davidson, W.S., Jonas, A., Clayton, D.F., et al., 1998. Stabilization of alpha-synuclein secondary structure upon binding to synthetic membranes. J. Biol. Chem. 273 (16), 9443–9449.

de Oliveira, R.M., Vicente Miranda, H., Francelle, L., et al., 2017. The mechanism of sirtuin 2-mediated exacerbation of alpha-synuclein toxicity in models of Parkinson disease. PLoS Biol. 15 (3), 1–27.

Del Tredici, K., Braak, H., 2012. Spinal cord lesions in sporadic Parkinson's disease. Acta Neuropathol. 124 (5), 643–664.

Desplats, P., Lee, H.J., Bae, E.J., et al., 2009. Inclusion formation and neuronal cell death through neuron-to-neuron transmission of alpha-synuclein. Proc. Natl. Acad. Sci. U. S. A. 106 (31), 13010–13015.

Devi, L., Anandatheerthavarada, H.K., 2010. Mitochondrial trafficking of APP and alpha synuclein: relevance to mitochondrial dysfunction in Alzheimer's and Parkinson's diseases. Biochim. Biophys. Acta (BBA) 1802 (1), 11–19.

Devi, L., Raghavendran, V., Prabhu, B.M., et al., 2008. Mitochondrial import and accumulation of α-synuclein impair complex I in human dopaminergic neuronal cultures and Parkinson disease brain. J. Biol. Chem. 283 (14), 9089–9100.

Devos, D., Lebouvier, T., Lardeux, B., et al., 2013. Colonic inflammation in Parkinson's disease. Neurobiol. Dis. 50 (1), 42–48.

Dieriks, B.V., Park, T.I., Fourie, C., et al., 2017. Alpha-synuclein transfer through tunneling nanotubes occurs in SH-SY5Y cells and primary brain pericytes from Parkinson's disease patients. Sci. Rep. 7, 42984.

Donadio, V., Incensi, A., Piccinini, C., et al., 2016. Skin nerve misfolded alpha-synuclein in pure autonomic failure and Parkinson disease. Ann. Neurol. 79 (2), 306–316.

Edwards, T.L., Scott, W.K., Almonte, C., et al., 2010. Genome-Wide association study confirms SNPs in SNCA and the MAPT region as common risk factors for Parkinson disease. Ann. Hum. Genet. 74 (2), 97–109.

Eliezer, D., Kutluay, E., Bussell, R., et al., 2001. Conformational properties of alpha-synuclein in its free and lipid-associated states. J. Mol. Biol. 307 (4), 1061–1073.

Emmanouilidou, E., Melachroinou, K., Roumeliotis, T., et al., 2010a. Cell-produced alpha-synuclein is secreted in a calcium-dependent manner by exosomes and impacts neuronal survival. J. Neurosci. 30 (20), 6838–6851.

Emmanouilidou, E., Stefanis, L., Vekrellis, K., 2010b. Cell-produced α-synuclein oligomers are targeted to, and impair, the 26S proteasome. Neurobiol. Aging 31 (6), 953–968.

Fanciulli, A., Wenning, G.K., 2015. Multiple-system atrophy. N. Engl. J. Med. 372 (14), 1375–1376.

Fares, M.B., Ait-Bouziad, N., Dikiy, I., et al., 2014. The novel Parkinson's disease linked mutation G51D attenuates in vitro aggregation and membrane binding of α-synuclein, and enhances its secretion and nuclear localization in cells. Hum. Mol. Genet. 23 (17), 4491–4509.

Fares, M.B., Maco, B., Oueslati, A., et al., 2016. Induction of de novo alpha-synuclein fibrillization in a neuronal model for Parkinson's disease. Proc. Natl. Acad. Sci. U. S. A. 113 (7), E912–E921.

Farrer, M., Kachergus, J., Forno, L., et al., 2004. Comparison of kindreds with parkinsonism and alpha-synuclein genomic multiplications. Ann. Neurol. 55 (2), 174–179.

Fauvet, B., Mbefo, M.K., Fares, M.-B., et al., 2012. α-Synuclein in central nervous system and from erythrocytes, mammalian cells, and *Escherichia coli* exists predominantly as disordered monomer. J. Biol. Chem. 287 (19), 15345–15364.

Ferreira, D.G., Temido-Ferreira, M., Miranda, H.V., et al., 2017. Alpha-synuclein interacts with PrP(C) to induce cognitive impairment through mGluR5 and NMDAR2B. Nat. Neurosci. 20 (11), 1569–1579.

Fitzgerald, E., Murphy, S., Martinson, H.A., 2019. Alpha-synuclein pathology and the role of the microbiota in Parkinson's disease. Front. Neurosci. 13, 369.

Flagmeier, P., Meisl, G., Vendruscolo, M., et al., 2016. Mutations associated with familial Parkinson's disease alter the initiation and amplification steps of α-synuclein aggregation. Proc. Natl. Acad. Sci. U. S. A. 113 (37), 10328–10333.

Flavin, W.P., Bousset, L., Green, Z.C., et al., 2017. Endocytic vesicle rupture is a conserved mechanism of cellular invasion by amyloid proteins. Acta Neuropathol. 134 (4), 629–653.

Fortin, D.L., 2004. Lipid rafts mediate the synaptic localization of α-synuclein. J. Neurosci. 24 (30), 6715–6723.

Freeman, D., Cedillos, R., Choyke, S., et al., 2013. Alpha-synuclein induces lysosomal rupture and cathepsin dependent reactive oxygen species following endocytosis. PLoS One 8 (4), e62143.

Fujiwara, H., Hasegawa, M., Dohmae, N., et al., 2003. α-Synuclein is phosphorylated in synucleinopathy lesions. Nat. Cell Biol. 4, 160–164, (February 2002).

Gan-Or, Z., Giladi, N., Rozovski, U., et al., 2008. Genotype-phenotype correlations between GBA mutations and Parkinson disease risk and onset. Neurology 70 (24), 2277–2283.

Gan-Or, Z., Amshalom, I., Kilarski, L.L., et al., 2015. Differential effects of severe vs mild GBA mutations on Parkinson disease. Neurology 84 (9), 880–887.

Garcia-Reitböck, P., Anichtchik, O., Bellucci, A., et al., 2010. SNARE protein redistribution and synaptic failure in a transgenic mouse model of Parkinson's disease. Brain 133 (7), 2032–2044.

Gegg, M.E., Burke, D., Heales, S.J., et al., 2012. Glucocerebrosidase deficiency in substantia nigra of Parkinson disease brains. Ann. Neurol. 72 (3), 455–463.

Giasson, B.I., 2000. Oxidative damage linked to neurodegeneration by selective alpha-synuclein nitration in synucleinopathy lesions. Science 290 (5493), 985–989.

Giasson, B.I., Murray, I.V.J., Trojanowski, J.Q., et al., 2001. A hydrophobic stretch of 12 amino acid residues in the middle of α-synuclein is essential for filament assembly. J. Biol. Chem. 276 (4), 2380–2386.

Gibb, W.R., Lees, A.J., 1988. The relevance of the Lewy body to the pathogenesis of idiopathic Parkinson's disease. J. Neurol. Neurosurg. Psychiatry 51 (6), 745–752.

Goedert, M., 2015. NEURODEGENERATION. Alzheimer's and Parkinson's diseases: the prion concept in relation to assembled Aβ, tau, and α-synuclein. Science 349 (6248), 1255555.

Goers, J., Manning-Bog, A.B., McCormack, A.L., et al., 2003. Nuclear localization of α-synuclein and its interaction with histones. Biochemistry 42 (28), 8465–8471.

Goker-Alpan, O., Schiffmann, R., LaMarca, M.E., et al., 2004. Parkinsonism among Gaucher disease carriers. J. Med. Genet. 41 (12), 937–940.

Gómez-Suaga, P., Bravo-San Pedro, J.M., González-Polo, R.A., et al., 2018. ER-mitochondria signaling in Parkinson's disease. Cell Death Dis. 9 (3), 337.

Gonçalves, S., Outeiro, T.F., 2013. Assessing the subcellular dynamics of alpha-synuclein using photoactivation microscopy. Mol. Neurobiol. 47 (3), 1081–1092.

Görlach, A., Klappa, P., Kietzmann, D.T., 2006. The endoplasmic reticulum: folding, calcium homeostasis, signaling, and redox control. Antioxid. Redox Signal. 8 (9–10), 1391–1418.

Grozdanov, V., Danzer, K.M., 2018. Release and uptake of pathologic alpha-synuclein. Cell Tissue Res. 373, 175–182.

Gu, M., Gash, M.T., Cooper, J.M., et al., 1997. Mitochondrial respiratory chain function in multiple system atrophy. Mov. Disord. 12 (3), 418–422.

Guardia-Laguarta, C., Area-Gomez, E., Rub, C., et al., 2014. Alpha-synuclein is localized to mitochondria-associated ER membranes. J. Neurosci. 34 (1), 249–259.

Guardia-Laguarta, C., Area-Gomez, E., Schon, E.A., et al., 2015. A new role for α-synuclein in Parkinson's disease: alteration of ER–mitochondrial communication. Mov. Disord. 30 (8), 1026–1033.

Hafner-Bratkovič, I., Benčina, M., Fitzgerald, K.A., et al., 2012. NLRP3 inflammasome activation in macrophage cell lines by prion protein fibrils as the source of IL-1β and neuronal toxicity. Cell. Mol. Life Sci. 69 (24), 4215–4228.

Hague, K., Lento, P., Morgello, S., et al., 1997. The distribution of Lewy bodies in pure autonomic failure: autopsy findings and review of the literature. Acta Neuropathol. 94 (2), 192–196.

Haj-Yahya, M., Fauvet, B., Herman-Bachinsky, Y., et al., 2013. Synthetic polyubiquitinated α-Synuclein reveals important insights into the roles of the ubiquitin chain in regulating its pathophysiology. Proc. Natl. Acad. Sci. U. S. A. 110 (44), 17726–17731.

Han, W., Liu, Y., Mi, Y., et al., 2015. Alpha-synuclein (SNCA) polymorphisms and susceptibility to Parkinson's disease: a meta-analysis. Am. J. Med. Genet. B Neuropsychiatr. Genet. 168B (2), 123–134.

Hansen, C., Angot, E., Bergstrom, A.L., et al., 2011. Alpha-synuclein propagates from mouse brain to grafted dopaminergic neurons and seeds aggregation in cultured human cells. J. Clin. Invest. 121 (2), 715–725.

Harms, A.S., Barnum, C.J., Ruhn, K.A., et al., 2011. Delayed dominant-negative TNF gene therapy halts progressive loss of nigral dopaminergic neurons in a rat model of Parkinson's disease. Mol. Ther. 19 (1), 46–52.

Hattori, N., Tanaka, M., Ozawa, T., et al., 1991. Immunohistochemical studies on complexes I, II, III, and IV of mitochondria in Parkinson's disease. Ann. Neurol. 30 (4), 563–571.

Hawkes, C.H., Del Tredici, K., Braak, H., 2009. Parkinson's disease: the dual hit theory revisited. Ann. N. Y. Acad. Sci. 1170, 615–622.

He, Y., Yu, Z., Chen, S., 2018. Alpha-synuclein nitration and its implications in Parkinson's disease. ACS Chem. Nerosci. 10, 777–782.

Hejjaoui, M., Haj-Yahya, M., Kumar, K.S.A., et al., 2011. Towards elucidation of the role of ubiquitination in the pathogenesis of Parkinson's disease with semisynthetic ubiquitinated α-synuclein. Angew. Chem. Int. Ed. 50 (2), 405–409.

Heman-Ackah, S.M., Manzano, R., Hoozemans, J.J.M., et al., 2017. Alpha-synuclein induces the unfolded protein response in Parkinson's disease SNCA triplication iPSC-derived neurons. Hum. Mol. Genet. 26 (22), 4441–4450.

Hodara, R., Norris, E.H., Giasson, B.I., et al., 2004. Functional consequences of alpha-synuclein tyrosine nitration: diminished binding to lipid vesicles and increased fibril formation. J. Biol. Chem. 279 (46), 47746–47753.

Hoffmann, A., Ettle, B., Bruno, A., et al., 2016. Alpha-synuclein activates BV2 microglia dependent on its aggregation state. Biochem. Biophys. Res. Commun. 479 (4), 881–886.

Holmes, B.B., DeVos, S.L., Kfoury, N., et al., 2013. Heparan sulfate proteoglycans mediate internalization and propagation of specific proteopathic seeds. Proc. Natl. Acad. Sci. U. S. A. 110 (33), E3138–E3147.

Holmqvist, S., Chutna, O., Bousset, L., et al., 2014. Direct evidence of Parkinson pathology spread from the gastrointestinal tract to the brain in rats. Acta Neuropathol. 128 (6), 805–820.

Holtz, W.A., O'Malley, K.L., 2003. Parkinsonian mimetics induce aspects of unfolded protein response in death of dopaminergic neurons. J. Biol. Chem. 278 (21), 19367–19377.

Houlden, H., Singleton, A.B., 2012. The genetics and neuropathology of Parkinson's disease. Acta Neuropathol. 124 (3), 325–338.

Hoyer, W., Cherny, D., Subramaniam, V., et al., 2004. Impact of the acidic C-terminal region comprising amino acids 109–140 on alpha-synuclein aggregation in vitro. Biochemistry 43 (51), 16233–16242.

Ibanez, P., Bonnet, A.M., Debarges, B., et al., 2004. Causal relation between alpha-synuclein gene duplication and familial Parkinson's disease. Lancet 364 (9440), 1169–1171.

Iranzo, A., 2018. The REM sleep circuit and how its impairment leads to REM sleep behavior disorder. Cell Tissue Res. 373 (1), 245–266.

Iranzo, A., Santamaria, J., Tolosa, E., 2016. Idiopathic rapid eye movement sleep behaviour disorder: diagnosis, management, and the need for neuroprotective interventions. Lancet Neurol. 15 (4), 405–419.

Irwin, D.J., Lee, V.M., Trojanowski, J.Q., 2013. Parkinson's disease dementia: convergence of alpha-synuclein, tau and amyloid-beta pathologies. Nat. Rev. Neurosci. 14 (9), 626–636.

Iwai, A., Masliah, E., Yoshimoto, M., et al., 1995. The precursor protein of non-Aβ component of Alzheimer's disease amyloid is a presynaptic protein of the central nervous system. Neuron 14 (2), 467–475.

Janezic, S., Threlfell, S., Dodson, P.D., et al., 2013. Deficits in dopaminergic transmission precede neuron loss and dysfunction in a new Parkinson model. Proc. Natl. Acad. Sci. U. S. A. 110 (42), E4016–E4025.

Jenner, P., Morris, H.R., Robbins, T.W., et al., 2013. Parkinson's disease—the debate on the clinical phenomenology, aetiology, pathology and pathogenesis. J. Park. Dis. 3 (1), 1–11.

Jensen, P.H., Hager, H., Nielsen, M.S., et al., 1999. α-Synuclein binds to Tau and stimulates the protein kinase A-catalyzed tau phosphorylation of serine residues 262 and 356. J. Biol. Chem. 274 (36), 25481–25489.

Kaufmann, H., 1996. Consensus statement on the definition of orthostatic hypotension, pure autonomic failure and multiple system atrophy. Clin. Auton. Res. 6 (2), 125–126.

Kaufmann, H., Goldstein, D.S., 2010. Pure autonomic failure: a restricted Lewy body synucleinopathy or early Parkinson disease? Neurology 74 (7), 536–537.

Keeney, P.M., 2006. Parkinson's disease brain mitochondrial complex I has oxidatively damaged subunits and is functionally impaired and misassembled. J. Neurosci. 26 (19), 5256–5264.

Keshavarzian, A., Green, S.J., Engen, P.A., et al., 2015. Colonic bacterial composition in Parkinson's disease. Mov. Disord. 30 (10), 1351–1360.

Kessler, J.C., Rochet, J.-C., Lansbury, P.T., 2003. The N-terminal repeat domain of alpha-synuclein inhibits beta-sheet and amyloid fibril formation. Biochemistry 42 (3), 672–678.

Khalaf, O., Fauvet, B., Oueslati, A., et al., 2014. The H50Q mutation enhances α-synuclein aggregation, secretion, and toxicity. J. Biol. Chem. 289 (32), 21856–21876.

Kim, M., Jung, W., Lee, I.H., et al., 2008. Impairment of microtubule system increases α-synuclein aggregation and toxicity. Biochem. Biophys. Res. Commun. 365 (4), 628–635.

Kim, C., Ho, D.H., Suk, J.E., et al., 2013. Neuron-released oligomeric alpha-synuclein is an endogenous agonist of TLR2 for paracrine activation of microglia. Nat. Commun. 4, 1562.

Kim, C., Lv, G., Lee, J.S., et al., 2016a. Exposure to bacterial endotoxin generates a distinct strain of alpha-synuclein fibril. Sci. Rep. 6, 30891.

Kim, Y.S., Anderson, M., Park, K., et al., 2016b. Coupled activation of primary sensory neurons contributes to chronic pain article coupled activation of primary sensory neurons contributes to chronic pain. Neuron 91 (5), 1085–1096.

Kirik, D., Rosenblad, C., Burger, C., et al., 2002. Parkinson-like neurodegeneration induced by targeted overexpression of α-synuclein in the nigrostriatal system. J. Neurosci. 22 (7), 2780–2791.

Klivenyi, P., Siwek, D., Gardian, G., et al., 2006. Mice lacking alpha-synuclein are resistant to mitochondrial toxins. Neurobiol. Dis. 21 (3), 541–548.

Kontopoulos, E., Parvin, J.D., Feany, M.B., 2006. Alpha-synuclein acts in the nucleus to inhibit histone acetylation and promote neurotoxicity. Hum. Mol. Genet. 15 (20), 3012–3023.

Kordower, J.H., Chu, Y., Hauser, R.A., et al., 2008. Lewy body-like pathology in long-term embryonic nigral transplants in Parkinson's disease. Nat. Med. 14 (5), 504–506.

Kovari, E., Horvath, J., Bouras, C., 2009. Neuropathology of Lewy body disorders. Brain Res. Bull. 80 (4–5), 203–210.

Kramer, M.L., Schulz-Schaeffer, W.J., 2007. Presynaptic alpha-synuclein aggregates, not Lewy bodies, cause neurodegeneration in dementia with Lewy bodies. J. Neurosci. 27 (6), 1405–1410.

Kruger, R., Kuhn, W., Muller, T., et al., 1998. Ala30Pro mutation in the gene encoding alpha-synuclein in Parkinson's disease. Nat. Genet. 18 (2), 106–108.

Krüger, R., Hardt, C., Tschentscher, F., et al., 2000. Genetic analysis of immunomodulating factors in sporadic Parkinson's disease. J. Neural Transm. 107, 553–562.

Krumova, P., Meulmeester, E., Garrido, M., et al., 2011. SUMOylation inhibits α-synuclein aggregation and toxicity. J. Cell Biol. 194 (1), 49–60.

Kubo, S., Nemani, V.M., Chalkley, R.J., et al., 2005. A combinatorial code for the interaction of α-synuclein with membranes. J. Biol. Chem. 280 (36), 31664–31672.

Kuusisto, E., Parkkinen, L., Alafuzoff, I., 2003. Morphogenesis of Lewy bodies: dissimilar incorporation of alpha-synuclein, ubiquitin, and p62. J. Neuropathol. Exp. Neurol. 62 (12), 1241–1253.

Larsen, K.E., Schmitz, Y., Troyer, M.D., et al., 2006. Alpha-synuclein overexpression in PC12 and chromaffin cells impairs catecholamine release by interfering with a late step in exocytosis. J. Neurosci. 26 (46), 11915–11922.

Lashuel, H.A., Petre, B.M., Wall, J., et al., 2002. Alpha-synuclein, especially the Parkinson's disease-associated mutants, forms pore-like annular and tubular protofibrils. J. Mol. Biol. 322 (5), 1089–1102.

Laugks, U., Taguchi, Y.V., Fernandez-Busnadiego, R., et al., 2017. Synucleins have multiple effects on presynaptic architecture. Cell Rep. 18 (1), 161–173.

Lázaro, D.F., Rodrigues, E.F., Langohr, R., et al., 2014. Systematic comparison of the effects of alpha-synuclein mutations on its oligomerization and aggregation. PLoS Genet. 10 (11), e1004741.

Leclair-Visonneau, L., Clairembault, T., Coron, E., et al., 2017. REM sleep behavior disorder is related to enteric neuropathology in Parkinson disease. Neurology 89 (15), 1612–1618.

Lee, H.-J., 2004. Clearance of alpha-synuclein oligomeric intermediates via the lysosomal degradation pathway. J. Neurosci. 24 (8), 1888–1896.

Lee, H.-J., Patel, S., Lee, S.-J., 2005. Intravesicular localization and exocytosis of α-synuclein and its aggregates. J. Neurosci. 25 (25), 6016–6024.

Lee, H.J., Khoshaghideh, F., Lee, S., et al., 2006. Impairment of microtubule-dependent trafficking by overexpression of α-synuclein. Eur. J. Neurosci. 24 (11), 3153–3162.

Lee, H.J., Suk, J.E., Bae, E.J., et al., 2008a. Assembly-dependent endocytosis and clearance of extracellular α-synuclein. Int. J. Biochem. Cell Biol. 40 (9), 1835–1849.

Lee, S.J., Jeon, H., Kandror, K.V., 2008b. Alpha-synuclein is localized in a subpopulation of rat brain synaptic vesicles. Acta Neurobiol. Exp. (Wars) 68 (4), 509–515.

Lee, J.G., Takahama, S., Zhang, G., et al., 2016. Unconventional secretion of misfolded proteins promotes adaptation to proteasome dysfunction in mammalian cells. Nat. Cell Biol. 18 (7), 765–776.

Lesage, S., Anheim, M., Letournel, F., et al., 2013. G51D alpha-synuclein mutation causes a novel parkinsonian-pyramidal syndrome. Ann. Neurol. 73 (4), 459–471.

Lewy, F., 1912. Paralysis agitans. In: Lewandowsky, M.A.G. (Ed.), Handbuch der Neurologie. Springer Verlag, Berlin, pp. 920–933.

Li, H.T., Du, H.N., Tang, L., et al., 2002. Structural transformation and aggregation of human α-synuclein in trifluoroethanol: non-amyloid component sequence is essential and β-sheet formation is prerequisite to aggregation. Biopolymers 64 (4), 221–226.

Li, W., West, N., Colla, E., et al., 2005. Aggregation promoting C-terminal truncation of alpha-synuclein is a normal cellular process and is enhanced by the familial Parkinson's disease-linked mutations. Proc. Natl. Acad. Sci. U. S. A. 102 (6), 2162–2167.

Li, W.W., Yang, R., Guo, J.C., et al., 2007. Localization of α-synuclein to mitochondria within midbrain of mice. Neuroreport 18 (15), 1543–1546.

Li, J.Y., Englund, E., Holton, J.L., et al., 2008. Lewy bodies in grafted neurons in subjects with Parkinson's disease suggest host-to-graft disease propagation. Nat. Med. 14 (5), 501–503.

Lindstrom, V., Gustafsson, G., Sanders, L.H., et al., 2017. Extensive uptake of alpha-synuclein oligomers in astrocytes results in sustained intracellular deposits and mitochondrial damage. Mol. Cell. Neurosci. 82, 143–156.

Loria, F., Vargas, J.Y., Bousset, L., et al., 2017. Alpha-synuclein transfer between neurons and astrocytes indicates that astrocytes play a role in degradation rather than in spreading. Acta Neuropathol. 134 (5), 789–808.

Lubomski, M., Tan, A.H., Lim, S.Y., et al., 2019. Parkinson's disease and the gastrointestinal microbiome. J. Neurol, 1–17.

Luk, K.C., Song, C., O'Brien, P., et al., 2009. Exogenous alpha-synuclein fibrils seed the formation of Lewy body-like intracellular inclusions in cultured cells. Proc. Natl. Acad. Sci. U. S. A. 106 (47), 20051–20056.

Ma, M.R., Hu, Z.W., Zhao, Y.F., et al., 2016. Phosphorylation induces distinct alpha-synuclein strain formation. Sci. Rep. 6, 37130.

Manfredsson, F.P., Luk, K.C., Benskey, M.J., et al., 2018. Induction of alpha-synuclein pathology in the enteric nervous system of the rat and non-human primate results in gastrointestinal dysmotility and transient CNS pathology. Neurobiol. Dis. 112, 106–118.

Mao, X., Ou, M.T., Karuppagounder, S.S., et al., 2016. Pathological alpha-synuclein transmission initiated by binding lymphocyte-activation gene 3. Science 353 (6307), aah3374.

Maroteaux, L., Campanelli, J.T., Scheller, R.H., 1988. Synuclein: a neuron-specific protein localized to the nucleus and presynaptic nerve terminal. J. Neurosci. 8 (8), 2804–2815.

Martin, L.J., 2006. Parkinson's disease α-synuclein transgenic mice develop neuronal mitochondrial degeneration and cell death. J. Neurosci. 26 (1), 41–50.

Marui, W., Iseki, E., Nakai, T., et al., 2002. Progression and staging of Lewy pathology in brains from patients with dementia with Lewy bodies. J. Neurol. Sci. 195 (2), 153–159.

Masaracchia, C., Hnida, M., Gerhardt, E., et al., 2018. Membrane binding, internalization, and sorting of alpha-synuclein in the cell. Acta Neuropathol. Commun. 6 (1), 79.

Mason, D.M., Nouraei, N., Pant, D.B., et al., 2016. Transmission of alpha-synucleinopathy from olfactory structures deep into the temporal lobe. Mol. Neurodegener. 11 (1), 49.

Mayo, M.C., Bordelon, Y., 2014. Dementia with Lewy bodies. Semin. Neurol. 34 (2), 182–188.

McGeer, P.L., Itagaki, S., Boyes, B.E., et al., 1988. Reactive microglia are positive for HLA-DR in the substantia nigra of Parkinson's and Alzheimer's disease brains. Neurology 38 (8), 1285.

McKeith, I.G., Boeve, B.F., Dickson, D.W., et al., 2017. Diagnosis and management of dementia with Lewy bodies. Neurology 89 (1), 88–100.

McKenna, D., Peever, J., 2017. Degeneration of rapid eye movement sleep circuitry underlies rapid eye movement sleep behavior disorder. Mov. Disord. 32 (5), 636–644.

McLean, P.J., Kawamata, H., Ribich, S., et al., 2000a. Membrane association and protein conformation of alpha-synuclein in intact neurons. Effect of Parkinson's disease-linked mutations. J. Biol. Chem. 275 (12), 8812–8816.

McLean, P.J., Ribich, S., Hyman, B.T., 2000b. Subcellular localization of alpha-synuclein in primary neuronal cultures: effect of missense mutations. J. Neural Transm. Suppl. (58), 53–63.

Miraglia, F., Ricci, A., Rota, L., et al., 2018. Subcellular localization of alpha-synuclein aggregates and their interaction with membranes. Neural Regen. Res. 13 (7), 1136–1144.

Mishizen-Eberz, A.J., Guttmann, R.P., Giasson, B.I., et al., 2003. Distinct cleavage patterns of normal and pathologic forms of α-synuclein by calpain I in vitro. J. Neurochem. 86 (4), 836–847.

Mishizen-Eberz, A.J., Norris, E.H., Giasson, B.I., et al., 2005. Cleavage of α-synuclein by calpain: potential role in degradation of fibrillized and nitrated species of α-synuclein. Biochemistry 44 (21), 7818–7829.

Mogi, M., Togari, A., Kondo, T., et al., 2000. Caspase activities and tumor necrosis factor receptor R1 (p55) level are elevated in the substantia nigra from parkinsonian brain. J. Neural Transm. (Vienna) 107 (3), 335–341.

Mor, D.E., Tsika, E., Mazzulli, J.R., et al., 2017. Dopamine induces soluble alpha-synuclein oligomers and nigrostriatal degeneration. Nat. Neurosci. 20, 1560–1568.

Mosharov, E.V., 2006. Alpha-synuclein overexpression increases cytosolic catecholamine concentration. J. Neurosci. 26 (36), 9304–9311.

Mullin, S., Schapira, A., 2015. The genetics of Parkinson's disease. Br. Med. Bull. 114 (1), 39–52.

Murray, I.V.J., Giasson, B.I., Quinn, S.M., et al., 2003. Role of α-synuclein Carboxy-terminus on fibril formation in vitro. Biochemistry 42, 8530–8540.

Nakamura, K., Nemani, V.M., Azarbal, F., et al., 2011. Direct membrane association drives mitochondrial fission by the Parkinson disease-associated protein α-synuclein. J. Biol. Chem. 286 (23), 20710–20726.

Nakamura, K., Mori, F., Kon, T., et al., 2015. Filamentous aggregations of phosphorylated alpha-synuclein in Schwann cells (Schwann cell cytoplasmic inclusions) in multiple system atrophy. Acta Neuropathol. Commun. 3, 29.

Nalls, M.A., Duran, R., Lopez, G., et al., 2013. A multicenter study of glucocerebrosidase mutations in dementia with Lewy bodies. JAMA Neurol. 70 (6), 727–735.

Narayanan, V., Scarlata, S., 2001. Membrane binding and self-association of α-synucleins. Biochemistry 40 (33), 9927–9934.

Nemani, V.M., Lu, W., Berge, V., et al., 2010. Increased expression of α-synuclein reduces neurotransmitter release by inhibiting synaptic vesicle reclustering after endocytosis. Neuron 65 (1), 66–79.

Ouberai, M.M., Wang, J., Swann, M.J., et al., 2013. α-Synuclein senses lipid packing defects and induces lateral expansion of lipids leading to membrane remodeling. J. Biol. Chem. 288 (29), 20883–20895.

Oueslati, A., Fournier, M., Lashuel, H.A., 2010. Role of post-translational modifications in modulating the structure, function and toxicity of α-synuclein. Prog. Brain Res. 183, 115–145.

Outeiro, T.F., Lindquist, S., 2003. Yeast cells provide insight into alpha-synuclein biology and pathobiology. Science 302 (5651), 1772–1775.

Outeiro, T.F., Putcha, P., Tetzlaff, J.E., et al., 2008. Formation of toxic oligomeric α-synuclein species in living cells. PLoS One 3 (4), e1867.

Outeiro, T.F., Klucken, J., Bercury, K., et al., 2009. Dopamine-Induced conformational changes in alpha-synuclein. PLoS One 4 (9), e6906.

Outeiro, T.F., Koss, D.J., Erskine, D., et al., 2019. Dementia with Lewy bodies: an update and outlook. Mol. Neurodegener. 14 (1), 1–18.

Ozawa, T., Healy, D.G., Abou-Sleiman, P.M., et al., 2006. The alpha-synuclein gene in multiple system atrophy. J. Neurol. Neurosurg. Psychiatry 77 (4), 464–467.

Paik, S.R., Shin, H., Lee, J., et al., 2004. Impact of the acidic C-terminal region comprising amino acids 109-140 on alpha-synuclein aggregation in vitro. Biochemistry 43 (51), 16233–16242.

Paiva, I., Jain, G., Lázaro, D.F., et al., 2018. Alpha-synuclein deregulates the expression of COL4A2 and impairs ER-Golgi function. Neurobiol. Dis. 119, 121–135.

Paleologou, K.E., Oueslati, A., Shakked, G., et al., 2010. Phosphorylation at S87 is enhanced in synucleinopathies, inhibits alpha-synuclein oligomerization, and influences synuclein-membrane interactions. J. Neurosci. 30 (9), 3184–3198.

Parker, W.D., Boyson, T.S.J., Parks, J.K., 1989. Abnormalities of the electron transport chain in idiopathic Parkinson's disease. Ann. Neurol. 26, 719–723.

Parkinson, J., 2002. An essay on the shaking palsy. 1817. J. Neuropsychiatry Clin. Neurosci. 14 (2), 223–236, (discussion 222).

Pasanen, P., Myllykangas, L., Siitonen, M., et al., 2014. Novel alpha-synuclein mutation A53E associated with atypical multiple system atrophy and Parkinson's disease-type pathology. Neurobiol. Aging 35 (9), 2180.e1-5.

Paumier, K.L., Luk, K.C., Manfredsson, F.P., et al., 2015. Intrastriatal injection of pre-formed mouse alpha-synuclein fibrils into rats triggers alpha-synuclein pathology and bilateral nigrostriatal degeneration. Neurobiol. Dis. 82, 185–199.

Pavlou, M.A.S., Pinho, R., Paiva, I., et al., 2016. The yin and yang of α-synuclein-associated epigenetics in Parkinson's disease. 140, 1–9.

Paxinou, E., Chen, Q., Weisse, M., et al., 2001. Induction of alpha-synuclein aggregation by intracellular nitrative insult. J. Neurosci. 21 (20), 8053–8061.

Peelaerts, W., Baekelandt, V., 2016. α-Synuclein strains and the variable pathologies of synucleinopathies. J. Neurochem. 139 (Suppl. 1), 256–274.

Peelaerts, W., Bousset, L., Van der Perren, A., et al., 2015. Alpha-synuclein strains cause distinct synucleinopathies after local and systemic administration. Nature 522 (7556), 340–344.

Peelaerts, W., Bousset, L., Baekelandt, V., et al., 2018. α-Synuclein strains and seeding in Parkinson's disease, incidental Lewy body disease, dementia with Lewy bodies and multiple system atrophy: similarities and differences. Cell Tissue Res. 373 (1), 195–212.

Peng, C., Gathagan, R.J., Covell, D.J., et al., 2018. Cellular milieu imparts distinct pathological alpha-synuclein strains in alpha-synucleinopathies. Nature 557 (7706), 558–563.

Perez, R.G., Lin, E., Liu, J.J., et al., 2002. A role for alpha-synuclein in the regulation of dopamine biosynthesis. J. Neurosci. 22 (8), 3090–3099.

Perez-Pardo, P., Hartog, M., Garssen, J., et al., 2017a. Microbes tickling your tummy: the importance of the gut-brain axis in Parkinson's disease. Curr. Behav. Neurosci. Rep. 4 (4), 361–368.

Perez-Pardo, P., Kliest, T., Dodiya, H.B., et al., 2017b. The gut-brain axis in Parkinson's disease: possibilities for food-based therapies. Eur. J. Pharmacol. 817, 86–95.

Perrin, R.J., Woods, W.S., Clayton, D.F., et al., 2001. Exposure to long chain polyunsaturated fatty acids triggers rapid multimerization of synucleins. J. Biol. Chem. 276 (45), 41958–41962.

Pieri, L., Madiona, K., Melki, R., 2016. Structural and functional properties of prefibrillar alpha-synuclein oligomers. Sci. Rep. 6, 24526.

Pinho, R., Paiva, I., Jerčić, K.G., et al., 2019. Nuclear localization and phosphorylation modulate pathological effects of alpha-synuclein. Hum. Mol. Genet. 28 (1), 31–50.

Plotegher, N., Bubacco, L., 2016. Lysines, Achilles' heel in alpha-synuclein conversion to a deadly neuronal endotoxin. Ageing Res. Rev. 26, 62–71.

Poewe, W., Seppi, K., Tanner, C.M., et al., 2017. Parkinson disease. Nat. Rev. Dis. Primers. 3, 17013.

Polinski, N.K., Volpicelli-Daley, L.A., Sortwell, C.E., et al., 2018. Best practices for generating and using alpha-synuclein pre-formed fibrils to model Parkinson's disease in rodents. J. Park. Dis. 8 (2), 303–322.

Polymeropoulos, M.H., Lavedan, C., Leroy, E., et al., 1997. Mutation in the alpha-synuclein gene identified in families with Parkinson's disease. Science 276 (5321), 2045–2047.

Postuma, R.B., Berg, D., Stern, M., et al., 2015a. MDS clinical diagnostic criteria for Parkinson's disease. Mov. Disord. 30 (12), 1591–1601.

Postuma, R.B., Gagnon, J.F., Bertrand, J.A., et al., 2015b. Parkinson risk in idiopathic REM sleep behavior disorder: preparing for neuroprotective trials. Neurology 84 (11), 1104–1113.

Quadri, M., Mandemakers, W., Grochowska, M.M., et al., 2018. LRP10 genetic variants in familial Parkinson's disease and dementia with Lewy bodies: a genome-wide linkage and sequencing study. Lancet Neurol. 17 (7), 597–608.

Quinn, N., 2015. A short clinical history of multiple system atrophy. Clin. Auton. Res. 25 (1), 3–7.

Recasens, A., Dehay, B., Bove, J., et al., 2014. Lewy body extracts from Parkinson disease brains trigger alpha-synuclein pathology and neurodegeneration in mice and monkeys. Ann. Neurol. 75 (3), 351–362.

Rey, N.L., Petit, G.H., Bousset, L., et al., 2013. Transfer of human alpha-synuclein from the olfactory bulb to interconnected brain regions in mice. Acta Neuropathol. 126 (4), 555–573.

Rey, N.L., Steiner, J.A., Maroof, N., et al., 2016. Widespread transneuronal propagation of alpha-synucleinopathy triggered in olfactory bulb mimics prodromal Parkinson's disease. J. Exp. Med. 213 (9), 1759–1778.

Reyes, J.F., Rey, N.L., Bousset, L., et al., 2014. Alpha-synuclein transfers from neurons to oligodendrocytes. Glia 62 (3), 387–398.

Roodveldt, C., Labrador-Garrido, A., Gonzalez-Rey, E., et al., 2013. Preconditioning of microglia by α-synuclein strongly affects the response induced by Toll-like receptor (TLR) stimulation. PLoS One 8 (11), 1–17.

Rostami, J., Holmqvist, S., Lindstrom, V., et al., 2017. Human astrocytes transfer aggregated alpha-synuclein via tunneling nanotubes. J. Neurosci. 37 (491), 11835–11853.

Rott, R., Szargel, R., Haskin, J., et al., 2008. Monoubiquitylation of α-Synuclein by seven in absentia homolog (SIAH) promotes its aggregation in dopaminergic cells. J. Biol. Chem. 283 (6), 3316–3328.

Rott, R., Szargel, R., Shani, V., et al., 2017. SUMOylation and ubiquitination reciprocally regulate α-synuclein degradation and pathological aggregation. Proc. Natl. Acad. Sci. U. S. A. 114 (50), 13176–13181.

Ryu, E.J., Harding, H.P., Angelastro, J.M., et al., 2018. Endoplasmic reticulum stress and the unfolded protein response in cellular models of Parkinson's disease. J. Neurosci. 22 (24), 10690–10698.

Sacino, A.N., Brooks, M., Thomas, M.A., et al., 2014. Amyloidogenic alpha-synuclein seeds do not invariably induce rapid, widespread pathology in mice. Acta Neuropathol. 127 (5), 645–665.

Saito, Y., Kawashima, A., Ruberu, N.N., et al., 2003. Accumulation of phosphorylated alpha-synuclein in aging human brain. J. Neuropathol. Exp. Neurol. 62 (6), 644–654.

Salganik, M., Sergeyev, V.G., Shinde, V., et al., 2015. The loss of glucose-regulated protein 78 (GRP78) during normal aging or from siRNA knockdown augments human alpha-synuclein (α-syn) toxicity to rat nigral neurons. Neurobiol. Aging 36 (6), 2213–2223.

Sampson, T.R., Debelius, J.W., Thron, T., et al., 2016. Gut microbiota regulate motor deficits and neuroinflammation in a model of Parkinson's disease. Cell 167 (6), 1469–1480.e12.

Satake, W., Nakabayashi, Y., Mizuta, I., et al., 2009. Genome-wide association study identifies common variants at four loci as genetic risk factors for Parkinson's disease. Nat. Genet. 41 (12), 1303–1307.

Schapira, A.H.V., Cooper, J.M., Dexter, D., et al., 1990. Mitochondrial complex I deficiency in Parkinson's disease. J. Neurochem. 54 (3), 823–827.

Schaser, A.J., Osterberg, V.R., Dent, S.E., et al., 2019. Alpha-synuclein is a DNA binding protein that modulates DNA repair with implications for Lewy body disorders. Sci. Rep. 9 (1), 10919.

Schenck, C.H., Bundlie, S.R., Ettinger, M.G., et al., 1986. Chronic behavioral disorders of human REM sleep: a new category of parasomnia. Sleep 9 (2), 293–308.

Scheperjans, F., Aho, V., Pereira, P.A., et al., 2015. Gut microbiota are related to Parkinson's disease and clinical phenotype. Mov. Disord. 30 (3), 350–358.

Scheperjans, F., Derkinderen, P., Borghammer, P., 2018. The gut and Parkinson's disease: hype or hope? J. Park. Dis. 8 (s1), S31–S39.

Schirinzi, T., Madeo, G., Martella, G., et al., 2016. Early synaptic dysfunction in Parkinson's disease: insights from animal models. Mov. Disord. 31 (6), 802–813.

Scholz, S.W., Houlden, H., Schulte, C., et al., 2009. SNCA variants are associated with increased risk for multiple system atrophy. Ann. Neurol. 65 (5), 610–614.

Schulz-Schaeffer, W.J., 2010. The synaptic pathology of α-synuclein aggregation in dementia with Lewy bodies, Parkinson's disease and Parkinson's disease dementia. Acta Neuropathol. 120 (2), 131–143.

Segrest, J.P., De Loof, H., Dohlman, J.G., et al., 1990. Amphipathic helix motif: classes and properties. Proteins Struct. Funct. Genet. 8 (2), 103–117.

Shannon, K., Vanden Berghe, P., 2018. The enteric nervous system in PD: gateway, bystander victim, or source of solutions. Cell Tissue Res. 373 (1), 313–326.

Sharon, R., Goldberg, M.S., Bar-Josef, I., et al., 2001. Alpha-Synuclein occurs in lipid-rich high molecular weight complexes, binds fatty acids, and shows homology to the fatty acid-binding proteins. Proc. Natl. Acad. Sci. U. S. A. 98 (16), 9110–9115.

Sharon, R., Bar-Joseph, I., Frosch, M.P., et al., 2003. The formation of highly soluble oligomers of alpha-synuclein is regulated by fatty acids and enhanced in Parkinson's disease. Neuron 37 (4), 583–595.

Shrivastava, A.N., Redeker, V., Fritz, N., et al., 2015. Alpha-synuclein assemblies sequester neuronal α3-Na+/K+-ATPase and impair Na+ gradient. EMBO J. 34 (19), 2408–2423.

Sidransky, E., Nalls, M.A., Aasly, J.O., et al., 2009. Multicenter analysis of glucocerebrosidase mutations in Parkinson's disease. N. Engl. J. Med. 361 (17), 1651–1661.

Simon-Sanchez, J., Schulte, C., Bras, J.M., et al., 2009. Genome-wide association study reveals genetic risk underlying Parkinson's disease. Nat. Genet. 41 (12), 1308–1312.

Singleton, A.B., Farrer, M., Johnson, J., et al., 2003. Alpha-synuclein locus triplication causes Parkinson's disease. Science 302 (5646), 841.

Smith, W.W., Jiang, H., Pei, Z., et al., 2005. Endoplasmic reticulum stress and mitochondrial cell death pathways mediate A53T mutant alpha-synuclein-induced toxicity. Hum. Mol. Genet. 14 (24), 3801–3811.

Smith, M.H., Ploegh, H.L., Weissman, J.S., 2011. Road to ruin: targeting proteins for degradation in the endoplasmic reticulum. Science 334 (6059), 1086–1090.

Soma, H., Yabe, I., Takei, A., et al., 2006. Heredity in multiple system atrophy. J. Neurol. Sci. 240 (1–2), 107–110.

Sousa, V.L., Bellani, S., Giannandrea, M., et al., 2009. α-Synuclein and Its A30P mutant affect actin cytoskeletal structure and dynamics. Mol. Biol. Cell 20 (16), 3725–3739.

Specht, C.G., Tigaret, C.M., Rast, G.F., et al., 2005. Subcellular localisation of recombinant α- and γ-synuclein. Mol. Cell. Neurosci. 28 (2), 326–334.

Spillantini, M.G., 1999. Parkinson's disease, dementia with Lewy bodies and multiple system atrophy are alpha-synucleinopathies. Parkinsonism Relat. Disord. 5 (4), 157–162.

Spillantini, M.G., Crowther, R.A., Jakes, R., et al., 1998a. Filamentous alpha-synuclein inclusions link multiple system atrophy with Parkinson's disease and dementia with Lewy bodies. Neurosci. Lett. 251 (3), 205–208.

Spillantini, M.G., Crowther, R.A., Jakes, R., et al., 1998b. α-Synuclein in filamentous inclusions of Lewy bodies from Parkinson's disease and dementia with Lewy bodies. Proc. Natl. Acad. Sci. U. S. A. 95 (11), 6469–6473.

Stefanis, L., Larsen, K.E., Rideout, H.J., et al., 2018. Expression of A53T mutant but not wild-type α-synuclein in PC12 cells induces alterations of the ubiquitin-dependent degradation system, loss of dopamine release, and autophagic cell death. J. Neurosci. 21 (24), 9549–9560.

Steiner, J.A., Quansah, E., Brundin, P., 2018. The concept of alpha-synuclein as a prion-like protein: ten years after. Cell Tissue Res. 373 (1), 161–173.

Stolzenberg, E., Berry, D., Yang, et al., 2017. A role for neuronal alpha-synuclein in gastro-intestinal immunity. J. Innate Immun. 9 (5), 456–463.

Sturm, E., Stefanova, N., 2014. Multiple system atrophy: genetic or epigenetic? Exp. Neurobiol. 23 (4), 277.

Surmeier, D.J., Obeso, J.A., Halliday, G.M., 2017. Parkinson's disease is not simply a prion disorder. J. Neurosci. 37 (41), 9799–9807.

Svensson, E., Horvath-Puho, E., Thomsen, R.W., et al., 2015. Vagotomy and subsequent risk of Parkinson's disease. Ann. Neurol. 78 (4), 522–529.

Szegő, É.M., Dominguez-Meijide, A., Gerhardt, E., et al., 2019. Cytosolic trapping of a mitochondrial heat shock protein is an early pathological event in synucleinopathies. Cell Rep. 28 (1), 65–77.e6.

Tanik, S.A., Schultheiss, C.E., Volpicelli-daley, L.A., et al., 2013. Lewy body-like α-synuclein aggregates resist degradation and impair macroautophagy. J. Biol. Chem. 288 (21), 15194–15210.

Tenreiro, S., Eckermann, K., Outeiro, T.F., 2014. Protein phosphorylation in neurodegeneration: friend or foe? Front. Mol. Neurosci. 7, 42.

Theillet, F.X., Binolfi, A., Bekei, B., et al., 2016. Structural disorder of monomeric α-synuclein persists in mammalian cells. Nature 530 (7588), 45–50.

Tilve, S., Difato, F., Chieregatti, E., 2015. Cofilin 1 activation prevents the defects in axon elongation and guidance induced by extracellular alpha-synuclein. Sci. Rep. 5 (November), 1–13.

Tofaris, G.K., 2017. A critical assessment of exosomes in the pathogenesis and stratification of Parkinson's disease. J. Park. Dis. 7 (4), 569–576.

Tofaris, G.K., Layfield, R., Spillantini, M.G., 2001. α-Synuclein metabolism and aggregation is linked to ubiquitin-independent degradation by the proteasome. FEBS Lett. 509 (1), 22–26.

Tofaris, G.K., Razzaq, A., Ghetti, B., et al., 2003. Ubiquitination of α-synuclein in Lewy bodies is a pathological event not associated with impairment of proteasome function. J. Biol. Chem. 278 (45), 44405–44411.

Tofaris, G.K., Kim, H.T., Hourez, R., et al., 2011. Ubiquitin ligase Nedd4 promotes alpha-synuclein degradation by the endosomal-lysosomal pathway. Proc. Natl. Acad. Sci. U. S. A. 108 (41), 17004–17009.

Tu, P.H., Galvin, J.E., Baba, M., et al., 1998. Glial cytoplasmic inclusions in white matter oligodendrocytes of multiple system atrophy brains contain insoluble alpha-synuclein. Ann. Neurol. 44 (3), 415–422.

Uéda, K., Fukushima, H., Masliah, E., et al., 1993. Molecular cloning of cDNA encoding an unrecognized component of amyloid in Alzheimer disease. Proc. Natl. Acad. Sci. U. S. A. 90 (23), 11282–11286.

Ulusoy, A., Rusconi, R., Perez-Revuelta, B.I., et al., 2013. Caudo-rostral brain spreading of alpha-synuclein through vagal connections. EMBO Mol. Med. 5 (7), 1119–1127.

Ulusoy, A., Musgrove, R.E., Rusconi, R., et al., 2015. Neuron-to-neuron alpha-synuclein propagation in vivo is independent of neuronal injury. Acta Neuropathol. Commun. 3, 13.

Unger, M.M., Spiegel, J., Dillmann, K.U., et al., 2016. Short chain fatty acids and gut microbiota differ between patients with Parkinson's disease and age-matched controls. Parkinsonism Relat. Disord. 32, 66–72.

Uversky, V.N., Li, J., Fink, A.L., 2001a. Evidence for a partially folded intermediate in alpha-synuclein fibril formation. J. Biol. Chem. 276 (14), 10737–10744.

Uversky, V.N., Li, J., Fink, A.L., 2001b. Metal-triggered structural transformations, aggregation, and fibrillation of human alpha-synuclein. A possible molecular NK between Parkinson's disease and heavy metal exposure. J. Biol. Chem. 276 (47), 44284–44296.

Uversky, V.N., Li, J., Bower, K., et al., 2002. Synergistic effects of pesticides and metals on the fibrillation of alpha-synuclein: implications for Parkinson's disease. Neurotoxicology 23, 527–536.

Vasconcellos, L.F., Pereira, J.S., 2015. Parkinson's disease dementia: diagnostic criteria and risk factor review. J. Clin. Exp. Neuropsychol. 37 (9), 988–993.

Vasili, E., Dominguez-Meijide, A., Outeiro, T.F., 2019. Spreading of alpha-synuclein and tau: a systematic comparison of the mechanisms involved. Front. Mol. Neurosci. 12, 107.

Vekrellis, K., Xilouri, M., Emmanouilidou, E., et al., 2009. Inducible over-expression of wild type alpha-synuclein in human neuronal cells leads to caspase-dependent non-apoptotic death. J. Neurochem. 109 (5), 1348–1362.

Vicente Miranda, H., Outeiro, T.F., 2010. The sour side of neurodegenerative disorders: the effects of protein glycation. J. Pathol. 221 (1), 13–25.

Vicente Miranda, H., Szego, É.M., Oliveira, L.M.A., et al., 2017. Glycation potentiates α-synuclein-associated neurodegeneration in synucleinopathies. Brain 140 (5), 1399–1419.

Vilar, M., Chou, H.-T., Lührs, T., et al., 2008. The fold of alpha-synuclein fibrils. Proc. Natl. Acad. Sci. U. S. A. 105 (25), 8637–8642.

Vilarino-Guell, C., Wider, C., Ross, O.A., et al., 2011. VPS35 mutations in Parkinson disease. Am. J. Hum. Genet. 89 (1), 162–167.

Vilas, D., Iranzo, A., Tolosa, E., et al., 2016. Assessment of alpha-synuclein in submandibular glands of patients with idiopathic rapid-eye-movement sleep behaviour disorder: a case-control study. Lancet Neurol. 15 (7), 708–718.

Villar-Piqué, A., Da Fonseca, T.L., Sant'Anna, R., et al., 2016. Environmental and genetic factors support the dissociation between α-synuclein aggregation and toxicity. Proc. Natl. Acad. Sci. U. S. A. 113 (42), E6506–E6515.

Villar-Piqué, A., Rossetti, G., Ventura, S., et al., 2017. Copper(II) and the pathological H50Q α-synuclein mutant: environment meets genetics. Commun. Integr. Biol. 10 (1), 1–4.

Visanji, N.P., Mollenhauer, B., Beach, T.G., et al., 2017. The systemic synuclein sampling study: toward a biomarker for Parkinson's disease. Biomark. Med. 11 (4), 359–368.

Vogiatzi, T., Xilouri, M., Vekrellis, K., et al., 2008. Wild type alpha-synuclein is degraded by chaperone-mediated autophagy and macroautophagy in neuronal cells. J. Biol. Chem. 283 (35), 23542–23556.

Volles, M.J., Lee, S.J., Rochet, J.C., et al., 2001. Vesicle permeabilization by protofibrillar α-synuclein: implications for the pathogenesis and treatment of Parkinson's disease. Biochemistry 40 (26), 7812–7819.

Volpicelli-Daley, L.A., Luk, K.C., Patel, T.P., et al., 2011. Exogenous α-synuclein fibrils induce Lewy body pathology leading to synaptic dysfunction and neuron death. Neuron 72 (1), 57–71.

Walsh, D.M., Selkoe, D.J., 2016. A critical appraisal of the pathogenic protein spread hypothesis of neurodegeneration. Nat. Rev. Neurosci. 17 (4), 251–260.

Wang, R., Brattain, M.G., 2007. The maximal size of protein to diffuse through the nuclear pore is larger than 60kDa. FEBS Lett. 581, 3164–3170.

Wang, W., Perovic, I., Chittuluru, J., et al., 2011. A soluble α-synuclein construct forms a dynamic tetramer. Proc. Natl. Acad. Sci. U. S. A. 108 (43), 17797–17802.

Wang, L., Das, U., Scott, D.A., et al., 2014. α-Synuclein multimers cluster synaptic vesicles and attenuate recycling. Curr. Biol. 24 (19), 2319–2326.

Webb, J.L., Ravikumar, B., Atkins, J., et al., 2003. α-Synuclein is degraded by both autophagy and the proteasome. J. Biol. Chem. 278 (27), 25009–25013.

Wersinger, C., Sidhu, A., 2003. Attenuation of dopamine transporter activity by α-synuclein. Neurosci. Lett. 340 (3), 189–192.

Winslow, A.R., Chen, C., Corrochano, S., et al., 2010. α-Synuclein impairs macroautophagy: implications for Parkinson's disease. J. Cell Biol. 190 (6), 1023–1037.

Wirdefeldt, K., Adami, H.O., Cole, P., et al., 2011. Epidemiology and etiology of Parkinson's disease: a review of the evidence. Eur. J. Epidemiol. 26 (Suppl. 1), S1–58.

Wu, Q., Takano, H., Riddle, D.M., et al., 2019. α-Synuclein (αSyn) preformed fibrils induce endogenous aSyn aggregation, compromise synaptic activity and enhance synapse loss in cultured excitatory hippocampal neurons. J. Neurosci. 39 (26), 5080–5094.

Yallapu, M.M., Jaggi, M., Chauhan, S.C., 2012. Curcumin nanoformulations: a future nano-medicine for cancer. Drug Discov. Today 17 (1–2), 71–80.

Yu, W.H., Dorado, B., Figueroa, H.Y., et al., 2009. Metabolic activity determines efficacy of macroautophagic clearance of pathological oligomeric α-synuclein. Am. J. Pathol. 175 (2), 736–747.

Zarranz, J.J., Alegre, J., Gómez-Esteban, J.C., et al., 2004. The new mutation, E46K, of α-synuclein causes Parkinson and Lewy body dementia. Ann. Neurol. 55 (2), 164–173.

Zhang, W., Wang, T., Pei, Z., et al., 2005. Aggregated alpha-synuclein activates microglia: a process leading to disease progression in Parkinson's disease. FASEB J. 19 (6), 533–542.

Zhang, L., Zhang, C., Zhu, Y., et al., 2008. Semi-quantitative analysis of α-synuclein in sub-cellular pools of rat brain neurons: an immunogold electron microscopic study using a C-terminal specific monoclonal antibody. Brain Res. 1244, 40–52.

Zhang, Q.S., Heng, Y., Yuan, Y.H., et al., 2017. Pathological α-synuclein exacerbates the progression of Parkinson's disease through microglial activation. Toxicol. Lett. 265, 30–37.

Zhu, M., Li, J., Fink, A.L., 2003. The Association of α-synuclein with membranes affects bilayer structure, stability, and fibril formation. J. Biol. Chem. 278 (41), 40186–40197.

Zigoneanu, I.G., Yang, Y.J., Krois, A.S., et al., 2012. Interaction of α-synuclein with vesicles that mimic mitochondrial membranes. Biochim. Biophys. Acta Biomembr. 1818 (3), 512–519.

The role of glia in Parkinson's disease: Emerging concepts and therapeutic applications

Katarzyna Z. Kuter[a,*], **M. Angela Cenci**[b], **Anna R. Carta**[c,*]

[a]Department of Neuropsychopharmacology, Maj Institute of Pharmacology, Polish Academy of Sciences, Krakow, Poland
[b]Basal Ganglia Pathophysiology Unit, Department of Experimental Medical Science, Lund University, Lund, Sweden
[c]Department of Biomedical Sciences, University of Cagliari, Cittadella Universitaria di Monserrato, Cagliari, Italy
*Corresponding authors: Tel.: +48-12-6623226 (K.Z.K.); +39-070-6758662 (A.R.C.),
e-mail address: kuter@if-pan.krakow.pl; acarta@unica.it

Abstract

Originally believed to primarily affect neurons, Parkinson's disease (PD) has recently been recognized to also affect the functions and integrity of microglia and astroglia, two cell categories of fundamental importance to brain tissue homeostasis, defense, and repair. Both a loss of glial supportive-defensive functions and a toxic gain of glial functions are implicated in the neurodegenerative process. Moreover, the chronic treatment with L-DOPA may cause maladaptive glial plasticity favoring a development of therapy complications. This chapter focuses on the pathophysiology of PD from a glial point of view, presenting this rapidly growing field from the first discoveries made to the most recent developments. We report and compare histopathological and molecular findings from experimental models of PD and human studies. We moreover discuss the important role played by astrocytes in compensatory adaptations taking place during presymptomatic disease stages. We finally describe examples of potential therapeutic applications stemming from an increased understanding of the important roles of glia in PD.

Keywords

Astrocytes, Microglia, Inflammation, α-Synuclein, Compensation, Energy metabolism, Pharmacology, Grafting, Dyskinesia, Levodopa

Progress in Brain Research, Volume 252, ISSN 0079-6123, https://doi.org/10.1016/bs.pbr.2020.02.004

1 Glial cells in PD: A brief introduction

In 1919, Kostantin Tretiakoff described a loss of pigmented neurons in the substantia nigra in patients affected by "paralysis agitans" (Lees et al., 2008). Since that first observation, most of the research on Parkinson's disease (PD) focused on documenting and understanding the pathology of neurons, first and foremost in the substantia nigra, then also in other parts of the brain. An involvement of glial cells started to be appreciated about 30 years ago thanks to the pioneering studies by the McGeer group. In 1988, a milestone paper from this group reported a large number of microglial cells positive for HLA-DR (a class II major histocompatibility antigen) in post-mortem substantia nigra from PD patients (McGeer et al., 1988). Subsequent studies confirmed and expanded these findings, demonstrating that microglial cells in PD-affected regions were positive for additional reactive markers (such as the phagocytosis-related receptor CD68) and moreover exhibited an activated morphology (Croisier et al., 2005; Imamura et al., 2003; Jyothi et al., 2015; Loeffler et al., 2006; McGeer et al., 1988). Reactive astrocytes were also reported, being recognized based on an increased immunostaining for glial fibrillar acidic protein (GFAP) (McGeer et al., 1988). Further post-mortem investigations of brain tissue from PD patients reported the presence of glial cells exhibiting markers of apoptosis in the substantia nigra and the occurrence of α-synuclein inclusions in astrocytes in many regions (substantia nigra, striatum, amygdala, thalamus, septum, claustrum, and cerebral cortex), whose abundance appeared to correlate with disease progression (Braak et al., 2007; Kösel et al., 1997; Wakabayashi et al., 2000).

In parallel with these histopathological investigations, positron emission tomography (PET) methods were developed to image microglia in the living human brain. The most widely used PET tracers utilized radioligands selective for the peripheral benzodiazepine receptors (PBR or TSPO), whose expression is enhanced in activated microglia (Banati et al., 1997; Papadopoulos et al., 2006). Neuroimaging studies using these tracers confirmed that microglial activation indeed occurs in both subcortical and cortical regions in PD (Edison et al., 2013; Gerhard et al., 2006), even though the levels of radioligand binding (reflecting the number of active microglial cells) often failed to correlate with indexes of disease severity or progression (Bartels et al., 2010; Gerhard et al., 2006; Ouchi et al., 2005). Based on these observations, it was suggested that microglial cells are activated early in the disease process and then maintain their reactive phenotype, contributing to neurodegeneration via release of pro-inflammatory cytokines (Gerhard et al., 2006).

Accordingly, increased levels of both pro-inflammatory and anti-inflammatory markers have been detected in the brain, cerebrospinal fluid (CSF) and blood samples from PD patients, indicating overactivation of glial cells associated with a dysregulation of cytokines production (Lopez Gonzalez et al., 2016; Mogi et al., 2007; Sawada et al., 2006). Notably, a recent study has reported a marked pro-inflammatory serum profile in PD patients presenting a faster progression of motor and cognitive symptoms (Williams-Gray et al., 2016).

Today, an involvement of both microglia and astroglia in the neurodegenerative process of PD can be considered as definitely ascertained (Braak et al., 2007; Capani et al., 2016; Kösel et al., 1997; Liddelow et al., 2017; Liu et al., 2017; McKenzie et al., 2017; Tang and Le, 2016; Wakabayashi et al., 2000), and several studies suggest that neuroinflammation plays an important role in both early and advanced stages of the disease (Booth et al., 2017; Braak et al., 2007; Cebrian et al., 2014; Marinova-Mutafchieva et al., 2009; Werner et al., 2008; Wissemann et al., 2013; Yasuda et al., 2007; Zhang et al., 2017). While microglia and astroglia normally cooperate to maintain brain tissue homeostasis and physiological neuronal transmission (Parpura et al., 2012; Pekny et al., 2016; Verkhratsky and Zorec, 2019), these glial cells appear to be pathologically affected in PD, becoming unable to fulfill their fundamental adaptive roles. Dynamic pathological changes in microglia and astroglia during PD progression result in unbalanced effector functions that convert these cells into active players in neurodegeneration (Joers et al., 2017).

In this article, we will first briefly review the main pathophysiological functions of microglia and astroglia and then comment on their possible contribution to different aspects of PD. In this regard, we will consider not only the primary neurodegenerative process but also the role of glia in presymptomatic compensation at early disease stages. Moreover, we will appraise the effects of chronic dopamine replacement therapy and the glial mechanisms implicated in the development of therapy complications. Finally, we will provide examples of novel therapeutic principles targeting glial mechanisms that have been successfully evaluated in preclinical models of PD.

1.1 Physiopathological functions of microglia

Microglia are the main immunocompetent macrophage cells in the central nervous system (CNS), in charge of immune surveillance, defense and maintenance of tissue homeostasis (Streit, 2002) (Fig. 1). As part of the innate immune system, in physiological conditions the so-called surveillant or homeostatic microglia continuously assay the surrounding microenvironment through their surface receptors, ready to rapidly respond to external insults such as pathogen invasion or internal signals released by damaged tissue (Davalos et al., 2005; Hanisch and Kettenmann, 2007; Nimmerjahn et al., 2005). Microglia activation in response to pathological insults is a complex and dynamic process that strictly depends on brain-launched signals (Gosselin et al., 2014). Activation entails both morphological and phenotypic changes that aim at protecting the CNS, resolving the damage and promoting tissue repair (Goldmann and Prinz, 2013). The morphology of active microglial cells gradually switches from a highly ramified to an amoeboid shape (Kreutzberg, 1996). An emerging concept is that different classes of transcription factors may be variably engaged in the microglial activation process in pathological conditions, conferring specific phenotypes that can be beneficial or deleterious (Holtman et al., 2017). Phenotypic changes in microglial cells may result in the expression of surface

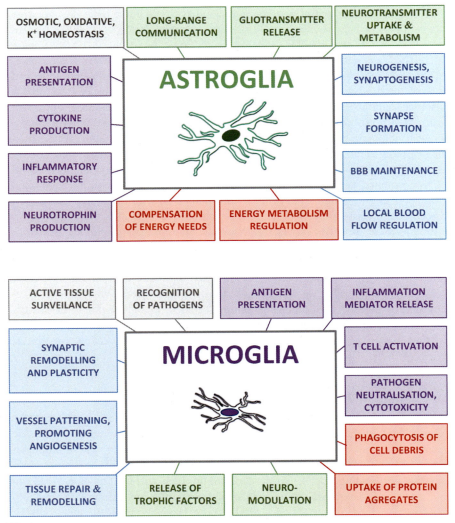

FIG. 1

The pathophysiological roles of glial cells. In astroglia colors of boxes indicate: gray—housekeeping functions, violet—immune and trophic roles, blue—structural, BBB (blood–brain barrier) and developmental roles, red—energy metabolism, green—cellular signaling related functions. In microglia colors of boxes indicate: gray—surveillance functions, violet—immune functions, blue—structural functions, red—phagocytic functions, green—regenerative and modulatory roles.

receptors, induction of phagocytic function, increase of pro-inflammatory or anti-inflammatory mediators (Hanisch and Kettenmann, 2007; Ransohoff and Perry, 2009). Classical definitions distinguish between two main microglial phenotypes, referred to as M1 and M2. The M1-like phenotype has been associated with increased production of pro-inflammatory cytokines TNF-α, IL-6, IL-12, and IL-1β and up-regulation of cell surface markers such as major histocompatibility complex-II (MHC-II) and CD86 (Martinez et al., 2006; Tang and Le, 2016). M2-like microglia is functionally further classified in the subtypes M2a (associated with the suppression of inflammation), M2b (with prevalent phagocytic activity), and M2c (involved in tissue remodeling) (Edwards et al., 2006; Martinez and Gordon, 2014; Melief et al., 2012; Wolf et al., 2017). Realistically, the activation process produces a heterogeneous population of phenotypes which may coexist within the same microenvironment or may consist of a continuum of intermediate phenotypes between pro- and anti-inflammatory functions. The critical concept is that, upon activation in response to pathogens or signals released by damaged cells, microglia may acquire a range of phenotypes which can either ameliorate a disease condition or contribute to disease progression promoting neurodegenerative changes.

Along with the identification of cell-specific regulatory mechanisms, the recognition of the highly dynamic nature of microglia is a pivotal concept for the development of targeted therapies aimed at modulating microglial activation in neurodegenerative diseases. In this regard, several transcription factors have been investigated as potential pharmacological targets, including: pro-inflammatory (nuclear factor NF-κB, STAT), anti-inflammatory/antioxidant (Nuclear receptor NURR1, estrogen receptors, ERs, Nuclear respiratory factor 2, NRF2), peroxisome proliferator-activated receptor gamma (PPAR-γ), and pleiotropic (interferon regulatory factors, IRF) mediators (Holtman et al., 2017). An issue crucial to drug development is that the expression of these transcription factors is not limited to microglia, which implies a risk of unwanted effects upon pharmacological manipulation. The first microglia-specific transcriptional regulator, *Sall1*, has been identified only 4 years ago. The *Sall1* gene is not expressed by other members of the mononuclear phagocyte system, nor by other CNS-resident cells (Buttgereit et al., 2016; Koso et al., 2016). *Sall1* has been involved in the maintenance of microglia homeostasis and its inactivation resulted in the conversion of microglia from resting tissue macrophages into inflammatory phagocytes (Buttgereit et al., 2016), opening the perspective for microglia-specific manipulation *in vivo*.

Regional differences have been reported in microglia distribution and phenotypes that may contribute to the different regional vulnerability in disease (de Haas et al., 2008; Grabert et al., 2016; Sharaf et al., 2013). For instance, microglia are denser in the telencephalon, and myelinated regions display higher microglia concentration than unmyelinated regions (Mittelbronn et al., 2001). Noteworthy, the substantia nigra is among the brain regions with highest microglia density (Kim et al., 2000; Lawson et al., 1990) and contains, in addition, high levels of tumor necrosis factor alpha (TNF-α) (Doorn et al., 2015). This aspect, together with the high concentration of iron and neuromelanin may contribute to the high vulnerability of the substantia nigra in PD (Walker et al., 2015). The different expression profiles of young and aged microglia across brain regions may aggravate the pathological scenario in age-related neurodegenerative diseases (Arcuri et al., 2017; Grabert et al., 2016; O'Neil et al., 2018).

Beyond the immune function, microglia play a prominent neuromodulatory role, influencing neuronal membrane properties and synaptic connectivity via the release of soluble signaling molecules or direct interaction with synaptic elements (Hong et al., 2016; Lewitus et al., 2016; Parkhurst et al., 2013; Sipe et al., 2016). The microglia-neuron crosstalk is anatomically and functionally embedded in the CNS cell-networks, representing a contributing factor in the pathophysiology of neurological disorders. This aspect is only recently gaining attention in relation to the pathophysiology of PD, where the chronic and unremittent microgliosis may profoundly affect basal ganglia synaptic connectivity (Carta et al., 2017). Microglia-released cytokines bind to neuronal and microglial receptors in a paracrine or autocrine manner, serving a critical neuromodulatory function (Marin and Kipnis, 2013; Salter and Beggs, 2014). Specifically, the pro-inflammatory cytokine TNF-α has been identified as an important regulator of neuronal excitability, synaptic strength, and plasticity (Beattie et al., 2002; Stellwagen and Malenka, 2006). In addition, microglial cells express many receptors for neurotransmitters and neuromodulators whereby sensing changes in neuronal activity (Carta et al., 2017; Pocock and Kettenmann, 2007; Tremblay et al., 2010). The microglial control of synaptic connectivity starts very early in life and continues throughout the adult age. During the prenatal period microglia shapes synaptic circuitries in the forebrain, while in the postnatal phase microglia-mediated synaptic pruning is necessary to remodel neural networks (Paolicelli et al., 2011; Schafer et al., 2013). In summary, microglia contributes to the homeostasis of neurons and synaptic circuits by regulating neuronal proliferation/differentiation, as well as the formation and remodeling of synaptic connections (Bialas and Stevens, 2013; Blank and Prinz, 2013; Hughes, 2012).

1.2 Physiopathological functions of astroglia

Astrocytes (collectively referred to as astroglia) have been classically attributed a supportive role toward neurons in the maintenance of osmotic, energetic, and structural tissue homeostasis (Fig. 1). The multiple active regulatory functions of these cells have been discovered only recently. Today we know that astroglia plays a role in the regulation of neurotransmitter clearance, extracellular ion balance, synaptogenesis, neurometabolic coupling, blood–brain barrier (BBB) permeability, cerebral blood flow, and glymphatic flow (Deitmer et al., 2019; Dienel and Carlson, 2019; Halassa and Haydon, 2010; Halassa et al., 2007; Haydon and Carmignoto, 2006; Kimelberg and Nedergaard, 2010; Seth and Koul, 2008). A still debated function of astrocytes is their direct modulation of neuronal excitability and synaptic plasticity via calcium-dependent neurotransmitter release (in particular, glutamate, ATP, D-serine), a process termed gliotransmission. While there is abundant evidence that gliotransmission can occur, its extent and physiological importance in the mature

CNS remains unclear, and it has been argued that some findings reported from *in vitro* preparations may reflect the more reactive phenotype exhibited by astrocytes under these experimental conditions (for review and discussion see Durkee and Araque, 2019, Savtchouk and Volterra, 2018). On the other hand, it is worth pointing out that astrocytes do acquire a reactive or altered phenotype in virtually all brain pathologies. There is therefore a need to define the physiopathological role of glio-transmission in specific disease models, including PD.

Astrocytes have extensive cellular processes forming complex three-dimensional structures with compartmentalized functions (Bindocci et al., 2017). Thanks to these processes, astrocytes can be physically apposed to synapses, on one hand, and to microvessels, on the other hand, acting as a functional bridge from neurons to blood flow (Zonta et al., 2003). Accordingly, a vast literature has revealed that astroglia can regulate acute and chronic responses of the microvasculature through release of vasoactive substances (e.g., prostanoids) and proangiogenic cytokines (Agulhon et al., 2012; Halassa and Haydon, 2010). In the CNS, astrocyte processes fill the local environment in non-overlapping domains, which appear to become enlarged in the basal ganglia after degeneration of the nigrostriatal DA pathway (Charron et al., 2014). Astrocytic domains exhibit a high degree of intracellular connectivity via gap junctions, enabling a prompt transfer of energetic substrates to neurons that may reside several cell layers away from the nearest blood capillary (Deitmer et al., 2019).

It is now well established that astrocytes provide neurons with both energy substrates (such as lactate, ketone bodies) and antioxidants (such as glutathione and ascorbic acid) required for scavenging reactive oxygen species (ROS) (Lian and Zheng, 2016; Parpura et al., 2012; Pekny et al., 2016; Verkhratsky and Zorec, 2019). Moreover, astrocytes release neurotrophic factors (NGF, BDNF, CNTF, GDNF, IGF-1, FGF-2, VEGF, ADNF) and thus promote neuronal survival and plasticity (Deitmer, 2001).

Astrocytes rapidly respond to a CNS insult by acquiring a reactive phenotype, a process referred to as astrogliosis that is induced by signals released from alarmed microglia or stressed neurons (Sofroniew, 2009). Most of the changes occurring in reactive astrocytes may have an adaptive value, increasing these cells' capacity for endogenous protection. However, strongly stimulated astrocytes can also exert harmful effects. Thus, severe astrogliosis can lead to a release of pro-inflammatory cytokines and trigger microglia-mediated neuroinflammation (Hung et al., 2016), or formation of a glial scar inhibiting axonal regeneration (Anderson et al., 2016). Recently, activated astrocytes have been classified into different subpopulations having opposite roles (Burda and Sofroniew, 2017). Reactive astrocytes that form in the presence of inflammatory mediators, called A1 type, exhibit a potentially neurotoxic, pro-inflammatory profile (Liddelow et al., 2017). In contrast, ischemia-reactive astrocytes, referred to as the A2 type, exhibit a supportive profile with enhanced expression of metabolism-associated genes and neurotrophic factors (Zamanian et al., 2012). Liddelow et al. (2017) have shown that A1 reactive astrocytes form when microglial cells secrete interleukin 1 alpha (Il-1α), TNF, and complement component 1, subcomponent q (C1q). These astrocytes not only lose the ability to promote neuron survival, neuritic outgrowth, synaptogenesis and phagocytosis, but can also induce the death of neurons (Liddelow et al., 2017). Using immunohistochemical markers, A1 astrocytes were found to occur in the brains of patients suffering from PD and other neurodegenerative diseases (Liddelow et al., 2017). Taken together, these important new findings shake the classical view of the astrocyte as a solely supportive cell. Instead, the occurrence of A1 astrocytes in human neurodegenerative diseases suggests that astroglia may aggravate, if not drive, the neurodegenerative process. In principle, astroglia could contribute to neurodegenerative disorders both via "loss of function" (i.e., a diminished capacity for trophic, antioxidant, bioenergetic support to neurons) and via toxic "gain of function" prompted by a neuroinflammatory milieu.

2 Role of glia in the neurodegenerative process

2.1 Microglia

Signs of microglial activation associated with a neuroinflammatory response have been reported in several preclinical models of PD. Toxin-based models, obtained by administration of methyl-4-phenyl-1,2,3,6-tetrahydropyridine (MPTP) or intracerebral injection of 6-hydroxydopamine (6-OHDA), show an early but possibly persistent microglial activation in the substantia nigra and the striatum (Armentero et al., 2006; Hurley et al., 2003; Maia et al., 2012; Schintu et al., 2009; Vazquez-Claverie et al., 2009; Walsh et al., 2011). Moreover, studies based on genetic models of PD have strongly supported the critical role of neuroinflammation in dopaminergic degeneration (Daher et al., 2014; Lin et al., 2009). More recently, neuroinflammation and elevated levels of inflammatory cytokines have been reported in α-synuclein-based models of PD (Su et al., 2009; Watson et al., 2012). Interestingly, a number of experimental studies have suggested a link between altered microglia phenotypes and dopaminergic degeneration. High levels of antigen-presentation markers such as MHC-I, MHC-II, and intercellular adhesion molecule 1 (ICAM-1) have been detected at early stages of nigrostriatal degeneration (Cebrian et al., 2014; Marinova-Mutafchieva et al., 2009; Yasuda et al., 2007), followed by an increase of pro-inflammatory microglia producing high levels of inflammatory factors in more advanced stages (Barcia et al., 2011; Bian et al., 2009; Liberatore et al., 1999; Lofrumento et al., 2011; Nagatsu et al., 2000; Pattarini et al., 2007; Pisanu et al., 2014; Yasuda et al., 2007). The latter response appears to coincide with a delayed downregulation of anti-inflammatory markers (Pisanu et al., 2014). All these experimental studies support the dynamic nature of microglia activation in PD and suggest that, in early disease stages, a mixed population of pro- and anti-inflammatory microglia may co-exist, whereas in advanced stages microglial cells lose their capability for tissue defense and repair, and their pro-inflammatory neurotoxic action prevails.

The toxin-based models have clearly demonstrated that dying dopamine neurons trigger microglia activation likely via the exposure of signaling molecules or the release of soluble activating factors (such as chemokine CX3CL1 and CCL21) that are recognized by microglial membrane receptors (Harrison et al., 1998; Wolf et al., 2017). Another major trigger of microglial activation are diffusible aggregates of α-synuclein, which have been found both in brain areas affected by neurodegeneration (Karpinar et al., 2009; Sharon et al., 2003) and in biological fluids of PD patients (Majbour et al., 2016; Tokuda et al., 2010). *In vitro* studies have shown that oligomeric α-synuclein species can directly interact not only with neurons but also with microglia and astrocytes, potentially driving both neuronal and glial pathology in PD (Lee et al., 2010a; Villar-Pique et al., 2016). The interaction of α-synuclein with microglia may represent a key event in PD pathogenesis, leading to an unremitting shift of microglia to pro-inflammatory phenotypes (Villar-Pique et al., 2016). Moreover, a structure-related toxicity of α-synuclein has been described whereby soluble protofibrils and oligomers would be toxic protein species driving neurodegeneration through several mechanisms, including a direct interaction with microglia (Fusco et al., 2017; Winner et al., 2011). In this scenario, toxic α-synuclein released by dying neurons behaves as a chemoattractant and establishes contacts with microglia (Emmanouilidou and Vekrellis, 2016; Kim et al., 2009; Yamada and Iwatsubo, 2018).

The interaction of microglia with neuronal signaling factors and α-synuclein released by dying neurons occurs mainly through Toll-like receptors (TLRs) TLR2 and TLR4, resulting in NF-κB nuclear translocation and driving the pro-inflammatory functions of these cells (Doorn et al., 2014; Fellner et al., 2013; Stefanova et al., 2011). Interestingly, several studies have suggested that the interaction of α-synuclein with TLRs is conformation-dependent, showing that protofibrillar/oligomeric forms display a greater inflammatory potential than the native monomeric protein (Daniele et al., 2015; Fellner et al., 2013; Kim et al., 2013; Klegeris et al., 2008; Lee et al., 2010a; Rojanathammanee et al., 2011; Zhang et al., 2005). In parallel with the inflammatory activity, deficits in microglial α-synuclein-clearing capacity may result in an increased extracellular concentration of this protein, thus favoring its neurotoxic activity (Stefanova et al., 2011; Venezia et al., 2017). The efficiency of microglial phagocytosis and clearance appears to depend on the protein structure, being stimulated by α-synuclein monomers but inhibited by oligomeric or mutated α-synuclein (Lee et al., 2008; Park et al., 2008; Roodveldt et al., 2010). In summary, current studies suggest that toxic α-synuclein species may alter the effector functions of microglial cells by affecting their phagocytic-proteostatic capacity and cytokine production. The imbalance between homeostatic and maladaptive responses may be further aggravated by age-related glial dysfunction (Arcuri et al., 2017; Cherry et al., 2014; Dominguez-Meijide et al., 2017; Labandeira-Garcia et al., 2017; Norden and Godbout, 2013; Tang and Le, 2016).

In addition to microglia-mediated innate immune responses, peripheral adaptive immunity plays a role in the neuropathological scenario of PD. The finding of

T-lymphocytes infiltration in the substantia nigra from PD patients suggested that these cells may contribute to microglia overactivation and dysregulation. Toxic α-synuclein species may enter brain microvessels and act as antigens to activate T-lymphocytes, triggering an adaptive immune response (Mosley and Gendelman, 2017). Inflammatory cytokines secreted by reactive microglia increase the BBB permeability, favoring an increased infiltration of activated T-lymphocytes into the brain (Farkas et al., 2000). Recently, a disbalance has been shown in CD4+ cell subpopulations in both PD patients and preclinical models of the disease, with a decreased number and activity of regulatory T cells (Treg) compared to effector T cells (Ambrosi et al., 2017; Kustrimovic et al., 2018; Solleiro-Villavicencio and Rivas-Arancibia, 2018). Once in the brain, dysregulated T-lymphocytes may intensify microglial activation and central inflammatory responses, thus contributing to neurodegeneration (Duffy et al., 2018; Mosley and Gendelman, 2017; Saunders et al., 2012).

In summary, as a consequence of pathological interactions with α-synuclein and other neuron-derived factors, microglial cells lose their capacity for self-regulation, leading to an imbalance between pro- and anti-inflammatory phenotypes and favoring a chronic inflammatory state that feeds neurodegeneration in PD (Hirsch and Hunot, 2009; Pisanu et al., 2014). The awareness that both innate and adaptive immune responses contribute to this scenario has widened the arsenal of possible therapeutic targets, explaining the current interest in pharmacological strategies targeting both central and peripheral immune systems.

2.2 Astroglia

Studies performed in various experimental models have now firmly established that, similar to microglia, astrocytes are able to internalize α-synuclein (Lee et al., 2010b). Astrocytes have been proposed to play an important role in the clearance of toxic α-synuclein species from the extracellular space. However, when their degrading capacity is overburdened, the intracellular accumulation of α-synuclein may cause proteostatic deficits and mitochondrial damage in these cells (Lindstrom et al., 2017; Rostami et al., 2017). Moreover, recent experimental data suggest that astrocytic networks may participate in the cell-to-cell transfer of α-synuclein aggregates (Cavaliere et al., 2017; Rostami et al., 2017), possibly mediating a spread of pathology within affected brain regions during the progression of PD. In addition, several of the proteins responsible for autosomal recessive forms of PD (Parkin, PINK-1, DJ-1) are highly expressed in astrocytes, suggesting that the "loss of function" associated with the corresponding genetic mutations has a direct deleterious impact on astroglia (Halliday and Stevens, 2011). As already mentioned, post-mortem histopathological studies have revealed that some astrocytes contain α-synuclein inclusions in PD (Braak et al., 2007; Wakabayashi et al., 2000). There is, however, uncertainty about the extent of astrocytosis, which seems to be variable and relatively modest in PD compared to other neurodegenerative diseases (Tong et al., 2015). The attenuated astrocytic reactivity has been attributed to a physical interaction of the internalized α-synuclein with proteins required for developing a reactive astrocytic phenotype (Halliday and Stevens, 2011). The lack of overt Lewy-like aggregates in astrocytes has instead been attributed to their high expression of β-synuclein, which is considered as a potent inhibitor of α-synuclein aggregation and toxicity (Israeli and Sharon, 2009).

An aspect of PD that seems to rely heavily on astrocytic mechanisms is the presymptomatic compensation occurring at early disease stages. It is well established that parkinsonian motor symptoms become manifest only when more than 70% of nigrostriatal DA neurons have degenerated (Blesa et al., 2017; Hornykiewicz, 1998). Likewise, very large dopaminergic lesions are required to induce permanent motor deficits in animal models of PD, while the deficits caused by smaller lesions spontaneously subside with time (Bezard and Gross, 1998; Cenci and Björklund, 2020; Kuter et al., 2016a, 2018; Robinson et al., 1994; Stanic et al., 2003; Zigmond, 1997). This is due to the compensatory mechanisms that occur at different levels, including an increased activity of remaining nigrostriatal neurons and synaptic adaptations of postsynaptic networks. The fact that compensatory mechanisms may mask the degeneration of up to 70% of nigral DA neurons gives a sense of the remarkable plastic potential that could be exploited for therapeutic purposes. The endogenous supportive and neuroprotective functions of astrocytes make them natural candidates to take part in the compensation processes. Accordingly, studies in neurotoxin-based models of PD have indicated that astrocytes play a key role in the compensatory adaptations that accompany partial nigrostriatal DA lesions. Using MPTP lesioned mice, L'Episcopo and colleagues found that reactive astrocytes in the ventral midbrain are a likely source of Wnt1, a protein involved in mesencephalic development, and that blocking Wnt signaling counteracts astrocyte-induced neuroprotection in primary mesencephalic astrocyte-neuron cultures. Moreover, lack of Wnt1 transcription in response to MPTP was causally linked with the compensation failure observed in middle-aged mice (L'Episcopo et al., 2011). Using a partial 6-OHDA lesion in rats, Kuter and colleagues found that fluorocitrate-induced death of 30% astrocytes in the substantia nigra blocked the compensation of locomotor deficits (Kuter et al., 2018). Supporting an involvement of astrocytes in compensatory adaptation, a proteomic study of nigral tissue from PD patients has revealed the upregulation of glial-specific proteins (glial fibrillary acidic protein, glial maturation factor beta, galectin 1, sorcin A), suggesting a mobilization of astrocyte-mediated neuroprotective pathways in the attempt to preserve remaining neurons (Werner et al., 2008).

In addition to being a source of trophic factors, astrocytes may support lesion-induced compensatory processes by providing energy substrates to neurons having increased metabolic demands, such as the remaining DA neurons in a degenerating substantia nigra. After partial 6-OHDA lesions in rats, the nigral tissue exhibits an increased expression and/or activity of carbohydrate metabolism proteins and mitochondrial proteins of the electron transport chain, associated with a rearrangement of their higher-order supercomplexes (Kuter et al., 2016a,b, 2019a,b). All these changes are predicted to lead to more efficient energy production via oxidative phosphorylation. Importantly, when the partially lesioned nigrostriatal system was deprived of astrocytic support, the above described energy metabolism adaptations were lost (Kuter et al., 2019a,b). More detailed studies of nigral tissue partially devoid of astrocytes revealed a general metabolic shift from glucose-based (glucose, lactate and glycogen) to fatty acid-based energy substrates, as indicated by an increased expression of carnitine palmitoyl-transferase 1c and 1a in neuronal and glial cells, respectively, along with an elevated production of ketone bodies in the substantia nigra (Kuter et al., 2019a). The lack of motor compensation in parkinsonian animals sustaining additional astrocytic damage, however, suggests that the observed adaptive metabolic changes in substantia nigra were not sufficient to achieve functional recovery without astrocytic support.

In summary, bioenergetic and metabolic adaptations in astrocytic-neuronal networks seem to provide a primary compensatory mechanism after nigrostriatal dopaminergic degeneration. Temporary or prolonged impairments of astroglial functioning may therefore result in an increased vulnerability of dopaminergic neurons, and may predispose to the development of idiopathic PD (Elstner et al., 2011).

3 Role of glia in dopamine replacement therapy
3.1 Role of glia in the uptake and metabolism of L-DOPA

Among all neurodegenerative diseases, PD is the only condition associated with lifelong DA replacement with L-DOPA, which is given as oral tablets several times per day, usually combined with other dopaminergic drugs in advanced disease stages (Cenci et al., 2011). Given that BBB microvessels are lined by astrocytic endfeet, it can be assumed that astroglia plays a key role in the transport of L-DOPA from the blood and its distribution over the brain parenchyma. Supporting this notion, rats receiving systemic injections of L-DOPA exhibit widespread histofluorescence for this compound within astrocytic cell bodies and endfeet apposed to microvessels (Inyushin et al., 2012). Astrocytes can internalize L-DOPA via the Large Neutral Amino Acids Transporter, also known as L-Type Amino Acid Transporter 1 or Sodium-Independent Neutral Amino Acid Transporter (LAT1) (reviewed in Verkhratsky and Zorec, 2019). Although earlier studies had entertained the possibility that astrocytes convert L-DOPA to DA in PD, the most recent investigations on the subject indicate that the efficiency of astrocytic L-DOPA conversion is very modest (Asanuma et al., 2014; De Deurwaerdère et al., 2017). Astrocytes may instead provide a temporary storage site for L-DOPA, and may internalize or release it depending on its extracellular concentrations (Asanuma et al., 2014). Moreover, reactive astrocytes in the DA-denervated striatum may express the DA transporter (DAT), thus being able to take up L-DOPA-derived DA once this is released from nerve terminals (for a review and discussion about the main sites of L-DOPA-derived DA release in PD see Cenci, 2014). Additionally, astrocytic DA uptake may involve Na$^+$-dependent monoamine transporters and low affinity transporters, such as the astrocytic organic cation transporter 3 (OCT-3) (Nishijima and Tomiyama, 2016).

DA is normally metabolized by monoamine oxidase (MAO)-A, MAO-B, aldehyde dehydrogenase, and catechol-O-methyltransferase (COMT). In the striatum, MAO-A is expressed in axon terminals and medium spiny neurons, while the majority of MAO-B as well as COMT are detected in astrocytes (Nishijima and Tomiyama, 2016). Therefore, astrocytes take part in the metabolism of DA produced from L-DOPA. Interestingly, the expression of MAO-B increases with aging, and its activity (which entails ROS production) correlates positively with the degree of cell loss in the substantia nigra. Conversely, smokers, who have lower risk of developing PD, have significantly decreased MAO-B activity (Kumar and Andersen, 2004). Transgenic mice overexpressing MAO-B in astrocytes were found to have age-related motor dysfunction that paralleled a degeneration of nigral neurons. The latter was associated with astrogliosis and microglial activation, suggesting that the oxidative stress caused by MAO-B overactivity in astrocytes leads to maladaptive glial reactions and nigrostriatal degeneration (Mallajosyula et al., 2008). Based on these and other data, MAO-B inhibitors that are routinely used for the symptomatic treatment of PD (such as selegiline and rasagiline) have been proposed to exert neuroprotective effects via the astroglia (Rappold and Tieu, 2010).

3.2 Role of glia in L-DOPA-induced dyskinesia

Although DA replacement therapy with L-DOPA represents the gold standard treatment for PD, this treatment leads almost unavoidably to a development of motor complications, such as motor fluctuations and L-DOPA-induced dyskinesia (LID) (Jankovic and Aguilar, 2008). The pathophysiology of LID is complex, involving both pre- and postsynaptic changes in DA transmission and secondary alterations to non-dopaminergic transmitter systems (Bastide et al., 2015). Classically, the attention of most investigators has focused on neuronal mechanisms involving dysregulation of striatal gene expression and abnormal synaptic plasticity in the corticostriatal pathway (Feyder et al., 2011; Picconi et al., 2003). During the past two decades a growing body of evidence has revealed the importance of non-neuronal mechanisms involving maladaptive glial and vascular responses (Carta et al., 2017).

A link between LID and neuroinflammation is supported by results from several research groups, showing intense inflammatory activation and release of pro-inflammatory cytokines in the DA-denervated striatum of L-DOPA-treated parkinsonian rats developing dyskinesia (Barnum et al., 2008; Bortolanza et al., 2015a,b; Mulas et al., 2016). These studies have also indicated that the inflammatory response to L-DOPA is amplified by the preexisting inflammatory milieu associated with the neurodegenerative process (Mulas et al., 2016). The mechanisms underlying an inflammatory response to L-DOPA in dyskinetic subjects are still elusive. It seems, however, likely that an increased DA metabolism and the associated oxidative load have a part to play (Carta et al., 2017; Zucca et al., 2017). Additional mechanisms may involve localized BBB leakage ensuing glial reactions (see below). In support of a causal link between LID and neuroinflammation, administration of lipopolysaccharide (LPS) has been shown to worsen the abnormal involuntary movements (Mulas et al., 2016), while treatment with corticosterone has been found to reduce their severity in rat models of LID (Barnum et al., 2008).

As mentioned above, microglia-released soluble cytokines function as neuromodulators in the brain. For instance, the pro-inflammatory cytokine TNF-α is a recognized regulator of synaptic plasticity and neuronal excitability via TNF receptor 1 and 2, (TNR1 and TNFR2), which are also expressed in neurons (Balosso et al., 2009; Beattie et al., 2002; Lewitus et al., 2014). The modulatory actions of cytokines at the glutamatergic synapses has been investigated in several studies showing, for instance, that the pro-inflammatory cytokine IL-1β affects postsynaptic NMDA receptor function leading to increased neurotransmitter release (Clark et al., 2015). In medium spiny neurons, TNF-α may cause a reduction of synaptic strength via the regulation of AMPA receptor trafficking (Lewitus et al., 2014). Therefore, the neuroinflammatory mechanism underpinning LID may involve microglia-secreted cytokines that, when released in excess, may contribute to the dysregulation of glutamatergic transmission (De Chiara et al., 2013) and corticostriatal synaptic plasticity (Centonze et al., 2009) driving the development of abnormal involuntary movements. Moreover, neuroinflammation is reciprocally linked with angiogenesis, which has been found to occur in specific basal ganglia regions in subjects affected by LIDs (see below). Inflammatory stimuli activate endothelial cells and increase microvascular permeability, and the pro-inflammatory cytokine TNF-α is a potent pro-angiogenic factor (Leibovich et al., 1987). Accordingly, in several pathological conditions treatment with anti-inflammatory drugs inhibits blood vessel formation (Monnier et al., 2005; Muller, 2014; Szade et al., 2015).

3.3 Gliovascular mechanisms in L-DOPA-induced dyskinesia

The first indication that LID may be linked with angiogenesis came from the report that several basal ganglia nuclei exhibit endothelial proliferation in 6-OHDA-lesioned rats developing dyskinesia upon repeated administration of L-DOPA (Westin et al., 2006). Regions showing endothelial proliferation were found to also exhibit focal increases in BBB permeability and astrocytic upregulation of vascular endothelial growth factor (VEGF-A) (Ohlin et al., 2011). In addition to L-DOPA, treatment with a selective D1-type receptor agonist was found to produce similar microvascular reactions correlating with the occurrence of dyskinesia (Lindgren et al., 2009). A causal involvement of angiogenesis in LID was demonstrated by co-administering L-DOPA with a small molecule inhibitor of VEGF receptor-signaling, which significantly attenuated the development of dyskinesia over a chronic course of L-DOPA treatment, while also suppressing the treatment-induced angiogenic activation and BBB hyperpermeability (Ohlin et al., 2011). Evidence of angiogenesis and VEGF upregulation was moreover found in post-mortem putaminal and pallidal tissues from PD patients with a history of LID (Ohlin et al., 2011). Congruent with these results, a recent brain imaging study examining the vasomotor response to hypercapnia showed that PD patients affected by LID have a larger capillary reserve in the putamen (Jourdain et al., 2017).

In parallel with these discoveries, studies in the rat model of LID revealed that the administration of L-DOPA is accompanied by large and transient increases in regional cerebral blood flow (rCBF) in both motor cortex, striatum, and deeper basal ganglia nuclei, and that such increases are accompanied by transient BBB leakage in the most affected regions (Ohlin et al., 2012). Furthermore, it was found that rCBF surges "on" L-DOPA are not necessarily paralleled by an increased rate of local glucose metabolism (Bimpisidis et al., 2017; Lerner et al., 2016; Ohlin et al., 2012), pointing to an abnormal dissociation between blood flow and metabolic activity in the affected regions similar to that reported in PD patients under the effect of L-DOPA (Hirano et al., 2008). Intriguingly, brain areas with severe flow-metabolism dissociation also exhibit immunohistochemical markers of angiogenesis (Lerner et al., 2017). Taken together, these and other findings suggest that angiogenesis, altered local blood flow regulation, and increased BBB permeability, all result from an abnormal response to L-DOPA that is primarily taking place at the gliovascular interface in DA-denervated brain regions. Since angiogenesis and inflammation are reciprocally linked under pathological conditions (Helmy et al., 2011; Landskron et al., 2014), this maladaptive gliovascular reactivity may feed the inflammatory mechanisms associated with LID (see above). In addition, foci of increased BBB permeability may cause an erratic uptake of medications within the brain parenchyma (Pisani et al., 2012), and favor fluctuations in L-DOPA and DA levels, which have in turn been linked with the development of LID (Cenci, 2014).

4 Therapeutic applications

4.1 Microglia as a therapeutic target for disease-modification

Based on the recognized role of both innate and adaptive immunity responses, and the improved understanding of microglia/lymphocytes dysregulation in PD pathogenesis, immune cells have become a compelling target for potential disease-modifying or antidyskinetic treatments in PD (Figs. 2 and 3).

In a first attempt to validate neuroinflammation as a mechanistic target for neuroprotective therapies, many studies have investigated the association between the use of non-steroidal anti-inflammatory drugs (NSAID) and risk of PD, yet reporting contrasting results, showing either a reduced risk of disease in individuals taking NSAID, or a lack of correlation (Becker et al., 2011; Chen et al., 2003; Poly et al., 2019; Ren et al., 2018). Among NSAID, ibuprofen was identified as a possible discriminating agent, since a reduced PD risk was reported in ibuprofen users (Gao et al., 2011).

At the preclinical level, different classes of clinically available drugs active on the immune system via an immunosuppressive or immunomodulatory mechanisms have been considered for repositioning in PD, and experimentation across various

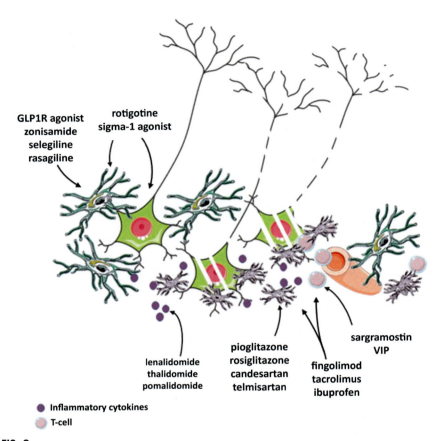

GLP1R agonist
zonisamide
selegiline
rasagiline

rotigotine
sigma-1 agonist

lenalidomide
thalidomide
pomalidomide

pioglitazone
rosiglitazone
candesartan
telmisartan

fingolimod
tacrolimus
ibuprofen

sargramostin
VIP

● Inflammatory cytokines
● T-cell

FIG. 2

Schematic representation of investigational disease-modifying therapies and therapeutic targets in glial and immune cells in the degenerating substantia nigra compacta. Both an intact neuron and degenerating neurons are depicted (white-dashed cells).
GLP1R—glucagon like-peptide-1 receptor, VIP—vasoactive intestinal protein.

models has suggested a benefit in slowing the neurodegenerative process and the related development of motor deficits (Martinez and Peplow, 2018). The immunosuppressant agent fingolimod, an oral drug approved for relapsing-remitting forms of multiple sclerosis, has displayed neuroprotective and anti-inflammatory effects in different PD models, including the 6-OHDA rat and mouse model, the rotenone model and the MPTP model, although with some contrasting results (Komnig et al., 2018; Motyl et al., 2018; Ren et al., 2017; Zhao et al., 2017). In the PD rat model of rAAV2/7-induced α-synuclein overexpression, the immunosuppressant tacrolimus (FK506) has been found to attenuate α-synuclein-induced neurodegeneration and microglial activation (Van der Perren et al., 2015), and similar results

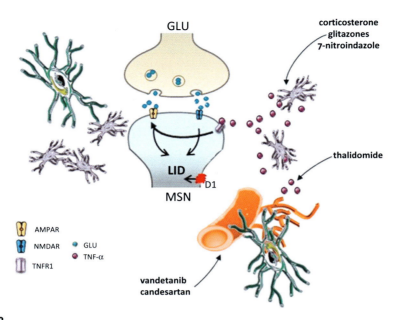

FIG. 3

Schematic representation of investigational anti-dyskinetic therapies and therapeutic targets in glial cells of the dyskinetic striatum. MSN—medium spiny neurons, GLU—glutamatergic neuron, LID—L-DOPA induced dyskinesias.

have been obtained in the acute MPTP mouse model (Manocha et al., 2017). However, the systemic adverse effects, narrow therapeutic window, and limited BBB permeability of these drugs limit their utility for the treatment of PD. More recently, agents acting through an immunomodulatory rather than immunosuppressive mechanism have started to gain interest in PD research. Besides the reduced risk of unwanted sides effects carried by this class of drugs, the recognition of coexistent yet dysregulated beneficial/harmful phenotypes of immune cells in PD suggests that the most effective neuroprotective strategy consists in modulating their activity by boosting reparative functions while suppressing the pro-inflammatory ones. When the immunomodulatory compounds lenalidomide and thalidomide were administered to transgenic mice overexpressing α-synuclein, an improvement in striatal dopaminergic fiber loss was reported that was associated with reduced levels of inflammatory cytokines (Valera et al., 2015). Accordingly, the last-generation thalidomide-derivative pomalidomide has recently been found to reduce dopaminergic cell loss in a transgenic leucin rich-repeat kinase 2 (LRRK2) drosophila model of PD (Casu et al., 2020).

In the attempt to modulate microglia toward a beneficial phenotype, transcription factors involved in microglia activation and polarization have been investigated as a therapeutic target of neuroprotective treatments. Among them, PPAR-γ signaling pathway plays a critical role in the polarization of microglia/macrophages in physiological and pathological conditions, acting as a master regulator that drives the acquisition of an anti-inflammatory activation state and promotes phagocytosis and tissue repair (Cai et al., 2018; Carta, 2013; Croasdell et al., 2015; Glass and Saijo, 2010; Lecca et al., 2018; Nagy et al., 2012; Odegaard et al., 2007). As such, PPAR-γ continues to be an attractive therapeutic target in a number of neurodegenerative diseases (Cai et al., 2018).

Several agonists of PPAR-γ, clinically prescribed as antidiabetic drugs, have been considered for repositioning in PD (Carta, 2013). Drugs belonging to glitazones such as pioglitazone, rosiglitazone and recently characterized derivatives have shown neuroprotective properties in experimental PD models, via the induction of anti-inflammatory and phagocytic functions in microglia (Lecca et al., 2018; Machado et al., 2019; Pinto et al., 2016; Pisanu et al., 2014). The disease-modifying efficacy of these drugs in human PD patients has not yet been confirmed. Thus far, one single clinical trial evaluating pioglitazone has been completed, showing no evidence of disease modulation (Simon et al., 2015). Nevertheless, a recent retrospective study has reported that glitazones use was associated with a significantly lower incidence of PD in diabetic patients, suggesting that further clinical studies are warranted to define the role of this category of drugs in PD (Brakedal et al., 2017).

The recently described local brain renin-angiotensin system (RAS) is also under investigation as a therapeutic target in PD. A dysregulation of the signaling activated by RAS receptors AT1 and AT2 in microglia may lead to pathological pro- or anti-inflammatory responses and may play a role in aging-related neurodegeneration and inflammation (Garrido-Gil et al., 2012; Labandeira-Garcia et al., 2017). The antagonists of angiotensin AT-1 receptor candesartan or telmisartan, widely used for the treatment of hypertension, have been proposed for repositioning in PD based on their demonstrated neuroprotective and anti-inflammatory properties in rats overexpressing mutated forms of α-synuclein (Labandeira-Garcia et al., 2017; Rodriguez-Perez et al., 2018).

In light of the recognized pathogenic role of adaptive immunity in PD, recent studies in experimental PD models have explored the use of immunomodulatory agents that may modulate T cells activation to dampen brain neuroinflammation and consequent neurodegeneration (Mosley et al., 2019). Among them, the human recombinant Granulocyte-macrophage colony-stimulating factor (GM-CSF) sargramostim, clinically used for cancer or post-transplantation therapy, was found to increase regulatory T cells (Treg) proliferation and is currently suggested for repositioning in PD (Gendelman et al., 2017; Hotta et al., 2019). In fact, sargramostim (Leukine) is now in early-phase clinical investigation (ClinicalTrials.gov Identifier: NCT03790670).

The vasoactive intestinal peptide (VIP) is a neuropeptide found in both the central and peripheral nervous system. Besides being a neuron-released neurotransmitter, VIP is released by immune cells and displays immune functions through VIP-receptors expressed in various cell types, including immune cells

(Delgado and Ganea, 2013). Among the multiple physiological functions, VIP is a potent anti-inflammatory mediator, whose actions include the modulation of T-lymphocytes toward the Th-2 phenotype, the increase of Tregs (Szema et al., 2011), and an inhibition of cytokine production by microglia/macrophages. VIP has shown promising neuroprotective effects in models of neurodegenerative disease, and VIP or VIP-receptor agonists have been found to exert significant dopaminergic neuroprotection in PD models (Delgado and Ganea, 2013; Mosley et al., 2019; Reynolds et al., 2010). Clinical testing of VIP has been proposed for several inflammatory diseases in the last decade, which may pave the way for a future potential application in PD.

4.2 Astroglia as an effector of disease-modifying treatments

Many studies using cultured astrocytes have reported expression of neurotransmitter receptors that could potentially be targeted for different therapeutic purposes, including neuroprotective ones. However, since cultured astrocytes may have immature or reactive properties, findings obtained *in vitro* need to be verified in the brain of adult animals in order to provide a solid basis for therapeutic development. In this section, we therefore focus on studies where the astrocytic expression of the therapeutic target has been verified *in vivo*.

In addition to being expressed in neurons, the serotonin 5-HT1A receptor has been found to be present on astrocytic cell bodies and processes in several brain regions (Azmitia et al., 1996). Stimulation of 5-HT_{1A} receptors has been shown to promote S100β secretion followed by astrocyte proliferation and activation of transcription factor Nrf2 (nuclear factor erythroid 2-related factor 2), along with an upregulation of antioxidant pathways (Isooka et al., 2020; Miyazaki et al., 2016). The anti-parkinsonian drug rotigotine can bind to both DA and 5-HT1A receptors. In a 6-OHDA-lesioned mouse model of PD, rotigotine administration has been found to increase the expression of the antioxidant molecule metallothionein in striatal astrocytes, partially counteracting nigrostriatal dopaminergic degeneration. These beneficial effects were blocked by co-administration of the 5-HT1A receptor antagonist WAY100635 (Isooka et al., 2020).

A potentially neuroprotective role of astrocytes has been shown upon treatment with zonisamide, a compound originally developed and used clinically as an antiepileptic agent and now approved for the treatment of PD in Japan (Miyazaki et al., 2016; Murata et al., 2018, 2019). Treatment with zonisamide has been found to increase cystine/glutamate exchange transporter expression and GSH production by astrocytes, reduce α-synuclein neurotoxicity and protect dopaminergic neurons (Finsterwald et al., 2015).

A neuroprotective approach that is currently being considered for clinical translation is the pharmacological stimulation of sigma-1 receptors. These are intracellular membrane-associated chaperones that positively regulate pathways of cells survival and neuroplasticity (Ruscher and Wieloch, 2015). Although sigma-1 receptors are ubiquitous, they exhibit relatively high expression levels in astrocytes within the nigrostriatal system (Francardo et al., 2014). Studies in 6-OHDA-lesioned mice have shown that high-affinity ligands of sigma-1 receptors exert both neuroprotective and neurorestorative effects on nigrostriatal DA neurons (Francardo et al., 2014, 2019). Treatment with neuroprotective doses of sigma-1 receptor agonists induced upregulation of GDNF and BDNF and dampened microglial activation in the substantia nigra and the striatum (Francardo et al., 2014, 2019). These results suggest that the beneficial effects of sigma-1 receptor agonists in this PD model involve a positive modulation of protective functions of glial cells.

Agonists of glucagon like-peptide-1 receptor (GLP1R) are currently regarded as a promising neuroprotective strategy for PD (Athauda and Foltynie, 2018), although the mechanisms underlying their neuroprotective action are not totally clear. Recently, a potent brain-penetrant GLP1R agonist has been reported to protects nigral DA neurons and yield motor improvement in a mouse model of nigrostriatal degeneration induced by α-synuclein fibril inoculation. Interestingly, the neuroprotective effect of GLP1R agonists could be attributed to the prevention of microglial-mediated conversion of astrocytes to an A1 neurotoxic phenotype (Yun et al., 2018).

4.3 Glial-based cell replacement therapies for PD

Cell therapy for PD has been considered as an approach to non-pharmacological DA replacement by transplanting DA neuroblasts that, once mature, will produce DA directly in the striatum. However, the implantation of cells into a disease-affected tissue faces many challenges. In particular, grafted DA neuroblasts need a permissive environment to differentiate into neurons, grow terminals and make functional connections with other cells. Because of the progressive neurodegenerative process in PD, the host tissue environment may become directly pathogenic to a transplant, a point that is vividly illustrated by the report of Lewy bodies within embryonic DA neurons grafted into the striatum in PD patients (Kordower et al., 2008; Li et al., 2008). Some attention may therefore be given to the possibility of co-transplanting astrocytes. In theory, such a strategy could replace potentially dysfunctional astrocytes in the host brain and thus support the differentiation and survival of the grafted neuroblasts, along with providing trophic support to the endogenous, residual dopaminergic innervation.

Astrocyte transplantation has thus far been considered mainly as an approach to raise brain levels of trophic factors. The main strategies so far evaluated in animal studies consist of implanting GDNF-secreting cells into the lesioned nigrostriatal system. In some studies, transplantation of DA-producing cells was combined with GDNF treatment to protect and increase survival of grafted DA neuroblasts. Some interesting results have been reported upon transplanting *in vitro*-differentiated glial precursors into the 6-OHDA-lesioned rat striatum. As a source of donor cells, Proschel et al. (2014) used a unique class of astrocytes termed "GDA[BMP]" obtained from a population of embryonic glial-restricted precursor cells by exposure to bone morphogenetic protein (BMP). These astrocytes exhibit increased levels of antioxidant pathway components (including glutathione) and trophic factors (including BDNF, GDNF, and neurturin). Transplantation of GDA^{BMP} into the 6-OHDA-lesioned rat striatum was found to restore the expression of tyrosine hydroxylase and promote behavioral recovery (Proschel et al., 2014).

In the studies by Song et al. (2018) astrocytes derived from ventral midbrain and transplanted to lesioned rat striatum showed increased expression of trophic factor genes, extracellular matrix and antioxidant proteins. The expression of anti-inflammatory phenotype markers prevailed over that of pro-inflammatory phenotype genes (Song et al., 2018). In the same study authors grafted also genetically engineered astrocytes with forced expression of Nurr1 and Foxa2, the transcription factors recently reported to polarize harmful activated glia into a neuroprotective phenotype (Song et al., 2018). Nurr1 is a transcription factor specific for developing and maintaining adult midbrain dopaminergic neurons (Saijo et al., 2009). Foxa2 on the other hand is a potent co-factor that synergizes the Nurr1-mediated anti-inflammatory roles in glia (Oh et al., 2015). Nurr1 + Foxa2 expression in experimentally transplanted astrocytes further improved astrocytic function to protect midbrain dopaminergic neurons against toxins, mainly by reducing inflammation (Song et al., 2018).

Astrocytes were also successfully co-grafted together with neural progenitor cells (NPCs) into the striatum in order to support their differentiation and survival rate (Song et al., 2018). Co-grafts improved NPC differentiation into midbrain DA neurons, promoting synaptic maturation, expression of midbrain-specific markers, presynaptic dopaminergic neuron function, and resistance against toxic stimuli. The effects of astrocyte co-grafting in this study showed almost complete behavioral restoration and extensive DA neuron engraftments in a rat PD model.

4.4 Targeting glial mechanisms to treat L-DOPA-induced dyskinesia

As mentioned above, neuroinflammation has recently gained attention as a possible target to reduce the motor complications associated with L-DOPA therapy, in particular dyskinesia (Carta et al., 2017) (Fig. 3). Accordingly, several studies have reported that the administration of anti-inflammatory or immunomodulatory compounds such as corticosterone, iNOS inhibitor 7-nitroindazole, PPAR-γ agonists or thalidomide, significantly attenuated the development of LID and reduced the microglia reactivity and inflammatory cytokines induced by L-DOPA (Barnum et al., 2008; Boi et al., 2019; Bortolanza et al., 2015a,b; Martinez et al., 2015). Of note, continuous L-DOPA delivery to parkinsonian rats, known to be associated with a low dyskinetic outcome, did not increase neuroinflammation in striatal motor regions (Mulas et al., 2016).

Cytokines secreted in excess upon L-DOPA treatment may contribute to synaptic maladaptive responses that drive the development of LID. Consistent with this hypothesis, the immunomodulatory compound 3,6-ditiothalidomide, which acts as a potent TNF-α inhibitor, significantly attenuated the severity of LID in a rat model of PD, and reduced the L-DOPA-induced increase of striatal AMPA receptor subunit GLUR1, which is attributed a role in the synaptic abnormalities at the basis of this movement disorder (Boi et al., 2019; Konitsiotis et al., 2000). It should be noted that drugs acting on the immune system have been consistently effective in preventing

LID but not in reducing the expression of already established dyskinesia on a short-term basis, suggesting that maladaptive plasticity, once established, is not easily amenable to cytokine re-modulation.

In light of the tight interconnection of neuroinflammation with angiogenesis (see above), it is intriguing that treatment with antiangiogenic compounds such as vandetanib and candesartan has antidyskinetic effects in parkinsonian rats (Munoz et al., 2014; Ohlin et al., 2011) (Fig. 3). Therefore, a pathological interplay exists between angiogenesis and L-DOPA-induced neuroinflammation, and interrupting it by TNF-α inhibitors or angiogenesis inhibitors may open a new therapeutic perspective to counteract LID. More specifically, targeting microglia or gliovascular mechanisms may offer a good approach to prevent the development of dyskinesia or perhaps "deprime" the abnormal plasticity at the basis of this movement disorder.

5 Concluding remarks

The recognized role played by microglial and astrocytic cells in multiple aspects of PD is profoundly changing the perspectives for therapeutic development. It is now ascertained that in PD the protective functions and integrity of glial cells are affected vis-á-vis those of neurons, leading to a loss of homeostatic, and restorative properties as well as a gain of neurotoxic roles. Dysfunctional glial cells play a key part in the neurodegenerative process, but recent research strongly suggests that they may also have a part to play in the development of L-DOPA-induced motor complications via aberrant modulation of synaptic plasticity and microvascular responses. Moreover, it has become clear that astrocytes play an important role in compensatory mechanisms occurring during the presymptomatic stage of PD that could be exploited for therapeutic purposes.

An increasing number of studies are proposing that new or old agents may modulate the phenotype of glial cells directly or indirectly via a control of cells involved in adaptive immune responses. Moreover, there is an increasing interest in agents that target glial-released soluble factors such as cytokines and growth factors. This growing area of research will likely lead to the clinical testing of glia-targeting therapeutics for disease-modification or dyskinesia prevention. In this area it is noteworthy that several clinically available agents have been proposed for repositioning in PD offering great opportunities to hasten the translational process.

References

Agulhon, C., Sun, M.Y., Murphy, T., Myers, T., Lauderdale, K., Fiacco, T.A., 2012. Calcium signaling and gliotransmission in normal vs. reactive astrocytes. Front. Pharmacol. 3, 139.

Ambrosi, G., Kustrimovic, N., Siani, F., Rasini, E., Cerri, S., et al., 2017. Complex changes in the innate and adaptive immunity accompany progressive degeneration of the nigrostriatal pathway induced by intrastriatal injection of 6-hydroxydopamine in the rat. Neurotox. Res. 32, 71–81.

Anderson, M.A., Burda, J.E., Ren, Y., Ao, Y., O'Shea, T.M., et al., 2016. Astrocyte scar formation aids central nervous system axon regeneration. Nature 532, 195–200.

Arcuri, C., Mecca, C., Bianchi, R., Giambanco, I., Donato, R., 2017. The pathophysiological role of microglia in dynamic surveillance, phagocytosis and structural remodeling of the developing CNS. Front. Mol. Neurosci. 10, 191.

Armentero, M.T., Levandis, G., Nappi, G., Bazzini, E., Blandini, F., 2006. Peripheral inflammation and neuroprotection: systemic pretreatment with complete Freund's adjuvant reduces 6-hydroxydopamine toxicity in a rodent model of Parkinson's disease. Neurobiol. Dis. 24, 492–505.

Asanuma, M., Miyazaki, I., Murakami, S., Diaz-Corrales, F.J., Ogawa, N., 2014. Striatal astrocytes act as a reservoir for L-DOPA. PLoS One 9, e106362.

Athauda, D., Foltynie, T., 2018. Protective effects of the GLP-1 mimetic exendin-4 in Parkinson's disease. Neuropharmacology 136, 260–270.

Azmitia, E.C., Gannon, P.J., Kheck, N.M., Whitaker-Azmitia, P.M., 1996. Cellular localization of the 5-HT1A receptor in primate brain neurons and glial cells. Neuropsychopharmacology 14, 35–46.

Balosso, S., Ravizza, T., Pierucci, M., Calcagno, E., Invernizzi, R., et al., 2009. Molecular and functional interactions between tumor necrosis factor-alpha receptors and the glutamatergic system in the mouse hippocampus: implications for seizure susceptibility. Neuroscience 161, 293–300.

Banati, R.B., Myers, R., Kreutzberg, G.W., 1997. PK ('peripheral benzodiazepine')—binding sites in the CNS indicate early and discrete brain lesions: microautoradiographic detection of [3H]PK11195 binding to activated microglia. J. Neurocytol. 26, 77–82.

Barcia, C., Ros, C.M., Annese, V., Gomez, A., Ros-Bernal, F., et al., 2011. IFN-gamma signaling, with the synergistic contribution of TNF-alpha, mediates cell specific microglial and astroglial activation in experimental models of Parkinson's disease. Cell Death Dis. 2, e142.

Barnum, C.J., Eskow, K.L., Dupre, K., Blandino P., Jr., Deak, T., Bishop, C., 2008. Exogenous corticosterone reduces L-DOPA-induced dyskinesia in the hemi-parkinsonian rat: role for interleukin-1beta. Neuroscience 156, 30–41.

Bartels, A.L., Willemsen, A.T., Doorduin, J., de Vries, E.F., Dierckx, R.A., Leenders, K.L., 2010. [11C]-PK11195 PET: quantification of neuroinflammation and a monitor of anti-inflammatory treatment in Parkinson's disease? Parkinsonism Relat. Disord. 16, 57–59.

Bastide, M.F., Meissner, W.G., Picconi, B., Fasano, S., Fernagut, P.O., et al., 2015. Pathophysiology of L-dopa-induced motor and non-motor complications in Parkinson's disease. Prog. Neurobiol. 132, 96–168.

Beattie, E.C., Stellwagen, D., Morishita, W., Bresnahan, J.C., Ha, B.K., et al., 2002. Control of synaptic strength by glial TNFalpha. Science 295, 2282–2285.

Becker, C., Jick, S.S., Meier, C.R., 2011. NSAID use and risk of Parkinson disease: a population-based case-control study. Eur. J. Neurol. 18, 1336–1342.

Bezard, E., Gross, C.E., 1998. Compensatory mechanisms in experimental and human parkinsonism: towards a dynamic approach. Prog. Neurobiol. 55, 93–116.

Bialas, A.R., Stevens, B., 2013. TGF-beta signaling regulates neuronal C1q expression and developmental synaptic refinement. Nat. Neurosci. 16, 1773–1782.

Bian, M.J., Li, L.M., Yu, M., Fei, J., Huang, F., 2009. Elevated interleukin-1beta induced by 1-methyl-4-phenyl-1,2,3,6-tetrahydropyridine aggravating dopaminergic neurodegeneration in old male mice. Brain Res. 1302, 256–264.

Bimpisidis, Z., Oberg, C.M., Maslava, N., Cenci, M.A., Lundblad, C., 2017. Differential effects of gaseous versus injectable anesthetics on changes in regional cerebral blood flow and metabolism induced by l-DOPA in a rat model of Parkinson's disease. Exp. Neurol. 292, 113–124.

Bindocci, E., Savtchouk, I., Liaudet, N., Becker, D., Carriero, G., Volterra, A., 2017. Three-dimensional Ca(2+) imaging advances understanding of astrocyte biology. Science 356, eaai8185.

Blank, T., Prinz, M., 2013. Microglia as modulators of cognition and neuropsychiatric disorders. Glia 61, 62–70.

Blesa, J., Trigo-Damas, I., Dileone, M., Del Rey, N.L., Hernandez, L.F., Obeso, J.A., 2017. Compensatory mechanisms in Parkinson's disease: circuits adaptations and role in disease modification. Exp. Neurol. 298, 148–161.

Boi, L., Pisanu, A., Greig, N.H., Scerba, M.T., Tweedie, D., et al., 2019. Immunomodulatory drugs alleviate l-dopa-induced dyskinesia in a rat model of Parkinson's disease. Mov. Disord. 34, 1818–1830.

Booth, H.D.E., Hirst, W.D., Wade-Martins, R., 2017. The role of astrocyte dysfunction in Parkinson's disease pathogenesis. Trends Neurosci. 40, 358–370.

Bortolanza, M., Cavalcanti-Kiwiatkoski, R., Padovan-Neto, F.E., da- Silva, C.A., Mitkovski, M., et al., 2015a. Glial activation is associated with l-DOPA induced dyskinesia and blocked by a nitric oxide synthase inhibitor in a rat model of Parkinson's disease. Neurobiol. Dis. 73, 377–387.

Bortolanza, M., Padovan-Neto, F.E., Cavalcanti-Kiwiatkoski, R., Dos Santos-Pereira, M., Mitkovski, M., et al., 2015b. Are cyclooxygenase-2 and nitric oxide involved in the dyskinesia of Parkinson's disease induced by L-DOPA? Philos. Trans. R. Soc. Lond. B Biol. Sci. 370, pii: 20140190.

Braak, H., Sastre, M., Del Tredici, K., 2007. Development of alpha-synuclein immunoreactive astrocytes in the forebrain parallels stages of intraneuronal pathology in sporadic Parkinson's disease. Acta Neuropathol. 114, 231–241.

Brakedal, B., Flones, I., Reiter, S.F., Torkildsen, O., Dolle, C., et al., 2017. Glitazone use associated with reduced risk of Parkinson's disease. Mov. Disord. 32, 1594–1599.

Burda, J.E., Sofroniew, M.V., 2017. Seducing astrocytes to the dark side. Cell Res. 27, 726–727.

Buttgereit, A., Lelios, I., Yu, X., Vrohlings, M., Krakoski, N.R., et al., 2016. Sall1 is a transcriptional regulator defining microglia identity and function. Nat. Immunol. 17, 1397–1406.

Cai, W., Yang, T., Liu, H., Han, L., Zhang, K., et al., 2018. Peroxisome proliferator-activated receptor gamma (PPARgamma): a master gatekeeper in CNS injury and repair. Prog. Neurobiol. 163–164, 27–58.

Capani, F., Quarracino, C., Caccuri, R., Sica, R.E., 2016. Astrocytes as the main players in primary degenerative disorders of the human central nervous system. Front. Aging Neurosci. 8, 45.

Carta, A.R., 2013. PPAR-gamma: therapeutic prospects in Parkinson's disease. Curr. Drug Targets 14, 743–751.

Carta, A.R., Mulas, G., Bortolanza, M., Duarte, T., Pillai, E., et al., 2017. l-DOPA-induced dyskinesia and neuroinflammation: do microglia and astrocytes play a role? Eur. J. Neurosci. 45, 73–91.

Casu, A.M., Mocci, I., Isola, R., Pisanu, A., Boi, L., et al., 2020. Neuroprotection by the immunomodulatory drug pomalidomide in the Drosophila LRRK2^{WD40} genetic model of Parkinson's disease. Front. Aging Neurosci. 12, 1–13.

Cavaliere, F., Cerf, L., Dehay, B., Ramos-Gonzalez, P., De Giorgi, F., et al., 2017. In vitro alpha-synuclein neurotoxicity and spreading among neurons and astrocytes using Lewy body extracts from Parkinson disease brains. Neurobiol. Dis. 103, 101–112.

Cebrian, C., Zucca, F.A., Mauri, P., Steinbeck, J.A., Studer, L., et al., 2014. MHC-I expression renders catecholaminergic neurons susceptible to T-cell-mediated degeneration. Nat. Commun. 5, 3633.

Cenci, M.A., 2014. Presynaptic mechanisms of l-DOPA-induced dyskinesia: the findings, the debate, and the therapeutic implications. Front. Neurol. 5, 242.

Cenci, M.A., Björklund, A., 2020. Animal models for preclinical Parkinson's research: an update. Prog. Brain Res. in press.

Cenci, M.A., Ohlin, K.E., Odin, P., 2011. Current options and future possibilities for the treatment of dyskinesia and motor fluctuations in Parkinson's disease. CNS Neurol. Disord. Drug Targets 10, 670–684.

Centonze, D., Muzio, L., Rossi, S., Cavasinni, F., De Chiara, V., et al., 2009. Inflammation triggers synaptic alteration and degeneration in experimental autoimmune encephalomyelitis. J. Neurosci. 29, 3442–3452.

Charron, G., Doudnikoff, E., Canron, M.H., Li, Q., Vega, C., et al., 2014. Astrocytosis in parkinsonism: considering tripartite striatal synapses in physiopathology? Front. Aging Neurosci. 6, 258.

Chen, H., Zhang, S.M., Hernan, M.A., Schwarzschild, M.A., Willett, W.C., et al., 2003. Nonsteroidal anti-inflammatory drugs and the risk of Parkinson disease. Arch. Neurol. 60, 1059–1064.

Cherry, J.D., Olschowka, J.A., O'Banion, M.K., 2014. Neuroinflammation and M2 microglia: the good, the bad, and the inflamed. J. Neuroinflammation 11, 98.

Clark, A.K., Gruber-Schoffnegger, D., Drdla-Schutting, R., Gerhold, K.J., Malcangio, M., Sandkühler, J., 2015. Selective activation of microglia facilitates synaptic strength. J. Neurosci. 35 (11), 4552–4570. https://doi.org/10.1523/JNEUROSCI.2061-14.2015.

Croasdell, A., Duffney, P.F., Kim, N., Lacy, S.H., Sime, P.J., Phipps, R.P., 2015. PPARgamma and the innate immune system mediate the resolution of inflammation. PPAR Res. 2015, 549691.

Croisier, E., Moran, L.B., Dexter, D.T., Pearce, R.K., Graeber, M.B., 2005. Microglial inflammation in the parkinsonian substantia nigra: relationship to alpha-synuclein deposition. J. Neuroinflammation 2, 14.

Daher, J.P., Volpicelli-Daley, L.A., Blackburn, J.P., Moehle, M.S., West, A.B., 2014. Abrogation of alpha-synuclein-mediated dopaminergic neurodegeneration in LRRK2-deficient rats. Proc. Natl. Acad. Sci. U. S. A. 111, 9289–9294.

Daniele, S.G., Beraud, D., Davenport, C., Cheng, K., Yin, H., Maguire-Zeiss, K.A., 2015. Activation of MyD88-dependent TLR1/2 signaling by misfolded alpha-synuclein, a protein linked to neurodegenerative disorders. Sci. Signal. 8, ra45.

Davalos, D., Grutzendler, J., Yang, G., Kim, J.V., Zuo, Y., et al., 2005. ATP mediates rapid microglial response to local brain injury in vivo. Nat. Neurosci. 8, 752–758.

De Chiara, V., Motta, C., Rossi, S., Studer, V., Barbieri, F., et al., 2013. Interleukin-1beta alters the sensitivity of cannabinoid CB1 receptors controlling glutamate transmission in the striatum. Neuroscience 250, 232–239.

De Deurwaerdère, P., Di Giovanni, G., Millan, M.J., 2017. Expanding the repertoire of L-DOPA's actions: a comprehensive review of its functional neurochemistry. Prog. Neurobiol. 151, 57–100.

de Haas, A.H., Boddeke, H.W., Biber, K., 2008. Region-specific expression of immunoregulatory proteins on microglia in the healthy CNS. Glia 56, 888–894.

Deitmer, J.W., 2001. Strategies for metabolic exchange between glial cells and neurons. Respir. Physiol. 129, 71–81.

Deitmer, J.W., Theparambil, S.M., Ruminot, I., Noor, S.I., Becker, H.M., 2019. Energy dynamics in the brain: contributions of astrocytes to metabolism and pH homeostasis. Front. Neurosci. 13, 1301.

Delgado, M., Ganea, D., 2013. Vasoactive intestinal peptide: a neuropeptide with pleiotropic immune functions. Amino Acids 45, 25–39.

Dienel, G.A., Carlson, G.M., 2019. Major advances in brain glycogen research: understanding of the roles of glycogen have evolved from emergency fuel reserve to dynamic, regulated participant in diverse brain functions. Adv. Neurobiol. 23, 1–16.

Dominguez-Meijide, A., Rodriguez-Perez, A.I., Diaz-Ruiz, C., Guerra, M.J., Labandeira-Garcia, J.L., 2017. Dopamine modulates astroglial and microglial activity via glial renin-angiotensin system in cultures. Brain Behav. Immun. 62, 277–290.

Doorn, K.J., Moors, T., Drukarch, B., van de Berg, W., Lucassen, P.J., van Dam, A.M., 2014. Microglial phenotypes and toll-like receptor 2 in the substantia nigra and hippocampus of incidental Lewy body disease cases and Parkinson's disease patients. Acta Neuropathol. Commun. 2, 90.

Doorn, K.J., Breve, J.J., Drukarch, B., Boddeke, H.W., Huitinga, I., et al., 2015. Brain region-specific gene expression profiles in freshly isolated rat microglia. Front. Cell. Neurosci. 9, 84.

Duffy, M.F., Collier, T.J., Patterson, J.R., Kemp, C.J., Luk, K.C., et al., 2018. Lewy body-like alpha-synuclein inclusions trigger reactive microgliosis prior to nigral degeneration. J. Neuroinflammation 15, 129.

Durkee, C.A., Araque, A., 2019. Diversity and specificity of astrocyte-neuron communication. Neuroscience 396, 73–78.

Edison, P., Ahmed, I., Fan, Z., Hinz, R., Gelosa, G., et al., 2013. Microglia, amyloid, and glucose metabolism in Parkinson's disease with and without dementia. Neuropsychopharmacology 38, 938–949.

Edwards, J.P., Zhang, X., Frauwirth, K.A., Mosser, D.M., 2006. Biochemical and functional characterization of three activated macrophage populations. J. Leukoc. Biol. 80, 1298–1307.

Elstner, M., Morris, C.M., Heim, K., Bender, A., Mehta, D., et al., 2011. Expression analysis of dopaminergic neurons in Parkinson's disease and aging links transcriptional dysregulation of energy metabolism to cell death. Acta Neuropathol. 122, 75–86.

Emmanouilidou, E., Vekrellis, K., 2016. Exocytosis and spreading of normal and aberrant alpha-synuclein. Brain Pathol. 26, 398–403.

Farkas, E., De Jong, G.I., Apro, E., De Vos, R.A., Steur, E.N., Luiten, P.G., 2000. Similar ultrastructural breakdown of cerebrocortical capillaries in Alzheimer's disease, Parkinson's disease, and experimental hypertension. What is the functional link? Ann. N. Y. Acad. Sci. 903, 72–82.

Fellner, L., Irschick, R., Schanda, K., Reindl, M., Klimaschewski, L., et al., 2013. Toll-like receptor 4 is required for alpha-synuclein dependent activation of microglia and astroglia. Glia 61, 349–360.

Feyder, M., Bonito-Oliva, A., Fisone, G., 2011. L-DOPA-induced dyskinesia and abnormal signaling in striatal medium spiny neurons: focus on dopamine D1 receptor-mediated transmission. Front. Behav. Neurosci. 5, 71.

Finsterwald, C., Magistretti, P.J., Lengacher, S., 2015. Astrocytes: new targets for the treatment of neurodegenerative diseases. Curr. Pharm. Des. 21, 3570–3581.

Francardo, V., Bez, F., Wieloch, T., Nissbrandt, H., Ruscher, K., Cenci, M.A., 2014. Pharmacological stimulation of sigma-1 receptors has neurorestorative effects in experimental parkinsonism. Brain 137, 1998–2014.

Francardo, V., Geva, M., Bez, F., Denis, Q., Steiner, L., et al., 2019. Pridopidine induces functional neurorestoration via the sigma-1 receptor in a mouse model of Parkinson's disease. Neurotherapeutics 16, 465–479.

Fusco, G., Chen, S.W., Williamson, P.T.F., Cascella, R., Perni, M., et al., 2017. Structural basis of membrane disruption and cellular toxicity by alpha-synuclein oligomers. Science 358, 1440–1443.

Gao, X., Chen, H., Schwarzschild, M.A., Ascherio, A., 2011. Use of ibuprofen and risk of Parkinson disease. Neurology 76, 863–869.

Garrido-Gil, P., Joglar, B., Rodriguez-Perez, A.I., Guerra, M.J., Labandeira-Garcia, J.L., 2012. Involvement of PPAR-gamma in the neuroprotective and anti-inflammatory effects of angiotensin type 1 receptor inhibition: effects of the receptor antagonist telmisartan and receptor deletion in a mouse MPTP model of Parkinson's disease. J. Neuroinflammation 9, 38.

Gendelman, H.E., Zhang, Y., Santamaria, P., Olson, K.E., Schutt, C.R., et al., 2017. Evaluation of the safety and immunomodulatory effects of sargramostim in a randomized, double-blind phase 1 clinical Parkinson's disease trial. NPJ Parkinsons Dis. 3, 10.

Gerhard, A., Pavese, N., Hotton, G., Turkheimer, F., Es, M., et al., 2006. In vivo imaging of microglial activation with [11C](R)-PK11195 PET in idiopathic Parkinson's disease. Neurobiol. Dis. 21, 404–412.

Glass, C.K., Saijo, K., 2010. Nuclear receptor transrepression pathways that regulate inflammation in macrophages and T cells. Nat. Rev. Immunol. 10, 365–376.

Goldmann, T., Prinz, M., 2013. Role of microglia in CNS autoimmunity. Clin. Dev. Immunol. 2013, 208093.

Gosselin, D., Link, V.M., Romanoski, C.E., Fonseca, G.J., Eichenfield, D.Z., et al., 2014. Environment drives selection and function of enhancers controlling tissue-specific macrophage identities. Cell 159, 1327–1340.

Grabert, K., Michoel, T., Karavolos, M.H., Clohisey, S., Baillie, J.K., et al., 2016. Microglial brain region-dependent diversity and selective regional sensitivities to aging. Nat. Neurosci. 19, 504–516.

Halassa, M.M., Haydon, P.G., 2010. Integrated brain circuits: astrocytic networks modulate neuronal activity and behavior. Annu. Rev. Physiol. 72, 335–355.

Halassa, M.M., Fellin, T., Haydon, P.G., 2007. The tripartite synapse: roles for gliotransmission in health and disease. Trends Mol. Med. 13, 54–63.

Halliday, G.M., Stevens, C.H., 2011. Glia: initiators and progressors of pathology in Parkinson's disease. Mov. Disord. 26, 6–17.

Hanisch, U.K., Kettenmann, H., 2007. Microglia: active sensor and versatile effector cells in the normal and pathologic brain. Nat. Neurosci. 10, 1387–1394.

Harrison, J.K., Jiang, Y., Chen, S., Xia, Y., Maciejewski, D., et al., 1998. Role for neuronally derived fractalkine in mediating interactions between neurons and CX3CR1-expressing microglia. Proc. Natl. Acad. Sci. U. S. A. 95, 10896–10901.

Haydon, P.G., Carmignoto, G., 2006. Astrocyte control of synaptic transmission and neurovascular coupling. Physiol. Rev. 86, 1009–1031.

Helmy, A., De Simoni, M.G., Guilfoyle, M.R., Carpenter, K.L., Hutchinson, P.J., 2011. Cytokines and innate inflammation in the pathogenesis of human traumatic brain injury. Prog. Neurobiol. 95, 352–372.

Hirano, S., Asanuma, K., Ma, Y., Tang, C., Feigin, A., et al., 2008. Dissociation of metabolic and neurovascular responses to levodopa in the treatment of Parkinson's disease. J. Neurosci. 28, 4201–4209.

Hirsch, E.C., Hunot, S., 2009. Neuroinflammation in Parkinson's disease: a target for neuroprotection? Lancet Neurol. 8, 382–397.

Holtman, I.R., Skola, D., Glass, C.K., 2017. Transcriptional control of microglia phenotypes in health and disease. J. Clin. Invest. 127, 3220–3229.

Hong, S., Dissing-Olesen, L., Stevens, B., 2016. New insights on the role of microglia in synaptic pruning in health and disease. Curr. Opin. Neurobiol. 36, 128–134.

Hornykiewicz, O., 1998. Biochemical aspects of Parkinson's disease. Neurology 51, S2–S9.

Hotta, M., Yoshimura, H., Satake, A., Tsubokura, Y., Ito, T., Nomura, S., 2019. GM-CSF therapy inhibits chronic graft-versus-host disease via expansion of regulatory T cells. Eur. J. Immunol. 49, 179–191.

Hughes, V., 2012. Microglia: the constant gardeners. Nature 485, 570–572.

Hung, C.C., Lin, C.H., Chang, H., Wang, C.Y., Lin, S.H., et al., 2016. Astrocytic GAP43 induced by the TLR4/NF-κB/STAT3 axis attenuates astrogliosis-mediated microglial activation and neurotoxicity. J. Neurosci. 36, 2027–2043.

Hurley, S.D., O'Banion, M.K., Song, D.D., Arana, F.S., Olschowka, J.A., Haber, S.N., 2003. Microglial response is poorly correlated with neurodegeneration following chronic, low-dose MPTP administration in monkeys. Exp. Neurol. 184, 659–668.

Imamura, K., Hishikawa, N., Sawada, M., Nagatsu, T., Yoshida, M., Hashizume, Y., 2003. Distribution of major histocompatibility complex class II-positive microglia and cytokine profile of Parkinson's disease brains. Acta Neuropathol. 106, 518–526.

Inyushin, M.Y., Huertas, A., Kucheryavykh, Y.V., Kucheryavykh, L.Y., Tsydzik, V., et al., 2012. L-DOPA uptake in astrocytic endfeet enwrapping blood vessels in rat brain. Parkinsons Dis. 2012, 321406.

Isooka, N., Miyazaki, I., Kikuoka, R., Wada, K., Nakayama, E., et al., 2020. Dopaminergic neuroprotective effects of rotigotine via 5-HT1A receptors: possibly involvement of metallothionein expression in astrocytes. Neurochem. Int. 132, 104608.

Israeli, E., Sharon, R., 2009. Beta-synuclein occurs in vivo in lipid-associated oligomers and forms hetero-oligomers with alpha-synuclein. J. Neurochem. 108, 465–474.

Jankovic, J., Aguilar, L.G., 2008. Current approaches to the treatment of Parkinson's disease. Neuropsychiatr. Dis. Treat. 4, 743–757.

Joers, V., Tansey, M.G., Mulas, G., Carta, A.R., 2017. Microglial phenotypes in Parkinson's disease and animal models of the disease. Prog. Neurobiol. 155, 57–75.

Jourdain, V.A., Schindlbeck, K.A., Tang, C.C., Niethammer, M., Choi, Y.Y., et al., 2017. Increased putamen hypercapnic vasoreactivity in levodopa-induced dyskinesia. JCI Insight, 2, e96411.

Jyothi, H.J., Vidyadhara, D.J., Mahadevan, A., Philip, M., Parmar, S.K., et al., 2015. Aging causes morphological alterations in astrocytes and microglia in human substantia nigra pars compacta. Neurobiol. Aging 36, 3321–3333.

Karpinar, D.P., Balija, M.B., Kugler, S., Opazo, F., Rezaei-Ghaleh, N., et al., 2009. Prefibrillar alpha-synuclein variants with impaired beta-structure increase neurotoxicity in Parkinson's disease models. EMBO J. 28, 3256–3268.

Kim, W.G., Mohney, R.P., Wilson, B., Jeohn, G.H., Liu, B., Hong, J.S., 2000. Regional difference in susceptibility to lipopolysaccharide-induced neurotoxicity in the rat brain: role of microglia. J. Neurosci. 20, 6309–6316.

Kim, S., Cho, S.H., Kim, K.Y., Shin, K.Y., Kim, H.S., et al., 2009. Alpha-synuclein induces migration of BV-2 microglial cells by up-regulation of CD44 and MT1-MMP. J. Neurochem. 109, 1483–1496.

Kim, C., Ho, D.H., Suk, J.E., You, S., Michael, S., et al., 2013. Neuron-released oligomeric alpha-synuclein is an endogenous agonist of TLR2 for paracrine activation of microglia. Nat. Commun. 4, 1562.

Kimelberg, H.K., Nedergaard, M., 2010. Functions of astrocytes and their potential as therapeutic targets. Neurotherapeutics 7, 338–353.

Klegeris, A., Pelech, S., Giasson, B.I., Maguire, J., Zhang, H., et al., 2008. Alpha-synuclein activates stress signaling protein kinases in THP-1 cells and microglia. Neurobiol. Aging 29, 739–752.

Komnig, D., Dagli, T.C., Habib, P., Zeyen, T., Schulz, J.B., Falkenburger, B.H., 2018. Fingolimod (FTY720) is not protective in the subacute MPTP mouse model of Parkinson's disease and does not lead to a sustainable increase of brain-derived neurotrophic factor. J. Neurochem. 147, 678–691.

Konitsiotis, S., Blanchet, P.J., Verhagen, L., Lamers, E., Chase, T.N., 2000. AMPA receptor blockade improves levodopa-induced dyskinesia in MPTP monkeys. Neurology 54, 1589–1595.

Kordower, J.H., Chu, Y., Hauser, R.A., Freeman, T.B., Olanow, C.W., 2008. Lewy body-like pathology in long-term embryonic nigral transplants in Parkinson's disease. Nat. Med. 14, 504–506.

Kösel, S., Egensperger, R., von Eitzen, U., Mehraein, P., Graeber, M.B., 1997. On the question of apoptosis in the parkinsonian substantia nigra. Acta Neuropathol. 93, 105–108.

Koso, H., Tsuhako, A., Lai, C.Y., Baba, Y., Otsu, M., et al., 2016. Conditional rod photoreceptor ablation reveals Sall1 as a microglial marker and regulator of microglial morphology in the retina. Glia 64, 2005–2024.

Kreutzberg, G.W., 1996. Microglia: a sensor for pathological events in the CNS. Trends Neurosci. 19, 312–318.

Kumar, M.J., Andersen, J.K., 2004. Perspectives on MAO-B in aging and neurological disease: where do we go from here? Mol. Neurobiol. 30, 77–89.

Kustrimovic, N., Comi, C., Magistrelli, L., Rasini, E., Legnaro, M., et al., 2018. Parkinson's disease patients have a complex phenotypic and functional Th1 bias: cross-sectional studies of CD4 + Th1/Th2/T17 and Treg in drug-naive and drug-treated patients. J. Neuroinflammation 15, 205.

Kuter, K., Kratochwil, M., Berghauzen-Maciejewska, K., Głowacka, U., Sugawa, M.D., et al., 2016a. Adaptation within mitochondrial oxidative phosphorylation supercomplexes and membrane viscosity during degeneration of dopaminergic neurons in an animal model of early Parkinson's disease. Biochim. Biophys. Acta 1862, 741–753.

Kuter, K., Kratochwil, M., Marx, S.H., Hartwig, S., Lehr, S., et al., 2016b. Native DIGE proteomic analysis of mitochondria from substantia nigra and striatum during neuronal degeneration and its compensation in an animal model of early Parkinson's disease. Arch. Physiol. Biochem. 122, 238–256.

Kuter, K., Olech, Ł., Głowacka, U., 2018. Prolonged dysfunction of astrocytes and activation of microglia accelerate degeneration of dopaminergic neurons in the rat substantia Nigra and block compensation of early motor dysfunction induced by 6-OHDA. Mol. Neurobiol. 55, 3049–3066.

Kuter, K., Olech, Ł., Głowacka, U., Paleczna, M., 2019a. Astrocyte support is important for the compensatory potential of the nigrostriatal system neurons during early neurodegeneration. J. Neurochem. 148, 63–79.

Kuter, K.Z., Olech, Ł., Dencher, N.A., 2019b. Increased energetic demand supported by mitochondrial electron transfer chain and astrocyte assistance is essential to maintain the compensatory ability of the dopaminergic neurons in an animal model of early Parkinson's disease. Mitochondrion 47, 227–237.

Labandeira-Garcia, J.L., Rodriguez-Perez, A.I., Garrido-Gil, P., Rodriguez-Pallares, J., Lanciego, J.L., Guerra, M.J., 2017. Brain renin-angiotensin system and microglial polarization: implications for aging and neurodegeneration. Front. Aging Neurosci. 9, 129.

Landskron, G., De la Fuente, M., Thuwajit, P., Thuwajit, C., Hermoso, M.A., 2014. Chronic inflammation and cytokines in the tumor microenvironment. J. Immunol. Res. 2014, 149185.

Lawson, L.J., Perry, V.H., Dri, P., Gordon, S., 1990. Heterogeneity in the distribution and morphology of microglia in the normal adult mouse brain. Neuroscience 39, 151–170.

Lecca, D., Janda, E., Mulas, G., Diana, A., Martino, C., et al., 2018. Boosting phagocytosis and anti-inflammatory phenotype in microglia mediates neuroprotection by PPARgamma agonist MDG548 in Parkinson's disease models. Br. J. Pharmacol. 175, 3298–3314.

Lee, H.J., Suk, J.E., Bae, E.J., Lee, S.J., 2008. Clearance and deposition of extracellular alpha-synuclein aggregates in microglia. Biochem. Biophys. Res. Commun. 372, 423–428.

Lee, E.J., Woo, M.S., Moon, P.G., Baek, M.C., Choi, I.Y., et al., 2010a. Alpha-synuclein activates microglia by inducing the expressions of matrix metalloproteinases and the subsequent activation of protease-activated receptor-1. J. Immunol. 185, 615–623.

Lee, H.J., Kim, C., Lee, S.J., 2010b. Alpha-synuclein stimulation of astrocytes: potential role for neuroinflammation and neuroprotection. Oxid. Med. Cell. Longev. 3, 283–287.

Lees, A.J., Selikhova, M., Andrade, L.A., Duyckaerts, C., 2008. The black stuff and Konstantin Nikolaevich Tretiakoff. Mov. Disord. 23 (6), 777–783. https://doi.org/10.1002/mds.21855.

Leibovich, S.J., Polverini, P.J., Shepard, H.M., Wiseman, D.M., Shively, V., Nuseir, N., 1987. Macrophage-induced angiogenesis is mediated by tumour necrosis factor-alpha. Nature 329, 630–632.

L'Episcopo, F., Tirolo, C., Testa, N., Caniglia, S., Morale, M.C., et al., 2011. Reactive astrocytes and Wnt/beta-catenin signaling link nigrostriatal injury to repair in 1-methyl-4-phenyl-1,2,3,6-tetrahydropyridine model of Parkinson's disease. Neurobiol. Dis. 41, 508–527.

Lerner, R.P., Bimpisidis, Z., Agorastos, S., Scherrer, S., Dewey, S.L., et al., 2016. Dissociation of metabolic and hemodynamic levodopa responses in the 6-hydroxydopamine rat model. Neurobiol. Dis. 96, 31–37.

Lerner, R.P., Francardo, V., Fujita, K., Bimpisidis, Z., Jourdain, V.A., et al., 2017. Levodopa-induced abnormal involuntary movements correlate with altered permeability of the blood-brain-barrier in the basal ganglia. Sci. Rep. 7, 16005.

Lewitus, G.M., Pribiag, H., Duseja, R., St-Hilaire, M., Stellwagen, D., 2014. An adaptive role of TNFalpha in the regulation of striatal synapses. J. Neurosci. 34, 6146–6155.

Lewitus, G.M., Konefal, S.C., Greenhalgh, A.D., Pribiag, H., Augereau, K., Stellwagen, D., 2016. Microglial TNF-alpha suppresses cocaine-induced plasticity and behavioral sensitization. Neuron 90, 483–491.

Li, J.Y., Englund, E., Holton, J.L., Soulet, D., Hagell, P., et al., 2008. Lewy bodies in grafted neurons in subjects with Parkinson's disease suggest host-to-graft disease propagation. Nat. Med. 14, 501–503.

Lian, H., Zheng, H., 2016. Signaling pathways regulating neuron-glia interaction and their implications in Alzheimer's disease. J. Neurochem. 136, 475–491.

Liberatore, G.T., Jackson-Lewis, V., Vukosavic, S., Mandir, A.S., Vila, M., et al., 1999. Inducible nitric oxide synthase stimulates dopaminergic neurodegeneration in the MPTP model of Parkinson disease. Nat. Med. 5, 1403–1409.

Liddelow, S.A., Guttenplan, K.A., Clarke, L.E., Bennett, F.C., Bohlen, C.J., et al., 2017. Neurotoxic reactive astrocytes are induced by activated microglia. Nature 541, 481–487.

Lin, X., Parisiadou, L., Gu, X.L., Wang, L., Shim, H., et al., 2009. Leucine-rich repeat kinase 2 regulates the progression of neuropathology induced by Parkinson's-disease-related mutant alpha-synuclein. Neuron 64, 807–827.

Lindgren, H.S., Ohlin, K.E., Cenci, M.A., 2009. Differential involvement of D1 and D2 dopamine receptors in L-DOPA-induced angiogenic activity in a rat model of Parkinson's disease. Neuropsychopharmacology 34, 2477–2488.

Lindstrom, V., Gustafsson, G., Sanders, L.H., Howlett, E.H., Sigvardson, J., et al., 2017. Extensive uptake of alpha-synuclein oligomers in astrocytes results in sustained intracellular deposits and mitochondrial damage. Mol. Cell. Neurosci. 82, 143–156.

Liu, B., Teschemacher, A.G., Kasparov, S., 2017. Astroglia as a cellular target for neuroprotection and treatment of neuro-psychiatric disorders. Glia 65, 1205–1226.

Loeffler, D.A., Camp, D.M., Conant, S.B., 2006. Complement activation in the Parkinson's disease substantia nigra: an immunocytochemical study. J. Neuroinflammation 3, 29.

Lofrumento, D.D., Saponaro, C., Cianciulli, A., De Nuccio, F., Mitolo, V., et al., 2011. MPTP-induced neuroinflammation increases the expression of pro-inflammatory cytokines and their receptors in mouse brain. Neuroimmunomodulation 18, 79–88.

Lopez Gonzalez, I., Garcia-Esparcia, P., Llorens, F., Ferrer, I., 2016. Genetic and transcriptomic profiles of inflammation in neurodegenerative diseases: Alzheimer, Parkinson, Creutzfeldt-Jakob and tauopathies. Int. J. Mol. Sci. 17, 206.

Machado, M.M.F., Bassani, T.B., Coppola-Segovia, V., Moura, E.L.R., Zanata, S.M., et al., 2019. PPAR-gamma agonist pioglitazone reduces microglial proliferation and NF-kappaB activation in the substantia nigra in the 6-hydroxydopamine model of Parkinson's disease. Pharmacol. Rep. 71, 556–564.

Maia, S., Arlicot, N., Vierron, E., Bodard, S., Vergote, J., et al., 2012. Longitudinal and parallel monitoring of neuroinflammation and neurodegeneration in a 6-hydroxydopamine rat model of Parkinson's disease. Synapse 66, 573–583.

Majbour, N.K., Vaikath, N.N., van Dijk, K.D., Ardah, M.T., Varghese, S., et al., 2016. Oligomeric and phosphorylated alpha-synuclein as potential CSF biomarkers for Parkinson's disease. Mol. Neurodegener. 11, 7.

Mallajosyula, J.K., Kaur, D., Chinta, S.J., Rajagopalan, S., Rane, A., et al., 2008. MAO-B elevation in mouse brain astrocytes results in Parkinson's pathology. PLoS One 3, e1616.

Manocha, G.D., Floden, A.M., Puig, K.L., Nagamoto-Combs, K., Scherzer, C.R., Combs, C.K., 2017. Defining the contribution of neuroinflammation to Parkinson's disease in humanized immune system mice. Mol. Neurodegener. 12, 17.

Marin, I., Kipnis, J., 2013. Learning and memory... and the immune system. Learn. Mem. 20, 601–606.

Marinova-Mutafchieva, L., Sadeghian, M., Broom, L., Davis, J.B., Medhurst, A.D., Dexter, D.T., 2009. Relationship between microglial activation and dopaminergic neuronal loss in the substantia nigra: a time course study in a 6-hydroxydopamine model of Parkinson's disease. J. Neurochem. 110, 966–975.

Martinez, F.O., Gordon, S., 2014. The M1 and M2 paradigm of macrophage activation: time for reassessment. F1000Prime Rep. 6, 13.

Martinez, B., Peplow, P.V., 2018. Neuroprotection by immunomodulatory agents in animal models of Parkinson's disease. Neural Regen. Res. 13, 1493–1506.

Martinez, F.O., Gordon, S., Locati, M., Mantovani, A., 2006. Transcriptional profiling of the human monocyte-to-macrophage differentiation and polarization: new molecules and patterns of gene expression. J. Immunol. 177, 7303–7311.

Martinez, A.A., Morgese, M.G., Pisanu, A., Macheda, T., Paquette, M.A., et al., 2015. Activation of PPAR gamma receptors reduces levodopa-induced dyskinesias in 6-OHDA-lesioned rats. Neurobiol. Dis. 74, 295–304.

McGeer, P.L., Itagaki, S., Boyes, B.E., McGeer, E.G., 1988. Reactive microglia are positive for HLA-DR in the substantia nigra of Parkinson's and Alzheimer's disease brains. Neurology 38, 1285–1291.

McKenzie, J.A., Spielman, L.J., Pointer, C.B., Lowry, J.R., Bajwa, E., et al., 2017. Neuroinflammation as a common mechanism associated with the modifiable risk factors for Alzheimer's and Parkinson's diseases. Curr. Aging Sci. 10, 158–176.

Melief, J., Koning, N., Schuurman, K.G., Van De Garde, M.D., Smolders, J., et al., 2012. Phenotyping primary human microglia: tight regulation of LPS responsiveness. Glia 60, 1506–1517.

Mittelbronn, M., Dietz, K., Schluesener, H.J., Meyermann, R., 2001. Local distribution of microglia in the normal adult human central nervous system differs by up to one order of magnitude. Acta Neuropathol. 101, 249–255.

Miyazaki, I., Murakami, S., Torigoe, N., Kitamura, Y., Asanuma, M., 2016. Neuroprotective effects of levetiracetam target xCT in astrocytes in parkinsonian mice. J. Neurochem. 136, 194–204.

Mogi, M., Kondo, T., Mizuno, Y., Nagatsu, T., 2007. p53 protein, interferon-gamma, and NF-kappaB levels are elevated in the parkinsonian brain. Neurosci. Lett. 414, 94–97.

Monnier, Y., Zaric, J., Ruegg, C., 2005. Inhibition of angiogenesis by non-steroidal anti-inflammatory drugs: from the bench to the bedside and back. Curr. Drug Targets Inflamm. Allergy 4, 31–38.

Mosley, R.L., Gendelman, H.E., 2017. T cells and Parkinson's disease. Lancet Neurol. 16, 769–771.

Mosley, R.L., Lu, Y., Olson, K.E., Machhi, J., Yan, W., et al., 2019. A synthetic agonist to vasoactive intestinal peptide receptor-2 Induces regulatory T cell neuroprotective activities in models of Parkinson's disease. Front. Cell. Neurosci. 13, 421.

Motyl, J., Przykaza, L., Boguszewski, P.M., Kosson, P., Strosznajder, J.B., 2018. Pramipexole and Fingolimod exert neuroprotection in a mouse model of Parkinson's disease by activation of sphingosine kinase 1 and Akt kinase. Neuropharmacology 135, 139–150.

Mulas, G., Espa, E., Fenu, S., Spiga, S., Cossu, G., et al., 2016. Differential induction of dyskinesia and neuroinflammation by pulsatile versus continuous l-DOPA delivery in the 6-OHDA model of Parkinson's disease. Exp. Neurol. 286, 83–92.

Muller, W.A., 2014. How endothelial cells regulate transmigration of leukocytes in the inflammatory response. Am. J. Pathol. 184, 886–896.

Munoz, A., Garrido-Gil, P., Dominguez-Meijide, A., Labandeira-Garcia, J.L., 2014. Angiotensin type 1 receptor blockage reduces l-dopa-induced dyskinesia in the 6-OHDA model of Parkinson's disease. Involvement of vascular endothelial growth factor and interleukin-1beta. Exp. Neurol. 261, 720–732.

Murata, M., Odawara, T., Hasegawa, K., Iiyama, S., Nakamura, M., et al., 2018. Adjunct zonisamide to levodopa for DLB parkinsonism: a randomized double-blind phase 2 study. Neurology 90, e664–e672.

Murata, M., Odawara, T., Hasegawa, K., Kajiwara, R., Takeuchi, H., et al., 2019. Effect of zonisamide on parkinsonism in patients with dementia with Lewy bodies: a phase 3 randomized clinical trial. Parkinsonism Relat. Disord. 12, pii: S1353-8020(19)30524-3.

Nagatsu, T., Mogi, M., Ichinose, H., Togari, A., 2000. Cytokines in Parkinson's disease. J. Neural Transm. Suppl. 143–151.

Nagy, L., Szanto, A., Szatmari, I., Szeles, L., 2012. Nuclear hormone receptors enable macrophages and dendritic cells to sense their lipid environment and shape their immune response. Physiol. Rev. 92, 739–789.

Nimmerjahn, A., Kirchhoff, F., Helmchen, F., 2005. Resting microglial cells are highly dynamic surveillants of brain parenchyma in vivo. Science 308, 1314–1318.

Nishijima, H., Tomiyama, M., 2016. What mechanisms are responsible for the reuptake of levodopa-derived dopamine in parkinsonian striatum? Front. Neurosci. 10, 575.

Norden, D.M., Godbout, J.P., 2013. Review: microglia of the aged brain: primed to be activated and resistant to regulation. Neuropathol. Appl. Neurobiol. 39, 19–34.

Odegaard, J.I., Ricardo-Gonzalez, R.R., Goforth, M.H., Morel, C.R., Subramanian, V., et al., 2007. Macrophage-specific PPARgamma controls alternative activation and improves insulin resistance. Nature 447, 1116–1120.

Oh, S.M., Chang, M.Y., Song, J.J., Rhee, Y.H., Joe, E.H., et al., 2015. Combined Nurr1 and Foxa2 roles in the therapy of Parkinson's disease. EMBO Mol. Med. 7, 510–525.

Ohlin, K.E., Francardo, V., Lindgren, H.S., Sillivan, S.E., O'Sullivan, S.S., et al., 2011. Vascular endothelial growth factor is upregulated by L-dopa in the parkinsonian brain: implications for the development of dyskinesia. Brain 134, 2339–2357.

Ohlin, K.E., Sebastianutto, I., Adkins, C.E., Lundblad, C., Lockman, P.R., Cenci, M.A., 2012. Impact of L-DOPA treatment on regional cerebral blood flow and metabolism in the basal ganglia in a rat model of Parkinson's disease. Neuroimage 61, 228–239.

O'Neil, S.M., Witcher, K.G., McKim, D.B., Godbout, J.P., 2018. Forced turnover of aged microglia induces an intermediate phenotype but does not rebalance CNS environmental cues driving priming to immune challenge. Acta Neuropathol. Commun. 6, 129.

Ouchi, Y., Yoshikawa, E., Sekine, Y., Futatsubashi, M., Kanno, T., et al., 2005. Microglial activation and dopamine terminal loss in early Parkinson's disease. Ann. Neurol. 57, 168–175.

Paolicelli, R.C., Bolasco, G., Pagani, F., Maggi, L., Scianni, M., et al., 2011. Synaptic pruning by microglia is necessary for normal brain development. Science 333, 1456–1458.

Papadopoulos, V., Baraldi, M., Guilarte, T.R., Knudsen, T.B., Lacapere, J.J., et al., 2006. Translocator protein (18kDa): new nomenclature for the peripheral-type benzodiazepine receptor based on its structure and molecular function. Trends Pharmacol. Sci. 27, 402–409.

Park, J.Y., Paik, S.R., Jou, I., Park, S.M., 2008. Microglial phagocytosis is enhanced by monomeric alpha-synuclein, not aggregated alpha-synuclein: implications for Parkinson's disease. Glia 56, 1215–1223.

Parkhurst, C.N., Yang, G., Ninan, I., Savas, J.N., Yates 3rd, J.R., et al., 2013. Microglia promote learning-dependent synapse formation through brain-derived neurotrophic factor. Cell 155, 1596–1609.

Parpura, V., Heneka, M.T., Montana, V., Oliet, S.H., Schousboe, A., et al., 2012. Glial cells in (patho)physiology. J. Neurochem. 121, 4–27.

Pattarini, R., Smeyne, R.J., Morgan, J.I., 2007. Temporal mRNA profiles of inflammatory mediators in the murine 1-methyl-4-phenyl-1,2,3,6-tetrahydropyrimidine model of Parkinson's disease. Neuroscience 145, 654–668.

Pekny, M., Pekna, M., Messing, A., Steinhäuser, C., Lee, J.M., et al., 2016. Astrocytes: a central element in neurological diseases. Acta Neuropathol. 131, 323–345.

Picconi, B., Centonze, D., Hakansson, K., Bernardi, G., Greengard, P., et al., 2003. Loss of bidirectional striatal synaptic plasticity in L-DOPA-induced dyskinesia. Nat. Neurosci. 6, 501–506.

Pinto, M., Nissanka, N., Peralta, S., Brambilla, R., Diaz, F., Moraes, C.T., 2016. Pioglitazone ameliorates the phenotype of a novel Parkinson's disease mouse model by reducing neuroinflammation. Mol. Neurodegener. 11, 25.

Pisani, V., Stefani, A., Pierantozzi, M., Natoli, S., Stanzione, P., et al., 2012. Increased blood-cerebrospinal fluid transfer of albumin in advanced Parkinson's disease. J. Neuroinflammation 9, 188.

Pisanu, A., Lecca, D., Mulas, G., Wardas, J., Simbula, G., et al., 2014. Dynamic changes in pro- and anti-inflammatory cytokines in microglia after PPAR-gamma agonist neuroprotective treatment in the MPTPp mouse model of progressive Parkinson's disease. Neurobiol. Dis. 71, 280–291.

Pocock, J.M., Kettenmann, H., 2007. Neurotransmitter receptors on microglia. Trends Neurosci. 30, 527–535.

Poly, T.N., Islam, M.M.R., Yang, H.C., Li, Y.J., 2019. Non-steroidal anti-inflammatory drugs and risk of Parkinson's disease in the elderly population: a meta-analysis. Eur. J. Clin. Pharmacol. 75, 99–108.

Proschel, C., Stripay, J.L., Shih, C.H., Munger, J.C., Noble, M.D., 2014. Delayed transplantation of precursor cell-derived astrocytes provides multiple benefits in a rat model of Parkinsons. EMBO Mol. Med. 6, 504–518.

Ransohoff, R.M., Perry, V.H., 2009. Microglial physiology: unique stimuli, specialized responses. Annu. Rev. Immunol. 27, 119–145.

Rappold, P.M., Tieu, K., 2010. Astrocytes and therapeutics for Parkinson's disease. Neurotherapeutics 7, 413–423.

Ren, M., Han, M., Wei, X., Guo, Y., Shi, H., et al., 2017. FTY720 attenuates 6-OHDA-associated dopaminergic degeneration in cellular and mouse parkinsonian models. Neurochem. Res. 42, 686–696.

Ren, L., Yi, J., Yang, J., Li, P., Cheng, X., Mao, P., 2018. Nonsteroidal anti-inflammatory drugs use and risk of Parkinson disease: a dose-response meta-analysis. Medicine (Baltimore) 97, e12172.

Reynolds, A.D., Stone, D.K., Hutter, J.A., Benner, E.J., Mosley, R.L., Gendelman, H.E., 2010. Regulatory T cells attenuate Th17 cell-mediated nigrostriatal dopaminergic neurodegeneration in a model of Parkinson's disease. J. Immunol. 184, 2261–2271.

Robinson, T.E., Mocsary, Z., Camp, D.M., Whishaw, I.Q., 1994. Time course of recovery of extracellular dopamine following partial damage to the nigrostriatal dopamine system. J. Neurosci. 14, 2687–2696.

Rodriguez-Perez, A.I., Sucunza, D., Pedrosa, M.A., Garrido-Gil, P., Kulisevsky, J., et al., 2018. Angiotensin type 1 receptor antagonists protect against alpha-synuclein-induced neuroinflammation and dopaminergic neuron death. Neurotherapeutics 15, 1063–1081.

Rojanathammanee, L., Murphy, E.J., Combs, C.K., 2011. Expression of mutant alpha-synuclein modulates microglial phenotype in vitro. J. Neuroinflammation 8, 44.

Roodveldt, C., Labrador-Garrido, A., Gonzalez-Rey, E., Fernandez-Montesinos, R., Caro, M., et al., 2010. Glial innate immunity generated by non-aggregated alpha-synuclein in mouse: differences between wild-type and Parkinson's disease-linked mutants. PLoS One 5, e13481.

Rostami, J., Holmqvist, S., Lindstrom, V., Sigvardson, J., Westermark, G.T., et al., 2017. Human astrocytes transfer aggregated alpha-synuclein via tunneling nanotubes. J. Neurosci. 37, 11835–11853.

Ruscher, K., Wieloch, T., 2015. The involvement of the sigma-1 receptor in neurodegeneration and neurorestoration. J. Pharmacol. Sci. 127, 30–35.

Saijo, K., Winner, B., Carson, C.T., Collier, J.G., Boyer, L., et al., 2009. A Nurr1/CoREST pathway in microglia and astrocytes protects dopaminergic neurons from inflammation-induced death. Cell 137, 47–59.

Salter, M.W., Beggs, S., 2014. Sublime microglia: expanding roles for the guardians of the CNS. Cell 158, 15–24.

Saunders, J.A., Estes, K.A., Kosloski, L.M., Allen, H.E., Dempsey, K.M., et al., 2012. CD4+ regulatory and effector/memory T cell subsets profile motor dysfunction in Parkinson's disease. J. Neuroimmune Pharmacol. 7, 927–938.

Savtchouk, I., Volterra, A., 2018. Gliotransmission: beyond black-and-white. J. Neurosci. 38, 14–25.

Sawada, M., Imamura, K., Nagatsu, T., 2006. Role of cytokines in inflammatory process in Parkinson's disease. J. Neural Transm. Suppl. 70, 373–381.

Schafer, D.P., Lehrman, E.K., Stevens, B., 2013. The "quad-partite" synapse: microglia-synapse interactions in the developing and mature CNS. Glia 61, 24–36.

Schintu, N., Frau, L., Ibba, M., Garau, A., Carboni, E., Carta, A.R., 2009. Progressive dopaminergic degeneration in the chronic MPTPp mouse model of Parkinson's disease. Neurotox. Res. 16, 127–139.

Seth, P., Koul, N., 2008. Astrocyte, the star avatar: redefined. J. Biosci. 33, 405–421.

Sharaf, A., Krieglstein, K., Spittau, B., 2013. Distribution of microglia in the postnatal murine nigrostriatal system. Cell Tissue Res. 351, 373–382.

Sharon, R., Bar-Joseph, I., Frosch, M.P., Walsh, D.M., Hamilton, J.A., Selkoe, D.J., 2003. The formation of highly soluble oligomers of alpha-synuclein is regulated by fatty acids and enhanced in Parkinson's disease. Neuron 37, 583–595.

Simon, D.K., Simuni, T., Elm, J., Clark-Matott, J., Graebner, A.K., et al., 2015. Peripheral biomarkers of Parkinson's disease progression and pioglitazone effects. J. Parkinsons Dis. 5, 731–736.

Sipe, G.O., Lowery, R.L., Tremblay, M.E., Kelly, E.A., Lamantia, C.E., Majewska, A.K., 2016. Microglial P2Y12 is necessary for synaptic plasticity in mouse visual cortex. Nat. Commun. 7, 10905.

Sofroniew, M.V., 2009. Molecular dissection of reactive astrogliosis and glial scar formation. Trends Neurosci. 32, 638–647.

Solleiro-Villavicencio, H., Rivas-Arancibia, S., 2018. Effect of chronic oxidative stress on neuroinflammatory response mediated by CD4(+)T cells in neurodegenerative diseases. Front. Cell. Neurosci. 12, 114.

Song, J.J., Oh, S.M., Kwon, O.C., Wulansari, N., Lee, H.S., et al., 2018. Cografting astrocytes improves cell therapeutic outcomes in a Parkinson's disease model. J. Clin. Invest. 128, 463–482.

Stanic, D., Finkelstein, D.I., Bourke, D.W., Drago, J., Horne, M.K., 2003. Timecourse of striatal re-innervation following lesions of dopaminergic SNpc neurons of the rat. Eur. J. Neurosci. 18, 1175–1188.

Stefanova, N., Fellner, L., Reindl, M., Masliah, E., Poewe, W., Wenning, G.K., 2011. Toll-like receptor 4 promotes alpha-synuclein clearance and survival of nigral dopaminergic neurons. Am. J. Pathol. 179, 954–963.

Stellwagen, D., Malenka, R.C., 2006. Synaptic scaling mediated by glial TNF-alpha. Nature 440, 1054–1059.

Streit, W.J., 2002. Microglia as neuroprotective, immunocompetent cells of the CNS. Glia 40, 133–139.

Su, X., Federoff, H.J., Maguire-Zeiss, K.A., 2009. Mutant alpha-synuclein overexpression mediates early proinflammatory activity. Neurotox. Res. 16, 238–254.

Szade, A., Grochot-Przeczek, A., Florczyk, U., Jozkowicz, A., Dulak, J., 2015. Cellular and molecular mechanisms of inflammation-induced angiogenesis. IUBMB Life 67, 145–159.

Szema, A.M., Hamidi, S.A., Golightly, M.G., Rueb, T.P., Chen, J.J., 2011. VIP regulates the development & proliferation of Treg in vivo in spleen. Allergy Asthma Clin. Immunol. 7, 19.

Tang, Y., Le, W., 2016. Differential roles of M1 and M2 microglia in neurodegenerative diseases. Mol. Neurobiol. 53, 1181–1194.

Tokuda, T., Qureshi, M.M., Ardah, M.T., Varghese, S., Shehab, S.A., et al., 2010. Detection of elevated levels of alpha-synuclein oligomers in CSF from patients with Parkinson disease. Neurology 75, 1766–1772.

Tong, J., Ang, L.C., Williams, B., Furukawa, Y., Fitzmaurice, P., et al., 2015. Low levels of astroglial markers in Parkinson's disease: relationship to alpha-synuclein accumulation. Neurobiol. Dis. 82, 243–253.

Tremblay, M.E., Lowery, R.L., Majewska, A.K., 2010. Microglial interactions with synapses are modulated by visual experience. PLoS Biol. 8, e1000527.

Valera, E., Mante, M., Anderson, S., Rockenstein, E., Masliah, E., 2015. Lenalidomide reduces microglial activation and behavioral deficits in a transgenic model of Parkinson's disease. J. Neuroinflammation 12, 93.

Van der Perren, A., Macchi, F., Toelen, J., Carlon, M.S., Maris, M., et al., 2015. FK506 reduces neuroinflammation and dopaminergic neurodegeneration in an alpha-synuclein-based rat model for Parkinson's disease. Neurobiol. Aging 36, 1559–1568.

Vazquez-Claverie, M., Garrido-Gil, P., San Sebastian, W., Izal-Azcarate, A., Belzunegui, S., et al., 2009. Acute and chronic 1-methyl-4-phenyl-1,2,3,6-tetrahydropyridine administrations elicit similar microglial activation in the substantia nigra of monkeys. J. Neuropathol. Exp. Neurol. 68, 977–984.

Venezia, S., Refolo, V., Polissidis, A., Stefanis, L., Wenning, G.K., Stefanova, N., 2017. Toll-like receptor 4 stimulation with monophosphoryl lipid A ameliorates motor deficits and nigral neurodegeneration triggered by extraneuronal alpha-synucleinopathy. Mol. Neurodegener. 12, 52.

Verkhratsky, A., Zorec, R., 2019. Astroglial signalling in health and disease. Neurosci. Lett. 689, 1–4.

Villar-Pique, A., Lopes da Fonseca, T., Outeiro, T.F., 2016. Structure, function and toxicity of alpha-synuclein: the Bermuda triangle in synucleinopathies. J. Neurochem. 139 (Suppl. 1), 240–255.

Wakabayashi, K., Hayashi, S., Yoshimoto, M., Kudo, H., Takahashi, H., 2000. NACP/alpha-synuclein-positive filamentous inclusions in astrocytes and oligodendrocytes of Parkinson's disease brains. Acta Neuropathol. 99, 14–20.

Walker, D.G., Lue, L.F., Serrano, G., Adler, C.H., Caviness, J.N., et al., 2015. Altered expression patterns of inflammation-associated and trophic molecules in substantia nigra and striatum brain samples from Parkinson's disease, incidental Lewy body disease and normal control cases. Front. Neurosci. 9, 507.

Walsh, S., Finn, D.P., Dowd, E., 2011. Time-course of nigrostriatal neurodegeneration and neuroinflammation in the 6-hydroxydopamine-induced axonal and terminal lesion models of Parkinson's disease in the rat. Neuroscience 175, 251–261.

Watson, M.B., Richter, F., Lee, S.K., Gabby, L., Wu, J., et al., 2012. Regionally-specific microglial activation in young mice over-expressing human wildtype alpha-synuclein. Exp. Neurol. 237, 318–334.

Werner, C.J., Heyny-von Haussen, R., Mall, G., Wolf, S., 2008. Proteome analysis of human substantia nigra in Parkinson's disease. Proteome Sci. 6, 8.

Westin, J.E., Lindgren, H.S., Gardi, J., Nyengaard, J.R., Brundin, P., et al., 2006. Endothelial proliferation and increased blood-brain barrier permeability in the basal ganglia in a rat model of 3,4-dihydroxyphenyl-L-alanine-induced dyskinesia. J. Neurosci. 26, 9448–9461.

Williams-Gray, C.H., Wijeyekoon, R., Yarnall, A.J., Lawson, R.A., Breen, D.P., et al., 2016. Serum immune markers and disease progression in an incident Parkinson's disease cohort (ICICLE-PD). Mov. Disord. 31, 995–1003.

Winner, B., Jappelli, R., Maji, S.K., Desplats, P.A., Boyer, L., et al., 2011. In vivo demonstration that alpha-synuclein oligomers are toxic. Proc. Natl. Acad. Sci. U. S. A. 108, 4194–4199.

Wissemann, W.T., Hill-Burns, E.M., Zabetian, C.P., Factor, S.A., Patsopoulos, N., et al., 2013. Association of Parkinson disease with structural and regulatory variants in the HLA region. Am. J. Hum. Genet. 93, 984–993.

Wolf, S.A., Boddeke, H.W., Kettenmann, H., 2017. Microglia in physiology and disease. Annu. Rev. Physiol. 79, 619–643.

Yamada, K., Iwatsubo, T., 2018. Extracellular alpha-synuclein levels are regulated by neuronal activity. Mol. Neurodegener. 13, 9.

Yasuda, Y., Shinagawa, R., Yamada, M., Mori, T., Tateishi, N., Fujita, S., 2007. Long-lasting reactive changes observed in microglia in the striatal and substantia nigral of mice after 1-methyl-4-phenyl-1,2,3,6-tetrahydropyridine. Brain Res. 1138, 196–202.

Yun, S.P., Kam, T.I., Panicker, N., Kim, S., Oh, Y., et al., 2018. Block of A1 astrocyte conversion by microglia is neuroprotective in models of Parkinson's disease. Nat. Med. 24, 931–938.

Zamanian, J.L., Xu, L., Foo, L.C., Nouri, N., Zhou, L., et al., 2012. Genomic analysis of reactive astrogliosis. J. Neurosci. 32, 6391–6410.

Zhang, W., Wang, T., Pei, Z., Miller, D.S., Wu, X., et al., 2005. Aggregated alpha-synuclein activates microglia: a process leading to disease progression in Parkinson's disease. FASEB J. 19, 533–542.

Zhang, Q.S., Heng, Y., Yuan, Y.H., Chen, N.H., 2017. Pathological α-synuclein exacerbates the progression of Parkinson's disease through microglial activation. Toxicol. Lett. 265, 30–37.

Zhao, P., Yang, X., Yang, L., Li, M., Wood, K., et al., 2017. Neuroprotective effects of fingolimod in mouse models of Parkinson's disease. FASEB J. 31, 172–179.

Zigmond, M.J., 1997. Do compensatory processes underlie the preclinical phase of neurodegenerative disease? Insights from an animal model of parkinsonism. Neurobiol. Dis. 4, 247–253.

Zonta, M., Angulo, M.C., Gobbo, S., Rosengarten, B., Hossmann, K.A., et al., 2003. Neuron-to-astrocyte signaling is central to the dynamic control of brain microcirculation. Nat. Neurosci. 6, 43–50.

Zucca, F.A., Segura-Aguilar, J., Ferrari, E., Munoz, P., Paris, I., et al., 2017. Interactions of iron, dopamine and neuromelanin pathways in brain aging and Parkinson's disease. Prog. Neurobiol. 155, 96–119.

Innate and adaptive immune responses in Parkinson's disease

Aubrey M. Schonhoff[†], Gregory P. Williams[†], Zachary D. Wallen, David G. Standaert, Ashley S. Harms[*]

Center for Neurodegeneration and Experimental Therapeutics, Department of Neurology, The University of Alabama at Birmingham, Birmingham, AL, United States
[*]Corresponding author: Tel.: +205-934-6142; Fax: +205-996-6580, e-mail address: anharms@uab.edu

Abstract

Parkinson's disease (PD) has classically been defined as a movement disorder, in which motor symptoms are explained by the aggregation of alpha-synuclein (α-syn) and subsequent death of dopaminergic neurons of the substantia nigra pars compacta (SNpc). More recently, the multisystem effects of the disease have been investigated, with the immune system being implicated in a number of these processes in the brain, the blood, and the gut. In this review, we highlight the dysfunctional immune system found in both human PD and animal models of the disease, and discuss how genetic risk factors and risk modifiers are associated with pro-inflammatory immune responses. Finally, we emphasize evidence that the immune response drives the pathogenesis and progression of PD, and discuss key questions that remain to be investigated in order to identify immunomodulatory therapies in PD.

Keywords

Immune system, Inflammation, Parkinson's disease, Alpha-synuclein, Microglia, T cells, Microbiota

1 Introduction

The involvement of the immune system in Parkinson's disease has been postulated since its initial description by James Parkinson in 1817. In his "An Essay on the Shaking Palsy," Parkinson suggests that attacks on the nervous system, "considered at the time merely as rheumatic affections, might lay the foundation of this lamentable disease, which might manifest itself at some distant period,

---

[†]These authors contributed equally.

Progress in Brain Research, Volume 252, ISSN 0079-6123, https://doi.org/10.1016/bs.pbr.2019.10.006

when the circumstances in which it had originated, had, perhaps, almost escaped the memory" (Parkinson, 2002). Now, more than 200 years after Parkinson's initial publication, the evidence has grown substantially to support the idea that inflammatory processes occur long before clinically apparent symptomatology, and initiate and/or drive the progression of PD.

Before addressing the research itself, it is helpful to clearly define what is meant by "inflammation," as the meaning of this word has expanded greatly in recent years. Classically the process of inflammation can be separated into three different components: (i) Responses from the host immune system that aid in the clearance of invading pathogens such as bacteria or viruses, (ii) responses to aid in the removal of dead or dying cells and wound repair, and (iii) responses that lead to excessive activation or misguided recognition of host antigens and which lead to damage of tissues. This review and others (Prinz and Priller, 2017) use a specific definition of inflammation encompassing all three of these responses. We consider inflammation to be a response that results in the infiltration of hematopoietic cells of a peripheral origin, such as monocytes, T cells, B cells, and neutrophils, into the affected tissue, along with the generation of cytokines and chemokines. Activation of only innate resident cells, such as microglia in "gliosis" or tissue resident macrophages in the absence of peripheral infiltration, should not be considered inflammation, even if this activation leads to cytokine production. While an innate response may result in the clearance of cellular debris, a true inflammatory response will engage the adaptive immune system for response escalation and resolution. Furthermore, the process of innate immune activation without adaptive involvement lacks specificity and memory. It is the interactions between innate (microglia, macrophages, neutrophils, monocytes, etc.) and adaptive cells (T and B cells) that carry out specific and targeted programs based upon the threat, and these interactions can occur at all sites throughout the body. It is with this definition that we propose Parkinson's disease be considered an inflammatory condition, as there is evidence for both innate and adaptive involvement in the pathogenesis of disease.

Historically, studies of PD have focused on the basal ganglia, where the hallmarks of the disease, including the dramatic death of dopaminergic neurons, the presence of α-syn rich inclusions, and the associated motor circuit dysfunction are most prominent. However, there is long-standing evidence, as well as more recent findings, that highlight early and persistent dysfunction within multiple systems, both within and outside the CNS in patients with PD. These findings of multisystem dysfunction are coupled with a growing awareness of non-motor dysfunction in patients with PD. This evidence supports the notion of PD as a systemic disease that involves multiple tissues and cell types, with the key players including the central nervous system (CNS), gastrointestinal tract, autonomic nervous system, and the innate and adaptive immune systems.

In this review of immune mechanisms in PD, we highlight evidence that the innate and adaptive responses to CNS antigens are involved in disease pathogenesis. We provide a basic overview of innate and adaptive immune system mechanisms, as well as an overview of the clinical and pathological manifestations of inflammation in PD, involving both the CNS and peripheral tissues. This is followed by an overview of the immune system's role in animal models of the disease, and how these models have identified key immune related pathways in PD pathogenesis. We conclude by identifying and discussing several key questions that we believe will drive future research and therapeutic discovery.

2 PD as a systemic and heterogeneous disease

Parkinson's disease has classically been considered a movement disorder, as the motor deficits are the most obvious symptoms of disease and are used in part for diagnosing the disease, although a definite diagnosis can only occur postmortem. Cardinal symptoms of PD include tremor, rigidity, akinesia or bradykinesia, and postural instability (Obeso et al., 2017). The symptoms themselves are associated with intraneuronal α-syn rich inclusions throughout the brain and the dramatic loss of dopamine-producing neurons in the substantia nigra pars compacta, leading to a deficit of dopaminergic signaling in the caudate and putamen within the basal ganglia (Obeso et al., 2017). Single-photon emission computerized tomography (SPECT) scans can be used to visualize this loss, through labeling of dopamine transporters (DAT) on the terminals of neurons projecting from the from the SNpc to the caudate and putamen (Benamer et al., 2000; Ichise et al., 1999) although this loss may not be a direct correlate of dopaminergic neuron loss (Honkanen et al., 2019). The loss of uptake is typically asymmetric (Benamer et al., 2000, Ichise et al., 1999), which is reflected in the asymmetric onset of the motor symptoms characteristic of PD.

A description of Parkinson's disease would be remiss without a brief discussion on the role of α-syn in the disease process. α-syn is typically found at the presynaptic terminals in neurons, but in PD, α-syn rich aggregates are found within the cell soma (Lewy bodies) and within neuronal processes (Lewy neurites) (Burre, 2015; Spillantini et al., 1997; Sulzer and Edwards, 2019). The normal function of α-syn remains somewhat of a mystery, although it has been implicated in processes such as synaptic vesicle release and recycling (Burre, 2015; Sulzer and Edwards, 2019), acting as a molecular chaperone for SNARE complex formation (Burre, 2015; Chandra et al., 2005), binding of dopamine and serotonin transporters (Burre, 2015), and regulating certain forms of synaptic plasticity (Sulzer and Edwards, 2019). Classically, the α-syn inclusions have been thought to be harmful, although there is more recent evidence that soluble, oligomeric forms of α-syn could be the more neurotoxic species (Cremades et al., 2012; Fusco et al., 2017; Luk et al., 2012). Regardless, α-syn pathology has been found throughout the brains of PD patients, and has been suggested to spread in a predictable prion-like manner between interconnected brain regions (Braak et al., 2003). However, it appears that α-syn inclusions alone cannot be blamed for PD, as there are cases of healthy people with high Lewy body load but without symptoms (Bengoa-Vergniory et al., 2017; Frigerio et al., 2011; Parkkinen et al., 2005) and Lewy body load does not necessarily correlate with the severity of symptoms (Bengoa-Vergniory et al., 2017), suggesting that there may be other mechanisms at work that explain the pattern of motor and non-motor symptoms in patients.

James Parkinson himself described many non-motor symptoms of PD including sleep disturbances, autonomic symptoms, and constipation (Parkinson, 2002). 200 years later, we now recognize that PD is a heterogeneous disease with widespread dysfunction and the involvement of multiple systems throughout the body—a disease in which each case has unique aspects. Indeed, it may be best to view PD as a syndrome with some core features, rather than a single disease (De Pablo-Fernandez et al., 2019). Anosmia (loss of smell), constipation, and rapid eye movement (REM) sleep disorder often begin prior to the appearance of motor symptoms, and when present together are highly predictive of later development of PD (Doty, 2012; Postuma et al., 2015). Other non-motor features appear later in the disease: approximately 20–40% of patients experience depression; and 40–56% experience anxiety (Barone et al., 2009). An even larger portion (30–80%) of PD patients will develop dementia, with half of non-demented patients still displaying mild cognitive impairment and cognitive decline over time (Aarsland et al., 2017). A neural correlate to this symptom could be the finding that PD patients have decreased cortical thickness compared to healthy controls (Deng et al., 2016; Sampedro et al., 2019; Uribe et al., 2018). Outside of the CNS, an exciting area of research is exploring the link between CNS and dysfunctions in the gastrointestinal tract, most notably preclinical constipation (Fasano et al., 2015) and associated intestinal inflammation in patients with PD (Devos et al., 2013). These motor and non-motor symptoms of PD and the associated pathology are indicative of a disease of global dysfunction.

3 Innate immune system

In order to understand inflammation as a process, it is important to distinguish between the two branches of the mammalian immune system. Most of our knowledge of the innate and adaptive immune systems comes from studies with mice, other mammals, or *in vitro* systems. The innate immune system is an ancient and highly conserved system that operates by nonspecific mechanisms and functions as a first line of defense that can later help to activate the adaptive system. Its purpose is to quickly resolve threats to the host, including clearance of invading pathogens such as bacteria or viruses, removal of dead/dying cells, and wound repair. However, under pathological conditions, the immune system can respond with excessive activation or misguided recognition of host antigens, leading to the damage of tissue.

The innate immune system is comprised of tissue resident macrophages that typically originate from the fetal yolk sac or fetal liver and self-renew (Goldmann et al., 2016; Hoeffel and Ginhoux, 2015), dendritic cells, neutrophils, circulating monocytes, granulocytes, and even some non-immune cells that adopt immunological functions as needed. The basic processes of innate immunity that will be discussed in this section are outlined in Fig. 1. Macrophages and dendritic cells, together referred to as antigen presenting cells (APCs), are thought to be the body's main

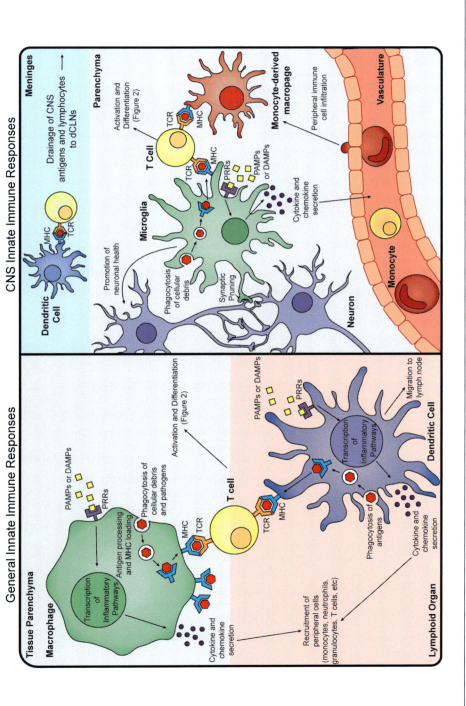

FIG. 1

Overview of Innate Immune Responses. General innate immune responses typically involve antigen presenting cell (APC) recognition of pathogen associated molecular patterns (PAMPs, yellow squares) or danger associated molecular patterns (DAMPs) by a pattern recognition receptor (PRR, purple receptor). Upon recognition, the APC will undergo transcription for inflammatory pathways, which will lead to secretion of cytokines and chemokines (purple circles) to recruit more immune cells to the site, and upregulation of surface molecules involved in antigen presentation (MHC and costimulatory molecules, blue receptor). APCs can also phagocytose cellular debris and pathogens, process them, and load the antigenic peptides onto an MHC to be presented to T cells. Macrophages (green) typically carry out these functions within a tissue, whereas dendritic cells (blue) are usually found at tissue boundaries, and may migrate to a lymph node upon antigen uptake. In the CNS, the predominant APCs are microglia (green), although dendritic cells (blue) are found in the leptomeninges. Antigen uptake, pattern recognition, and cytokine secretion are thought to occur similar to general innate immune responses. Microglia also have tissue-specific homeostatic functions, such as synapse pruning and support of neuronal (blue) health. Infiltrating cells such as monocytes (red) and T cells (yellow), however, can be neuroprotective or neurotoxic, depending on the inflammatory stimulus. While there is some evidence that monocyte-derived macrophages (red) can play a role in antigen presentation during an inflammatory response, there is little known of their longevity in the parenchyma after resolution of the immune response.

sensors of danger and initiator of an immune response; they constantly sample their environment, phagocytosing any debris, presenting findings on their surface via major histocompatibility complex (MHC) molecules, and sending out signals or interacting with other immune cells when they encounter danger. The MHC has two classes, MHC class I (MHCI) and MHC class II (MHCII). MHCI is expressed by all nucleated cells whereas MHCII is mainly expressed by the APCs described above. The ability of the innate immune system to present antigen is integral to mounting an effective immune response and communicating with other immune cells. It is important to note that macrophages are generally believed to be less competent at presenting antigens to T cells than dendritic cells, although both are capable of interacting with and activating T cells (Mildner and Jung, 2014). It is generally thought that dendritic cells or macrophages begin any immune response, and lead to its amplification through the recruitment of other cells such as monocytes, granulocytes, and neutrophils.

Within the brain parenchyma itself, there are few, if any, dendritic cells. Most are found within the leptomeninges, though in very small populations (Mrdjen et al., 2018). Despite their relatively small numbers, their contribution to PD inflammation cannot be ruled out, as meningeal DCs are implicated in antigen presentation and subsequent neuroinflammation in models of multiple sclerosis (MS) (Mundt et al., 2019). However, the role of meningeal immunity and trafficking of immune cells in this space is poorly understood and remains understudied in the context of PD. The largest population of immune cells in the healthy CNS is microglia, the main tissue resident macrophage of the brain (Mrdjen et al., 2018). These cells arise early in development from the fetal yolk sac, and self-renew throughout life without major contribution from peripheral hematopoietic sources (Ginhoux et al., 2010; Goldmann et al., 2016; Nayak et al., 2014). They are thought of as first responders to any injury, modulators of homeostasis, and mediators of neuroinflammation. In homeostatic conditions, microglia express low to undetectable levels of MHCII, and have a generally anti-inflammatory phenotype (Colonna and Butovsky, 2017; Mrdjen et al., 2018; Nayak et al., 2014). However, during inflammation, they can upregulate MHCII and produce cytokines and chemokines. Additionally, peripheral myeloid cells such as monocytes can be recruited to the parenchyma during CNS inflammation, and further contribute to the inflammatory processes (Prinz and Priller, 2017). Whether or not microglia are pathogenic or protective, promoting either damaging inflammation or resolution and tissue repair in disease states is hotly debated, and seems to depend on the exact immune stimulus (Colonna and Butovsky, 2017; Nayak et al., 2014). The differential role between microglia and infiltrating peripheral myeloid cells is also a current area of research; with some hypothesizing that infiltrating cells could prove to be the more pathogenic population.

APCs mount immune responses following the activation of pattern recognition receptors (PRRs) by pathogen associated molecular patterns (PAMPs) or damage associated molecular patterns (DAMPs), activation of the complement cascade, or through antigen presentation to T cells (Fig. 1). Examples of pattern molecules that activate PRRs in these cells are: lipopolysaccharide (LPS) activating toll-like receptor 4 (TLR4) (Qureshi et al., 1999) and UV damaged self-RNA activating TLR3 (Bernard et al., 2012). Recognition of these PAMPs or DAMPs leads to the induction of signaling pathways to produce antimicrobial genes and inflammatory cytokines, depending on the specific immunological challenge (Lamkanfi and Dixit, 2014). Furthermore, these processes can amplify the ability of innate cells to actively present antigen, the process by which peptides (from host cells or from microbes) are loaded and presented on the MHC and interrogated by T cells. Dendritic cells normally take this antigen and migrate to the lymph nodes, where they will then direct T cell activation. Macrophages, on the other hand, typically secrete cytokines and chemokines from their origin tissue to recruit other cells such as monocytes and T cells, to the site of damage (Fig. 1). This initiates an immune response that will be further amplified and then resolved by incoming immune cells (Rankin and Artis, 2018). While innate immune responses are typically nonspecific, they are rapid and therefore key in determining the subsequent immunological response program that is initiated (Rivera et al., 2016). In many cases, the innate immune system alone can ward off pathogens and retain a rudimentary memory of the invader (Netea et al., 2016). However, without the other half of the system, true memory and robust inflammatory reactions would not occur.

4 Adaptive immune system

In contrast to the innate immune system, the adaptive immune system is highly specific and able to remember and effectively mount responses against previously encountered immunological threats. The specificity of the adaptive immune system is achieved due to the ability of T and B cells, collectively called lymphocytes, to rearrange their genomes and create unique antigen specific receptors: T cell receptors (TCRs) and B cell receptors (BCRs) (Bassing et al., 2002). The number of possible unique TCRs or BCRs is estimated to be in the 10^{13} range (Calis and Rosenberg, 2014; Laydon et al., 2015) and allows recognition of a wide range of bacterial, viral, or fungal antigens in a highly specific manner. Accompanying this vast antigen detection capacity is the ability of a small subset of pathogen-responsive T and B cells to differentiate into a long-lived population that can rapidly respond to re-exposures to the initial pathogen, a process collectively referred to as "memory" (Vitetta et al., 1991). While both T and B cells possess these unique sensing and recall abilities, their downstream immune effector functions are markedly different. The basic processes of the adaptive immunity that will be discussed in this section are outlined in Fig. 2.

In a process similar to that in innate immune cells, B cells have the ability to sample soluble antigen directly via their BCR and can differentiate, proliferate, and secrete antibodies specific to that particular antigen (Fig. 2). B cells can also obtain help from T cells to produce high-affinity antibodies (also called immunoglobulins, Igs) that rapidly recognize and clear immunological threats (Chaplin, 2010). However, in the setting of autoimmunity (harmful immune reactions to the body's own

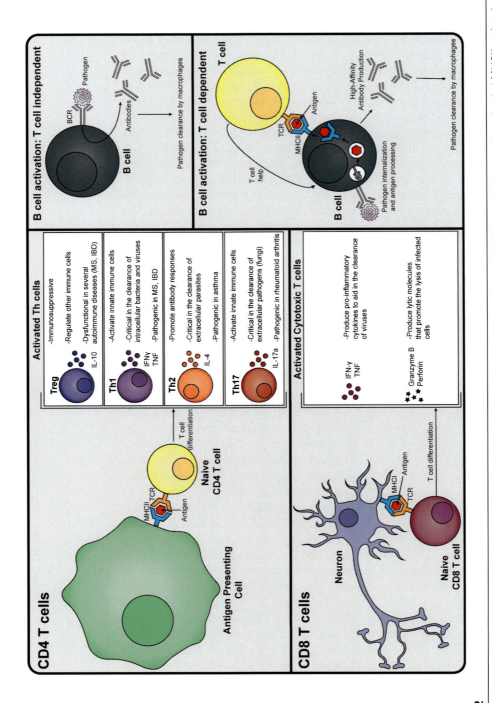

FIG. 2

Overview of Adaptive Immune Responses. CD4 T cells (yellow) interact with antigen presenting cells (APCs, green) via an antigen-loaded MHCII molecule, and costimulatory molecules (not shown). Upon antigen recognition by their T cell receptor (TCR), a CD4 T cell can differentiate into a Treg, Th1, Th2, or Th17 cell type, which have different roles in inflammation and produce signature cytokines. A CD8 T cell (maroon) interacts with antigen-loaded onto an MHCI molecule, which can be displayed on the surface of any cell, including neurons (blue). Upon antigen recognition by the TCR, a CD8 T cell will differentiate and produce inflammatory cytokines and lytic molecules. B cells (dark green) can act with or without the help of T cells. The B cell receptor (BCR) can recognize extracellular pathogens (purple), leading to the production and secretion of antibodies targeted toward that pathogen. A B cell can also recognize a pathogen via its BCR, process it, and load an antigenic peptide onto an MHCII molecule, which can then interact with the TCR of a CD4 T cell (yellow). This results in the production of high affinity antibodies targeted toward the pathogen that will promote that pathogen's clearance.

tissues), B cells produce autoantibodies to self-peptides that lead to tissue dysfunction and destruction by the body's own immune system, such as in the disease systemic lupus erythematosus.

T cells, on the other hand, detect antigens that have been loaded onto the MHC of other cells via their TCR. The MHC has two separate classes that dictate the nature of the antigens as well as the subsequent T cell response. MHC class I is expressed by all nucleated cells in the mammalian body and is responsible for antigen sampling within the intracellular space. CD8 T cells, often referred to as cytotoxic T cells (Tc), are one of the two major subsets of T cells and function mainly via their TCR in recognizing foreign antigens (e.g., viruses) on MHCI. When activated, they proliferate and release cytokines and lytic molecules (e.g., granzyme and perforin) that promote the lysis of the infected cell (Lieberman, 2003) (Fig. 2). In contrast, CD4 T cells recognize antigen-loaded onto MHC class II, which is expressed by antigen presenting cells like macrophages, monocytes, dendritic cells, and B cells. Upon TCR recognition of an MHC-bound antigen, CD4 T cells produce effector cytokines that "help" mediate pathogen clearance by boosting the function of the innate immune system and antibody-producing B cells (Fig. 2).

The cytokines produced by helper CD4 T cells (Th) are pathogen and tissue dependent, and have been generally categorized as having Th1, Th2, Th17, or T regulatory (Treg) type responses (Fig. 2) (Leung et al., 2010). Generally, Th1 immune responses are directed toward intracellular bacteria or viruses and they help clear these immunological threats in part by secreting the cytokine interferon gamma (IFN-γ) which serves to amplify the innate immune system's antimicrobial and antiviral capabilities (Leung et al., 2010; Zhu and Paul, 2008) (Fig. 2). However, there are instances where autoimmune Th1 responses occur, as is the case in patients with MS (McFarland and Martin, 2007) or inflammatory bowel disease (IBD) (Neurath, 2014), where they can promote destructive tissue damage through excessive cytokine release and innate immune system activation. Th2 immune responses are typically directed toward extracellular parasites and promote clearance of these pathogens through the secretion of the cytokine IL-4, which helps to support the humoral B cell response (Leung et al., 2010; Zhu and Paul, 2008) (Fig. 2). The overactive allergic responses observed in patients with asthma are an example of aberrant Th2 responses (Lambrecht and Hammad, 2015). Th17 responses are critical in the response to extracellular pathogens such as fungi, in part through the secretion of IL-17a, a cytokine that enhances the effectiveness of the innate immune system— particularly neutrophils (Leung et al., 2010; Zhu and Paul, 2008) (Fig. 2). However, in some autoinflammatory contexts, Th17 responses can become aberrant and promote excessive inflammation and tissue damage, as is the case with rheumatoid arthritis (Leipe et al., 2010). Lastly, Tregs play a regulatory role that serves to facilitate the process of resolving inflammation during pro-inflammatory events as well as preventing misguided host-tissue reactions in part through the innate and adaptive modulating cytokine IL-10 (Vignali et al., 2008). However, breakdowns in the function of Tregs can lead to the loss of regulation of pro-inflammatory immune reactions and can lead to damage of tissues. In fact, nearly all autoimmune diseases have evidence of dysfunctional Tregs that contribute to the break of tolerance and inability to resolve inflammatory events (Dominguez-Villar and Hafler, 2018; Leung et al., 2010) (Fig. 2).

Typically, the innate and adaptive arms of the immune system function in harmony, mounting and resolving effective and protective immune responses. However, as in the examples of autoimmunity cited above, these normal functions can turn aberrant and pathological, and become damaging toward a self-antigen, or something endogenously produced by the body. Typically, self-reactive T cells are deleted during development, but some escape and normally remain inert without the presence of additional signals (Bouneaud et al., 2000; Nemazee, 2017; Theofilopoulos et al., 2017). However, upon failure of central and peripheral tolerance of reactive T cells, an aberrant immune response may be mounted as T cells recognize self-antigens that are presented on MHC molecules of APCs. This response often fails to resolve itself, especially since IL-10 producing Tregs are impaired in either number or function in many autoimmune conditions (Dominguez-Villar and Hafler, 2018; Miyara et al., 2011). This could also be affected by the altered microbiome found in autoimmune diseases, as the gut microbiome strongly influences the development of Tregs (Honda and Littman, 2016; Theofilopoulos et al., 2017). Much of what we know of aberrant, harmful inflammatory processes comes from the study of autoimmune disease, and will likely inform future research in PD. Specifically, the role of APCs and T cells in the pathogenesis of PD should be investigated, as well as the target antigen of the inflammatory response observed.

5 CNS inflammation in human PD

5.1 CNS parenchyma

One of the first findings that connected the immune system to the pathobiology of PD was the observation of an increased number of reactive microglia, originally defined by morphology (Foix and Nicolesco, 1925) and later human leukocyte antigen-DR isotype (HLA-DR, a component of the human MHCII) expression, in the postmortem substantia nigra of PD patients (Mcgeer et al., 1988b). This microgliosis and other immune-PD phenotypes are summarized in Fig. 3. These activated microglia have been observed in direct proximity to Lewy pathology and free melanin in the brain (Mcgeer et al., 1988b). However, that same study acknowledged that the assumption of a microglial identity for these cells was based off their shared reactive morphology to the glioma associated microglia originally described by Penfield (Penfield, 1925). The exact identity of these reactive "microglia" still remains to be determined, as it is possible that some may be peripherally derived macrophages of a monocyte lineage rather than true microglia (Hoeffel and Ginhoux, 2015) (Fig. 1). Regardless of their exact origin, multiple groups have provided further evidence of a pro-inflammatory phenotype (HLA-DR[+], CD68[+], ICAM-1[+]) or innate activation (increased TLR2 expression) through histological studies (Croisier et al., 2005; Dzamko et al., 2017;

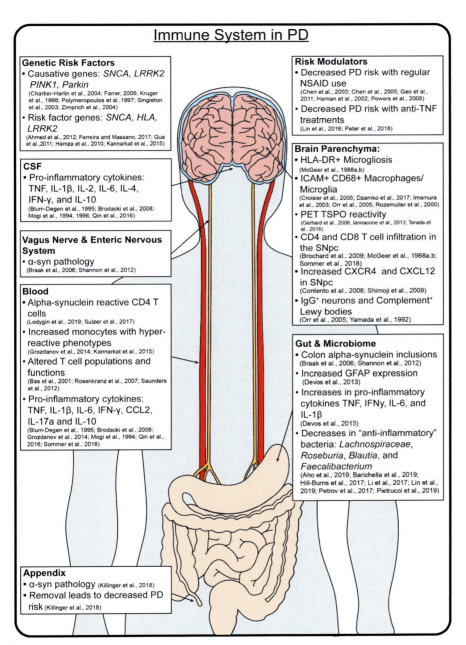

Immune System in PD

Genetic Risk Factors
- Causative genes: *SNCA, LRRK2 PINK1, Parkin*
  (Chartier-Harlin et al., 2004; Farrer, 2006; Kruger et al., 1998; Polymeropoulos et al.,1997; Singleton et al., 2003; Zimprich et al., 2004)
- Risk factor genes: *SNCA, HLA, LRRK2*
  (Ahmed et al., 2012; Ferreira and Massano, 2017; Gua et al.,2011; Hamza et al., 2010; Kannarkat et al., 2015)

CSF
- Pro-inflammatory cytokines: TNF, IL-1β, IL-2, IL-6, IL-4, IFN-γ, and IL-10
  (Blum-Degen et al., 1995; Brodacki et al., 2008; Mogi et al., 1994, 1996; Qin et al., 2016)

Vagus Nerve & Enteric Nervous System
- α-syn pathology
  (Braak et al., 2006; Shannon et al., 2012)

Blood
- Alpha-synuclein reactive CD4 T cells
  (Lodygin et al., 2019; Sulzer et al., 2017)
- Increased monocytes with hyper-reactive phenotypes
  (Grozdanov et al., 2014; Kannarkat et al., 2015)
- Altered T cell populations and functions
  (Bas et al., 2001; Rosenkranz et al., 2007; Saunders et al., 2012)
- Pro-inflammatory cytokines: TNF, IL-1β, IL-6, IFN-γ, CCL2, IL-17a and IL-10
  (Blum-Degen et al., 1995; Brodacki et al., 2008; Grozdanov et al., 2014; Mogi et al., 1994; Qin et al., 2016; Sommer et al., 2018)

Risk Modulators
- Decreased PD risk with regular NSAID use
  (Chen et al., 2003; Chen et al., 2005; Gao et al., 2011; Hernan et al., 2002; Powers et al., 2008)
- Decreased PD risk with anti-TNF treatments
  (Lin et al., 2016; Peter et al., 2018)

Brain Parenchyma:
- HLA-DR+ Microgliosis
  (McGeer et al., 1988a,b)
- ICAM+ CD68+ Macrophages/Microglia
  (Croisier et al., 2005; Dzamko et al., 2017; Imamura et al., 2003; Orr et al., 2005; Rozemuller et al., 2000)
- PET TSPO reactivity
  (Gerhard et al., 2006; Iannacone et al., 2013; Terada et al., 2016)
- CD4 and CD8 T cell infiltration in the SNpc
  (Brochard et al., 2009; McGeer et al., 1988a,b; Sommer et al., 2018)
- Increased CXCR4 and CXCL12 in SNpc
  (Contento et al., 2008; Shimoji et al., 2009)
- IgG⁺ neurons and Complement⁺ Lewy bodies
  (Orr et al., 2005; Yamada et al., 1992)

Gut & Microbiome
- Colon alpha-synuclein inclusions
  (Braak et al., 2006; Shannon et al., 2012)
- Increased GFAP expression
  (Devos et al., 2013)
- Increases in pro-inflammatory cytokines TNF, IFNγ, IL-6, and IL-1β
  (Devos et al., 2013)
- Decreases in "anti-inflammatory" bacteria: *Lachnospiraceae, Roseburia, Blautia,* and *Faecalibacterium*
  (Aho et al., 2019; Barichella et al., 2019; Hill-Burns et al., 2017; Li et al., 2017; Lin et al., 2019; Petrov et al., 2017; Pietrucci et al., 2019)

Appendix
- α-syn pathology (Killinger et al., 2018)
- Removal leads to decreased PD risk (Killinger et al., 2018)

FIG. 3

The Immune System in Human PD. PD is a disease of global dysfunction, with inflammation found in multiple tissues. Risk is conferred by specific genes, and modulated by certain anti-inflammatory treatments. Notable inflammatory markers are found throughout the brain, CSF, blood, gut, enteric nervous system, and microbiome.

Imamura et al., 2003; Orr et al., 2005; Rozemuller et al., 2000), as well as through the use of positron emission tomography (PET) imaging studies (Gerhard et al., 2006; Iannaccone et al., 2013; Terada et al., 2016). In the PET studies, ligands are specific to the translocator protein, TSPO (previously referred to as the "peripheral benzodiazepine receptor"), a protein that is expressed on the mitochondrial membrane of activated myeloid cells and astrocytes (Cosenza-Nashat et al., 2009; Lavisse et al., 2012). Chronic, increased signal from the TSPO ligands were found in PD patients, suggesting a chronic neuroinflammatory condition of patients that had only previously been observed in postmortem tissue (Fig. 3). This use of noninvasive imaging to detect and track brain inflammation in PD patients could be very useful in determining the effectiveness of immunotherapies for the disease, however, early studies using this method to test anti-inflammatory treatments on reducing TSPO binding (i.e., microglia inflammation) in PD patients have produced mixed results (Bartels et al., 2010; Jucaite et al., 2015). In addition to the evidence of cellular morphology indicative of inflammation in the brains of PD patients, there is also evidence of increased pro-inflammatory cytokine and chemokine molecules in the brain parenchyma. These findings include increased levels of both the pro-inflammatory cytokine tumor necrosis factor (TNF, previously TNF-α) and the T cell associated chemokine CXCL12 in the caudate, putamen, and substantia nigra of PD postmortem brain, respectively (Mogi et al., 1994; Shimoji et al., 2009).

While earlier findings suggested the presence of Tc (cytotoxic) CD8 T cells in the postmortem substantia nigra of PD patients (Mcgeer et al., 1988a), Brochard's (Brochard et al., 2009) discovery of increased CD4$^+$ and CD8$^+$ T cells surrounding neuromelanin positive neurons in the PD postmortem brain provided more definitive evidence that the adaptive immune system was involved in disease and has been since replicated (Sommer et al., 2018). This finding is further supported by the observation that levels of CXCR4, a chemokine receptor expressed by T cells (Contento et al., 2008), and its reciprocal ligand CXCL12 are both increased in postmortem brains of PD patients compared to healthy controls (Shimoji et al., 2009) (Fig. 3). While B cells have never been reported to be within patient brain samples directly, there are increases in the number of IgG positive neurons (Orr et al., 2005) and complement positive Lewy bodies (Yamada et al., 1992) in the substantia nigra of patients with PD. These data would suggest that B cells may in fact contribute to the pathobiology of Parkinson's disease, though likely from a primed secondary lymphoid organ. Taken together, there is a substantial amount of data to suggest that cells of both the innate (microglia or macrophages) and the adaptive immune system (T cells) are in close proximity to each other and neuromelanin containing neurons in the PD brain.

5.2 Cerebrospinal fluid

Cytokine and chemokine molecules are crucial signal transducers in the recruitment, stimulation, and regulation of multiple cell types in an inflammatory response (Fig. 2). In PD, there is substantial evidence that signaling and effector molecules are secreted in both the brain and cerebrospinal fluid (CSF). One of the first results to substantiate this idea was the observation that protein levels of the cytokine TNF are increased in both the caudate and putamen as well as the CSF of PD patients compared to controls (Mogi et al., 1994). Other early observations in the CSF of PD patients showed increased levels of IL-1β, IL-2, IL-6, and IL-4 (Blum-Degen et al., 1995; Mogi et al., 1996; Qin et al., 2016). The combination of IL-1β, IL-2, and IL-6 suggests the presence of a pro-inflammatory process as all three cytokines have been shown to have detrimental effects on neurons and other vulnerable cell types in other inflammatory diseases (Filiano et al., 2017; Neurath, 2014). IL-4, however, has been linked with neuroprotective and neuroregenerative responses in the CNS (Filiano et al., 2017; Walsh et al., 2015) along with non-CNS allergic or autoimmune responses that have heavy humoral influence, such as asthma (Lambrecht and Hammad, 2015). Another more recent study has recapitulated these above findings but also found increased levels of TNF, IFN-γ, and IL-10 in the CSF of PD patients (Brodacki et al., 2008). The presence of TNF and IFN-γ provides further evidence of a pro-inflammatory process, as both are potent immune activators (O'Shea et al., 2002) and detrimental to neuron health (Cebrian et al., 2014; Filiano et al., 2017). Interestingly, IL-10 is typically associated with the regulatory responses (Fig. 2) produced by Tregs to modulate the innate immune system. It is possible that a compensatory and regulatory process occurs alongside the pro-inflammatory response observed in PD.

6 Peripheral inflammation in PD

In addition to the evidence of an activated innate and adaptive immune response in the CNS of PD patients, there is also ample data supporting the idea that a similar inflammatory response occurs throughout the body. This peripheral PD immune phenotype is consistent with the overarching concept that Parkinson's disease is a systemic, multisystem disease affecting the whole body, and is summarized in Fig. 3.

6.1 Blood

One possible mechanism that can connect the systemic inflammatory phenotypes observed in PD is the bloodstream, the major pathway for immune cell trafficking throughout the body. Furthermore, due to the challenge of detecting inflammation in the brains of living PD patients, the study of blood in individuals with PD has produced much of our understanding of immune responses in the disease. Higher levels of inflammatory cytokines and chemokines are found in the blood and CSF, including: TNF, IL-1β, IL-2, IL-6, IFN-γ, and CCL2 (Blum-Degen et al., 1995; Grozdanov et al., 2014; Mogi et al., 1994, 1996; Qin et al., 2016). IL-17a and IL-10 are also elevated in the blood of PD patients (Brodacki et al., 2008; Sommer et al., 2018). One study correlated higher serum levels of TNF at the start of the study with faster decline of motor function over the following 3 years, and higher IL-1β and IL-2 with faster cognitive decline (Williams-Gray et al., 2016). However, these cytokine measurements are snapshots, taken at one time point, and it is therefore difficult to determine if these inflammatory markers in the bloodstream remain stable over time. Regardless, these findings mirror both the pro-inflammatory and regulatory phenotypes that have been measured in patient brain, indicating that the inflammatory process is potentially one of global coordination and involvement.

Furthermore, there are changes in both the number and activation profiles of immune cells in PD blood. In general, classical monocytes (CD14$^+$CD16$^-$) are enriched in PD patients compared to controls and are hyper-reactive (Grozdanov et al., 2014). Specifically, monocytes derived from patients produce more IL-6 to an LPS stimulus compared to those from healthy controls, and this cytokine production is positively correlated with the Hoehn and Yahr stage of disease severity. Additionally, both healthy controls and PD patients that carry a single nucleotide polymorphism (SNP) in HLA-DR that confers a higher PD risk have higher baseline MHCII expression on blood B cells and monocytes (Kannarkat et al., 2015), indicating that a SNP in HLA-DR could predispose individuals to a more inflammatory phenotype. This could indicate a lower threshold for activation, or a sensitized response to inflammatory stimuli that is present in PD immune cells and predisposes patients to stronger and more damaging immune reactions. PD patients also have lower lymphocyte numbers in their blood compared to controls, an effect which is mostly driven by a decrease in subsets of CD4$^+$ T cells, specifically naïve and memory subsets, whereas numbers of CD8 cytotoxic T cells are unchanged. Although naïve and memory subsets decrease, there is an overall increase in the number of activated, antigen-experienced CD4$^+$ T cells (CD4$^+$ CD25$^+$) (Bas et al., 2001) (Fig. 3). These findings indicate that there is an overall shift to an activated, antigen-experienced T cell phenotype in the blood. Additionally, one study found an increase in T regulatory cells in PD patients, while another reported decreased ability of PD Tregs to suppress the activity of effector T cells *in vitro* (Rosenkranz et al., 2007; Saunders et al., 2012). These data could be indicative of an attempt to regulate the inflammatory response, but a decreased ability to do so effectively. The idea of T cell driven neuroinflammation is further supported by evidence from (Sulzer et al., 2017), who exposed patient PBMCs to various α-syn peptides, and found cytokine responses from CD4 and CD8 T cells. The predominant responses were IL-5 or IFN-γ, indicating CD4 Th2-MHCII and CD8-MHCI responses to α-syn, with the highest responses to two specific α-syn peptide sequences. One peptide included cleavage sites, and the other included the Ser129 region that is often phosphorylated (pSer129) in abnormal α-syn species. These data have been recapitulated in part by Lodygin et al. (2019), who found that blood circulating α-*syn*-reactive CD4 T cells were expanded in PD patients. Further studies have implicated a role for Th17 cells in neuron death through IL-17/IL-17R signaling using PD T cells and human iPSC derived midbrain neurons (Sommer et al., 2018). These findings indicate the presence of systemic immune activation, possibly directed toward abnormal α-syn, in human PD.

6.2 Gastrointestinal tract, enteric nervous system, and the microbiome

It has been observed that constipation in PD patients begins long before the onset of motor symptoms, and that fewer bowel movements per day are associated with a higher risk for PD (Abbott et al., 2001). While α-syn pathology is typically discussed in relation to the CNS, inclusions are also found in the colon, in neurons of the enteric nervous system, and within the vagus nerve itself (Braak et al., 2006; Shannon et al., 2012). Additionally, a recent study reported the appendix to be a rich source of α-syn pathology and found an association of reduced PD risk in individuals whose appendix was removed early in life (Killinger et al., 2018). These findings of α-syn GI system are accompanied by increases in markers of inflammation, such as increased glial fibrillary acidic protein (GFAP) and increases in a number of pro-inflammatory cytokines including TNF, IFN-γ, IL-6, and IL-1β (Devos et al., 2013). Interestingly, these cytokines in the gut are higher earlier in disease and decrease with time, perhaps indicating gut inflammation could be an early event in the pathogenesis of Parkinson's disease. It is difficult to say whether gut inflammation directly contributes to disease pathogenesis, especially since a bona fide innate and adaptive immune response has not yet been reported in PD gut, and additionally it is not known if α-syn inclusions or gut inflammation occurs first. Nevertheless, this hypothesis is supported by findings that a diagnosis of inflammatory bowel disease, in particular Crohn's disease, increases the risk of developing PD in certain populations (Lin et al., 2016).

Given the observations of both gastrointestinal dysfunction and inflammation in individuals with PD, it is logical to investigate changes in the gut microbiome of these individuals. Not only is the microbiome important in overall gut function, it is also now better appreciated for its role in shaping as well as being shaped by the host's immune system (Hooper et al., 2012). Therefore, it is not surprising that multiple studies have observed alterations in the microbiomes of diverse cohorts of people with PD across the world (Aho et al., 2019; Barichella et al., 2019; Hill-Burns et al., 2017; Li et al., 2017; Lin et al., 2019; Petrov et al., 2017; Pietrucci et al., 2019; Scheperjans et al., 2015). More specifically, individual bacterial taxa have been implicated in the disease state from the aforementioned studies. Certain bacterial taxa associated with an "anti-inflammatory" environment have consistently found to be reduced in PD stool. These include the bacterial family *Lachnospiraceae*, and some of its genera *Roseburia*, *Blautia*, and *Faecalibacterium*. *Roseburia* spp. have been shown experimentally *in vitro* and *in vivo* to promote anti-inflammatory processes through a variety of potential mechanisms which include enhancement of gut lining health via increased expression of tight junction proteins (Tan et al., 2019) and modulation of immune responses via downregulation of pro-inflammatory cytokines (e.g., IL-17) (Zhu et al., 2018) and induction of anti-inflammatory cytokines (e.g., IL-10, TGFβ) (Patterson et al., 2017; Shen et al., 2018). Additionally, a species of *Faecalibacterium* and its cellular byproducts has been shown experimentally to reduce inflammation through multiple potential mechanisms including regulation of Th17 and Treg cell differentiation (Zhou et al., 2018), inhibition of pro-inflammatory pathways and cytokine production (Martin et al., 2014; Sokol et al., 2008), induction of anti-inflammatory molecules (Breyner et al., 2017; Sokol et al., 2008), and promotion of gut barrier health (Carlsson et al., 2013; Martin et al., 2015). With decreased abundance of these species, it is possible that the gut becomes a site predisposed to inflammation.

Overall, the composition of the gut microbiota is clearly altered in PD, as this finding is replicated across numerous studies from multiple geographical populations. How this dysbiosis of the gut microbiota in PD relates to inflammation and immunity in PD is still under investigation. The picture should become clearer as additional research is conducted to investigate how changes in the gut microbiota, at the global and individual microorganisms level, influence, or are affected by, PD pathogenesis.

7 The immuno-genetics of PD

Although a majority of Parkinson's disease cases appear to be idiopathic, there are multiple gene mutations or SNPs (especially related to the α-syn locus) that can promote familial forms or significantly increase one's risk to develop Parkinson's disease. In this section, we will overview some of those PD genes and their links to the inflammatory immune response observed in PD.

7.1 SNCA

Since the initial discovery of genetic variations in *SNCA*, the gene encoding α-syn, as the cause of certain familial forms of PD (Polymeropoulos et al., 1997), years of research have only strengthened the findings that the hallmark PD protein and the genetic control of its expression are a major determining factor for the development of familial or idiopathic Parkinson's disease (Chartier-Harlin et al., 2004; Edwards et al., 2010; Kruger et al., 1998; Singleton et al., 2003). These studies and others support the idea that high levels of α-syn can lead to the neurotoxic phenotype observed in human PD. The exact mechanism of how this increased expression leads to neurodegeneration is still unclear, but defects in autophagy, vesicle trafficking, and the generation of toxic, oligomerized species due to α-syn accumulation have been proposed (Lashuel et al., 2013). Another possible mechanism, and one that involves the immune system, is that pathogenic forms of α-syn due to mutation, overaccumulation, or oligomerization can directly activate the immune system. This idea is supported by the finding that pathogenic forms of α-syn can directly induce toll-like receptor 4 (TLR4) mediated microglial cytokine production, ROS production, and phagocytic activity (Fellner et al., 2013). α-syn may also be capable of activating the adaptive immune system, as (Sulzer et al., 2017) identified two antigenic regions of α-syn capable of eliciting inflammatory IFN-γ and IL-5 cytokine responses from PD patient derived CD4 and CD8 T cells. The antigenic Y39 region identified is near several of the PD-associated mutations (A30P, E46K, H50Q, G51D, A53E/T) (Hernandez et al., 2016), while the second antigenic region contains the Ser129 amino acid and requires it's phosphorylation to activate T cells—further supporting the link of mutated, pathogenic forms of α-syn promoting the inflammatory phenotype observed in PD.

7.2 HLA

HLA is a large, highly polymorphic gene set located on human chromosome 6 and is subdivided into class I and class II regions (MHCI and MHCII, respectively) (Mosaad, 2015). The control of these key immune system genes is essential to the selection of T cells, antigen sampling, and the induction of an immune response. In the context of disease, variations of certain alleles in the *HLA* region have been associated with the development of a wide range of diseases from rheumatoid arthritis (Stastny, 1978) to narcolepsy (Mignot et al., 1997) and PD (Hamza et al., 2010). The initial findings of (Hamza et al., 2010) which observed associations between SNPs in the *HLA-DRA* region with late-onset idiopathic Parkinson's disease, have since been independently replicated (Guo et al., 2011; Kannarkat et al., 2015) and expanded to include the *HLA-DRB5* and *HLA-DRB1* gene loci as well (Ahmed et al., 2012; International Parkinson Disease Genomics Consortium et al., 2011). Interestingly, in the case of *HLA-DRA* PD-associated polymorphisms, there is evidence to suggest that this risk may be mediated by higher baseline and induced levels of MHCII proteins on blood monocytes, B cells, and a skew of their phenotype toward inflammatory responses (Kannarkat et al., 2015). Most recently, a class I allele (*HLA-A*) has also been associated with PD and the IFN-γ response from blood derived CD8 T cells to α-syn (Sulzer et al., 2017), possibly providing an insight to their function in the brain parenchyma of patient. Overall, the exact mechanisms that underlie the PD-associated *HLA* alleles effect on disease pathology remain unclear, but the most straightforward implication is that they may potentiate the pro-inflammatory microglial and T cell responses observed in PD patients.

7.3 LRRK2

Mutations in leucine-rich repeat kinase (*LRRK2*) are a genetic cause of PD that appears to act in part by influencing the immune system (Zimprich et al., 2004). Mutations in *LRRK2* that cause PD are associated with increased kinase activity, and these increases are also observed in PD patients without a *LRRK2* mutation (Di Maio et al., 2018; Ferreira and Massano, 2017). LRRK2 protein is thought to phosphorylate Rab GTPases (Steger et al., 2016) and modulate intracellular vesicle trafficking. Though LRRK2's expression is mainly thought of in the context of neurons, it is also found to be expressed in high levels by immune cells such as macrophages, monocytes, and B cells (Gardet et al., 2010; Hakimi et al., 2011), where Rab GTPase mediated vesicle trafficking is crucial to the initiation of their immune responses (Prashar et al., 2017). Moreover, LRRK2 alterations are consistently linked to other, classical inflammatory diseases such as leprosy and Crohn's disease (Bae and Lee, 2015; Hui et al., 2018), and increased kinase activity could amplify the already pro-inflammatory functions of LRRK2 in inflammasome activation (Liu et al., 2017) and nuclear factor of activated T cells (NFAT) activation (Liu et al., 2011). However, further study of LRRK2 and how it functions in the innate and adaptive immune compartments within the context of PD is required to better understand its contribution to potential disease driving neuroinflammation in PD.

7.4 PINK1 and Parkin

Mutations in the genes encoding PTEN-induced kinase 1 (PINK1) and E3 ubiquitin ligase (Parkin) can result in an early onset, slowly progressing, and L-DOPA responsive form of Parkinson's disease (Farrer, 2006). Furthermore, SNPs in the promoter and coding regions of Parkin have been associated with the development of late-onset PD (Mata et al., 2004). The two proteins have recently been shown to play important roles in the process of mitophagy, or the removal of damaged mitochondria within the cell (Pickrell and Youle, 2015), a process that is important in immune system responses.

It has been shown that individuals with mono or biallelic mutations in parkin, regardless of PD status, have higher amounts of pro-inflammatory cytokines and chemokines (IL-6, IL-1β, CCL2, and CCL4) in their serum compared to controls (Sliter et al., 2018). PINK1 and Parkin knockout mice display a similar inflammatory response in their serum. In both of these knockout mice, the inflammatory response is the result of an accumulation of mitochondrial DNA and its subsequent activation of the innate immune system's type I interferon response through the stimulator of interferon genes (STING) pathway. Additionally, two recent studies performed in mice reported activation of the adaptive immune system, through the induction of pro-inflammatory cytotoxic CD8 T cells targeting host mitochondrial antigens in PINK1 or Parkin deficient mice subjected to bacterial exposure (Matheoud et al., 2016, 2019). B cells may also play a role in disease as increased amounts of anti-nuclear/dsDNA antibodies can be detected in the serum of oxidatively stressed PINK1 or Parkin deficient mice (Sliter et al., 2018). Broadly, given that the dopamine neurons in the substantia nigra have been shown to have a higher basal rate in mitophagy than other dopamine producing neurons (Guzman et al., 2018), it is logical that a breakdown in the mitophagy machinery, either by dysfunction in PINK1, Parkin, or α-syn pathology could contribute to the neurodegeneration observed in PD. Interestingly, these findings suggest that one consequence of the breakdown of mitophagy machinery is the production of cytosolic DNA and mitochondrial damage that can lead to the activation of both the innate and adaptive immune systems to produce tissue destructive inflammation.

Taken together, there is ample evidence of multiple PD-associated genes having various effects on the immune system. These effects seem to promote a pro-inflammatory response, which coincides with what is observed globally in the immune systems of individuals with PD. Interestingly, this potential PD-variant genetic effect on immune cells may be more direct than we think as (Raj et al., 2014) observed an overrepresentation of immune-specific expression quantitative trait loci (eQTL) overlapping with PD susceptibility loci (e.g., *SNCA*, *LRRK2*) from blood monocytes of healthy controls. These data suggest there may be a cell-autonomous effect of PD-associated genes within immune cells themselves (especially innate immune cells) that could contribute to the pathobiology of PD (Harms and Standaert, 2014). However, much more work is needed to better characterize these PD-gene and immune system interactions.

8 Immune modulation and risk of Parkinson's disease

A number of population studies support the idea that the immune system has an important role in the development of PD. These studies found associations between the chronic use of anti-inflammatory drugs and reduced risk of developing Parkinson's disease. Nonsteroidal anti-inflammatory drugs (NSAIDs) were originally found to be neuroprotective against dopamine cell loss in the 1-methyl-4-phenyl-1,2,3,6-tetrahydropyridine (MPTP) mouse model of PD (reviewed in next section) (Aubin et al., 1998; Teismann and Ferger, 2001). Additionally, (Chen et al., 2003) observed that men and women who had reported regular NSAID use had a significantly lower risk of later developing Parkinson's disease. These same general findings have been observed in subsequent studies using a variety of different cohorts (Chen et al., 2005; Gao et al., 2011; Hernan et al., 2002; Powers et al., 2008). The mechanism by which NSAIDs favor PD protection is unclear, but their ability to inhibit the aberrant COX-2 and NF-κB signaling observed in PD may be relevant (Asanuma and Miyazaki, 2007).

Another anti-inflammatory therapy that is associated with reduced PD risk is the use of anti-TNF therapy (Peter et al., 2018). Inflammatory bowel disease (IBD) patients, a population who are at higher risk of developing PD (Lin et al., 2016; Peter et al., 2018), who were on anti-TNF treatments had a 78% reduction in their PD incidence rate compared to IBD patients not on anti-TNF treatment. Elevated TNF levels in PD brain, CSF, and blood have been reported by several groups (Brodacki et al., 2008; Mogi et al., 1994; Qin et al., 2016), suggesting that TNF may have a key role in promoting the pathophysiology of PD.

9 Inflammation in murine models of PD

It is increasingly clear that findings in human PD have strongly implicated a pro-inflammatory response in the pathogenesis of disease. It has also highlighted that PD is a systemic disease, in which symptoms are found in multiple organ systems. However, it is difficult to parse the initiating events or mechanistic processes in humans, and thus the field has turned to animal models. Models of PD have evolved rapidly over time, beginning with neurotoxin-based models that focused on neuronal death to models that highlight the importance of α-syn in disease processes. Here, we will summarize some of the evidence for inflammation in both neurotoxin and α-syn models of PD.

9.1 Neurotoxin-based animal models of PD

9.1.1 6-OHDA

The 6-hydroxy-dopamine (6-OHDA) model is a neurotoxin model of PD that has been widely used and characterized since its initial development for studying the death of monoamine neurons in the CNS of rats (Ungerstedt, 1968). The mechanism of 6-OHDA's neurotoxic effect is through the production of excessive reactive oxygen species in neurons after entering through their dopamine transporter (Soto-Otero et al., 2000). The universal features of the model include the loss of nigrostriatal neurons, the development of L-Dopa responsive bradykinetic motor symptoms, and the marked activation of the innate immune system in the CNS (Bove and Perier, 2012).

In this model, activation of the CNS innate immune system, first described by (Akiyama and Mcgeer, 1989), has been shown to precede overt neuron loss (Cicchetti et al., 2002; Depino et al., 2003; Marinova-Mutafchieva et al., 2009), suggesting that the priming of these inflammatory myeloid cells may help drive this neuron loss via phagocytosis or enhanced inflammatory cytokine production. This notion is supported by the fact that administration of CX3CL1 (a neuron-derived microglial suppressor molecule) (Pabon et al., 2011), iNOS inhibitors (Broom et al., 2011), or dominant negative anti-TNF therapy (Harms et al., 2011) are all neuroprotective in the 6-OHDA model. While the innate immune response in the 6-OHDA model is thought to be neurotoxic, the role of the adaptive immune system is less studied and less clear. T (both CD4 and CD8) and B cells have been found to infiltrate the CNS after 6-OHDA administration at a similar time point as myeloid activation (Theodore and Maragos, 2015). However, these infiltrating lymphocytes may have a net positive neurotrophic effect as one study found that 6-OHDA treated athymic RNU$^{-/-}$ rats lacking T cells displayed exacerbated motor deficits compared to 6-OHDA treated controls (Wheeler et al., 2014). Perhaps the neuroregenerative anti-inflammatory effects of 6-OHDA responding Tregs outweighs any other pro-inflammatory cytokine producing T cells, but no follow-up experiments have been performed to support or reject this hypothesis.

9.1.2 MPTP

1-methyl-4-phenyl-1,2,3,6-tetrahydropyridine (MPTP) readily crosses the blood brain barrier, where it is metabolized by glial cells to its final neurotoxic form of MPP$^+$ (Meredith and Rademacher, 2011), a potent mitochondrial complex I inhibitor that leads to rapid dopaminergic neuron death and the development of Parkinsonian symptoms in humans (Langston et al., 1983), nonhuman primates (Tetrud and Langston, 1989), and mice (Sonsalla and Heikkila, 1986).

In mice and nonhuman primates, MPTP treatment leads to robust MHCII$^+$ microglial activation (Czlonkowska et al., 1996; Kohutnicka et al., 1998; Mcgeer et al., 2003). Similar MHCII$^+$ microglial activation has been found in human postmortem tissue of individuals who had been exposed to MPTP and developed Parkinsonism (Langston et al., 1999). Similar to the 6-OHDA model of PD, efforts to reduce microglial activation or associated downstream signaling effects (NF-κB, IL-1β, IL-6) via treatment with anti-inflammatory iNOS inhibitors (Du et al., 2001; Wu et al., 2002) or COX-1/2 inhibitors (Teismann and Ferger, 2001) have proven neuroprotective against the effects of MPTP in mice. Additionally, there is ample evidence of an adaptive immune response to MPTP. This includes the infiltration of CD4 and CD8 T cells (Benner et al., 2008; Brochard et al., 2009; Kurkowska-Jastrzebska et al., 1999; Reynolds et al., 2010) as well as B cell mediated IgG deposition (Benner et al., 2008). Additionally, Reynolds et al. (2010) showed that if MPTP intoxication is combined with nitrated α-syn immunizations, α-syn responding CD4 T cells produced multiple pro-inflammatory cytokines including IFN- γ, TNF, IL-17a, and IL-2. Moreover, Th1 (IFN-γ) and Th17 (IL-17a) polarized T cells adoptively transferred into MPTP intoxicated mice exacerbated SN TH$^+$ neuron loss, while Th2 (IL-4), Treg (IL-10), and vasoactive intestinal peptide (VIP, a neuropeptide associated with Treg responses) treated mice did not. Further evidence of the inflammatory role of CD4 T cells in the MPTP model comes from the findings that SCID, Tcrb$^{-/-}$, Cd4$^{-/-}$, but not Cd8$^{-/-}$ mice administered MPTP are protected from neurodegeneration (Benner et al., 2008; Brochard et al., 2009). Taken together, these data support the hypothesis that MPTP induces an activation of both the innate and adaptive immune that promotes further pro-inflammatory and neurotoxic effects.

9.1.3 Rotenone and paraquat

Rotenone is a plant derived insecticide and pesticide whose mechanism of action is mainly through binding to complex I in the electron transport chain, and thus disrupting mitochondrial oxidative phosphorylation (Schuler and Casida, 2001). On the other hand, paraquat, a popular herbicide, is thought to be deleterious through promoting excessive ROS production (Day et al., 1999). Similar to 6-OHDA and MPTP, both rotenone and paraquat disrupt mitochondria and promote oxidative stress in the cell.

Both mouse and rat rotenone models of PD display loss of dopaminergic neurons, α-syn pathology, and motor deficits (Bove and Perier, 2012). In one study, rotenone administration in rats recapitulated the hallmark microgliosis that is observed in human PD (Sherer et al., 2003). Another study showed that rotenone administration induced the production of multiple pro-inflammatory signaling mediators including IL-1β, TNF, IL-6, NF-κB, and that these inflammatory molecules, as well as iNOS, could be attenuated with treatment of a heat shock protein inducer (Thakur and Nehru, 2015). However, while there is evidence of an innate response to the rotenone model of PD, to date there is no clear evidence of a role for the adaptive immune system. This absence of data on the adaptive immune system's role in rotenone models appears to be due to lack of direct study as opposed to any evidence that the PD pathologies produced by the model to be independent of T or B cell responses. Similar to the rotenone model, the paraquat mouse model of PD recapitulates the human hallmarks of disease, including the loss of dopamine neurons in the SN (<25%) (Bove and Perier, 2012) and the presence of α-syn pathology (Manning-Bog et al., 2002). In this model, microglia become activated (Peng et al., 2009; Purisai et al., 2007) and are thought to be crucial in mediating ROS-induced neuro-degeneration. Like the rotenone model, there is a clear role for the CNS innate immune system in the paraquat model, but no clear evidence of the involvement of the adaptive arm. Again, this is most likely due to the lack of studies as opposed to the adaptive immune system playing no role in the pathophysiology of the paraquat model.

9.2 α-syn based animal models of PD

9.2.1 Human α-syn transgenics

As previously mentioned, mutations in the α-syn gene (*SNCA*), or polymorphisms increasing or enhancing α-syn expression in human PD are associated with increased risk for PD (Fuchs et al., 2008; Maraganore et al., 2006). Numerous animal models have been created that express either normal or familial mutations of α-syn. There are many non-murine models that utilize human α-syn, but other reviews have covered this topic extensively (Maulik et al., 2017; Vanhauwaert and Verstreken, 2015; Visanji et al., 2016), and only murine models will be discussed here. Mice that over-express full length human α-syn under the Thy1, PDGF-β, or even the rat TH promoters develop intracellular inclusions (Fleming et al., 2004, 2011; Masliah et al., 2000; Rockenstein et al., 2002), and some develop decreases in striatal dopa-mine levels (Lam et al., 2011; Masliah et al., 2000). It should be noted that these models do not present with overt neurodegeneration, and thus often do not display motor deficits (Antony et al., 2011) with the exception of the PDGF-β α-syn mice at 12 months (Masliah et al., 2000) and the Thy1 α-syn mice. Thy1 α-syn mice develop progressive α-syn inclusions, loss of striatal dopamine, and notably, mild motor def-icits in the absence of overt cell loss (Chesselet et al., 2012). These mice have been the most extensively characterized α-syn based transgenic model with regards to inflammation. Thy1-α-syn mice develop increased expression of TLR 1, 2, 4, and 8 in the substantia nigra with age, and TNF mRNA increases in both the striatum and substantia nigra (Watson et al., 2012). Furthermore, these same mice have in-creased nigral MHCII expression at 14 months, and increased CD4 and CD8 T cells in the blood at 22 months. This suggests that the increasing α-syn burden found with age could be related to progressive innate and adaptive immune system involvement in these mice.

Two autosomal dominant mutations, the A53T (Polymeropoulos et al., 1997) that results in an alanine to threonine substitution and A30P (Kruger et al., 1998) that results in an alanine to proline substitution, have been identified in multiple human cohorts. While many of these mutated forms have been made into mouse models, most research has not investigated inflammation. The few studies that have will be discussed here. Mice overexpressing a double mutated form of α-syn at A53T and A30P from the rat TH promoter develop increased ionized calcium binding adaptor molecule 1 (IBA1, which increases on myeloid cells during inflammation) immunostaining in the substantia nigra and striatum, increased numbers of IBA1$^+$ cells (which could represent microglia or infiltrating myeloid cells), increased levels of TNF in both the substantia nigra and striatum, increased cytokine and ROS production, as well as decreased IL-10 levels (Su et al., 2009). The mouse model with PrP (mouse prion protein) driven A53T expression has decreased stool frequency and gastric emptying, although no pure inflammatory markers have been reported (Vidal-Martinez et al., 2016). However, the administration of a drug targeting sphingosine-1-phosphate receptor (fingolimod) attenuates the problems with gut motility, decreases both gut and CNS α-syn pathology, and is neuroprotective (Vidal-Martinez et al., 2016). This is especially interesting, as it strengthens the argument that gut pathology and possibly inflammation contribute to disease progression. Thus, it could be the case that a predisposition to abnormal α-syn, combined with inflammatory processes, leads to further immune system involvement and disease.

9.2.2 Viral mediated α-syn expression

Some PD models utilize adeno-associated viral vectors (AAVs) to induce overexpression of normal or familial forms of mutated human α-syn in neurons. These models are helpful in studying the effect of excess α-syn in neurons of specific brain regions, as they are temporally and spatially restricted depending on where the viral vector is injected. AAVs also preferentially infect postmitotic cells, a mostly neuronal population in the brain, although the type of neuron varies depending on viral serotype. AAV serotype 2 (AAV2) seems to infect dopaminergic neurons of the substantia nigra most effectively, and is used either alone or as a hybrid vector to express α-syn in neurons. These models are especially useful, as they can be used in any transgenic mouse and allow the study of non-cell-autonomous mechanisms of neurodegeneration.

Numerous AAV serotypes have been used to induce human α-syn expression, and lead to intraneuronal α-syn pathology (Lindgren et al., 2012; Volpicelli-Daley, 2017). Just as the case with other α-syn based models of PD, the inflammatory response has not been investigated in depth in a majority of the AAV models. Two exceptions are the AAV2/5 model and AAV2 models. AAV2/5 is used in rats and leads to increased expression of MHCII and CD68, and infiltration CD4 and CD8 T cells in the midbrain (Sanchez-Guajardo et al., 2010). AAV2 based overexpression of human α-syn results in production of pSer129$^+$ intraneuronal α-syn inclusions and death of ~30% of dopaminergic neurons (Harms et al., 2013; Theodore et al., 2008). This model is especially interesting in the context of inflammation, as neuroinflammation is a driving force in the observed neurodegeneration rather than just an effect of the disease process (Harms et al., 2013, 2018; Theodore et al., 2008; Thome et al., 2016; Williams et al., 2018). The neuroinflammatory responses reported in the aforementioned AAV2 studies include: microglial upregulation of MHCII, infiltration of inflammatory monocytes, infiltration of T cells, deposition of IgG in the midbrain, and the production of inflammatory cytokines and chemokines (i.e., IL-6, TNF, and iNOS at early time points). Disruption of many of these processes, such as MHCII knockout (Harms et al., 2013), targeted CIITA (the transcriptional coactivator for inducible MHCII expression) knockdown (Williams et al., 2018), or CCR2 knockout (Harms et al., 2018), which prevents CCR2$^+$ inflammatory monocyte migration to tissues, are protective against neurodegeneration. Additionally, as only a portion of microglia begin expressing MHCII, this could indicate the presence of a group of disease associated microglia (DAMs), as has been identified in other neurodegenerative diseases such as Alzheimer's disease. Furthermore, increased levels of B cells infiltrate in response to α-syn, and FcγR knockout mice are protected from inflammation and neurodegeneration as well (Cao et al., 2010; Theodore et al., 2008). Viral models have been crucial in delineating the role of an immune response mouse models of PD, and have highlighted a role for inflammation in driving α-syn induced neurodegeneration.

9.2.3 α-syn fibrils

Pathological α-syn can corrupt endogenous protein and lead to the further spread of pathological species (Luk et al., 2009, 2012), a finding that has led to the development of a model of α-syn propagation: the pre-formed fibril (PFF) model. This model uses human or mouse α-syn monomer, which alone is non-pathological (Volpicelli-Daley et al., 2011), to create small fibrils. These fibrils are injected into the brain, and pSer129 positive inclusions appear throughout interconnected regions, depending on the location of injection, and are comprised of both injected and endogenous α-syn protein. In rats and mice, motor deficits appear over time (Luk et al., 2012; Paumier et al., 2015). *In vitro*, these fibrils are neurotoxic, pro-inflammatory and can activate cultured microglia inducing MHCII and iNOS expression (Cremades et al., 2012; Harms et al., 2013; Williams et al., 2018). Human PFFs injected into M20 human α-*syn*-expressing mice leads to increased GFAP and IBA1 immunostaining, indicative of astrocytosis and a myeloid response (Sacino et al., 2014; Sorrentino et al., 2017). Additionally, injection of mouse PFFs into WT Sprague Dawley rats led to increased MHCII and IBA1 expression in the midbrain, accompanied by the infiltration of peripheral myeloid cells and CD4 T cells prior to measurable neurodegeneration (Duffy et al., 2018; Harms et al., 2017). However, while there are no reports to date of peripheral cell infiltration in mice given PFFs, robust astrocytosis has been reported. Specifically, the emergence of A1 astrocytes (nomenclature mirrors the M1/M2 phenotypes of macrophages), or astrocytes that are associated with disease states and have a pro-inflammatory skew are found in the fibril model (Yun et al., 2018). However, our picture of the contribution of inflammation to this model remains incomplete, as manipulations aimed at halting an inflammatory response to fibrillar α-syn have not been performed *in vivo*.

10 Conclusions and key questions

Taken together, the data presented in this chapter indicate that the immune system plays a driving role in both the pathogenesis and progression of PD. As reviewed earlier, these findings harken back to the original description of inflammation in the disease by James Parkinson. The evidence consists of multiple observations of an abnormal innate and adaptive immune system in the blood, gut, and CNS of patients with PD, as well as several genetic and environmental associations linking the immune system and PD risk. This idea has been supported by numerous findings in both humans and animal models. Both genetic familial forms of the disease and genetic risk factors, such as *HLA* and *SNCA*, are associated with immune dysfunction. Additionally, PD patients display signatures of global inflammation in the blood, gut, and CNS. This includes elevated blood and CSF cytokine levels, infiltration of peripheral immune cells into the CNS, hyperreactive circulating immune cells, and a dysregulated gut microbiome. There is also evidence that blocking inflammatory events leads to a reduced risk for later developing PD. While there is no evidence to conclude that inflammation alone causes PD itself, many studies in animal models suggest that interrupting the inflammatory response, such as blocking T cell responses, MHCII signaling, CCR2 signaling, use of iNOS inhibitors, or anti-TNF therapy is a neuroprotective strategy, leading us to believe that inflammation is a main driver of disease pathogenesis. Based on the data discussed here, we hypothesize that PD begins and progresses in an autoinflammatory manner, where abnormal α-syn and genetic/environmental risks together lead to disease, as outlined in Fig. 4.

While recent research has identified roles for the innate and adaptive immune systems in human disease and animal models, much more work is needed to integrate these findings into a mechanistic understanding of the disease process. Here, we have highlighted four key questions that, if addressed, would provide a clearer picture of how the immune system contributes to the progression of PD (identified in Fig. 4).

10.1 Key question 1: What is the role of α-syn in activating the immune system?

There is ample evidence for immune activation within the CNS in human PD (Fig. 3), although the signal initiating the inflammatory response remains unknown. One likely candidate is the neuronal postsynaptic protein α-syn. α-syn inclusions are found throughout the brain, and are theorized to spread in a prion-like manner through retrograde transport (Henderson et al., 2019; Ma et al., 2019), although this is currently a hotly debated topic. Much of the research on α-syn propagation comes from *in vitro* experiments, where only monosynaptic transfer is demonstrated (Grozdanov and Danzer, 2018; Mao et al., 2016), or from *in vivo* models where the extent of spread, and whether it is across only one synapse or multiple, even spreading to neuroanatomical regions that are not directly connected, has been a source of much disagreement within the field (Grozdanov and Danzer, 2018; Kim et al., 2019; Sorrentino et al., 2017). However, in humans, brain derived exosomes

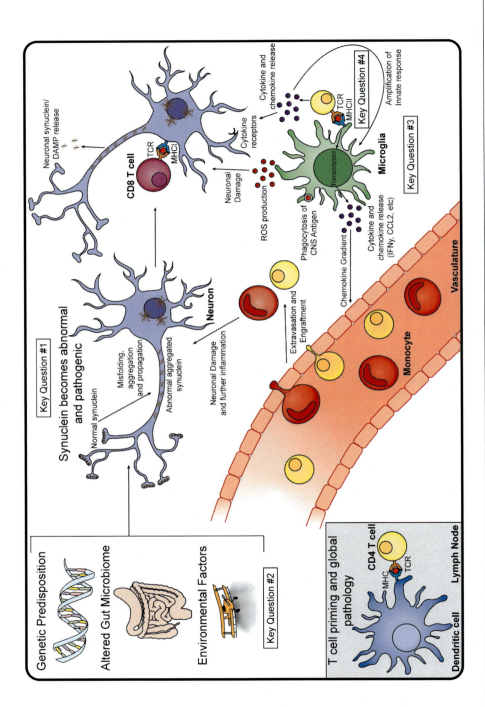

FIG. 4

Proposed Mechanism of Immune Involvement in PD and Key Questions. We hypothesize that PD begins due to genetic predispositions, alterations in the gut microbiome, and the influence of external, environmental factors. It is possible that a CNS antigen drains to the lymph node (bottom left), where a dendritic cell (blue) can present it to a CD4 T cell (yellow), priming the T cell for an inflammatory response. Within the CNS, we believe that normal α-syn (brown) becomes misfolded, and begins to propagate and aggregate in neurons (blue). This can lead to the release of toxic α-syn species from neurons, and neuronal presentation of antigen to CD8 T cells (maroon) via MHCI. Microglia take up CNS antigen and present this on their MHCII to CD4 T cells, leading to T cell differentiation and cytokine release. The MHCII-TCR interaction and subsequent cytokines can amplify microglial activation, leading to further production of cytokines and chemokines. Peripheral immune cells, such as monocytes (red) and additional T cells will home to the site of chemokine production, and extravagate from the vasculature or meningeal lymphatics (not shown), and cause further inflammation and neuronal damage in the CNS. As much of this is speculative, and ties together many disparate pieces of data, several key questions remain. These include: #1 "What is the role of α-syn in activating the immune system?" (top middle) #2 "Is Parkinson disease caused by an immune response that originates in the gut?" (left) #3 "What is the role of microglia in Parkinson's disease?" (bottom right) and #4 "How do T cells contribute to the pathobiology of Parkinson disease?" (bottom right).

have been found to contain elevated levels of α-syn that correlate with disease sever-ity (more α-syn correlated with worse disease measures) (Shi et al., 2014). Therefore, there is evidence of α-syn release from the CNS in human disease, and this α-syn is found throughout the body. The enteric nervous system actually expresses high levels of α-syn as well, and α-syn pathology can be found throughout the enteric neu-rons in PD (Braak et al., 2006; Shannon et al., 2012). Widespread α-syn pathology, if it is indeed an autoantigen, could help to explain the global symptoms exhibited by PD patients.

Although α-syn pathology is widespread, this does not directly implicate α-syn as an antigen. Multiple experiments have demonstrated that α-syn is able to activate myeloid cells and cultured microglia. Particularly, toll-like receptors on microglia can be activated by α-syn, similar to a PAMP or DAMP (Daniele et al., 2015; Fellner et al., 2013; Kim et al., 2013, 2016; Sanchez-Guajardo et al., 2015; Yun et al., 2018). Interestingly, TLR2 expression is increased in postmortem tissue of PD patients on both neurons and IBA1[+] cells, which could represent microglia or other infiltrating myeloid cells (Dzamko et al., 2017). When investigated closely *in vitro*, neurons secrete oligomeric α-syn, which then activates microglia via TLR2 signaling (Kim et al., 2013, 2016). The secretion of α-syn from neurons has been proposed to occur through the release of exosomes as a way to rid the cell of excess protein (Stefanis et al., 2019). While it is controversial whether neurons *in vivo* can secrete α-syn, the external application of α-syn seems to reliably activate microglia through TLR1/2 heterodimers, TLR2, or TLR4 (Daniele et al., 2015; Fellner et al., 2013; Kim et al., 2013). In Thy1-asyn mice, TLR 1, 2, 4, and 8 are increased in the substantia nigra, indicating that this process may also occur *in vivo* (Watson et al., 2012). Activation of TLRs by α-syn can also lead to the production of pro-inflammatory cytokines and chemokines, such as TNF, IL-1β, IL-6, Cox2, IL-1α, and iNOS and reactive oxygen species (Fellner et al., 2013; Russo et al., 2019; Sanchez-Guajardo et al., 2015; Yun et al., 2018). Therefore, in these ways, α-syn acts similar to a foreign antigen, PAMP, or DAMP that would typically activate an APC.

There is also evidence that some T cells in PD patients may be particularly reactive to specific fragments of α-syn. T cells isolated from PD patients contain TCRs that respond to native, fibrillar, and antigenic peptides of α-syn. Additionally, specific alleles of HLA (part of the MHCII complex) bind these peptides with a much higher affinity (Sulzer et al., 2017). It has also been shown in both the pre-formed fibril model in rats and the AAV2 model in mice that abnormal α-syn in the brain can lead to MHCII expression and T cell infiltration (see Sections 9.2.2 and 9.2.3). As α-syn is the common denominator and single manipulation here, it is logical that this response would be directed against some form of α-syn.

This idea of α-syn as the antigen begs the question of: what form of α-syn could be antigenic? There has been much speculation on which form is toxic to neurons (nitrated, phosphorylated, fibrillar, etc.), but much less speculation on which of these species could be antigenic. Proteins present in homeostasis can become post-translationally modified under conditions of cellular stress or inflammation, leading to their recognition by autoreactive T cells that were not deleted in the thymus (Doyle and Mamula, 2012). In the case of α-syn, it seems as if overabundance of α-syn is enough to lead to hyperphosphorylation, ubiquitylation, and aggregation (Chesselet et al., 2012; Fleming et al., 2004; Masliah et al., 2000). As stated above, PD patients have autoreactive T cells to specific fragments of α-syn, including at pSer129, a site that becomes phosphorylated in overabundance and fibril models of PD. Therefore, it could be that cellular stress or genetic predispositions could lead to the generation and recognition of antigenic forms of α-syn, which then lead to inflammation and neurodegeneration. However the question of how native α-syn becomes modified in the first place remains unanswered.

Additionally, there remains the possibility that there is a yet unidentified antigen in disease, as other autoimmune diseases such as multiple sclerosis and rheumatoid arthritis have multiple known autoantigens. It is also possible that α-syn becomes recognized through epitope spreading or molecular mimicry—where a different, but similarly structured antigen becomes recognized first, followed by the recognition of α-syn. α-syn is widespread throughout the body, and this provides many opportunities for interaction with the immune system.

10.2 Key question 2: Is Parkinson's disease caused by an immune response that originates in the gut?

Gastrointestinal dysfunction was initially remarked upon in the original "An Essay on the Shaking Palsy" and has continued to be an important disease symptom, as most patients with PD suffer from constipation (Fasano et al., 2015). Furthermore, there is evidence of increased intestinal epithelial barrier permeability (Forsyth et al., 2011) and pro-inflammatory cytokines in the GI tract of PD patients (Devos et al., 2013). Abnormal α-syn species have also been detected throughout the whole GI tract, including in the submandibular gland, the colon and the innervating enteric nervous system (Fasano et al., 2015). Braak's hypothesis attempts to tie these two phenotypes together by suggesting that a pathogen could pass through the leaky intestinal wall of presymptomatic PD patients and promote the formation of abnormal α-syn species across the GI tract and enteric nervous system—a hypothesis that has only grown more feasible with the discovery of direct connections between gut epithelial and vagal neurons (Kaelberer et al., 2018). This reinforces the Braak hypothesis, in which α-syn could be a prion, with the initial seeding event occurring in the enteric nervous system and then propagating up the vagus nerve into the CNS proper. Now, with the better understanding and appreciation of the gut-immune-brain axis (Fung et al., 2017), the evidence of an abnormal microbiota signature in PD patients, and the consistent findings of an overactive immune system in PD, it may be time to update Braak's hypothesis to reflect this.

The immune system is crucial in directing the composition and location of microbiota in mammals, while microbiota are simultaneously critical in the development and future responses of the immune system in both health and disease (Hooper et al., 2012). In the case of PD, there is ample evidence that the gut microbiota is altered in

PD patients, which includes the loss of anti-inflammatory bacteria that could be important in controlling GI inflammation (see Section 6.2). Additionally, there is evidence of innate and adaptive immune activation in the gut, periphery, and central nervous system associated with PD (Fig. 3). Braak initially suggested that a pathogen could be involved in PD, and perhaps the pathogen(s) are particular gut bacteria that elicit damaging, pro-inflammatory responses from the innate and adaptive immune systems, including the production of reactive oxygen species.

It is known that one of the primary biological tools of the inflammatory immune response is the production of ROS to diminish bacterial or viral survival as well as mediate several signaling pathways including the activation of the inflammasome pathway (Yang et al., 2013) which is upregulated in PD (Gordon et al., 2018). Oxidative stress has been shown in multiple PD models to promote the oligomerization and pathogenicity of α-syn (Dias et al., 2013). It is possible then that this pro-inflammatory, ROS rich environment produced by the microbiota and immune system could lead to the formation or propagation of pathogenic α-syn species in the gut and the innervating nervous system. These pathogenic α-syn species, now residing in an inflammatory environment, may also serve as a new target for the immune reaction in the gut, which would help explain the evidence of a α-syn specific cytokine recall response in PD patient derived T cells (Sulzer et al., 2017). Following this, it is possible that the resident and circulating immune system could react to the CNS α-syn pathology in a similar way that it may have in the gut. Whatever the exact mechanisms may be, the idea that the initiation of the pathobiology of PD could be initiated in the gut due to a combination of abnormal bacteria, an overactive immune system, and the propensity of α-syn to oligomerize in oxidative stress conditions is one that warrants consideration in the field of PD.

10.3 Key question 3: What is the role of microglia in Parkinson's disease?

Microglia have long been the scapegoat for neuroinflammation. They are the tissue resident macrophages of the CNS parenchyma, and are thought of as first responders to any injury, modulators of homeostasis, and mediators of neuroinflammation. Across neurodegenerative diseases, however, there is an ongoing debate of whether the activity of microglia can be harmful or helpful, promoting either damaging inflammation or resolution and tissue repair.

There is extensive data demonstrating that microglia react to abnormal α-syn, as discussed in the Section 9.2 and Key Question 1. These phenotypic changes may be similar to what is seen in other models of neurodegeneration, and microglia may in fact exhibit a mixed phenotype, with signatures characteristic of both damaging inflammation and resolution. There is evidence from models of acute neuron death demonstrating that microglia expand clonally (Tay et al., 2017), suggesting that specific subsets of microglia respond to cell death and general threats to the environment. Recent studies have also identified a subset of disease associated macrophages (DAMs) that contribute to neuroinflammation and cell death (Keren-Shaul et al., 2017).

In fact, these DAMs have been found to be consistently present across many models of neurodegeneration, including models of Alzheimer's disease, ALS, and multiple sclerosis (Song and Colonna, 2018). While no studies have yet investigated whether DAMs are present in models of PD, it is likely that they play a role, as the phenotype can be induced with aging, neurodegeneration, or through acute models that lead to apoptosis and cell death. Interestingly, DAMs express both pro- and anti-inflammatory molecules, making their exact effect on the inflammatory process unclear (Keren-Shaul et al., 2017; Song and Colonna, 2018). Upon exposure to abnormal α-syn, or any robust inflammatory stimulus, a portion of the microglia begins to upregulate MHCII on their surface and become activated. This has been found in the AAV2 overexpression mouse model (Harms et al., 2013, 2018), an α-syn fibril rat model (Harms et al., 2017), as well as neurotoxin models such as paraquat (Peng et al., 2009; Purisai et al., 2007). This would seem to indicate that the portion of microglia upregulating MHCII may indeed be involved in antigen presentation to CD4 T cells and act as DAMs sensing alterations in the local environment. Media analyzed from co-culture of microglia and CD4 T cells exposed to α-syn shows increases in both pro- and anti-inflammatory cytokines, such as TNF, IFN-γ, and Il-10 (Harms et al., 2013). While this is by no means conclusive, it demonstrates that microglia in models of PD may adopt a mixed, DAM-like phenotype.

One caveat to these findings is the issue that, until recently, it has been incredibly difficult to differentiate between infiltrating monocytes and macrophages and tissue resident microglia due to overlap in surface markers between the two populations. Most *in vivo* studies report all IBA1$^+$ cells as microglia, although many different cell types express this marker promiscuously. This is also a key issue with the TSPO based PET scanning, as this ligand is not specific to microglia or even immune cells—indeed it has been shown to bind to astrocytes (Kuhlmann and Guilarte, 2000; Pannell et al., 2019). In other disease models, such as stroke, it has been shown that infiltrating monocytes are actually protective, and dampen a harmful microglial response (Greenhalgh et al., 2018), and it is therefore integral that the two populations be studied independently. However, based on the current available data, it is clear that there is an inflammatory response ongoing in the brains of mouse models and humans, even if we do not yet know the involved cell types.

10.4 Key question 4: How do t cells contribute to the pathobiology of Parkinson's disease?

Interaction between antigen presenting cells (whether MHCI or MHCII) and T cells likely accounts for the elevated levels of T cell associated cytokines in the brain, CSF, and blood of PD patients. Sulzer et al. (2017) proposed that "altered degradation of proteins including α-syn could produce antigenic epitopes that trigger immune reactions during aging and Parkinson's disease." How exactly are these T cell derived pro-inflammatory cytokines contributing to the α-syn pathology and overt neurodegeneration seen in Parkinson's disease? Activation of T cells via their TCR can directly lead to the robust production of ROS as well as T cell derived cytokines such as IFN-γ promoting further ROS signaling in macrophages and monocytes (Yang et al., 2013). Additionally, the pro-inflammatory cytokines associated with PD have been shown to be able to skew microglia to be more phagocytic, which could be detrimental to neuron survival (Merson et al., 2010). Another potential mechanism for T cell driven neurodegeneration in PD is through the direct interaction of T cell derived cytokines and their cognate receptors on neurons. It has been recently demonstrated in an iPSC based preclinical model of PD that the Th17 associated cytokine IL-17a could bind to the IL-17R on PD derived midbrain neurons and lead to their apoptosis (Sommer et al., 2018). Lastly, in another preclinical model, it has been suggested that CD8 cytotoxic T cells could directly induce neuronal apoptosis via their TCR interaction with MHCI on dopaminergic neurons in PD (Cebrian et al., 2014). It is likely that a combination of these proposed mechanisms including excess ROS production from the innate and adaptive immune system, direct cytokine signaling on neuronal populations, and T cell directed phagocytosis are contributing to the Parkinson's disease state (Fig. 4).

10.5 Conclusion

In conclusion, human PD and rodent models of the disease are consistently associated with a pro-inflammatory innate and adaptive immune phenotype. Work in several animal models of the disease have demonstrated the effectiveness of targeting the innate and adaptive immune compartments as a neuroprotective strategy. As the field continues to elucidate the origin and the exact nature of these potentially disease driving immune responses, it will be important to translate these findings into creating and employing much needed disease-modifying immunotherapies.

References

Aarsland, D., Creese, B., Politis, M., Chaudhuri, K.R., Ffytche, D.H., Weintraub, D., Ballard, C., 2017. Cognitive decline in Parkinson disease. Nat. Rev. Neurol. 13, 217–231.

Abbott, R.D., Petrovitch, H., White, L.R., Masaki, K.H., Tanner, C.M., Curb, J.D., Grandinetti, A., Blanchette, P.L., Popper, J.S., Ross, G.W., 2001. Frequency of bowel movements and the future risk of Parkinson's disease. Neurology 57, 456–462.

Ahmed, I., Tamouza, R., Delord, M., Krishnamoorthy, R., Tzourio, C., Mulot, C., Nacfer, M., Lambert, J.C., Beaune, P., Laurent-Puig, P., Loriot, M.A., Charron, D., Elbaz, A., 2012. Association between Parkinson's disease and the HLA-DRB1 locus. Mov. Disord. 27, 1104–1110.

Aho, V.T.E., Pereira, P.A.B., Voutilainen, S., Paulin, L., Pekkonen, E., Auvinen, P., Scheperjans, F., 2019. Gut microbiota in Parkinson's disease: temporal stability and relations to disease progression. EBioMedicine 44, 691–707.

Akiyama, H., Mcgeer, P.L., 1989. Microglial response to 6-hydroxydopamine-induced substantia nigra lesions. Brain Res. 489, 247–253.

Antony, P.M., Diederich, N.J., Balling, R., 2011. Parkinson's disease mouse models in translational research. Mamm. Genome 22, 401–419.

Asanuma, M., Miyazaki, I., 2007. Common anti-inflammatory drugs are potentially therapeutic for Parkinson's disease? Exp. Neurol. 206, 172–178.

Aubin, N., Curet, O., Deffois, A., Carter, C., 1998. Aspirin and salicylate protect against MPTP-induced dopamine depletion in mice. J. Neurochem. 71, 1635–1642.

Bae, J.R., Lee, B.D., 2015. Function and dysfunction of leucine-rich repeat kinase 2 (LRRK2): Parkinson's disease and beyond. BMB Rep. 48, 243–248.

Barichella, M., Severgnini, M., Cilia, R., Cassani, E., Bolliri, C., Caronni, S., Ferri, V., Cancello, R., Ceccarani, C., Faierman, S., Pinelli, G., De Bellis, G., Zecca, L., Cereda, E., Consolandi, C., Pezzoli, G., 2019. Unraveling gut microbiota in Parkinson's disease and atypical parkinsonism. Mov. Disord. 34, 396–405.

Barone, P., Antonini, A., Colosimo, C., Marconi, R., Morgante, L., Avarello, T.P., Bottacchi, E., Cannas, A., Ceravolo, G., Ceravolo, R., Cicarelli, G., Gaglio, R.M., Giglia, R.M., Iemolo, F., Manfredi, M., Meco, G., Nicoletti, A., Pederzoli, M., Petrone, A., Pisani, A., Pontieri, F.E., Quatrale, R., Ramat, S., Scala, R., Volpe, G., Zappulla, S., Bentivoglio, A.R., Stocchi, F., Trianni, G., Dotto, P.D., PRIAMO Study Group, 2009. The Priamo study: a multicenter assessment of nonmotor symptoms and their impact on quality of life in Parkinson's disease. Mov. Disord. 24, 1641–1649.

Bartels, A.L., Willemsen, A.T., Doorduin, J., De Vries, E.F., Dierckx, R.A., Leenders, K.L., 2010. [11C]-PK11195 PET: quantification of neuroinflammation and a monitor of anti-inflammatory treatment in Parkinson's disease? Parkinsonism Relat. Disord. 16, 57–59.

Bas, J., Calopa, M., Mestre, M., Mollevi, D.G., Cutillas, B., Ambrosio, S., Buendia, E., 2001. Lymphocyte populations in Parkinson's disease and in rat models of parkinsonism. J. Neuroimmunol. 113, 146–152.

Bassing, C.H., Swat, W., Alt, F.W., 2002. The mechanism and regulation of chromosomal V(D)J recombination. Cell 109 (Suppl), S45–S55.

Benamer, H.T., Patterson, J., Wyper, D.J., Hadley, D.M., Macphee, G.J., Grosset, D.G., 2000. Correlation of Parkinson's disease severity and duration with 123I-FP-CIT SPECT striatal uptake. Mov. Disord. 15, 692–698.

Bengoa-Vergniory, N., Roberts, R.F., Wade-Martins, R., Alegre-Abarrategui, J., 2017. Alpha-synuclein oligomers: a new hope. Acta Neuropathol. 134, 819–838.

Benner, E.J., Banerjee, R., Reynolds, A.D., Sherman, S., Pisarev, V.M., Tsiperson, V., Nemachek, C., Ciborowski, P., Przedborski, S., Mosley, R.L., Gendelman, H.E., 2008. Nitrated alpha-synuclein immunity accelerates degeneration of nigral dopaminergic neurons. PLoS One 3, e1376.

Bernard, J.J., Cowing-Zitron, C., Nakatsuji, T., Muehleisen, B., Muto, J., Borkowski, A.W., Martinez, L., Greidinger, E.L., Yu, B.D., Gallo, R.L., 2012. Ultraviolet radiation damages self noncoding RNA and is detected by TLR3. Nat. Med. 18, 1286–1290.

Blum-Degen, D., Muller, T., Kuhn, W., Gerlach, M., Przuntek, H., Riederer, P., 1995. Interleukin-1 beta and interleukin-6 are elevated in the cerebrospinal fluid of Alzheimer's and de novo Parkinson's disease patients. Neurosci. Lett. 202, 17–20.

Bouneaud, C., Kourilsky, P., Bousso, P., 2000. Impact of negative selection on the T cell repertoire reactive to a self-peptide: a large fraction of T cell clones escapes clonal deletion. Immunity 13, 829–840.

Bove, J., Perier, C., 2012. Neurotoxin-based models of Parkinson's disease. Neuroscience 211, 51–76.

Braak, H., Del Tredici, K., Rub, U., De Vos, R.A., Jansen Steur, E.N., Braak, E., 2003. Staging of brain pathology related to sporadic Parkinson's disease. Neurobiol. Aging 24, 197–211.

Braak, H., De Vos, R.A., Bohl, J., Del Tredici, K., 2006. Gastric alpha-synuclein immunoreactive inclusions in Meissner's and Auerbach's plexuses in cases staged for Parkinson's disease-related brain pathology. Neurosci. Lett. 396, 67–72.

Breyner, N.M., Michon, C., De Sousa, C.S., Vilas Boas, P.B., Chain, F., Azevedo, V.A., Langella, P., Chatel, J.M., 2017. Microbial anti-inflammatory molecule (MAM) from Faecalibacterium prausnitzii shows a protective effect on DNBS and DSS-induced colitis model in mice through inhibition of NF-kappaB pathway. Front. Microbiol. 8, 114.

Brochard, V., Combadiere, B., Prigent, A., Laouar, Y., Perrin, A., Beray-Berthat, V., Bonduelle, O., Alvarez-Fischer, D., Callebert, J., Launay, J.M., Duyckaerts, C., Flavell, R.A., Hirsch, E.C., Hunot, S., 2009. Infiltration of CD4+ lymphocytes into the brain contributes to neurodegeneration in a mouse model of Parkinson disease. J. Clin. Invest. 119, 182–192.

Brodacki, B., Staszewski, J., Toczylowska, B., Kozlowska, E., Drela, N., Chalimoniuk, M., Stepien, A., 2008. Serum interleukin (IL-2, IL-10, IL-6, IL-4), TNFalpha, and INFgamma concentrations are elevated in patients with atypical and idiopathic parkinsonism. Neurosci. Lett. 441, 158–162.

Broom, L., Marinova-Mutafchieva, L., Sadeghian, M., Davis, J.B., Medhurst, A.D., Dexter, D.T., 2011. Neuroprotection by the selective iNOS inhibitor GW274150 in a model of Parkinson disease. Free Radic. Biol. Med. 50, 633–640.

Burre, J., 2015. The synaptic function of alpha-synuclein. J. Park. Dis. 5, 699–713.

Calis, J.J., Rosenberg, B.R., 2014. Characterizing immune repertoires by high throughput sequencing: strategies and applications. Trends Immunol. 35, 581–590.

Cao, S., Theodore, S., Standaert, D.G., 2010. Fcgamma receptors are required for NF-kappaB signaling, microglial activation and dopaminergic neurodegeneration in an AAV-synuclein mouse model of Parkinson's disease. Mol. Neurodegener. 5, 42.

Carlsson, A.H., Yakymenko, O., Olivier, I., Hakansson, F., Postma, E., Keita, A.V., Soderholm, J.D., 2013. Faecalibacterium prausnitzii supernatant improves intestinal barrier function in mice DSS colitis. Scand. J. Gastroenterol. 48, 1136–1144.

Cebrian, C., Zucca, F.A., Mauri, P., Steinbeck, J.A., Studer, L., Scherzer, C.R., Kanter, E., Budhu, S., Mandelbaum, J., Vonsattel, J.P., Zecca, L., Loike, J.D., Sulzer, D., 2014. MHC-I expression renders catecholaminergic neurons susceptible to T-cell-mediated degeneration. Nat. Commun. 5, 3633.

Chandra, S., Gallardo, G., Fernandez-Chacon, R., Schluter, O.M., Sudhof, T.C., 2005. Alpha-synuclein cooperates with CSPalpha in preventing neurodegeneration. Cell 123, 383–396.

Chaplin, D.D., 2010. Overview of the immune response. J. Allergy Clin. Immunol. 125, S3–23.

Chartier-Harlin, M.C., Kachergus, J., Roumier, C., Mouroux, V., Douay, X., Lincoln, S., Levecque, C., Larvor, L., Andrieux, J., Hulihan, M., Waucquier, N., Defebvre, L., Amouyel, P., Farrer, M., Destee, A., 2004. Alpha-synuclein locus duplication as a cause of familial Parkinson's disease. Lancet 364, 1167–1169.

Chen, H., Zhang, S.M., Hernan, M.A., Schwarzschild, M.A., Willett, W.C., Colditz, G.A., Speizer, F.E., Ascherio, A., 2003. Nonsteroidal anti-inflammatory drugs and the risk of Parkinson disease. Arch. Neurol. 60, 1059–1064.

Chen, H., Jacobs, E., Schwarzschild, M.A., McCullough, M.L., Calle, E.E., Thun, M.J., Ascherio, A., 2005. Nonsteroidal antiinflammatory drug use and the risk for Parkinson's disease. Ann. Neurol. 58, 963–967.

Chesselet, M.F., Richter, F., Zhu, C., Magen, I., Watson, M.B., Subramaniam, S.R., 2012. A progressive mouse model of Parkinson's disease: the Thy1-aSyn ("Line 61") mice. Neurotherapeutics 9, 297–314.

Cicchetti, F., Brownell, A.L., Williams, K., Chen, Y.I., Livni, E., Isacson, O., 2002. Neuroinflammation of the nigrostriatal pathway during progressive 6-OHDA dopamine degeneration in rats monitored by immunohistochemistry and PET imaging. Eur. J. Neurosci. 15, 991–998.

Colonna, M., Butovsky, O., 2017. Microglia function in the central nervous system during health and neurodegeneration. Annu. Rev. Immunol. 35, 441–468.

Contento, R.L., Molon, B., Boularan, C., Pozzan, T., Manes, S., Marullo, S., Viola, A., 2008. CXCR4-CCR5: a couple modulating T cell functions. Proc. Natl. Acad. Sci. U. S. A. 105, 10101–10106.

Cosenza-Nashat, M., Zhao, M.L., Suh, H.S., Morgan, J., Natividad, R., Morgello, S., Lee, S.C., 2009. Expression of the translocator protein of 18 kDa by microglia, macrophages and astrocytes based on immunohistochemical localization in abnormal human brain. Neuropathol. Appl. Neurobiol. 35, 306–328.

Cremades, N., Cohen, S.I., Deas, E., Abramov, A.Y., Chen, A.Y., Orte, A., Sandal, M., Clarke, R.W., Dunne, P., Aprile, F.A., Bertoncini, C.W., Wood, N.W., Knowles, T.P., Dobson, C.M., Klenerman, D., 2012. Direct observation of the interconversion of normal and toxic forms of alpha-synuclein. Cell 149, 1048–1059.

Croisier, E., Moran, L.B., Dexter, D.T., Pearce, R.K., Graeber, M.B., 2005. Microglial inflammation in the parkinsonian substantia nigra: relationship to alpha-synuclein deposition. J. Neuroinflammation 2, 14.

Czlonkowska, A., Kohutnicka, M., Kurkowska-Jastrzebska, I., Czlonkowski, A., 1996. Microglial reaction in MPTP (1-methyl-4-phenyl-1,2,3,6-tetrahydropyridine) induced Parkinson's disease mice model. Neurodegeneration 5, 137–143.

Daniele, S.G., Beraud, D., Davenport, C., Cheng, K., Yin, H., Maguire-Zeiss, K.A., 2015. Activation of MyD88-dependent TLR1/2 signaling by misfolded alpha-synuclein, a protein linked to neurodegenerative disorders. Sci. Signal. 8, ra45.

Day, B.J., Patel, M., Calavetta, L., Chang, L.Y., Stamler, J.S., 1999. A mechanism of paraquat toxicity involving nitric oxide synthase. Proc. Natl. Acad. Sci. U. S. A. 96, 12760–12765.

De Pablo-Fernandez, E., Lees, A.J., Holton, J.L., Warner, T.T., 2019. Prognosis and neuropathologic correlation of clinical subtypes of Parkinson disease. JAMA Neurol. 76, 470–479.

Deng, X., Tang, C.Y., Zhang, J., Zhu, L., Xie, Z.C., Gong, H.H., Xiao, X.Z., Xu, R.S., 2016. The cortical thickness correlates of clinical manifestations in the mid-stage sporadic Parkinson's disease. Neurosci. Lett. 633, 279–289.

Depino, A.M., Earl, C., Kaczmarczyk, E., Ferrari, C., Besedovsky, H., Del Rey, A., Pitossi, F.J., Oertel, W.H., 2003. Microglial activation with atypical proinflammatory cytokine expression in a rat model of Parkinson's disease. Eur. J. Neurosci. 18, 2731–2742.

Devos, D., Lebouvier, T., Lardeux, B., Biraud, M., Rouaud, T., Pouclet, H., Coron, E., Bruley Des Varannes, S., Naveilhan, P., Nguyen, J.M., Neunlist, M., Derkinderen, P., 2013. Colonic inflammation in Parkinson's disease. Neurobiol. Dis. 50, 42–48.

Di Maio, R., Hoffman, E.K., Rocha, E.M., Keeney, M.T., Sanders, L.H., De Miranda, B.R., Zharikov, A., Van Laar, A., Stepan, A.F., Lanz, T.A., Kofler, J.K., Burton, E.A., Alessi, D.R., Hastings, T.G., Greenamyre, J.T., 2018. LRRK2 activation in idiopathic Parkinson's disease. Sci. Transl. Med. 10, 451.

Dias, V., Junn, E., Mouradian, M.M., 2013. The role of oxidative stress in Parkinson's disease. J. Park. Dis. 3, 461–491.

Dominguez-Villar, M., Hafler, D.A., 2018. Regulatory T cells in autoimmune disease. Nat. Immunol. 19, 665–673.

Doty, R.L., 2012. Olfaction in Parkinson's disease and related disorders. Neurobiol. Dis. 46, 527–552.

Doyle, H.A., Mamula, M.J., 2012. Autoantigenesis: the evolution of protein modifications in autoimmune disease. Curr. Opin. Immunol. 24, 112–118.

Du, Y., Ma, Z., Lin, S., Dodel, R.C., Gao, F., Bales, K.R., Triarhou, L.C., Chernet, E., Perry, K.W., Nelson, D.L., Luecke, S., Phebus, L.A., Bymaster, F.P., Paul, S.M., 2001. Minocycline prevents nigrostriatal dopaminergic neurodegeneration in the MPTP model of Parkinson's disease. Proc. Natl. Acad. Sci. U. S. A. 98, 14669–14674.

Duffy, M.F., Collier, T.J., Patterson, J.R., Kemp, C.J., Luk, K.C., Tansey, M.G., Paumier, K.L., Kanaan, N.M., Fischer, D.L., Polinski, N.K., Barth, O.L., Howe, J.W., Vaikath, N.N., Majbour, N.K., El-Agnaf, O.M.A., Sortwell, C.E., 2018. Lewy body-like alpha-synuclein inclusions trigger reactive microgliosis prior to nigral degeneration. J. Neuroinflammation 15, 129.

Dzamko, N., Gysbers, A., Perera, G., Bahar, A., Shankar, A., Gao, J., Fu, Y., Halliday, G.M., 2017. Toll-like receptor 2 is increased in neurons in Parkinson's disease brain and may contribute to alpha-synuclein pathology. Acta Neuropathol. 133, 303–319.

Edwards, T.L., Scott, W.K., Almonte, C., Burt, A., Powell, E.H., Beecham, G.W., Wang, L., Zuchner, S., Konidari, I., Wang, G., Singer, C., Nahab, F., Scott, B., Stajich, J.M., Pericak-Vance, M., Haines, J., Vance, J.M., Martin, E.R., 2010. Genome-wide association study confirms SNPs in SNCA and the MAPT region as common risk factors for Parkinson disease. Ann. Hum. Genet. 74, 97–109.

Farrer, M.J., 2006. Genetics of Parkinson disease: paradigm shifts and future prospects. Nat. Rev. Genet. 7, 306–318.

Fasano, A., Visanji, N.P., Liu, L.W., Lang, A.E., Pfeiffer, R.F., 2015. Gastrointestinal dysfunction in Parkinson's disease. Lancet Neurol. 14, 625–639.

Fellner, L., Irschick, R., Schanda, K., Reindl, M., Klimaschewski, L., Poewe, W., Wenning, G.K., Stefanova, N., 2013. Toll-like receptor 4 is required for alpha-synuclein dependent activation of microglia and astroglia. Glia 61, 349–360.

Ferreira, M., Massano, J., 2017. An updated review of Parkinson's disease genetics and clinicopathological correlations. Acta Neurol. Scand. 135, 273–284.

Filiano, A.J., Gadani, S.P., Kipnis, J., 2017. How and why do T cells and their derived cytokines affect the injured and healthy brain? Nat. Rev. Neurosci. 18, 375–384.

Fleming, S.M., Salcedo, J., Fernagut, P.O., Rockenstein, E., Masliah, E., Levine, M.S., Chesselet, M.F., 2004. Early and progressive sensorimotor anomalies in mice overexpressing wild-type human alpha-synuclein. J. Neurosci. 24, 9434–9440.

Fleming, S.M., Mulligan, C.K., Richter, F., Mortazavi, F., Lemesre, V., Frias, C., Zhu, C., Stewart, A., Gozes, I., Morimoto, B., Chesselet, M.F., 2011. A pilot trial of the microtubule-interacting peptide (NAP) in mice overexpressing alpha-synuclein shows improvement in motor function and reduction of alpha-synuclein inclusions. Mol. Cell. Neurosci. 46, 597–606.

Foix, C., Nicolesco, J., 1925. Anatomie cérébrale. In: Les noyaux gris centraux et la région mésencéphalo-sous-optique, suivi d'un appendice sur l'anatomie pathologique de la maladie de Parkinson. Masson et Cie; Paris, pp. 508–539.

Forsyth, C.B., Shannon, K.M., Kordower, J.H., Voigt, R.M., Shaikh, M., Jaglin, J.A., Estes, J.D., Dodiya, H.B., Keshavarzian, A., 2011. Increased intestinal permeability correlates with sigmoid mucosa alpha-synuclein staining and endotoxin exposure markers in early Parkinson's disease. PLoS One 6, e28032.

Frigerio, R., Fujishiro, H., Ahn, T.B., Josephs, K.A., Maraganore, D.M., Delledonne, A., Parisi, J.E., Klos, K.J., Boeve, B.F., Dickson, D.W., Ahlskog, J.E., 2011. Incidental Lewy body disease: do some cases represent a preclinical stage of dementia with Lewy bodies? Neurobiol. Aging 32, 857–863.

Fuchs, J., Tichopad, A., Golub, Y., Munz, M., Schweitzer, K.J., Wolf, B., Berg, D., Mueller, J.C., Gasser, T., 2008. Genetic variability in the SNCA gene influences alpha-synuclein levels in the blood and brain. FASEB J. 22, 1327–1334.

Fung, T.C., Olson, C.A., Hsiao, E.Y., 2017. Interactions between the microbiota, immune and nervous systems in health and disease. Nat. Neurosci. 20, 145–155.

Fusco, G., Chen, S.W., Williamson, P.T.F., Cascella, R., Perni, M., Jarvis, J.A., Cecchi, C., Vendruscolo, M., Chiti, F., Cremades, N., Ying, L., Dobson, C.M., De Simone, A., 2017. Structural basis of membrane disruption and cellular toxicity by alpha-synuclein oligomers. Science 358, 1440–1443.

Gao, X., Chen, H., Schwarzschild, M.A., Ascherio, A., 2011. Use of ibuprofen and risk of Parkinson disease. Neurology 76, 863–869.

Gardet, A., Benita, Y., Li, C., Sands, B.E., Ballester, I., Stevens, C., Korzenik, J.R., Rioux, J.D., Daly, M.J., Xavier, R.J., Podolsky, D.K., 2010. LRRK2 is involved in the IFN-gamma response and host response to pathogens. J. Immunol. 185, 5577–5585.

Gerhard, A., Pavese, N., Hotton, G., Turkheimer, F., Es, M., Hammers, A., Eggert, K., Oertel, W., Banati, R.B., Brooks, D.J., 2006. In vivo imaging of microglial activation with [11C](R)-PK11195 PET in idiopathic Parkinson's disease. Neurobiol. Dis. 21, 404–412.

Ginhoux, F., Greter, M., Leboeuf, M., Nandi, S., See, P., Gokhan, S., Mehler, M.F., Conway, S.J., Ng, L.G., Stanley, E.R., Samokhvalov, I.M., Merad, M., 2010. Fate mapping analysis reveals that adult microglia derive from primitive macrophages. Science 330, 841–845.

Goldmann, T., Wieghofer, P., Jordao, M.J., Prutek, F., Hagemeyer, N., Frenzel, K., Amann, L., Staszewski, O., Kierdorf, K., Krueger, M., Locatelli, G., Hochgerner, H., Zeiser, R., Epelman, S., Geissmann, F., Priller, J., Rossi, F.M., Bechmann, I., Kerschensteiner, M., Linnarsson, S., Jung, S., Prinz, M., 2016. Origin, fate and dynamics of macrophages at central nervous system interfaces. Nat. Immunol. 17, 797–805.

Gordon, R., Albornoz, E.A., Christie, D.C., Langley, M.R., Kumar, V., Mantovani, S., Robertson, A.A.B., Butler, M.S., Rowe, D.B., O'neill, L.A., Kanthasamy, A.G., Schroder, K., Cooper, M.A., Woodruff, T.M., 2018. Inflammasome inhibition prevents alpha-synuclein pathology and dopaminergic neurodegeneration in mice. Sci. Transl. Med. 10, 465.

Greenhalgh, A.D., Zarruk, J.G., Healy, L.M., Baskar Jesudasan, S.J., Jhelum, P., Salmon, C.K., Formanek, A., Russo, M.V., Antel, J.P., Mcgavern, D.B., McColl, B.W., David, S., 2018. Peripherally derived macrophages modulate microglial function to reduce inflammation after CNS injury. PLoS Biol. 16, e2005264.

Grozdanov, V., Danzer, K.M., 2018. Release and uptake of pathologic alpha-synuclein. Cell Tissue Res. 373, 175–182.

Grozdanov, V., Bliederhaeuser, C., Ruf, W.P., Roth, V., Fundel-Clemens, K., Zondler, L., Brenner, D., Martin-Villalba, A., Hengerer, B., Kassubek, J., Ludolph, A.C., Weishaupt, J.H., Danzer, K.M., 2014. Inflammatory dysregulation of blood monocytes in Parkinson's disease patients. Acta Neuropathol. 128, 651–663.

Guo, Y., Deng, X., Zheng, W., Xu, H., Song, Z., Liang, H., Lei, J., Jiang, X., Luo, Z., Deng, H., 2011. HLA rs3129882 variant in Chinese Han patients with late-onset sporadic Parkinson disease. Neurosci. Lett. 501, 185–187.

Guzman, J.N., Ilijic, E., Yang, B., Sanchez-Padilla, J., Wokosin, D., Galtieri, D., Kondapalli, J., Schumacker, P.T., Surmeier, D.J., 2018. Systemic isradipine treatment diminishes calcium-dependent mitochondrial oxidant stress. J. Clin. Invest. 128, 2266–2280.

Hakimi, M., Selvanantham, T., Swinton, E., Padmore, R.F., Tong, Y., Kabbach, G., Venderova, K., Girardin, S.E., Bulman, D.E., Scherzer, C.R., Lavoie, M.J., Gris, D., Park, D.S., Angel, J.B., Shen, J., Philpott, D.J., Schlossmacher, M.G., 2011. Parkinson's disease-linked LRRK2 is expressed in circulating and tissue immune cells and upregulated following recognition of microbial structures. J. Neural Transm. (Vienna) 118, 795–808.

Hamza, T.H., Zabetian, C.P., Tenesa, A., Laederach, A., Montimurro, J., Yearout, D., Kay, D.M., Doheny, K.F., Paschall, J., Pugh, E., Kusel, V.I., Collura, R., Roberts, J., Griffith, A., Samii, A., Scott, W.K., Nutt, J., Factor, S.A., Payami, H., 2010. Common genetic variation in the HLA region is associated with late-onset sporadic Parkinson's disease. Nat. Genet. 42, 781–785.

Harms, A.S., Standaert, D.G., 2014. Monocytes and Parkinson's disease: invaders from outside? Mov. Disord. 29, 1242.

Harms, A.S., Barnum, C.J., Ruhn, K.A., Varghese, S., Trevino, I., Blesch, A., Tansey, M.G., 2011. Delayed dominant-negative TNF gene therapy halts progressive loss of nigral dopaminergic neurons in a rat model of Parkinson's disease. Mol. Ther. 19, 46–52.

Harms, A.S., Cao, S., Rowse, A.L., Thome, A.D., Li, X., Mangieri, L.R., Cron, R.Q., Shacka, J.J., Raman, C., Standaert, D.G., 2013. MHCII is required for alpha-synuclein-induced activation of microglia, CD4 T cell proliferation, and dopaminergic neurodegeneration. J. Neurosci. 33, 9592–9600.

Harms, A.S., Delic, V., Thome, A.D., Bryant, N., Liu, Z., Chandra, S., Jurkuvenaite, A., West, A.B., 2017. Alpha-synuclein fibrils recruit peripheral immune cells in the rat brain prior to neurodegeneration. Acta Neuropathol. Commun. 5, 85.

Harms, A.S., Thome, A.D., Yan, Z., Schonhoff, A.M., Williams, G.P., Li, X., Liu, Y., Qin, H., Benveniste, E.N., Standaert, D.G., 2018. Peripheral monocyte entry is required for alpha-synuclein induced inflammation and neurodegeneration in a model of Parkinson disease. Exp. Neurol. 300, 179–187.

Henderson, M.X., Trojanowski, J.Q., Lee, V.M., 2019. alpha-synuclein pathology in Parkinson's disease and related alpha-synucleinopathies. Neurosci. Lett. 709, 134316.

Hernan, M.A., Takkouche, B., Caamano-Isorna, F., Gestal-Otero, J.J., 2002. A meta-analysis of coffee drinking, cigarette smoking, and the risk of Parkinson's disease. Ann. Neurol. 52, 276–284.

Hernandez, D.G., Reed, X., Singleton, A.B., 2016. Genetics in Parkinson disease: Mendelian versus non-Mendelian inheritance. J. Neurochem. 139 (Suppl. 1), 59–74.

Hill-Burns, E.M., Debelius, J.W., Morton, J.T., Wissemann, W.T., Lewis, M.R., Wallen, Z.D., Peddada, S.D., Factor, S.A., Molho, E., Zabetian, C.P., Knight, R., Payami, H., 2017. Parkinson's disease and Parkinson's disease medications have distinct signatures of the gut microbiome. Mov. Disord. 32, 739–749.

Hoeffel, G., Ginhoux, F., 2015. Ontogeny of tissue-resident macrophages. Front. Immunol. 6, 486.

Honda, K., Littman, D.R., 2016. The microbiota in adaptive immune homeostasis and disease. Nature 535, 75–84.

Honkanen, E.A., Saari, L., Orte, K., Gardberg, M., Noponen, T., Joutsa, J., Kaasinen, V., 2019. No link between striatal dopaminergic axons and dopamine transporter imaging in Parkinson's disease. Mov. Disord. 34, 1562–1566.

Hooper, L.V., Littman, D.R., Macpherson, A.J., 2012. Interactions between the microbiota and the immune system. Science 336, 1268–1273.

Hui, K.Y., Fernandez-Hernandez, H., Hu, J., Schaffner, A., Pankratz, N., Hsu, N.Y., Chuang, L.S., Carmi, S., Villaverde, N., Li, X., Rivas, M., Levine, A.P., Bao, X., Labrias, P.R., Haritunians, T., Ruane, D., Gettler, K., Chen, E., Li, D., Schiff, E.R., Pontikos, N., Barzilai, N., Brant, S.R., Bressman, S., Cheifetz, A.S., Clark, L.N., Daly, M.J., Desnick, R.J., Duerr, R.H., Katz, S., Lencz, T., Myers, R.H., Ostrer, H., Ozelius, L., Payami, H., Peter, Y., Rioux, J.D., Segal, A.W., Scott, W.K., Silverberg, M.S., Vance, J.M., Ubarretxena-Belandia, I., Foroud, T., Atzmon, G., Pe'er, I., Ioannou, Y., McGovern, D.P.B., Yue, Z., Schadt, E.E., Cho, J.H., Peter, I., 2018. Functional variants in the LRRK2 gene confer shared effects on risk for Crohn's disease and Parkinson's disease. Sci. Transl. Med. 10, 423.

Iannaccone, S., Cerami, C., Alessio, M., Garibotto, V., Panzacchi, A., Olivieri, S., Gelsomino, G., Moresco, R.M., Perani, D., 2013. In vivo microglia activation in very early dementia with Lewy bodies, comparison with Parkinson's disease. Parkinsonism Relat. Disord. 19, 47–52.

Ichise, M., Kim, Y.J., Ballinger, J.R., Vines, D., Erami, S.S., Tanaka, F., Lang, A.E., 1999. Spect imaging of pre- and postsynaptic dopaminergic alterations in L-dopa-untreated PD. Neurology 52, 1206–1214.

Imamura, K., Hishikawa, N., Sawada, M., Nagatsu, T., Yoshida, M., Hashizume, Y., 2003. Distribution of major histocompatibility complex class II-positive microglia and cytokine profile of Parkinson's disease brains. Acta Neuropathol. 106, 518–526.

International Parkinson Disease Genomics Consortium, Nalls, M.A., Plagnol, V., Hernandez, D.G., Sharma, M., Sheerin, U.M., Saad, M., Simon-Sanchez, J., Schulte, C., Lesage, S., Sveinbjornsdottir, S., Stefansson, K., Martinez, M., Hardy, J., Heutink, P., Brice, A., Gasser, T., Singleton, A.B., Wood, N.W., 2011. Imputation of sequence variants for identification of genetic risks for Parkinson's disease: a meta-analysis of genome-wide association studies. Lancet 377, 641–649.

Jucaite, A., Svenningsson, P., Rinne, J.O., Cselenyi, Z., Varnas, K., Johnstrom, P., Amini, N., Kirjavainen, A., Helin, S., Minkwitz, M., Kugler, A.R., Posener, J.A., Budd, S., Halldin, C., Varrone, A., Farde, L., 2015. Effect of the myeloperoxidase inhibitor AZD3241 on microglia: a PET study in Parkinson's disease. Brain 138, 2687–2700.

Kaelberer, M.M., Buchanan, K.L., Klein, M.E., Barth, B.B., Montoya, M.M., Shen, X., Bohorquez, D.V., 2018. A gut-brain neural circuit for nutrient sensory transduction. Science 361, 6408.

Kannarkat, G.T., Cook, D.A., Lee, J.K., Chang, J., Chung, J., Sandy, E., Paul, K.C., Ritz, B., Bronstein, J., Factor, S.A., Boss, J.M., Tansey, M.G., 2015. Common genetic variant association with altered HLA expression, synergy with pyrethroid exposure, and risk for Parkinson's disease: an observational and case-control study. NPJ Parkinsons Dis. 1, 15002.

Keren-Shaul, H., Spinrad, A., Weiner, A., Matcovitch-Natan, O., Dvir-Szternfeld, R., Ulland, T.K., David, E., Baruch, K., Lara-Astaiso, D., Toth, B., Itzkovitz, S., Colonna, M., Schwartz, M., Amit, I., 2017. A unique microglia type associated with restricting development of Alzheimer's disease. Cell 169, 1276–1290 e17.

Killinger, B.A., Madaj, Z., Sikora, J.W., Rey, N., Haas, A.J., Vepa, Y., Lindqvist, D., Chen, H., Thomas, P.M., Brundin, P., Brundin, L., Labrie, V., 2018. The vermiform appendix impacts the risk of developing Parkinson's disease. Sci. Transl. Med. 10, 465.

Kim, C., Ho, D.H., Suk, J.E., You, S., Michael, S., Kang, J., Joong Lee, S., Masliah, E., Hwang, D., Lee, H.J., Lee, S.J., 2013. Neuron-released oligomeric alpha-synuclein is an endogenous agonist of TLR2 for paracrine activation of microglia. Nat. Commun. 4, 1562.

Kim, C., Lee, H.J., Masliah, E., Lee, S.J., 2016. Non-cell-autonomous neurotoxicity of alpha-synuclein through microglial toll-like receptor 2. Exp. Neurobiol. 25, 113–119.

Kim, S., Kwon, S.H., Kam, T.I., Panicker, N., Karuppagounder, S.S., Lee, S., Lee, J.H., Kim, W.R., Kook, M., Foss, C.A., Shen, C., Lee, H., Kulkarni, S., Pasricha, P.J., Lee, G., Pomper, M.G., Dawson, V.L., Dawson, T.M., Ko, H.S., 2019. Transneuronal propagation of pathologic alpha-synuclein from the gut to the brain models Parkinson's disease. Neuron 103, 627–641 e7.

Kohutnicka, M., Lewandowska, E., Kurkowska-Jastrzebska, I., Czlonkowski, A., Czlonkowska, A., 1998. Microglial and astrocytic involvement in a murine model of Parkinson's disease induced by 1-methyl-4-phenyl-1,2,3,6-tetrahydropyridine (MPTP). Immunopharmacology 39, 167–180.

Kruger, R., Kuhn, W., Muller, T., Woitalla, D., Graeber, M., Kosel, S., Przuntek, H., Epplen, J.T., Schols, L., Riess, O., 1998. Ala30Pro mutation in the gene encoding alpha-synuclein in Parkinson's disease. Nat. Genet. 18, 106–108.

Kuhlmann, A.C., Guilarte, T.R., 2000. Cellular and subcellular localization of peripheral benzodiazepine receptors after trimethyltin neurotoxicity. J. Neurochem. 74, 1694–1704.

Kurkowska-Jastrzebska, I., Wronska, A., Kohutnicka, M., Czlonkowski, A., Czlonkowska, A., 1999. The inflammatory reaction following 1-methyl-4-phenyl-1,2,3,6-tetrahydropyridine intoxication in mouse. Exp. Neurol. 156, 50–61.

Lam, H.A., Wu, N., Cely, I., Kelly, R.L., Hean, S., Richter, F., Magen, I., Cepeda, C., Ackerson, L.C., Walwyn, W., Masliah, E., Chesselet, M.F., Levine, M.S., Maidment, N.T., 2011. Elevated tonic extracellular dopamine concentration and altered dopamine modulation of synaptic activity precede dopamine loss in the striatum of mice overexpressing human alpha-synuclein. J. Neurosci. Res. 89, 1091–1102.

Lambrecht, B.N., Hammad, H., 2015. The immunology of asthma. Nat. Immunol. 16, 45–56.

Lamkanfi, M., Dixit, V.M., 2014. Mechanisms and functions of inflammasomes. Cell 157, 1013–1022.

Langston, J.W., Ballard, P., Tetrud, J.W., Irwin, I., 1983. Chronic Parkinsonism in humans due to a product of meperidine-analog synthesis. Science 219, 979–980.

Langston, J.W., Forno, L.S., Tetrud, J., Reeves, A.G., Kaplan, J.A., Karluk, D., 1999. Evidence of active nerve cell degeneration in the substantia nigra of humans years after 1-methyl-4-phenyl-1,2,3,6-tetrahydropyridine exposure. Ann. Neurol. 46, 598–605.

Lashuel, H.A., Overk, C.R., Oueslati, A., Masliah, E., 2013. The many faces of alpha-synuclein: from structure and toxicity to therapeutic target. Nat. Rev. Neurosci. 14, 38–48.

Lavisse, S., Guillermier, M., Herard, A.S., Petit, F., Delahaye, M., Van Camp, N., Ben Haim, L., Lebon, V., Remy, P., Dolle, F., Delzescaux, T., Bonvento, G., Hantraye, P., Escartin, C., 2012. Reactive astrocytes overexpress TSPO and are detected by TSPO positron emission tomography imaging. J. Neurosci. 32, 10809–10818.

Laydon, D.J., Bangham, C.R., Asquith, B., 2015. Estimating T-cell repertoire diversity: limitations of classical estimators and a new approach. Philos. Trans. R. Soc. Lond. Ser. B Biol. Sci. 370, 1675.

Leipe, J., Grunke, M., Dechant, C., Reindl, C., Kerzendorf, U., Schulze-Koops, H., Skapenko, A., 2010. Role of Th17 cells in human autoimmune arthritis. Arthritis Rheum. 62, 2876–2885.

Leung, S., Liu, X., Fang, L., Chen, X., Guo, T., Zhang, J., 2010. The cytokine milieu in the interplay of pathogenic Th1/Th17 cells and regulatory T cells in autoimmune disease. Cell. Mol. Immunol. 7, 182–189.

Li, W., Wu, X., Hu, X., Wang, T., Liang, S., Duan, Y., Jin, F., Qin, B., 2017. Structural changes of gut microbiota in Parkinson's disease and its correlation with clinical features. Sci. China Life Sci. 60, 1223–1233.

Lieberman, J., 2003. The ABCS of granule-mediated cytotoxicity: new weapons in the arsenal. Nat. Rev. Immunol. 3, 361–370.

Lin, J.C., Lin, C.S., Hsu, C.W., Lin, C.L., Kao, C.H., 2016. Association between Parkinson's disease and inflammatory bowel disease: a nationwide Taiwanese retrospective cohort study. Inflamm. Bowel Dis. 22, 1049–1055.

Lin, C.H., Chen, C.C., Chiang, H.L., Liou, J.M., Chang, C.M., Lu, T.P., Chuang, E.Y., Tai, Y.C., Cheng, C., Lin, H.Y., Wu, M.S., 2019. Altered gut microbiota and inflammatory cytokine responses in patients with Parkinson's disease. J. Neuroinflammation 16, 129.

Lindgren, H.S., Lelos, M.J., Dunnett, S.B., 2012. Do alpha-synuclein vector injections provide a better model of Parkinson's disease than the classic 6-hydroxydopamine model? Exp. Neurol. 237, 36–42.

Liu, Z., Lee, J., Krummey, S., Lu, W., Cai, H., Lenardo, M.J., 2011. The kinase LRRK2 is a regulator of the transcription factor NFAT that modulates the severity of inflammatory bowel disease. Nat. Immunol. 12, 1063–1070.

Liu, W., Liu, X., Li, Y., Zhao, J., Liu, Z., Hu, Z., Wang, Y., Yao, Y., Miller, A.W., Su, B., Cookson, M.R., Li, X., Kang, Z., 2017. LRRK2 promotes the activation of NLRC4 inflammasome during Salmonella typhimurium infection. J. Exp. Med. 214, 3051–3066.

Lodygin, D., Hermann, M., Schweingruber, N., Flugel-Koch, C., Watanabe, T., Schlosser, C., Merlini, A., Korner, H., Chang, H.F., Fischer, H.J., Reichardt, H.M., Zagrebelsky, M., Mollenhauer, B., Kugler, S., Fitzner, D., Frahm, J., Stadelmann, C., Haberl, M., Odoardi, F., Flugel, A., 2019. beta-Synuclein-reactive T cells induce autoimmune CNS grey matter degeneration. Nature 566, 503–508.

Luk, K.C., Song, C., O'brien, P., Stieber, A., Branch, J.R., Brunden, K.R., Trojanowski, J.Q., Lee, V.M., 2009. Exogenous alpha-synuclein fibrils seed the formation of Lewy body-like intracellular inclusions in cultured cells. Proc. Natl. Acad. Sci. U. S. A. 106, 20051–20056.

Luk, K.C., Kehm, V., Carroll, J., Zhang, B., O'brien, P., Trojanowski, J.Q., Lee, V.M., 2012. Pathological alpha-synuclein transmission initiates Parkinson-like neurodegeneration in nontransgenic mice. Science 338, 949–953.

Ma, J., Gao, J., Wang, J., Xie, A., 2019. Prion-like mechanisms in Parkinson's disease. Front. Neurosci. 13, 552.

Manning-Bog, A.B., McCormack, A.L., Li, J., Uversky, V.N., Fink, A.L., Di Monte, D.A., 2002. The herbicide paraquat causes up-regulation and aggregation of alpha-synuclein in mice: paraquat and alpha-synuclein. J. Biol. Chem. 277, 1641–1644.

Mao, X., Ou, M.T., Karuppagounder, S.S., Kam, T.I., Yin, X., Xiong, Y., Ge, P., Umanah, G.E., Brahmachari, S., Shin, J.H., Kang, H.C., Zhang, J., Xu, J., Chen, R., Park, H., Andrabi, S.A., Kang, S.U., Goncalves, R.A., Liang, Y., Zhang, S., Qi, C., Lam, S., Keiler, J.A., Tyson, J., Kim, D., Panicker, N., Yun, S.P., Workman, C.J., Vignali, D.A., Dawson, V.L., Ko, H.S., Dawson, T.M., 2016. Pathological alpha-synuclein transmission initiated by binding lymphocyte-activation gene 3. Science 353, 6307.

Maraganore, D.M., De Andrade, M., Elbaz, A., Farrer, M.J., Ioannidis, J.P., Kruger, R., Rocca, W.A., Schneider, N.K., Lesnick, T.G., Lincoln, S.J., Hulihan, M.M., Aasly, J.O., Ashizawa, T., Chartier-Harlin, M.C., Checkoway, H., Ferrarese, C., Hadjigeorgiou, G., Hattori, N., Kawakami, H., Lambert, J.C., Lynch, T., Mellick, G.D.,

Papapetropoulos, S., Parsian, A., Quattrone, A., Riess, O., Tan, E.K., Van Broeckhoven, C., Genetic Epidemiology Of Parkinson'S Disease Consortium, 2006. Collaborative analysis of alpha-synuclein gene promoter variability and Parkinson disease. JAMA 296, 661–670.

Marinova-Mutafchieva, L., Sadeghian, M., Broom, L., Davis, J.B., Medhurst, A.D., Dexter, D.T., 2009. Relationship between microglial activation and dopaminergic neuronal loss in the substantia nigra: a time course study in a 6-hydroxydopamine model of Parkinson's disease. J. Neurochem. 110, 966–975.

Martin, R., Chain, F., Miquel, S., Lu, J., Gratadoux, J.J., Sokol, H., Verdu, E.F., Bercik, P., Bermudez-Humaran, L.G., Langella, P., 2014. The commensal bacterium Faecalibacterium prausnitzii is protective in DNBS-induced chronic moderate and severe colitis models. Inflamm. Bowel Dis. 20, 417–430.

Martin, R., Miquel, S., Chain, F., Natividad, J.M., Jury, J., Lu, J., Sokol, H., Theodorou, V., Bercik, P., Verdu, E.F., Langella, P., Bermudez-Humaran, L.G., 2015. Faecalibacterium prausnitzii prevents physiological damages in a chronic low-grade inflammation murine model. BMC Microbiol. 15, 67.

Masliah, E., Rockenstein, E., Veinbergs, I., Mallory, M., Hashimoto, M., Takeda, A., Sagara, Y., Sisk, A., Mucke, L., 2000. Dopaminergic loss and inclusion body formation in alpha-synuclein mice: implications for neurodegenerative disorders. Science 287, 1265–1269.

Mata, I.F., Lockhart, P.J., Farrer, M.J., 2004. Parkin genetics: one model for Parkinson's disease. Hum. Mol. Genet. 13, R127–R133. Spec No 1.

Matheoud, D., Sugiura, A., Bellemare-Pelletier, A., Laplante, A., Rondeau, C., Chemali, M., Fazel, A., Bergeron, J.J., Trudeau, L.E., Burelle, Y., Gagnon, E., McBride, H.M., Desjardins, M., 2016. Parkinson's disease-related proteins PINK1 and parkin repress mitochondrial antigen presentation. Cell 166, 314–327.

Matheoud, D., Cannon, T., Voisin, A., Penttinen, A.M., Ramet, L., Fahmy, A.M., Ducrot, C., Laplante, A., Bourque, M.J., Zhu, L., Cayrol, R., Le Campion, A., McBride, H.M., Gruenheid, S., Trudeau, L.E., Desjardins, M., 2019. Intestinal infection triggers Parkinson's disease-like symptoms in PINK1(-/-) mice. Nature 571, 565–569.

Maulik, M., Mitra, S., Bult-Ito, A., Taylor, B.E., Vayndorf, E.M., 2017. Behavioral phenotyping and pathological indicators of Parkinson's disease in C. elegans models. Front. Genet. 8, 77.

McFarland, H.F., Martin, R., 2007. Multiple sclerosis: a complicated picture of autoimmunity. Nat. Immunol. 8, 913–919.

Mcgeer, P.L., Itagaki, S., Akiyama, H., Mcgeer, E.G., 1988a. Rate of cell death in parkinsonism indicates active neuropathological process. Ann. Neurol. 24, 574–576.

Mcgeer, P.L., Itagaki, S., Boyes, B.E., Mcgeer, E.G., 1988b. Reactive microglia are positive for HLA-DR in the substantia nigra of Parkinson's and Alzheimer's disease brains. Neurology 38, 1285–1291.

Mcgeer, P.L., Schwab, C., Parent, A., Doudet, D., 2003. Presence of reactive microglia in monkey substantia nigra years after 1-methyl-4-phenyl-1,2,3,6-tetrahydropyridine administration. Ann. Neurol. 54, 599–604.

Meredith, G.E., Rademacher, D.J., 2011. MPTP mouse models of Parkinson's disease: an update. J. Park. Dis. 1, 19–33.

Merson, T.D., Binder, M.D., Kilpatrick, T.J., 2010. Role of cytokines as mediators and regulators of microglial activity in inflammatory demyelination of the CNS. NeuroMol. Med. 12, 99–132.

Mignot, E., Hayduk, R., Black, J., Grumet, F.C., Guilleminault, C., 1997. HLA DQB1*0602 is associated with cataplexy in 509 narcoleptic patients. Sleep 20, 1012–1020.

Mildner, A., Jung, S., 2014. Development and function of dendritic cell subsets. Immunity 40, 642–656.

Miyara, M., Gorochov, G., Ehrenstein, M., Musset, L., Sakaguchi, S., Amoura, Z., 2011. Human FoxP3+ regulatory T cells in systemic autoimmune diseases. Autoimmun. Rev. 10, 744–755.

Mogi, M., Harada, M., Riederer, P., Narabayashi, H., Fujita, K., Nagatsu, T., 1994. Tumor necrosis factor-alpha (TNF-alpha) increases both in the brain and in the cerebrospinal fluid from parkinsonian patients. Neurosci. Lett. 165, 208–210.

Mogi, M., Harada, M., Narabayashi, H., Inagaki, H., Minami, M., Nagatsu, T., 1996. Interleukin (Il)-1 beta, IL-2, IL-4, IL-6 and transforming growth factor-alpha levels are elevated in ventricular cerebrospinal fluid in juvenile parkinsonism and Parkinson's disease. Neurosci. Lett. 211, 13–16.

Mosaad, Y.M., 2015. Clinical role of human leukocyte antigen in health and disease. Scand. J. Immunol. 82, 283–306.

Mrdjen, D., Pavlovic, A., Hartmann, F.J., Schreiner, B., Utz, S.G., Leung, B.P., Lelios, I., Heppner, F.L., Kipnis, J., Merkler, D., Greter, M., Becher, B., 2018. High-dimensional single-cell mapping of central nervous system immune cells reveals distinct myeloid subsets in health, aging, and disease. Immunity 48, 380–395 e6.

Mundt, S., Mrdjen, D., Utz, S.G., Greter, M., Schreiner, B., Becher, B., 2019. Conventional DCs sample and present myelin antigens in the healthy CNS and allow parenchymal T cell entry to initiate neuroinflammation. Sci. Immunol. 4, 31.

Nayak, D., Roth, T.L., Mcgavern, D.B., 2014. Microglia development and function. Annu. Rev. Immunol. 32, 367–402.

Nemazee, D., 2017. Mechanisms of central tolerance for B cells. Nat. Rev. Immunol. 17, 281–294.

Netea, M.G., Joosten, L.A., Latz, E., Mills, K.H., Natoli, G., Stunnenberg, H.G., O'neill, L.A., Xavier, R.J., 2016. Trained immunity: a program of innate immune memory in health and disease. Science 352, aaf1098.

Neurath, M.F., 2014. Cytokines in inflammatory bowel disease. Nat. Rev. Immunol. 14, 329–342.

Obeso, J.A., Stamelou, M., Goetz, C.G., Poewe, W., Lang, A.E., Weintraub, D., Burn, D., Halliday, G.M., Bezard, E., Przedborski, S., Lehericy, S., Brooks, D.J., Rothwell, J.C., Hallett, M., Delong, M.R., Marras, C., Tanner, C.M., Ross, G.W., Langston, J.W., Klein, C., Bonifati, V., Jankovic, J., Lozano, A.M., Deuschl, G., Bergman, H., Tolosa, E., Rodriguez-Violante, M., Fahn, S., Postuma, R.B., Berg, D., Marek, K., Standaert, D.G., Surmeier, D.J., Olanow, C.W., Kordower, J.H., Calabresi, P., Schapira, A.H.V., Stoessl, A.J., 2017. Past, present, and future of Parkinson's disease: a special essay on the 200th anniversary of the shaking palsy. Mov. Disord. 32, 1264–1310.

Orr, C.F., Rowe, D.B., Mizuno, Y., Mori, H., Halliday, G.M., 2005. A possible role for humoral immunity in the pathogenesis of Parkinson's disease. Brain 128, 2665–2674.

O'shea, J.J., Ma, A., Lipsky, P., 2002. Cytokines and autoimmunity. Nat. Rev. Immunol. 2, 37–45.

Pabon, M.M., Bachstetter, A.D., Hudson, C.E., Gemma, C., Bickford, P.C., 2011. CX3CL1 reduces neurotoxicity and microglial activation in a rat model of Parkinson's disease. J. Neuroinflammation 8, 9.

Pannell, M., Economopoulos, V., Wilson, T.C., Kersemans, V., Isenegger, P.G., Larkin, J.R., Smart, S., Gilchrist, S., Gouverneur, V., Sibson, N.R., 2019. Imaging of Translocator Protein Upregulation is Selective for Pro-Inflammatory Polarized Astrocytes and Microglia. Glia.

Parkinson, J., 2002. An essay on the shaking palsy. 1817. J. Neuropsychiatr. Clin. Neurosci. 14, 223–236. discussion 222.

Parkkinen, L., Pirttila, T., Tervahauta, M., Alafuzoff, I., 2005. Widespread and abundant alpha-synuclein pathology in a neurologically unimpaired subject. Neuropathology 25, 304–314.

Patterson, A.M., Mulder, I.E., Travis, A.J., Lan, A., Cerf-Bensussan, N., Gaboriau-Routhiau, V., Garden, K., Logan, E., Delday, M.I., Coutts, A.G.P., Monnais, E., Ferraria, V.C., Inoue, R., Grant, G., Aminov, R.I., 2017. Human gut symbiont Roseburia hominis promotes and regulates innate immunity. Front. Immunol. 8, 1166.

Paumier, K.L., Luk, K.C., Manfredsson, F.P., Kanaan, N.M., Lipton, J.W., Collier, T.J., Steece-Collier, K., Kemp, C.J., Celano, S., Schulz, E., Sandoval, I.M., Fleming, S., Dirr, E., Polinski, N.K., Trojanowski, J.Q., Lee, V.M., Sortwell, C.E., 2015. Intrastriatal injection of pre-formed mouse alpha-synuclein fibrils into rats triggers alpha-synuclein pathology and bilateral nigrostriatal degeneration. Neurobiol. Dis. 82, 185–199.

Penfield, W., 1925. Microglia and the process of phagocytosis in gliomas. Am. J. Pathol. 1, 77–90 15.

Peng, J., Stevenson, F.F., Oo, M.L., Andersen, J.K., 2009. Iron-enhanced paraquat-mediated dopaminergic cell death due to increased oxidative stress as a consequence of microglial activation. Free Radic. Biol. Med. 46, 312–320.

Peter, I., Dubinsky, M., Bressman, S., Park, A., Lu, C., Chen, N., Wang, A., 2018. Anti-tumor necrosis factor therapy and incidence of Parkinson disease among patients with inflammatory bowel disease. JAMA Neurol. 75, 939–946.

Petrov, V.A., Saltykova, I.V., Zhukova, I.A., Alifirova, V.M., Zhukova, N.G., Dorofeeva, Y.B., Tyakht, A.V., Kovarsky, B.A., Alekseev, D.G., Kostryukova, E.S., Mironova, Y.S., Izhboldina, O.P., Nikitina, M.A., Perevozchikova, T.V., Fait, E.A., Babenko, V.V., Vakhitova, M.T., Govorun, V.M., Sazonov, A.E., 2017. Analysis of gut microbiota in patients with Parkinson's disease. Bull. Exp. Biol. Med. 162, 734–737.

Pickrell, A.M., Youle, R.J., 2015. The roles of PINK1, parkin, and mitochondrial fidelity in Parkinson's disease. Neuron 85, 257–273.

Pietrucci, D., Cerroni, R., Unida, V., Farcomeni, A., Pierantozzi, M., Mercuri, N.B., Biocca, S., Stefani, A., Desideri, A., 2019. Dysbiosis of gut microbiota in a selected population of Parkinson's patients. Parkinsonism Relat. Disord. 65, 124–130.

Polymeropoulos, M.H., Lavedan, C., Leroy, E., Ide, S.E., Dehejia, A., Dutra, A., Pike, B., Root, H., Rubenstein, J., Boyer, R., Stenroos, E.S., Chandrasekharappa, S., Athanassiadou, A., Papapetropoulos, T., Johnson, W.G., Lazzarini, A.M., Duvoisin, R.C., Di Iorio, G., Golbe, L.I., Nussbaum, R.L., 1997. Mutation in the alpha-synuclein gene identified in families with Parkinson's disease. Science 276, 2045–2047.

Postuma, R.B., Adler, C.H., Dugger, B.N., Hentz, J.G., Shill, H.A., Driver-Dunckley, E., Sabbagh, M.N., Jacobson, S.A., Belden, C.M., Sue, L.I., Serrano, G., Beach, T.G., 2015. REM sleep behavior disorder and neuropathology in Parkinson's disease. Mov. Disord. 30, 1413–1417.

Powers, K.M., Kay, D.M., Factor, S.A., Zabetian, C.P., Higgins, D.S., Samii, A., Nutt, J.G., Griffith, A., Leis, B., Roberts, J.W., Martinez, E.D., Montimurro, J.S., Checkoway, H., Payami, H., 2008. Combined effects of smoking, coffee, and NSAIDs on Parkinson's disease risk. Mov. Disord. 23, 88–95.

Prashar, A., Schnettger, L., Bernard, E.M., Gutierrez, M.G., 2017. Rab GTPases in immunity and inflammation. Front. Cell. Infect. Microbiol. 7, 435.

Prinz, M., Priller, J., 2017. The role of peripheral immune cells in the CNS in steady state and disease. Nat. Neurosci. 20, 136–144.

Purisai, M.G., McCormack, A.L., Cumine, S., Li, J., Isla, M.Z., Di Monte, D.A., 2007. Microglial activation as a priming event leading to paraquat-induced dopaminergic cell degeneration. Neurobiol. Dis. 25, 392–400.

Qin, X.Y., Zhang, S.P., Cao, C., Loh, Y.P., Cheng, Y., 2016. Aberrations in peripheral inflammatory cytokine levels in Parkinson disease: a systematic review and meta-analysis. JAMA Neurol. 73, 1316–1324.

Qureshi, S.T., Lariviere, L., Leveque, G., Clermont, S., Moore, K.J., Gros, P., Malo, D., 1999. Endotoxin-tolerant mice have mutations in toll-like receptor 4 (Tlr4). J. Exp. Med. 189, 615–625.

Raj, T., Rothamel, K., Mostafavi, S., Ye, C., Lee, M.N., Replogle, J.M., Feng, T., Lee, M., Asinovski, N., Frohlich, I., Imboywa, S., Von Korff, A., Okada, Y., Patsopoulos, N.A., Davis, S., McCabe, C., Paik, H.I., Srivastava, G.P., Raychaudhuri, S., Hafler, D.A., Koller, D., Regev, A., Hacohen, N., Mathis, D., Benoist, C., Stranger, B.E., De Jager, P.L., 2014. Polarization of the effects of autoimmune and neurodegenerative risk alleles in leukocytes. Science 344, 519–523.

Rankin, L.C., Artis, D., 2018. Beyond host defense: emerging functions of the immune system in regulating complex tissue physiology. Cell 173, 554–567.

Reynolds, A.D., Stone, D.K., Hutter, J.A., Benner, E.J., Mosley, R.L., Gendelman, H.E., 2010. Regulatory T cells attenuate Th17 cell-mediated nigrostriatal dopaminergic neurodegeneration in a model of Parkinson's disease. J. Immunol. 184, 2261–2271.

Rivera, A., Siracusa, M.C., Yap, G.S., Gause, W.C., 2016. Innate cell communication kickstarts pathogen-specific immunity. Nat. Immunol. 17, 356–363.

Rockenstein, E., Mallory, M., Hashimoto, M., Song, D., Shults, C.W., Lang, I., Masliah, E., 2002. Differential neuropathological alterations in transgenic mice expressing alpha-synuclein from the platelet-derived growth factor and Thy-1 promoters. J. Neurosci. Res. 68, 568–578.

Rosenkranz, D., Weyer, S., Tolosa, E., Gaenslen, A., Berg, D., Leyhe, T., Gasser, T., Stoltze, L., 2007. Higher frequency of regulatory T cells in the elderly and increased suppressive activity in neurodegeneration. J. Neuroimmunol. 188, 117–127.

Rozemuller, A.J., Eikelenboom, P., Theeuwes, J.W., Jansen Steur, E.N., De Vos, R.A., 2000. Activated microglial cells and complement factors are unrelated to cortical Lewy bodies. Acta Neuropathol. 100, 701–708.

Russo, I., Kaganovich, A., Ding, J., Landeck, N., Mamais, A., Varanita, T., Biosa, A., Tessari, I., Bubacco, L., Greggio, E., Cookson, M.R., 2019. Transcriptome analysis of LRRK2 knock-out microglia cells reveals alterations of inflammatory- and oxidative stress-related pathways upon treatment with alpha-synuclein fibrils. Neurobiol. Dis. 129, 67–78.

Sacino, A.N., Brooks, M., McKinney, A.B., Thomas, M.A., Shaw, G., Golde, T.E., Giasson, B.I., 2014. Brain injection of alpha-synuclein induces multiple proteinopathies, gliosis, and a neuronal injury marker. J. Neurosci. 34, 12368–12378.

Sampedro, F., Marin-Lahoz, J., Martinez-Horta, S., Pagonabarraga, J., Kulisevsky, J., 2019. Dopaminergic degeneration induces early posterior cortical thinning in Parkinson's disease. Neurobiol. Dis. 124, 29–35.

Sanchez-Guajardo, V., Febbraro, F., Kirik, D., Romero-Ramos, M., 2010. Microglia acquire distinct activation profiles depending on the degree of alpha-synuclein neuropathology in a rAAV based model of Parkinson's disease. PLoS One 5, e8784.

Sanchez-Guajardo, V., Tentillier, N., Romero-Ramos, M., 2015. The relation between alpha-synuclein and microglia in Parkinson's disease: recent developments. Neuroscience 302, 47–58.

Saunders, J.A., Estes, K.A., Kosloski, L.M., Allen, H.E., Dempsey, K.M., Torres-Russotto, D.R., Meza, J.L., Santamaria, P.M., Bertoni, J.M., Murman, D.L., Ali, H.H., Standaert, D.G., Mosley, R.L., Gendelman, H.E., 2012. CD4+ regulatory and effector/memory T cell subsets profile motor dysfunction in Parkinson's disease. J. NeuroImmune Pharmacol. 7, 927–938.

Scheperjans, F., Aho, V., Pereira, P.A., Koskinen, K., Paulin, L., Pekkonen, E., Haapaniemi, E., Kaakkola, S., Eerola-Rautio, J., Pohja, M., Kinnunen, E., Murros, K., Auvinen, P., 2015. Gut microbiota are related to Parkinson's disease and clinical phenotype. Mov. Disord. 30, 350–358.

Schuler, F., Casida, J.E., 2001. Functional coupling of PSST and ND1 subunits in NADH:ubiquinone oxidoreductase established by photoaffinity labeling. Biochim. Biophys. Acta 1506, 79–87.

Shannon, K.M., Keshavarzian, A., Mutlu, E., Dodiya, H.B., Daian, D., Jaglin, J.A., Kordower, J.H., 2012. Alpha-synuclein in colonic submucosa in early untreated Parkinson's disease. Mov. Disord. 27, 709–715.

Shen, Z., Zhu, C., Quan, Y., Yang, J., Yuan, W., Yang, Z., Wu, S., Luo, W., Tan, B., Wang, X., 2018. Insights into Roseburia intestinalis which alleviates experimental colitis pathology by inducing anti-inflammatory responses. J. Gastroenterol. Hepatol. 33, 1751–1760.

Sherer, T.B., Betarbet, R., Kim, J.H., Greenamyre, J.T., 2003. Selective microglial activation in the rat rotenone model of Parkinson's disease. Neurosci. Lett. 341, 87–90.

Shi, M., Liu, C., Cook, T.J., Bullock, K.M., Zhao, Y., Ginghina, C., Li, Y., Aro, P., Dator, R., He, C., Hipp, M.J., Zabetian, C.P., Peskind, E.R., Hu, S.C., Quinn, J.F., Galasko, D.R., Banks, W.A., Zhang, J., 2014. Plasma exosomal alpha-synuclein is likely CNS-derived and increased in Parkinson's disease. Acta Neuropathol. 128, 639–650.

Shimoji, M., Pagan, F., Healton, E.B., Mocchetti, I., 2009. CXCR4 and CXCL12 expression is increased in the nigro-striatal system of Parkinson's disease. Neurotox. Res. 16, 318–328.

Singleton, A.B., Farrer, M., Johnson, J., Singleton, A., Hague, S., Kachergus, J., Hulihan, M., Peuralinna, T., Dutra, A., Nussbaum, R., Lincoln, S., Crawley, A., Hanson, M., Maraganore, D., Adler, C., Cookson, M.R., Muenter, M., Baptista, M., Miller, D., Blancato, J., Hardy, J., Gwinn-Hardy, K., 2003. alpha-Synuclein locus triplication causes Parkinson's disease. Science 302, 841.

Sliter, D.A., Martinez, J., Hao, L., Chen, X., Sun, N., Fischer, T.D., Burman, J.L., Li, Y., Zhang, Z., Narendra, D.P., Cai, H., Borsche, M., Klein, C., Youle, R.J., 2018. Parkin and PINK1 mitigate sting-induced inflammation. Nature 561, 258–262.

Sokol, H., Pigneur, B., Watterlot, L., Lakhdari, O., Bermudez-Humaran, L.G., Gratadoux, J.J., Blugeon, S., Bridonneau, C., Furet, J.P., Corthier, G., Grangette, C., Vasquez, N., Pochart, P., Trugnan, G., Thomas, G., Blottiere, H.M., Dore, J., Marteau, P., Seksik, P., Langella, P., 2008. Faecalibacterium prausnitzii is an anti-inflammatory commensal bacterium identified by gut microbiota analysis of Crohn disease patients. Proc. Natl. Acad. Sci. U. S. A. 105, 16731–16736.

Sommer, A., Marxreiter, F., Krach, F., Fadler, T., Grosch, J., Maroni, M., Graef, D., Eberhardt, E., Riemenschneider, M.J., Yeo, G.W., Kohl, Z., Xiang, W., Gage, F.H., Winkler, J., Prots, I., Winner, B., 2018. Th17 lymphocytes induce neuronal cell death in a human iPSC-based model of Parkinson's disease. Cell Stem Cell 23, 123–131 e6.

Song, W.M., Colonna, M., 2018. The identity and function of microglia in neurodegeneration. Nat. Immunol. 19, 1048–1058.

Sonsalla, P.K., Heikkila, R.E., 1986. The influence of dose and dosing interval on MPTP-induced dopaminergic neurotoxicity in mice. Eur. J. Pharmacol. 129, 339–345.

Sorrentino, Z.A., Brooks, M.M.T., Hudson 3rd, V., Rutherford, N.J., Golde, T.E., Giasson, B.I., Chakrabarty, P., 2017. Intrastriatal injection of alpha-synuclein can lead to widespread synucleinopathy independent of neuroanatomic connectivity. Mol. Neurodegener. 12, 40.

Soto-Otero, R., Mendez-Alvarez, E., Hermida-Ameijeiras, A., Munoz-Patino, A.M., Labandeira-Garcia, J.L., 2000. Autoxidation and neurotoxicity of 6-hydroxydopamine in the presence of some antioxidants: potential implication in relation to the pathogenesis of Parkinson's disease. J. Neurochem. 74, 1605–1612.

Spillantini, M.G., Schmidt, M.L., Lee, V.M., Trojanowski, J.Q., Jakes, R., Goedert, M., 1997. Alpha-synuclein in Lewy bodies. Nature 388, 839–840.

Stastny, P., 1978. Association of the B-cell alloantigen DRw4 with rheumatoid arthritis. N. Engl. J. Med. 298, 869–871.

Stefanis, L., Emmanouilidou, E., Pantazopoulou, M., Kirik, D., Vekrellis, K., Tofaris, G.K., 2019. How is alpha-synuclein cleared from the cell? J. Neurochem. 150, 577–590.

Steger, M., Tonelli, F., Ito, G., Davies, P., Trost, M., Vetter, M., Wachter, S., Lorentzen, E., Duddy, G., Wilson, S., Baptista, M.A., Fiske, B.K., Fell, M.J., Morrow, J.A., Reith, A.D., Alessi, D.R., Mann, M., 2016. Phosphoproteomics reveals that Parkinson's disease kinase LRRK2 regulates a subset of Rab GTPases. elife 5, e12813.

Su, X., Federoff, H.J., Maguire-Zeiss, K.A., 2009. Mutant alpha-synuclein overexpression mediates early proinflammatory activity. Neurotox. Res. 16, 238–254.

Sulzer, D., Edwards, R.H., 2019. The physiological role of alpha-synuclein and its relationship to Parkinson's disease. J. Neurochem. 150, 475–486.

Sulzer, D., Alcalay, R.N., Garretti, F., Cote, L., Kanter, E., Agin-Liebes, J., Liong, C., Mcmurtrey, C., Hildebrand, W.H., Mao, X., Dawson, V.L., Dawson, T.M., Oseroff, C., Pham, J., Sidney, J., Dillon, M.B., Carpenter, C., Weiskopf, D., Phillips, E., Mallal, S., Peters, B., Frazier, A., Lindestam Arlehamn, C.S., Sette, A., 2017. T cells from patients with Parkinson's disease recognize alpha-synuclein peptides. Nature 546, 656–661.

Tan, B., Luo, W., Shen, Z., Xiao, M., Wu, S., Meng, X., Wu, X., Yang, Z., Tian, L., Wang, X., 2019. Roseburia intestinalis inhibits oncostatin M and maintains tight junction integrity in a murine model of acute experimental colitis. Scand. J. Gastroenterol. 54, 432–440.

Tay, T.L., Mai, D., Dautzenberg, J., Fernandez-Klett, F., Lin, G., Sagar, Datta, M., Drougard, A., Stempfl, T., Ardura-Fabregat, A., Staszewski, O., Margineanu, A., Sporbert, A., Steinmetz, L.M., Pospisilik, J.A., Jung, S., Priller, J., Grun, D., Ronneberger, O., Prinz, M., 2017. A new fate mapping system reveals context-dependent random or clonal expansion of microglia. Nat. Neurosci. 20, 793–803.

Teismann, P., Ferger, B., 2001. Inhibition of the cyclooxygenase isoenzymes COX-1 and COX-2 provide neuroprotection in the MPTP-mouse model of Parkinson's disease. Synapse 39, 167–174.

Terada, T., Yokokura, M., Yoshikawa, E., Futatsubashi, M., Kono, S., Konishi, T., Miyajima, H., Hashizume, T., Ouchi, Y., 2016. Extrastriatal spreading of microglial activation in Parkinson's disease: a positron emission tomography study. Ann. Nucl. Med. 30, 579–587.

Tetrud, J.W., Langston, J.W., 1989. MPTP-induced parkinsonism as a model for Parkinson's disease. Acta Neurol. Scand. Suppl. 126, 35–40.

Thakur, P., Nehru, B., 2015. Inhibition of neuroinflammation and mitochondrial dysfunctions by carbenoxolone in the rotenone model of Parkinson's disease. Mol. Neurobiol. 51, 209–219.

Theodore, S., Maragos, W., 2015. 6-Hydroxydopamine as a tool to understand adaptive immune system-induced dopamine neurodegeneration in Parkinson's disease. Immunopharmacol. Immunotoxicol. 37, 393–399.

Theodore, S., Cao, S., Mclean, P.J., Standaert, D.G., 2008. Targeted overexpression of human alpha-synuclein triggers microglial activation and an adaptive immune zresponse in a mouse model of Parkinson disease. J. Neuropathol. Exp. Neurol. 67, 1149–1158.

Theofilopoulos, A.N., Kono, D.H., Baccala, R., 2017. The multiple pathways to autoimmunity. Nat. Immunol. 18, 716–724.

Thome, A.D., Harms, A.S., Volpicelli-Daley, L.A., Standaert, D.G., 2016. microRNA-155 regulates alpha-synuclein-induced inflammatory responses in models of Parkinson disease. J. Neurosci. 36, 2383–2390.

Ungerstedt, U., 1968. 6-Hydroxy-dopamine induced degeneration of central monoamine neurons. Eur. J. Pharmacol. 5, 107–110.

Uribe, C., Segura, B., Baggio, H.C., Abos, A., Garcia-Diaz, A.I., Campabadal, A., Marti, M.J., Valldeoriola, F., Compta, Y., Bargallo, N., Junque, C., 2018. Gray/white matter contrast in Parkinson's disease. Front. Aging Neurosci. 10, 89.

Vanhauwaert, R., Verstreken, P., 2015. Flies with Parkinson's disease. Exp. Neurol. 274, 42–51.

Vidal-Martinez, G., Vargas-Medrano, J., Gil-Tommee, C., Medina, D., Garza, N.T., Yang, B., Segura-Ulate, I., Dominguez, S.J., Perez, R.G., 2016. FTY720/fingolimod reduces synucleinopathy and improves gut motility in A53T Mice: CONTRIBUTIONS OF PRO-BRAIN-DERIVED NEUROTROPHIC FACTOR (PRO-BDNF) AND MATURE BDNF. J. Biol. Chem. 291, 20811–20821.

Vignali, D.A., Collison, L.W., Workman, C.J., 2008. How regulatory T cells work. Nat. Rev. Immunol. 8, 523–532.

Visanji, N.P., Brotchie, J.M., Kalia, L.V., Koprich, J.B., Tandon, A., Watts, J.C., Lang, A.E., 2016. Alpha-Synuclein-based animal models of Parkinson's disease: challenges and opportunities in a new era. Trends Neurosci. 39, 750–762.

Vitetta, E.S., Berton, M.T., Burger, C., Kepron, M., Lee, W.T., Yin, X.M., 1991. Memory B and T cells. Annu. Rev. Immunol. 9, 193–217.

Volpicelli-Daley, L.A., 2017. Effects of alpha-synuclein on axonal transport. Neurobiol. Dis. 105, 321–327.

Volpicelli-Daley, L.A., Luk, K.C., Patel, T.P., Tanik, S.A., Riddle, D.M., Stieber, A., Meaney, D.F., Trojanowski, J.Q., Lee, V.M., 2011. Exogenous alpha-synuclein fibrils induce Lewy body pathology leading to synaptic dysfunction and neuron death. Neuron 72, 57–71.

Walsh, J.T., Hendrix, S., Boato, F., Smirnov, I., Zheng, J., Lukens, J.R., Gadani, S., Hechler, D., Golz, G., Rosenberger, K., Kammertons, T., Vogt, J., Vogelaar, C., Siffrin, V., Radjavi, A., Fernandez-Castaneda, A., Gaultier, A., Gold, R., Kanneganti, T.D., Nitsch, R., Zipp, F., Kipnis, J., 2015. MHCII-independent CD4+ T cells protect injured CNS neurons via IL-4. J. Clin. Invest. 125, 699–714.

Watson, M.B., Richter, F., Lee, S.K., Gabby, L., Wu, J., Masliah, E., Effros, R.B., Chesselet, M.F., 2012. Regionally-specific microglial activation in young mice overexpressing human wildtype alpha-synuclein. Exp. Neurol. 237, 318–334.

Wheeler, C.J., Seksenyan, A., Koronyo, Y., Rentsendorj, A., Sarayba, D., Wu, H., Gragg, A., Siegel, E., Thomas, D., Espinosa, A., Thompson, K., Black, K., Koronyo-Hamaoui, M.,

Pechnick, R., Irvin, D.K., 2014. T-Lymphocyte deficiency exacerbates behavioral deficits in the 6-OHDA unilateral lesion rat model for Parkinson's disease. J. Neurol. Neurophysiol. 5, 3.

Williams, G.P., Schonhoff, A.M., Jurkuvenaite, A., Thome, A.D., Standaert, D.G., Harms, A.S., 2018. Targeting of the class II transactivator attenuates inflammation and neurodegeneration in an alpha-synuclein model of Parkinson's disease. J. Neuroinflammation 15, 244.

Williams-Gray, C.H., Wijeyekoon, R., Yarnall, A.J., Lawson, R.A., Breen, D.P., Evans, J.R., Cummins, G.A., Duncan, G.W., Khoo, T.K., Burn, D.J., Barker, R.A., ICICLE-PD Study Group, 2016. Serum immune markers and disease progression in an incident Parkinson's disease cohort (ICICLE-PD). Mov. Disord. 31, 995–1003.

Wu, D.C., Jackson-Lewis, V., Vila, M., Tieu, K., Teismann, P., Vadseth, C., Choi, D.K., Ischiropoulos, H., Przedborski, S., 2002. Blockade of microglial activation is neuroprotective in the 1-methyl-4-phenyl-1,2,3,6-tetrahydropyridine mouse model of Parkinson disease. J. Neurosci. 22, 1763–1771.

Yamada, T., Mcgeer, P.L., Mcgeer, E.G., 1992. Lewy bodies in Parkinson's disease are recognized by antibodies to complement proteins. Acta Neuropathol. 84, 100–104.

Yang, Y., Bazhin, A.V., Werner, J., Karakhanova, S., 2013. Reactive oxygen species in the immune system. Int. Rev. Immunol. 32, 249–270.

Yun, S.P., Kam, T.I., Panicker, N., Kim, S., Oh, Y., Park, J.S., Kwon, S.H., Park, Y.J., Karuppagounder, S.S., Park, H., Kim, S., Oh, N., Kim, N.A., Lee, S., Brahmachari, S., Mao, X., Lee, J.H., Kumar, M., An, D., Kang, S.U., Lee, Y., Lee, K.C., Na, D.H., Kim, D., Lee, S.H., Roschke, V.V., Liddelow, S.A., Mari, Z., Barres, B.A., Dawson, V.L., Lee, S., Dawson, T.M., Ko, H.S., 2018. Block of A1 astrocyte conversion by microglia is neuroprotective in models of Parkinson's disease. Nat. Med. 24, 931–938.

Zhou, L., Zhang, M., Wang, Y., Dorfman, R.G., Liu, H., Yu, T., Chen, X., Tang, D., Xu, L., Yin, Y., Pan, Y., Zhou, Q., Zhou, Y., Yu, C., 2018. Faecalibacterium prausnitzii produces butyrate to maintain Th17/Treg balance and to ameliorate colorectal colitis by inhibiting histone deacetylase 1. Inflamm. Bowel Dis. 24 (9), 1926–1940. https://doi.org/10.1093/ibd/izy182.

Zhu, J., Paul, W.E., 2008. CD4 T cells: fates, functions, and faults. Blood 112, 1557–1569.

Zhu, C., Song, K., Shen, Z., Quan, Y., Tan, B., Luo, W., Wu, S., Tang, K., Yang, Z., Wang, X., 2018. Roseburia intestinalis inhibits interleukin17 excretion and promotes regulatory T cells differentiation in colitis. Mol. Med. Rep. 17, 7567–7574.

Zimprich, A., Biskup, S., Leitner, P., Lichtner, P., Farrer, M., Lincoln, S., Kachergus, J., Hulihan, M., Uitti, R.J., Calne, D.B., Stoessl, A.J., Pfeiffer, R.F., Patenge, N., Carbajal, I.C., Vieregge, P., Asmus, F., Muller-Myhsok, B., Dickson, D.W., Meitinger, T., Strom, T.M., Wszolek, Z.K., Gasser, T., 2004. Mutations in LRRK2 cause autosomal-dominant parkinsonism with pleomorphic pathology. Neuron 44, 601–607.

Pathways of protein synthesis and degradation in PD pathogenesis

7

Rebekah G. Langston, Mark R. Cookson[*]

Laboratory of Neurogenetics, National Institute on Aging, National Institutes of Health, Bethesda, MD, United States

[*]*Corresponding author: Tel.: +1-301-451-3870, e-mail address: cookson@mail.nih.gov*

Abstract

Since the discovery of protein aggregates in the brains of individuals with Parkinson's disease (PD) in the early 20th century, the scientific community has been interested in the role of dysfunctional protein metabolism in PD etiology. Recent advances in the field have implicated defective protein handling underlying PD through genetic, *in vitro*, and *in vivo* studies incorporating many disease models alongside neuropathological evidence. Here, we discuss the existing body of research focused on understanding cellular pathways of protein synthesis and degradation, and how aberrations in either system could engender PD pathology with special attention to α-synuclein-related consequences. We consider transcription, translation, and post-translational modification to constitute protein synthesis, and protein degradation to encompass proteasome-, lysosome- and endoplasmic reticulum-dependent mechanisms. Novel findings connecting each of these steps in protein metabolism to development of PD indicate that deregulation of protein production and turnover remains an exciting area in PD research.

Keywords

Synthesis, Degradation, α-Synuclein, Transcription, Translation, Autophagy, Proteasome

1 Introduction

Protein inclusions named Lewy bodies were an early piece of evidence that pointed to dysregulation of protein metabolism as a key part of the pathophysiological process underlying Parkinson's disease (PD; reviewed in Engelhardt, 2017). Initially, Lewy bodies were identified in surviving neurons in various brain regions by classic histological staining methods. Subsequently, it was demonstrated that a major component of Lewy bodies is the normally synaptic protein α-synuclein, although other proteins and lipids are also present in these structures (Baba et al., 1998; Mezey et al., 1998; Spillantini et al., 1997). Staining with α-synuclein antibodies revealed a range

Progress in Brain Research, Volume 252, ISSN 0079-6123, https://doi.org/10.1016/bs.pbr.2020.01.002

of structures in neurons of the PD brain, including Lewy bodies in the perikaryon and Lewy neurites in axons and dendrites. It is known that α-synuclein is aggregated within Lewy bodies and post-translationally modified, leading to relative insolubility of the protein after extraction from brains.

However, it still is not clear how Lewy bodies arise and it remains controversial as to the role they play in the neurodegeneration that leads to PD. In general, protein aggregates in a cell are thought to form due to increased levels of substrate overwhelming the degradation machinery or decreased proteolytic activity of a dysfunctional degradation pathway. Here, we will discuss pathways of protein synthesis and degradation relevant to PD in general and to α-synuclein metabolism in particular with a view to understanding how Lewy bodies and Lewy neurites form and may contribute to the disease process.

2 Protein synthesis

Protein synthesis is the culmination of the "central dogma of molecular biology" where genomic DNA is transcribed to RNA, then translated to protein (Crick, 1970). In order to achieve their proper conformations and fulfill their intended cellular functions, proteins undergo post-translational modifications such as glycosylation, phosphorylation, and methylation (Cooper, 2000; Khoury et al., 2011). Cells therefore require careful regulation of transcription, translation, and post-translational modification in order to regulate protein amount and function. This section discusses disruptions in each of these three protein synthesis steps that have been implicated in PD pathophysiology, as illustrated in Fig. 1.

2.1 Transcription

As well as being a marker of Lewy bodies, point mutations in the gene that encodes α-synuclein can cause familial forms of PD (Krüger et al., 1998; Polymeropoulos, 1997). Identification of these *SNCA* mutations lead to several suggestions for how mutations may change function to cause pathogenesis, including hypotheses suggesting protein aggregation as a mechanism (reviewed in Nussbaum, 2017). However, the subsequent discovery that triplication of the *SNCA* locus caused an autosomal dominant form of Lewy body disease suggested that altered gene dosage might contribute to disease risk (Singleton et al., 2003). *SNCA* triplication on one allele, plus one copy on the other allele, gave affected individuals four functional copies of *SNCA* rather than the usual two effectively doubling the gene dosage (Singleton et al., 2003). Doubled gene dosage corresponded to a twofold increase in *SNCA* mRNA in human brain in a small post-mortem analysis of frontal cortex and cerebellum of *SNCA* triplication carriers compared to age-matched controls, indicating that extra copies of *SNCA* results in increased *SNCA* transcription (Miller et al., 2004). *SNCA* gene dosage has been shown to correlate with age of disease onset,

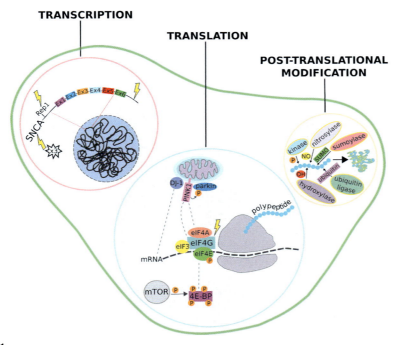

TRANSCRIPTION

TRANSLATION

**POST-TRANSLATIONAL
MODIFICATION**

FIG. 1

Illustration of protein synthesis pathways relevant to Parkinson's disease pathogenesis. Transcription of the α-synuclein gene *SNCA* is highlighted, occurring in the nucleus of the cell. Three genetic insults associated with PD are indicated by lightning bolts: locus triplication, Rep1 dinucleotide repeat expansion, and 3′ variation. The ribosome is depicted with translation initiation factors (eIFs) shown to be mechanistically involved in PD, along with mitochondrial and mammalian target of rapamycin (mTOR) cascades that may influence the translation process. PD-associated genetic mutations in mitochondrial proteins are indicated by lightning bolts. Post-translational modification enzymes represent the numerous alterations made to polypeptides and proteins during their lifetime in the cell, which may be deregulated in PD.

severity of Parkinsonian symptoms, and disease progression in parkinsonism families harboring *SNCA* multiplications (Chartier-Harlin et al., 2004; Farrer et al., 2004; Fuchs et al., 2007; Ibáñez et al., 2004; Ross et al., 2008). The correlation with disease onset was validated in a meta-analysis of 59 families with regions of multiplication encompassing *SNCA* and a range of other genes in the genomic locus (Book et al., 2018). Though there is not necessarily a linear relationship between gene copy number and gene expression, these studies suggest that chronic overexpression of wild-type α-synuclein protein following from *SNCA* multiplication can be as deleterious as mutations in α-synuclein.

In addition to rare *SNCA* multiplication (Johnson et al., 2004) underlying transcriptional dysregulation in PD, common variation in the promoter region of *SNCA* has also been shown to influence α-synuclein expression level. Rep1, a polymorphic dinucleotide repeat element about 9 kb upstream of the transcriptional start site of *SNCA*, was predicted to be functionally important by comparative sequence analysis of the *SNCA* locus in human and mouse and found to be required for normal basal *SNCA* expression *in vitro* (Touchman et al., 2001) and *in vivo* (Cronin et al., 2009). Five Rep1 alleles each differing in length by two nucleotides have been observed within study populations, with three of the five commonly occurring (Farrer et al., 2001; Hellman et al., 1998; Krüger et al., 1999; Xia et al., 1996). An increased number of Rep1 dinucleotide repeats has been associated with increased risk of sporadic PD in some studies (Farrer et al., 2001; Krüger et al., 1999; Maraganore et al., 2006; Tan et al., 2000), but not all (Izumi et al., 2001; Parsian et al., 1998; Soldner et al., 2016). This disparity might be partially explained by co-inheritance with other PD susceptibility single nucleotide polymorphisms (SNPs) (Mizuta et al., 2006; Pals et al., 2004) or by sequence variability in alleles of the same length (Farrer et al., 2001), though one study demonstrated that sequence variation between same-sized Rep1 alleles did not significantly affect promoter activity *in vitro* (Chiba-Falek et al., 2003). Supporting the hypothesis that increased Rep1 length enhances *SNCA* transcription, human *SNCA* locus transgenic mice homozygous for the longer PD risk-associated Rep1 allele expressed higher brain levels of both human *SNCA* mRNA and α-synuclein protein compared to mice homozygous for the shorter protective Rep1 allele (Cronin et al., 2009). This effect was also observed in neurologically normal post-mortem human brain including substantia nigra, with lower *SNCA* mRNA levels observed in individuals homozygous for the shorter Rep1 allele compared to individuals carrying at least one longer PD risk-conferring Rep1 allele (Linnertz et al., 2009). Although the precise impact of Rep1 on *SNCA* transcription remains somewhat uncertain, collectively these results show that PD-associated variation in the *SNCA* promoter region may facilitate increased α-synuclein expression.

Interrogation of PD-associated genetic variation beyond Rep1 has offered further evidence for contribution of abnormal *SNCA* transcription to Parkinson's disease pathogenesis. Genome-wide association studies have nominated variants upstream, downstream, and within the *SNCA* gene as contributors to PD risk (Nalls et al., 2014), with three independent signals at the locus, one 5' and two 3', driving the disease association (Pihlstrøm et al., 2018). The SNP most highly associated with PD risk is located downstream of the *SNCA* gene, and may be functional as it resides within a brain-specific regulatory element and the risk allele was associated with higher levels of α-synuclein in CSF (Pihlstrøm et al., 2018). Variation in the 3' region modulates *SNCA* transcription in PD brain as well (Fuchs et al., 2008; Linnertz et al., 2009; Rhinn et al., 2012). PD-associated variation within the *SNCA* gene also affects α-synuclein expression. *SNCA* mRNA levels in human frontal cortex samples from both cases (PD, DLB) and controls were positively correlated with number of PD susceptibility SNPs located primarily within *SNCA* that were in high linkage disequilibrium with a PD risk variant identified via a screen of 121 candidate genes

(Mizuta et al., 2006). Using CRISPR/Cas9-mediated genome editing in induced pluripotent stem cell (iPSC)-derived neuronal cells, Soldner et al. demonstrated that a PD risk SNP in a distal enhancer element in *SNCA* intron 4 increased *SNCA* expression through altered binding of brain-specific transcription factors (Soldner et al., 2016). Epigenetic modification of the *SNCA* locus has been recognized as an important regulator of α-synuclein expression as well (Labbé et al., 2016). One example is detection of significant PD-specific demethylation of the promoter region of *SNCA*, shown to activate *SNCA* expression *in vitro*, in substantia nigra pars compacta (SNc) DNA of patients compared to controls (Jowaed et al., 2010; Matsumoto et al., 2010).

These results predict genetic variants associated with PD at the *SNCA* locus impart disease risk by increased transcription. However, measurements of *SNCA* mRNA in the SNc of PD brain which would test this hypothesis in a simple manner have been inconsistent. Some investigations of *SNCA* mRNA in PD brain have reported decreased α-synuclein expression in SNc dopaminergic neurons of PD patients compared to controls as determined by *in situ* hybridization (Kingsbury et al., 2004), ribonuclease protection assay (Neystat et al., 1999), or quantitative real-time PCR (qPCR) (Dächsel et al., 2007; Papapetropoulos et al., 2007). Another report found no difference between *SNCA* mRNA levels in control and PD nigral samples by qPCR (Fuchs et al., 2008). However, some studies have measured increased *SNCA* mRNA in PD substantia nigra *versus* control by ribonuclease protection assay (Rockenstein et al., 2001), qPCR (Chiba-Falek et al., 2006), and UV-laser microdissection of individual neuromelanin- and TH-positive SNc neurons followed by qPCR (Gründemann et al., 2008), indicating that elevated α-synuclein expression may foster disease in a specific cell type rather than at the tissue level. Through examination of more readily accessible blood cells, individuals carrying *SNCA* triplication or a greater number of Rep1 dinucleotide repeats were shown to have elevated levels of α-synuclein protein levels in peripheral blood mononuclear cells (PBMCs) in addition to increased *SNCA* mRNA (Fuchs et al., 2008; Gardai et al., 2013; Miller et al., 2004) indicating that transcriptional changes impact α-synuclein translation, potentially overloading cells with α-synuclein protein. Taken together, these reports suggest that disruption of normal transcription may be an underlying mechanism in sporadic and some familial forms of PD, although some results still need to be confirmed or refuted by additional studies.

2.2 Translation

Protein translation occurs in three phases, namely initiation, elongation, and termination. Multiple eukaryotic initiation factors (eIFs) coordinate translation initiation, with GTP-bound eIF-2 required to recruit tRNAs while eIF-3 and eIF-4 associate with the ribosome to together form a complex that can bind mRNA (Gingras et al., 1999). Cap-dependent translation relies on the binding of the 5′ cap of a mature mRNA by eIF4E and the interaction of cap-bound eIF4E with eIF4G, the "backbone" of the translation initiation complex, at the small ribosomal subunit (Richter and Sonenberg, 2005). Availability of eIF4E is controlled by 4E-binding proteins (4E-BPs) which bind and sequester eIF4E when hypophosphorylated and dissociate when phosphorylated, promoting translation initiation. Phosphorylation status of 4E-BP is regulated canonically by the kinase FRAP/mTOR (Richter and Sonenberg, 2005). Another key posttranscriptional regulatory mechanism influencing the initiation step of translation is driven by the RNA-induced silencing complex (RISC). Composed of regulatory RNA such as microRNA (miRNA) or small interfering RNA (siRNA) and a member of the Argonaute protein family, RISC can repress translation through several methods including by cleaving mRNA (Hammond et al., 2000; Pratt and MacRae, 2009). Once translation initiation is accomplished, elongation is performed by a cohort of elongation factors that coordinate tRNA positioning at the ribosome and addition of new amino acids carried by tRNAs, which recognize each codon in the mRNA, to the growing peptide chain through peptide bond formation. Termination is mediated by release factors when the translation complex recognizes a stop codon, resulting in dissociation of the ribosome from the mRNA (Cooper, 2000; Taymans et al., 2015). As for transcription, interest in the contribution of aberrant translation to Parkinson's disease pathogenesis has arisen largely through elucidation of the genetic underpinnings of PD as several nominated PD genes may directly or indirectly affect protein translation: *parkin* (*PARK2*), *PINK1* (*PARK6*), *DJ-1* (*PARK7*) and *LRRK2* (*PARK8*) (Chang et al., 2017; Klein and Westenberger, 2012). Loss of function mutations in parkin, PTEN-induced kinase 1 (PINK1), and DJ-1 underlie autosomal recessive forms of familial parkinsonism, while variants in leucine-rich repeat kinase 2 (LRRK2) are associated with autosomal dominant inherited PD as well as sporadic PD (Hauser et al., 2017).

Interaction of the mitochondrial modulators PINK1 and parkin with protein translation has been demonstrated in *Drosophila melanogaster* models of PD. Activation of translation through the target of rapamycin (TOR) pathway resulted in exacerbated pathogenic motor phenotypes and dopaminergic neuron loss in PINK1/parkin deficient *Drosophila* that could be rescued by blocking TOR signaling with either rapamycin, leading to hypophosphorylation and activation of 4E-BP, or by knockdown of downstream effector proteins (Liu and Lu, 2010; Tain et al., 2009). PINK1 and parkin were found to locally regulate translation of respiratory chain complex mRNAs at the mitochondria in PINK1-mutant *Drosophila* and in PINK1-mutant dopaminergic neurons induced from patient fibroblasts, with PINK1 observed to associate with eIF4A and eIF4G in an RNA-dependent manner potentially displacing translational repressors *in vitro* (Gehrke et al., 2015).

In transgenic *Drosophila*, human LRRK2 was shown to phosphorylate and inactivate 4E-BP, and increased LRRK2 activity led to increased eIF4E-mediated translation and reduced dopaminergic neuron survival (Imai et al., 2008). LRRK2 harboring PD-associated gain-of-function mutation G2019S activated translation by phosphorylating and activating ribosomal protein s15 (Rps15) also in transgenic *Drosophila* (Martin et al., 2014). Of note, neither 4E-BP nor Rps15 are likely to be direct physiological substrates of human LRRK2 (Kumar et al., 2010; Steger et al., 2016) warranting cautious interpretation of these results in the context of human PD.

In another *Drosophila* study, LRRK2-G2019S stimulated protein synthesis by decreasing Argonaute protein availability and suppressing miRNAs to antagonize RISC function, resulting in increased synthesis of cell cycle protein E2F1 (Gehrke et al., 2010). Suggesting some relevance to the PD disease process in humans, E2F1 protein was detected in dopaminergic neurons of post-mortem SNc tissue from PD patients but not controls by immunohistochemistry (Höglinger et al., 2007). Additionally, increased phosphorylation of Rps15 was measured in ribosomal fractions of lysates of post-mortem human cortex from G2019S carriers compared to wild-type LRRK2 samples, and associated with increased toxicity in embryonic stem cell-derived human neuron cultures (Martin et al., 2014). Pathologic phenotypes in both G2019S-LRRK2 transgenic flies and PINK1/parkin mutant flies were ameliorated by inhibition of translation either pharmacologically with anisomycin (Martin et al., 2014) or by overexpression of 4E-BP (Tain et al., 2009), further supporting the role of dysregulated protein synthesis in PD pathophysiology.

In *Escherichia coli*, the bacterial homolog of DJ-1 (YajL) was found to interact with bacterial ribosomal proteins, and YajL deficiency led to defective translation (Kthiri et al., 2010). Validated in mammalian cells, ribosomal proteins and other proteins involved in protein synthesis made up a significant proportion of all proteins determined to associate with both α-synuclein and DJ-1 in dopaminergic MES cells via proteomics analysis (Jin et al., 2007). DJ-1 was shown to modulate translation by binding mRNA in human dopaminergic neuroblastoma cells and in mouse brain, promoting synthesis of members of the PTEN/Akt survival pathway in aged animals likely as a response to oxidative stress (van der Brug et al., 2008). This response was absent in DJ-1 knockout mice, again indicating that disruption of normal protein translation is an outcome of PD-associated protein dysfunction.

Along with elucidation of the interplay between established PD genes and protein translation pathways, additional studies have directly implicated protein synthesis machinery in PD pathogenesis. Genome-wide analysis of a French family with members affected by late-onset parkinsonism and DLB revealed a missense mutation in eIF4G1 that segregated with disease and was also found in an additional nine PD patients and no controls in a larger screen for the putative pathogenic mutation (Chartier-Harlin et al., 2011). *In vitro* the mutant protein was unable to interact normally with binding partners eIF3e and eIF4E suggesting that the eIF4G1 mutation could cause disease by disrupting formation of the translation initiation complex (Chartier-Harlin et al., 2011). Offering further evidence that abnormal eIF4G can contribute to disease states, inhibition of eIF4G worsened the degenerative phenotypes of PINK1-deficient *Drosophila* (Gehrke et al., 2015) and a genetic interaction between eIF4G1 and another PD gene *VPS35* (see Section 3.2) was observed in yeast and nematodes (Dhungel et al., 2015). Though additional missense mutations in eIF4G1 were found in other familial and sporadic PD cases (Chartier-Harlin et al., 2011), later studies have concluded that eIF4G1 variants are likely benign polymorphisms not causative of PD (Lesage et al., 2012; Nichols et al., 2015; Schulte et al., 2012; Tucci et al., 2012). Finally, eIF2 signaling was identified as being deregulated in a comparison of transcriptome profiles of PBMCs from sporadic and LRRK2 mutation carrier PD patients with those of controls (Mutez et al., 2014). As eIF2 acts to decrease protein synthesis in response to cell stress as part of the unfolded protein response (Holcik and Sonenberg, 2005), its deregulation signifies heightened stress and a disturbance in control of protein translation in PD patient cells. These studies together build a case in support of the hypothesis that dysregulation of protein synthesis contributes to the development of Parkinson's disease.

2.3 Post-translational modification

After mRNA is translated into an amino acid chain, additional modifications to the amino acids are required for the production of a functional protein. These post-translational modifications (PTMs) are typically reversible enzymatic reactions that add or remove a functional group, such as a phosphate group. PTMs are often required for proper protein folding to occur and they can control the activity of a folded protein by changing subcellular localization or inducing a switch between active and inactive conformations. Glycosylation and phosphorylation are two of the most common post-translational modifications (Khoury et al., 2011), but amino acids can undergo a wide variety of modifications including linkage to other proteins, e.g., ubiquitin, and those that cause structural alterations such as disulfide bridge formation (Cooper, 2000). Because of their extensive influence on protein function, differences in PTMs between a healthy and a disease state may provide clues about the underlying pathogenic mechanism. Here, we offer a brief overview of the effects of post-translational modification on two major players in PD pathophysiology, α-synuclein and parkin. For more in depth review, please refer to Barrett and Timothy Greenamyre (2015), Chakraborty et al. (2017), and Stefanis et al. (2019).

The study of PTMs in the PD field was motivated in part by efforts to understand how and why α-synuclein protein came to be a predominant constituent of Lewy bodies. The most prevalent and selective PTM of α-synuclein in synucleinopathy lesions was shown to be phosphorylation at serine 129 (Anderson et al., 2006; Fujiwara et al., 2002). G protein-coupled receptor kinases, the polo-like kinase family (Mbefo et al., 2010) and casein kinase 2 can modify this phosphorylation site (Ishii et al., 2007; Mbefo et al., 2010; Pronin et al., 2000; Wang et al., 2019). Phosphorylation status of Ser^{129} has been observed to modulate α-synuclein clearance, directing it to the proteasome (Arawaka et al., 2017; Machiya et al., 2010) or to the lysosome (Oueslati et al., 2013), and to affect its propensity to aggregate. Effects of Ser^{129} phosphorylation on α-synuclein aggregation and toxicity *in vivo* are not yet clarified (Oueslati, 2016).

α-Synuclein present in Lewy bodies is often ubiquitinated in addition to being phosphorylated (Anderson et al., 2006; Tofaris et al., 2003) though ubiquitination is not required for deposition of α-synuclein (Sampathu et al., 2003). Ubiquitylation by Nedd4 was found to target α-synuclein to the endolysosomal pathway (Sugeno et al., 2014; Tofaris et al., 2011), and to reduce its accumulation and toxicity in *Drosophila* and in rat substantia nigra (Davies et al., 2014). Nedd4 did not ubiquitylate pathogenic A53T α-synuclein as efficiently as wild-type *in vitro*, linking disruption in Nedd4 regulation to the human PD process (Mund et al., 2018). Ubiquitylation of α-synuclein by E3 ligase SIAH1 as well as its deubiquitylation by USP8 and USP9X deubiquitinases also regulate α-synuclein turnover and toxicity (Alexopoulou et al., 2016; Lee et al., 2007; Rott et al., 2008, 2011).

Post-translational nitration modifications of α-synuclein observed in Lewy bodies (Giasson, 2000) have been suggested to promote aggregation and to interfere with the ability of α-synuclein to bind membrane lipids (He et al., 2019). Post-translational addition of the small ubiquitin-like modifier (SUMO) family of proteins, known as SUMOylation, may promote (Kim et al., 2011; Oh et al., 2011; Rott et al., 2017) or inhibit (Krumova et al., 2011) α-synuclein accumulation depending on site and SUMO isoform that is SUMOylated (Abeywardana and Pratt, 2015). Supporting a detrimental effect of increased SUMOylation, familial PD mutations A30P and A53T rendered α-synuclein more susceptible to SUMOylation than wild-type protein *in vitro*, and elevated levels of SUMOylated α-synuclein were measured in post-mortem PD substantia nigra and cortex samples compared to control samples (Rott et al., 2017). α-Synuclein function may be governed by the complex interplay between different post-translational modifications, rather than being determined by a single phosphorylation, ubiquitylation, nitration, or SUMOy-lation event (Haj-Yahya et al., 2013; Shahpasandzadeh et al., 2014). However, at this time it is not firmly established if such PTMs alter normal function or if they are simply adventitious events after deposition of the protein, perhaps even as markers of attempted clearance of aggregates within neurons.

Parkin is an E3 ubiquitin ligase, an enzyme that covalently modifies proteins by the addition of ubiquitin and participates in different cellular processes as directed by post-translational events at certain residues (Chakraborty et al., 2017). Parkin serves a neuroprotective function by promoting clearance of superfluous or damaged proteins and organelles, particularly mitochondria (Pickrell and Youle, 2015). Parkin is itself regulated by post-translational phosphorylation, ubiquitylation, nitro-sylation, and sulfhydration. Phosphorylation of parkin by nonreceptor tyrosine kinase c-Abl (Ko et al., 2010), casein kinase-1 (Rubio de la Torre et al., 2009; Yamamoto et al., 2005), and cyclin-dependent kinase 5 (Avraham et al., 2007; Rubio de la Torre et al., 2009) was shown to impair parkin activity and reduce its neuroprotective effects *in vitro*. An increase in phosphorylated parkin relative to total parkin was observed in sporadic PD caudate nucleus (Rubio de la Torre et al., 2009), suggesting phosphorylation-induced parkin deficit could foster disease in the same manner as loss-of-function parkin mutations in familial parkinsonism.

Phosphorylation of parkin by the mitochondrial kinase PINK1 is required for par-kin activation and recruitment to the mitochondria (Kim et al., 2008; Kondapalli et al., 2012; Pao et al., 2016; Sha et al., 2010; Shiba-Fukushima et al., 2012). PD-associated PINK1 mutations result in loss of these phosphorylation events impairing parkin ac-tivation (Pao et al., 2016; Sha et al., 2010). Phosphorylation of ubiquitin by PINK1 followed by binding of serine 65-phosphorylated ubiquitin was found to enhance par-kin activation (Kane et al., 2014; Kazlauskaite et al., 2015; Koyano et al., 2014; Pao et al., 2016), and PD-linked parkin mutations permitting parkin auto-ubiquitylation increased parkin activation as well (Chaugule et al., 2011). Conversely, ubiquitylation of parkin at another residue prevented its recruitment to the mitochondria, inhibiting its function in mitochondrial maintenance (Durcan et al., 2014). As with phosphorylation and ubiquitylation, S-nitrosylation at one residue suppressed parkin activity (Chung, 2004; Sunico et al., 2013; Yao et al., 2004) but at a different residue elevated parkin activity (Ozawa et al., 2013). Levels of S-nitrosylated parkin were significantly higher in PD and DLB brain samples compared to controls (Chung, 2004; Sunico et al., 2013; Vandiver et al., 2013), supporting a loss of parkin-mediated neuroprotection occurring with an overall increase in nitrosylated parkin. Finally, parkin E3 ligase activity was shown to be increased by both sulfhydration (Vandiver et al., 2013) and neddylation, the covalent linkage of ubiquitin-like NEDD8 protein (Choo et al., 2012; Um et al., 2012). Parkin sulfhydration was reduced in brain samples from PD patients *versus* control samples (Vandiver et al., 2013). In sum, changes in parkin PTM status observed in PD dysregulate parkin activity thereby threatening neuronal health, likely by preventing proper regulation of mitochondrial integrity, which is also impacted by PINK1.

3 Protein degradation

Protein levels within a cell at any given time are dictated by both the rate of protein synthesis and the rate of protein degradation. In addition to providing a means to clear defective or redundant components to maintain cell health, degradation pathways also permit differential regulation of protein digestion (Cooper, 2000). Modulating turnover speed of proteins is an important method of controlling cell fate, as a protein's half-life determines its availability to perform its intended role as well as detrimental activities. The two major mechanisms of protein degradation are the ubiquitin-proteasome pathway and the autophagy-lysosome pathway, with the endoplasmic reticulum governing a minor mechanism. This section reviews how malfunction of these systems can contribute to Parkinson's disease pathobiology, as depicted in Fig. 2.

3.1 Proteasome

The ubiquitin-proteasome system (UPS) is responsible for turnover of many proteins in the cell, typically those with relatively short half-lives. The proteasome is a large cylindrical protease complex made up of a proteolytic core known as the 20S proteasome and a 19S regulatory cap also called PA700, which together form the active 26S proteasome (Bochtler et al., 1999). Proteins destined for proteasomal degradation undergo polyubiquitylation, a procedure that is tightly controlled by three enzymes: E1 activates and transfers ubiquitin to E2, a carrier enzyme that conjugates ubiquitin to the target protein together with a specific E3 ubiquitin ligase. There are hundreds of E3 ligases with affinities for distinct substrates, conferring selectivity to the process. PA700 recognizes a ubiquitin-tagged protein and channels it into the central lumen of the proteasome, where it is then disassembled into small peptides that

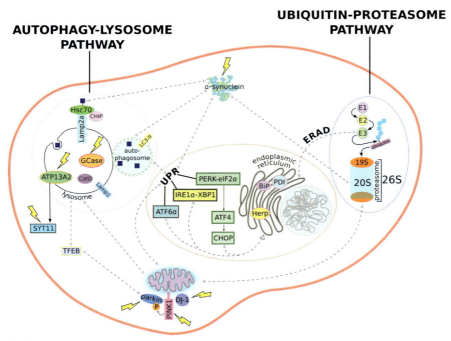

FIG. 2

Illustration of protein degradation pathways relevant to Parkinson's disease pathogenesis. Genetic perturbations associated with PD are indicated by lightning bolts. On the right, the ubiquitin-proteasome system is summarized as the three ligases (E1, E2, and E3) that control polyubiquitylation of proteins targeted for proteasomal degradation together with the active 26S proteasome, composed of a 20S core and 19S regulatory caps. On the left, a simplified version of the autophagy-lysosome pathway is portrayed. Microautophagy, chaperone-mediated autophagy (CMA), and macroautophagy are depicted clockwise from the top left of the lysosome. In the center, the unfolded protein response (UPR) with its three described branches is shown activating endoplasmic reticulum-associated degradation (ERAD), and also autophagy. Proposed interactions of both α-synuclein protein and the mitochondria with each of these pathways are indicated by dashed lines.

can be reused to construct other proteins (reviewed in Nandi et al., 2006). As evidenced by the extensive neurodegeneration and Lewy-like intraneuronal aggregations observed in a mouse model with 26S proteasomal dysfunction caused by knockout of 26S subunit PSMC1 (Bedford et al., 2008), proper function of the UPS is required for neuronal homeostasis.

Dysregulation of the UPS has long been considered a potential cause of Parkinson's disease pathology. UPS dysfunction in PD was nominated first by observations of ubiquitin in Lewy bodies (Bancher et al., 1989; Iwatsubo et al., 1996; Kuzuhara et al., 1988; Lennox et al., 1989; Love et al., 1988; Lowe et al., 1988), then by the presence of UPS machinery including ubiquitin hydrolase PGP 9.5 (Lowe et al., 1990), a multi-catalytic proteinase (Kwak et al., 1991), and proteasome subunits (Ii et al., 1997) in these aggregates. Deletions in parkin E3 ligase, also found in Lewy bodies, were shown to underlie autosomal recessive juvenile parkinsonism (AR-JP) (Imai et al., 2000; Kitada et al., 1998; Schlossmacher et al., 2002; Shimura et al., 2000). Soon after, McNaught et al. published a series of experiments that demonstrated decreased protein levels of the proteasome α-subunit as well as deficits in proteasomal enzymatic activities in post-mortem substantia nigra pars compacta (SNc) of sporadic PD patients compared to age-matched controls (McNaught et al., 2002a, 2003; McNaught and Jenner, 2001). Similar results were observed in other studies as well (Bukhatwa et al., 2010; Grünblatt et al., 2004; Tofaris et al., 2003). Additionally, an α-synuclein species present in Lewy bodies in PD and DLB brain tissue was shown to be aberrantly ubiquitinated (Tofaris et al., 2003). These discoveries set the stage for an interest in the role of UPS dysfunction in Parkinson's disease pathogenesis.

3.1.1 α-Synuclein and the UPS

Although synuclein pathology clearly marks the brain regions that are affected in PD and related disorders, whether Lewy bodies are causal for neurodegeneration remains unclear. It is known that protein aggregates are generally toxic *in vitro* (Bucciantini et al., 2002). One potential mechanism by which this occurs is that protein aggregates impair the UPS, disrupting cell homeostasis and potentially initiating a positive feedback mechanism of more aggregation ultimately leading to cell death (Bence et al., 2001). Therefore, a key question becomes how α-synuclein turnover can be regulated.

It has been claimed that turnover of α-synuclein involves both proteasomal and lysosomal pathways (Webb et al., 2003), and occurs in both ubiquitin-dependent (Haj-Yahya et al., 2013; Rott et al., 2011; Stefanis et al., 2001) and ubiquitin-independent fashions (Liu et al., 2003; Tofaris et al., 2001) *in vitro*. Of note, α-synuclein is not a proteasomal substrate in all cell types or under all experimental conditions (Ancolio et al., 2000; Paxinou et al., 2001). Therefore, whether the proteasome is a major pathway for α-synuclein degradation is controversial. The contrasting lysosomal degradation pathways are discussed below. However, disease-relevant α-synuclein can impair proteasomal function (Emmanouilidou et al., 2010; Zondler et al., 2017), potentially by directly binding the proteasome (Ghee et al., 2000; Lindersson et al., 2004; Snyder et al., 2003; Zhang et al., 2008). PD-associated A53T and A30P mutant α-synuclein have been shown to negatively affect the UPS to a greater extent than the wild-type protein. In transiently transfected SH-SY5Y human neuroblastoma cells, mutant A53T α-synuclein was degraded by the proteasome at half the rate of wild type suggesting the mutation may increase the proclivity of the protein to accumulate and aggregate (Bennett et al., 1999). Overexpression of A53T and A30P α-synuclein in cells decreased proteasome activity (Petrucelli et al., 2002; Smith et al., 2005; Stefanis et al., 2001; Tanaka et al., 2001) and increased sensitivity to proteasome inhibitor treatment (MG132 or lactacystin) as evidenced by increased cell death (Lee et al., 2001a; Petrucelli et al., 2002;

Tanaka et al., 2001; Zondler et al., 2017). Reduced baseline proteasome activity was observed in human α-synuclein transgenic mouse cortex (Ebrahimi-Fakhari et al., 2011; Emmanouilidou et al., 2010), with soluble α-synuclein oligomers measured in the proteasomal fraction of cortical homogenate (Emmanouilidou et al., 2010). Interestingly, Petrucelli et al. found TH-positive neurons in mouse primary midbrain cultures to be selectively vulnerable to both overexpression of mutant α-synuclein and to pharmacological proteasome inhibition, measuring reduced TH-positive cell numbers while TH-negative cell numbers remained stable (Petrucelli et al., 2002). Collectively, these results suggest that even if the proteasome is not a major degradation site for α-synuclein, it may be inhibited as a consequence of α-synuclein buildup in neurons.

3.1.2 Proteasome inhibitor models of PD

Given the robust observation of impaired proteasome function in both post-mortem PD brain tissue and in various models of PD, several studies have employed pharmacological inhibition of the UPS to interrogate the mechanism of PD pathophysiology. Selective proteasomal inhibition leads to aggregation of protein and inclusion body formation (Johnston et al., 1998). β-Lactone, a selective proteasome inhibitor, reduced degradation of recombinant and endogenous α-synuclein in neural SH-SY5Y cells (Bennett et al., 1999). Chronic low-level inhibition of proteasome activity with MG115 was also shown to cause increased protein aggregation and oxidation in SH-SY5Y cells (Ding et al., 2003). Application of lactacystin to dopaminergic PC12 rat cells induced death of both naïve and neuronally differentiated cells, and formation of ubiquitinated inclusions in surviving cells (Rideout et al., 2001). Proteasome inhibitors, but not lysosomal protease inhibitors, promoted apoptosis in Rat-1 and PC12 cells as well as in cultured rat cortical neurons (Lopes et al., 1997; Qiu et al., 2000; Rideout and Stefanis, 2002). Both proteasome inhibition by lactacystin or epoxomicin and ubiquitin C-terminal hydrolase inhibition by ubiquitin aldehyde, led to selective degeneration of dopaminergic neurons along with accumulation of ubiquitin and α-synuclein in embryonic rat ventral midbrain neuron cultures (McNaught et al., 2002c; Rideout et al., 2005).

Similarly, in an *in vivo* rat model, stereotaxic infusion of lactacystin into the SNc resulted in dose-dependent loss of dopaminergic cells and the formation of cytoplasmic α-synuclein aggregates along with motor abnormalities including bradykinesia (McNaught et al., 2002b). Infusion of lactacystin or epoxomicin into rat striatum also resulted in loss of nigrostriatal DAn and accumulation of parkin-, α-synuclein-, and ubiquitin-positive protein aggregates in remaining neurons in both regions (Fornai et al., 2003; Miwa et al., 2005). Topical application of selective proteasome inhibitor clasto-lactacystin-β-lactone to mouse cortex also led to elevated levels of both endogenous α-synuclein in non-transgenic mice and overexpressed human wild-type α-synuclein in transgenic mice, with significantly more accumulation observed in older mice compared to younger mice indicating a role for the UPS in α-synuclein degradation in both normal and disease states influenced by age (Ebrahimi-Fakhari et al., 2011). Systemic administration of proteasome inhibitors epoxomicin and synthetic peptide aldehyde PSI (Z-Ile-Glu(O*t*Bu)-Ala-Leu-al) to rats via injections over a 2-week course induced the development of a parkinsonian phenotype that appeared to recapitulate key disease features including motor dysfunction (e.g., bradykinesia, tremor), nigral neurodegeneration, and formation of Lewy body-like intraneuronal inclusions in rat brain (McNaught et al., 2004). However, reproducibility of this model has come into question as the results of proteasome inhibition in rodents have been inconsistent and the findings in the original rat model were not reliably replicable (Beal and Lang, 2006; Bové et al., 2006; Kordower et al., 2006; Manning-Boğ et al., 2006; McNaught and Olanow, 2006; Schapira et al., 2006; Zeng et al., 2006a). For more extensive review of proteasome inhibitor models of PD we direct the interested reader to Bentea et al. (2017).

3.1.3 Interaction of the UPS and oxidative stress

Although there are therefore significant concerns as to whether UPS inhibition can be used to model PD, there are reasons to think that some of the potential downstream effects of UPS dysfunction might interact with other pathways relevant to PD. First, the proteasomal degradation cycle requires ATP, with both construction and destruction of ubiquitin-protein conjugates involving ATP (Hershko et al., 1984); thus, decreased production of ATP by the mitochondria would have a negative impact on the proteasome. Indeed, rotenone, an inhibitor of mitochondrial complex I, inhibits the UPS, presumably as a consequence of altered energy metabolism (Höglinger et al., 2003; Shamoto-Nagai et al., 2003) and PD-associated mitochondrial deficit was shown to trigger UPS impairment (Martins-Branco et al., 2012) *in vitro*. 1-Methyl-4-phenyl-1,2,5,6-tetrahydropyridine (MPTP), a mitochondrial toxin classically used to model PD *in vivo* (Heikkila et al., 1984), decreased proteasomal function when administered to both mice (Fornai et al., 2005) and marmosets (Zeng et al., 2006b). Second, as the proteasome is normally responsible for clearing oxidatively damaged proteins from within a cell (Grune et al., 1995), excessive oxidized substrate could overwhelm the UPS (Guo et al., 2013). Using the potent oxidizer hydrogen peroxide (Reinheckel et al., 1998) or the lipid peroxidation product 4-hydroxy-2-nonenal (Friguet and Szweda, 1997; Okada et al., 1999) in model systems, the production of reactive oxygen species (ROS) leading to oxidative stress has been shown to impair the proteasome.

Normal catabolism and auto-oxidation of endogenous dopamine yields ROS that could create oxidative stress within a cell (Stokes et al., 1999), potentially contributing to the selective vulnerability of dopaminergic neurons in PD. Proteasome inhibition by dopamine and its oxidative metabolites has been demonstrated *in vitro* (Keller et al., 2000; Zafar et al., 2007; Zhou and Lim, 2009) and exposure to oxidizing agents promoted α-synuclein aggregation and inclusion body formation *in vitro* (Paxinou et al., 2001) and *in vivo* (Fornai et al., 2005). Conversely, pharmacological proteasome inhibition by lactacystin caused increased oxidative stress in human cancer cell lines (Lee et al., 2001b) and in rat SNc neurons (Miwa et al., 2005), with the neurotoxic effects of lactacystin found to depend on endogenous dopamine in rat nigrostriatal neurons (Fornai et al., 2003). Further evidence for a link between

PD-associated mitochondrial deficit and UPS dysfunction has been provided by the discovery that DJ-1, a mitochondrial maintenance protein in which loss-of-function mutations cause autosomal recessive parkinsonism (Cookson, 2012), modulates the 20S proteasome (Moscovitz et al., 2015; Saito et al., 2016).

3.2 Lysosome

Along with the ubiquitin-proteasome system, the other major pathway for degradation of proteins within a cell is lysosomal proteolysis. The lysosome is a membrane-bound organelle with an acidic lumen containing many catabolic enzymes that can degrade a variety of cellular materials, including nucleic acids, carbohydrates, lipids, and proteins. Both extracellular proteins captured by endocytosis and intracellular proteins engulfed via autophagy can be directed to the lysosome for digestion (Cooper, 2000). The endocytic pathway involves transfer of a cargo through a series of vesicles and membrane-bound organelles, which are defined in part by their association with certain regulatory Rab GTPases (Rink et al., 2005). The process begins with invagination of the plasma membrane around an external material to form an intracellular membrane-limited endocytic vesicle, which then fuses with an early endosome (Rab5-positive). After acquiring various other components within intraluminal vesicles, the endosome morphs into a multivesicular body called a late endosome (Rab7-positive) that fuses with a lysosome thereby releasing its cargo into the lysosomal lumen (Langemeyer et al., 2018; Roosen and Cookson, 2016). The term "autophagy" describes generally the uptake of cytoplasmic cargo for transport to the lysosome, and encompasses three different routes of delivery: microautophagy, chaperone-mediated autophagy (CMA), and macroautophagy (Levine and Kroemer, 2019). Microautophagy refers to the direct uptake of substrate immediately surrounding a lysosome by invagination of the lysosomal membrane (Li et al., 2012). CMA depends on recognition of a substrate by chaperone protein Hsc70, followed by binding of the chaperone-substrate complex to a specific receptor on the lysosomal membrane, Lamp2a, for translocation of the substrate into the lysosome (Maria Cuervo, 2004). Finally, macroautophagy (hereafter referred to as "autophagy") occurs when a double-layered membrane termed a phagophore expands to encircle a cytosolic substrate, producing an autophagosome that later fuses with a lysosome to submit its contents for degradation (Feng et al., 2014).

As lysosomes are essential for maintenance of protein homeostasis, or proteostasis, within a cell, the presence of Lewy body protein aggregations in PD brain may point to possible disruption of this degradative organelle. Accumulation of autophagic vacuoles and other features of autophagic degeneration have been observed in dopaminergic neurons of post-mortem PD patient substantia nigra (Anglade et al., 1997; Chu, 2006), further implicating dysfunction of the autophagy-lysosome pathway. Autophagy is important in the maintenance of neuronal health, but becomes impaired in normal aging (Komatsu et al., 2006; Rubinsztein et al., 2011). Decreasing efficiency of lysosomal degradation pathways with increasing age may induce cellular stress that leads to neurodegeneration. Consistent with this idea, buildup of ubiquitin-protein conjugates within the lysosomal system was found to contribute to intraneuronal inclusions occurring with neurodegenerative diseases (Lowe et al., 1993; Mayer et al., 1991, 1996). Polyubiquitin chains do not target substrate proteins exclusively to the proteasome as the PD-associated E3 ligase parkin was shown to direct misfolded proteins, including UCH-L1 and DJ-1, into the autophagy pathway (McKeon et al., 2015; Olzmann et al., 2007; Olzmann and Chin, 2008) via ligation of ubiquitin to lysine residue 63 (K63) within the preceding ubiquitin rather than to K48 as required for canonical proteasome delivery (Pickart and Fushman, 2004). Another ubiquitin ligase, Nedd4, was similarly found to direct α-synuclein to the lysosome by attachment of K63-linked polyubiquitin chains (Davies et al., 2014; Tofaris et al., 2011). Also demonstrating lysosomal deficiency, decreased levels of lysosomal proteins glucocerebrosidase, lysosome-associated membrane protein 1 (Lamp1), and cathepsin D (CatD) were measured along with increased amounts of α-synuclein in post-mortem sporadic PD brain compared to age-matched controls (Chu et al., 2009; Gegg et al., 2012; Murphy et al., 2014).

3.2.1 Lysosomal degradation of α-synuclein

Following from the hypothesis that lysosomal deficiency leads to accumulation of disease-relevant protein in the brain, many studies have investigated the role of the lysosome in degradation of α-synuclein in normal and disease states. Pharmacological inhibition of lysosomal activity has been shown to increase intracellular α-synuclein levels in cell models (Lee et al., 2004; Paxinou et al., 2001; Vogiatzi et al., 2008; Webb et al., 2003) and in transgenic mice expressing human α-synuclein (Ebrahimi-Fakhari et al., 2011), while pharmacological stimulation of autophagy with rapamycin reduced α-synuclein levels under some experimental conditions (Webb et al., 2003) but not all situations (Lee et al., 2004). Both *in vitro* and *in vivo*, transcription factor EB (TFEB)-mediated lysosomal stimulation protected against α-synuclein toxicity while autophagy inhibition led to increased cell death (Decressac et al., 2013; Dehay et al., 2010; Kilpatrick et al., 2015; Napolitano and Ballabio, 2016). α-Synuclein aggregates blocked autophagy in HEK cells and primary mouse neurons, causing accumulation of LC3-II-positive autophagosomes and depletion of Lamp1-positive lysosomes also demonstrating an interaction between α-synuclein and the autophagy-lysosome pathway (Tanik et al., 2013).

In addition to the autophagy pathway, the CMA pathway is also important for α-synuclein clearance. Cuervo et al. determined that the α-synuclein amino acid sequence contains a CMA recognition motif and observed translocation of wild-type α-synuclein into lysosomes for degradation. Interestingly, PD-associated A53T and A30P mutant α-synuclein blocked Lamp2a-controlled lysosomal uptake, inhibiting their own degradation and that of other CMA substrates (Cuervo et al., 2004). Knockdown of Lamp2a led to increased levels of α-synuclein in dopaminergic SH-SY5Y cells (Alvarez-Erviti et al., 2010) and in primary rat neuron cultures (Vogiatzi et al., 2008) providing further evidence of α-synuclein turnover by CMA. Considering these reports in light of the many studies of proteasomal degradation of α-synuclein, Hsp70-interacting protein (CHIP) has been suggested to act as a "molecular switch" that can direct α-synuclein to either the proteasome or the lysosome for degradation offering a potential mechanism for reconciliation of these disparate cellular processes (Shin et al., 2005). Examination of CMA proteins in post-mortem human brain samples revealed decreased levels of Lamp2a and Hsc70 in SNc, amygdala, and anterior cingulate cortex of PD patients *versus* age-matched control subjects (Alvarez-Erviti et al., 2010; Murphy et al., 2015). In anterior cingulate cortex, loss of the Lamp2a isoform was selectively reduced and correlated with increased levels of α-synuclein protein in PD samples (Murphy et al., 2015). Lysosomal degradation through either autophagy or CMA thus appears to modulate intracellular levels of α-synuclein protein, and disruption of these pathways may contribute to pathogenic accumulation of α-synuclein in PD.

3.2.2 *Genetic framework of lysosomal dysfunction in PD*

Interest in involvement of the lysosome in Parkinson's disease pathophysiology further grew from reports of a subset of patients with Gaucher's disease (GD), an inherited lysosomal storage disorder (LSD), who presented with Parkinsonian symptoms including bradykinesia, tremor, and rigidity (Machaczka et al., 1999; Neudorfer et al., 1996; Tayebi et al., 2001). Additionally, neuropathological exam of Gaucher's patients with parkinsonism features revealed synuclein-positive Lewy bodies (Tayebi et al., 2003). GD is caused by autosomal recessive loss-of-function mutations in the gene *GBA1*, which encodes the lysosomal enzyme glucocerebrosidase (GCase). While GD arises due to mutations in both *GBA1* alleles, individuals having just one affected allele do not experience symptoms of Gaucher's but do have increased incidence of PD, suggesting that *GBA1* mutations predispose to the development of PD (Goker-Alpan, 2004; Halperin et al., 2006; Lwin et al., 2004). A large international study of 5691 patients with PD and 4898 controls found a significantly higher rate of *GBA1* mutations in the PD group, affirming an association between deficiency of *GBA1* and Parkinson's disease (Sidransky et al., 2009). Additional studies of this association have conclusively nominated mutations in *GBA1* as the most common risk factor for PD known to date (Alcalay et al., 2014; Mitsui et al., 2009; Rana et al., 2013; Ryan et al., 2019). Though the mechanism linking GCase and PD is not yet fully understood, the observation of a reciprocal relationship between GCase levels and α-synuclein levels in cell and animal models has tied loss of normal GCase function to pathological Lewy body formation (Aflaki et al., 2017; Cleeter et al., 2013; Manning-Boğ et al., 2009; Mazzulli et al., 2011; Rockenstein et al., 2016; Sardi et al., 2013). In midbrain dopaminergic neurons differentiated from iPSC lines derived from both GD and PD patients harboring *GBA1* mutations, impaired autophagosome-lysosome fusion and accumulation of enlarged lysosomes were observed in addition to increased glucocerebroside and α-synuclein, indicating that altered α-synuclein metabolism as a result of GCase deficiency is due to broad disruption of the autophagy-lysosome pathway (Schöndorf et al., 2014). Interestingly, inhibition of glucocerebroside synthesis ameliorated α-synuclein pathology in hippocampi of both GBA1-deficient mice and A53T-SNCA transgenic mice, further evidencing biochemical interplay between GCase and α-synuclein

(Sardi et al., 2017). GCase protein and enzymatic activity levels were significantly lower in post-mortem brain tissue samples from idiopathic PD subjects compared to age-matched controls (Gegg et al., 2012; Murphy et al., 2014; Rocha et al., 2015), indicating that lysosomal deficit may be a factor in PD pathogenesis even in the absence of a mutation in *GBA1*.

Genetic studies of PD have increasingly highlighted the contribution of variation in lysosomal genes beyond *GBA1* to disease risk. Examining the influence of genetic variants in other LSD genes on PD susceptibility, analysis of whole exome sequencing data from PD cases *versus* controls revealed that over half of PD cases in the study population carried at least one LSD-associated variant and that variants in lysosomal genes *CTSD*, *SLC17A5*, and *ASAH1* were linked to increased PD risk (Robak et al., 2017). Heterozygous mutations in *NPC1*, causative of the LSD Niemann-Pick Disease type C, were also found in PD patients (Josephs et al., 2004; Kluenemann et al., 2013) and a variant in *SMPD1*, responsible for Niemann-Pick Disease types A and B, was nominated as a risk factor for PD (Gan-Or et al., 2013). Variants in arylsulfatase A (ARSA), underlying the LSD metachromatic leukodystrophy, were shown to modify PD risk and ARSA was proposed to serve as a chaperone of α-synuclein as these molecules physically interacted, and ARSA deficiency promoted α-synuclein aggregation *in vitro* (Lee et al., 2019). Along with these convincing genetic connections between lysosomal disorders and PD, a meta-analysis of genome-wide association studies incorporating over 26,000 PD cases and 400,000 controls identified novel risk loci encompassing several autophagy-lysosomal genes including *KAT8*, *CTSB*, *ATP6V0A1*, and *GALC* broadly implicating disruption of lysosome function in PD risk (Chang et al., 2017).

The PD gene *ATP13A2*, encoding a lysosomal ATPase, has been subjected to functional as well as genetic analyses. A loss-of-function mutation in *ATP13A2* (*PARK9*) was identified in a family affected by an autosomal recessive early onset parkinsonism syndrome (Ramirez et al., 2006), and later confirmed in additional families (Park et al., 2015). ATP13A2 mutations have been found to cause lysosomal deficits and disturbances in autophagy in several PD models. Fibroblasts derived from patients bearing ATP13A2 mutations had reduced capacity for lysosomal degradation due to impaired acidification of lysosomes and aberrant processing of lysosomal enzymes (Dehay et al., 2012; Usenovic et al., 2012b). Knockdown of ATP13A2 in HeLa cells and in primary mouse cortical neurons caused toxic accumulation of α-synuclein that was mitigated by α-synuclein silencing (Bento et al., 2016; Usenovic et al., 2012b). *In vivo*, loss of ATP13A2 resulted in motor deficits and increased amounts of autophagy substrate p62, decreased mature CatD, and ubiquitin-positive inclusions in mouse brain indicating dysregulation of the autophagy-lysosome pathway. In these mice, and in contrast to cell culture models, neurotoxic effects of ATP13A2 deficiency were not dependent on α-synuclein (Kett et al., 2015). ATP13A2 was also determined to interact with another PD-associated protein, SYT11 or synaptotagmin XI, which is involved in vesicular trafficking and regulated by parkin (Huynh et al., 2003; IPDGC, 2011; Usenovic et al., 2012a). SYT11 has been shown to be a downstream effector of ATP13A2, with loss of

ATP13A2 resulting in decreased SYT11 expression via reduction in nuclear TFEB levels. SYT11 knockdown and ATP13A2 knockdown caused similar disturbances of the autophagy-lysosome pathway and the impaired autophagy phenotype observed following ATP13A2 knockdown was rescued by SYT11 overexpression (Bento et al., 2016). This pathway may indeed be relevant to the human disease process as dopaminergic SNc neurons in post-mortem sporadic PD samples contained less ATP13A2 protein compared to control samples, and most LBs were ATP13A2 positive by IHC (Dehay et al., 2012). Reduced function of ATP13A2 may contribute to PD pathophysiology by impeding autophagy.

Leucine-rich repeat kinase 2 is a relatively common cause of inherited PD, and has also been found to influence lysosomal biology through functional assays. LRRK2 can indirectly modulate phosphorylation status of p62, and reciprocally, p62 was found to regulate LRRK2 turnover in its role as a selective autophagy receptor (Park et al., 2016), and to modulate LRRK2 toxicity (Kalogeropulou et al., 2018) *in vitro*. LRRK2 was shown to induce autophagy through activation of the CAMKK-β/AMPK pathway in HEK293 cells (Gómez-Suaga et al., 2012) and through phosphorylation of MAPK/ERK kinases in PD patient fibroblasts (Bravo-San Pedro et al., 2013). LRRK2 silencing as well as LRRK2 kinase inhibition suppressed autophagy in immune cells and activated LRRK2 re-localized to autophagosome membranes (Schapansky et al., 2014). In *Drosophila* follicle cells, *Drosophila* Lrrk protein localized to late endosomal/lysosomal membranes and was found to modulate lysosomal positioning through negative regulation of Rab7 (Dodson et al., 2012). In Lrrk-deficient fly follicle cells, disruption of endolysosomal and autophagic pathways exhibited by the presence of enlarged lysosomes and accumulation of autophagosomes could be ameliorated by constitutively active Rab9, suggesting Lrrk plays a part in Rab9-dependent vesicular transport (Dodson et al., 2014). Studies of LRRK2 knockout mouse kidneys, a tissue in which LRRK2 is normally highly expressed, demonstrated age-dependent alterations in LC3-II and p62 levels as well as accumulation of α-synuclein and autophagolysosomes, reinforcing the idea that LRRK2 regulates autophagy *in vivo* (Tong et al., 2010, 2012). These reports support the conclusion that pathogenic LRRK2 activity may deregulate the autophagy-lysosomal pathway as well as the endolysosomal pathway by perturbing normal protein trafficking.

3.2.3 Mitochondria and the lysosome

While a precise mechanism to describe the pathophysiology underlying Parkinson's disease has not yet been detailed, it has become clear that there is an important intersection between dysregulation of α-synuclein, lysosomes, and mitochondria (Lin et al., 2019). Autophagy serves as an organellar quality control pathway, with lysosomal digestion of unneeded or malfunctioning mitochondria referred to as "mitophagy" (Levine and Kroemer, 2019). Disruption of either the mitochondria or the lysosome can thus disturb homeostasis of both organelles, resulting in release of ROS that can cause oxidative damage throughout the cell. For example, lysosomal depletion was observed in the ventral midbrain of mice treated with mitochondrial toxin MPTP likely due to damage of lysosomal membranes by mitochondrial ROS (Dehay et al., 2010). Conversely, lysosomal deficiency due to knockdown of ATP13A2 led to mitochondrial fragmentation and oxidative stress *in vitro* (Gusdon et al., 2012). The interplay may be complex, with mitophagy shown to modulate the lysosome via PINK1- and parkin-mediated activation of TFEB promoting lysosomal biogenesis (Nezich et al., 2015). The link between mitochondrial and lysosomal function is also being explored as a means of combating PD-associated pathobiology. Induction of autophagy by overexpression of autophagy protein Atg1 ameliorated dopaminergic neuron loss and muscle degeneration in PINK1-deficient *Drosophila* (Liu and Lu, 2010). Similarly, in cells treated with MPTP, TFEB overexpression rescued lysosome levels and decreased cell death (Dehay et al., 2010). Oxidation of dopamine has also been proposed to be a point of convergence between established PD pathways. Mitochondrial deficiency and a concomitant accumulation of oxidized dopamine was found to disrupt activity of lysosomal enzyme glucocerebrosidase, increasing oxidation and aggregation of α-synuclein in PD patient iPSC-derived dopaminergic neurons (Burbulla et al., 2017). Further, dopamine-modified α-synuclein blocked the CMA pathway *in vitro* (Martinez-Vicente et al., 2008). These results support the hypothesis that cell stress caused by lysosomal and/or mitochondrial dysfunction yields abnormal ROS production that amplifies basal dopamine oxidation contributing to the selective death of dopaminergic neurons in PD.

3.2.4 Vesicular trafficking defects underlying PD-associated lysosome dysfunction

As with many other topics within the realm of PD research, investigation of the contribution of aberrant protein transport to lysosomal insufficiency in PD began with α-synuclein. Cooper et al. found that expression of human α-synuclein in yeast impeded anterograde ER-to-Golgi trafficking though inhibition of the yeast homolog of Rab1. Overexpression of Rab1 protected against α-synuclein-induced toxicity in yeast, *C. elegans, Drosophila*, and in rat dopaminergic neurons, confirming that proper function Rab1 in the early secretory pathway is vital to cell health (Cooper et al., 2006). Highlighting the interconnectedness of autophagy to other cellular pathways, proper function of the early secretory pathway is required for normal formation of autophagosomes in yeast (Hamasaki et al., 2003). Overexpression of the post-Golgi Rab8a and presynaptic Rab3a, also rescued α-synuclein-induced cell death in *C. elegans* and rat primary midbrain culture indicating that α-synuclein can disrupt multiple intracellular trafficking events (Gitler et al., 2008). Overexpression of human α-synuclein in transgenic mice was demonstrated to impair autophagy by hindering Rab1 function, causing mislocalization of autophagy protein Atg9, and thereby inhibiting autophagosome production. These results were recapitulated with Rab1a knockdown, and reversed by Rab1a overexpression *in vitro* (Winslow et al., 2010). Examining axonal transport in primary mouse neurons, α-synuclein aggregates were shown to alter trafficking of autophagosomes and Rab7-positive endosomes thus impairing autophagosome fusion with late endosome/lysosomes

(Volpicelli-Daley et al., 2014). Altogether, these experiments point to perturbation of vesicular trafficking as a mechanism of autophagy-lysosome disruption by pathogenic levels of intracellular α-synuclein.

The genetic architecture of PD has further implicated malfunction of protein transport pathways as the abnormality underlying lysosomal malfunction in the disease process. A missense mutation in the *VPS35* gene, which encodes vacuolar protein sorting protein 35 (Vps35), was determined to be a cause of autosomal dominant PD (Sharma et al., 2012; Vilariño-Güell et al., 2011; Zimprich et al., 2011). Because Vps35 is a core subunit of the retromer complex that mediates endosome-to-Golgi complex and endosome-to-plasma membrane transport to facilitate movement of many proteins including lysosomal acid hydrolases, neurodegeneration incited by mutant Vps35 is likely due to defective intracellular trafficking (Williams et al., 2017). The PD-associated mutation D620N has been shown to impair autophagy by disrupting trafficking of Atg9a (Zavodszky et al., 2014), endosomes (Follett et al., 2014), Lamp2a (Tang et al., 2015), and CatD (Follett et al., 2014; Miura et al., 2014) *in vitro*. As CatD is the primary protease involved in α-synuclein degradation within the lysosome (Cullen et al., 2009; Sevlever et al., 2008) and Lamp2a is required for clearance of α-synuclein by CMA, Vps35 deficiency precipitated α-synuclein accumulation *in vitro* and *in vivo* in flies and mice (Miura et al., 2014; Tang et al., 2015). Measurement of reduced Vps35 mRNA in the substantia nigra of PD patients compared to unaffected controls affirms the relevance of these studies (MacLeod et al., 2013). Elucidation of Vps35 interactions with other PD proteins further supports deregulation of protein trafficking in PD pathogenesis. LRRK2 has been separately linked to protein trafficking through its regulation of Rab GTPases (Beilina et al., 2014; MacLeod et al., 2013; Steger et al., 2016). Additionally, LRRK2 has been suggested to influence Vps35 function in synaptic vesicle endocytosis (Inoshita et al., 2017) and Vps35-D620N was shown to enhance LRRK2-mediated phosphorylation of Rab substrates potentially initiating a toxic gain-of-function cascade (Mir et al., 2018). Finally, parkin was found to ubiquitylate Vps35 (Martinez et al., 2017) and modulate its role in the endolysosomal pathway (Song et al., 2016). Therefore, protein trafficking along with protein synthesis and protein degradation appears to be important in the development of PD.

3.3 Endoplasmic reticulum

The endoplasmic reticulum (ER) is a cellular organelle responsible for proper folding of newly synthesized proteins, by way of numerous resident chaperones and post-translational modification enzymes, before packaging of the proteins in membrane-bound vesicles for transport. As an early location for proteins on the secretory pathway, the ER coordinates assembly and trafficking of proteins destined for both intracellular and extracellular locations (Lehtonen et al., 2019; Wang and Kaufman, 2016). Briefly, the secretory pathway describes the anterograde journey of a protein from ribosome to ER, to Golgi apparatus, to secretory vesicles for export. However, not all proteins that enter the ER are secreted. Constituent ER and Golgi proteins, as well as lysosomal and plasma membrane proteins also traverse the initial steps of this pathway (Cooper, 2000). When the folding process goes awry within the ER and a protein does not attain its correct conformation, the misfolded polypeptides become substrates for ER-associated protein degradation (ERAD) which acts as a quality control mechanism and prevents buildup of misfolded protein (Ruggiano et al., 2014). ERAD encompasses several mechanisms distinguished by their reliance on different E3 ubiquitin ligases that serve to transfer misfolded conformations to the cytoplasm and on to the proteasome for degradation (Smith et al., 2011). Disruption of these normal processes, e.g., by proteasome inhibition, aberrant increase in protein synthesis, or malfunction of ER machinery due to environmental insult, causes ER stress and activation of the unfolded protein response (UPR) (Kaufman, 1999). The UPR involves three parallel signaling pathways, headed by (1) PRKR-like ER kinase-eukaryotic translation initiation factor 2α (PERK-eIF2α), (2) inositol-requiring protein 1α-X-box-binding protein 1 (IRE1α-XBP1), and (3) activating transcription factor (ATF)6α (Wang and Kaufman, 2016). These pathways orchestrate a transient decrease in protein synthesis together with increases of folding and transport of proteins as well as ERAD and autophagy in an attempt to clear the ER. Failure of the UPR to mitigate protein misfolding leads to cell death by apoptosis (Lin et al., 2007; Xu et al., 2005).

Immunohistochemical studies of post-mortem substantia nigral tissue obtained from PD patients have provided evidence for contribution of UPR activation to disease pathophysiology. The protein folding chaperone pancreatic protein disulfide isomerase (PDI) was found to be expressed in SNc neurons of PD and DLB subjects but not control subjects by immunohistochemistry (IHC), with some but not all LBs being stained for PDI (Conn et al., 2004). Phosphorylated (active) PERK, eIF2α, and IRE1α were observed in nigral dopaminergic neurons of PD cases but not controls, and phospho-PERK and phospho-IRE1α were present only in neurons containing α-synuclein indicating that α-synuclein accumulation may trigger the UPR (Heman-Ackah et al., 2017; Hoozemans et al., 2007). Intensity of phospho-PERK staining was also higher in prefrontal cortex samples from PD and DLB patients compared to controls (Baek et al., 2016). Consistent with this observation, expression of activating transcription factor 4 (ATF4), a downstream mediator of PERK-eIF2α signaling, was increased in neuromelanin-positive neurons of a subset of PD patients relative to controls with some LBs displaying ATF4-positivity (Sun et al., 2013). Another ER-associated protein known as Herp (homocysteine-inducible ER stress protein) showed stronger immunoreactivity in the substantia nigra of sporadic PD patients compared to controls, especially in glial cells, and Herp deposits were seen in the core of all measured LBs (Slodzinski et al., 2009). Herp is involved in ERAD and in maintenance of calcium homeostasis, playing a key role in neuroprotection during ER stress (Chan et al., 2004; Schulze et al., 2005); therefore, its increased expression level implies increased ER stress in PD SNc. This hypothesis is supported by an *in vitro* model in which Herp knockdown resulted in increased ER stress-associated cell death in PC12 cells stably overexpressing mutant A53T α-synuclein (Belal et al., 2012). Finally, increased protein level of ER chaperone

GRP78/BiP was measured by western blot in cingulate gyrus, but not in parietal, prefrontal or temporal cortex, of PD and DLB patients compared to controls along with α-synuclein colocalization in some neurons observed by IHC (Baek et al., 2016). These analyses of UPR-related findings in PD-relevant regions of the human brain suggest that ER stress occurs in PD. However, and in contrast to lysosomal enzymes, none of the genes encoding the proteins discussed above are risk factors for PD, making it more difficult to assign a causal role for ER stress in PD pathogenesis at this time.

3.3.1 α-Synuclein and the unfolded protein response

The importance of the interplay between α-synuclein and UPR activation in PD pathogenesis has been highlighted through studies of patient-derived induced pluripotent stem cell (iPSC) models. Following an unbiased screen in yeast that yielded predictions of pathogenic cellular phenotypes evoked by α-synuclein toxicity, Chung et al. demonstrated increased ER stress in cortical neurons differentiated from patient iPSCs carrying the A53T mutation in α-synuclein compared to mutation-corrected cells. A state of ER stress in A53T neurons was quantified by elevated levels of ER chaperones PDI and BiP as well as ER accumulation of ERAD substrates GCase and nicastrin that could be partially rescued by lentiviral expression of E3 ubiquitin ligases Nedd4 or synoviolin-1 (Chung et al., 2013). Using comparative transcriptomic analysis, Heman-Ackah et al. showed upregulation of all three UPR branches in cortical neurons produced from iPSCs carrying triplication of *SNCA versus* iPSCs edited to correct gene dosage, with "protein processing at the ER" identified as the most highly enriched pathway between genotypes. Validation of the RNA sequencing results by quantitative PCR confirmed activation of the IRE1α-XBP1 axis in neurons with *SNCA* triplication that was partially reversed with triplication correction (Heman-Ackah et al., 2017). Genetic perturbations in α-synuclein known to cause PD thus appear to disrupt ER function in these models employing iPSC-derived neurons as surrogates of PD patient brain cells.

Additional details about the mechanism by which α-synuclein impairs normal ER processes have been elucidated by other disease models. α-Synuclein was shown to interact with ER chaperone GRP78/BiP (glucose-regulated protein/binding immunoglobulin protein) *in vitro* and *in vivo* in transgenic mice overexpressing aggregation-prone truncated human α-synuclein, with increased levels of GRP78/BiP and ATF4 measured in the substantia nigra compared to that in wild-type mice indicating activation of the PERK-eIF2α arm of the UPR in response to ER accumulation of α-synuclein (Bellucci et al., 2011). GRP78/BiP has been posited to be a master regulator of the UPR, dissociating from UPR branch leaders PERK, ATF6α and IRE1α upon sensing misfolded protein to allow their autophosphorylation and activation (Bellucci et al., 2011; Wang and Kaufman, 2016). Although this mechanism has been challenged in recent years (Carrara et al., 2015; Gardner and Walter, 2011), the response of GRP78/BiP to aggregated α-synuclein remains of interest in the context of PD pathogenesis. Activation of the PERK-eIF2α-ATF4-CHOP cascade by α-synuclein aggregates also occurred in rotenone-treated SH-SY5Y cells overexpressing wild-type α-synuclein, a toxic effect that was found to be dependent on phosphorylation of α-synuclein at serine 129 (Sugeno et al., 2008). In contrast, the PERK-eIF2α axis was not activated in spinal cord of transgenic mice expressing human A53T α-synuclein but increased levels of activated UPR mediator XBP1 and of ER chaperones including GRP78/BiP and PDI were detected (Colla et al., 2012). Neurotoxic effects of A53T α-synuclein were ameliorated *in vitro* and in rodent models by treatment with salubrinal, an ER stress inhibitor that acts on eIF2α (Boyce et al., 2005), substantiating the role of aberrant UPR activation and ER stress in PD pathophysiology (Colla et al., 2012; Smith et al., 2005).

In cell models of α-synuclein-dependent degeneration, deregulation of calcium signaling was shown to be caused by α-synuclein aggregates binding and stimulating ER calcium pump SERCA (sarco/endoplasmic reticulum calcium ATPase) (Betzer et al., 2018). Similarly, overexpression of human A53T α-synuclein in PC12 cells and in transgenic mice induced ER stress by deregulating ER calcium release via increased protein levels of ER calcium channels and ERAD substrates IP$_3$R (inositol triphosphate receptor) and RYR (ryanodine receptor), possibly due to binding of A53T α-synuclein to ERAD facilitator Herp which modulates turnover of these calcium channels (Belal et al., 2012) to help maintain calcium homeostasis and stabilize both ER and mitochondria (Chan et al., 2004). Further suggesting that mutant A53T α-synuclein impedes both ER and mitochondrial function, increased ROS production and mitochondrial release of cytochrome *c* was observed in PC12 cells with inducible expression of A53T α-synuclein as well as suppressed proteasomal activity and increased ER stress (elevated levels of phospho-eIF2α) underlying higher rates of apoptotic cell death compared to non-induced cells (Smith et al., 2005). Overall, these studies suggest that accumulation of PD-relevant forms of α-synuclein instigates chronic ER stress that contributes to the neurotoxicity of α-synuclein aggregates.

3.3.2 Link between oxidative stress and ER stress

Another key area of study into how the UPR is involved in PD pathogenesis has been exploration of the interaction between oxidative stress and ER stress. Mitochondrial toxins such as MPTP and 6-hydroxydopamine (6-OHDA) have been employed *in vitro* and *in vivo* to investigate ER-mitochondria crosstalk. Application of 6-OHDA or the MPTP metabolite 1-methyl-4-phenyl-pyridinium (MPP+) was shown to upregulate ER stress genes including chaperones PDI and BiP, and UPR branch cascade initiators IRE1α-XBP1 and PERK-eIF2α in dopaminergic cells (Conn et al., 2004; Holtz and O'Malley, 2003; Ryu et al., 2002). Additionally, primary sympathetic neurons derived from PERK-knockout mice that thus had limited ability to rectify ER stress were more sensitive to 6-OHDA-induced toxicity (Ryu et al., 2002). Likewise, acute silencing of XBP1 in the SNc of adult mice incited persistent ER stress and dopaminergic neurodegeneration while local delivery of active XBP1 to mouse SNc reduced dopaminergic cell death in animals subjected to intrastriatal injection of 6-OHDA (Valdés et al., 2014) suggesting that an intact UPR network is vital for the health of dopaminergic neurons. Exogenous expression of active XBP1 was also found to be neuroprotective in both cells and in mice treated with MPTP (Sado et al., 2009). Knockout of CHOP/GADD153, a transcription factor downstream of PERK-eIF2α that can induce apoptosis if ER stress is not resolved (Wang and Kaufman, 2016), reduced nigral dopamine neuron death following 6-OHDA intrastriatal injection in mice as well (Silva et al., 2005). Upregulation of ATF4, which is also downstream of PERK-eIF2α, was observed in cells exposed to 6-OHDA or MPP+, with silencing of ATF4 exacerbating and overexpression of ATF4 ameliorating cell death caused by these toxins (Sun et al., 2013).

Further indicating a neuroprotective role of the UPR in the cellular response to mitochondrial stress, systemic administration of MPTP to mice triggered activation of ATF6α, the initiator of the third major UPR division, along with upregulation of ER chaperone proteins and ERAD participant Derlin-3 in midbrain and striatum. ATF6α-knockout mice had decreased levels of these downstream effectors and increased ubiquitin-positive inclusions and dopaminergic neuron death in the SNc following MPTP challenge compared to wild-type mice (Egawa et al., 2011). Investigation into the mechanism by which MPTP precipitates ER stress revealed that MPTP disrupts calcium homeostasis through downregulation of transient receptor potential channel 1 (TRPC1), resulting in a disturbance of ER calcium levels triggering the UPR (Selvaraj et al., 2012). Analysis of the SNc of Trpc1-knockout mice showed elevation of UPR markers including phospho-eIF2α and CHOP and reduced dopaminergic neuron number *versus* wild-type mice. Interestingly, decreased TRPC1 and increased UPR protein levels were also measured in post-mortem SNc lysates of PD patients compared to age-matched controls (Selvaraj et al., 2012). Altogether, these studies build a compelling case for the hypothesis that ER and oxidative stress pathways converge and contribute to neurodegeneration.

Examining parkin function in the context of ER stress is relevant in the consideration of this hypothesis. As mentioned previously, parkin and PINK1 collectively promote mitophagy and hence deficiency of either of these proteins results in accumulation of damaged mitochondria and increased reactive oxygen species within the cell (Barodia et al., 2017). Parkin was first tied to the UPR with *in vitro* observations of parkin upregulation in response to ER stress induced by either β-mercaptoethanol or tunicamycin, compounds that prevent normal protein folding by inhibiting disulfide bridge formation and *N*-glycosylation, respectively (Imai et al., 2000; Ledesma et al., 2002). Parkin was shown to attenuate UPR-induced cell death through identification of a putative physiological substrate, Pael receptor, that became unfolded and insoluble with either overexpression or application of an ER stressor requiring parkin-mediated ubiquitylation to promote ERAD clearance of the misfolded protein (Imai et al., 2001). Accumulation of insoluble Pael receptor was seen in frontal cortex of AR-JP patients but not in controls, supporting the relevance of this substrate to PD pathophysiology (Imai et al., 2001). Bouman et al. demonstrated that parkin upregulation in response to UPR activation is mediated by ATF4 of the PERK-eIF2α pathway binding the parkin promoter to activate transcription (Bouman et al., 2011), and reciprocally parkin was determined to be required for ATF4 to diminish the toxic effects of 6-OHDC and MPP+ (Sun et al., 2013). Elucidation of this mechanism therefore directly linked parkin to the UPR. Finally, parkin was found to modulate ER stress through interaction with UPR branch IRE1α-XBP1 and DJ-1.

Investigating the ramifications of ER stress in a cell model, Duplan et al. described a
signaling cascade in which parkin upregulation in response to ER stress represses
p53, resulting in upregulation of active XBP1 which acts on the DJ-1 promoter to
enhance transcription (Duplan et al., 2013). In sum, loss-of-function mutations in
any of three parkinsonism-associated proteins, parkin, PINK1, or DJ-1, could disrupt
parkin-regulated cellular processes initiated by ER stress and fuel an aberrant UPR.

4 Summary

Understanding the complex pathways of protein synthesis and degradation as they
occur normally and in the context of PD is an active area of research. We have
reviewed the current evidence demonstrating how disease may arise due to
disruption of any one of the stages in the life cycle of a protein, from synthesis
via transcription, translation, and post-translational modification, to degradation
by proteasome, lysosome, or endoplasmic reticulum pathways, with intermediate
protein trafficking. Recent work in this field has highlighted the genetic component
of PD and the many ways in which it points to dysfunctional protein metabolism,
especially as pertains to α-synuclein homeostasis and involvement of the lysosome
in pathogenic processes. Future studies will shed even brighter light on the functional
implications of genetic clues to PD pathophysiology as capacity for genomic studies
and personalized medicine strategies continues to build. With a mind to identification of disease-modifying treatment targets, rather than symptomatic therapeutics alone, interrogation of protein deregulation underlying Parkinson's disease
will progress.

Acknowledgment

This research was supported by the Intramural Research Program of the NIH, National
Institute on Aging.

References

Abeywardana, T., Pratt, M.R., 2015. Extent of inhibition of α-synuclein aggregation in vitro by
    SUMOylation is conjugation site- and SUMO isoform-selective. Biochemistry 54, 959–961.
    https://doi.org/10.1021/bi501512m.
Aflaki, E., Westbroek, W., Sidransky, E., 2017. The complicated relationship between
    Gaucher disease and parkinsonism: insights from a rare disease. Neuron 93, 737–746.
    https://doi.org/10.1016/j.neuron.2017.01.018.
Alcalay, R.N., Dinur, T., Quinn, T., Sakanaka, K., Levy, O., Waters, C., Fahn, S., Dorovski, T.,
    Chung, W.K., Pauciulo, M., Nichols, W., Rana, H.Q., Balwani, M., Bier, L., Elstein, D.,
    Zimran, A., 2014. Comparison of Parkinson risk in Ashkenazi Jewish patients with
    Gaucher disease and GBA heterozygotes. JAMA Neurol. 71, 752–757. https://doi.org/
    10.1001/jamaneurol.2014.313.

Alexopoulou, Z., Lang, J., Perrett, R.M., Elschami, M., Hurry, M.E.D., Kim, H.T., Mazaraki, D., Szabo, A., Kessler, B.M., Goldberg, A.L., Ansorge, O., Fulga, T.A., Tofaris, G.K., 2016. Deubiquitinase Usp8 regulates α-synuclein clearance and modifies its toxicity in Lewy body disease. Proc. Natl. Acad. Sci. U. S. A. 113, E4688–E4697. https://doi.org/10.1073/pnas.1523597113.

Alvarez-Erviti, L., Rodriguez-Oroz, M.C., Cooper, J.M., Caballero, C., Ferrer, I., Obeso, J.A., Schapira, A.H.V., 2010. Chaperone-mediated autophagy markers in Parkinson disease brains. Arch. Neurol. 67, 1464–1472. https://doi.org/10.1001/archneurol.2010.198.

Ancolio, K., Alves da Costa, C., Uéda, K., Checler, F., 2000. α-Synuclein and the Parkinson's disease-related mutant Ala53Thr-α-synuclein do not undergo proteasomal degradation in HEK293 and neuronal cells. Neurosci. Lett. 285, 79–82. https://doi.org/10.1016/S0304-3940(00)01049-1.

Anderson, J.P., Walker, D.E., Goldstein, J.M., de Laat, R., Banducci, K., Caccavello, R.J., Barbour, R., Huang, J., Kling, K., Lee, M., Diep, L., Keim, P.S., Shen, X., Chataway, T., Schlossmacher, M.G., Seubert, P., Schenk, D., Sinha, S., Gai, W.P., Chilcote, T.J., 2006. Phosphorylation of Ser-129 is the dominant pathological modification of alpha-synuclein in familial and sporadic Lewy body disease. J. Biol. Chem. 281, 29739–29752. https://doi.org/10.1074/jbc.M600933200.

Anglade, P., Vyas, S., Javoy-Agid, F., Herrero, M.T., Michel, P.P., Marquez, J., Mouatt-Prigent, A., Ruberg, M., Hirsch, E.C., Agid, Y., 1997. Apoptosis and autophagy in nigral neurons of patients with Parkinson's disease. Histol. Histopathol. 12, 25–31.

Arawaka, S., Sato, H., Sasaki, A., Koyama, S., Kato, T., 2017. Mechanisms underlying extensive Ser129-phosphorylation in α-synuclein aggregates. Acta Neuropathol. Commun. 5, 48. https://doi.org/10.1186/s40478-017-0452-6.

Avraham, E., Rott, R., Liani, E., Szargel, R., Engelender, S., 2007. Phosphorylation of Parkin by the cyclin-dependent kinase 5 at the linker region modulates its ubiquitin-ligase activity and aggregation. J. Biol. Chem. 282, 12842–12850. https://doi.org/10.1074/jbc.M608243200.

Baba, M., Nakajo, S., Tu, P.H., Tomita, T., Nakaya, K., Lee, V.M., Trojanowski, J.Q., Iwatsubo, T., 1998. Aggregation of alpha-synuclein in Lewy bodies of sporadic Parkinson's disease and dementia with Lewy bodies. Am. J. Pathol. 152, 879–884.

Baek, J.-H., Whitfield, D., Howlett, D., Francis, P., Bereczki, E., Ballard, C., Hortobágyi, T., Attems, J., Aarsland, D., 2016. Unfolded protein response is activated in Lewy body dementias: UPR activation in Lewy body dementias. Neuropathol. Appl. Neurobiol. 42, 352–365. https://doi.org/10.1111/nan.12260.

Bancher, C., Lassmann, H., Budka, H., Jellinger, K., Grundke-Iqbal, I., Iqbal, K., Wiche, G., Seitelberger, F., Wisniewski, H.M., 1989. An antigenic profile of Lewy bodies: immunocytochemical indication for protein phosphorylation and ubiquitination. J. Neuropathol. Exp. Neurol. 48, 81–93.

Barodia, S.K., Creed, R.B., Goldberg, M.S., 2017. Parkin and PINK1 functions in oxidative stress and neurodegeneration. Brain Res. Bull. 133, 51–59. https://doi.org/10.1016/j.brainresbull.2016.12.004.

Barrett, P.J., Timothy Greenamyre, J., 2015. Post-translational modification of α-synuclein in Parkinson's disease. Brain Res. 1628, 247–253. https://doi.org/10.1016/j.brainres.2015.06.002.

Beal, F., Lang, A., 2006. The proteasomal inhibition model of Parkinson's disease: "Boon or bust"? Ann. Neurol. 60, 158–161. https://doi.org/10.1002/ana.20939.

Bedford, L., Hay, D., Devoy, A., Paine, S., Powe, D.G., Seth, R., Gray, T., Topham, I., Fone, K., Rezvani, N., Mee, M., Soane, T., Layfield, R., Sheppard, P.W., Ebendal, T.,

Usoskin, D., Lowe, J., Mayer, R.J., 2008. Depletion of 26S proteasomes in mouse brain neurons causes neurodegeneration and Lewy-like inclusions resembling human pale bodies. J. Neurosci. 28, 8189–8198. https://doi.org/10.1523/JNEUROSCI.2218-08.2008.

Beilina, A., Rudenko, I.N., Kaganovich, A., Civiero, L., Chau, H., Kalia, S.K., Kalia, L.V., Lobbestael, E., Chia, R., Ndukwe, K., Ding, J., Nalls, M.A., International Parkinson's Disease Genomics Consortium, North American Brain Expression Consortium, Olszewski, M., Hauser, D.N., Kumaran, R., Lozano, A.M., Baekelandt, V., Greene, L.E., Taymans, J.-M., Greggio, E., Cookson, M.R., 2014. Unbiased screen for interactors of leucine-rich repeat kinase 2 supports a common pathway for sporadic and familial Parkinson disease. Proc. Natl. Acad. Sci. U. S. A. 111, 2626–2631. https://doi.org/10.1073/pnas.1318306111.

Belal, C., Ameli, N.J., El Kommos, A., Bezalel, S., Al'Khafaji, A.M., Mughal, M.R., Mattson, M.P., Kyriazis, G.A., Tyrberg, B., Chan, S.L., 2012. The homocysteine-inducible endoplasmic reticulum (ER) stress protein Herp counteracts mutant α-synuclein-induced ER stress via the homeostatic regulation of ER-resident calcium release channel proteins. Hum. Mol. Genet. 21, 963–977. https://doi.org/10.1093/hmg/ddr502.

Bellucci, A., Navarria, L., Zaltieri, M., Falarti, E., Bodei, S., Sigala, S., Battistin, L., Spillantini, M., Missale, C., Spano, P., 2011. Induction of the unfolded protein response by α-synuclein in experimental models of Parkinson's disease: α-synuclein accumulation induces the UPR. J. Neurochem. 116, 588–605. https://doi.org/10.1111/j.1471-4159.2010.07143.x.

Bence, N.F., Sampat, R.M., Kopito, R.R., 2001. Impairment of the ubiquitin-proteasome system by protein aggregation. Science 292, 1552–1555. https://doi.org/10.1126/science.292.5521.1552.

Bennett, M.C., Bishop, J.F., Leng, Y., Chock, P.B., Chase, T.N., Mouradian, M.M., 1999. Degradation of alpha-synuclein by proteasome. J. Biol. Chem. 274, 33855–33858.

Bentea, E., Verbruggen, L., Massie, A., 2017. The proteasome inhibition model of Parkinson's disease. J. Park. Dis. 7, 31–63. https://doi.org/10.3233/JPD-160921.

Bento, C.F., Ashkenazi, A., Jimenez-Sanchez, M., Rubinsztein, D.C., 2016. The Parkinson's disease-associated genes ATP13A2 and SYT11 regulate autophagy via a common pathway. Nat. Commun. 7, 11803. https://doi.org/10.1038/ncomms11803.

Betzer, C., Lassen, L.B., Olsen, A., Kofoed, R.H., Reimer, L., Gregersen, E., Zheng, J., Calì, T., Gai, W.-P., Chen, T., Moeller, A., Brini, M., Fu, Y., Halliday, G., Brudek, T., Aznar, S., Pakkenberg, B., Andersen, J.P., Jensen, P.H., 2018. Alpha-synuclein aggregates activate calcium pump SERCA leading to calcium dysregulation. EMBO Rep. 19, e44617. https://doi.org/10.15252/embr.201744617.

Bochtler, M., Ditzel, L., Groll, M., Hartmann, C., Huber, R., 1999. The proteasome. Annu. Rev. Biophys. Biomol. Struct. 28, 295–317. https://doi.org/10.1146/annurev.biophys.28.1.295.

Book, A., Guella, I., Candido, T., Brice, A., Hattori, N., Jeon, B., Farrer, M.J., SNCA Multiplication Investigators of the GEoPD Consortium, 2018. A meta-analysis of α-synuclein multiplication in familial parkinsonism. Front. Neurol. 9, 1021. https://doi.org/10.3389/fneur.2018.01021.

Bouman, L., Schlierf, A., Lutz, A.K., Shan, J., Deinlein, A., Kast, J., Galehdar, Z., Palmisano, V., Patenge, N., Berg, D., Gasser, T., Augustin, R., Trümbach, D., Irrcher, I., Park, D.S., Wurst, W., Kilberg, M.S., Tatzelt, J., Winklhofer, K.F., 2011. Parkin is transcriptionally regulated by ATF4: evidence for an interconnection between mitochondrial stress and ER stress. Cell Death Differ. 18, 769–782. https://doi.org/10.1038/cdd.2010.142.

Bové, J., Zhou, C., Jackson-Lewis, V., Taylor, J., Chu, Y., Rideout, H.J., Wu, D.-C., Kordower, J.H., Petrucelli, L., Przedborski, S., 2006. Proteasome inhibition and Parkinson's disease modeling. Ann. Neurol. 60, 260–264. https://doi.org/10.1002/ana.20937.

Boyce, M., Bryant, K.F., Jousse, C., Long, K., Harding, H.P., Scheuner, D., Kaufman, R.J., Ma, D., Coen, D.M., Ron, D., Yuan, J., 2005. A selective inhibitor of eIF2alpha dephosphorylation protects cells from ER stress. Science 307, 935–939. https://doi.org/10.1126/science.1101902.

Bravo-San Pedro, J.M., Niso-Santano, M., Gómez-Sánchez, R., Pizarro-Estrella, E., Aiastui-Pujana, A., Gorostidi, A., Climent, V., López de Maturana, R., Sanchez-Pernaute, R., López de Munain, A., Fuentes, J.M., González-Polo, R.A., 2013. The LRRK2 G2019S mutant exacerbates basal autophagy through activation of the MEK/ERK pathway. Cell. Mol. Life Sci. 70, 121–136. https://doi.org/10.1007/s00018-012-1061-y.

Bucciantini, M., Giannoni, E., Chiti, F., Baroni, F., Formigli, L., Zurdo, J., Taddei, N., Ramponi, G., Dobson, C.M., Stefani, M., 2002. Inherent toxicity of aggregates implies a common mechanism for protein misfolding diseases. Nature 416, 507–511. https://doi.org/10.1038/416507a.

Bukhatwa, S., Zeng, B.-Y., Rose, S., Jenner, P., 2010. A comparison of changes in proteasomal subunit expression in the substantia nigra in Parkinson's disease, multiple system atrophy and progressive supranuclear palsy. Brain Res. 1326, 174–183. https://doi.org/10.1016/j.brainres.2010.02.045.

Burbulla, L.F., Song, P., Mazzulli, J.R., Zampese, E., Wong, Y.C., Jeon, S., Santos, D.P., Blanz, J., Obermaier, C.D., Strojny, C., Savas, J.N., Kiskinis, E., Zhuang, X., Krüger, R., Surmeier, D.J., Krainc, D., 2017. Dopamine oxidation mediates mitochondrial and lysosomal dysfunction in Parkinson's disease. Science 357, 1255–1261. https://doi.org/10.1126/science.aam9080.

Carrara, M., Prischi, F., Nowak, P.R., Kopp, M.C., Ali, M.M., 2015. Noncanonical binding of BiP ATPase domain to Ire1 and Perk is dissociated by unfolded protein CH1 to initiate ER stress signaling. eLife 4, e03522. https://doi.org/10.7554/eLife.03522.

Chakraborty, J., Basso, V., Ziviani, E., 2017. Post translational modification of Parkin. Biol. Direct 12, 6. https://doi.org/10.1186/s13062-017-0176-3.

Chan, S.L., Fu, W., Zhang, P., Cheng, A., Lee, J., Kokame, K., Mattson, M.P., 2004. Herp stabilizes neuronal Ca2+ homeostasis and mitochondrial function during endoplasmic reticulum stress. J. Biol. Chem. 279, 28733–28743. https://doi.org/10.1074/jbc.M404272200.

Chang, D., Nalls, M.A., Hallgrímsdóttir, I.B., Hunkapiller, J., van der Brug, M., Cai, F., International Parkinson's Disease Genomics Consortium, 23andMe Research Team, Kerchner, G.A., Ayalon, G., Bingol, B., Sheng, M., Hinds, D., Behrens, T.W., Singleton, A.B., Bhangale, T.R., Graham, R.R., 2017. A meta-analysis of genome-wide association studies identifies 17 new Parkinson's disease risk loci. Nat. Genet. 49, 1511.

Chartier-Harlin, M.-C., Kachergus, J., Roumier, C., Mouroux, V., Douay, X., Lincoln, S., Levecque, C., Larvor, L., Andrieux, J., Hulihan, M., Waucquier, N., Defebvre, L., Amouyel, P., Farrer, M., Destée, A., 2004. Alpha-synuclein locus duplication as a cause of familial Parkinson's disease. Lancet Lond. Engl. 364, 1167–1169. https://doi.org/10.1016/S0140-6736(04)17103-1.

Chartier-Harlin, M.-C., Dachsel, J.C., Vilariño-Güell, C., Lincoln, S.J., Leprêtre, F., Hulihan, M.M., Kachergus, J., Milnerwood, A.J., Tapia, L., Song, M.-S., Le Rhun, E., Mutez, E., Larvor, L., Duflot, A., Vanbesien-Mailliot, C., Kreisler, A., Ross, O.A., Nishioka, K., Soto-Ortolaza, A.I., Cobb, S.A., Melrose, H.L., Behrouz, B., Keeling, B.H., Bacon, J.A., Hentati, E., Williams, L., Yanagiya, A., Sonenberg, N.,

Lockhart, P.J., Zubair, A.C., Uitti, R.J., Aasly, J.O., Krygowska-Wajs, A., Opala, G., Wszolek, Z.K., Frigerio, R., Maraganore, D.M., Gosal, D., Lynch, T., Hutchinson, M., Bentivoglio, A.R., Valente, E.M., Nichols, W.C., Pankratz, N., Foroud, T., Gibson, R.A., Hentati, F., Dickson, D.W., Destée, A., Farrer, M.J., 2011. Translation initiator EIF4G1 mutations in familial Parkinson disease. Am. J. Hum. Genet. 89, 398–406. https://doi.org/10.1016/j.ajhg.2011.08.009.

Chaugule, V.K., Burchell, L., Barber, K.R., Sidhu, A., Leslie, S.J., Shaw, G.S., Walden, H., 2011. Autoregulation of Parkin activity through its ubiquitin-like domain. EMBO J. 30, 2853–2867. https://doi.org/10.1038/emboj.2011.204.

Chiba-Falek, O., Touchman, J.W., Nussbaum, R.L., 2003. Functional analysis of intra-allelic variation at NACP-Rep1 in the alpha-synuclein gene. Hum. Genet. 113, 426–431. https://doi.org/10.1007/s00439-003-1002-9.

Chiba-Falek, O., Lopez, G.J., Nussbaum, R.L., 2006. Levels of alpha-synuclein mRNA in sporadic Parkinson disease patients. Mov. Disord. 21, 1703–1708. https://doi.org/10.1002/mds.21007.

Choo, Y.S., Vogler, G., Wang, D., Kalvakuri, S., Iliuk, A., Tao, W.A., Bodmer, R., Zhang, Z., 2012. Regulation of parkin and PINK1 by neddylation. Hum. Mol. Genet. 21, 2514–2523. https://doi.org/10.1093/hmg/dds070.

Chu, C.T., 2006. Autophagic stress in neuronal injury and disease. J. Neuropathol. Exp. Neurol. 65, 423–432. https://doi.org/10.1097/01.jnen.0000229233.75253.be.

Chu, Y., Dodiya, H., Aebischer, P., Olanow, C.W., Kordower, J.H., 2009. Alterations in lysosomal and proteasomal markers in Parkinson's disease: relationship to alpha-synuclein inclusions. Neurobiol. Dis. 35, 385–398. https://doi.org/10.1016/j.nbd.2009.05.023.

Chung, K.K.K., 2004. S-nitrosylation of parkin regulates ubiquitination and compromises parkin's protective function. Science 304, 1328–1331. https://doi.org/10.1126/science.1093891.

Chung, C.Y., Khurana, V., Auluck, P.K., Tardiff, D.F., Mazzulli, J.R., Soldner, F., Baru, V., Lou, Y., Freyzon, Y., Cho, S., Mungenast, A.E., Muffat, J., Mitalipova, M., Pluth, M.D., Jui, N.T., Schüle, B., Lippard, S.J., Tsai, L.-H., Krainc, D., Buchwald, S.L., Jaenisch, R., Lindquist, S., 2013. Identification and rescue of α-synuclein toxicity in Parkinson patient-derived neurons. Science 342, 983–987. https://doi.org/10.1126/science.1245296.

Cleeter, M.W.J., Chau, K.-Y., Gluck, C., Mehta, A., Hughes, D.A., Duchen, M., Wood, N.W., Hardy, J., Mark Cooper, J., Schapira, A.H., 2013. Glucocerebrosidase inhibition causes mitochondrial dysfunction and free radical damage. Neurochem. Int. 62, 1–7. https://doi.org/10.1016/j.neuint.2012.10.010.

Colla, E., Coune, P., Liu, Y., Pletnikova, O., Troncoso, J.C., Iwatsubo, T., Schneider, B.L., Lee, M.K., 2012. Endoplasmic reticulum stress is important for the manifestations of α-synucleinopathy in vivo. J. Neurosci. 32, 3306–3320. https://doi.org/10.1523/JNEUROSCI.5367-11.2012.

Conn, K.J., Gao, W., McKee, A., Lan, M.S., Ullman, M.D., Eisenhauer, P.B., Fine, R.E., Wells, J.M., 2004. Identification of the protein disulfide isomerase family member PDIp in experimental Parkinson's disease and Lewy body pathology. Brain Res. 1022, 164–172. https://doi.org/10.1016/j.brainres.2004.07.026.

Cookson, M.R., 2012. Parkinsonism due to mutations in PINK1, parkin, and DJ-1 and oxidative stress and mitochondrial pathways. Cold Spring Harb. Perspect. Med. 2, a009415. https://doi.org/10.1101/cshperspect.a009415.

Cooper, G.M., 2000. The Cell: A Molecular Approach, second ed. Sinauer Associates, Sunderland, MA.

Cooper, A.A., Gitler, A.D., Cashikar, A., Haynes, C.M., Hill, K.J., Bhullar, B., Liu, K., Xu, K., Strathearn, K.E., Liu, F., Cao, S., Caldwell, K.A., Caldwell, G.A., Marsischky, G., Kolodner, R.D., Labaer, J., Rochet, J.-C., Bonini, N.M., Lindquist, S., 2006. Alpha-synuclein blocks ER-Golgi traffic and Rab1 rescues neuron loss in Parkinson's models. Science 313, 324–328. https://doi.org/10.1126/science.1129462.

Crick, F., 1970. Central dogma of molecular biology. Nature 227, 561–563. https://doi.org/10.1038/227561a0.

Cronin, K.D., Ge, D., Manninger, P., Linnertz, C., Rossoshek, A., Orrison, B.M., Bernard, D.J., El-Agnaf, O.M.A., Schlossmacher, M.G., Nussbaum, R.L., Chiba-Falek, O., 2009. Expansion of the Parkinson disease-associated SNCA-Rep1 allele upregulates human alpha-synuclein in transgenic mouse brain. Hum. Mol. Genet. 18, 3274–3285. https://doi.org/10.1093/hmg/ddp265.

Cuervo, A.M., Stefanis, L., Fredenburg, R., Lansbury, P.T., Sulzer, D., 2004. Impaired degradation of mutant alpha-synuclein by chaperone-mediated autophagy. Science 305, 1292–1295. https://doi.org/10.1126/science.1101738.

Cullen, V., Lindfors, M., Ng, J., Paetau, A., Swinton, E., Kolodziej, P., Boston, H., Saftig, P., Woulfe, J., Feany, M.B., Myllykangas, L., Schlossmacher, M.G., Tyynelä, J., 2009. Cathepsin D expression level affects alpha-synuclein processing, aggregation, and toxicity in vivo. Mol. Brain 2, 5. https://doi.org/10.1186/1756-6606-2-5.

Dächsel, J.C., Lincoln, S.J., Gonzalez, J., Ross, O.A., Dickson, D.W., Farrer, M.J., 2007. The ups and downs of alpha-synuclein mRNA expression. Mov. Disord. 22, 293–295. https://doi.org/10.1002/mds.21223.

Davies, S.E., Hallett, P.J., Moens, T., Smith, G., Mangano, E., Kim, H.T., Goldberg, A.L., Liu, J.-L., Isacson, O., Tofaris, G.K., 2014. Enhanced ubiquitin-dependent degradation by Nedd4 protects against α-synuclein accumulation and toxicity in animal models of Parkinson's disease. Neurobiol. Dis. 64, 79–87. https://doi.org/10.1016/j.nbd.2013.12.011.

Decressac, M., Mattsson, B., Weikop, P., Lundblad, M., Jakobsson, J., Björklund, A., 2013. TFEB-mediated autophagy rescues midbrain dopamine neurons from α-synuclein toxicity. Proc. Natl. Acad. Sci. U. S. A. 110, E1817–E1826. https://doi.org/10.1073/pnas.1305623110.

Dehay, B., Bové, J., Rodríguez-Muela, N., Perier, C., Recasens, A., Boya, P., Vila, M., 2010. Pathogenic lysosomal depletion in Parkinson's disease. J. Neurosci. 30, 12535–12544. https://doi.org/10.1523/JNEUROSCI.1920-10.2010.

Dehay, B., Ramirez, A., Martinez-Vicente, M., Perier, C., Canron, M.-H., Doudnikoff, E., Vital, A., Vila, M., Klein, C., Bezard, E., 2012. Loss of P-type ATPase ATP13A2/PARK9 function induces general lysosomal deficiency and leads to Parkinson disease neurodegeneration. Proc. Natl. Acad. Sci. U. S. A. 109, 9611–9616. https://doi.org/10.1073/pnas.1112368109.

Dhungel, N., Eleuteri, S., Li, L., Kramer, N.J., Chartron, J.W., Spencer, B., Kosberg, K., Fields, J.A., Stafa, K., Adame, A., Lashuel, H., Frydman, J., Shen, K., Masliah, E., Gitler, A.D., 2015. Parkinson's disease genes VPS35 and EIF4G1 interact genetically and converge on α-synuclein. Neuron 85, 76–87. https://doi.org/10.1016/j.neuron.2014.11.027.

Ding, Q., Dimayuga, E., Martin, S., Bruce-Keller, A.J., Nukala, V., Cuervo, A.M., Keller, J.N., 2003. Characterization of chronic low-level proteasome inhibition on neural homeostasis. J. Neurochem. 86, 489–497.

Dodson, M.W., Zhang, T., Jiang, C., Chen, S., Guo, M., 2012. Roles of the Drosophila LRRK2 homolog in Rab7-dependent lysosomal positioning. Hum. Mol. Genet. 21, 1350–1363. https://doi.org/10.1093/hmg/ddr573.

Dodson, M.W., Leung, L.K., Lone, M., Lizzio, M.A., Guo, M., 2014. Novel ethyl methane-sulfonate (EMS)-induced null alleles of the Drosophila homolog of LRRK2 reveal a crucial role in endolysosomal functions and autophagy in vivo. Dis. Model. Mech. 7, 1351–1363. https://doi.org/10.1242/dmm.017020.

Duplan, E., Giaime, E., Viotti, J., Sévalle, J., Corti, O., Brice, A., Ariga, H., Qi, L., Checler, F., Alves da Costa, C., 2013. ER-stress-associated functional link between Parkin and DJ-1 via a transcriptional cascade involving the tumor suppressor p53 and the spliced X-box binding protein XBP-1. J. Cell Sci. 126, 2124–2133. https://doi.org/10.1242/jcs.127340.

Durcan, T.M., Tang, M.Y., Pérusse, J.R., Dashti, E.A., Aguileta, M.A., McLelland, G.-L., Gros, P., Shaler, T.A., Faubert, D., Coulombe, B., Fon, E.A., 2014. USP8 regulates mito-phagy by removing K6-linked ubiquitin conjugates from parkin. EMBO J. 33, 2473–2491. https://doi.org/10.15252/embj.201489729.

Ebrahimi-Fakhari, D., Cantuti-Castelvetri, I., Fan, Z., Rockenstein, E., Masliah, E., Hyman, B.T., McLean, P.J., Unni, V.K., 2011. Distinct roles in vivo for the ubiquitin-proteasome system and the autophagy-lysosomal pathway in the degradation of α-synu-clein. J. Neurosci. 31, 14508–14520. https://doi.org/10.1523/JNEUROSCI.1560-11.2011.

Egawa, N., Yamamoto, K., Inoue, H., Hikawa, R., Nishi, K., Mori, K., Takahashi, R., 2011. The endoplasmic reticulum stress sensor, ATF6α, protects against neurotoxin-induced do-paminergic neuronal death. J. Biol. Chem. 286, 7947–7957. https://doi.org/10.1074/jbc.M110.156430.

Emmanouilidou, E., Stefanis, L., Vekrellis, K., 2010. Cell-produced alpha-synuclein oligo-mers are targeted to, and impair, the 26S proteasome. Neurobiol. Aging 31, 953–968. https://doi.org/10.1016/j.neurobiolaging.2008.07.008.

Engelhardt, E., 2017. Lafora and Trétiakoff: the naming of the inclusion bodies discovered by Lewy. Arq. Neuropsiquiatr. 75, 751–753. https://doi.org/10.1590/0004-282X20170116.

Farrer, M., Maraganore, D.M., Lockhart, P., Singleton, A., Lesnick, T.G., de Andrade, M., West, A., de Silva, R., Hardy, J., Hernandez, D., 2001. alpha-Synuclein gene haplotypes are associated with Parkinson's disease. Hum. Mol. Genet. 10, 1847–1851. https://doi.org/10.1093/hmg/10.17.1847.

Farrer, M., Kachergus, J., Forno, L., Lincoln, S., Wang, D.-S., Hulihan, M., Maraganore, D., Gwinn-Hardy, K., Wszolek, Z., Dickson, D., Langston, J.W., 2004. Comparison of kin-dreds with parkinsonism and alpha-synuclein genomic multiplications. Ann. Neurol. 55, 174–179. https://doi.org/10.1002/ana.10846.

Feng, Y., He, D., Yao, Z., Klionsky, D.J., 2014. The machinery of macroautophagy. Cell Res. 24, 24–41. https://doi.org/10.1038/cr.2013.168.

Follett, J., Norwood, S.J., Hamilton, N.A., Mohan, M., Kovtun, O., Tay, S., Zhe, Y., Wood, S.A., Mellick, G.D., Silburn, P.A., Collins, B.M., Bugarcic, A., Teasdale, R.D., 2014. The Vps35 D620N mutation linked to Parkinson's disease disrupts the cargo sorting function of retromer: Parkinson's disease causing mutation alters retromer's function. Traffic 15, 230–244. https://doi.org/10.1111/tra.12136.

Fornai, F., Lenzi, P., Gesi, M., Ferrucci, M., Lazzeri, G., Busceti, C.L., Ruffoli, R., Soldani, P., Ruggieri, S., Alessandri, M.G., Paparelli, A., 2003. Fine structure and biochemical mech-anisms underlying nigrostriatal inclusions and cell death after proteasome inhibition. J. Neurosci. 23, 8955–8966.

Fornai, F., Schlüter, O.M., Lenzi, P., Gesi, M., Ruffoli, R., Ferrucci, M., Lazzeri, G., Busceti, C.L., Pontarelli, F., Battaglia, G., Pellegrini, A., Nicoletti, F., Ruggieri, S., Paparelli, A., Südhof, T.C., 2005. Parkinson-like syndrome induced by continuous MPTP infusion: convergent roles of the ubiquitin-proteasome system and alpha-synuclein. Proc. Natl. Acad. Sci. U. S. A. 102, 3413–3418. https://doi.org/10.1073/pnas.0409713102.

Friguet, B., Szweda, L.I., 1997. Inhibition of the multicatalytic proteinase (proteasome) by 4-hydroxy-2-nonenal cross-linked protein. FEBS Lett. 405, 21–25. https://doi.org/10.1016/S0014-5793(97)00148-8.

Fuchs, J., Nilsson, C., Kachergus, J., Munz, M., Larsson, E.-M., Schüle, B., Langston, J.W., Middleton, F.A., Ross, O.A., Hulihan, M., Gasser, T., Farrer, M.J., 2007. Phenotypic variation in a large Swedish pedigree due to SNCA duplication and triplication. Neurology 68, 916–922. https://doi.org/10.1212/01.wnl.0000254458.17630.c5.

Fuchs, J., Tichopad, A., Golub, Y., Munz, M., Schweitzer, K.J., Wolf, B., Berg, D., Mueller, J.C., Gasser, T., 2008. Genetic variability in the SNCA gene influences alpha-synuclein levels in the blood and brain. FASEB J. 22, 1327–1334. https://doi.org/10.1096/fj.07-9348com.

Fujiwara, H., Hasegawa, M., Dohmae, N., Kawashima, A., Masliah, E., Goldberg, M.S., Shen, J., Takio, K., Iwatsubo, T., 2002. alpha-Synuclein is phosphorylated in synucleinopathy lesions. Nat. Cell Biol. 4, 160–164. https://doi.org/10.1038/ncb748.

Gan-Or, Z., Ozelius, L.J., Bar-Shira, A., Saunders-Pullman, R., Mirelman, A., Kornreich, R., Gana-Weisz, M., Raymond, D., Rozenkrantz, L., Deik, A., Gurevich, T., Gross, S.J., Schreiber-Agus, N., Giladi, N., Bressman, S.B., Orr-Urtreger, A., 2013. The p.L302P mutation in the lysosomal enzyme gene SMPD1 is a risk factor for Parkinson disease. Neurology 80, 1606–1610. https://doi.org/10.1212/WNL.0b013e31828f180e.

Gardai, S.J., Mao, W., Schüle, B., Babcock, M., Schoebel, S., Lorenzana, C., Alexander, J., Kim, S., Glick, H., Hilton, K., Fitzgerald, J.K., Buttini, M., Chiou, S.-S., McConlogue, L., Anderson, J.P., Schenk, D.B., Bard, F., Langston, J.W., Yednock, T., Johnston, J.A., 2013. Elevated alpha-synuclein impairs innate immune cell function and provides a potential peripheral biomarker for Parkinson's disease. PLoS One 8, e71634. https://doi.org/10.1371/journal.pone.0071634.

Gardner, B.M., Walter, P., 2011. Unfolded proteins are Ire1-activating ligands that directly induce the unfolded protein response. Science 333, 1891–1894. https://doi.org/10.1126/science.1209126.

Gegg, M.E., Burke, D., Heales, S.J.R., Cooper, J.M., Hardy, J., Wood, N.W., Schapira, A.H. V., 2012. Glucocerebrosidase deficiency in substantia nigra of Parkinson disease brains. Ann. Neurol. 72, 455–463. https://doi.org/10.1002/ana.23614.

Gehrke, S., Imai, Y., Sokol, N., Lu, B., 2010. Pathogenic LRRK2 negatively regulates microRNA-mediated translational repression. Nature 466, 637–641. https://doi.org/10.1038/nature09191.

Gehrke, S., Wu, Z., Klinkenberg, M., Sun, Y., Auburger, G., Guo, S., Lu, B., 2015. PINK1 and parkin control localized translation of respiratory chain component mRNAs on mitochondria outer membrane. Cell Metab. 21, 95–108. https://doi.org/10.1016/j.cmet.2014.12.007.

Ghee, M., Fournier, A., Mallet, J., 2000. Rat alpha-synuclein interacts with Tat binding protein 1, a component of the 26S proteasomal complex. J. Neurochem. 75, 2221–2224. https://doi.org/10.1046/j.1471-4159.2000.0752221.x.

Giasson, B.I., 2000. Oxidative damage linked to neurodegeneration by selective alpha-synuclein nitration in synucleinopathy lesions. Science 290, 985–989. https://doi.org/10.1126/science.290.5493.985.

Gingras, A.-C., Raught, B., Sonenberg, N., 1999. eIF4 initiation factors: effectors of mRNA recruitment to ribosomes and regulators of translation. Annu. Rev. Biochem. 68, 913–963. https://doi.org/10.1146/annurev.biochem.68.1.913.

Gitler, A.D., Bevis, B.J., Shorter, J., Strathearn, K.E., Hamamichi, S., Su, L.J., Caldwell, K.A., Caldwell, G.A., Rochet, J.-C., McCaffery, J.M., Barlowe, C., Lindquist, S., 2008. The Parkinson's disease protein-synuclein disrupts cellular Rab homeostasis. Proc. Natl. Acad. Sci. U. S. A. 105, 145–150. https://doi.org/10.1073/pnas.0710685105.

Goker-Alpan, O., 2004. Parkinsonism among Gaucher disease carriers. J. Med. Genet. 41, 937–940. https://doi.org/10.1136/jmg.2004.024455.

Gómez-Suaga, P., Luzón-Toro, B., Churamani, D., Zhang, L., Bloor-Young, D., Patel, S., Woodman, P.G., Churchill, G.C., Hilfiker, S., 2012. Leucine-rich repeat kinase 2 regulates autophagy through a calcium-dependent pathway involving NAADP. Hum. Mol. Genet. 21, 511–525. https://doi.org/10.1093/hmg/ddr481.

Grünblatt, E., Mandel, S., Jacob-Hirsch, J., Zeligson, S., Amariglo, N., Rechavi, G., Li, J., Ravid, R., Roggendorf, W., Riederer, P., Youdim, M.B.H., 2004. Gene expression profiling of parkinsonian substantia nigra pars compacta; alterations in ubiquitin-proteasome, heat shock protein, iron and oxidative stress regulated proteins, cell adhesion/cellular matrix and vesicle trafficking genes. J. Neural Transm. (Vienna) 111, 1543–1573. https://doi.org/10.1007/s00702-004-0212-1.

Gründemann, J., Schlaudraff, F., Haeckel, O., Liss, B., 2008. Elevated alpha-synuclein mRNA levels in individual UV-laser-microdissected dopaminergic substantia nigra neurons in idiopathic Parkinson's disease. Nucleic Acids Res. 36, e38. https://doi.org/10.1093/nar/gkn084.

Grune, T., Reinheckel, T., Joshi, M., Davies, K.J., 1995. Proteolysis in cultured liver epithelial cells during oxidative stress. Role of the multicatalytic proteinase complex, proteasome. J. Biol. Chem. 270, 2344–2351. https://doi.org/10.1074/jbc.270.5.2344.

Guo, C., Sun, L., Chen, X., Zhang, D., 2013. Oxidative stress, mitochondrial damage and neurodegenerative diseases. Neural Regen. Res. 8, 2003–2014. https://doi.org/10.3969/j.issn.1673-5374.2013.21.009.

Gusdon, A.M., Zhu, J., Van Houten, B., Chu, C.T., 2012. ATP13A2 regulates mitochondrial bioenergetics through macroautophagy. Neurobiol. Dis. 45, 962–972. https://doi.org/10.1016/j.nbd.2011.12.015.

Haj-Yahya, M., Fauvet, B., Herman-Bachinsky, Y., Hejjaoui, M., Bavikar, S.N., Karthikeyan, S.V., Ciechanover, A., Lashuel, H.A., Brik, A., 2013. Synthetic polyubiquitinated α-synuclein reveals important insights into the roles of the ubiquitin chain in regulating its pathophysiology. Proc. Natl. Acad. Sci. U. S. A. 110, 17726–17731. https://doi.org/10.1073/pnas.1315654110.

Halperin, A., Elstein, D., Zimran, A., 2006. Increased incidence of Parkinson disease among relatives of patients with Gaucher disease. Blood Cells. Mol. Dis. 36, 426–428. https://doi.org/10.1016/j.bcmd.2006.02.004.

Hamasaki, M., Noda, T., Ohsumi, Y., 2003. The early secretory pathway contributes to autophagy in yeast. Cell Struct. Funct. 28, 49–54. https://doi.org/10.1247/csf.28.49.

Hammond, S.M., Bernstein, E., Beach, D., Hannon, G.J., 2000. An RNA-directed nuclease mediates post-transcriptional gene silencing in Drosophila cells. Nature 404, 293–296. https://doi.org/10.1038/35005107.

Hauser, D.N., Primiani, C.T., Cookson, M.R., 2017. The effects of variants in the parkin, PINK1, and DJ-1 genes along with evidence for their pathogenicity. Curr. Protein Pept. Sci. 18, 702–714. https://doi.org/10.2174/1389203717666160311121954.

He, Y., Yu, Z., Chen, S., 2019. Alpha-synuclein nitration and its implications in Parkinson's disease. ACS Chem. Neurosci. 10, 777–782. https://doi.org/10.1021/acschemneuro.8b00288.

Heikkila, R.E., Hess, A., Duvoisin, R.C., 1984. Dopaminergic neurotoxicity of 1-methyl-4-phenyl-1,2,5,6-tetrahydropyridine in mice. Science 224, 1451–1453.

Hellman, N.E., Grant, E.A., Goate, A.M., 1998. Failure to replicate a protective effect of allele 2 of NACP/alpha-synuclein polymorphism in Alzheimer's disease: an association study. Ann. Neurol. 44, 278–281. https://doi.org/10.1002/ana.410440223.

Heman-Ackah, S.M., Manzano, R., Hoozemans, J.J.M., Scheper, W., Flynn, R., Haerty, W., Cowley, S.A., Bassett, A.R., Wood, M.J.A., 2017. Alpha-synuclein induces the unfolded protein response in Parkinson's disease SNCA triplication iPSC-derived neurons. Hum. Mol. Genet. 26, 4441–4450. https://doi.org/10.1093/hmg/ddx331.

Hershko, A., Leshinsky, E., Ganoth, D., Heller, H., 1984. ATP-dependent degradation of ubiquitin-protein conjugates. Proc. Natl. Acad. Sci. U. S. A. 81, 1619–1623. https://doi.org/10.1073/pnas.81.6.1619.

Höglinger, G.U., Carrard, G., Michel, P.P., Medja, F., Lombès, A., Ruberg, M., Friguet, B., Hirsch, E.C., 2003. Dysfunction of mitochondrial complex I and the proteasome: interactions between two biochemical deficits in a cellular model of Parkinson's disease. J. Neurochem. 86, 1297–1307.

Höglinger, G.U., Breunig, J.J., Depboylu, C., Rouaux, C., Michel, P.P., Alvarez-Fischer, D., Boutillier, A.-L., Degregori, J., Oertel, W.H., Rakic, P., Hirsch, E.C., Hunot, S., 2007. The pRb/E2F cell-cycle pathway mediates cell death in Parkinson's disease. Proc. Natl. Acad. Sci. U. S. A. 104, 3585–3590. https://doi.org/10.1073/pnas.0611671104.

Holcik, M., Sonenberg, N., 2005. Translational control in stress and apoptosis. Nat. Rev. Mol. Cell Biol. 6, 318–327. https://doi.org/10.1038/nrm1618.

Holtz, W.A., O'Malley, K.L., 2003. Parkinsonian mimetics induce aspects of unfolded protein response in death of dopaminergic neurons. J. Biol. Chem. 278, 19367–19377. https://doi.org/10.1074/jbc.M211821200.

Hoozemans, J.J.M., van Haastert, E.S., Eikelenboom, P., de Vos, R.A., Rozemuller, J.M., Scheper, W., 2007. Activation of the unfolded protein response in Parkinson's disease. Biochem. Biophys. Res. Commun. 354, 707–711. https://doi.org/10.1016/j.bbrc.2007.01.043.

Huynh, D.P., Scoles, D.R., Nguyen, D., Pulst, S.M., 2003. The autosomal recessive juvenile Parkinson disease gene product, parkin, interacts with and ubiquitinates synaptotagmin XI. Hum. Mol. Genet. 12, 2587–2597. https://doi.org/10.1093/hmg/ddg269.

Ibáñez, P., Bonnet, A.-M., Débarges, B., Lohmann, E., Tison, F., Pollak, P., Agid, Y., Dürr, A., Brice, A., 2004. Causal relation between alpha-synuclein gene duplication and familial Parkinson's disease. Lancet Lond. Engl. 364, 1169–1171. https://doi.org/10.1016/S0140-6736(04)17104-3.

Ii, K., Ito, H., Tanaka, K., Hirano, A., 1997. Immunocytochemical co-localization of the proteasome in ubiquitinated structures in neurodegenerative diseases and the elderly. J. Neuropathol. Exp. Neurol. 56, 125–131.

Imai, Y., Soda, M., Takahashi, R., 2000. Parkin suppresses unfolded protein stress-induced cell death through its E3 ubiquitin-protein ligase activity. J. Biol. Chem. 275, 35661–35664. https://doi.org/10.1074/jbc.C000447200.

Imai, Y., Soda, M., Inoue, H., Hattori, N., Mizuno, Y., Takahashi, R., 2001. An unfolded putative transmembrane polypeptide, which can lead to endoplasmic reticulum stress, is a substrate of Parkin. Cell 105, 891–902.

Imai, Y., Gehrke, S., Wang, H.-Q., Takahashi, R., Hasegawa, K., Oota, E., Lu, B., 2008. Phosphorylation of 4E-BP by LRRK2 affects the maintenance of dopaminergic neurons in Drosophila. EMBO J. 27, 2432–2443. https://doi.org/10.1038/emboj.2008.163.

Inoshita, T., Arano, T., Hosaka, Y., Meng, H., Umezaki, Y., Kosugi, S., Morimoto, T., Koike, M., Chang, H.-Y., Imai, Y., Hattori, N., 2017. Vps35 in cooperation with LRRK2 regulates synaptic vesicle endocytosis through the endosomal pathway in Drosophila. Hum. Mol. Genet. 26, 2933–2948. https://doi.org/10.1093/hmg/ddx179.

IPDGC, 2011. Imputation of sequence variants for identification of genetic risks for Parkinson's disease: a meta-analysis of genome-wide association studies. Lancet 377, 641–649. https://doi.org/10.1016/S0140-6736(10)62345-8.

Ishii, A., Nonaka, T., Taniguchi, S., Saito, T., Arai, T., Mann, D., Iwatsubo, T., Hisanaga, S.-I., Goedert, M., Hasegawa, M., 2007. Casein kinase 2 is the major enzyme in brain that phosphorylates Ser129 of human alpha-synuclein: Implication for alpha-synucleinopathies. FEBS Lett. 581, 4711–4717. https://doi.org/10.1016/j.febslet.2007.08.067.

Iwatsubo, T., Yamaguchi, H., Fujimuro, M., Yokosawa, H., Ihara, Y., Trojanowski, J.Q., Lee, V.M., 1996. Purification and characterization of Lewy bodies from the brains of patients with diffuse Lewy body disease. Am. J. Pathol. 148, 1517–1529.

Izumi, Y., Morino, H., Oda, M., Maruyama, H., Udaka, F., Kameyama, M., Nakamura, S., Kawakami, H., 2001. Genetic studies in Parkinson's disease with an alpha-synuclein/NACP gene polymorphism in Japan. Neurosci. Lett. 300, 125–127. https://doi.org/10.1016/s0304-3940(01)01557-9.

Jin, J., Li, G.J., Davis, J., Zhu, D., Wang, Y., Pan, C., Zhang, J., 2007. Identification of novel proteins associated with both α-synuclein and DJ-1. Mol. Cell. Proteomics 6, 845–859. https://doi.org/10.1074/mcp.M600182-MCP200.

Johnson, J., Hague, S.M., Hanson, M., Gibson, A., Wilson, K.E., Evans, E.W., Singleton, A.A., McInerney-Leo, A., Nussbaum, R.L., Hernandez, D.G., Gallardo, M., McKeith, I.G., Burn, D.J., Ryu, M., Hellstrom, O., Ravina, B., Eerola, J., Perry, R.H., Jaros, E., Tienari, P., Weiser, R., Gwinn-Hardy, K., Morris, C.M., Hardy, J., Singleton, A.B., 2004. SNCA multiplication is not a common cause of Parkinson disease or dementia with Lewy bodies. Neurology 63, 554–556. https://doi.org/10.1212/01.wnl.0000133401.09043.44.

Johnston, J.A., Ward, C.L., Kopito, R.R., 1998. Aggresomes: a cellular response to misfolded proteins. J. Cell Biol. 143, 1883–1898. https://doi.org/10.1083/jcb.143.7.1883.

Josephs, K.A., Matsumoto, J.Y., Lindor, N.M., 2004. Heterozygous niemann-pick disease type C presenting with tremor. Neurology 63, 2189–2190. https://doi.org/10.1212/01.WNL.0000145710.25588.2F.

Jowaed, A., Schmitt, I., Kaut, O., Wüllner, U., 2010. Methylation regulates alpha-synuclein expression and is decreased in Parkinson's disease patients' brains. J. Neurosci. 30, 6355–6359. https://doi.org/10.1523/JNEUROSCI.6119-09.2010.

Kalogeropulou, A.F., Zhao, J., Bolliger, M.F., Memou, A., Narasimha, S., Molitor, T.P., Wilson, W.H., Rideout, H.J., Nichols, R.J., 2018. P62/SQSTM1 is a novel leucine-rich repeat kinase 2 (LRRK2) substrate that enhances neuronal toxicity. Biochem. J. 475, 1271–1293. https://doi.org/10.1042/BCJ20170699.

Kane, L.A., Lazarou, M., Fogel, A.I., Li, Y., Yamano, K., Sarraf, S.A., Banerjee, S., Youle, R.J., 2014. PINK1 phosphorylates ubiquitin to activate parkin E3 ubiquitin ligase activity. J. Cell Biol. 205, 143–153. https://doi.org/10.1083/jcb.201402104.

Kaufman, R.J., 1999. Stress signaling from the lumen of the endoplasmic reticulum: coordination of gene transcriptional and translational controls. Genes Dev. 13, 1211–1233. https://doi.org/10.1101/gad.13.10.1211.

Kazlauskaite, A., Martínez-Torres, R.J., Wilkie, S., Kumar, A., Peltier, J., Gonzalez, A., Johnson, C., Zhang, J., Hope, A.G., Peggie, M., Trost, M., van Aalten, D.M.F., Alessi, D.R., Prescott, A.R., Knebel, A., Walden, H., Muqit, M.M.K., 2015. Binding to serine 65-phosphorylated ubiquitin primes Parkin for optimal PINK1-dependent phosphorylation and activation. EMBO Rep. 16, 939–954.https://doi.org/10.15252/embr.201540352.

Keller, J.N., Huang, F.F., Dimayuga, E.R., Maragos, W.F., 2000. Dopamine induces proteasome inhibition in neural PC12 cell line. Free Radic. Biol. Med. 29, 1037–1042. https://doi.org/10.1016/S0891-5849(00)00412-3.

Kett, L.R., Stiller, B., Bernath, M.M., Tasset, I., Blesa, J., Jackson-Lewis, V., Chan, R.B., Zhou, B., Di Paolo, G., Przedborski, S., Cuervo, A.M., Dauer, W.T., 2015. α-Synuclein-independent histopathological and motor deficits in mice lacking the endolysosomal parkinsonism protein Atp13a2. J. Neurosci. 35, 5724–5742. https://doi.org/10.1523/JNEUROSCI.0632-14.2015.

Khoury, G.A., Baliban, R.C., Floudas, C.A., 2011. Proteome-wide post-translational modification statistics: frequency analysis and curation of the swiss-prot database. Sci. Rep. 1, https://doi.org/10.1038/srep00090.

Kilpatrick, K., Zeng, Y., Hancock, T., Segatori, L., 2015. Genetic and chemical activation of TFEB mediates clearance of aggregated α-synuclein. PLoS One 10, e0120819. https://doi.org/10.1371/journal.pone.0120819.

Kim, Y., Park, J., Kim, S., Song, S., Kwon, S.-K., Lee, S.-H., Kitada, T., Kim, J.-M., Chung, J., 2008. PINK1 controls mitochondrial localization of parkin through direct phosphorylation. Biochem. Biophys. Res. Commun. 377, 975–980. https://doi.org/10.1016/j.bbrc.2008.10.104.

Kim, Y.M., Jang, W.H., Quezado, M.M., Oh, Y., Chung, K.C., Junn, E., Mouradian, M.M., 2011. Proteasome inhibition induces α-synuclein SUMOylation and aggregate formation. J. Neurol. Sci. 307, 157–161. https://doi.org/10.1016/j.jns.2011.04.015.

Kingsbury, A.E., Daniel, S.E., Sangha, H., Eisen, S., Lees, A.J., Foster, O.J.F., 2004. Alteration in alpha-synuclein mRNA expression in Parkinson's disease. Mov. Disord. 19, 162–170. https://doi.org/10.1002/mds.10683.

Kitada, T., Asakawa, S., Hattori, N., Matsumine, H., Yamamura, Y., Minoshima, S., Yokochi, M., Mizuno, Y., Shimizu, N., 1998. Mutations in the parkin gene cause autosomal recessive juvenile parkinsonism. Nature 392, 605–608. https://doi.org/10.1038/33416.

Klein, C., Westenberger, A., 2012. Genetics of Parkinson's disease. Cold Spring Harb. Perspect. Med. 2, a008888. https://doi.org/10.1101/cshperspect.a008888.

Kluenemann, H.H., Nutt, J.G., Davis, M.Y., Bird, T.D., 2013. Parkinsonism syndrome in heterozygotes for Niemann–Pick C1. J. Neurol. Sci. 335, 219–220. https://doi.org/10.1016/j.jns.2013.08.033.

Ko, H.S., Lee, Y., Shin, J.-H., Karuppagounder, S.S., Gadad, B.S., Koleske, A.J., Pletnikova, O., Troncoso, J.C., Dawson, V.L., Dawson, T.M., 2010. Phosphorylation by the c-Abl protein tyrosine kinase inhibits parkin's ubiquitination and protective function. Proc. Natl. Acad. Sci. U. S. A. 107, 16691–16696. https://doi.org/10.1073/pnas.1006083107.

Komatsu, M., Waguri, S., Chiba, T., Murata, S., Iwata, J., Tanida, I., Ueno, T., Koike, M., Uchiyama, Y., Kominami, E., Tanaka, K., 2006. Loss of autophagy in the central nervous system causes neurodegeneration in mice. Nature 441, 880–884. https://doi.org/10.1038/nature04723.

Kondapalli, C., Kazlauskaite, A., Zhang, N., Woodroof, H.I., Campbell, D.G., Gourlay, R., Burchell, L., Walden, H., Macartney, T.J., Deak, M., Knebel, A., Alessi, D.R., Muqit, M.M.K., 2012. PINK1 is activated by mitochondrial membrane potential depolarization and stimulates parkin E3 ligase activity by phosphorylating serine 65. Open Biol. 2, 120080. https://doi.org/10.1098/rsob.120080.

Kordower, J.H., Kanaan, N.M., Chu, Y., Suresh Babu, R., Stansell, J., Terpstra, B.T., Sortwell, C.E., Steece-Collier, K., Collier, T.J., 2006. Failure of proteasome inhibitor administration to provide a model of Parkinson's disease in rats and monkeys. Ann. Neurol. 60, 264–268. https://doi.org/10.1002/ana.20935.

Koyano, F., Okatsu, K., Kosako, H., Tamura, Y., Go, E., Kimura, M., Kimura, Y., Tsuchiya, H., Yoshihara, H., Hirokawa, T., Endo, T., Fon, E.A., Trempe, J.-F., Saeki, Y., Tanaka, K., Matsuda, N., 2014. Ubiquitin is phosphorylated by PINK1 to activate parkin. Nature 510, 162–166. https://doi.org/10.1038/nature13392.

Krüger, R., Kuhn, W., Müller, T., Woitalla, D., Graeber, M., Kösel, S., Przuntek, H., Epplen, J.T., Schöls, L., Riess, O., 1998. Ala30Pro mutation in the gene encoding alpha-synuclein in Parkinson's disease. Nat. Genet. 18, 106–108. https://doi.org/10.1038/ng0298-106.

Krüger, R., Vieira-Saecker, A.M., Kuhn, W., Berg, D., Müller, T., Kühnl, N., Fuchs, G.A., Storch, A., Hungs, M., Woitalla, D., Przuntek, H., Epplen, J.T., Schöls, L., Riess, O., 1999. Increased susceptibility to sporadic Parkinson's disease by a certain combined alpha-synuclein/apolipoprotein E genotype. Ann. Neurol. 45, 611–617.

Krumova, P., Meulmeester, E., Garrido, M., Tirard, M., Hsiao, H.-H., Bossis, G., Urlaub, H., Zweckstetter, M., Kügler, S., Melchior, F., Bähr, M., Weishaupt, J.H., 2011. Sumoylation inhibits α-synuclein aggregation and toxicity. J. Cell Biol. 194, 49–60. https://doi.org/10.1083/jcb.201010117.

Kthiri, F., Gautier, V., Le, H.-T., Prere, M.-F., Fayet, O., Malki, A., Landoulsi, A., Richarme, G., 2010. Translational defects in a mutant deficient in YajL, the bacterial homolog of the parkinsonism-associated protein DJ-1. J. Bacteriol. 192, 6302–6306. https://doi.org/10.1128/JB.01077-10.

Kumar, A., Greggio, E., Beilina, A., Kaganovich, A., Chan, D., Taymans, J.-M., Wolozin, B., Cookson, M.R., 2010. The Parkinson's disease associated LRRK2 exhibits weaker in vitro phosphorylation of 4E-BP compared to autophosphorylation. PLoS One 5, e8730. https://doi.org/10.1371/journal.pone.0008730.

Kuzuhara, S., Mori, H., Izumiyama, N., Yoshimura, M., Ihara, Y., 1988. Lewy bodies are ubiquitinated. A light and electron microscopic immunocytochemical study. Acta Neuropathol. 75, 345–353.

Kwak, S., Masaki, T., Ishiura, S., Sugita, H., 1991. Multicatalytic proteinase is present in Lewy bodies and neurofibrillary tangles in diffuse Lewy body disease brains. Neurosci. Lett. 128, 21–24.

Labbé, C., Lorenzo-Betancor, O., Ross, O.A., 2016. Epigenetic regulation in Parkinson's disease. Acta Neuropathol. 132, 515–530. https://doi.org/10.1007/s00401-016-1590-9.

Langemeyer, L., Fröhlich, F., Ungermann, C., 2018. Rab GTPase function in endosome and lysosome biogenesis. Trends Cell Biol. 28, 957–970. https://doi.org/10.1016/j.tcb.2018.06.007.

Ledesma, M.D., Galvan, C., Hellias, B., Dotti, C., Jensen, P.H., 2002. Astrocytic but not neuronal increased expression and redistribution of parkin during unfolded protein stress: parkin in astrocytes during unfolded protein stress. J. Neurochem. 83, 1431–1440. https://doi.org/10.1046/j.1471-4159.2002.01253.x.

Lee, M., Hyun, D.-H., Halliwell, B., Jenner, P., 2001a. Effect of the overexpression of wild-type or mutant α-synuclein on cell susceptibility to insult: mutant α-synuclein, oxidative stress and cell apoptosis. J. Neurochem. 76, 998–1009. https://doi.org/10.1046/j.1471-4159.2001.00149.x.

Lee, M.H., Hyun, D.H., Jenner, P., Halliwell, B., 2001b. Effect of proteasome inhibition on cellular oxidative damage, antioxidant defences and nitric oxide production. J. Neurochem. 78, 32–41.

Lee, H.-J., Khoshaghideh, F., Patel, S., Lee, S.-J., 2004. Clearance of alpha-synuclein oligomeric intermediates via the lysosomal degradation pathway. J. Neurosci. 24, 1888–1896. https://doi.org/10.1523/JNEUROSCI.3809-03.2004.

Lee, J.T., Wheeler, T.C., Li, L., Chin, L.-S., 2007. Ubiquitination of α-synuclein by Siah-1 promotes α-synuclein aggregation and apoptotic cell death. Hum. Mol. Genet. 17, 906–917. https://doi.org/10.1093/hmg/ddm363.

Lee, J.S., Kanai, K., Suzuki, M., Kim, W.S., Yoo, H.S., Fu, Y., Kim, D.-K., Jung, B.C., Choi, M., Oh, K.W., Li, Y., Nakatani, M., Nakazato, T., Sekimoto, S., Funayama, M., Yoshino, H., Kubo, S.-I., Nishioka, K., Sakai, R., Ueyama, M., Mochizuki, H., Lee, H.-J., Sardi, S.P., Halliday, G.M., Nagai, Y., Lee, P.H., Hattori, N., Lee, S.-J., 2019. Arylsulfatase A, a genetic modifier of Parkinson's disease, is an α-synuclein chaperone. Brain J. Neurol. 142 (9), 2845–2859. https://doi.org/10.1093/brain/awz205.

Lehtonen, Š., Sonninen, T.-M., Wojciechowski, S., Goldsteins, G., Koistinaho, J., 2019. Dysfunction of cellular proteostasis in Parkinson's disease. Front. Neurosci. 13, 457. https://doi.org/10.3389/fnins.2019.00457.

Lennox, G., Lowe, J., Morrell, K., Landon, M., Mayer, R.J., 1989. Anti-ubiquitin immunocytochemistry is more sensitive than conventional techniques in the detection of diffuse Lewy body disease. J. Neurol. Neurosurg. Psychiatry 52, 67–71.

Lesage, S., Condroyer, C., Klebe, S., Lohmann, E., Durif, F., Damier, P., Tison, F., Anheim, M., Honoré, A., Viallet, F., Bonnet, A.-M., Ouvrard-Hernandez, A.-M., Vidailhet, M., Durr, A., Brice, A., French Parkinson's Disease Genetics Study Group, 2012. EIF4G1 in familial Parkinson's disease: pathogenic mutations or rare benign variants? Neurobiol. Aging 33, 2233. e1–2233.e5. https://doi.org/10.1016/j.neurobiolaging.2012.05.006.

Levine, B., Kroemer, G., 2019. Biological functions of autophagy genes: a disease perspective. Cell 176, 11–42. https://doi.org/10.1016/j.cell.2018.09.048.

Li, W., Li, J., Bao, J., 2012. Microautophagy: lesser-known self-eating. Cell. Mol. Life Sci. 69, 1125–1136. https://doi.org/10.1007/s00018-011-0865-5.

Lin, J.H., Li, H., Yasumura, D., Cohen, H.R., Zhang, C., Panning, B., Shokat, K.M., Lavail, M.M., Walter, P., 2007. IRE1 signaling affects cell fate during the unfolded protein response. Science 318, 944–949. https://doi.org/10.1126/science.1146361.

Lin, K.-J., Lin, K.-L., Chen, S.-D., Liou, C.-W., Chuang, Y.-C., Lin, H.-Y., Lin, T.-K., 2019. The overcrowded crossroads: mitochondria, alpha-synuclein, and the endo-lysosomal system interaction in Parkinson's disease. Int. J. Mol. Sci. 20, 5312. https://doi.org/10.3390/ijms20215312.

Lindersson, E., Beedholm, R., Højrup, P., Moos, T., Gai, W., Hendil, K.B., Jensen, P.H., 2004. Proteasomal inhibition by alpha-synuclein filaments and oligomers. J. Biol. Chem. 279, 12924–12934. https://doi.org/10.1074/jbc.M306390200.

Linnertz, C., Saucier, L., Ge, D., Cronin, K.D., Burke, J.R., Browndyke, J.N., Hulette, C.M., Welsh-Bohmer, K.A., Chiba-Falek, O., 2009. Genetic regulation of alpha-synuclein mRNA expression in various human brain tissues. PLoS One 4, e7480. https://doi.org/10.1371/journal.pone.0007480.

Liu, S., Lu, B., 2010. Reduction of protein translation and activation of autophagy protect against PINK1 pathogenesis in Drosophila melanogaster. PLoS Genet. 6, e1001237. https://doi.org/10.1371/journal.pgen.1001237.

Liu, C.-W., Corboy, M.J., DeMartino, G.N., Thomas, P.J., 2003. Endoproteolytic activity of the proteasome. Science 299, 408–411. https://doi.org/10.1126/science.1079293.

Lopes, U.G., Erhardt, P., Yao, R., Cooper, G.M., 1997. p53-Dependent induction of apoptosis by proteasome inhibitors. J. Biol. Chem. 272, 12893–12896. https://doi.org/10.1074/jbc.272.20.12893.

Love, S., Saitoh, T., Quijada, S., Cole, G.M., Terry, R.D., 1988. Alz-50, ubiquitin and tau immunoreactivity of neurofibrillary tangles, Pick bodies and Lewy bodies. J. Neuropathol. Exp. Neurol. 47, 393–405.

Lowe, J., Blanchard, A., Morrell, K., Lennox, G., Reynolds, L., Billett, M., Landon, M., Mayer, R.J., 1988. Ubiquitin is a common factor in intermediate filament inclusion bodies of diverse type in man, including those of Parkinson's disease, Pick's disease, and Alzheimer's disease, as well as Rosenthal fibres in cerebellar astrocytomas, cytoplasmic bodies in muscle, and mallory bodies in alcoholic liver disease. J. Pathol. 155, 9–15. https://doi.org/10.1002/path.1711550105.

Lowe, J., McDermott, H., Landon, M., Mayer, R.J., Wilkinson, K.D., 1990. Ubiquitin carboxyl-terminal hydrolase (PGP 9.5) is selectively present in ubiquitinated inclusion bodies characteristic of human neurodegenerative diseases. J. Pathol. 161, 153–160. https://doi.org/10.1002/path.1711610210.

Lowe, J., Mayer, R.J., Landon, M., 1993. Ubiquitin in neurodegenerative diseases. Brain Pathol. Zurich Switz. 3, 55–65.

Lwin, A., Orvisky, E., Goker-Alpan, O., LaMarca, M.E., Sidransky, E., 2004. Glucocerebrosidase mutations in subjects with parkinsonism. Mol. Genet. Metab. 81, 70–73.

Machaczka, M., Rucinska, M., Skotnicki, A.B., Jurczak, W., 1999. Parkinson's syndrome preceding clinical manifestation of Gaucher's disease. Am. J. Hematol. 61, 216–217. https://doi.org/10.1002/(SICI)1096-8652(199907)61:3<216::AID-AJH12>3.0.CO;2-B.

Machiya, Y., Hara, S., Arawaka, S., Fukushima, S., Sato, H., Sakamoto, M., Koyama, S., Kato, T., 2010. Phosphorylated alpha-synuclein at Ser-129 is targeted to the proteasome pathway in a ubiquitin-independent manner. J. Biol. Chem. 285, 40732–40744. https://doi.org/10.1074/jbc.M110.141952.

MacLeod, D.A., Rhinn, H., Kuwahara, T., Zolin, A., Di Paolo, G., McCabe, B.D., MacCabe, B.D., Marder, K.S., Honig, L.S., Clark, L.N., Small, S.A., Abeliovich, A., 2013. RAB7L1 interacts with LRRK2 to modify intraneuronal protein sorting and Parkinson's disease risk. Neuron 77, 425–439. https://doi.org/10.1016/j.neuron.2012.11.033.

Manning-Boğ, A.B., Reaney, S.H., Chou, V.P., Johnston, L.C., McCormack, A.L., Johnston, J., Langston, J.W., Di Monte, D.A., 2006. Lack of nigrostriatal pathology in a rat model of proteasome inhibition. Ann. Neurol. 60, 256–260. https://doi.org/10.1002/ana.20938.

Manning-Boğ, A.B., Schüle, B., Langston, J.W., 2009. Alpha-synuclein-glucocerebrosidase interactions in pharmacological Gaucher models: a biological link between Gaucher disease and parkinsonism. Neurotoxicology 30, 1127–1132. https://doi.org/10.1016/j.neuro.2009.06.009.

Maraganore, D.M., de Andrade, M., Elbaz, A., Farrer, M.J., Ioannidis, J.P., Krüger, R., Rocca, W.A., Schneider, N.K., Lesnick, T.G., Lincoln, S.J., Hulihan, M.M., Aasly, J.O., Ashizawa, T., Chartier-Harlin, M.-C., Checkoway, H., Ferrarese, C.,

Hadjigeorgiou, G., Hattori, N., Kawakami, H., Lambert, J.-C., Lynch, T., Mellick, G.D., Papapetropoulos, S., Parsian, A., Quattrone, A., Riess, O., Tan, E.-K., Van Broeckhoven, C., Genetic Epidemiology of Parkinson's Disease (GEO-PD) Consortium, 2006. Collaborative analysis of alpha-synuclein gene promoter variability and Parkinson disease. JAMA 296, 661–670. https://doi.org/10.1001/jama.296.6.661.

Maria Cuervo, A., 2004. Autophagy: in sickness and in health. Trends Cell Biol. 14, 70–77. https://doi.org/10.1016/j.tcb.2003.12.002.

Martin, I., Kim, J.W., Lee, B.D., Kang, H.C., Xu, J.-C., Jia, H., Stankowski, J., Kim, M.-S., Zhong, J., Kumar, M., Andrabi, S.A., Xiong, Y., Dickson, D.W., Wszolek, Z.K., Pandey, A., Dawson, T.M., Dawson, V.L., 2014. Ribosomal protein s15 phosphorylation mediates LRRK2 neurodegeneration in Parkinson's disease. Cell 157, 472–485. https://doi.org/10.1016/j.cell.2014.01.064.

Martinez, A., Lectez, B., Ramirez, J., Popp, O., Sutherland, J.D., Urbé, S., Dittmar, G., Clague, M.J., Mayor, U., 2017. Quantitative proteomic analysis of Parkin substrates in Drosophila neurons. Mol. Neurodegener. 12, 29. https://doi.org/10.1186/s13024-017-0170-3.

Martinez-Vicente, M., Talloczy, Z., Kaushik, S., Massey, A.C., Mazzulli, J., Mosharov, E.V., Hodara, R., Fredenburg, R., Wu, D.-C., Follenzi, A., Dauer, W., Przedborski, S., Ischiropoulos, H., Lansbury, P.T., Sulzer, D., Cuervo, A.M., 2008. Dopamine-modified alpha-synuclein blocks chaperone-mediated autophagy. J. Clin. Invest. 118, 777–788. https://doi.org/10.1172/JCI32806.

Martins-Branco, D., Esteves, A.R., Santos, D., Arduino, D.M., Swerdlow, R.H., Oliveira, C.R., Januario, C., Cardoso, S.M., 2012. Ubiquitin proteasome system in Parkinson's disease: a keeper or a witness? Exp. Neurol. 238, 89–99. https://doi.org/10.1016/j.expneurol.2012.08.008.

Matsumoto, L., Takuma, H., Tamaoka, A., Kurisaki, H., Date, H., Tsuji, S., Iwata, A., 2010. CpG demethylation enhances alpha-synuclein expression and affects the pathogenesis of Parkinson's disease. PLoS One 5, e15522. https://doi.org/10.1371/journal.pone.0015522.

Mayer, R.J., Lowe, J., Landon, M., McDermott, H., László, L., 1991. The role of protein ubiquitination in neurodegenerative disease. Acta Biol. Hung. 42, 21–26.

Mayer, R.J., Tipler, C., Arnold, J., Laszlo, L., Al-Khedhairy, A., Lowe, J., Landon, M., 1996. Endosome-lysosomes, ubiquitin and neurodegeneration. Adv. Exp. Med. Biol. 389, 261–269. https://doi.org/10.1007/978-1-4613-0335-0_33.

Mazzulli, J.R., Xu, Y.-H., Sun, Y., Knight, A.L., McLean, P.J., Caldwell, G.A., Sidransky, E., Grabowski, G.A., Krainc, D., 2011. Gaucher disease glucocerebrosidase and α-synuclein form a bidirectional pathogenic loop in synucleinopathies. Cell 146, 37–52. https://doi.org/10.1016/j.cell.2011.06.001.

Mbefo, M.K., Paleologou, K.E., Boucharaba, A., Oueslati, A., Schell, H., Fournier, M., Olschewski, D., Yin, G., Zweckstetter, M., Masliah, E., Kahle, P.J., Hirling, H., Lashuel, H.A., 2010. Phosphorylation of synucleins by members of the polo-like kinase family. J. Biol. Chem. 285, 2807–2822. https://doi.org/10.1074/jbc.M109.081950.

McKeon, J.E., Sha, D., Li, L., Chin, L.-S., 2015. Parkin-mediated K63-polyubiquitination targets ubiquitin C-terminal hydrolase L1 for degradation by the autophagy-lysosome system. Cell. Mol. Life Sci. 72, 1811–1824. https://doi.org/10.1007/s00018-014-1781-2.

McNaught, K.S., Jenner, P., 2001. Proteasomal function is impaired in substantia nigra in Parkinson's disease. Neurosci. Lett. 297, 191–194.

McNaught, K.S.P., Olanow, C.W., 2006. Proteasome inhibitor-induced model of Parkinson's disease. Ann. Neurol. 60, 243–247. https://doi.org/10.1002/ana.20936.

McNaught, K.S.P., Belizaire, R., Jenner, P., Olanow, C.W., Isacson, O., 2002a. Selective loss of 20S proteasome alpha-subunits in the substantia nigra pars compacta in Parkinson's disease. Neurosci. Lett. 326, 155–158.

McNaught, K.S.P., Björklund, L.M., Belizaire, R., Isacson, O., Jenner, P., Olanow, C.W., 2002b. Proteasome inhibition causes nigral degeneration with inclusion bodies in rats. Neuroreport 13, 1437–1441.

McNaught, K.S.P., Mytilineou, C., Jnobaptiste, R., Yabut, J., Shashidharan, P., Jennert, P., Olanow, C.W., 2002c. Impairment of the ubiquitin-proteasome system causes dopaminergic cell death and inclusion body formation in ventral mesencephalic cultures. J. Neurochem. 81, 301–306.

McNaught, K.S.P., Belizaire, R., Isacson, O., Jenner, P., Olanow, C.W., 2003. Altered proteasomal function in sporadic Parkinson's disease. Exp. Neurol. 179, 38–46.

McNaught, K.S.P., Perl, D.P., Brownell, A.-L., Olanow, C.W., 2004. Systemic exposure to proteasome inhibitors causes a progressive model of Parkinson's disease. Ann. Neurol. 56, 149–162. https://doi.org/10.1002/ana.20186.

Mezey, E., Dehejia, A.M., Harta, G., Tresser, N., Suchy, S.F., Nussbaum, R.L., Brownstein, M.J., Polymeropoulos, M.H., 1998. Alpha synuclein is present in Lewy bodies in sporadic Parkinson's disease. Mol. Psychiatry 3, 493–499.

Miller, D.W., Hague, S.M., Clarimon, J., Baptista, M., Gwinn-Hardy, K., Cookson, M.R., Singleton, A.B., 2004. α-Synuclein in blood and brain from familial Parkinson disease with SNCA locus triplication. Neurology 62, 1835–1838. https://doi.org/10.1212/01.WNL.0000127517.33208.F4.

Mir, R., Tonelli, F., Lis, P., Macartney, T., Polinski, N.K., Martinez, T.N., Chou, M.-Y., Howden, A.J.M., König, T., Hotzy, C., Milenkovic, I., Brücke, T., Zimprich, A., Sammler, E., Alessi, D.R., 2018. The Parkinson's disease VPS35[D620N] mutation enhances LRRK2-mediated Rab protein phosphorylation in mouse and human. Biochem. J. 475, 1861–1883. https://doi.org/10.1042/BCJ20180248.

Mitsui, J., Mizuta, I., Toyoda, A., Ashida, R., Takahashi, Y., Goto, J., Fukuda, Y., Date, H., Iwata, A., Yamamoto, M., Hattori, N., Murata, M., Toda, T., Tsuji, S., 2009. Mutations for Gaucher disease confer high susceptibility to Parkinson disease. Arch. Neurol. 66, 571–576. https://doi.org/10.1001/archneurol.2009.72.

Miura, E., Hasegawa, T., Konno, M., Suzuki, M., Sugeno, N., Fujikake, N., Geisler, S., Tabuchi, M., Oshima, R., Kikuchi, A., Baba, T., Wada, K., Nagai, Y., Takeda, A., Aoki, M., 2014. VPS35 dysfunction impairs lysosomal degradation of α-synuclein and exacerbates neurotoxicity in a Drosophila model of Parkinson's disease. Neurobiol. Dis. 71, 1–13. https://doi.org/10.1016/j.nbd.2014.07.014.

Miwa, H., Kubo, T., Suzuki, A., Nishi, K., Kondo, T., 2005. Retrograde dopaminergic neuron degeneration following intrastriatal proteasome inhibition. Neurosci. Lett. 380, 93–98. https://doi.org/10.1016/j.neulet.2005.01.024.

Mizuta, I., Satake, W., Nakabayashi, Y., Ito, C., Suzuki, S., Momose, Y., Nagai, Y., Oka, A., Inoko, H., Fukae, J., Saito, Y., Sawabe, M., Murayama, S., Yamamoto, M., Hattori, N., Murata, M., Toda, T., 2006. Multiple candidate gene analysis identifies alpha-synuclein as a susceptibility gene for sporadic Parkinson's disease. Hum. Mol. Genet. 15, 1151–1158. https://doi.org/10.1093/hmg/ddl030.

Moscovitz, O., Ben-Nissan, G., Fainer, I., Pollack, D., Mizrachi, L., Sharon, M., 2015. The Parkinson's-associated protein DJ-1 regulates the 20S proteasome. Nat. Commun. 6, 6609. https://doi.org/10.1038/ncomms7609.

Mund, T., Masuda-Suzukake, M., Goedert, M., Pelham, H.R., 2018. Ubiquitination of alpha-synuclein filaments by Nedd4 ligases. PLoS One 13, e0200763. https://doi.org/10.1371/journal.pone.0200763.

Murphy, K.E., Gysbers, A.M., Abbott, S.K., Tayebi, N., Kim, W.S., Sidransky, E., Cooper, A., Garner, B., Halliday, G.M., 2014. Reduced glucocerebrosidase is associated with increased α-synuclein in sporadic Parkinson's disease. Brain J. Neurol. 137, 834–848. https://doi.org/10.1093/brain/awt367.

Murphy, K.E., Gysbers, A.M., Abbott, S.K., Spiro, A.S., Furuta, A., Cooper, A., Garner, B., Kabuta, T., Halliday, G.M., 2015. Lysosomal-associated membrane protein 2 isoforms are differentially affected in early Parkinson's disease: early loss of LAMP2A protein in PD. Mov. Disord. 30, 1639–1647. https://doi.org/10.1002/mds.26141.

Mutez, E., Nkiliza, A., Belarbi, K., de Broucker, A., Vanbesien-Mailliot, C., Bleuse, S., Duflot, A., Comptdaer, T., Semaille, P., Blervaque, R., Hot, D., Leprêtre, F., Figeac, M., Destée, A., Chartier-Harlin, M.-C., 2014. Involvement of the immune system, endocytosis and EIF2 signaling in both genetically determined and sporadic forms of Parkinson's disease. Neurobiol. Dis. 63, 165–170. https://doi.org/10.1016/j.nbd.2013.11.007.

Nalls, M.A., Pankratz, N., Lill, C.M., Do, C.B., Hernandez, D.G., Saad, M., DeStefano, A.L., Kara, E., Bras, J., Sharma, M., Schulte, C., Keller, M.F., Arepalli, S., Letson, C., Edsall, C., Stefansson, H., Liu, X., Pliner, H., Lee, J.H., Cheng, R., International Parkinson's Disease Genomics Consortium (IPDGC), Parkinson's Study Group (PSG) Parkinson's Research: The Organized GENetics Initiative (PROGENI), 23andMe, GenePD, NeuroGenetics Research Consortium (NGRC), Hussman Institute of Human Genomics (HIHG), Ashkenazi Jewish Dataset Investigator, Cohorts for Health and Aging Research in Genetic Epidemiology (CHARGE), North American Brain Expression Consortium (NABEC), United Kingdom Brain Expression Consortium (UKBEC), Greek Parkinson's Disease Consortium, Alzheimer Genetic Analysis Group, Ikram, M.A., Ioannidis, J.P.A., Hadjigeorgiou, G.M., Bis, J.C., Martinez, M., Perlmutter, J.S., Goate, A., Marder, K., Fiske, B., Sutherland, M., Xiromerisiou, G., Myers, R.H., Clark, L.N., Stefansson, K., Hardy, J.A., Heutink, P., Chen, H., Wood, N.W., Houlden, H., Payami, H., Brice, A., Scott, W.K., Gasser, T., Bertram, L., Eriksson, N., Foroud, T., Singleton, A.B., 2014. Large-scale meta-analysis of genome-wide association data identifies six new risk loci for Parkinson's disease. Nat. Genet. 46, 989–993. https://doi.org/10.1038/ng.3043.

Nandi, D., Tahiliani, P., Kumar, A., Chandu, D., 2006. The ubiquitin-proteasome system. J. Biosci. 31, 137–155.

Napolitano, G., Ballabio, A., 2016. TFEB at a glance. J. Cell Sci. 129, 2475–2481. https://doi.org/10.1242/jcs.146365.

Neudorfer, O., Giladi, N., Elstein, D., Abrahamov, A., Turezkite, T., Aghai, E., Reches, A., Bembi, B., Zimran, A., 1996. Occurrence of Parkinson's syndrome in type I Gaucher disease. QJM Mon. J. Assoc. Physicians 89, 691–694. https://doi.org/10.1093/qjmed/89.9.691.

Neystat, M., Lynch, T., Przedborski, S., Kholodilov, N., Rzhetskaya, M., Burke, R.E., 1999. Alpha-synuclein expression in substantia nigra and cortex in Parkinson's disease. Mov. Disord. 14, 417–422.

Nezich, C.L., Wang, C., Fogel, A.I., Youle, R.J., 2015. MiT/TFE transcription factors are activated during mitophagy downstream of parkin and Atg5. J. Cell Biol. 210, 435–450. https://doi.org/10.1083/jcb.201501002.

Nichols, N., Bras, J.M., Hernandez, D.G., Jansen, I.E., Lesage, S., Lubbe, S., Singleton, A.B., International Parkinson's Disease Genomics Consortium, 2015. EIF4G1 mutations do not cause Parkinson's disease. Neurobiol. Aging 36 (2444), e1–e4. https://doi.org/10.1016/j.neurobiolaging.2015.04.017.

Nussbaum, R.L., 2017. The identification of alpha-synuclein as the first Parkinson disease gene. J. Park. Dis. 7, S43–S49. https://doi.org/10.3233/JPD-179003.

Oh, Y., Kim, Y.M., Mouradian, M.M., Chung, K.C., 2011. Human polycomb protein 2 promotes α-synuclein aggregate formation through covalent SUMOylation. Brain Res. 1381, 78–89. https://doi.org/10.1016/j.brainres.2011.01.039.

Okada, K., Wangpoengtrakul, C., Osawa, T., Toyokuni, S., Tanaka, K., Uchida, K., 1999. 4-Hydroxy-2-nonenal-mediated impairment of intracellular proteolysis during oxidative stress. Identification of proteasomes as target molecules. J. Biol. Chem. 274, 23787–23793. https://doi.org/10.1074/jbc.274.34.23787.

Olzmann, J.A., Chin, L.-S., 2008. Parkin-mediated K63-linked polyubiquitination: a signal for targeting misfolded proteins to the aggresome-autophagy pathway. Autophagy 4, 85–87. https://doi.org/10.4161/auto.5172.

Olzmann, J.A., Li, L., Chudaev, M.V., Chen, J., Perez, F.A., Palmiter, R.D., Chin, L.-S., 2007. Parkin-mediated K63-linked polyubiquitination targets misfolded DJ-1 to aggresomes via binding to HDAC6. J. Cell Biol. 178, 1025–1038. https://doi.org/10.1083/jcb.200611128.

Oueslati, A., 2016. Implication of alpha-synuclein phosphorylation at S129 in synucleinopathies: what have we learned in the last decade? J. Park. Dis. 6, 39–51. https://doi.org/10.3233/JPD-160779.

Oueslati, A., Schneider, B.L., Aebischer, P., Lashuel, H.A., 2013. Polo-like kinase 2 regulates selective autophagic α-synuclein clearance and suppresses its toxicity in vivo. Proc. Natl. Acad. Sci. U. S. A. 110, E3945–E3954. https://doi.org/10.1073/pnas.1309991110.

Ozawa, K., Komatsubara, A.T., Nishimura, Y., Sawada, T., Kawafune, H., Tsumoto, H., Tsuji, Y., Zhao, J., Kyotani, Y., Tanaka, T., Takahashi, R., Yoshizumi, M., 2013. S-nitrosylation regulates mitochondrial quality control via activation of parkin. Sci. Rep. 3, 2202. https://doi.org/10.1038/srep02202.

Pals, P., Lincoln, S., Manning, J., Heckman, M., Skipper, L., Hulihan, M., Van den Broeck, M., De Pooter, T., Cras, P., Crook, J., Van Broeckhoven, C., Farrer, M.J., 2004. α-Synuclein promoter confers susceptibility to Parkinson's disease. Ann. Neurol. 56, 591–595. https://doi.org/10.1002/ana.20268.

Pao, K.-C., Stanley, M., Han, C., Lai, Y.-C., Murphy, P., Balk, K., Wood, N.T., Corti, O., Corvol, J.-C., Muqit, M.M.K., Virdee, S., 2016. Probes of ubiquitin E3 ligases enable systematic dissection of parkin activation. Nat. Chem. Biol. 12, 324–331. https://doi.org/10.1038/nchembio.2045.

Papapetropoulos, S., Adi, N., Mash, D.C., Shehadeh, L., Bishopric, N., Shehadeh, L., 2007. Expression of α-synuclein mRNA in Parkinson's disease. Mov. Disord. 22, 1057–1059. https://doi.org/10.1002/mds.21466.

Park, J.-S., Blair, N.F., Sue, C.M., 2015. The role of ATP13A2 in Parkinson's disease: clinical phenotypes and molecular mechanisms: ATP13A2 in Parkinson's disease. Mov. Disord. 30, 770–779. https://doi.org/10.1002/mds.26243.

Park, S., Han, S., Choi, I., Kim, B., Park, S.P., Joe, E.-H., Suh, Y.H., 2016. Interplay between leucine-rich repeat kinase 2 (LRRK2) and p62/SQSTM-1 in selective autophagy. PLoS One 11, e0163029. https://doi.org/10.1371/journal.pone.0163029.

Parsian, A., Racette, B., Zhang, Z.H., Chakraverty, S., Rundle, M., Goate, A., Perlmutter, J.S., 1998. Mutation, sequence analysis, and association studies of alpha-synuclein in Parkinson's disease. Neurology 51, 1757–1759. https://doi.org/10.1212/wnl.51.6.1757.

Paxinou, E., Chen, Q., Weisse, M., Giasson, B.I., Norris, E.H., Rueter, S.M., Trojanowski, J.Q., Lee, V.M., Ischiropoulos, H., 2001. Induction of alpha-synuclein aggregation by intracellular nitrative insult. J. Neurosci. 21, 8053–8061.

Petrucelli, L., O'Farrell, C., Lockhart, P.J., Baptista, M., Kehoe, K., Vink, L., Choi, P., Wolozin, B., Farrer, M., Hardy, J., Cookson, M.R., 2002. Parkin protects against the toxicity associated with mutant alpha-synuclein: proteasome dysfunction selectively affects catecholaminergic neurons. Neuron 36, 1007–1019.

Pickart, C.M., Fushman, D., 2004. Polyubiquitin chains: polymeric protein signals. Curr. Opin. Chem. Biol. 8, 610–616. https://doi.org/10.1016/j.cbpa.2004.09.009.

Pickrell, A.M., Youle, R.J., 2015. The roles of PINK1, parkin, and mitochondrial fidelity in Parkinson's disease. Neuron 85, 257–273. https://doi.org/10.1016/j.neuron.2014.12.007.

Pihlstrøm, L., Blauwendraat, C., Cappelletti, C., Berge-Seidl, V., Langmyhr, M., Henriksen, S.P., van de Berg, W.D.J., Gibbs, J.R., Cookson, M.R., International Parkinson Disease Genomics Consortium, North American Brain Expression Consortium, Singleton, A.B., Nalls, M.A., Toft, M., 2018. A comprehensive analysis of SNCA-related genetic risk in sporadic Parkinson disease. Ann. Neurol. 84, 117–129. https://doi.org/10.1002/ana.25274.

Polymeropoulos, M.H., 1997. Mutation in the α-synuclein gene identified in families with Parkinson's disease. Science 276, 2045–2047. https://doi.org/10.1126/science.276.5321.2045.

Pratt, A.J., MacRae, I.J., 2009. The RNA-induced silencing complex: a versatile gene-silencing machine. J. Biol. Chem. 284, 17897–17901. https://doi.org/10.1074/jbc.R900012200.

Pronin, A.N., Morris, A.J., Surguchov, A., Benovic, J.L., 2000. Synucleins are a novel class of substrates for G protein-coupled receptor kinases. J. Biol. Chem. 275, 26515–26522. https://doi.org/10.1074/jbc.M003542200.

Qiu, J.H., Asai, A., Chi, S., Saito, N., Hamada, H., Kirino, T., 2000. Proteasome inhibitors induce cytochrome c-caspase-3-like protease-mediated apoptosis in cultured cortical neurons. J. Neurosci. 20, 259–265.

Ramirez, A., Heimbach, A., Gründemann, J., Stiller, B., Hampshire, D., Cid, L.P., Goebel, I., Mubaidin, A.F., Wriekat, A.-L., Roeper, J., Al-Din, A., Hillmer, A.M., Karsak, M., Liss, B., Woods, C.G., Behrens, M.I., Kubisch, C., 2006. Hereditary parkinsonism with dementia is caused by mutations in ATP13A2, encoding a lysosomal type 5 P-type ATPase. Nat. Genet. 38, 1184–1191. https://doi.org/10.1038/ng1884.

Rana, H.Q., Balwani, M., Bier, L., Alcalay, R.N., 2013. Age-specific Parkinson disease risk in GBA mutation carriers: information for genetic counseling. Genet. Med. 15, 146–149. https://doi.org/10.1038/gim.2012.107.

Reinheckel, T., Sitte, N., Ullrich, O., Kuckelkorn, U., Davies, K.J., Grune, T., 1998. Comparative resistance of the 20S and 26S proteasome to oxidative stress. Biochem. J. 335 (Pt. 3), 637–642.

Rhinn, H., Qiang, L., Yamashita, T., Rhee, D., Zolin, A., Vanti, W., Abeliovich, A., 2012. Alternative α-synuclein transcript usage as a convergent mechanism in Parkinson's disease pathology. Nat. Commun. 3, 1084. https://doi.org/10.1038/ncomms2032.

Richter, J.D., Sonenberg, N., 2005. Regulation of cap-dependent translation by eIF4E inhibitory proteins. Nature 433, 477–480. https://doi.org/10.1038/nature03205.

Rideout, H.J., Stefanis, L., 2002. Proteasomal inhibition-induced inclusion formation and death in cortical neurons require transcription and ubiquitination. Mol. Cell. Neurosci. 21, 223–238.

Rideout, H.J., Larsen, K.E., Sulzer, D., Stefanis, L., 2001. Proteasomal inhibition leads to formation of ubiquitin/alpha-synuclein-immunoreactive inclusions in PC12 cells. J. Neurochem. 78, 899–908.

Rideout, H.J., Lang-Rollin, I.C.J., Savalle, M., Stefanis, L., 2005. Dopaminergic neurons in rat ventral midbrain cultures undergo selective apoptosis and form inclusions, but do not up-regulate iHSP70, following proteasomal inhibition. J. Neurochem. 93, 1304–1313. https://doi.org/10.1111/j.1471-4159.2005.03124.x.

Rink, J., Ghigo, E., Kalaidzidis, Y., Zerial, M., 2005. Rab conversion as a mechanism of progression from early to late endosomes. Cell 122, 735–749. https://doi.org/10.1016/j.cell.2005.06.043.

Robak, L.A., Jansen, I.E., van Rooij, J., Uitterlinden, A.G., Kraaij, R., Jankovic, J., International Parkinson's Disease Genomics Consortium (IPDGC), Heutink, P., Shulman, J.M., 2017. Excessive burden of lysosomal storage disorder gene variants in Parkinson's disease. Brain J. Neurol. 140, 3191–3203. https://doi.org/10.1093/brain/awx285.

Rocha, E.M., Smith, G.A., Park, E., Cao, H., Brown, E., Hallett, P., Isacson, O., 2015. Progressive decline of glucocerebrosidase in aging and Parkinson's disease. Ann. Clin. Transl. Neurol. 2, 433–438. https://doi.org/10.1002/acn3.177.

Rockenstein, E., Hansen, L.A., Mallory, M., Trojanowski, J.Q., Galasko, D., Masliah, E., 2001. Altered expression of the synuclein family mRNA in Lewy body and Alzheimer's disease. Brain Res. 914, 48–56. https://doi.org/10.1016/S0006-8993(01)02772-X.

Rockenstein, E., Clarke, J., Viel, C., Panarello, N., Treleaven, C.M., Kim, C., Spencer, B., Adame, A., Park, H., Dodge, J.C., Cheng, S.H., Shihabuddin, L.S., Masliah, E., Sardi, S.P., 2016. Glucocerebrosidase modulates cognitive and motor activities in murine models of Parkinson's disease. Hum. Mol. Genet. 25, 2645–2660. https://doi.org/10.1093/hmg/ddw124.

Roosen, D.A., Cookson, M.R., 2016. LRRK2 at the interface of autophagosomes, endosomes and lysosomes. Mol. Neurodegener. 11, 73. https://doi.org/10.1186/s13024-016-0140-1.

Ross, O.A., Braithwaite, A.T., Skipper, L.M., Kachergus, J., Hulihan, M.M., Middleton, F.A., Nishioka, K., Fuchs, J., Gasser, T., Maraganore, D.M., Adler, C.H., Larvor, L., Chartier-Harlin, M.-C., Nilsson, C., Langston, J.W., Gwinn, K., Hattori, N., Farrer, M.J., 2008. Genomic investigation of alpha-synuclein multiplication and parkinsonism. Ann. Neurol. 63, 743–750. https://doi.org/10.1002/ana.21380.

Rott, R., Szargel, R., Haskin, J., Shani, V., Shainskaya, A., Manov, I., Liani, E., Avraham, E., Engelender, S., 2008. Monoubiquitylation of α-synuclein by seven in absentia homolog (SIAH) promotes its aggregation in dopaminergic cells. J. Biol. Chem. 283, 3316–3328. https://doi.org/10.1074/jbc.M704809200.

Rott, R., Szargel, R., Haskin, J., Bandopadhyay, R., Lees, A.J., Shani, V., Engelender, S., 2011. α-Synuclein fate is determined by USP9X-regulated monoubiquitination. Proc. Natl. Acad. Sci. U. S. A. 108, 18666–18671. https://doi.org/10.1073/pnas.1105725108.

Rott, R., Szargel, R., Shani, V., Hamza, H., Savyon, M., Abd Elghani, F., Bandopadhyay, R., Engelender, S., 2017. SUMOylation and ubiquitination reciprocally regulate α-synuclein degradation and pathological aggregation. Proc. Natl. Acad. Sci. U. S. A. 114, 13176–13181. https://doi.org/10.1073/pnas.1704351114.

Rubinsztein, D.C., Mariño, G., Kroemer, G., 2011. Autophagy and aging. Cell 146, 682–695. https://doi.org/10.1016/j.cell.2011.07.030.

Rubio de la Torre, E., Luzón-Toro, B., Forte-Lago, I., Minguez-Castellanos, A., Ferrer, I., Hilfiker, S., 2009. Combined kinase inhibition modulates parkin inactivation. Hum. Mol. Genet. 18, 809–823. https://doi.org/10.1093/hmg/ddn407.

Ruggiano, A., Foresti, O., Carvalho, P., 2014. ER-associated degradation: protein quality control and beyond. J. Cell Biol. 204, 869–879. https://doi.org/10.1083/jcb.201312042.

Ryan, E., Seehra, G., Sharma, P., Sidransky, E., 2019. GBA1-associated parkinsonism: new insights and therapeutic opportunities. Curr. Opin. Neurol. (4), 589–596. https://doi.org/10.1097/WCO.0000000000000715.

Ryu, E.J., Harding, H.P., Angelastro, J.M., Vitolo, O.V., Ron, D., Greene, L.A., 2002. Endoplasmic reticulum stress and the unfolded protein response in cellular models of Parkinson's disease. J. Neurosci. 22, 10690–10698.

Sado, M., Yamasaki, Y., Iwanaga, T., Onaka, Y., Ibuki, T., Nishihara, S., Mizuguchi, H., Momota, H., Kishibuchi, R., Hashimoto, T., Wada, D., Kitagawa, H., Watanabe, T.K., 2009. Protective effect against Parkinson's disease-related insults through the activation of XBP1. Brain Res. 1257, 16–24. https://doi.org/10.1016/j.brainres.2008.11.104.

Saito, Y., Akazawa-Ogawa, Y., Matsumura, A., Saigoh, K., Itoh, S., Sutou, K., Kobayashi, M., Mita, Y., Shichiri, M., Hisahara, S., Hara, Y., Fujimura, H., Takamatsu, H., Hagihara, Y., Yoshida, Y., Hamakubo, T., Kusunoki, S., Shimohama, S., Noguchi, N., 2016. Oxidation and interaction of DJ-1 with 20S proteasome in the erythrocytes of early stage Parkinson's disease patients. Sci. Rep. 6, 30793. https://doi.org/10.1038/srep30793.

Sampathu, D.M., Giasson, B.I., Pawlyk, A.C., Trojanowski, J.Q., Lee, V.M.-Y., 2003. Ubiquitination of alpha-synuclein is not required for formation of pathological inclusions in alpha-synucleinopathies. Am. J. Pathol. 163, 91–100. https://doi.org/10.1016/s0002-9440(10)63633-4.

Sardi, S.P., Clarke, J., Viel, C., Chan, M., Tamsett, T.J., Treleaven, C.M., Bu, J., Sweet, L., Passini, M.A., Dodge, J.C., Yu, W.H., Sidman, R.L., Cheng, S.H., Shihabuddin, L.S., 2013. Augmenting CNS glucocerebrosidase activity as a therapeutic strategy for parkinsonism and other Gaucher-related synucleinopathies. Proc. Natl. Acad. Sci. U. S. A. 110, 3537–3542. https://doi.org/10.1073/pnas.1220464110.

Sardi, S.P., Viel, C., Clarke, J., Treleaven, C.M., Richards, A.M., Park, H., Olszewski, M.A., Dodge, J.C., Marshall, J., Makino, E., Wang, B., Sidman, R.L., Cheng, S.H., Shihabuddin, L.S., 2017. Glucosylceramide synthase inhibition alleviates aberrations in synucleinopathy models. Proc. Natl. Acad. Sci. U. S. A. 114, 2699–2704. https://doi.org/10.1073/pnas.1616152114.

Schapansky, J., Nardozzi, J.D., Felizia, F., LaVoie, M.J., 2014. Membrane recruitment of endogenous LRRK2 precedes its potent regulation of autophagy. Hum. Mol. Genet. 23, 4201–4214. https://doi.org/10.1093/hmg/ddu138.

Schapira, A.H.V., Cleeter, M.W.J., Muddle, J.R., Workman, J.M., Cooper, J.M., King, R.H.M., 2006. Proteasomal inhibition causes loss of nigral tyrosine hydroxylase neurons. Ann. Neurol. 60, 253–255. https://doi.org/10.1002/ana.20934.

Schlossmacher, M.G., Frosch, M.P., Gai, W.P., Medina, M., Sharma, N., Forno, L., Ochiishi, T., Shimura, H., Sharon, R., Hattori, N., Langston, J.W., Mizuno, Y., Hyman, B.T., Selkoe, D.J., Kosik, K.S., 2002. Parkin localizes to the Lewy bodies of Parkinson disease and dementia with Lewy bodies. Am. J. Pathol. 160, 1655–1667. https://doi.org/10.1016/S0002-9440(10)61113-3.

Schöndorf, D.C., Aureli, M., McAllister, F.E., Hindley, C.J., Mayer, F., Schmid, B., Sardi, S.P., Valsecchi, M., Hoffmann, S., Schwarz, L.K., Hedrich, U., Berg, D., Shihabuddin, L.S., Hu, J., Pruszak, J., Gygi, S.P., Sonnino, S., Gasser, T., Deleidi, M., 2014. iPSC-derived neurons from GBA1-associated Parkinson's disease patients show autophagic defects and impaired calcium homeostasis. Nat. Commun. 5, 4028. https://doi.org/10.1038/ncomms5028.

Schulte, E.C., Mollenhauer, B., Zimprich, A., Bereznai, B., Lichtner, P., Haubenberger, D., Pirker, W., Brücke, T., Molnar, M.J., Peters, A., Gieger, C., Trenkwalder, C., Winkelmann, J., 2012. Variants in eukaryotic translation initiation factor 4G1 in sporadic Parkinson's disease. Neurogenetics 13, 281–285. https://doi.org/10.1007/s10048-012-0334-9.

Schulze, A., Standera, S., Buerger, E., Kikkert, M., van Voorden, S., Wiertz, E., Koning, F., Kloetzel, P.-M., Seeger, M., 2005. The ubiquitin-domain protein HERP forms a complex with components of the endoplasmic reticulum associated degradation pathway. J. Mol. Biol. 354, 1021–1027. https://doi.org/10.1016/j.jmb.2005.10.020.

Selvaraj, S., Sun, Y., Watt, J.A., Wang, S., Lei, S., Birnbaumer, L., Singh, B.B., 2012. Neurotoxin-induced ER stress in mouse dopaminergic neurons involves downregulation of TRPC1 and inhibition of AKT/mTOR signaling. J. Clin. Invest. 122, 1354–1367. https://doi.org/10.1172/JCI61332.

Sevlever, D., Jiang, P., Yen, S.-H.C., 2008. Cathepsin D is the main lysosomal enzyme involved in the degradation of α-synuclein and generation of its carboxy-terminally truncated species. Biochemistry 47, 9678–9687. https://doi.org/10.1021/bi800699v.

Sha, D., Chin, L.-S., Li, L., 2010. Phosphorylation of parkin by Parkinson disease-linked kinase PINK1 activates parkin E3 ligase function and NF-kappaB signaling. Hum. Mol. Genet. 19, 352–363. https://doi.org/10.1093/hmg/ddp501.

Shahpasandzadeh, H., Popova, B., Kleinknecht, A., Fraser, P.E., Outeiro, T.F., Braus, G.H., 2014. Interplay between sumoylation and phosphorylation for protection against α-synuclein inclusions. J. Biol. Chem. 289, 31224–31240. https://doi.org/10.1074/jbc.M114.559237.

Shamoto-Nagai, M., Maruyama, W., Kato, Y., Isobe, K., Tanaka, M., Naoi, M., Osawa, T., 2003. An inhibitor of mitochondrial complex I, rotenone, inactivates proteasome by oxidative modification and induces aggregation of oxidized proteins in SH-SY5Y cells. J. Neurosci. Res. 74, 589–597. https://doi.org/10.1002/jnr.10777.

Sharma, M., Ioannidis, J.P.A., Aasly, J.O., Annesi, G., Brice, A., Bertram, L., Bozi, M., Barcikowska, M., Crosiers, D., Clarke, C.E., Facheris, M.F., Farrer, M., Garraux, G., Gispert, S., Auburger, G., Vilariño-Güell, C., Hadjigeorgiou, G.M., Hicks, A.A., Hattori, N., Jeon, B.S., Jamrozik, Z., Krygowska-Wajs, A., Lesage, S., Lill, C.M., Lin, J.-J., Lynch, T., Lichtner, P., Lang, A.E., Libioulle, C., Murata, M., Mok, V., Jasinska-Myga, B., Mellick, G.D., Morrison, K.E., Meitnger, T., Zimprich, A., Opala, G., Pramstaller, P.P., Pichler, I., Park, S.S., Quattrone, A., Rogaeva, E., Ross, O.A., Stefanis, L., Stockton, J.D., Satake, W., Silburn, P.A., Strom, T.M., Theuns, J., Tan, E.-K., Toda, T., Tomiyama, H., Uitti, R.J., Van Broeckhoven, C., Wirdefeldt, K., Wszolek, Z., Xiromerisiou, G., Yomono, H.S., Yueh, K.-C., Zhao, Y., Gasser, T., Maraganore, D., Krüger, R., on behalf of GEOPD Consortium, 2012. A multi-centre clinico-genetic analysis of the VPS35 gene in Parkinson disease indicates reduced penetrance for disease-associated variants. J. Med. Genet. 49, 721–726. https://doi.org/10.1136/jmedgenet-2012-101155.

Shiba-Fukushima, K., Imai, Y., Yoshida, S., Ishihama, Y., Kanao, T., Sato, S., Hattori, N., 2012. PINK1-mediated phosphorylation of the Parkin ubiquitin-like domain primes mitochondrial translocation of Parkin and regulates mitophagy. Sci. Rep. 2, 1002. https://doi.org/10.1038/srep01002.

Shimura, H., Hattori, N., Kubo, S.i., Mizuno, Y., Asakawa, S., Minoshima, S., Shimizu, N., Iwai, K., Chiba, T., Tanaka, K., Suzuki, T., 2000. Familial Parkinson disease gene product, parkin, is a ubiquitin-protein ligase. Nat. Genet. 25, 302–305. https://doi.org/10.1038/77060.

Shin, Y., Klucken, J., Patterson, C., Hyman, B.T., McLean, P.J., 2005. The co-chaperone carboxyl terminus of Hsp70-interacting protein (CHIP) mediates α-synuclein degradation decisions between proteasomal and lysosomal pathways. J. Biol. Chem. 280, 23727–23734. https://doi.org/10.1074/jbc.M503326200.

Sidransky, E., Nalls, M.A., Aasly, J.O., Aharon-Peretz, J., Annesi, G., Barbosa, E.R., Bar-Shira, A., Berg, D., Bras, J., Brice, A., Chen, C.-M., Clark, L.N., Condroyer, C., De Marco, E.V., Dürr, A., Eblan, M.J., Fahn, S., Farrer, M.J., Fung, H.-C., Gan-Or, Z., Gasser, T., Gershoni-Baruch, R., Giladi, N., Griffith, A., Gurevich, T., Januario, C., Kropp, P., Lang, A.E., Lee-Chen, G.-J., Lesage, S., Marder, K., Mata, I.F., Mirelman, A., Mitsui, J., Mizuta, I., Nicoletti, G., Oliveira, C., Ottman, R., Orr-Urtreger, A., Pereira, L.V., Quattrone, A., Rogaeva, E., Rolfs, A., Rosenbaum, H., Rozenberg, R., Samii, A., Samaddar, T., Schulte, C., Sharma, M., Singleton, A., Spitz, M., Tan, E.-K., Tayebi, N., Toda, T., Troiano, A.R., Tsuji, S., Wittstock, M., Wolfsberg, T.G., Wu, Y.-R., Zabetian, C.P., Zhao, Y., Ziegler, S.G., 2009. Multicenter analysis of glucocerebrosidase mutations in Parkinson's disease. N. Engl. J. Med. 361, 1651–1661. https://doi.org/10.1056/NEJMoa0901281.

Silva, R.M., Ries, V., Oo, T.F., Yarygina, O., Jackson-Lewis, V., Ryu, E.J., Lu, P.D., Marciniak, S.J., Ron, D., Przedborski, S., Kholodilov, N., Greene, L.A., Burke, R.E., 2005. CHOP/GADD153 is a mediator of apoptotic death in substantia nigra dopamine neurons in an in vivo neurotoxin model of parkinsonism. J. Neurochem. 95, 974–986. https://doi.org/10.1111/j.1471-4159.2005.03428.x.

Singleton, A.B., Farrer, M., Johnson, J., Singleton, A., Hague, S., Kachergus, J., Hulihan, M., Peuralinna, T., Dutra, A., Nussbaum, R., Lincoln, S., Crawley, A., Hanson, M., Maraganore, D., Adler, C., Cookson, M.R., Muenter, M., Baptista, M., Miller, D., Blancato, J., Hardy, J., Gwinn-Hardy, K., 2003. alpha-Synuclein locus triplication causes Parkinson's disease. Science 302, 841. https://doi.org/10.1126/science.1090278.

Slodzinski, H., Moran, L.B., Michael, G.J., Wang, B., Novoselov, S., Cheetham, M.E., Pearce, R.K.B., Graeber, M.B., 2009. Homocysteine-induced endoplasmic reticulum protein (herp) is up-regulated in parkinsonian substantia nigra and present in the core of Lewy bodies. Clin. Neuropathol. 28, 333–343.

Smith, W.W., Jiang, H., Pei, Z., Tanaka, Y., Morita, H., Sawa, A., Dawson, V.L., Dawson, T.M., Ross, C.A., 2005. Endoplasmic reticulum stress and mitochondrial cell death pathways mediate A53T mutant alpha-synuclein-induced toxicity. Hum. Mol. Genet. 14, 3801–3811. https://doi.org/10.1093/hmg/ddi396.

Smith, M.H., Ploegh, H.L., Weissman, J.S., 2011. Road to ruin: targeting proteins for degradation in the endoplasmic reticulum. Science 334, 1086–1090. https://doi.org/10.1126/science.1209235.

Snyder, H., Mensah, K., Theisler, C., Lee, J., Matouschek, A., Wolozin, B., 2003. Aggregated and monomeric alpha-synuclein bind to the S6' proteasomal protein and inhibit proteasomal function. J. Biol. Chem. 278, 11753–11759. https://doi.org/10.1074/jbc.M208641200.

Soldner, F., Stelzer, Y., Shivalila, C.S., Abraham, B.J., Latourelle, J.C., Barrasa, M.I., Goldmann, J., Myers, R.H., Young, R.A., Jaenisch, R., 2016. Parkinson-associated risk variant in distal enhancer of α-synuclein modulates target gene expression. Nature 533, 95–99. https://doi.org/10.1038/nature17939.

Song, P., Trajkovic, K., Tsunemi, T., Krainc, D., 2016. Parkin modulates endosomal organization and function of the endo-lysosomal pathway. J. Neurosci. 36, 2425–2437. https://doi.org/10.1523/JNEUROSCI.2569-15.2016.

Spillantini, M.G., Schmidt, M.L., Lee, V.M., Trojanowski, J.Q., Jakes, R., Goedert, M., 1997. Alpha-synuclein in Lewy bodies. Nature 388, 839–840. https://doi.org/10.1038/42166.

Stefanis, L., Larsen, K.E., Rideout, H.J., Sulzer, D., Greene, L.A., 2001. Expression of A53T mutant but not wild-type alpha-synuclein in PC12 cells induces alterations of the ubiquitin-dependent degradation system, loss of dopamine release, and autophagic cell death. J. Neurosci. 21, 9549–9560.

Stefanis, L., Emmanouilidou, E., Pantazopoulou, M., Kirik, D., Vekrellis, K., Tofaris, G.K., 2019. How is alpha-synuclein cleared from the cell? J. Neurochem. 150 (5), 577–590. https://doi.org/10.1111/jnc.14704.

Steger, M., Tonelli, F., Ito, G., Davies, P., Trost, M., Vetter, M., Wachter, S., Lorentzen, E., Duddy, G., Wilson, S., Baptista, M.A., Fiske, B.K., Fell, M.J., Morrow, J.A., Reith, A.D., Alessi, D.R., Mann, M., 2016. Phosphoproteomics reveals that Parkinson's disease kinase LRRK2 regulates a subset of Rab GTPases. eLife 5, e12813. https://doi.org/10.7554/eLife.12813.

Stokes, A.H., Hastings, T.G., Vrana, K.E., 1999. Cytotoxic and genotoxic potential of dopamine. J. Neurosci. Res. 55, 659–665. https://doi.org/10.1002/(SICI)1097-4547(19990315)55:6<659::AID-JNR1>3.0.CO;2-C.

Sugeno, N., Takeda, A., Hasegawa, T., Kobayashi, M., Kikuchi, A., Mori, F., Wakabayashi, K., Itoyama, Y., 2008. Serine 129 phosphorylation of alpha-synuclein induces unfolded protein response-mediated cell death. J. Biol. Chem. 283, 23179–23188. https://doi.org/10.1074/jbc.M802223200.

Sugeno, N., Hasegawa, T., Tanaka, N., Fukuda, M., Wakabayashi, K., Oshima, R., Konno, M., Miura, E., Kikuchi, A., Baba, T., Anan, T., Nakao, M., Geisler, S., Aoki, M., Takeda, A., 2014. Lys-63-linked ubiquitination by E3 ubiquitin ligase Nedd4-1 facilitates endosomal sequestration of internalized α-synuclein. J. Biol. Chem. 289, 18137–18151. https://doi.org/10.1074/jbc.M113.529461.

Sun, X., Liu, J., Crary, J.F., Malagelada, C., Sulzer, D., Greene, L.A., Levy, O.A., 2013. ATF4 protects against neuronal death in cellular Parkinson's disease models by maintaining levels of parkin. J. Neurosci. 33, 2398–2407. https://doi.org/10.1523/JNEUROSCI.2292-12.2013.

Sunico, C.R., Nakamura, T., Rockenstein, E., Mante, M., Adame, A., Chan, S.F., Newmeyer, T.F., Masliah, E., Nakanishi, N., Lipton, S.A., 2013. S-nitrosylation of parkin as a novel regulator of p53-mediated neuronal cell death in sporadic Parkinson's disease. Mol. Neurodegener. 8, 29. https://doi.org/10.1186/1750-1326-8-29.

Tain, L.S., Mortiboys, H., Tao, R.N., Ziviani, E., Bandmann, O., Whitworth, A.J., 2009. Rapamycin activation of 4E-BP prevents parkinsonian dopaminergic neuron loss. Nat. Neurosci. 12, 1129–1135. https://doi.org/10.1038/nn.2372.

Tan, E.K., Matsuura, T., Nagamitsu, S., Khajavi, M., Jankovic, J., Ashizawa, T., 2000. Polymorphism of NACP-Rep1 in Parkinson's disease: an etiologic link with essential tremor? Neurology 54, 1195–1198. https://doi.org/10.1212/wnl.54.5.1195.

Tanaka, Y., Engelender, S., Igarashi, S., Rao, R.K., Wanner, T., Tanzi, R.E., Sawa, A., L Dawson, V., Dawson, T.M., Ross, C.A., 2001. Inducible expression of mutant alpha-synuclein decreases proteasome activity and increases sensitivity to mitochondria-dependent apoptosis. Hum. Mol. Genet. 10, 919–926.

Tang, F.-L., Erion, J.R., Tian, Y., Liu, W., Yin, D.-M., Ye, J., Tang, B., Mei, L., Xiong, W.-C., 2015. VPS35 in dopamine neurons is required for endosome-to-golgi retrieval of Lamp2a, a receptor of chaperone-mediated autophagy that is critical for α-synuclein degradation and prevention of pathogenesis of Parkinson's disease. J. Neurosci. 35, 10613–10628. https://doi.org/10.1523/JNEUROSCI.0042-15.2015.

Tanik, S.A., Schultheiss, C.E., Volpicelli-Daley, L.A., Brunden, K.R., Lee, V.M.Y., 2013. Lewy body-like α-synuclein aggregates resist degradation and impair macroautophagy. J. Biol. Chem. 288, 15194–15210. https://doi.org/10.1074/jbc.M113.457408.

Tayebi, N., Callahan, M., Madike, V., Stubblefield, B.K., Orvisky, E., Krasnewich, D., Fillano, J.J., Sidransky, E., 2001. Gaucher disease and parkinsonism: a phenotypic and genotypic characterization. Mol. Genet. Metab. 73, 313–321. https://doi.org/10.1006/mgme.2001.3201.

Tayebi, N., Walker, J., Stubblefield, B., Orvisky, E., LaMarca, M.E., Wong, K., Rosenbaum, H., Schiffmann, R., Bembi, B., Sidransky, E., 2003. Gaucher disease with parkinsonian manifestations: does glucocerebrosidase deficiency contribute to a vulnerability to parkinsonism? Mol. Genet. Metab. 79, 104–109.

Taymans, J.-M., Nkiliza, A., Chartier-Harlin, M.-C., 2015. Deregulation of protein translation control, a potential game-changing hypothesis for Parkinson's disease pathogenesis. Trends Mol. Med. 21, 466–472. https://doi.org/10.1016/j.molmed.2015.05.004.

Tofaris, G.K., Layfield, R., Spillantini, M.G., 2001. alpha-Synuclein metabolism and aggregation is linked to ubiquitin-independent degradation by the proteasome. FEBS Lett. 509, 22–26.

Tofaris, G.K., Razzaq, A., Ghetti, B., Lilley, K.S., Spillantini, M.G., 2003. Ubiquitination of alpha-synuclein in Lewy bodies is a pathological event not associated with impairment of proteasome function. J. Biol. Chem. 278, 44405–44411. https://doi.org/10.1074/jbc.M308041200.

Tofaris, G.K., Kim, H.T., Hourez, R., Jung, J.-W., Kim, K.P., Goldberg, A.L., 2011. Ubiquitin ligase Nedd4 promotes alpha-synuclein degradation by the endosomal-lysosomal pathway. Proc. Natl. Acad. Sci. U. S. A. 108, 17004–17009. https://doi.org/10.1073/pnas.1109356108.

Tong, Y., Yamaguchi, H., Giaime, E., Boyle, S., Kopan, R., Kelleher, R.J., Shen, J., 2010. Loss of leucine-rich repeat kinase 2 causes impairment of protein degradation pathways, accumulation of α-synuclein, and apoptotic cell death in aged mice. Proc. Natl. Acad. Sci. U. S. A. 107, 9879–9884. https://doi.org/10.1073/pnas.1004676107.

Tong, Y., Giaime, E., Yamaguchi, H., Ichimura, T., Liu, Y., Si, H., Cai, H., Bonventre, J.V., Shen, J., 2012. Loss of leucine-rich repeat kinase 2 causes age-dependent bi-phasic alterations of the autophagy pathway. Mol. Neurodegener. 7, 2. https://doi.org/10.1186/1750-1326-7-2.

Touchman, J.W., Dehejia, A., Chiba-Falek, O., Cabin, D.E., Schwartz, J.R., Orrison, B.M., Polymeropoulos, M.H., Nussbaum, R.L., 2001. Human and mouse alpha-synuclein genes: comparative genomic sequence analysis and identification of a novel gene regulatory element. Genome Res. 11, 78–86. https://doi.org/10.1101/gr.165801.

Tucci, A., Charlesworth, G., Sheerin, U.-M., Plagnol, V., Wood, N.W., Hardy, J., 2012. Study of the genetic variability in a Parkinson's disease gene: EIF4G1. Neurosci. Lett. 518, 19–22. https://doi.org/10.1016/j.neulet.2012.04.033.

Um, J.W., Han, K.A., Im, E., Oh, Y., Lee, K., Chung, K.C., 2012. Neddylation positively regulates the ubiquitin E3 ligase activity of parkin. J. Neurosci. Res. 90, 1030–1042. https://doi.org/10.1002/jnr.22828.

Usenovic, M., Knight, A.L., Ray, A., Wong, V., Brown, K.R., Caldwell, G.A., Caldwell, K.A., Stagljar, I., Krainc, D., 2012a. Identification of novel ATP13A2 interactors and their role in α-synuclein misfolding and toxicity. Hum. Mol. Genet. 21, 3785–3794. https://doi.org/10.1093/hmg/dds206.

Usenovic, M., Tresse, E., Mazzulli, J.R., Taylor, J.P., Krainc, D., 2012b. Deficiency of ATP13A2 leads to lysosomal dysfunction, α-synuclein accumulation, and neurotoxicity. J. Neurosci. 32, 4240–4246. https://doi.org/10.1523/JNEUROSCI.5575-11.2012.

Valdés, P., Mercado, G., Vidal, R.L., Molina, C., Parsons, G., Court, F.A., Martinez, A., Galleguillos, D., Armentano, D., Schneider, B.L., Hetz, C., 2014. Control of dopaminergic neuron survival by the unfolded protein response transcription factor XBP1. Proc. Natl. Acad. Sci. U. S. A. 111, 6804–6809. https://doi.org/10.1073/pnas.1321845111.

van der Brug, M.P., Blackinton, J., Chandran, J., Hao, L.-Y., Lal, A., Mazan-Mamczarz, K., Martindale, J., Xie, C., Ahmad, R., Thomas, K.J., Beilina, A., Gibbs, J.R., Ding, J., Myers, A.J., Zhan, M., Cai, H., Bonini, N.M., Gorospe, M., Cookson, M.R., 2008. RNA binding activity of the recessive parkinsonism protein DJ-1 supports involvement in multiple cellular pathways. Proc. Natl. Acad. Sci. U. S. A. 105, 10244–10249. https://doi.org/10.1073/pnas.0708518105.

Vandiver, M.S., Paul, B.D., Xu, R., Karuppagounder, S., Rao, F., Snowman, A.M., Ko, H.S., Lee, Y.I., Dawson, V.L., Dawson, T.M., Sen, N., Snyder, S.H., 2013. Sulfhydration mediates neuroprotective actions of parkin. Nat. Commun. 4, 1626. https://doi.org/10.1038/ncomms2623.

Vilariño-Güell, C., Wider, C., Ross, O.A., Dachsel, J.C., Kachergus, J.M., Lincoln, S.J., Soto-Ortolaza, A.I., Cobb, S.A., Wilhoite, G.J., Bacon, J.A., Behrouz, B., Melrose, H.L., Hentati, E., Puschmann, A., Evans, D.M., Conibear, E., Wasserman, W.W., Aasly, J.O., Burkhard, P.R., Djaldetti, R., Ghika, J., Hentati, F., Krygowska-Wajs, A., Lynch, T., Melamed, E., Rajput, A., Rajput, A.H., Solida, A., Wu, R.-M., Uitti, R.J., Wszolek, Z.K., Vingerhoets, F., Farrer, M.J., 2011. VPS35 mutations in Parkinson disease. Am. J. Hum. Genet. 89, 162–167. https://doi.org/10.1016/j.ajhg.2011.06.001.

Vogiatzi, T., Xilouri, M., Vekrellis, K., Stefanis, L., 2008. Wild type alpha-synuclein is degraded by chaperone-mediated autophagy and macroautophagy in neuronal cells. J. Biol. Chem. 283, 23542–23556. https://doi.org/10.1074/jbc.M801992200.

Volpicelli-Daley, L.A., Gamble, K.L., Schultheiss, C.E., Riddle, D.M., West, A.B., Lee, V.M.-Y., 2014. Formation of α-synuclein Lewy neurite-like aggregates in axons impedes the transport of distinct endosomes. Mol. Biol. Cell 25, 4010–4023. https://doi.org/10.1091/mbc.E14-02-0741.

Wang, M., Kaufman, R.J., 2016. Protein misfolding in the endoplasmic reticulum as a conduit to human disease. Nature 529, 326–335. https://doi.org/10.1038/nature17041.

Wang, R., Wang, Y., Qu, L., Chen, B., Jiang, H., Song, N., Xie, J., 2019. Iron-induced oxidative stress contributes to α-synuclein phosphorylation and up-regulation via polo-like kinase 2 and casein kinase 2. Neurochem. Int. 125, 127–135. https://doi.org/10.1016/j.neuint.2019.02.016.

Webb, J.L., Ravikumar, B., Atkins, J., Skepper, J.N., Rubinsztein, D.C., 2003. Alpha-synuclein is degraded by both autophagy and the proteasome. J. Biol. Chem. 278, 25009–25013. https://doi.org/10.1074/jbc.M300227200.

Williams, E.T., Chen, X., Moore, D.J., 2017. VPS35, the retromer complex and Parkinson's disease. J. Park. Dis. 7, 219–233. https://doi.org/10.3233/JPD-161020.

Winslow, A.R., Chen, C.-W., Corrochano, S., Acevedo-Arozena, A., Gordon, D.E., Peden, A.A., Lichtenberg, M., Menzies, F.M., Ravikumar, B., Imarisio, S., Brown, S., O'Kane, C.J., Rubinsztein, D.C., 2010. α-synuclein impairs macroautophagy: implications for Parkinson's disease. J. Cell Biol. 190, 1023–1037. https://doi.org/10.1083/jcb.201003122.

Xia, Y., Rohan de Silva, H.A., Rosi, B.L., Yamaoka, L.H., Rimmler, J.B., Pericak-Vance, M.A., Roses, A.D., Chen, X., Masliah, E., DeTeresa, R., Iwai, A., Sundsmo, M., Thomas, R.G., Hofstetter, C.R., Gregory, E., Hansen, L.A., Katzman, R., Thal, L.J., Saitoh, T., 1996. Genetic studies in Alzheimer's disease with an NACP/alpha-synuclein polymorphism. Ann. Neurol. 40, 207–215. https://doi.org/10.1002/ana.410400212.

Xu, C., Bailly-Maitre, B., Reed, J.C., 2005. Endoplasmic reticulum stress: cell life and death decisions. J. Clin. Invest. 115, 2656–2664. https://doi.org/10.1172/JCI26373.

Yamamoto, A., Friedlein, A., Imai, Y., Takahashi, R., Kahle, P.J., Haass, C., 2005. Parkin phosphorylation and modulation of its E3 ubiquitin ligase activity. J. Biol. Chem. 280, 3390–3399. https://doi.org/10.1074/jbc.M407724200.

Yao, D., Gu, Z., Nakamura, T., Shi, Z.-Q., Ma, Y., Gaston, B., Palmer, L.A., Rockenstein, E.M., Zhang, Z., Masliah, E., Uehara, T., Lipton, S.A., 2004. Nitrosative stress linked to sporadic Parkinson's disease: S-nitrosylation of parkin regulates its E3 ubiquitin ligase activity. Proc. Natl. Acad. Sci. U. S. A. 101, 10810–10814. https://doi.org/10.1073/pnas.0404161101.

Zafar, K.S., Inayat-Hussain, S.H., Ross, D., 2007. A comparative study of proteasomal inhibition and apoptosis induced in N27 mesencephalic cells by dopamine and MG132. J. Neurochem. 102, 913–921. https://doi.org/10.1111/j.1471-4159.2007.04637.x.

Zavodszky, E., Seaman, M.N.J., Moreau, K., Jimenez-Sanchez, M., Breusegem, S.Y., Harbour, M.E., Rubinsztein, D.C., 2014. Mutation in VPS35 associated with Parkinson's disease impairs WASH complex association and inhibits autophagy. Nat. Commun. 5, 3828. https://doi.org/10.1038/ncomms4828.

Zeng, B.-Y., Bukhatwa, S., Hikima, A., Rose, S., Jenner, P., 2006a. Reproducible nigral cell loss after systemic proteasomal inhibitor administration to rats. Ann. Neurol. 60, 248–252. https://doi.org/10.1002/ana.20932.

Zeng, B.-Y., Iravani, M.M., Lin, S.-T., Irifune, M., Kuoppamäki, M., Al-Barghouthy, G., Smith, L., Jackson, M.J., Rose, S., Medhurst, A.D., Jenner, P., 2006b. MPTP treatment of common marmosets impairs proteasomal enzyme activity and decreases expression of structural and regulatory elements of the 26S proteasome. Eur. J. Neurosci. 23, 1766–1774. https://doi.org/10.1111/j.1460-9568.2006.04718.x.

Zhang, N.-Y., Tang, Z., Liu, C.-W., 2008. alpha-synuclein protofibrils inhibit 26S proteasome-mediated protein degradation: understanding the cytotoxicity of protein protofibrils in neurodegenerative disease pathogenesis. J. Biol. Chem. 283, 20288–20298. https://doi.org/10.1074/jbc.M710560200.

Zhou, Z.D., Lim, T.M., 2009. Dopamine (DA) induced irreversible proteasome inhibition via DA derived quinones. Free Radic. Res. 43, 417–430. https://doi.org/10.1080/10715760902801533.

Zimprich, A., Benet-Pagès, A., Struhal, W., Graf, E., Eck, S.H., Offman, M.N., Haubenberger, D., Spielberger, S., Schulte, E.C., Lichtner, P., Rossle, S.C., Klopp, N., Wolf, E., Seppi, K., Pirker, W., Presslauer, S., Mollenhauer, B., Katzenschlager, R., Foki, T., Hotzy, C., Reinthaler, E., Harutyunyan, A., Kralovics, R., Peters, A., Zimprich, F., Brücke, T., Poewe, W., Auff, E., Trenkwalder, C., Rost, B.,

Ransmayr, G., Winkelmann, J., Meitinger, T., Strom, T.M., 2011. A mutation in VPS35,
encoding a subunit of the retromer complex, causes late-onset Parkinson disease. Am. J.
Hum. Genet. 89, 168–175. https://doi.org/10.1016/j.ajhg.2011.06.008.

Zondler, L., Kostka, M., Garidel, P., Heinzelmann, U., Hengerer, B., Mayer, B.,
Weishaupt, J.H., Gillardon, F., Danzer, K.M., 2017. Proteasome impairment by α-synuclein. PLoS One 12, e0184040. https://doi.org/10.1371/journal.pone.0184040.

Endosomal sorting pathways in the pathogenesis of Parkinson's disease

Lindsey A. Cunningham[a,b], Darren J. Moore[b,*]

[a]*Van Andel Institute Graduate School, Grand Rapids, MI, United States*
[b]*Center for Neurodegenerative Science, Van Andel Institute, Grand Rapids, MI, United States*
[*]*Corresponding author: Tel.: +1-6162345346, e-mail address: darren.moore@vai.org*

Abstract

The identification of Parkinson's disease (PD)-associated genes has created a powerful platform to begin to understand and nominate pathophysiological disease mechanisms. Herein, we discuss the genetic and experimental evidence supporting endolysosomal dysfunction as a major pathway implicated in PD. Well-studied familial PD-linked gene products, including LRRK2, VPS35, and α-synuclein, demonstrate how disruption of different aspects of endolysosomal sorting pathways by disease-causing mutations may manifest into PD-like phenotypes in many disease models. Newly-identified PD-linked genes, including auxilin, synaptojanin-1 and Rab39b, as well as putative risk genes for idiopathic PD (endophilinA1, Rab29, GAK), further support endosomal sorting deficits as being central to PD. LRRK2 may represent a nexus by regulating many distinct features of endosomal sorting, potentially via phosphorylation of key endocytosis machinery (i.e., auxilin, synaptojanin-1, endoA1) and Rab GTPases (i.e., Rab29, Rab8A, Rab10) that function within these pathways. In turn, LRRK2 kinase activity is critically regulated by Rab29 at the Golgi complex and retromer-associated VPS35 at endosomes. Taken together, the known functions of PD-associated gene products, the impact of disease-linked mutations, and the emerging functional interactions between these proteins points to endosomal sorting pathways as a key point of convergence in the pathogenesis of PD.

Keywords

Parkinson's disease, Endolysosomal sorting, VPS35, LRRK2, α-Synuclein, Synaptic vesicle endocytosis, Auxilin, GAK, Synaptojanin-1, Endophilin, Rab GTPases

Progress in Brain Research, Volume 252, ISSN 0079-6123, https://doi.org/10.1016/bs.pbr.2020.02.001

1 Genetics lead the way to understanding Parkinson's disease pathogenesis

As the global population grows and ages, neurological disorders have become the largest source of disability and the second largest cause of death worldwide (GBD 2016 Neurology Collaborators, 2019). Of these disorders Parkinson's disease (PD) is one of the fastest growing with nearly 22% increase in age-standardized prevalence between 1990 and 2016 (GBD 2016 Parkinson's Disease Collaborators, 2018). PD is the most common neurodegenerative movement disorder affecting 10 million people worldwide and projected to affect nearly 1 million in North America alone by 2020 (Dorsey and Bloem, 2018; Dorsey et al., 2018; Marras et al., 2018). Clinically, PD is largely diagnosed via cardinal motor symptoms, including resting tremor, bradykinesia, rigidity and postural instability (Lang and Lozano, 1998a,b). Pathologically, PD is characterized by the relatively selective loss of dopaminergic neurons in the *substantia nigra pars compacta* (SNpc) resulting in dopamine depletion of the nigrostriatal pathway and brainstem Lewy pathology (Lang and Lozano, 1998a,b). Despite urgent clinical need and impending societal health burden, current therapies are purely palliative and often lose efficacy as disease progresses. The lack of tractable therapeutic success is likely due to our nascent understanding of disease etiology. Over the past 200 years since PD was originally described, many pathogenic mechanisms have been proposed based on genetic, epidemiological, experimental and pathological evidence including; endolysosomal dysfunction, mitochondrial deficits, synaptic dysfunction, protein aggregation, and inflammation, to name a few (Meissner et al., 2011; Moore et al., 2005). Yet today, it remains uncertain what causes the selective loss of nigral dopaminergic neurons and other neuronal populations in people with PD.

Although the majority of PD is idiopathic, PD can also occur in a familial manner due to the Mendelian inheritance of distinct genetic mutations, with the identification of at least 15 genes to date. While familial PD due to autosomal dominant inheritance (i.e., *LRRK2*, *VPS35*, *SNCA*) is usually similar to idiopathic disease with typical late-onset, clinical symptoms and neuropathology, autosomal recessive PD (i.e., *Parkin*, *PINK1*, *DJ-1*) tends to be early-onset, slowly progressive, and most often lacks Lewy pathology (Blauwendraat et al., 2020; Hernandez et al., 2016). Genome-wide association studies (GWAS) for idiopathic PD have also highlighted common genetic variants (~90 SNPs) in loci involved in PD risk, that include genes linked to familial PD (i.e., *SNCA*, *LRRK2*) as well as within candidate genes that cluster in similar biological pathways (Chang et al., 2017; Nalls et al., 2014, 2019). The genes linked to familial PD or nominated by GWAS provide key insight into the molecular and cellular mechanisms that regulate age-related neuronal vulnerability and Lewy pathology in PD, and collectively implicate the endolysosomal system as an important subcellular location and pathway for disease pathogenesis (Table 1). In this review, we will discuss the most well-studied familial PD genes involved in the endolysosomal system; LRRK2, VPS35 and α-synuclein, and their connections to other, recently identified PD-linked endosomal genes. Currently, human genetic and cell biological studies provide the strongest evidence for disease pathogenesis through endolysosomal dysfunction (Abeliovich and Gitler, 2016).

Table 1 PD-associated genes and their functions within endosomal pathways.

Gene/locus name	Protein name	Phenotype	Function	References
SNCA (*PARK1*)	α-Synuclein	AD LO PD	Synaptic vesicle exocytosis; SNARE chaperone	Polymeropoulos et al. (1997), Kruger et al. (1998), Zarranz et al. (2004), Appel-Cresswell et al. (2013), and Lesage et al. (2013)
SNCA (*PARK4*)	α-Synuclein	AD EO PD		Singleton et al. (2003) and Chartier-Harlin et al. (2004)
LRRK2 (*PARK8*)	LRRK2	AD LO PD	Kinase, GTPase	Paisan-Ruiz et al. (2013) and Zimprich et al. (2004)
PARK16	Rab29 (Rab7L1)	PD risk, GWAS	Rab GTPase	Nalls et al. (2014), Tucci et al. (2010), and MacLeod et al. (2013)
VPS35 (*PARK17*)	VPS35	AD LO PD	Endosomal cargo sorting—retromer	Deng et al. (2012), Vilarino-Guell et al. (2011), Sharma et al. (2012), and Zimprich et al. (2011)
DNAJC6 (*PARK19*)	Auxilin	AR Juvenile PD	Clathrin uncoating	Edvardson et al. (2012) and Song et al. (2017)
SYNJ1 (*PARK20*)	Synaptojanin-1	AR EO PD	Lipid phosphatase	Krebs et al. (2013), Quadri et al. (2013), Olgiati et al. (2014), Chen et al. (2015), and Kirola et al. (2016)
RAB32	Rab32	AD, LO PD	Rab GTPase	Gustavsson et al. (2017)
RAB39B	Rab39b	X-linked EO PD	Rab GTPase	Lesage et al. (2015) and Wilson et al. (2014)
GAK (*DNAJC26*)	GAK	PD risk, GWAS	Clathrin uncoating	Chen et al. (2013) and Beilina et al. (2014)

Key: AD, autosomal dominant; AR, autosomal recessive; LO, late-onset; EO, early-onset; PD, Parkinson's disease; GWAS, genome-wide association study.

---

2 Role of familial PD genes in endosomal sorting pathways

2.1 The endolysosomal system

In order to understand how alterations in familial PD genes cause disease, an understanding of how they play roles within, or interact with, the endosomal system is required. Briefly, proteins enter the endosomal system from two main directions, either via endocytosis at the plasma membrane or from the *trans*-Golgi network (TGN) (Fig. 1). Endocytic vesicles from either direction fuse with endosomes, adding their contents and/or membrane-bound "cargo" proteins into the endosomal lumen. Once

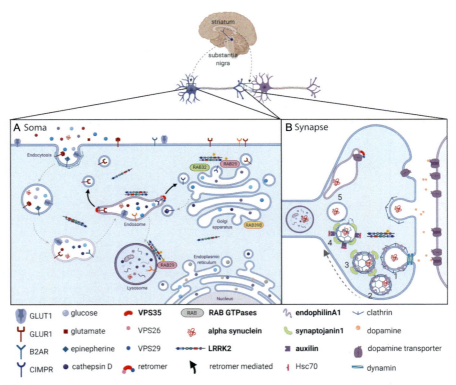

FIG. 1

PD-related proteins (bolded) in endolysosomal sorting pathways. (A) In the neuronal soma ligands, receptors and transporters are taken up via endocytosis at the plasma membrane. As endosomes mature, lysosomal hydrolases, including cathepsin D, are delivered to the late endosome and lysosome from the *trans*-Golgi (TGN) as vesicles become more acidic (indicated by deepening purple). The retromer associates with early endosomes to identify cargo for retrograde transport (bold arrows) to the TGN or plasma membrane. Cargo that fail to be retrieved from endosomes accumulate and eventually undergo degradation in lysosomes. LRRK2 is cytoplasmic but is enriched upon different vesicular membranes in a dimeric, active conformation. Certain Rab GTPases are phosphorylated by LRRK2 with Rab29 regulating LRRK2 recruitment to the TGN and stressed lysosomes. PD-linked Rab32 and Rab39b may play a role within the Golgi complex. (B) At the synapse, PD-associated proteins are responsible for clathrin-mediated synaptic vesicle endocytosis. (1) EndophilinA1 regulates membrane curvature and recruits dynamin-1 for constriction of plasma membrane buds and (2) eventual fission of clathrin-coated vesicles (CCV). (3) Synaptojanin-1 is recruited by endophilinA1 to dephosphorylate membrane lipids and release adaptor proteins (not shown), allowing auxilin to bind. (4) Auxilin and its cofactor Hsc70 remove the clathrin coat to produce synaptic vesicles that can fuse with endosomes or be loaded with neurotransmitters. Endocytosis represents one potential route for intake of misfolded or aggregated α-synuclein. Created using BioRender.com.

in the endosomal system, cargo can be recycled back to the plasma membrane or TGN, transported to a new destination or remain in the endosome as it acidifies and matures into a lysosome for eventual degradation. At this stage the lysosome may fuse with other organelles such as autophagosomes for organelle or protein degradation and clearance. Perturbations to this system can result in cargo that abnormally remains in the maturing endosome and is eventually degraded in the lysosome. It is hypothesized that the accumulation of protein cargo or pathogenic proteins with time may eventually contribute to lysosome dysfunction and cell death. A number of PD models support this concept as perturbation of PD-linked proteins have been shown to consistently disrupt endolysosomal functions resulting in the accumulation and/or swelling of different vesicular structures.

For retrieval and recycling from the endosome, cargo must be recognized and sorted to other compartments. Recognition of cargo in the endosomal membrane is performed by three main complexes, including the retromer, retriever and ESCRT (endosomal sorting complex required for transport) (McNally and Cullen, 2018; Seaman, 2012; Vietri et al., 2020). Briefly, both retromer and retriever select cargo from early and recycling endosomes for retrograde transport, subsequently preventing cargo from undergoing lysosomal degradation. Although retromer and retriever are functionally and structurally similar (i.e., heterotrimers that include a VPS29 subunit), each complex recognizes a specific set of protein cargo for recycling (McNally and Cullen, 2018; McNally et al., 2017; Seaman, 2012). In contrast, the ESCRT complex is less cargo-selective and instead gathers ubiquitinated proteins upon late endosomes for inclusion into intra-luminal vesicles (ILV). Cargo-enriched ILVs are contained in multivesicular bodies for future degradation by the lysosome (Christ et al., 2017; Wollert and Hurley, 2010). Here, we focus on VPS35 and the retromer due to their importance to endosomal sorting and familial PD (Williams et al., 2017).

2.2 Vacuolar protein sorting-associated protein 35 (VPS35)

Retromer is a heteropentameric complex responsible for selecting cargo in the early endosome for retrograde transport to the plasma membrane or TGN. VPS35, VPS26, and VPS29 form the cargo-selective complex (CSC) which is required for the identification of cargo for recycling (Seaman, 2012). VPS35 is the largest component of the CSC consisting of an α-solenoid that makes up a central backbone that binds to VPS26 at its N-terminus and VPS29 at its C-terminus (Hierro et al., 2007; Kovtun et al., 2018). The CSC trimer associates with a sorting nexin (SNX) dimer, consisting of either SNX1 or SNX2 and SNX5 or SNX6, to form a pentamer (Cullen and Korswagen, 2011). Although the CSC is required for cargo recognition, it is unable to associate with the endosomal membrane on its own and requires membrane recruitment via the SNX dimer and Rab7A (Seaman, 2012). A proportion of the retromer associates with FAM21, a component of the pentameric Wiscott–Aldrich syndrome protein and SCAR homolog (WASH) complex, to recruit WASH to the endosomal membrane where it mediates F-actin polymerization and the formation of patches required for the partitioning of cargo into endosomal tubules for sorting (Seaman and Freeman, 2014; Seaman et al., 2013).

One classical retromer cargo is the cation-independent mannose-6 phosphate receptor (CIMPR) that binds newly synthesized acid hydrolases in the TGN, including cathepsin D, for transport to the maturing endolysosome (Arighi et al., 2004; Seaman, 2004). Note that the mechanism involved in sorting proteins to the endosome does not involve the retromer. As the endolysosome matures and acidifies, the cathepsin D ligand is released into the lumen. CIMPR is subsequently identified by the retromer and recycled to the TGN for reuse. Similarly, cargo that undergo endocytosis at the plasma membrane often consist of receptors and ion channels, including the glucose transporter (GLUT1), AMPA receptor subunits (GluR1/R2), or the β2-adrenergic receptor, although in neurons there is also a large degree of specialized endocytosis at synapses including pre- and post-synaptic compartments (Choy et al., 2014; Steinberg et al., 2013; Temkin et al., 2011). In both locations, clathrin-coated pits form at the plasma membrane to remove receptors and transporters following ligand binding and/or activation. Following endocytosis, receptor cargo are sequestered into early endosomes where they remain until being recycled back to the cell surface through recognition and sorting by the retromer and WASH complexes, or sorted for degradation via the lysosome. These examples indicate that cargo retrieval is important for protein recycling and efficient sorting but also to maintain normal lysosomal health.

Mutations in *VPS35* were first identified to cause autosomal dominant, familial PD less than a decade ago, but this and other discoveries have promoted great interest into the role of endosomal sorting in PD (Vilarino-Guell et al., 2011; Zimprich et al., 2011). Several putative mutations in *VPS35* have been identified to date, but only a single missense mutation (p.Asp620Asn, D620N, c.1858G > A) has been confirmed as pathogenic based upon segregation with disease in multiple PD families (Deng et al., 2012; Kumar et al., 2012; Sharma et al., 2012; Vilarino-Guell et al., 2011; Williams et al., 2017; Zimprich et al., 2011). Familial PD due to *VPS35* mutations is clinically similar to idiopathic disease with late-onset, slow progression, and a good response to levodopa therapy. Only a single *VPS35*-linked PD subject has been assessed at autopsy albeit incompletely without assessment of the brainstem, and therefore the neuropathological hallmarks associated with *VPS35* mutations are not yet clear (Struhal et al., 2014; Wider et al., 2008). PET imaging of the caudate putamen in *VPS35*-linked PD subjects is consistent with degeneration of the nigrostriatal dopaminergic pathway (Deng et al., 2013; Wider et al., 2008). A dominant pattern of inheritance combined with a lack of familial mutations that produce deleterious truncations, rearrangements or exonic deletions, may suggest that heterozygous *VPS35* mutations most likely cause disease through a toxic gain-of-function or dominant-negative mechanism (Williams et al., 2017). However, the pathogenic effects of the D620N mutation in different cells and models are still being investigated.

At the present time, the only known molecular defect induced by the D620N mutation is a reduced interaction of VPS35 with the WASH complex via impaired binding to FAM21 (McGough et al., 2014; Zavodszky et al., 2014). The D620N mutation does not alter the protein stability of VPS35, alter retromer assembly or distribution in cells or brain, and does not induce global defects in retromer cargo sorting including sortilin or sorLA (Tsika et al., 2014). However, altered sorting of CIMPR, AMPA receptors or ATG9A have been reported in certain cell types expressing D620N VPS35 (Follett et al., 2016; MacLeod et al., 2013; Munsie et al., 2015; Zavodszky et al., 2014). The effects of the D620N mutation on the interaction of VPS35 with other retromer cargo or accessory proteins is not yet clear, whereas it is not known whether the reduction in WASH complex recruitment contributes to cellular damage (Zavodszky et al., 2014). While D620N VPS35 retains many of the functions of its wild-type counterpart (Williams et al., 2017), there are examples where the D620N protein may act in a loss-of-function manner depending on the specific phenotypic or cellular context (Linhart et al., 2014; MacLeod et al., 2013; Zavodszky et al., 2014). Together, these studies indicate that D620N VPS35 may produce subtle functional deficits most consistent with a partial loss-of-function effect but how or whether these effects are sufficient to manifest neurodegeneration remains unknown.

Rodent models have further helped to understand the pathogenic mechanism of the D620N mutation. The viral-mediated overexpression of human D620N VPS35 within the nigrostriatal pathway of adult rats is sufficient to induce dopaminergic neurodegeneration and axonal pathology (Tsika et al., 2014), consistent with a gain-of-function or partial dominant-negative mechanism. Furthermore, the recent development of *D620N VPS35* knockin mice indicate that homozygous D620N mice exhibit normal development, survival and motor function (Cataldi et al., 2018; Chen et al., 2019; Ishizu et al., 2016). This contrasts with the embryonic lethality of *VPS35* knockout mice (Chen et al., 2019; Wen et al., 2011), demonstrating that the D620N VPS35 protein is largely functional and unlikely to manifest via a loss-of-function. In addition, both heterozygous and homozygous *D620N VPS35* knockin mice display equivalent and progressive dopaminergic neuronal degeneration, axonal damage and somatodendritic tau pathology with advanced age (Chen et al., 2019). A recent study provides evidence for LRRK2 hyperactivation in tissues from *D620N VPS35* knockin mice as indicated by increased phosphorylation of the LRRK2 substrate Rab10 as well as LRRK2 autophosphorylation (i.e., at Ser1292) (Mir et al., 2018). LRRK2 hyperactivation is equivalent between heterozygous and homozygous knockin mice (Mir et al., 2018), which together with neuropathological observations in these mice (Chen et al., 2019), support a toxic gain-of-function mechanism for the D620N mutation and exclude a loss-of-function mechanism.

In general, the loss of *VPS35* expression is known to be detrimental in different cells and animal models (Williams et al., 2017), indicating a critical requirement of VPS35 function for normal cellular health and viability. Earlier studies have focused on the role of VPS35 loss-of-function in Alzheimer's disease (AD) since vulnerable brain regions of AD subjects exhibit decreased VPS35 protein levels (Muhammad et al., 2008; Small et al., 2005; Wen et al., 2011). In PD, studies have explored a potential functional interaction between PD-linked LRRK2 and VPS35, and have shown that PD-linked LRRK2 mutants can phenocopy the effects of *VPS35* depletion in *Drosophila* and primary cortical neuronal models (Linhart et al., 2014; MacLeod et al., 2013). Further evidence that these two proteins may operate in a common pathway is supported by observations that the overexpression of wild-type VPS35 (but not the D620N mutant) can rescue mutant LRRK2-induced neuronal phenotypes in these models (Linhart et al., 2014; MacLeod et al., 2013). Moreover, endogenous VPS35 protein is reduced by PD-linked LRRK2 mutants in cells or brains of LRRK2 transgenic mice (MacLeod et al., 2013); however, there are conflicting data concerning whether VPS35 levels are also reduced in human brains from idiopathic or *LRRK2* mutant PD subjects (Tsika et al., 2014; Zhao et al., 2018). The concept that PD-linked *LRRK2* mutations may exert their pathogenic effects in PD by decreasing VPS35 expression and impairing retromer function is attractive since, (1) this would indicate that LRRK2 and VPS35 interact and operate in a common cellular pathway relevant to PD that manifests in retromer deficiency, and (2) that targeting this interaction could provide a single therapeutic strategy for different familial forms of PD, and potentially be relevant to idiopathic PD as well as AD. Additional studies are now warranted to further validate this potential LRRK2/VPS35 disease mechanism, especially using rodent models of PD and human brain tissue.

Due to the relevance of VPS35 loss to AD pathophysiology (Li et al., 2019; Small and Petsko, 2015; Small et al., 2005), there have been significant efforts to develop therapeutic compounds that can boost VPS35 levels and/or stabilize the retromer (Berman et al., 2015). While such compounds have been developed with AD in mind, it is not yet clear whether this approach could also be used in PD subjects, especially those harboring *VPS35* mutations. R33 and R55 are small molecule chemical chaperones that stabilize the interaction between VPS35 and VPS29 to increase retromer stability and boost its expression (Berman et al., 2015; Mecozzi et al., 2014). While these compounds are effective at increasing retromer and reducing Aβ peptide accumulation in cultured hippocampal neurons (Mecozzi et al., 2014), their pharmacokinetic properties and brain permeability for neuroprotective studies in rodent models of AD or PD are being evaluated (Li et al., 2020). While PD-linked *LRRK2* mutations may drive retromer deficiency in some simple models (Linhart et al., 2014; MacLeod et al., 2013), studies in *D620N VPS35* knockin rodent models do not necessarily suggest a loss-of-function mechanism (Chen et al., 2019). Therefore, the utility of retromer-stabilizing compounds for PD awaits further evaluation and mechanistic studies in PD-linked *VPS35* rodent models.

The homozygous deletion of *VPS35* in mice results in early embryonic lethality whereas heterozygous null ($VPS35^{+/-}$) mice exhibit normal development and lifespan (Chen et al., 2019; Tang et al., 2015a; Wen et al., 2011). Therefore, VPS35 is critically required for normal embryonic development and viability. $VPS35^{+/-}$ mice develop chronic and progressive nigrostriatal degeneration resulting in the modest loss of dopaminergic neurons (~20% loss) by 12 months of age but without overt neuronal loss in the hippocampus (Tang et al., 2015a). Furthermore, the conditional homozygous deletion of *VPS35* selectively in DAT-positive dopaminergic neurons ($VPS35^{DAT-Cre}$ mice) induces early dopaminergic neuronal and axonal loss

(~30%) by 2–3 months of age (Tang et al., 2015b), although it is unclear whether neuronal loss is progressive in this model. These studies indicate that *VPS35* depletion in mice is sufficient to produce age-dependent PD-like neuropathology, and demonstrate that VPS35 is critically required for the normal maintenance and survival of dopaminergic neurons. However, what remains uncertain is whether *VPS35* deletion selectively impacts dopaminergic neurons of the substantia nigra (A9 population), versus other dopaminergic (i.e., A10 ventral tegmental area) or non-dopaminergic neuronal populations, and how these effects specifically relate to the pathogenic mechanisms induced by the PD-linked D620N mutation. Recent studies have reported the conditional deletion of *VPS35* selectively in microglia or forebrain neurons of mice. Deletion of *VPS35* in microglial cells (*VPS35^{CX3CR1-CreER}*) results in normal survival but increased microglial density and activity in the hippocampus, increased neuronal progenitor proliferation but decreased neuronal differentiation (Appel et al., 2018). Curiously, *VPS35* deletion in CamKIIα-positive forebrain neurons results in glial activation (Qureshi et al., 2019), whereas deletion in embryonic pyramidal neurons (*VPS35^{neurod6-Cre}*) impairs axonal and dendritic terminal differentiation, and induces cortical atrophy and reactive gliosis (Tang et al., 2020). Together, these studies suggest that different brain cells are vulnerable to *VPS35* loss, including midbrain and forebrain neurons, although selective deletion in microglia is not sufficient alone to produce PD-like pathology implying that dopaminergic neuronal loss is likely to be a cell-autonomous mechanism (Tang et al., 2015a,b). Whether *VPS35* depletion is relevant to the mechanisms underlying PD-linked mutations requires additional investigation, especially given the difference in survival between homozygous *VPS35* knockout and *D620N VPS35* knockin mice (Chen et al., 2019; Wen et al., 2011), and the limited evidence for reduced VPS35 in PD brains (MacLeod et al., 2013; Tsika et al., 2014; Zhao et al., 2018).

While VPS35 plays a distinct role in retromer-dependent endosomal sorting and is directly linked to familial PD, it is also known to physically or functionally interact with proteins implicated in endosomal pathways and PD pathophysiology including LRRK2, α-synuclein, tau, parkin and Rab29, which will be addressed in later sections (Chen et al., 2019; Dhungel et al., 2015; Linhart et al., 2014; MacLeod et al., 2013; Miura et al., 2014; Williams et al., 2018).

2.3 Leucine-rich repeat kinase 2 (LRRK2)

LRRK2 is reported to influence a number of cellular processes or pathways and its precise function within the endosomal system is incompletely understood. LRRK2 is enriched upon multiple intracellular membranes and vesicular compartments including early and late endosomal membranes (Alegre-Abarrategui et al., 2009; Berger et al., 2010; Biskup et al., 2006; Stafa et al., 2014). Furthermore, LRRK2 is reported to interact with a number of PD-linked gene products that can reside and/or function within the endosomal system, including VPS35, α-synuclein, auxilin/DNAJC6, synaptojanin-1 (SYNJ1), endophilinA1 (EndoA1) and Rab GTPases (i.e., Rab29,

Rab32). LRRK2 can also regulate endocytosis and the sorting of receptors or cargo for recycling or lysosomal degradation in different mammalian cells (Gomez-Suaga et al., 2014; MacLeod et al., 2013; Wallings et al., 2019; Xiong et al., 2010). Together, these observations suggest LRRK2 may serve as a nexus in the endosomal system in PD, although the specific molecular mechanisms involved have not been elucidated.

Mutations in the *LRRK2* gene cause late-onset, autosomal dominant PD and represent the most common cause of familial PD (Biskup and West, 2009; Healy et al., 2008; Paisan-Ruiz et al., 2004; Zimprich et al., 2004). Common genetic variation at the *LRRK2* locus is also associated with increased risk for idiopathic PD (Nalls et al., 2014), indicating that *LRRK2* can serve as a pleomorphic risk locus. Similar to *VPS35*-linked PD, familial PD due to *LRRK2* mutations is clinically indistinguishable from idiopathic disease although mutations can produce pleomorphic neuropathology (Biskup and West, 2009; Healy et al., 2008; Zimprich et al., 2004). LRRK2 is a large multi-domain protein that is notable for containing two distinct enzymatic domains, including a Ras-of-Complex (Roc) GTPase domain and a serine/threonine-directed protein kinase domain (Islam and Moore, 2017). LRRK2 belongs to the ROCO protein family that is defined by containing a Roc domain in tandem with a C-terminal-of-Roc (COR) linker domain. At least seven missense mutations in *LRRK2* have been identified that segregate with disease in families with PD, including N1437H, R1441C/G/H, Y1699C, G2019S, and I2020T (Hernandez et al., 2016; Islam and Moore, 2017). These pathogenic mutations cluster within the central Roc-COR-kinase catalytic domains and share the capacity to enhance LRRK2 kinase activity in cells via direct effects on the kinase activation loop (G2019S, I2020T) or indirectly by impairing GTP hydrolysis and prolonging the GTP-bound "on" state (Islam and Moore, 2017; Sheng et al., 2012; Steger et al., 2016, 2017; West et al., 2005, 2007; Xiong et al., 2010). These observations have supported the development of LRRK2 kinase inhibitors as a potential therapeutic strategy for PD. Structurally-distinct LRRK2 kinase inhibitors can produce on-target pathology in rodents, resembling the phenotypes of *LRRK2* knockout models (Baptista et al., 2013; Herzig et al., 2011; Tong et al., 2010, 2012), including pigmented and enlarged kidneys with lysosomal accumulation or lamellar bodies in lung pneumocytes (Andersen et al., 2018; Fuji et al., 2015). Such pathology is dose-dependent, reversible after a short washout period, and does not impact kidney or lung function following chronic inhibitor administration (Andersen et al., 2018).

Despite a general consensus that familial LRRK2 mutations increase kinase activity in different cells (Steger et al., 2016, 2017), the mechanisms by which LRRK2 activation induces neuronal damage are still being explored (Islam and Moore, 2017). Originally, most studies relied upon in vitro kinase assays to identify LRRK2 substrates, which include ArfGAP1, RPS15, MARK1 and β-tubulin (Gillardon, 2009; Islam and Moore, 2017; Krumova et al., 2015; Martin et al., 2014; Stafa et al., 2012; Xiong et al., 2012), in addition to LRRK2 autophosphorylation at Ser1292 (Kluss et al., 2018; Sheng et al., 2012). With the exception of pSer1292-LRRK2, many of these earlier substrates have not been extensively validated in mammalian cells or brain tissue and await the development of phospho-specific substrate antibodies. Recent proteome-wide mass spectrometry studies using embryonic fibroblast cells derived from *G2019S LRRK2* knockin mice have identified at least 14 Rab GTPases as the first bona fide cellular substrates of LRRK2, including Rab8A, Rab10, Rab12 and Rab29 (Steger et al., 2016, 2017). Rab phosphorylation occurs on a highly conserved threonine or serine residue with the Switch II catalytic motif (i.e., Thr73 in Rab10). These Rabs are discreetly distributed across intracellular membranes within the endolysosomal system where they are thought to play roles in vesicular membrane recognition, sorting and fission/fusion events (Bonet-Ponce and Cookson, 2019). Phospho-specific Rab antibodies and PhosTag methodology have been developed as important tools for monitoring LRRK2 activity in different cells and tissues and for evaluating the efficacy of LRRK2 kinase inhibitors (Ito et al., 2016; Kluss et al., 2018; Lis et al., 2018). While increased LRRK2-dependent Rab phosphorylation has been detected in different tissues from PD-linked mutant *LRRK2* knockin and transgenic rodent models (Ito et al., 2016; Lis et al., 2018; Steger et al., 2016, 2017), the significance of Rab phosphorylation for mediating cellular processes such as endolysosomal sorting as well as neuronal damage are still emerging. Whether Rabs represent the only or most relevant substrates of LRRK2 kinase activity, especially within brain cells, is not yet known as detection of robust Rab phosphorylation in the brain has proven challenging (Kelly et al., 2018; Steger et al., 2016). Nevertheless, roles for the LRRK2-mediated phosphorylation of certain Rabs (i.e., Rab8A, Rab10, Rab29) have recently emerged in mediating ciliogenesis and centrosomal cohesion deficits, the response to lysosomal stress, and the recruitment and activation of LRRK2 at the TGN (Beilina et al., 2014; Dhekne et al., 2018; Eguchi et al., 2018; Gomez et al., 2019; Lara Ordonez et al., 2019; Liu et al., 2018; Purlyte et al., 2018; Steger et al., 2017).

Although LRRK2 has been implicated in myriad cellular processes, the removal of LRRK2 does not clearly indicate a required function in vivo. *LRRK2* knockout rodents are viable and do not exhibit PD-related neuropathology or behavioral deficits (Andres-Mateos et al., 2009; Baptista et al., 2013; Herzig et al., 2011; Hinkle et al., 2012; Pellegrini et al., 2018; Tong et al., 2010). This is consistent with the idea that familial mutations act via a gain-of-function mechanism, given that PD-linked mutations commonly enhance kinase activity and mutant LRRK2 overexpression promotes neuronal damage in primary culture models and certain transgenic or viral-mediated rodent models (Blauwendraat et al., 2018; Dusonchet et al., 2011; Greggio et al., 2006; Lee et al., 2010; Ramonet et al., 2011; Smith et al., 2006; Steger et al., 2016; Tsika et al., 2015; West et al., 2005, 2007). However, *LRRK2* knockout rodents consistently develop enlarged, darkly pigmented kidneys with lysosomal accumulation, and type II pneumocytes in lung bearing lamellar bodies, though without an obvious impact on organ function (Andersen et al., 2018; Baptista et al., 2013; Boddu et al., 2015; Herzig et al., 2011; Hinkle et al., 2012; Ness et al., 2013; Tong et al., 2010, 2012). The accumulation of lysosomal-like vesicles in knockout rodents supports an important role for LRRK2 in regulating the endolysosomal pathway.

Further cementing a role for LRRK2 in endosomal sorting pathways is the intriguing connection with the PD-related proteins, Rab29 and VPS35. Rab29 (a.k.a. Rab7L1) is normally enriched upon the TGN where it plays a role in the recruitment of LRRK2 (Beilina et al., 2014; Liu et al., 2018; Purlyte et al., 2018). This recruitment is dependent on LRRK2 kinase activity, with increased relocalization of PD-linked LRRK2 mutants that are kinase hyperactive, and requires membrane- and GTP-bound Rab29 (Gomez et al., 2019; Liu et al., 2018). In turn, Rab29 binding to the N-terminal region of LRRK2 is required for kinase activation including Ser1292 autophosphorylation and Rab phosphorylation (Purlyte et al., 2018). Rab29 itself also serves as a substrate of LRRK2 (Liu et al., 2018, Purlyte et al., 2018). Rab29 recruitment and activation of LRRK2 to the TGN may serve to stabilize membrane and GTP-bound Rabs and regulate their interactions with effector proteins (Liu et al., 2018). This process has been implicated in the clearance of Golgi-derived vesicles via the autophagy-lysosomal system (Beilina et al., 2014). A similar Rab29-dependent mechanism has also been suggested to occur at the lysosomal membrane that recruits and activates LRRK2 in response to lysosomal stress to maintain their homeostasis via lysosomal secretion (Eguchi et al., 2018). A recent study suggests that Rab29-mediated membrane association is the most important determinant for LRRK2 recruitment and activation rather than the membrane type or subcellular location per se (Gomez et al., 2019).

Distinct from Rab29 function, VPS35 and the retromer mediate endosomal cargo recycling to the TGN (Seaman, 2012). As discussed earlier, familial LRRK2 mutants are reported to induce a retromer deficiency by interacting with and decreasing VPS35 protein levels in cell lines or brains of R1441C LRRK2 transgenic mice via an unknown mechanism (MacLeod et al., 2013). G2019S LRRK2 can disrupt endosomal sorting of the retromer cargo CIMPR in primary neurons, and this can be rescued by VPS35 overexpression (MacLeod et al., 2013). Similarly, VPS35 overexpression can rescue the pathogenic effects of G2019S LRRK2 on neurite outgrowth in primary cultures and dopaminergic neuronal loss in *Drosophila* (MacLeod et al., 2013). In *Drosophila* models, VPS35 or VPS26 overexpression can rescue eye phenotypes, reduced lifespan and locomotor deficits induced by I2020T LRRK2, whereas *VPS35* gene silencing can phenocopy I2020T LRRK2 in flies (Linhart et al., 2014). These observations support the idea that PD-linked *LRRK2* mutations and *VPS35* loss-of-function may act through similar mechanisms. Lastly, PD brains harboring *G2019S LRRK2* mutations reveal decreased VPS35 protein levels in frontal and occipital cortex relative to idiopathic PD or control brains (Zhao et al., 2018), although an earlier study found no difference in VPS35 levels in frontal cortex of *G2019S* PD brains (Tsika et al., 2014). These studies provide initial support for a functional interaction of LRRK2 and VPS35 in vivo and further suggest that PD-linked *LRRK2* mutations may exert their pathogenic effects via retromer disruption. Additional validation of this mechanism is now required in rodent models of PD, potentially including neuroprotection studies in mutant LRRK2 models with retromer-stabilizing chemical chaperones (i.e., R33/R55) (Mecozzi et al., 2014).

Further complicating our understanding of a LRRK2/VPS35 molecular pathway is the recent observation that *D620N VPS35* knockin mice unexpectedly display robust LRRK2 hyperactivation in cells and tissues as indicated by increased Rab phosphorylation (Rab8a, Rab10, Rab12) and LRRK2 Ser1292 autophosphorylation (Mir et al., 2018). Notably, LRRK2 activation in these *D620N VPS35* mice is greater (~6-fold) than that produced in *R1441C* or *G2019S LRRK2* knockin mice (~4- or 2-fold, respectively), and surprisingly endogenous VPS35 is even required in part for the increased LRRK2 kinase activity in *R1441C LRRK2* mice (Mir et al., 2018). Rab phosphorylation in *VPS35* knockin mice was shown to be dependent on LRRK2 using the potent and selective kinase inhibitor MLi-2 (Mir et al., 2018). LRRK2 hyperactivation could also be detected in human primary neutrophils and monocytes derived from PD subjects harboring a heterozygous *D620N VPS35* mutation (Mir et al., 2018). This new study supports a previously unappreciated role for the retromer in critically regulating LRRK2 activity and supports a putative model where VPS35 may lie upstream of LRRK2. The mechanism by which D620N VPS35 activates LRRK2 is not yet known, and whether this occurs through a direct interaction, via an interaction with an intermediate regulatory protein or complex, and/or by altering LRRK2 membrane occupancy or subcellular localization, awaits further study. Strategies to correct the pathogenic effects associated with the D620N VPS35 protein or LRRK2 kinase inhibition may provide therapeutic opportunities for treating both *VPS35-* and *LRRK2*-linked familial PD. It will be important in future studies to confirm and reconcile the observations that familial *LRRK2* mutations can induce retromer deficiency whereas *VPS35* mutations can induce LRRK2 activation (MacLeod et al., 2013; Mir et al., 2018), although both potential mechanisms are not necessarily mutually exclusive. Also, it will be important to address how the LRRK2/VPS35 pathway impacts upon specific endolysosomal functions and its relevance to disease pathogenesis.

2.4 **SNCA/α-synuclein**

Missense mutations and multiplications of the *SNCA* gene encoding the α-synuclein protein cause autosomal dominant PD (Appel-Cresswell et al., 2013; Chartier-Harlin et al., 2004; Kruger et al., 1998; Lesage et al., 2013; Polymeropoulos et al., 1997; Singleton et al., 2003; Zarranz et al., 2004). α-Synuclein fibrils are a major component of Lewy bodies and neurites, one of the pathological hallmarks of PD (Spillantini et al., 1997, 1998). GWAS have identified common variation at the *SNCA* locus as a risk factor for idiopathic PD (Nalls et al., 2014; Simon-Sanchez et al., 2009; Zhang et al., 2018), indicating that *SNCA* serves as a pleomorphic risk gene similar to *LRRK2*. α-Synuclein mutations were first linked to familial PD in 1997 (Polymeropoulos et al., 1997), making α-synuclein the most widely studied PD-related protein, yet the primary function of α-synuclein and how it contributes to the molecular pathogenesis of PD is still a matter of some debate. While the process of α-synuclein fibrillization and aggregation into Lewy pathology, and the propagation of α-synuclein species between cells, are intrinsically linked to disease (Burre et al., 2018), interference with these events in humans is eagerly awaited to confirm their importance as key pathogenic mechanisms (McFarthing and Simuni, 2019).

α-Synuclein is a small 140 residue protein lacking enzymatic domains that is enriched at presynaptic nerve terminals and interacts with proteins involved in the synaptic vesicle machinery suggesting a role in synaptic vesicle trafficking (Burre, 2015; Burre et al., 2018). Overexpression of α-synuclein reduces neurotransmitter release by inhibiting synaptic vesicle reclustering after endocytosis leading to a reduction in the synaptic vesicle recycling pool (Nemani et al., 2010), whereas α-synuclein knockout mice have limited alterations in synaptic transmission (Abeliovich et al., 2000). Multimeric forms of α-synuclein have been reported to assemble upon membranes where they can act as a SNARE complex chaperone at the presynaptic terminal (Burre et al., 2014). Familial PD-linked mutations in α-synuclein result in an increased propensity to aggregate into oligomeric and/or fibrillar species that promote neurotoxicity (Burre et al., 2018). Therefore, in terms of endosomal sorting, α-synuclein aggregation impairs its normal endocytic function at synapses while aggregates themselves can serve as endocytic cargo and in turn interfere with endosomal sorting (Burre, 2015). α-Synuclein species can propagate from neuron-to-neuron in a "prion-like" manner in experimental models, and this process is reported to involve both exocytosis and endocytosis (Ma et al., 2019).

Intriguingly, recent studies have linked α-synuclein-induced neuronal degeneration to both LRRK2 and VPS35. The genetic deletion or pharmacological kinase inhibition of LRRK2 was shown to rescue dopaminergic neuronal loss induced by the AAV-mediated delivery of human α-synuclein in rats (Daher et al., 2014, 2015). These studies suggest that endogenous LRRK2 is critical for mediating the pathogenic effects of α-synuclein in this model. It is less clear whether endogenous LRRK2 plays a similar role in α-synuclein pathology induced by preformed fibrils in rodents, although G2019S LRRK2 overexpression is reported to worsen α-synuclein pathology and propagation in these models (Bieri et al., 2019; Henderson et al., 2019; Volpicelli-Daley et al., 2016). VPS35 is also reported to genetically interact with α-synuclein where *VPS35* deletion in yeast and *C. elegans* models exacerbates the toxic effects of human α-synuclein (Dhungel et al., 2015). In addition, the viral-mediated overexpression of wild-type VPS35 was shown to attenuate hippocampal neuronal loss, gliosis and α-synuclein pathology that develop in a well-established transgenic mouse model expressing human wild-type α-synuclein (Dhungel et al., 2015). These observations suggest a potentially critical role for VPS35 and the retromer in mediating α-synuclein pathology and neurotoxicity, although it is not yet clear whether α-synuclein-induced toxicity manifests via a retromer deficiency similar to the effects of familial *LRRK2* mutations. It is of interest that *D620N VPS35* knockin mice do not develop α-synuclein pathology within their lifespan, but instead develop somatodendritic tau pathology (Chen et al., 2019), and it will be important to determine whether Lewy pathology is a feature of *VPS35*-linked PD brains (Wider et al., 2008). Furthermore, the *D620N VPS35* mutation is not sufficient to alter the lethal neurodegenerative phenotype that develops in human A53T α-synuclein transgenic mice (Chen et al., 2019). These studies suggest that the interaction between α-synuclein and VPS35 is likely complex and/or context-dependent and now awaits additional studies further clarifying the nature of their interaction. In addition to interactions within the endosomal pathway, α-synuclein is also reported to be selectively translocated into lysosomes for degradation by the chaperone-mediated autophagy (CMA) pathway (Cuervo et al., 2004). Aggregation-prone mutant α-synuclein can bind to the LAMP2a receptor on lysosomal membranes and serve to block uptake by CMA and inhibit degradation (Cuervo et al., 2004). LRRK2 is also reported to inhibit CMA by interfering with the LAMP2a translocation complex (Orenstein et al., 2013). Together, these studies suggest roles for α-synuclein in endolysosomal pathways, including interactions with PD-linked LRRK2 and VPS35, and such pathways are important for mediating the pathogenic effects of α-synuclein aggregates.

3 Emerging endosomal evidence in PD: Synaptic vesicle endocytosis and Rab GTPases

3.1 Synaptic vesicle endocytosis (SVE)

Synaptic vesicle endocytosis (SVE) plays an important role in the uptake of clathrin-coated vesicles at the presynaptic terminal for delivery to the endosomal system (Dittman and Ryan, 2009; Saheki and De Camilli, 2012). In neurons, this is critical for recycling neurotransmitter receptors and other proteins required for synaptic transmission, for example, the dopamine receptor. After SVE these proteins are sorted and trafficked by the endosomal system in a process identical to that in the soma (Fig. 1). First, endophilin is recruited to the plasma membrane by adaptor proteins to regulate membrane curvature and for recruitment of the membrane scission protein, dynamin. Once vesicle fission occurs, endophilin recruits the dual lipid phosphatase synaptojanin-1 to clathrin-coated vesicles (CCVs) to dephosphorylate membrane lipids (i.e., phosphoinositides), which serves to facilitate the release of adaptor proteins and allows auxilin/Hsc70 to bind to CCVs and mediate clathrin uncoating. Notably, a number of PD-linked mutations have recently been identified in genes involved in SVE and endosomal sorting, including auxilin (*DNAJC6*) and synaptojanin-1 (*SYNJ1*), whereas cyclin-G-associated kinase (*GAK*) and endophilinA1 (*EndoA1*) have been nominated as PD risk loci by GWAS (Table 1) (Nguyen et al., 2019).

3.1.1 Auxilin/DNAJC6

Mutations in the *DNAJC6* gene encoding auxilin were first discovered in two Palestinian brothers resulting in juvenile-onset, autosomal recessive PD (Edvardson et al., 2012). Additional mutations have since been identified in other PD families that cause mRNA splicing defects or exonic deletions that consequently decrease auxilin protein levels and act via a loss-of-function mechanism. While PD patients with *DNAJC6* mutations display cardinal motor symptoms, some also develop epilepsy and intellectual disability (Koroglu et al., 2013; Olgiati et al., 2016). The phenotypic variation indicates widespread neurological problems beyond the nigrostriatal pathway indicating that *DNAJC6* mutations can manifest atypical forms of PD.

Auxilin is part of the evolutionarily conserved DNAJ/Hsp40 family that stimulates the ATPase activity of Hsc70 to mediate the uncoating of CCVs following endocytosis (Saheki and De Camilli, 2012). Clathrin coat removal is required for vesicles to fuse and enter the endosomal pathway for cargo sorting. Auxilin knockout models display defects in clathrin-mediated endocytosis, including the accumulation of CCVs and empty clathrin cages, and exhibit dopaminergic neurodegeneration (Hirst et al., 2008; Song et al., 2017; Yim et al., 2010). Auxilin reduction in *Drosophila* results in climbing deficits, reduced lifespan, and dopaminergic neurodegeneration that phenocopies, and is exacerbated by, α-synuclein overexpression (Song et al., 2017). Interestingly, auxilin knockout mice display a threefold increase in the auxilin homolog GAK, that also serves as a co-chaperone for Hsc70 in clathrin uncoating (Yim et al., 2010). Auxilin is further tied to PD pathogenesis via its interaction with LRRK2. Auxilin can be phosphorylated by LRRK2 at Ser627 that results in differential binding to clathrin and impaired SVE (Nguyen and Krainc, 2018). This results in the accumulation of oxidized dopamine in iPSC-derived dopaminergic neurons from PD subjects harboring *LRRK2 R1441C, R1441G* or *G2019S* mutations (Nguyen and Krainc, 2018). *Auxilin R857G* knockin mice have recently been developed as a model of PD that develop motor deficits, impaired synaptic recycling, dystrophic Golgi morphology and lipid accumulation (Roosen et al., 2019). Similar to the effects of LRRK2 phosphorylation, PD-linked mutations in auxilin selectively impair the interaction with clathrin but not clathrin adaptor proteins (Roosen et al., 2019).

3.1.2 Cyclin-G-associated kinase (GAK)

GAK is an auxilin homolog and performs similar functions in SVE, including recruitment of Hsc70 to mediate clathrin uncoating. In contrast to auxilin, which is primarily expressed in neurons, GAK is ubiquitously expressed and its deletion selectively in the brain or germline is lethal (Lee et al., 2008). In auxilin knockout mice, GAK can compensate for auxilin in some capacities indicating some functional redundancy (Yim et al., 2010), but auxilin does not appear to be sufficient to rescue GAK knockout in vivo. *GAK* has gained attention in PD since common variation at the *GAK/TMEM175* locus has been linked to increased PD risk by GWAS (Nalls et al., 2014, 2019; Yu et al., 2015). GAK can form a functional complex with LRRK2 and Rab29 in cells and mouse brain, that participates in the clearance of Golgi-derived vesicles (Beilina et al., 2014). Carriers of PD risk variants at the *GAK* locus are associated with the increased expression of α-synuclein in the frontal cortex (Dumitriu et al., 2011). Although *GAK* represents an interesting and compelling risk gene for PD directly involved in the endosomal system, it is not clear whether PD risk variants at this locus impact the expression of *GAK*, *TMEM175* or both genes (Nalls et al., 2014, 2019). *TMEM175* encodes a lysosomal potassium pump that would also fit with pathogenic mechanisms implicated in PD (Jinn et al., 2019; Krohn et al., 2020). Future studies of risk-associated variants will attempt to understand how the *GAK/TMEM175* locus contributes to PD.

3.1.3 Synaptojanin-1 (SYNJ1)

Similar to auxilin, *SYNJ1* mutations cause early-onset, autosomal recessive Parkinsonism with some cases also displaying seizures or other atypical symptoms (Chen et al., 2015; Kirola et al., 2016; Krebs et al., 2013; Quadri et al., 2013). Synaptojanin-1 is a lipid phosphatase that dephosphorylates membrane lipids (i.e., phosphatidylinositol-4,5-bisphosphate) to initiate the release of adaptor proteins from CCVs (Drouet and Lesage, 2014). Once adaptor proteins are removed from CCVs, auxilin/Hsc70 can bind and mediate clathrin uncoating. Mutations in *SYNJ1* (i.e., R258Q, R459P) are rare with those located in the Sac 1 domain leading to decreased phosphatase activity (Drouet and Lesage, 2014). Although synaptojanin-1 and auxilin have different functions, their sequential roles in SVE may explain why knockout of either gene causes similar phenotypes. Both *SYNJ1* and *auxilin* knockout mice similarly display neurological deficits, motor abnormalities and accumulate CCVs at the synaptic terminal (Cremona et al., 1999; Kim et al., 2002; Yim et al., 2010). *R258Q SYNJ1* knockin mice also display SVE deficits, the accumulation of clathrin-coated intermediates, a compensatory elevation of auxilin, parkin and endoA1 proteins, and dystrophic dopaminergic axonal terminals (Cao et al., 2017). Alternatively, *Drosophila* models harboring an *R258Q SYNJ1* knockin display normal synaptic vesicle cycling but instead exhibit deficits in synaptic autophagosome maturation and modest dopaminergic neuronal loss (Vanhauwaert et al., 2017), indicating that *SYNJ1* mutations may impact both endocytosis and autophagy pathways. Synaptojanin-1 was first identified by phospho-proteome profiling as a LRRK2 substrate in *Drosophila* brain and human synaptojanin-1 can be phosphorylated by LRRK2 at Thr1173 in vitro (Islam et al., 2016). Functionally, synaptojanin-1 phosphorylation by LRRK2 at a second site (Thr1205) was reported to disrupt its interaction with endoA1 that is required for SVE (Pan et al., 2017). Furthermore, *SYNJ1* haploinsufficiency in G2019S LRRK2 transgenic mice impaired sustained exocytosis in midbrain neurons and caused mild alterations in motor function (Pan et al., 2017). Therefore, similar to auxilin, synaptojanin-1 interactions and function can also be modulated by LRRK2. Auxilin and SYNJ1 further highlight the importance of SVE and the endosomal pathway in the development of early-onset PD, whereas the interaction with LRRK2 also links both SVE proteins to typical late-onset PD.

3.1.4 EndophilinA1 (EndoA1)

Common variation at the *SH3GL2* locus that encodes endoA1 has been identified as a risk factor for idiopathic PD by GWAS (Chang et al., 2017). Prior to this, endoA1 has long been of interest in PD due to its proximity to other PD-linked proteins including synaptojanin-1, LRRK2, and parkin (Islam et al., 2016; Matta et al., 2012; Trempe et al., 2009). In SVE, endoA1 plays an intermediary role between initial vesicle invagination and dynamin-dependent scission, and the recruitment of synaptojanin-1 and auxilin to initiate CCV uncoating (Fig. 1). Similar to auxilin and synaptojanin-1, endoA1 can also be phosphorylated by LRRK2 at Ser75, which alters its association with and the tubulation of vesicle membranes that result in SVE deficits in

Drosophila (Ambroso et al., 2014; Matta et al., 2012). The connection of endoA1 to PD further supports the concept that multiple steps in the SVE pathway are important to the pathophysiological mechanisms underlying PD.

3.2 Rab GTPases

The role of Rab GTPases in PD has gained recent interest following their identification as LRRK2 kinase substrates, human genetic evidence linking certain Rabs to PD, and due to their roles in the endolysosomal pathway (Bonet-Ponce and Cookson, 2019). Mutations in *RAB32* and *RAB39B* have recently been identified to cause familial PD whereas common variation at or near the *RAB29* (*PARK16*) locus may contribute to the risk of idiopathic PD. There is also emerging evidence that these Rabs may functionally interact with other PD-linked gene products, namely LRRK2, VPS35 or α-synuclein. Herein, we briefly summarize evidence for a role of Rab GTPases in PD pathophysiology, but for a more thorough review please refer to (Bonet-Ponce and Cookson, 2019; Gao et al., 2018).

3.2.1 Rab39B

The function of Rab39B is largely uncharacterized, although it is enriched in the brain and highly expressed in neurons, and localizes to the secretory network including the endoplasmic reticulum and *cis*-Golgi interface, the TGN and recycling endosomes (Gambarte Tudela et al., 2019; Giannandrea et al., 2010). Mutations in *RAB39B* were first identified to cause early-onset, X-linked dominant PD in Australian and Wisconsin kindreds due to a ∼45 kb deletion encompassing the entire *RAB39B* gene or a T168K mutation, respectively (Wilson et al., 2014). Although affected family members exhibit typical PD symptoms and Lewy pathology, they also develop intellectual disability. These findings were later confirmed through the identification of a nonsense mutation (Trp186Stop) in *RAB39B* in a French subject, and a missense mutation (G192R) in seven affected members of a US family with PD (Lesage et al., 2015; Mata et al., 2015). While most mutations decrease Rab39B levels due to reduced protein stability, gene deletions or truncations (Gao et al., 2020), the G192R mutation induces Rab39B mislocalization in cells (Mata et al., 2015). PD-linked mutations in Rab39B therefore manifest disease through a loss-of-function mechanism. *Rab39B* gene silencing in cultured primary neurons decreased neurite branching and the number of growth cones (Giannandrea et al., 2010), and surprisingly reduced the steady-state levels of α-synuclein and the density of α-synuclein-positive puncta in dendritic processes (Wilson et al., 2014). These findings suggest that Rab39B depletion can dysregulate α-synuclein homeostasis, and this effect now awaits confirmation in rodent models with Rab39B deletion.

3.2.2 Rab32

A missense mutation (S71R) in *RAB32* has been suggested to cause familial PD (Gustavsson et al., 2017). Rab32 is also connected to PD via its direct interaction with the N-terminal armadillo domain of LRRK2, where they colocalize upon recycling endosomes and transport vesicles (McGrath et al., 2019; Waschbusch et al., 2014). The putative PD-linked S71R mutation in Rab32 disrupts a highly conserved residue within the catalytic Switch II region, that in related Rabs is known to serve as a robust site of phosphorylation by LRRK2 (Steger et al., 2016). Rab32 was recently shown to interact with SNX6, one of the components of the retromer-associated SNX dimer, to influence retromer-dependent sorting of CIMPR to the TGN (Waschbusch et al., 2019). Rab32 therefore provides an interesting new link to PD that may cooperate with LRRK2 and the retromer to regulate endosomal sorting.

3.2.3 Rab29 (Rab7L1)

Common variation at the *PARK16* locus is associated with an increased risk for idiopathic PD (Chang et al., 2017; Nalls et al., 2014; Satake et al., 2009; Simon-Sanchez et al., 2009). Although five genes are nominated within the *PARK16* locus, *RAB29* is of special interest due to its genetic and functional interaction with LRRK2 and its known role in endosomal sorting (Beilina et al., 2014; Liu et al., 2018; MacLeod et al., 2013; Purlyte et al., 2018). Genetic studies initially suggested that PD risk variants at the *PARK16* and *LRRK2* loci are highly interrelated in terms of their transcriptional impact in brain and risk associations (MacLeod et al., 2013). *RAB29* was nominated as the critical risk gene at the *PARK16* locus with PD risk variants suggested to alter mRNA splicing and lower Rab29 expression (MacLeod et al., 2013). Accordingly, Rab29 silencing in cultured primary neurons was shown to phenocopy the effects of G2019S LRRK2 on inhibition of neurite outgrowth, the accumulation of enlarged lysosomes, and altered sorting of CIMPR (MacLeod et al., 2013; Wang et al., 2014). Rab29 overexpression could oppositely rescue these G2019S LRRK2-induced phenotypes in neurons (MacLeod et al., 2013). Similarly, in *Drosophila*, G2019S LRRK2 overexpression and silencing of the Rab29 ortholog *Lightoid* produce similar phenotypes, including reduced survival and dopaminergic neuronal loss, whereas Rab29 overexpression can rescue the effects of G2019S LRRK2 in flies (MacLeod et al., 2013). Lastly, mice lacking *LRRK2*, *RAB29* or both genes exhibit similar phenotypes including enlarged and pigmented kidneys and the accumulation of lysosome-related pathology in kidneys and lungs (Kuwahara et al., 2016). These phenotypes were slightly less severe in *RAB29* knockout animals but were not additive in the *LRRK2/RAB29* double knockout mice, suggesting that LRRK2 and Rab29 may function in the same pathway (Kuwahara et al., 2016).

In agreement with a common LRRK2/Rab29 pathway, Rab29 is now known to interact with and recruit LRRK2 to the TGN where it becomes phosphorylated, a process that is enhanced by PD-linked *LRRK2* mutations and requires membrane- and GTP-bound Rab29 (Beilina et al., 2014; Liu et al., 2018; MacLeod et al., 2013). The Rab29 interaction with LRRK2 stimulates its kinase activity including LRRK2 autophosphorylation and Rab phosphorylation (i.e., Rab8A, Rab10) (Liu et al., 2018; Purlyte et al., 2018). Rab29 therefore serves as a master regulator of LRRK2 activity and localization in cells (Purlyte et al., 2018). Rab29 and LRRK2 function together in the endolysosomal pathway that may be important for understanding the pathophysiological mechanisms driving PD.

4 Conclusion and future directions

Substantial human genetic and experimental evidence now points to endosomal sorting pathways as important to the pathophysiology of PD (Abeliovich and Gitler, 2016; Bandres-Ciga et al., 2019; Bonet-Ponce and Cookson, 2019; Gao et al., 2018; Shahmoradian et al., 2019). As discussed, LRRK2 may function as a nexus within these pathways, where it can influence many distinct functions of endosomal sorting via protein interactions with the key endosomal machinery and phosphorylation of Rabs that decorate these intracellular vesicles and membranes (Islam et al., 2016; Matta et al., 2012; Nguyen and Krainc, 2018; Steger et al., 2016, 2017). LRRK2 may also regulate α-synuclein-induced neurotoxicity (Bieri et al., 2019; Daher et al., 2014; Daniel et al., 2015; Volpicelli-Daley et al., 2016), potentially via the convergence of these proteins upon lysosomal function (Cuervo et al., 2004; Orenstein et al., 2013), and may regulate VPS35 levels and retromer function (MacLeod et al., 2013). LRRK2 activity can in turn be regulated by these sorting factors, with Rab29 serving as a master regulator to recruit and activate LRRK2 at the TGN and potentially lysosomes (Beilina et al., 2014; Eguchi et al., 2018; Liu et al., 2018; Purlyte et al., 2018), and PD-linked mutations in VPS35 leading to LRRK2 hyperactivation (Mir et al., 2018). It is unclear at this juncture how these functional interactions will translate into identifying new therapeutic targets and strategies, since endosomal sorting pathways are dynamic, multifaceted and increasingly complex. It may not be sufficient to simply improve bulk output, such as by improving lysosomal function, as many cargo require sorting and recycling from the endosome to other compartments to prevent their inappropriate lysosomal degradation. Such a strategy may be appropriate for the removal of neurotoxic proteins, such as oligomeric or aggregated α-synuclein, as suggested by studies that improve CMA by LAMP2a overexpression (Xilouri et al., 2013). Restoring or improving the function of single endosomal protein components, such as auxilin, synaptojanin-1 or endoA1, may not lead to general improvements in endosomal function.

What is perhaps needed is a key focus on the consequences of disease-causing mutations in PD-linked gene products that are likely to impact endosomal function more broadly. For example, in the case of LRRK2, this could generally involve inhibiting or normalizing overactive kinase activity, or interfering with key downstream kinase substrates (i.e., Rab8A or Rab10) or master regulators of LRRK2 activity (i.e., Rab29) (Liu et al., 2018; Purlyte et al., 2018; Steger et al., 2016, 2017). Additional general strategies might include inhibition of LRRK2 activity via modulation of GTPase activity, disrupting dimerization or downregulation by antisense oligonucleotides (Nguyen and Moore, 2017). Likewise, approaches for normalizing VPS35 and retromer function might include retromer-stabilizing chemical chaperones (i.e., R33/R55), improving retromer membrane recruitment (by inhibition of TBC1D5 that leads to enhanced GTP- and membrane-bound Rab7A) or by restoration of specific retromer cargo depletion or accumulation (Follett et al., 2016; Mecozzi et al., 2014; Seaman et al., 2018; Zavodszky et al., 2014). Some of these retromer-targeted approaches may also serve to attenuate LRRK2 kinase activity (Mir et al., 2018). Much will now depend on establishing whether PD-linked VPS35 mutations act via a loss- or gain-of-function mechanism to induce neurodegeneration in PD (Chen et al., 2019; Williams et al., 2017). An improved molecular and cellular understanding of how PD-linked gene products regulate endosomal sorting pathways in different brain and peripheral cells, and the functional interactions or points of convergence between these proteins, will prove pivotal in the design of effective therapeutics and for biomarker development for PD. Endosomal sorting pathways are likely to provide a critical point of convergence for understanding the pathophysiological mechanisms driving familial and sporadic PD.

Acknowledgments

The authors work is supported by funding from the National Institutes of Health (R01 NS091719, R01 NS105432, R21 AG058241), Michael J. Fox Foundation for Parkinson's Research, and the Van Andel Institute.

References

Abeliovich, A., Gitler, A.D., 2016. Defects in trafficking bridge Parkinson's disease pathology and genetics. Nature 539, 207–216.

Abeliovich, A., Schmitz, Y., Farinas, I., Choi-Lundberg, D., Ho, W.H., Castillo, P.E., Shinsky, N., Verdugo, J.M., Armanini, M., Ryan, A., Hynes, M., Phillips, H., Sulzer, D., Rosenthal, A., 2000. Mice lacking alpha-synuclein display functional deficits in the nigrostriatal dopamine system. Neuron 25, 239–252.

Alegre-Abarrategui, J., Christian, H., Lufino, M.M., Mutihac, R., Venda, L.L., Ansorge, O., Wade-Martins, R., 2009. LRRK2 regulates autophagic activity and localizes to specific membrane microdomains in a novel human genomic reporter cellular model. Hum. Mol. Genet. 18, 4022–4034.

Ambroso, M.R., Hegde, B.G., Langen, R., 2014. Endophilin A1 induces different membrane shapes using a conformational switch that is regulated by phosphorylation. Proc. Natl. Acad. Sci. U. S. A. 111, 6982–6987.

Andersen, M.A., Wegener, K.M., Larsen, S., Badolo, L., Smith, G.P., Jeggo, R., Jensen, P.H., Sotty, F., Christensen, K.V., Thougaard, A., 2018. PFE-360-induced LRRK2 inhibition induces reversible, non-adverse renal changes in rats. Toxicology 395, 15–22.

Andres-Mateos, E., Mejias, R., Sasaki, M., Li, X., Lin, B.M., Biskup, S., Zhang, L., Banerjee, R., Thomas, B., Yang, L., Liu, G., Beal, M.F., Huso, D.L., Dawson, T.M., Dawson, V.L., 2009. Unexpected lack of hypersensitivity in LRRK2 knock-out mice to MPTP (1-methyl-4-phenyl-1,2,3,6-tetrahydropyridine). J. Neurosci. 29, 15846–15850.

Appel, J.R., Ye, S., Tang, F., Sun, D., Zhang, H., Mei, L., Xiong, W.C., 2018. Increased microglial activity, impaired adult hippocampal neurogenesis, and depressive-like behavior in microglial VPS35-depleted mice. J. Neurosci. 38, 5949–5968.

Appel-Cresswell, S., Vilarino-Guell, C., Encarnacion, M., Sherman, H., Yu, I., Shah, B., Weir, D., Thompson, C., Szu-Tu, C., Trinh, J., Aasly, J.O., Rajput, A., Rajput, A.H., Jon Stoessl, A., Farrer, M.J., 2013. Alpha-synuclein p.H50Q, a novel pathogenic mutation for Parkinson's disease. Mov. Disord. 28, 811–813.

Arighi, C.N., Hartnell, L.M., Aguilar, R.C., Haft, C.R., Bonifacino, J.S., 2004. Role of the mammalian retromer in sorting of the cation-independent mannose 6-phosphate receptor. J. Cell Biol. 165, 123–133.

Bandres-Ciga, S., Saez-Atienzar, S., Bonet-Ponce, L., Billingsley, K., Vitale, D., Blauwendraat, C., Gibbs, J.R., Pihlstrom, L., Gan-Or, Z., Cookson, M.R., Nalls, M.A., Singleton, A.B., 2019. The endocytic membrane trafficking pathway plays a major role in the risk of Parkinson's disease. Mov. Disord. 34, 460–468.

Baptista, M.A., Dave, K.D., Frasier, M.A., Sherer, T.B., Greeley, M., Beck, M.J., Varsho, J.S., Parker, G.A., Moore, C., Churchill, M.J., Meshul, C.K., Fiske, B.K., 2013. Loss of leucine-rich repeat kinase 2 (LRRK2) in rats leads to progressive abnormal phenotypes in peripheral organs. PLoS One 8, e80705.

Beilina, A., Rudenko, I.N., Kaganovich, A., Civiero, L., Chau, H., Kalia, S.K., Kalia, L.V., Lobbestael, E., Chia, R., Ndukwe, K., Ding, J., Nalls, M.A., International Parkinson'S Disease Genomics, C, North American Brain Expression, C, Olszewski, M., Hauser, D.N., Kumaran, R., Lozano, A.M., Baekelandt, V., Greene, L.E., Taymans, J.M., Greggio, E., Cookson, M.R., 2014. Unbiased screen for interactors of leucine-rich repeat kinase 2 supports a common pathway for sporadic and familial Parkinson disease. Proc. Natl. Acad. Sci. U. S. A. 111, 2626–2631.

Berger, Z., Smith, K.A., Lavoie, M.J., 2010. Membrane localization of LRRK2 is associated with increased formation of the highly active LRRK2 dimer and changes in its phosphorylation. Biochemistry 49, 5511–5523.

Berman, D.E., Ringe, D., Petsko, G.A., Small, S.A., 2015. The use of pharmacological retromer chaperones in Alzheimer's disease and other endosomal-related disorders. Neurotherapeutics 12, 12–18.

Bieri, G., Brahic, M., Bousset, L., Couthouis, J., Kramer, N.J., Ma, R., Nakayama, L., Monbureau, M., Defensor, E., Schule, B., Shamloo, M., Melki, R., Gitler, A.D., 2019. LRRK2 modifies alpha-syn pathology and spread in mouse models and human neurons. Acta Neuropathol. 137, 961–980.

Biskup, S., West, A.B., 2009. Zeroing in on LRRK2-linked pathogenic mechanisms in Parkinson's disease. Biochim. Biophys. Acta 1792, 625–633.

Biskup, S., Moore, D.J., Celsi, F., Higashi, S., West, A.B., Andrabi, S.A., Kurkinen, K., Yu, S.W., Savitt, J.M., Waldvogel, H.J., Faull, R.L., Emson, P.C., Torp, R., Ottersen, O.P., Dawson, T.M., Dawson, V.L., 2006. Localization of LRRK2 to membranous and vesicular structures in mammalian brain. Ann. Neurol. 60, 557–569.

Blauwendraat, C., Reed, X., Kia, D.A., Gan-Or, Z., Lesage, S., Pihlstrom, L., Guerreiro, R., Gibbs, J.R., Sabir, M., Ahmed, S., Ding, J., Alcalay, R.N., Hassin-Baer, S., Pittman, A.M., Brooks, J., Edsall, C., Hernandez, D.G., Chung, S.J., Goldwurm, S., Toft, M., Schulte, C., Bras, J., Wood, N.W., Brice, A., Morris, H.R., Scholz, S.W., Nalls, M.A., Singleton, A.B., Cookson, M.R., 2018. Frequency of loss of function variants in LRRK2 in Parkinson disease. JAMA Neurol. 75, 1416–1422.

Blauwendraat, C., Nalls, M.A., Singleton, A.B., 2020. The genetic architecture of Parkinson's disease. Lancet Neurol. 19, 170–178.

Boddu, R., Hull, T.D., Bolisetty, S., Hu, X., Moehle, M.S., Daher, J.P., Kamal, A.I., Joseph, R., George, J.F., Agarwal, A., Curtis, L.M., West, A.B., 2015. Leucine-rich repeat kinase 2 deficiency is protective in rhabdomyolysis-induced kidney injury. Hum. Mol. Genet. 24, 4078–4093.

Bonet-Ponce, L., Cookson, M.R., 2019. The role of Rab GTPases in the pathobiology of Parkinson' disease. Curr. Opin. Cell Biol. 59, 73–80.

Burre, J., 2015. The synaptic function of alpha-synuclein. J. Parkinsons Dis. 5, 699–713.

Burre, J., Sharma, M., Sudhof, T.C., 2014. alpha-Synuclein assembles into higher-order multi-mers upon membrane binding to promote SNARE complex formation. Proc. Natl. Acad. Sci. U. S. A. 111, E4274–E4283.

Burre, J., Sharma, M., Sudhof, T.C., 2018. Cell biology and pathophysiology of alpha-synuclein. Cold Spring Harb. Perspect. Med. 8, a024091

Cao, M., Wu, Y., Ashrafi, G., McCartney, A.J., Wheeler, H., Bushong, E.A., Boassa, D., Ellisman, M.H., Ryan, T.A., De Camilli, P., 2017. Parkinson sac domain mutation in synaptojanin 1 impairs clathrin uncoating at synapses and triggers dystrophic changes in dopaminergic axons. Neuron 93, 882–896.e5.

Cataldi, S., Follett, J., Fox, J.D., Tatarnikov, I., Kadgien, C., Gustavsson, E.K., Khinda, J., Milnerwood, A.J., Farrer, M.J., 2018. Altered dopamine release and monoamine trans-porters in Vps35 p.D620N knock-in mice. NPJ Parkinsons Dis. 4, 27.

Chang, D., Nalls, M.A., Hallgrimsdottir, I.B., Hunkapiller, J., Van Der Brug, M., Cai, F., Kerchner, G.A., Ayalon, G., Bingol, B., Sheng, M., Hinds, D., Behrens, T.W., Singleton, A.B., Bhangale, T.R., Graham, R.R., 2017. A meta-analysis of genome-wide asso-ciation studies identifies 17 new Parkinson's disease risk loci. Nat. Genet. 49, 1511–1516.

Chartier-Harlin, M.C., Kachergus, J., Roumier, C., Mouroux, V., Douay, X., Lincoln, S., Levecque, C., Larvor, L., Andrieux, J., Hulihan, M., Waucquier, N., Defebvre, L., Amouyel, P., Farrer, M., Destee, A., 2004. Alpha-synuclein locus duplication as a cause of familial Parkinson's disease. Lancet 364, 1167–1169.

Chen, Y.P., Song, W., Huang, R., Chen, K., Zhao, B., Li, J., Yang, Y., Shang, H.F., 2013. GAK rs1564282 and DGKQ rs11248060 increase the risk for Parkinson's disease in a Chinese population. J. Clin. Neurosci. 20, 880–883.

Chen, K.H., Wu, R.M., Lin, H.I., Tai, C.H., Lin, C.H., 2015. Mutational analysis of SYNJ1 gene (PARK20) in Parkinson's disease in a Taiwanese population. Neurobiol. Aging 36, 2905.e7-8.

Chen, X., Kordich, J.K., Williams, E.T., Levine, N., Cole-Strauss, A., Marshall, L., Labrie, V., Ma, J., Lipton, J.W., Moore, D.J., 2019. Parkinson's disease-linked D620N VPS35 knockin mice manifest tau neuropathology and dopaminergic neurodegeneration. Proc. Natl. Acad. Sci. U. S. A. 116, 5765–5774.

Choy, R.W., Park, M., Temkin, P., Herring, B.E., Marley, A., Nicoll, R.A., Von Zastrow, M., 2014. Retromer mediates a discrete route of local membrane delivery to dendrites. Neuron 82, 55–62.

Christ, L., Raiborg, C., Wenzel, E.M., Campsteijn, C., Stenmark, H., 2017. Cellular functions and molecular mechanisms of the ESCRT membrane-scission machinery. Trends Bio-chem. Sci. 42, 42–56.

Cremona, O., Di Paolo, G., Wenk, M.R., Luthi, A., Kim, W.T., Takei, K., Daniell, L., Nemoto, Y., Shears, S.B., Flavell, R.A., McCormick, D.A., De Camilli, P., 1999. Essential role of phosphoinositide metabolism in synaptic vesicle recycling. Cell 99, 179–188.

Cuervo, A.M., Stefanis, L., Fredenburg, R., Lansbury, P.T., Sulzer, D., 2004. Impaired degradation of mutant alpha-synuclein by chaperone-mediated autophagy. Science 305, 1292–1295.

Cullen, P.J., Korswagen, H.C., 2011. Sorting nexins provide diversity for retromer-dependent trafficking events. Nat. Cell Biol. 14, 29–37.

Daher, J.P., Volpicelli-Daley, L.A., Blackburn, J.P., Moehle, M.S., West, A.B., 2014. Abrogation of alpha-synuclein-mediated dopaminergic neurodegeneration in LRRK2-deficient rats. Proc. Natl. Acad. Sci. U. S. A. 111, 9289–9294.

Daher, J.P., Abdelmotilib, H.A., Hu, X., Volpicelli-Daley, L.A., Moehle, M.S., Fraser, K.B., Needle, E., Chen, Y., Steyn, S.J., Galatsis, P., Hirst, W.D., West, A.B., 2015. Leucine-rich repeat kinase 2 (LRRK2) pharmacological inhibition abates alpha-synuclein gene-induced neurodegeneration. J. Biol. Chem. 290, 19433–19444.

Daniel, G., Musso, A., Tsika, E., Fiser, A., Glauser, L., Pletnikova, O., Schneider, B.L., Moore, D.J., 2015. alpha-Synuclein-induced dopaminergic neurodegeneration in a rat model of Parkinson's disease occurs independent of ATP13A2 (PARK9). Neurobiol. Dis. 73, 229–243.

Deng, H., Xu, H., Deng, X., Song, Z., Zheng, W., Gao, K., Fan, X., Tang, J., 2012. VPS35 mutation in Chinese Han patients with late-onset Parkinson's disease. Eur. J. Neurol. 19, e96–e97.

Deng, H., Gao, K., Jankovic, J., 2013. The VPS35 gene and Parkinson's disease. Mov. Disord. 28, 569–575.

Dhekne, H.S., Yanatori, I., Gomez, R.C., Tonelli, F., Diez, F., Schule, B., Steger, M., Alessi, D.R., Pfeffer, S.R., 2018. A pathway for Parkinson's disease LRRK2 kinase to block primary cilia and Sonic hedgehog signaling in the brain. Elife, 7, e40202

Dhungel, N., Eleuteri, S., Li, L.B., Kramer, N.J., Chartron, J.W., Spencer, B., Kosberg, K., Fields, J.A., Stafa, K., Adame, A., Lashuel, H., Frydman, J., Shen, K., Masliah, E., Gitler, A.D., 2015. Parkinson's disease genes VPS35 and EIF4G1 interact genetically and converge on alpha-synuclein. Neuron 85, 76–87.

Dittman, J., Ryan, T.A., 2009. Molecular circuitry of endocytosis at nerve terminals. Annu. Rev. Cell Dev. Biol. 25, 133–160.

Dorsey, E.R., Bloem, B.R., 2018. The parkinson pandemic-A call to action. JAMA Neurol. 75, 9–10.

Dorsey, E.R., Sherer, T., Okun, M.S., Bloem, B.R., 2018. The emerging evidence of the parkinson pandemic. J. Parkinsons Dis. 8, S3–s8.

Drouet, V., Lesage, S., 2014. Synaptojanin 1 mutation in Parkinson's disease brings further insight into the neuropathological mechanisms. Biomed. Res. Int. 2014, 289728.

Dumitriu, A., Pacheco, C.D., Wilk, J.B., Strathearn, K.E., Latourelle, J.C., Goldwurm, S., Pezzoli, G., Rochet, J.C., Lindquist, S., Myers, R.H., 2011. Cyclin-G-associated kinase modifies alpha-synuclein expression levels and toxicity in Parkinson's disease: results from the GenePD Study. Hum. Mol. Genet. 20, 1478–1487.

Dusonchet, J., Kochubey, O., Stafa, K., Young, S.M., Jr., Zufferey, R., Moore, D.J., Schneider, B.L., Aebischer, P., 2011. A rat model of progressive nigral neurodegeneration induced by the Parkinson's disease-associated G2019S mutation in LRRK2. J. Neurosci. 31, 907–912.

Edvardson, S., Cinnamon, Y., Ta-Shma, A., Shaag, A., Yim, Y.I., Zenvirt, S., Jalas, C., Lesage, S., Brice, A., Taraboulos, A., Kaestner, K.H., Greene, L.E., Elpeleg, O., 2012. A deleterious mutation in DNAJC6 encoding the neuronal-specific clathrin-uncoating co-chaperone auxilin, is associated with juvenile parkinsonism. PLoS One 7, e36458.

Eguchi, T., Kuwahara, T., Sakurai, M., Komori, T., Fujimoto, T., Ito, G., Yoshimura, S.I., Harada, A., Fukuda, M., Koike, M., Iwatsubo, T., 2018. LRRK2 and its substrate Rab GTPases are sequentially targeted onto stressed lysosomes and maintain their homeostasis. Proc. Natl. Acad. Sci. U. S. A. 115, E9115–e9124.

Follett, J., Bugarcic, A., Yang, Z., Ariotti, N., Norwood, S.J., Collins, B.M., Parton, R.G., Teasdale, R.D., 2016. Parkinson disease-linked Vps35 R524W mutation impairs the endosomal association of retromer and induces alpha-synuclein aggregation. J. Biol. Chem. 291, 18283–18298.

Fuji, R.N., Flagella, M., Baca, M., Baptista, M.A., Brodbeck, J., Chan, B.K., Fiske, B.K., Honigberg, L., Jubb, A.M., Katavolos, P., Lee, D.W., Lewin-Koh, S.C., Lin, T., Liu, X., Liu, S., Lyssikatos, J.P., O'mahony, J., Reichelt, M., Roose-Girma, M., Sheng, Z., Sherer, T., Smith, A., Solon, M., Sweeney, Z.K., Tarrant, J., Urkowitz, A., Warming, S., Yaylaoglu, M., Zhang, S., Zhu, H., Estrada, A.A., Watts, R.J., 2015. Effect of selective LRRK2 kinase inhibition on nonhuman primate lung. Sci. Transl. Med. 7, 273ra15.

Gambarte Tudela, J., Buonfigli, J., Lujan, A., Alonso Bivou, M., Cebrian, I., Capmany, A., Damiani, M.T., 2019. Rab39a and Rab39b display different intracellular distribution and function in sphingolipids and phospholipids transport. Int. J. Mol. Sci. 20, E1688.

Gao, Y., Wilson, G.R., Stephenson, S.E.M., Bozaoglu, K., Farrer, M.J., Lockhart, P.J., 2018. The emerging role of Rab GTPases in the pathogenesis of Parkinson's disease. Mov. Disord. 33, 196–207.

Gao, Y., Martinez-Cerdeno, V., Hogan, K.J., Mclean, C.A., Lockhart, P.J., 2020. Clinical and neuropathological features associated with loss of RAB39B. Mov. Disord. https://doiorg/10.1002/mds.27951.

GBD 2016 Neurology Collaborators. 2019. Global, regional, and national burden of neurological disorders, 1990-2016: a systematic analysis for the Global Burden of Disease Study 2016. Lancet Neurol. 18, 459–480.

GBD 2016 Parkinson's Disease Collaborators. 2018. Global, regional, and national burden of Parkinson's disease, 1990–2016: a systematic analysis for the Global Burden of Disease Study 2016. Lancet Neurol. 17, 939–953.

Giannandrea, M., Bianchi, V., Mignogna, M.L., Sirri, A., Carrabino, S., D'elia, E., Vecellio, M., Russo, S., Cogliati, F., Larizza, L., Ropers, H.H., Tzschach, A., Kalscheuer, V., Oehl-Jaschkowitz, B., Skinner, C., Schwartz, C.E., Gecz, J., Van Esch, H., Raynaud, M., Chelly, J., De Brouwer, A.P., Toniolo, D., D'adamo, P., 2010. Mutations in the small GTPase gene RAB39B are responsible for X-linked mental retardation associated with autism, epilepsy, and macrocephaly. Am. J. Hum. Genet. 86, 185–195.

Gillardon, F., 2009. Leucine-rich repeat kinase 2 phosphorylates brain tubulin-beta isoforms and modulates microtubule stability–a point of convergence in parkinsonian neurodegeneration? J. Neurochem. 110, 1514–1522.

Gomez, R.C., Wawro, P., Lis, P., Alessi, D.R., Pfeffer, S.R., 2019. Membrane association but not identity is required for LRRK2 activation and phosphorylation of Rab GTPases. J. Cell Biol. 218, 4157–4170.

Gomez-Suaga, P., Rivero-Rios, P., Fdez, E., Blanca Ramirez, M., Ferrer, I., Aiastui, A., Lopez De Munain, A., Hilfiker, S., 2014. LRRK2 delays degradative receptor trafficking by impeding late endosomal budding through decreasing Rab7 activity. Hum. Mol. Genet. 23, 6779–6796.

Greggio, E., Jain, S., Kingsbury, A., Bandopadhyay, R., Lewis, P., Kaganovich, A., Van Der Brug, M.P., Beilina, A., Blackinton, J., Thomas, K.J., Ahmad, R., Miller, D.W., Kesavapany, S., Singleton, A., Lees, A., Harvey, R.J., Harvey, K., Cookson, M.R., 2006. Kinase activity is required for the toxic effects of mutant LRRK2/dardarin. Neurobiol. Dis. 23, 329–341.

Gustavsson, E., Follett, J., Ramirez, C.M., Trinh, J., Fox, J., Aasly, E., Pourcher, E., Hentati, F., Farrer, M., 2017. RAB32 as a cause for familial Parkinson's disease.

In: International Congress of Parkinson's Disease and Movement Disorders, British Columbia, Canada.

Healy, D.G., Falchi, M., O'sullivan, S.S., Bonifati, V., Durr, A., Bressman, S., Brice, A., Aasly, J., Zabetian, C.P., Goldwurm, S., Ferreira, J.J., Tolosa, E., Kay, D.M., Klein, C., Williams, D.R., Marras, C., Lang, A.E., Wszolek, Z.K., Berciano, J., Schapira, A.H., Lynch, T., Bhatia, K.P., Gasser, T., Lees, A.J., Wood, N.W., International, L. C, 2008. Phenotype, genotype, and worldwide genetic penetrance of LRRK2-associated Parkinson's disease: a case-control study. Lancet Neurol. 7, 583–590.

Henderson, M.X., Sengupta, M., Mcgeary, I., Zhang, B., Olufemi, M.F., Brown, H., Trojanowski, J.Q., Lee, V.M.Y., 2019. LRRK2 inhibition does not impart protection from alpha-synuclein pathology and neuron death in non-transgenic mice. Acta Neuropathol. Commun. 7, 28.

Hernandez, D.G., Reed, X., Singleton, A.B., 2016. Genetics in Parkinson disease: mendelian versus non-mendelian inheritance. J. Neurochem. 139 (Suppl. 1), 59–74.

Herzig, M.C., Kolly, C., Persohn, E., Theil, D., Schweizer, T., Hafner, T., Stemmelen, C., Troxler, T.J., Schmid, P., Danner, S., Schnell, C.R., Mueller, M., Kinzel, B., Grevot, A., Bolognani, F., Stirn, M., Kuhn, R.R., Kaupmann, K., Van Der Putten, P.H., Rovelli, G., Shimshek, D.R., 2011. LRRK2 protein levels are determined by kinase function and are crucial for kidney and lung homeostasis in mice. Hum. Mol. Genet. 20, 4209–4223.

Hierro, A., Rojas, A.L., Rojas, R., Murthy, N., Effantin, G., Kajava, A.V., Steven, A.C., Bonifacino, J.S., Hurley, J.H., 2007. Functional architecture of the retromer cargo-recognition complex. Nature 449, 1063–1067.

Hinkle, K.M., Yue, M., Behrouz, B., Dachsel, J.C., Lincoln, S.J., Bowles, E.E., Beevers, J.E., Dugger, B., Winner, B., Prots, I., Kent, C.B., Nishioka, K., Lin, W.L., Dickson, D.W., Janus, C.J., Farrer, M.J., Melrose, H.L., 2012. LRRK2 knockout mice have an intact dopaminergic system but display alterations in exploratory and motor co-ordination behaviors. Mol. Neurodegener. 7, 25.

Hirst, J., Sahlender, D.A., Li, S., Lubben, N.B., Borner, G.H., Robinson, M.S., 2008. Auxilin depletion causes self-assembly of clathrin into membraneless cages in vivo. Traffic 9, 1354–1371.

Ishizu, N., Yui, D., Hebisawa, A., Aizawa, H., Cui, W., Fujita, Y., Hashimoto, K., Ajioka, I., Mizusawa, H., Yokota, T., Watase, K., 2016. Impaired striatal dopamine release in homozygous Vps35 D620N knock-in mice. Hum. Mol. Genet. 25, 4507–4517.

Islam, M.S., Moore, D.J., 2017. Mechanisms of LRRK2-dependent neurodegeneration: role of enzymatic activity and protein aggregation. Biochem. Soc. Trans. 45, 163–172.

Islam, M.S., Nolte, H., Jacob, W., Ziegler, A.B., Putz, S., Grosjean, Y., Szczepanowska, K., Trifunovic, A., Braun, T., Heumann, H., Heumann, R., Hovemann, B., Moore, D.J., Kruger, M., 2016. Human R1441C LRRK2 regulates the synaptic vesicle proteome and phosphoproteome in a Drosophila model of Parkinson's disease. Hum. Mol. Genet. 25, 5365–5382.

Ito, G., Katsemonova, K., Tonelli, F., Lis, P., Baptista, M.A., Shpiro, N., Duddy, G., Wilson, S., Ho, P.W., Ho, S.L., Reith, A.D., Alessi, D.R., 2016. Phos-tag analysis of Rab10 phosphorylation by Lrrk2: a powerful assay for assessing kinase function and inhibitors. Biochem. J. 473, 2671–2685.

Jinn, S., Blauwendraat, C., Toolan, D., Gretzula, C.A., Drolet, R.E., Smith, S., Nalls, M.A., Marcus, J., Singleton, A.B., Stone, D.J., 2019. Functionalization of the TMEM175 p.M393T variant as a risk factor for Parkinson disease. Hum. Mol. Genet. 28, 3244–3254.

Kelly, K., Wang, S., Boddu, R., Liu, Z., Moukha-Chafiq, O., Augelli-Szafran, C., West, A.B., 2018. The G2019S mutation in LRRK2 imparts resiliency to kinase inhibition. Exp. Neurol. 309, 1–13.

Kim, W.T., Chang, S., Daniell, L., Cremona, O., Di Paolo, G., De Camilli, P., 2002. Delayed reentry of recycling vesicles into the fusion-competent synaptic vesicle pool in synaptojanin 1 knockout mice. Proc. Natl. Acad. Sci. U. S. A. 99, 17143–17148.

Kirola, L., Behari, M., Shishir, C., Thelma, B.K., 2016. Identification of a novel homozygous mutation Arg459Pro in SYNJ1 gene of an Indian family with autosomal recessive juvenile Parkinsonism. Parkinsonism Relat. Disord. 31, 124–128.

Kluss, J.H., Conti, M.M., Kaganovich, A., Beilina, A., Melrose, H.L., Cookson, M.R., Mamais, A., 2018. Detection of endogenous S1292 LRRK2 autophosphorylation in mouse tissue as a readout for kinase activity. NPJ Parkinsons Dis. 4, 13.

Koroglu, C., Baysal, L., Cetinkaya, M., Karasoy, H., Tolun, A., 2013. DNAJC6 is responsible for juvenile parkinsonism with phenotypic variability. Parkinsonism Relat. Disord. 19, 320–324.

Kovtun, O., Leneva, N., Bykov, Y.S., Ariotti, N., Teasdale, R.D., Schaffer, M., Engel, B.D., Owen, D.J., Briggs, J.A.G., Collins, B.M., 2018. Structure of the membrane-assembled retromer coat determined by cryo-electron tomography. Nature 561, 561–564.

Krebs, C.E., Karkheiran, S., Powell, J.C., Cao, M., Makarov, V., Darvish, H., Di Paolo, G., Walker, R.H., Shahidi, G.A., Buxbaum, J.D., De Camilli, P., Yue, Z., Paisan-Ruiz, C., 2013. The Sac1 domain of SYNJ1 identified mutated in a family with early-onset progressive Parkinsonism with generalized seizures. Hum. Mutat. 34, 1200–1207.

Krohn, L., Ozturk, T.N., Vanderperre, B., Ouled Amar Bencheikh, B., Ruskey, J.A., Laurent, S.B., Spiegelman, D., Postuma, R.B., Arnulf, I., Hu, M.T.M., Dauvilliers, Y., Hogl, B., Stefani, A., Monaca, C.C., Plazzi, G., Antelmi, E., Ferini-Strambi, L., Heidbreder, A., Rudakou, U., Cochen De Cock, V., Young, P., Wolf, P., Oliva, P., Zhang, X.K., Greenbaum, L., Liong, C., Gagnon, J.F., Desautels, A., Hassin-Baer, S., Montplaisir, J.Y., Dupre, N., Rouleau, G.A., Fon, E.A., Trempe, J.F., Lamoureux, G., Alcalay, R.N., Gan-Or, Z., 2020. Genetic, structural, and functional evidence link TMEM175 to synucleinopathies. Ann. Neurol. 87, 139–153.

Kruger, R., Kuhn, W., Muller, T., Woitalla, D., Graeber, M., Kosel, S., Przuntek, H., Epplen, J.T., Schols, L., Riess, O., 1998. Ala30Pro mutation in the gene encoding alpha-synuclein in Parkinson's disease. Nat. Genet. 18, 106–108.

Krumova, P., Reyniers, L., Meyer, M., Lobbestael, E., Stauffer, D., Gerrits, B., Muller, L., Hoving, S., Kaupmann, K., Voshol, J., Fabbro, D., Bauer, A., Rovelli, G., Taymans, J.M., Bouwmeester, T., Baekelandt, V., 2015. Chemical genetic approach identifies microtubule affinity-regulating kinase 1 as a leucine-rich repeat kinase 2 substrate. FASEB J. 29, 2980–2992.

Kumar, K.R., Weissbach, A., Heldmann, M., Kasten, M., Tunc, S., Sue, C.M., Svetel, M., Kostic, V.S., Segura-Aguilar, J., Ramirez, A., Simon, D.K., Vieregge, P., Munte, T.F., Hagenah, J., Klein, C., Lohmann, K., 2012. Frequency of the D620N mutation in VPS35 in Parkinson disease. Arch. Neurol. 69, 1360–1364.

Kuwahara, T., Inoue, K., D'agati, V.D., Fujimoto, T., Eguchi, T., Saha, S., Wolozin, B., Iwatsubo, T., Abeliovich, A., 2016. LRRK2 and RAB7L1 coordinately regulate axonal morphology and lysosome integrity in diverse cellular contexts. Sci. Rep. 6, 29945.

Lang, A.E., Lozano, A.M., 1998a. Parkinson's disease. First of two parts. N. Engl. J. Med. 339, 1044–1053.

Lang, A.E., Lozano, A.M., 1998b. Parkinson's disease. Second of two parts. N. Engl. J. Med. 339, 1130–1143.

Lara Ordonez, A.J., Fernandez, B., Fdez, E., Romo-Lozano, M., Madero-Perez, J., Lobbestael, E., Baekelandt, V., Aiastui, A., Lopez De Munain, A., Melrose, H.L., Civiero, L., Hilfiker, S., 2019. RAB8, RAB10 and RILPL1 contribute to both LRRK2 kinase-mediated centrosomal cohesion and ciliogenesis deficits. Hum. Mol. Genet. 28, 3552–3568.

Lee, D.W., Zhao, X., Yim, Y.I., Eisenberg, E., Greene, L.E., 2008. Essential role of cyclin-G-associated kinase (Auxilin-2) in developing and mature mice. Mol. Biol. Cell 19, 2766–2776.

Lee, B.D., Shin, J.H., Vankampen, J., Petrucelli, L., West, A.B., Ko, H.S., Lee, Y.I., Maguire-Zeiss, K.A., Bowers, W.J., Federoff, H.J., Dawson, V.L., Dawson, T.M., 2010. Inhibitors of leucine-rich repeat kinase-2 protect against models of Parkinson's disease. Nat. Med. 16, 998–1000.

Lesage, S., Anheim, M., Letournel, F., Bousset, L., Honore, A., Rozas, N., Pieri, L., Madiona, K., Durr, A., Melki, R., Verny, C., Brice, A., 2013. G51D alpha-synuclein mutation causes a novel parkinsonian-pyramidal syndrome. Ann. Neurol. 73, 459–471.

Lesage, S., Bras, J., Cormier-Dequaire, F., Condroyer, C., Nicolas, A., Darwent, L., Guerreiro, R., Majounie, E., Federoff, M., Heutink, P., Wood, N.W., Gasser, T., Hardy, J., Tison, F., Singleton, A., Brice, A., 2015. Loss-of-function mutations in RAB39B are associated with typical early-onset Parkinson disease. Neurol. Genet. 1, e9.

Li, J.G., Chiu, J., Pratico, D., 2019. Full recovery of the Alzheimer's disease phenotype by gain of function of vacuolar protein sorting 35. Mol. Psychiatry. https://doi.org/10.1038/s41380-019-0364-x.

Li, J.G., Chiu, J., Ramanjulu, M., Blass, B.E., Pratico, D., 2020. A pharmacological chaperone improves memory by reducing Abeta and tau neuropathology in a mouse model with plaques and tangles. Mol. Neurodegener. 15, 1.

Linhart, R., Wong, S.A., Cao, J., Tran, M., Huynh, A., Ardrey, C., Park, J.M., Hsu, C., Taha, S., Peterson, R., Shea, S., Kurian, J., Venderova, K., 2014. Vacuolar protein sorting 35 (Vps35) rescues locomotor deficits and shortened lifespan in Drosophila expressing a Parkinson's disease mutant of Leucine-Rich Repeat Kinase 2 (LRRK2). Mol. Neurodegener. 9, 23.

Lis, P., Burel, S., Steger, M., Mann, M., Brown, F., Diez, F., Tonelli, F., Holton, J.L., Ho, P.W., Ho, S.L., Chou, M.Y., Polinski, N.K., Martinez, T.N., Davies, P., Alessi, D.R., 2018. Development of phospho-specific Rab protein antibodies to monitor in vivo activity of the LRRK2 Parkinson's disease kinase. Biochem. J. 475, 1–22.

Liu, Z., Bryant, N., Kumaran, R., Beilina, A., Abeliovich, A., Cookson, M.R., West, A.B., 2018. LRRK2 phosphorylates membrane-bound Rabs and is activated by GTP-bound Rab7L1 to promote recruitment to the trans-Golgi network. Hum. Mol. Genet. 27, 385–395.

Ma, J., Gao, J., Wang, J., Xie, A., 2019. Prion-like mechanisms in Parkinson's disease. Front. Neurosci. 13, 552.

MacLeod, D.A., Rhinn, H., Kuwahara, T., Zolin, A., Di Paolo, G., McCabe, B.D., Marder, K.S., Honig, L.S., Clark, L.N., Small, S.A., Abeliovich, A., 2013. RAB7L1 interacts with LRRK2 to modify intraneuronal protein sorting and Parkinson's disease risk. Neuron 77, 425–439.

Marras, C., Beck, J.C., Bower, J.H., Roberts, E., Ritz, B., Ross, G.W., Abbott, R.D., Savica, R., Van Den Eeden, S.K., Willis, A.W., Tanner, C.M., 2018. Prevalence of Parkinson's disease across North America. NPJ Parkinsons Dis. 4, 21.

Martin, I., Kim, J.W., Lee, B.D., Kang, H.C., Xu, J.C., Jia, H., Stankowski, J., Kim, M.S., Zhong, J., Kumar, M., Andrabi, S.A., Xiong, Y., Dickson, D.W., Wszolek, Z.K., Pandey, A., Dawson, T.M., Dawson, V.L., 2014. Ribosomal protein s15 phosphorylation mediates LRRK2 neurodegeneration in Parkinson's disease. Cell 157, 472–485.

Mata, I.F., Jang, Y., Kim, C.H., Hanna, D.S., Dorschner, M.O., Samii, A., Agarwal, P., Roberts, J.W., Klepitskaya, O., Shprecher, D.R., Chung, K.A., Factor, S.A., Espay, A.J., Revilla, F.J., Higgins, D.S., Litvan, I., Leverenz, J.B., Yearout, D., Inca-Martinez, M., Martinez, E., Thompson, T.R., Cholerton, B.A., Hu, S.C., Edwards, K.L., Kim, K.S., Zabetian, C.P., 2015. The RAB39B p.G192R mutation causes X-linked dominant Parkinson's disease. Mol. Neurodegener. 10, 50.

Matta, S., Van Kolen, K., Da Cunha, R., Van Den Bogaart, G., Mandemakers, W., Miskiewicz, K., De Bock, P.J., Morais, V.A., Vilain, S., Haddad, D., Delbroek, L., Swerts, J., Chavez-Gutierrez, L., Esposito, G., Daneels, G., Karran, E., Holt, M., Gevaert, K., Moechars, D.W., De Strooper, B., Verstreken, P., 2012. LRRK2 controls an EndoA phosphorylation cycle in synaptic endocytosis. Neuron 75, 1008–1021.

McFarthing, K., Simuni, T., 2019. Clinical trial highlights: targetting alpha-synuclein. J. Parkinsons Dis. 9, 5–16.

McGough, I.J., Steinberg, F., Jia, D., Barbuti, P.A., McMillan, K.J., Heesom, K.J., Whone, A.L., Caldwell, M.A., Billadeau, D.D., Rosen, M.K., Cullen, P.J., 2014. Retromer binding to FAM21 and the WASH complex is perturbed by the Parkinson disease-linked VPS35(D620N) mutation. Curr. Biol. 24, 1670–1676.

McGrath, E., Waschbusch, D., Baker, B.M., Khan, A.R., 2019. LRRK2 binds to the Rab32 subfamily in a GTP-dependent manner via its armadillo domain. Small GTPases 1–14. https://doi.org/10.1080/21541248.2019.1666623.

McNally, K.E., Cullen, P.J., 2018. Endosomal retrieval of cargo: retromer is not alone. Trends Cell Biol. 28, 807–822.

McNally, K.E., Faulkner, R., Steinberg, F., Gallon, M., Ghai, R., Pim, D., Langton, P., Pearson, N., Danson, C.M., Nagele, H., Morris, L.L., Singla, A., Overlee, B.L., Heesom, K.J., Sessions, R., Banks, L., Collins, B.M., Berger, I., Billadeau, D.D., Burstein, E., Cullen, P.J., 2017. Retriever is a multiprotein complex for retromer-independent endosomal cargo recycling. Nat. Cell Biol. 19, 1214–1225.

Mecozzi, V.J., Berman, D.E., Simoes, S., Vetanovetz, C., Awal, M.R., Patel, V.M., Schneider, R.T., Petsko, G.A., Ringe, D., Small, S.A., 2014. Pharmacological chaperones stabilize retromer to limit APP processing. Nat. Chem. Biol. 10, 443–449.

Meissner, W.G., Frasier, M., Gasser, T., Goetz, C.G., Lozano, A., Piccini, P., Obeso, J.A., Rascol, O., Schapira, A., Voon, V., Weiner, D.M., Tison, F., Bezard, E., 2011. Priorities in Parkinson's disease research. Nat. Rev. Drug Discov. 10, 377–393.

Mir, R., Tonelli, F., Lis, P., Macartney, T., Polinski, N.K., Martinez, T.N., Chou, M.Y., Howden, A.J.M., Konig, T., Hotzy, C., Milenkovic, I., Brucke, T., Zimprich, A., Sammler, E., Alessi, D.R., 2018. The Parkinson's disease VPS35[D620N] mutation enhances LRRK2-mediated Rab protein phosphorylation in mouse and human. Biochem. J. 475, 1861–1883.

Miura, E., Hasegawa, T., Konno, M., Suzuki, M., Sugeno, N., Fujikake, N., Geisler, S., Tabuchi, M., Oshima, R., Kikuchi, A., Baba, T., Wada, K., Nagai, Y., Takeda, A., Aoki, M., 2014. VPS35 dysfunction impairs lysosomal degradation of alpha-synuclein and exacerbates neurotoxicity in a Drosophila model of Parkinson's disease. Neurobiol. Dis. 71, 1–13.

Moore, D.J., West, A.B., Dawson, V.L., Dawson, T.M., 2005. Molecular pathophysiology of Parkinson's disease. Annu. Rev. Neurosci. 28, 57–87.

Muhammad, A., Flores, I., Zhang, H., Yu, R., Staniszewski, A., Planel, E., Herman, M., Ho, L., Kreber, R., Honig, L.S., Ganetzky, B., Duff, K., Arancio, O., Small, S.A., 2008. Retromer deficiency observed in Alzheimer's disease causes hippocampal dysfunction, neurodegeneration, and Abeta accumulation. Proc. Natl. Acad. Sci. U. S. A. 105, 7327–7332.

Munsie, L.N., Milnerwood, A.J., Seibler, P., Beccano-Kelly, D.A., Tatarnikov, I., Khinda, J., Volta, M., Kadgien, C., Cao, L.P., Tapia, L., Klein, C., Farrer, M.J., 2015. Retromer-dependent neurotransmitter receptor trafficking to synapses is altered by the Parkinson's disease VPS35 mutation p.D620N. Hum. Mol. Genet. 24, 1691–1703.

Nalls, M.A., Pankratz, N., Lill, C.M., Do, C.B., Hernandez, D.G., Saad, M., Destefano, A.L., Kara, E., Bras, J., Sharma, M., Schulte, C., Keller, M.F., Arepalli, S., Letson, C., Edsall, C., Stefansson, H., Liu, X., Pliner, H., Lee, J.H., Cheng, R., Ikram, M.A., Ioannidis, J.P., Hadjigeorgiou, G.M., Bis, J.C., Martinez, M., Perlmutter, J.S., Goate, A., Marder, K., Fiske, B., Sutherland, M., Xiromerisiou, G., Myers, R.H., Clark, L.N., Stefansson, K., Hardy, J.A., Heutink, P., Chen, H., Wood, N.W., Houlden, H., Payami, H., Brice, A., Scott, W.K., Gasser, T., Bertram, L., Eriksson, N., Foroud, T., Singleton, A.B., 2014. Large-scale meta-analysis of genome-wide association data identifies six new risk loci for Parkinson's disease. Nat. Genet. 46, 989–993.

Nalls, M.A., Blauwendraat, C., Vallerga, C.L., Heilbron, K., Bandres-Ciga, S., Chang, D., Tan, M., Kia, D.A., Noyce, A.J., Xue, A., Bras, J., Young, E., Von Coelln, R., Simon-Sanchez, J., Schulte, C., Sharma, M., Krohn, L., Pihlstrom, L., Siitonen, A., Iwaki, H., Leonard, H., Faghri, F., Gibbs, J.R., Hernandez, D.G., Scholz, S.W., Botia, J.A., Martinez, M., Corvol, J.C., Lesage, S., Jankovic, J., Shulman, L.M., Sutherland, M., Tienari, P., Majamaa, K., Toft, M., Andreassen, O.A., Bangale, T., Brice, A., Yang, J., Gan-Or, Z., Gasser, T., Heutink, P., Shulman, J.M., Wood, N.W., Hinds, D.A., Hardy, J.A., Morris, H.R., Gratten, J., Visscher, P.M., Graham, R.R., Singleton, A.B., Andme Research, T., System Genomics Of Parkinson'S Disease, C, International Parkinson'S Disease Genomics, C, 2019. Identification of novel risk loci, causal insights, and heritable risk for Parkinson's disease: a meta-analysis of genome-wide association studies. Lancet Neurol. 18, 1091–1102.

Nemani, V.M., Lu, W., Berge, V., Nakamura, K., Onoa, B., Lee, M.K., Chaudhry, F.A., Nicoll, R.A., Edwards, R.H., 2010. Increased expression of alpha-synuclein reduces neurotransmitter release by inhibiting synaptic vesicle reclustering after endocytosis. Neuron 65, 66–79.

Ness, D., Ren, Z., Gardai, S., Sharpnack, D., Johnson, V.J., Brennan, R.J., Brigham, E.F., Olaharski, A.J., 2013. Leucine-rich repeat kinase 2 (LRRK2)-deficient rats exhibit renal tubule injury and perturbations in metabolic and immunological homeostasis. PLoS One 8, e66164.

Nguyen, M., Krainc, D., 2018. LRRK2 phosphorylation of auxilin mediates synaptic defects in dopaminergic neurons from patients with Parkinson's disease. Proc. Natl. Acad. Sci. U. S. A. 115, 5576–5581.

Nguyen, A.P., Moore, D.J., 2017. Understanding the Gtpase activity of LRRK2: regulation, function, and neurotoxicity. Adv. Neurobiol. 14, 71–88.

Nguyen, M., Wong, Y.C., Ysselstein, D., Severino, A., Krainc, D., 2019. Synaptic, mitochondrial, and lysosomal dysfunction in Parkinson's disease. Trends Neurosci. 42, 140–149.

Olgiati, S., De Rosa, A., Quadri, M., Criscuolo, C., Breedveld, G.J., Picillo, M., Pappata, S., Quarantelli, M., Barone, P., De Michele, G., Bonifati, V., 2014. PARK20 caused by SYNJ1 homozygous Arg258Gln mutation in a new Italian family. Neurogenetics 15, 183–188.

Olgiati, S., Quadri, M., Fang, M., Rood, J.P., Saute, J.A., Chien, H.F., Bouwkamp, C.G., Graafland, J., Minneboo, M., Breedveld, G.J., Zhang, J., Verheijen, F.W., Boon, A.J., Kievit, A.J., Jardim, L.B., Mandemakers, W., Barbosa, E.R., Rieder, C.R., Leenders, K.L., Wang, J., Bonifati, V., 2016. DNAJC6 mutations associated with early-onset Parkinson's disease. Ann. Neurol. 79, 244–256.

Orenstein, S.J., Kuo, S.H., Tasset, I., Arias, E., Koga, H., Fernandez-Carasa, I., Cortes, E., Honig, L.S., Dauer, W., Consiglio, A., Raya, A., Sulzer, D., Cuervo, A.M., 2013. Interplay of LRRK2 with chaperone-mediated autophagy. Nat. Neurosci. 16, 394–406.

Paisan-Ruiz, C., Jain, S., Evans, E.W., Gilks, W.P., Simon, J., Van Der Brug, M., Lopez De Munain, A., Aparicio, S., Gil, A.M., Khan, N., Johnson, J., Martinez, J.R., Nicholl, D., Marti Carrera, I., Pena, A.S., De Silva, R., Lees, A., Marti-Masso, J.F., Perez-Tur, J., Wood, N.W., Singleton, A.B., 2004. Cloning of the gene containing mutations that cause PARK8-linked Parkinson's disease. Neuron 44, 595–600.

Paisan-Ruiz, C., Lewis, P.A., Singleton, A.B., 2013. LRRK2: cause, risk, and mechanism. J. Parkinsons Dis. 3, 85–103.

Pan, P.Y., Li, X., Wang, J., Powell, J., Wang, Q., Zhang, Y., Chen, Z., Wicinski, B., Hof, P., Ryan, T.A., Yue, Z., 2017. Parkinson's disease-associated LRRK2 hyperactive kinase mutant disrupts synaptic vesicle trafficking in ventral midbrain neurons. J. Neurosci. 37, 11366–11376.

Pellegrini, L., Hauser, D.N., Li, Y., Mamais, A., Beilina, A., Kumaran, R., Wetzel, A., Nixon-Abell, J., Heaton, G., Rudenko, I., Alkaslasi, M., Ivanina, N., Melrose, H.L., Cookson, M.R., Harvey, K., 2018. Proteomic analysis reveals co-ordinated alterations in protein synthesis and degradation pathways in LRRK2 knockout mice. Hum. Mol. Genet. 27, 3257–3271.

Polymeropoulos, M.H., Lavedan, C., Leroy, E., Ide, S.E., Dehejia, A., Dutra, A., Pike, B., Root, H., Rubenstein, J., Boyer, R., Stenroos, E.S., Chandrasekharappa, S., Athanassiadou, A., Papapetropoulos, T., Johnson, W.G., Lazzarini, A.M., Duvoisin, R.C., Di Iorio, G., Golbe, L.I., Nussbaum, R.L., 1997. Mutation in the alpha-synuclein gene identified in families with Parkinson's disease. Science 276, 2045–2047.

Purlyte, E., Dhekne, H.S., Sarhan, A.R., Gomez, R., Lis, P., Wightman, M., Martinez, T.N., Tonelli, F., Pfeffer, S.R., Alessi, D.R., 2018. Rab29 activation of the Parkinson's disease-associated LRRK2 kinase. EMBO J. 37, 1–18.

Quadri, M., Fang, M., Picillo, M., Olgiati, S., Breedveld, G.J., Graafland, J., Wu, B., Xu, F., Erro, R., Amboni, M., Pappata, S., Quarantelli, M., Annesi, G., Quattrone, A., Chien, H.F., Barbosa, E.R., Oostra, B.A., Barone, P., Wang, J., Bonifati, V., 2013. Mutation in the SYNJ1 gene associated with autosomal recessive, early-onset Parkinsonism. Hum. Mutat. 34, 1208–1215.

Qureshi, Y.H., Berman, D.E., Klein, R.L., Patel, V.M., Simoes, S., Kannan, S., Cox, R., Waksal, S.D., Stevens, B., Petsko, G.A., Small, S.A., 2019. Retromer repletion with AAV9-VPS35 restores endosomal function in the mouse hippocampus. bioRxiv, 618496. https://doi.org/10.1101/618496.

Ramonet, D., Daher, J.P., Lin, B.M., Stafa, K., Kim, J., Banerjee, R., Westerlund, M., Pletnikova, O., Glauser, L., Yang, L., Liu, Y., Swing, D.A., Beal, M.F., Troncoso, J.C., Mccaffery, J.M., Jenkins, N.A., Copeland, N.G., Galter, D., Thomas, B., Lee, M.K., Dawson, T.M., Dawson, V.L., Moore, D.J., 2011. Dopaminergic neuronal loss, reduced neurite complexity and autophagic abnormalities in transgenic mice expressing G2019S mutant LRRK2. PLoS One 6, e18568.

Roosen, D.A., Landeck, N., Conti, M., Smith, N., Saez-Atienzar, S., Ding, J., Beilina, A., Kumaran, R., Kaganovich, A., Du Hoffmann, J., Williamson, C.D., Gershlick, D.C.,

Bonet-Ponce, L., Sampieri, L., Bleck, C.K.E., Liu, C., Bonifacino, J.S., Li, Y., Lewis, P.A., Cookson, M.R., 2019. Mutations in Auxilin cause parkinsonism via impaired clathrin-mediated trafficking at the Golgi apparatus and synapse. bioRxiv, 830802.

Saheki, Y., De Camilli, P., 2012. Synaptic vesicle endocytosis. Cold Spring Harb. Perspect. Biol. 4, a005645.

Satake, W., Nakabayashi, Y., Mizuta, I., Hirota, Y., Ito, C., Kubo, M., Kawaguchi, T., Tsunoda, T., Watanabe, M., Takeda, A., Tomiyama, H., Nakashima, K., Hasegawa, K., Obata, F., Yoshikawa, T., Kawakami, H., Sakoda, S., Yamamoto, M., Hattori, N., Murata, M., Nakamura, Y., Toda, T., 2009. Genome-wide association study identifies common variants at four loci as genetic risk factors for Parkinson's disease. Nat. Genet. 41, 1303–1307.

Seaman, M.N., 2004. Cargo-selective endosomal sorting for retrieval to the Golgi requires retromer. J. Cell Biol. 165, 111–122.

Seaman, M.N., 2012. The retromer complex—endosomal protein recycling and beyond. J. Cell Sci. 125, 4693–4702.

Seaman, M., Freeman, C.L., 2014. Analysis of the Retromer complex-WASH complex interaction illuminates new avenues to explore in Parkinson disease. Commun. Integr. Biol. 7, e29483.

Seaman, M.N., Gautreau, A., Billadeau, D.D., 2013. Retromer-mediated endosomal protein sorting: all WASHed up!. Trends Cell Biol. 23, 522–528.

Seaman, M.N.J., Mukadam, A.S., Breusegem, S.Y., 2018. Inhibition of TBC1D5 activates Rab7a and can enhance the function of the retromer cargo-selective complex. J. Cell Sci. 131, jcs217398

Shahmoradian, S.H., Lewis, A.J., Genoud, C., Hench, J., Moors, T.E., Navarro, P.P., Castano-Diez, D., Schweighauser, G., Graff-Meyer, A., Goldie, K.N., Sutterlin, R., Huisman, E., Ingrassia, A., Gier, Y., Rozemuller, A.J.M., Wang, J., Paepe, A., Erny, J., Staempfli, A., Hoernschemeyer, J., Grosseruschkamp, F., Niedieker, D., El-Mashtoly, S.F., Quadri, M., Van, I.W.F.J., Bonifati, V., Gerwert, K., Bohrmann, B., Frank, S., Britschgi, M., Stahlberg, H., Van De Berg, W.D.J., Lauer, M.E., 2019. Lewy pathology in Parkinson's disease consists of crowded organelles and lipid membranes. Nat. Neurosci. 22, 1099–1109.

Sharma, M., Ioannidis, J.P., Aasly, J.O., Annesi, G., Brice, A., Bertram, L., Bozi, M., Barcikowska, M., Crosiers, D., Clarke, C.E., Facheris, M.F., Farrer, M., Garraux, G., Gispert, S., Auburger, G., Vilarino-Guell, C., Hadjigeorgiou, G.M., Hicks, A.A., Hattori, N., Jeon, B.S., Jamrozik, Z., Krygowska-Wajs, A., Lesage, S., Lill, C.M., Lin, J.J., Lynch, T., Lichtner, P., Lang, A.E., Libioulle, C., Murata, M., Mok, V., Jasinska-Myga, B., Mellick, G.D., Morrison, K.E., Meitnger, T., Zimprich, A., Opala, G., Pramstaller, P.P., Pichler, I., Park, S.S., Quattrone, A., Rogaeva, E., Ross, O.A., Stefanis, L., Stockton, J.D., Satake, W., Silburn, P.A., Strom, T.M., Theuns, J., Tan, E.K., Toda, T., Tomiyama, H., Uitti, R.J., Van Broeckhoven, C., Wirdefeldt, K., Wszolek, Z., Xiromerisiou, G., Yomono, H.S., Yueh, K.C., Zhao, Y., Gasser, T., Maraganore, D., Kruger, R., 2012. A multi-centre clinico-genetic analysis of the VPS35 gene in Parkinson disease indicates reduced penetrance for disease-associated variants. J. Med. Genet. 49, 721–726.

Sheng, Z., Zhang, S., Bustos, D., Kleinheinz, T., Le Pichon, C.E., Dominguez, S.L., Solanoy, H.O., Drummond, J., Zhang, X., Ding, X., Cai, F., Song, Q., Li, X., Yue, Z., Van Der Brug, M.P., Burdick, D.J., Gunzner-Toste, J., Chen, H., Liu, X.,

Estrada, A.A., Sweeney, Z.K., Scearce-Levie, K., Moffat, J.G., Kirkpatrick, D.S., Zhu, H., 2012. Ser1292 autophosphorylation is an indicator of LRRK2 kinase activity and contributes to the cellular effects of PD mutations. Sci. Transl. Med. 4, 164ra161.

Simon-Sanchez, J., Schulte, C., Bras, J.M., Sharma, M., Gibbs, J.R., Berg, D., Paisan-Ruiz, C., Lichtner, P., Scholz, S.W., Hernandez, D.G., Kruger, R., Federoff, M., Klein, C., Goate, A., Perlmutter, J., Bonin, M., Nalls, M.A., Illig, T., Gieger, C., Houlden, H., Steffens, M., Okun, M.S., Racette, B.A., Cookson, M.R., Foote, K.D., Fernandez, H.H., Traynor, B.J., Schreiber, S., Arepalli, S., Zonozi, R., Gwinn, K., Van Der Brug, M., Lopez, G., Chanock, S.J., Schatzkin, A., Park, Y., Hollenbeck, A., Gao, J., Huang, X., Wood, N.W., Lorenz, D., Deuschl, G., Chen, H., Riess, O., Hardy, J.A., Singleton, A.B., Gasser, T., 2009. Genome-wide association study reveals genetic risk underlying Parkinson's disease. Nat. Genet. 41, 1308–1312.

Singleton, A.B., Farrer, M., Johnson, J., Singleton, A., Hague, S., Kachergus, J., Hulihan, M., Peuralinna, T., Dutra, A., Nussbaum, R., Lincoln, S., Crawley, A., Hanson, M., Maraganore, D., Adler, C., Cookson, M.R., Muenter, M., Baptista, M., Miller, D., Blancato, J., Hardy, J., Gwinn-Hardy, K., 2003. alpha-Synuclein locus triplication causes Parkinson's disease. Science 302, 841.

Small, S.A., Petsko, G.A., 2015. Retromer in Alzheimer disease, Parkinson disease and other neurological disorders. Nat. Rev. Neurosci. 16, 126–132.

Small, S.A., Kent, K., Pierce, A., Leung, C., Kang, M.S., Okada, H., Honig, L., Vonsattel, J.P., Kim, T.W., 2005. Model-guided microarray implicates the retromer complex in Alzheimer's disease. Ann. Neurol. 58, 909–919.

Smith, W.W., Pei, Z., Jiang, H., Dawson, V.L., Dawson, T.M., Ross, C.A., 2006. Kinase activity of mutant LRRK2 mediates neuronal toxicity. Nat. Neurosci. 9, 1231–1233.

Song, L., He, Y., Ou, J., Zhao, Y., Li, R., Cheng, J., Lin, C.H., Ho, M.S., 2017. Auxilin underlies progressive locomotor deficits and dopaminergic neuron loss in a drosophila model of Parkinson's disease. Cell Rep. 18, 1132–1143.

Spillantini, M.G., Schmidt, M.L., Lee, V.M., Trojanowski, J.Q., Jakes, R., Goedert, M., 1997. Alpha-synuclein in Lewy bodies. Nature 388, 839–840.

Spillantini, M.G., Crowther, R.A., Jakes, R., Hasegawa, M., Goedert, M., 1998. alpha-Synuclein in filamentous inclusions of Lewy bodies from Parkinson's disease and dementia with lewy bodies. Proc. Natl. Acad. Sci. U. S. A. 95, 6469–6473.

Stafa, K., Trancikova, A., Webber, P.J., Glauser, L., West, A.B., Moore, D.J., 2012. GTPase activity and neuronal toxicity of Parkinson's disease-associated LRRK2 is regulated by ArfGAP1. PLoS Genet. 8, e1002526.

Stafa, K., Tsika, E., Moser, R., Musso, A., Glauser, L., Jones, A., Biskup, S., Xiong, Y., Bandopadhyay, R., Dawson, V.L., Dawson, T.M., Moore, D.J., 2014. Functional interaction of Parkinson's disease-associated LRRK2 with members of the dynamin GTPase superfamily. Hum. Mol. Genet. 23, 2055–2077.

Steger, M., Tonelli, F., Ito, G., Davies, P., Trost, M., Vetter, M., Wachter, S., Lorentzen, E., Duddy, G., Wilson, S., Baptista, M.A., Fiske, B.K., Fell, M.J., Morrow, J.A., Reith, A.D., Alessi, D.R., Mann, M., 2016. Phosphoproteomics reveals that Parkinson's disease kinase LRRK2 regulates a subset of Rab GTPases. Elife, 5, e12813

Steger, M., Diez, F., Dhekne, H.S., Lis, P., Nirujogi, R.S., Karayel, O., Tonelli, F., Martinez, T.N., Lorentzen, E., Pfeffer, S.R., Alessi, D.R., Mann, M., 2017. Systematic proteomic analysis of LRRK2-mediated Rab GTPase phosphorylation establishes a connection to ciliogenesis. Elife, 6, e31012

Steinberg, F., Gallon, M., Winfield, M., Thomas, E.C., Bell, A.J., Heesom, K.J., Tavare, J.M., Cullen, P.J., 2013. A global analysis of SNX27-retromer assembly and cargo specificity reveals a function in glucose and metal ion transport. Nat. Cell Biol. 15, 461–471.

Struhal, W., Presslauer, S., Spielberger, S., Zimprich, A., Auff, E., Bruecke, T., Poewe, W., Ransmayr, G., 2014. VPS35 Parkinson's disease phenotype resembles the sporadic disease. J. Neural Transm. (Vienna) 121, 755–759.

Tang, F.L., Erion, J.R., Tian, Y., Liu, W., Yin, D.M., Ye, J., Tang, B., Mei, L., Xiong, W.C., 2015a. VPS35 in dopamine neurons is required for endosome-to-Golgi retrieval of Lamp2a, a receptor of chaperone-mediated autophagy that is critical for alpha-synuclein degradation and prevention of pathogenesis of Parkinson's disease. J. Neurosci. 35, 10613–10628.

Tang, F.L., Liu, W., Hu, J.X., Erion, J.R., Ye, J., Mei, L., Xiong, W.C., 2015b. VPS35 deficiency or mutation causes dopaminergic neuronal loss by impairing mitochondrial fusion and function. Cell Rep. 12, 1631–1643.

Tang, F.L., Zhao, L., Zhao, Y., Sun, D., Zhu, X.J., Mei, L., Xiong, W.C., 2020. Coupling of terminal differentiation deficit with neurodegenerative pathology in Vps35-deficient pyramidal neurons. Cell Death Differ. https://doi.org/10.1038/s41418-019-0487-2

Temkin, P., Lauffer, B., Jager, S., Cimermancic, P., Krogan, N.J., Von Zastrow, M., 2011. SNX27 mediates retromer tubule entry and endosome-to-plasma membrane trafficking of signalling receptors. Nat. Cell Biol. 13, 715–721.

Tong, Y., Yamaguchi, H., Giaime, E., Boyle, S., Kopan, R., Kelleher 3rd, R.J., Shen, J., 2010. Loss of leucine-rich repeat kinase 2 causes impairment of protein degradation pathways, accumulation of alpha-synuclein, and apoptotic cell death in aged mice. Proc. Natl. Acad. Sci. U. S. A. 107, 9879–9884.

Tong, Y., Giaime, E., Yamaguchi, H., Ichimura, T., Liu, Y., Si, H., Cai, H., Bonventre, J.V., Shen, J., 2012. Loss of leucine-rich repeat kinase 2 causes age-dependent bi-phasic alterations of the autophagy pathway. Mol. Neurodegener. 7, 2.

Trempe, J.F., Chen, C.X., Grenier, K., Camacho, E.M., Kozlov, G., McPherson, P.S., Gehring, K., Fon, E.A., 2009. SH3 domains from a subset of BAR proteins define a Ubl-binding domain and implicate parkin in synaptic ubiquitination. Mol. Cell 36, 1034–1047.

Tsika, E., Glauser, L., Moser, R., Fiser, A., Daniel, G., Sheerin, U.M., Lees, A., Troncoso, J.C., Lewis, P.A., Bandopadhyay, R., Schneider, B.L., Moore, D.J., 2014. Parkinson's disease-linked mutations in VPS35 induce dopaminergic neurodegeneration. Hum. Mol. Genet. 23, 4621–4638.

Tsika, E., Nguyen, A.P., Dusonchet, J., Colin, P., Schneider, B.L., Moore, D.J., 2015. Adeno-viral-mediated expression of G2019S LRRK2 induces striatal pathology in a kinase-dependent manner in a rat model of Parkinson's disease. Neurobiol. Dis. 77, 49–61.

Tucci, A., Nalls, M.A., Houlden, H., Revesz, T., Singleton, A.B., Wood, N.W., Hardy, J., Paisan-Ruiz, C., 2010. Genetic variability at the PARK16 locus. Eur. J. Hum. Genet. 18, 1356–1359.

Vanhauwaert, R., Kuenen, S., Masius, R., Bademosi, A., Manetsberger, J., Schoovaerts, N., Bounti, L., Gontcharenko, S., Swerts, J., Vilain, S., Picillo, M., Barone, P., Munshi, S.T., De Vrij, F.M., Kushner, S.A., Gounko, N.V., Mandemakers, W., Bonifati, V., Meunier, F.A., Soukup, S.F., Verstreken, P., 2017. The SAC1 domain in synaptojanin is required for autophagosome maturation at presynaptic terminals. EMBO J. 36, 1392–1411.

Vietri, M., Radulovic, M., Stenmark, H., 2020. The many functions of ESCRTs. Nat. Rev. Mol. Cell Biol. 21, 25–42.

Vilarino-Guell, C., Wider, C., Ross, O.A., Dachsel, J.C., Kachergus, J.M., Lincoln, S.J., Soto-Ortolaza, A.I., Cobb, S.A., Wilhoite, G.J., Bacon, J.A., Behrouz, B., Melrose, H.L., Hentati, E., Puschmann, A., Evans, D.M., Conibear, E., Wasserman, W.W., Aasly, J.O., Burkhard, P.R., Djaldetti, R., Ghika, J., Hentati, F., Krygowska-Wajs, A., Lynch, T., Melamed, E., Rajput, A., Rajput, A.H., Solida, A., Wu, R.M., Uitti, R.J., Wszolek, Z.K., Vingerhoets, F., Farrer, M.J., 2011. VPS35 mutations in Parkinson disease. Am. J. Hum. Genet. 89, 162–167.

Volpicelli-Daley, L.A., Abdelmotilib, H., Liu, Z., Stoyka, L., Daher, J.P., Milnerwood, A.J., Unni, V.K., Hirst, W.D., Yue, Z., Zhao, H.T., Fraser, K., Kennedy, R.E., West, A.B., 2016. G2019S-LRRK2 expression augments alpha-synuclein sequestration into inclusions in neurons. J. Neurosci. 36, 7415–7427.

Wallings, R., Connor-Robson, N., Wade-Martins, R., 2019. LRRK2 interacts with the vacuolar-type H+-ATPase pump a1 subunit to regulate lysosomal function. Hum. Mol. Genet. 28, 2696–2710.

Wang, S., Ma, Z., Xu, X., Wang, Z., Sun, L., Zhou, Y., Lin, X., Hong, W., Wang, T., 2014. A role of Rab29 in the integrity of the trans-Golgi network and retrograde trafficking of mannose-6-phosphate receptor. PLoS One 9, e96242.

Waschbusch, D., Michels, H., Strassheim, S., Ossendorf, E., Kessler, D., Gloeckner, C.J., Barnekow, A., 2014. LRRK2 transport is regulated by its novel interacting partner Rab32. PLoS One 9, e111632.

Waschbusch, D., Hubel, N., Ossendorf, E., Lobbestael, E., Baekelandt, V., Lindsay, A.J., McCaffrey, M.W., Khan, A.R., Barnekow, A., 2019. Rab32 interacts with SNX6 and affects retromer-dependent Golgi trafficking. PLoS One 14, e0208889.

Wen, L., Tang, F.L., Hong, Y., Luo, S.W., Wang, C.L., He, W., Shen, C., Jung, J.U., Xiong, F., Lee, D.H., Zhang, Q.G., Brann, D., Kim, T.W., Yan, R., Mei, L., Xiong, W.C., 2011. VPS35 haploinsufficiency increases Alzheimer's disease neuropathology. J. Cell Biol. 195, 765–779.

West, A.B., Moore, D.J., Biskup, S., Bugayenko, A., Smith, W.W., Ross, C.A., Dawson, V.L., Dawson, T.M., 2005. Parkinson's disease-associated mutations in leucine-rich repeat kinase 2 augment kinase activity. Proc. Natl. Acad. Sci. U. S. A. 102, 16842–16847.

West, A.B., Moore, D.J., Choi, C., Andrabi, S.A., Li, X., Dikeman, D., Biskup, S., Zhang, Z., Lim, K.L., Dawson, V.L., Dawson, T.M., 2007. Parkinson's disease-associated mutations in LRRK2 link enhanced GTP-binding and kinase activities to neuronal toxicity. Hum. Mol. Genet. 16, 223–232.

Wider, C., Skipper, L., Solida, A., Brown, L., Farrer, M., Dickson, D., Wszolek, Z.K., Vingerhoets, F.J., 2008. Autosomal dominant dopa-responsive parkinsonism in a multigenerational Swiss family. Parkinsonism Relat. Disord. 14, 465–470.

Williams, E.T., Chen, X., Moore, D.J., 2017. VPS35, the retromer complex and Parkinson's disease. J. Parkinsons Dis. 7, 219–233.

Williams, E.T., Glauser, L., Tsika, E., Jiang, H., Islam, S., Moore, D.J., 2018. Parkin mediates the ubiquitination of VPS35 and modulates retromer-dependent endosomal sorting. Hum. Mol. Genet. 27, 3189–3205.

Wilson, G.R., Sim, J.C., Mclean, C., Giannandrea, M., Galea, C.A., Riseley, J.R., Stephenson, S.E., Fitzpatrick, E., Haas, S.A., Pope, K., Hogan, K.J., Gregg, R.G., Bromhead, C.J., Wargowski, D.S., Lawrence, C.H., James, P.A., Churchyard, A., Gao, Y., Phelan, D.G., Gillies, G., Salce, N., Stanford, L., Marsh, A.P., Mignogna, M.L., Hayflick, S.J., Leventer, R.J., Delatycki, M.B., Mellick, G.D., Kalscheuer, V.M., D'adamo, P., Bahlo, M., Amor, D.J., Lockhart, P.J., 2014. Mutations in RAB39B cause X-linked intellectual disability and early-onset Parkinson disease with alpha-synuclein pathology. Am. J. Hum. Genet. 95, 729–735.

Wollert, T., Hurley, J.H., 2010. Molecular mechanism of multivesicular body biogenesis by ESCRT complexes. Nature 464, 864.

Xilouri, M., Brekk, O.R., Landeck, N., Pitychoutis, P.M., Papasilekas, T., Papadopoulou-Daifoti, Z., Kirik, D., Stefanis, L., 2013. Boosting chaperone-mediated autophagy in vivo mitigates alpha-synuclein-induced neurodegeneration. Brain 136, 2130–2146.

Xiong, Y., Coombes, C.E., Kilaru, A., Li, X., Gitler, A.D., Bowers, W.J., Dawson, V.L., Dawson, T.M., Moore, D.J., 2010. GTPase activity plays a key role in the pathobiology of LRRK2. PLoS Genet. 6, e1000902.

Xiong, Y., Yuan, C., Chen, R., Dawson, T.M., Dawson, V.L., 2012. ArfGAP1 is a GTPase activating protein for LRRK2: reciprocal regulation of ArfGAP1 by LRRK2. J. Neurosci. 32, 3877–3886.

Yim, Y.I., Sun, T., Wu, L.G., Raimondi, A., De Camilli, P., Eisenberg, E., Greene, L.E., 2010. Endocytosis and clathrin-uncoating defects at synapses of auxilin knockout mice. Proc. Natl. Acad. Sci. U. S. A. 107, 4412–4417.

Yu, W.J., Cheng, L., Li, N.N., Wang, L., Tan, E.K., Peng, R., 2015. Interaction between SNCA, LRRK2 and GAK increases susceptibility to Parkinson's disease in a Chinese population. eNeurologicalSci. 1, 3–6.

Zarranz, J.J., Alegre, J., Gomez-Esteban, J.C., Lezcano, E., Ros, R., Ampuero, I., Vidal, L., Hoenicka, J., Rodriguez, O., Atares, B., Llorens, V., Gomez Tortosa, E., Del Ser, T., Munoz, D.G., De Yebenes, J.G., 2004. The new mutation, E46K, of alpha-synuclein causes Parkinson and Lewy body dementia. Ann. Neurol. 55, 164–173.

Zavodszky, E., Seaman, M.N., Moreau, K., Jimenez-Sanchez, M., Breusegem, S.Y., Harbour, M.E., Rubinsztein, D.C., 2014. Mutation in VPS35 associated with Parkinson's disease impairs WASH complex association and inhibits autophagy. Nat. Commun. 5, 3828.

Zhang, Y., Shu, L., Sun, Q., Pan, H., Guo, J., Tang, B., 2018. A comprehensive analysis of the association between SNCA polymorphisms and the risk of Parkinson's disease. Front. Mol. Neurosci. 11, 391.

Zhao, Y., Perera, G., Takahashi-Fujigasaki, J., Mash, D.C., Vonsattel, J.P.G., Uchino, A., Hasegawa, K., Jeremy Nichols, R., Holton, J.L., Murayama, S., Dzamko, N., Halliday, G.M., 2018. Reduced LRRK2 in association with retromer dysfunction in post-mortem brain tissue from LRRK2 mutation carriers. Brain 141, 486–495.

Zimprich, A., Biskup, S., Leitner, P., Lichtner, P., Farrer, M., Lincoln, S., Kachergus, J., Hulihan, M., Uitti, R.J., Calne, D.B., Stoessl, A.J., Pfeiffer, R.F., Patenge, N., Carbajal, I.C., Vieregge, P., Asmus, F., Muller-Myhsok, B., Dickson, D.W., Meitinger, T., Strom, T.M., Wszolek, Z.K., Gasser, T., 2004. Mutations in LRRK2 cause autosomal-dominant parkinsonism with pleomorphic pathology. Neuron 44, 601–607.

Zimprich, A., Benet-Pages, A., Struhal, W., Graf, E., Eck, S.H., Offman, M.N., Haubenberger, D., Spielberger, S., Schulte, E.C., Lichtner, P., Rossle, S.C., Klopp, N., Wolf, E., Seppi, K., Pirker, W., Presslauer, S., Mollenhauer, B., Katzenschlager, R., Foki, T., Hotzy, C., Reinthaler, E., Harutyunyan, A., Kralovics, R., Peters, A., Zimprich, F., Brucke, T., Poewe, W., Auff, E., Trenkwalder, C., Rost, B., Ransmayr, G., Winkelmann, J., Meitinger, T., Strom, T.M., 2011. A mutation in VPS35, encoding a subunit of the retromer complex, causes late-onset Parkinson disease. Am. J. Hum. Genet. 89, 168–175.

New players in basal ganglia dysfunction in Parkinson's disease

9

Sara Meoni[a,b], Rubens Gisbert Cury[c], Elena Moro[a,b,*]

[a]*Movement Disorders Unit, Division of Neurology, CHU of Grenoble, Grenoble Alpes University, Grenoble, France*
[b]*INSERM U1216, Grenoble Institute of Neurosciences, Grenoble, France*
[c]*Movement Disorders Center, Department of Neurology, School of Medicine, University of São Paulo, São Paulo, Brazil*
Corresponding author: Tel.: +33-4-76-76-57-91, e-mail address: elenamfmoro@gmail.com

Abstract

The classical model of the basal ganglia (BG) circuit has been recently revised with the identification of other structures that play an increasing relevant role especially in the pathophysiology of Parkinson's disease (PD). Numerous studies have supported the spreading of the alpha-synuclein pathology to several areas beyond the BG and likely even before their involvement.

With the aim of better understanding PD pathophysiology and finding new targets for treatment, the spinal cord, the pedunculopontine nucleus, the substantia nigra pars reticulata, the retina, the superior colliculus, the cerebellum, the nucleus parabrachialis and the Meynert's nucleus have been investigated both in animal and human studies.

In this chapter, we describe the main anatomical and functional connections between the above structures and the BG, the relationship between their pathology and PD features, and the rational of applying neuromodulation treatment to improve motor and non-motor symptoms in PD. Some of these new players in the BG circuits might also have a potential intriguing role as early biomarkers of PD.

Keywords

Basal ganglia, Dysfunction, Parkinson's disease, Pathophysiology, Retina, Superior colliculus

1 Introduction

Since the basal ganglia (BG) are involved in both voluntary and automatic movements, their dysfunction can result in hypokinetic or hyperkinetic disorders (Obeso et al., 2002). Among the hypokinetic disorders, Parkinson's disease (PD)

Progress in Brain Research, Volume 252, ISSN 0079-6123, https://doi.org/10.1016/bs.pbr.2020.01.001

symptoms are mainly related to the loss of dopaminergic neurons in the substantia nigra pars compacta (SNc) which translates in loss of dopamine (Lotharius and Brundin, 2002; Schroll and Hamker, 2013). However, over the last years, advances in the PD pathophysiology research have introduced new players in the BG dysfunction. Indeed, according to the Braak's hypothesis of PD (Braak et al., 2003), alpha-synuclein pathology spreads from the periphery (gut and/or olfactory bulb) to the central nervous system (CNS), thus involving several other structures such as the spinal cord, the pedunculopontine nucleus (PPN), the substantia nigra pars reticulate (SNr), the retina, the superior colliculus (SC), the cerebellum, the parabrachialis nucleus (PBN), and the Meynert's nucleus (NBM) (Braak et al., 2003; Rietdijk et al., 2017; Visanji et al., 2013).

Here, we will illustrate the anatomofunctional networks linking the above CNS structures between them and with the BG (Fig. 1). Moreover, we will discuss how pathological changes in their anatomy and/or physiology might contribute to BG dysfunction, and how some surgical treatments may help to restore motor impairment in PD patients. These new players in the BG dysfunction provide new insights with better understanding the pathogenesis underpinning PD motor and non-motor features, and can help to find new useful biomarkers of early diagnosis, disease severity and progression of PD.

2 The spinal cord

There is evidence supporting presence and progression of the alpha-synuclein pathology in the spinal cord early during the course of the disease (Raudino and Leva, 2012). The spinal cord projects to several supraspinal structures through the spinothalamic, spinoreticular and spinocerebellar tracts. Conversely, multiple supraspinal structures are connected directly and indirectly with the spinal cord (Rossignol et al., 2006).

Concerning physiology of gait, the spinal circuitry is involved in the automatic locomotor network, in the control of muscle tone during locomotion, and stretch and flexion reflexes involved in the control of posture (Takakusaki, 2017). Therefore, spinal cord has been investigated as a potential therapeutic target for gait disturbances in PD (Yadav and Nicolelis, 2017).

2.1 Stimulation of the spinal cord

The effects of spinal cord stimulation (SCS) on bradykinesia and gait were initially studied in rodents (Fuentes et al., 2009) and in nonhuman primate models (Santana et al., 2014). Later, several independent reports have demonstrated positive results of short-term SCS on gait disturbances in small samples of PD patients (Agari and Date, 2012; Fénelon et al., 2012; Pinto de Souza et al., 2017). The mechanisms of SCS in ameliorating gait in PD are still unclear. In animal studies, analysis of local field potential (LFP) recordings from both motor cortex and striatum during SCS have

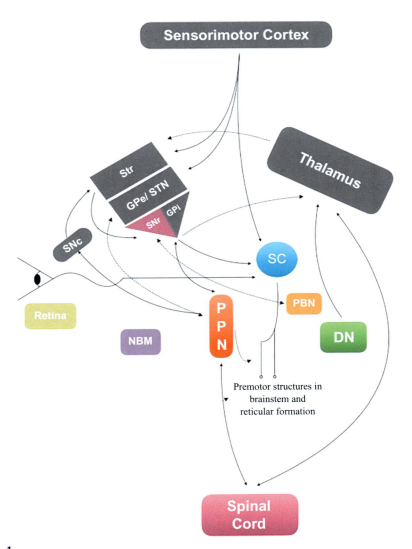

FIG. 1

Interconnected pathways between sensorimotor cortex, basal ganglia, brainstem structures and spinal cord in Parkinson's disease. DN, dentate nucleus; GPe, globus pallidus externus; GPi, globus pallidus internus; NBM, nucleus basalis of Meynert; PPN, peduncolopontine nucleus; SC, superior colliculus; SNc, substantia nigra compacta; SNr, substantia nigra reticulata; STN, subthalamic nucleus; Str, striatum nuclei; PBN, parabrachialis nucleus.

revealed changes in spectral power frequencies (Fuentes et al., 2009). Moreover, motor improvement was associated with disruption of aberrant low-frequency synchronous corticostriatal oscillations (beta-band power) (Petersson et al., 2019), leading to the appearance of neuronal activity that resembles to the state normally preceding spontaneous initiation of locomotion. SCS might also modulate specific somatosensory pathways and recruit brainstem arousal systems, favoring movement and locomotion (Fanselow et al., 2000).

Another possible mechanism of SCS in PD involves the BG. The mesencephalic locomotor region receives tonic inhibition fibers from the output nuclei of the BG, which in turn receive excitatory inputs from the cortex (Deniau et al., 2007). In PD, the lack of dopamine would finally induce tonic inhibition of the mesencephalic locomotor region thus compromising the locomotory system (Fuentes et al., 2009). SCS may work by activating cortical and thalamic inputs to the striatum, leading to depolarization and, subsequently, facilitating the activation of striatal projection neurons. Based on the current knowledge that the pedunculopontine nucleus (PPN) plays a key role in the initiation and modulation of gait, modulation of the PPN by either ascending pathways from the spinal cord or by retrograde descending projections to spinal neurons may explain improvements in motor symptoms in PD patients following SCS (de Andrade et al., 2016; de Lima-Pardini et al., 2018; Fonoff et al., 2019; Hubsch et al., 2019; Samotus et al., 2018). Nevertheless, efficacy of SCS in PD gait needs to be supported by larger randomized double-blinded studies.

3 The pedunculopontine nucleus

The pedunculopontine nucleus (PPN) is a collection of heterogeneous neurons at the junction of the midbrain and pons; it hosts cholinergic, glutamatergic, gamma-aminobutyric acid (GABAergic), and probably glycinergic neurons (Jenkinson et al., 2009). The PPN has an impressive array of reciprocal connections with the BG, motor cortex and spinal cord motor neurons (Jenkinson et al., 2009). Ascending connections are mostly concentrated on BG and thalamus, and descending fibers target the spinal cord and the medullary and pontine reticular formations (Pahapill and Lozano, 2000). The PPN also communicates with the contralateral PPN. There are bidirectional connection between the globus pallidus internus (GPi) and the PPN (Jenkinson et al., 2009). Small ipsilateral fibers from the PPN also reach the striatum, with an even smaller contralateral component (Lavoie and Parent, 1994).

Besides the pallidum, the PPN has strong glutaminergic and cholinergic connections with the SNc, whereas the SNr sends GABAergic projections to the PPN (Childs and Gale, 1983). The PPN innervates the subthalamic nucleus (STN) via cholinergic, glutamatergic, and GABAergic fibers (Bevan et al., 1995; Jenkinson et al., 2009). As a result, the STN sends a smaller glutamatergic projection to the PPN (Jenkinson et al., 2009).

Taken together, anatomical connections and physiological properties of the PPN exert a profound influence on the BG. These connections are mainly with the SN and STN, and to a lesser extend with the pallidum. Consequently, it is likely that PPN dysfunction caused by degeneration of the nucleus or indirectly due to BG abnormalities may play a major role in gait and balance in PD and related BG disorders.

3.1 Stimulation of the pedunculopontine nucleus area

There has been great hope in PPN stimulation for gait and balance disturbances in the last two decades (Pahapill and Lozano, 2000). Preclinical deep brain stimulation (DBS) experiments (Nandi et al., 2002) were followed by several clinical trials of PPN DBS in PD (Garcia-Rill et al., 2019). These investigations have proposed that unilateral or bilateral PPN DBS can improve gait freezing in both off and on medication states in the first 2 years after surgery (Ferraye et al., 2010; Moro et al., 2010). One long-term evaluation demonstrated persistent benefit in four out of six patients in terms of gait freezing and falls 4 years after unilateral PPN DBS (Mestre et al., 2016). Overall, the majority of the available studies have found that medication refractory gait freezing and falls can be ameliorated by PPN DBS (Thevathasan et al., 2018). However, strong conclusions cannot be drawn due to the small number of patients, the variability of patients' clinical characteristics and electrodes position (Thevathasan et al., 2018).

In tandem with freezing of gait, PPN stimulation has also been linked to improvements in other aspects of PD, including balance (Perera et al., 2018) and posture (Shih et al., 2013).

4 The substantia nigra pars reticulata

Together with the GPi, the SNr is a primary output nucleus of the BG (Deniau et al., 2007). The SNr sends GABAergic projections to the mesencephalic locomotor region, particularly the PPN (Childs and Gale, 1983; Grofova and Zhou, 1998). In PD, the SNr is pathologically overactivated (Breit et al., 2006), leading to inhibition of the locomotor network and contributing to the axial problems typical of PD progression. The integrative role of the SNr during locomotion has been recently emphasized by several studies (Deniau et al., 2007; Heilbronn et al., 2019; Lafreniere-Roula et al., 2010).

4.1 Stimulation of the substantia nigra pars reticulata

Whereas the STN DBS is effective for the management of segmental motor symptoms and fluctuations, axial motor symptoms show limited therapeutic response as the disease progresses. Combined stimulation of the SNr (using caudal contacts) and the STN (using rostral contacts) has been attempted to modulate locomotor integration (Chastan et al., 2009; Weiss et al., 2013). SNr stimulation at 130 Hz improved forelimb akinesia in a rat model of PD (Sutton et al., 2013). In a first double-blind randomized controlled clinical trial, combined stimulation (STN + SNr) was superior in controlling freezing of gait comparing to STN stimulation alone, whereas balance impairment remained unchanged. Another study showed that neuromodulation of the SNr but not of the STN improved the control of anticipatory postural adjustments in PD (Heilbronn et al., 2019). The authors hypothesized that STN stimulation would act predominantly by modulating the STN–thalamo–cortical circuit, while SNr stimulation would impact predominantly the descending nigro-pontine pathway, culminating to modulate distinct anticipatory postural adjustments modulation.

A more recently cross-over, randomized trial investigated the effects of simultaneous stimulation in both the STN and SNr at different frequencies in PD (126 Hz in STN and 63 Hz in SNr). This study compared the combined stimulation with the STN or the SNr stimulation alone. For most patients, the combined stimulation achieved the best freezing and balance control, suggesting that the add-on SNr stimulation to STN DBS alone could improve PD-associated gait disorders (Valldeoriola et al., 2019).

Freezing of gait has also been reported to be improved by combined STN and SNr stimulation (Weiss et al., 2019). Larger randomized controlled trials are needed to assess the efficacy of combined stimulation on gait disorders in PD.

5 The retina

The mammalian retina is a neural structure belonging to the CNS, composed by multiple layers of five types of neurons: photoreceptors in the outermost layer; bipolar, horizontal and dopaminergic amacrine cells in the middle layer, and retinal ganglion cells (RGCs) in the innermost layer (Dowling, 2002). The retina is responsible for the transduction of light into electrical signals, which are transmitted to the brain via the optic nerve, the lateral geniculate nucleus and the SC, to reach the primary visual cortex (V1). The retina has also indirect connections with the SNc and with other midbrain structures as well as the BG loop through the SC, as illustrated further below.

In PD patients, several oculo-visual abnormalities have been described, including abnormalities in primary vision (visual acuity, contrast sensitivity, color vision), in retinal morphology and in visuospatial attention (Guo et al., 2018).

Several studies have reported an impaired contrast sensitivity in PD patients, occurring early in the disease (Štenc Bradvica et al., 2015), in both foveal and peripheral retinal regions (Bodis-Wollner et al., 1987; Bulens et al., 1987; Delalande et al., 1996; Langheinrich et al., 2000; Štenc Bradvica et al., 2015). These visual defects have been related to progressive neurodegeneration of the retinal cells and the consequent dopaminergic deficiency, as L-dopa treatment has been showed to improve contrast sensitivity (Bulens et al., 1987; Hutton et al., 1993). Moreover, the loss in contrast sensitivity in PD seems to correlate with disease severity and cognitive impairment so that it could be a potential early diagnostic biomarker of PD (Archibald et al., 2011; Miri et al., 2016; Ridder et al., 2017).

A dysfunction in color vision has also been described in PD patients. An impaired color discrimination has been found in several tasks during clinical and psychophysical assessments (Büttner et al., 1995; Pieri et al., 2000; Regan et al., 1998), with changes in both chromatic (parvocellular for red-green and koniocellular for blue-yellow) and achromatic (magnicellular) pathways (Oh et al., 2011). These color contrast defects can occur early in PD (Birch et al., 1998; Piro et al., 2014), and correlate with the disease progression (Diederich et al., 2002; Müller et al., 2002; Oh et al., 2011) and improvement with dopaminergic treatment (Büttner et al., 1995).

Indeed, several retinal electrophysiological abnormalities, such as a decreased electrical activity at the fovea in electroretinogram (ERG) (Kaur et al., 2015; Moschos et al., 2011) and increased visual evoked potential (VEP) latency (Miri et al., 2016) have been reported in PD patients with impaired visual function. These functional alterations have been associated to dopaminergic loss, as they are improved by dopaminergic treatment in patients (Peppe et al., 1995) and worsened by dopaminergic receptor blocking agents in PD animal models (Bodis-Wollner and Tzelepi, 1998).

The underlying mechanisms of contrast and color dysfunction in PD are still under debate. It is likely that a retinal dopaminergic synaptic dysfunction, as described in ERG patterns (Price et al., 1994; Sartucci and Porciatti, 2006), coexists with the disease-related dysfunction of the other cortical and subcortical structures involved in the visual pathway.

Recent advances in imaging technologies, such as the optical coherence tomography (OCT), have allowed a direct investigation of the morphology of the retina. There is evidence of significant thinning of the peripapillary retinal nerve fiber layer (RNFL) and ganglion cell layer (GCL) abnormalities in PD patients. A recent meta-analysis of spectral-domain OCT studies (Chrysou et al., 2019) shows that PD patients have significantly thinner retinas compared to age- and gender-matched controls. The inner retinal layers with dopaminergic amacrine neurons, consisting of the RNFL and the ganglion cell layer-inner plexiform layer (GCL-IPL), are mostly affected. Moreover, it seems to exist a spatial distribution in the RNFL abnormalities, with a sparing of the nasal sector of the retina. Moreover, the retinal thinning has been shown to be present in the early stages of PD, and to correlate with disease severity, thus suggesting a link with nigral dopaminergic degeneration (Ahn et al., 2018).

Indeed, as a SNC structure, the retina shares with the brain and the spinal cord several molecular and cellular pathologies (London et al., 2013). Immunohistological studies in the postmortem retinas of PD patients have shown diffuse alpha-synuclein depositions and Lewy bodies in the inner plexiform layer and the RNFL, corresponding to the sites of retina thinning observed with OCT (Beach et al., 2014; Bodis-Wollner et al., 1987).

Taken all together, these imaging and histological findings suggest that alpha-synuclein pathology in the retina may mirror—and even precede—PD brain pathology. As such, OCT parameters as the retinal thickness could be useful as an early diagnostic/disease progression biomarkers of PD.

6 The superior colliculus

The superior colliculus (SC) is a complex mesencephalic structure involved in the integration of multimodal stimuli (visual, tactile, auditory) and the rapid generation of motor responses adapted to the gaze, head and limbs (Fecteau and Munoz, 2006). It is composed of superficial sensory neurons (SLSC) that receive direct visual input from retina by the retinotectal pathway in response to changes in luminance and movement in the visual field and indirect input from visual cortex. The SC includes also sensory and pre-motor neurons in the deep-layers (DLSC) that project to the brainstem burst generator for ocular movements and to the tectospinal and tecto-reticulospinal pathways. The SLSC and the DLSC are modulated by a population of GABA (gamma-aminobutyric acid)-ergic interneurons. The SC is deeply integrated in the BG loop; it receives afferences from the PPN and the SNpr, and it projects to the SNpc, the STN and the thalamus (Redgrave et al., 2010).

Although based on few neuro-anatomical studies in mammalian models, a network of connections between the cortical and subcortical visual structures, including the SC, and the BG has been demonstrated (López-Figueroa et al., 1995; McHaffie et al., 2006; Saint-Cyr et al., 2000; Sil'kis, 2007).

Using neuroanatomical tracing and electrophysiology techniques in animal models, the SC has been shown to project toward the main output structures of the BG, such as the STN (Coizet et al., 2009; Comoli et al., 2003; McHaffie et al., 2006) allowing direct transmission of visual information (Coizet et al., 2009; Dommett et al., 2005).

Recently, a SC dysfunction has been found in parkinsonian animal models (Coizet et al., 2009), likely due to an abnormal SC plasticity secondary to an increased inhibition from the SNr. A similar SC dysfunction in early PD patients has been recently discovered (Moro et al., 2020). This functional MRI study has reported no modulation of the SC responses to the increase of luminance contrast in PD patients compared to controls. These findings differ from the alteration of luminance contrast processing with normal aging (Bellot et al., 2016) and seem to be related to PD pathology. Indeed, the α-synuclein accumulation in the human SC have been reported in previous studies, with a more severe pathology in the SC layers connected to midbrain (Erskine et al., 2017). Moreover, alpha-synuclein spreading to the SC may occur via the vagal nerve through the parabrachial nucleus (Gauriau and Bernard, 2002), and via the retina (Beach et al., 2014). Therefore, the pathological process is likely to affect the SC before the SNc so that the SC visual dysfunction may be an early diagnostic biomarker of PD.

Because of these close anatomical connections between the BG and visual structures such as visual cortex, SC and retina, the alpha-synuclein pathology in PD could interfere with the transmission and processing of visual information at several levels of the visual pathway, explaining the different visual disturbances, including visual hallucinations, observed in parkinsonian patients. A recent functional MRI study has reported that in PD patients the visual hallucinations are associated with functional changes in associative visual cortices (V2, V3 and the fusiform gyri bilaterally), possibly related to an abnormal strengthened stability of resting-state networks (Dujardin et al., 2019).

7 The cerebellum

The cerebellum influences several cortical areas involved in the BG loops through projections to distinct thalamic nuclei (Bostan and Strick, 2018). Animal studies have demonstrated substantial interactions between the BG and the cerebellum. There is evidence that the output of the dentate is linked to the striatum via a disynaptic connection and to the GPe via a trisynaptic connection (Hoshi et al., 2005). It is likely that these connections are mediated by intralaminar nuclei and the ventrolateral thalamus (Hoshi et al., 2005). In humans, probabilistic diffusion tractography has confirmed the presence of dentato-thalamo-striato-pallidal and subthalamo-cerebellar connections (Pelzer et al., 2013). Consequently, abnormal cerebellar activity may alter the BG function and cause or worsen movement disorders (Tewari et al., 2017).

Recent studies have provided compelling evidence supporting a role of cerebellar-BG pathway dysfunction in the pathophysiology of parkinsonian resting tremor (França et al., 2018; Helmich et al., 2012). Using a functional MRI protocol, tremor amplitude-related activity was localized in the cerebello-thalamo-cortical circuit (Helmich et al., 2012). Tremor-dominant PD patients showed increased functional connectivity between the BG and the cerebello-thalamo-cortical circuit compared with non-tremor PD patients and healthy subjects (Helmich et al., 2012).

Functional and metabolic changes in the cerebellum have been observed also in PD patients treated with surgeries known to reduce levodopa-induced dyskinesia (LID) (Cenci et al., 2011, 2018), such as GPi DBS or pallidotomy (Fukuda et al., 2001). The binding potential of cerebellar sigma-receptor (related to LID) in dyskinetic PD patients was found markedly elevated compared with healthy volunteers (Nimura et al., 2004). After pallidotomy, a reduction in the binding potentials was associated to an almost complete amelioration of dyskinesia, supporting the role of the cerebellum in the LID pathophysiology.

Due its significant connections with subcortical and cortical regions, the cerebellum could be a promising target for neuromodulation in movement disorders (França et al., 2018). Indeed, cerebellar transcranial magnetic stimulation might have an antidyskinetic effect in PD patients with LID (Koch et al., 2009; Sanna et al., 2019), and also be an effective therapy in PD patients with tremor (Bologna et al., 2015).

8 The parabrachialis nucleus

The parabrachialis nucleus (PBN) is a structure located at the pons-midbrain boundary, laterally to the locus coeruleus, and dissected by the superior cerebellar peduncle into a medial and a lateral region (Hashimoto et al., 2009).

The PBN receives afferent connections from the dorsal laminae of the spinal cord, the nucleus tractus solitarius, and the forebrain regions (Tokita et al., 2009). It projects to several forebrain nuclei (thalamus, amygdala, bed nucleus of the stria terminalis, hypothalamus), to the midbrain and to the cortex (Fulwiler and Saper, 1984).

The PBN is primarily involved in the transmission of sensory stimuli (visceral malaise, taste, temperature, pain, and itch) from periphery and spinal cord to forebrain structures, which in turn modulate the PBN activity in a feedback loop (Palmiter, 2018).

Evidence from animal studies suggests that the PBN and the BG (i.e., the STN) are linked in a nociceptive network (Pautrat et al., 2018). Specifically, the PBN projects pain information to both STN (Klop et al., 2005) and SNc (Coizet et al., 2010). The STN is well known to largely project to BG output structures such as the SNr (Alexander et al., 1986; Gurney et al., 2004), which in turn projects to the SC and PBN (Deniau and Chevalier, 1992; Schneider, 1986).

In PD, pain and variable burning or stabbing sensations without any physical cause are frequently described (Ha and Jankovic, 2012). Pain threshold is likely reduced in PD patients, as they seem to be more sensitive to painful stimuli than healthy population (Brefel-Courbon et al., 2013). As such, brain circuits involved in pain perception and processing may be altered in PD. Indeed, the abnormal activity observed in the STN (Albin et al., 1995; Bergman et al., 1994) could contribute to some of the pain symptoms reported by parkinsonian patients, due to the involvement of the STN in the nociceptive network connecting the SC and the PBN. The evidence that STN DBS can improve pain in parkinsonian patients (Jung et al., 2015) further supports the hypothesis of a direct modulation of the STN pain network (Pautrat et al., 2018).

9 The nucleus basalis of Meynert

The nucleus basalis of Meynert (NBM) is a structure in the basal forebrain, located inferior to the GPi, which provides the major source of cholinergic innervation to the cortex (Mesulam et al., 1983).

This nuclei is largely involved in cognitive and behavioral functions, including arousal, attention, perception, and memory (Goard and Dan, 2009; Kalmbach et al., 2012).

The degeneration of this structure has long been implicated in the pathophysiology of PD dementia (PDD), as well as Alzheimer's disease (AD) (Candy et al., 1983; Gratwicke and Foltynie, 2018).

More recently, increasing attention has been given to the involvement of this nuclei in the early stage of PD. Indeed, it has been hypothesized that PD pathology in the NBM begins early in the anterior portion, which has connections with the olfactory bulb, and then progresses caudally (Liu et al., 2015). This anatomical progression supports the prion-like propagation hypothesis and the dual-hit hypothesis of alpha-synuclein pathology (Hawkes et al., 2009).

The NBM degenerates considerably in patients with PDD, correlating with cortical cholinergic deficits and cognitive impairment (Choi et al., 2012; Shimada et al., 2009). Consequently, structural imaging of the basal forebrain region has been investigated as a biomarker of cognitive decline in PD (Choi et al., 2012). In a recent MRI study, NBM atrophy seems to predict cognitive decline in de novo PD at 2 years (Ray et al., 2018).

9.1 Stimulation of the nucleus basalis of Meynert

Cognitive impairment and dementia are an important source of disability and reduction in quality of life for both patients and caregiver (Aarsland et al., 2009; Rosenthal et al., 2010). As the current treatment of PDD with acetylcholinesterase inhibitors and N-methyl D-aspartate–receptor antagonists moderately improve at best cognitive function, new therapeutic options are under investigation, such as NBM stimulation. Based on animal models (Goard and Dan, 2009), both the NBM and the fornix (belonging to the Papez's circuit for memory processes) have been hypothesized to be a key DBS target to enhance cognitive function in patients with cognitive impairment. To date, three DBS clinical trials have been performed in AD (Kuhn et al., 2015; Laxton et al., 2010; Lozano et al., 2016). Although DBS seems to be safe, no significant improvement of cognitive scores has been reported so far. Only one small, randomized, clinical trial, investigating the effect of low frequency (20 Hz) DBS of NBM on cognitive function is available in PDD (Gratwicke and Foltynie, 2018). Similarly to the trials in AD, no improvement was observed in cognitive outcomes. Larger trials with longer follow-up are ongoing to better investigated utility and long-term safety of NBM DBS in PD with cognitive impairment.

10 Conclusions

There is increasing evidence that the prion-like spreading of alpha-synuclein pathology in PD, likely associated to several other pathological processes (neuroinflammation, mitochondrial dysfunction, etc.), drives a degeneration of several non-dopaminergic pathways (noradrenergic, cholinergic, serotoninergic) in addition to the dopaminergic nigrostriatal pathway (Titova et al., 2017). This complex multisystem pathology results in the clinical heterogeneity of motor and non-motor features observed from the prodromal to the advanced stage of PD.

Several new structures connected to the BG are becoming very relevant for understanding PD pathophysiology. Pathological changes in morphology or function of non-dopaminergic circuits and their connections to BG network may represent potential biomarkers for early PD diagnosis (such as the visual dysfunction of the SC or the thinning of the retina), patient phenotyping, prediction of disease progression and therapeutic interventions.

References

Aarsland, D., Marsh, L., Schrag, A., 2009. Neuropsychiatric symptoms in Parkinson's disease. Mov. Disord. 24, 2175–2186. https://doi.org/10.1002/mds.22589.

Agari, T., Date, I., 2012. Spinal cord stimulation for the treatment of abnormal posture and gait disorder in patients with Parkinson's disease. Neurol. Med. Chir. 52, 470–474.

Ahn, J., Lee, J.-Y., Kim, T.W., Yoon, E.J., Oh, S., Kim, Y.K., Kim, J.-M., Woo, S.J., Kim, K.W., Jeon, B., 2018. Retinal thinning associates with nigral dopaminergic loss in de novo Parkinson disease. Neurology 91, e1003–e1012. https://doi.org/10.1212/WNL.0000000000006157.

Albin, R.L., Young, A.B., Penney, J.B., 1995. The functional anatomy of disorders of the basal ganglia. Trends Neurosci. 18, 63–64.

Alexander, G.E., DeLong, M.R., Strick, P.L., 1986. Parallel organization of functionally segregated circuits linking basal ganglia and cortex. Annu. Rev. Neurosci. 9, 357–381. https://doi.org/10.1146/annurev.ne.09.030186.002041.

Archibald, N.K., Clarke, M.P., Mosimann, U.P., Burn, D.J., 2011. Visual symptoms in Parkinson's disease and Parkinson's disease dementia. Mov. Disord. 26, 2387–2395. https://doi.org/10.1002/mds.23891.

Beach, T.G., Carew, J., Serrano, G., Adler, C.H., Shill, H.A., Sue, L.I., Sabbagh, M.N., Akiyama, H., Cuenca, N., Arizona Parkinson's Disease Consortium, 2014. Phosphorylated α-synuclein-immunoreactive retinal neuronal elements in Parkinson's disease subjects. Neurosci. Lett. 571, 34–38. https://doi.org/10.1016/j.neulet.2014.04.027.

Bellot, E., Coizet, V., Warnking, J., Knoblauch, K., Moro, E., Dojat, M., 2016. Effects of aging on low luminance contrast processing in humans. Neuroimage 139, 415–426. https://doi.org/10.1016/j.neuroimage.2016.06.051.

Bergman, H., Wichmann, T., Karmon, B., DeLong, M.R., 1994. The primate subthalamic nucleus. II. Neuronal activity in the MPTP model of parkinsonism. J. Neurophysiol. 72, 507–520. https://doi.org/10.1152/jn.1994.72.2.507.

Bevan, M.D., Francis, C.M., Bolam, J.P., 1995. The glutamate-enriched cortical and thalamic input to neurons in the subthalamic nucleus of the rat: convergence with GABA-positive terminals. J. Comp. Neurol. 361, 491–511. https://doi.org/10.1002/cne.903610312.

Birch, J., Kolle, R.U., Kunkel, M., Paulus, W., Upadhyay, P., 1998. Acquired colour deficiency in patients with Parkinson's disease. Vision Res. 38, 3421–3426. https://doi.org/10.1016/s0042-6989(97)00398-2.

Bodis-Wollner, I., Tzelepi, A., 1998. The push-pull action of dopamine on spatial tuning of the monkey retina: the effects of dopaminergic deficiency and selective D1 and D2 receptor ligands on the pattern electroretinogram. Vision Res. 38, 1479–1487. https://doi.org/10.1016/s0042-6989(98)00028-5.

Bodis-Wollner, I., Marx, M.S., Mitra, S., Bobak, P., Mylin, L., Yahr, M., 1987. Visual dysfunction in Parkinson's disease. Loss in spatiotemporal contrast sensitivity. Brain 110 (Pt. 6), 1675–1698. https://doi.org/10.1093/brain/110.6.1675.

Bologna, M., Di Biasio, F., Conte, A., Iezzi, E., Modugno, N., Berardelli, A., 2015. Effects of cerebellar continuous theta burst stimulation on resting tremor in Parkinson's disease. Parkinsonism Relat. Disord. 21, 1061–1066. https://doi.org/10.1016/j.parkreldis.2015.06.015.

Bostan, A.C., Strick, P.L., 2018. The basal ganglia and the cerebellum: nodes in an integrated network. Nat. Rev. Neurosci. 19, 338–350. https://doi.org/10.1038/s41583-018-0002-7.

Braak, H., Del Tredici, K., Rüb, U., de Vos, R.A.I., Jansen Steur, E.N.H., Braak, E., 2003. Staging of brain pathology related to sporadic Parkinson's disease. Neurobiol. Aging 24, 197–211.

Brefel-Courbon, C., Ory-Magne, F., Thalamas, C., Payoux, P., Rascol, O., 2013. Nociceptive brain activation in patients with neuropathic pain related to Parkinson's disease. Parkinsonism Relat. Disord. 19, 548–552. https://doi.org/10.1016/j.parkreldis.2013.02.003.

Breit, S., Lessmann, L., Unterbrink, D., Popa, R.C., Gasser, T., Schulz, J.B., 2006. Lesion of the pedunculopontine nucleus reverses hyperactivity of the subthalamic nucleus and substantia nigra pars reticulata in a 6-hydroxydopamine rat model. Eur. J. Neurosci. 24, 2275–2282. https://doi.org/10.1111/j.1460-9568.2006.05106.x.

Bulens, C., Meerwaldt, J.D., Van der Wildt, G.J., Van Deursen, J.B., 1987. Effect of levodopa treatment on contrast sensitivity in Parkinson's disease. Ann. Neurol. 22, 365–369. https://doi.org/10.1002/ana.410220313.

Büttner, T., Kuhn, W., Müller, T., Patzold, T., Przuntek, H., 1995. Color vision in Parkinson's disease: missing influence of amantadine sulphate. Clin. Neuropharmacol. 18, 458–463. https://doi.org/10.1097/00002826-199510000-00009.

Candy, J.M., Perry, R.H., Perry, E.K., Irving, D., Blessed, G., Fairbairn, A.F., Tomlinson, B.E., 1983. Pathological changes in the nucleus of Meynert in Alzheimer's and Parkinson's diseases. J. Neurol. Sci. 59, 277–289. https://doi.org/10.1016/0022-510x(83)90045-x.

Cenci, M.A., Ohlin, K.E., Odin, P., 2011. Current options and future possibilities for the treatment of dyskinesia and motor fluctuations in Parkinson's disease. CNS Neurol. Disord. Drug Targets 10, 670–684.

Cenci, M.A., Jorntell, H., Petersson, P., 2018. On the neuronal circuitry mediating L-DOPA-induced dyskinesia. J. Neural Transm. 125, 115–769.

Chastan, N., Westby, G.W.M., Yelnik, J., Bardinet, E., Do, M.C., Agid, Y., Welter, M.L., 2009. Effects of nigral stimulation on locomotion and postural stability in patients with Parkinson's disease. Brain 132, 172–184. https://doi.org/10.1093/brain/awn294.

Childs, J.A., Gale, K., 1983. Neurochemical evidence for a nigrotegmental GABAergic projection. Brain Res. 258, 109–114. https://doi.org/10.1016/0006-8993(83)91233-7.

Choi, S.H., Jung, T.M., Lee, J.E., Lee, S.-K., Sohn, Y.H., Lee, P.H., 2012. Volumetric analysis of the substantia innominata in patients with Parkinson's disease according to cognitive status. Neurobiol. Aging 33, 1265–1272. https://doi.org/10.1016/j.neurobiolaging.2010.11.015.

Chrysou, A., Jansonius, N.M., van Laar, T., 2019. Retinal layers in Parkinson's disease: a meta-analysis of spectral-domain optical coherence tomography studies. Parkinsonism Relat. Disord. 64, 40–49. https://doi.org/10.1016/j.parkreldis.2019.04.023.

Coizet, V., Graham, J.H., Moss, J., Bolam, J.P., Savasta, M., McHaffie, J.G., Redgrave, P., Overton, P.G., 2009. Short-latency visual input to the subthalamic nucleus is provided by the midbrain superior colliculus. J. Neurosci. 29, 5701–5709. https://doi.org/10.1523/JNEUROSCI.0247-09.2009.

Coizet, V., Dommett, E.J., Klop, E.M., Redgrave, P., Overton, P.G., 2010. The parabrachial nucleus is a critical link in the transmission of short latency nociceptive information to midbrain dopaminergic neurons. Neuroscience 168, 263–272. https://doi.org/10.1016/j.neuroscience.2010.03.049.

Comoli, E., Coizet, V., Boyes, J., Bolam, J.P., Canteras, N.S., Quirk, R.H., Overton, P.G., Redgrave, P., 2003. A direct projection from superior colliculus to substantia nigra for detecting salient visual events. Nat. Neurosci. 6, 974–980. https://doi.org/10.1038/nn1113.

de Andrade, E.M., Ghilardi, M.G., Cury, R.G., Barbosa, E.R., Fuentes, R., Teixeira, M.J., Fonoff, E.T., 2016. Spinal cord stimulation for Parkinson's disease: a systematic review. Neurosurg. Rev. 39, 27–35. discussion 35. https://doi.org/10.1007/s10143-015-0651-1.

de Lima-Pardini, A.C., Coelho, D.B., Souza, C.P., Souza, C.O., Ghilardi, M.G.D.S., Garcia, T., Voos, M., Milosevic, M., Hamani, C., Teixeira, L.A., Fonoff, E.T., 2018. Effects of spinal cord stimulation on postural control in Parkinson's disease patients with freezing of gait. Elife 7, e37727. https://doi.org/10.7554/eLife.37727.

Delalande, I., Destée, A., Hache, J.C., Forzy, G., Bughin, M., Benhadjali, J., 1996. Visual evoked potentials and spatiotemporal contrast sensitivity changes in idiopathic Parkinson's disease and multiple system atrophy. Adv. Neurol. 69, 319–325.

Deniau, J.M., Chevalier, G., 1992. The lamellar organization of the rat substantia nigra pars reticulata: distribution of projection neurons. Neuroscience 46, 361–377. https://doi.org/10.1016/0306-4522(92)90058-a.

Deniau, J.M., Mailly, P., Maurice, N., Charpier, S., 2007. The pars reticulata of the substantia nigra: a window to basal ganglia output. Prog. Brain Res. 160, 151–172. https://doi.org/10.1016/S0079-6123(06)60009-5.

Diederich, N.J., Raman, R., Leurgans, S., Goetz, C.G., 2002. Progressive worsening of spatial and chromatic processing deficits in Parkinson disease. Arch. Neurol. 59, 1249–1252. https://doi.org/10.1001/archneur.59.8.1249.

Dommett, E., Coizet, V., Blaha, C.D., Martindale, J., Lefebvre, V., Walton, N., Mayhew, J.E.W., Overton, P.G., Redgrave, P., 2005. How visual stimuli activate dopaminergic neurons at short latency. Science 307, 1476–1479. https://doi.org/10.1126/science.1107026.

Dowling, J.E., 2002. In: Ramachandran, V.S. (Ed.), Encyclopedia of the Human Brain. Academic Press, pp. 217–235.

Dujardin, K., Roman, D., Baille, G., Pins, D., Lefebvre, S., Delmaire, C., Defebvre, L., Jardri, R., 2019. What can we learn from fMRI capture of visual hallucinations in Parkinson's disease? Brain Imaging Behav. 13 (4), 1–7. https://doi.org/10.1007/s11682-019-00185-6.

Erskine, D., Thomas, A.J., Taylor, J.-P., Savage, M.A., Attems, J., McKeith, I.G., Morris, C.M., Khundakar, A.A., 2017. Neuronal loss and A-synuclein pathology in the superior colliculus and its relationship to visual hallucinations in dementia with Lewy bodies. Am. J. Geriatr. Psychiatry 25, 595–604. https://doi.org/10.1016/j.jagp.2017.01.005.

Fanselow, E.E., Reid, A.P., Nicolelis, M.A., 2000. Reduction of pentylenetetrazole-induced seizure activity in awake rats by seizure-triggered trigeminal nerve stimulation. J. Neurosci. 20, 8160–8168.

Fecteau, J.H., Munoz, D.P., 2006. Salience, relevance, and firing: a priority map for target selection. Trends Cogn. Sci. (Regul. Ed.) 10, 382–390. https://doi.org/10.1016/j.tics.2006.06.011.

Fénelon, G., Goujon, C., Gurruchaga, J.-M., Cesaro, P., Jarraya, B., Palfi, S., Lefaucheur, J.-P., 2012. Spinal cord stimulation for chronic pain improved motor function in a patient with Parkinson's disease. Parkinsonism Relat. Disord. 18, 213–214. https://doi.org/10.1016/j.parkreldis.2011.07.015.

Ferraye, M.U., Debû, B., Fraix, V., Goetz, L., Ardouin, C., Yelnik, J., Henry-Lagrange, C., Seigneuret, E., Piallat, B., Krack, P., Le Bas, J.-F., Benabid, A.-L., Chabardès, S., Pollak, P., 2010. Effects of pedunculopontine nucleus area stimulation on gait disorders in Parkinson's disease. Brain 133, 205–214. https://doi.org/10.1093/brain/awp229.

Fonoff, E.T., de Lima-Pardini, A.C., Coelho, D.B., Monaco, B.A., Machado, B., Pinto de Souza, C., Dos Santos Ghilardi, M.G., Hamani, C., 2019. Spinal cord stimulation for freezing of gait: from bench to bedside. Front. Neurol. 10, 905. https://doi.org/10.3389/fneur.2019.00905.

França, C., de Andrade, D.C., Teixeira, M.J., Galhardoni, R., Silva, V., Barbosa, E.R., Cury, R.G., 2018. Effects of cerebellar neuromodulation in movement disorders: a systematic review. Brain Stimul. 11, 249–260. https://doi.org/10.1016/j.brs.2017.11.015.

Fuentes, R., Petersson, P., Siesser, W.B., Caron, M.G., Nicolelis, M.A.L., 2009. Spinal cord stimulation restores locomotion in animal models of Parkinson's disease. Science 323, 1578–1582. https://doi.org/10.1126/science.1164901.

Fukuda, M., Mentis, M.J., Ma, Y., Dhawan, V., Antonini, A., Lang, A.E., Lozano, A.M., Hammerstad, J., Lyons, K., Koller, W.C., Moeller, J.R., Eidelberg, D., 2001. Networks mediating the clinical effects of pallidal brain stimulation for Parkinson's disease: a PET study of resting-state glucose metabolism. Brain 124, 1601–1609. https://doi.org/10.1093/brain/124.8.1601.

Fulwiler, C.E., Saper, C.B., 1984. Subnuclear organization of the efferent connections of the parabrachial nucleus in the rat. Brain Res. 319, 229–259. https://doi.org/10.1016/0165-0173(84)90012-2.

Garcia-Rill, E., Saper, C.B., Rye, D.B., Kofler, M., Nonnekes, J., Lozano, A., Valls-Solé, J., Hallett, M., 2019. Focus on the pedunculopontine nucleus. Consensus review from the May 2018 brainstem society meeting in Washington, DC, USA. Clin. Neurophysiol. 130, 925–940. https://doi.org/10.1016/j.clinph.2019.03.008.

Gauriau, C., Bernard, J.-F., 2002. Pain pathways and parabrachial circuits in the rat. Exp. Physiol. 87, 251–258. https://doi.org/10.1113/eph8702357.

Goard, M., Dan, Y., 2009. Basal forebrain activation enhances cortical coding of natural scenes. Nat. Neurosci. 12, 1444–1449. https://doi.org/10.1038/nn.2402.

Gratwicke, J.P., Foltynie, T., 2018. Early nucleus basalis of Meynert degeneration predicts cognitive decline in Parkinson's disease. Brain 141, 7–10. https://doi.org/10.1093/brain/awx333.

Grofova, I., Zhou, M., 1998. Nigral innervation of cholinergic and glutamatergic cells in the rat mesopontine tegmentum: light and electron microscopic anterograde tracing and immunohistochemical studies. J. Comp. Neurol. 395, 359–379.

Guo, L., Normando, E.M., Shah, P.A., De Groef, L., Cordeiro, M.F., 2018. Oculo-visual abnormalities in Parkinson's disease: possible value as biomarkers. Mov. Disord. 33, 1390–1406. https://doi.org/10.1002/mds.27454.

Gurney, K.N., Humphries, M., Wood, R., Prescott, T.J., Redgrave, P., 2004. Testing computational hypotheses of brain systems function: a case study with the basal ganglia. Network 15, 263–290.

Ha, A.D., Jankovic, J., 2012. Pain in Parkinson's disease. Mov. Disord. 27, 485–491. https://doi.org/10.1002/mds.23959.

Hashimoto, K., Obata, K., Ogawa, H., 2009. Characterization of parabrachial subnuclei in mice with regard to salt tastants: possible independence of taste relay from visceral processing. Chem. Senses 34, 253–267. https://doi.org/10.1093/chemse/bjn085.

Hawkes, C.H., Del Tredici, K., Braak, H., 2009. Parkinson's disease: the dual hit theory revisited. Ann. N. Y. Acad. Sci. 1170, 615–622. https://doi.org/10.1111/j.1749-6632.2009.04365.x.

Heilbronn, M., Scholten, M., Schlenstedt, C., Mancini, M., Schöllmann, A., Cebi, I., Pötter-Nerger, M., Gharabaghi, A., Weiss, D., 2019. Anticipatory postural adjustments are modulated by substantia nigra stimulation in people with Parkinson's disease and freezing of gait. Parkinsonism Relat. Disord. 66, 34–39. https://doi.org/10.1016/j.parkreldis.2019.06.023.

Helmich, R.C., Hallett, M., Deuschl, G., Toni, I., Bloem, B.R., 2012. Cerebral causes and consequences of parkinsonian resting tremor: a tale of two circuits? Brain 135, 3206–3226. https://doi.org/10.1093/brain/aws023.

Hoshi, E., Tremblay, L., Féger, J., Carras, P.L., Strick, P.L., 2005. The cerebellum communicates with the basal ganglia. Nat. Neurosci. 8, 1491–1493. https://doi.org/10.1038/nn1544.

Hubsch, C., D'Hardemare, V., Ben Maacha, M., Ziegler, M., Patte-Karsenti, N., Thiebaut, J.B., Gout, O., Brandel, J.P., 2019. Tonic spinal cord stimulation as therapeutic option in Parkinson disease with axial symptoms: effects on walking and quality of life. Parkinsonism Relat. Disord. 63, 235–237. https://doi.org/10.1016/j.parkreldis.2019.02.044.

Hutton, J.T., Morris, J.L., Elias, J.W., 1993. Levodopa improves spatial contrast sensitivity in Parkinson's disease. Arch. Neurol. 50, 721–724. https://doi.org/10.1001/archneur.1993.00540070041012.

Jenkinson, N., Nandi, D., Muthusamy, K., Ray, N.J., Gregory, R., Stein, J.F., Aziz, T.Z., 2009. Anatomy, physiology, and pathophysiology of the pedunculopontine nucleus. Mov. Disord. 24, 319–328. https://doi.org/10.1002/mds.22189.

Jung, Y.J., Kim, H.-J., Jeon, B.S., Park, H., Lee, W.-W., Paek, S.H., 2015. An 8-year follow-up on the effect of subthalamic nucleus deep brain stimulation on pain in Parkinson disease. JAMA Neurol. 72 (5), 504–510. https://doi.org/10.1001/jamaneurol.2015.8.

Kalmbach, A., Hedrick, T., Waters, J., 2012. Selective optogenetic stimulation of cholinergic axons in neocortex. J. Neurophysiol. 107, 2008–2019. https://doi.org/10.1152/jn.00870.2011.

Kaur, M., Saxena, R., Singh, D., Behari, M., Sharma, P., Menon, V., 2015. Correlation between structural and functional retinal changes in Parkinson disease. J. Neuroophthalmol. 35, 254–258. https://doi.org/10.1097/WNO.0000000000000240.

Klop, E.M., Mouton, L.J., Hulsebosch, R., Boers, J., Holstege, G., 2005. In cat four times as many lamina I neurons project to the parabrachial nuclei and twice as many to the periaqueductal gray as to the thalamus. Neuroscience 134, 189–197. https://doi.org/10.1016/j.neuroscience.2005.03.035.

Koch, G., Brusa, L., Carrillo, F., Lo Gerfo, E., Torriero, S., Oliveri, M., Mir, P., Caltagirone, C., Stanzione, P., 2009. Cerebellar magnetic stimulation decreases levodopa-induced dyskinesias in Parkinson disease. Neurology 73, 113–119. https://doi.org/10.1212/WNL.0b013e3181ad5387.

Kuhn, J., Hardenacke, K., Shubina, E., Lenartz, D., Visser-Vandewalle, V., Zilles, K., Sturm, V., Freund, H.-J., 2015. Deep brain stimulation of the nucleus basalis of Meynert in early stage of Alzheimer's dementia. Brain Stimul. 8, 838–839. https://doi.org/10.1016/j.brs.2015.04.002.

Lafreniere-Roula, M., Kim, E., Hutchison, W.D., Lozano, A.M., Hodaie, M., Dostrovsky, J.O., 2010. High-frequency microstimulation in human globus pallidus and substantia nigra. Exp. Brain Res. 205, 251–261. https://doi.org/10.1007/s00221-010-2362-8.

Langheinrich, T., Tebartz van Elst, L., Lagrèze, W.A., Bach, M., Lücking, C.H., Greenlee, M.W., 2000. Visual contrast response functions in Parkinson's disease: evidence from electroretinograms, visually evoked potentials and psychophysics. Clin. Neurophysiol. 111, 66–74. https://doi.org/10.1016/s1388-2457(99)00223-0.

Lavoie, B., Parent, A., 1994. Pedunculopontine nucleus in the squirrel monkey: projections to the basal ganglia as revealed by anterograde tract-tracing methods. J. Comp. Neurol. 344, 210–231. https://doi.org/10.1002/cne.903440204.

Laxton, A.W., Tang-Wai, D.F., McAndrews, M.P., Zumsteg, D., Wennberg, R., Keren, R., Wherrett, J., Naglie, G., Hamani, C., Smith, G.S., Lozano, A.M., 2010. A phase I trial of deep brain stimulation of memory circuits in Alzheimer's disease. Ann. Neurol. 68, 521–534. https://doi.org/10.1002/ana.22089.

Liu, A.K.L., Chang, R.C.-C., Pearce, R.K.B., Gentleman, S.M., 2015. Nucleus basalis of Meynert revisited: anatomy, history and differential involvement in Alzheimer's and Parkinson's disease. Acta Neuropathol. 129, 527–540. https://doi.org/10.1007/s00401-015-1392-5.

London, A., Benhar, I., Schwartz, M., 2013. The retina as a window to the brain-from eye research to CNS disorders. Nat. Rev. Neurol. 9, 44–53. https://doi.org/10.1038/nrneurol.2012.227.

López-Figueroa, M.O., Ramirez-Gonzalez, J.A., Divac, I., 1995. Projections from the visual areas to the neostriatum in rats. A re-examination. Acta Neurobiol. Exp. (Wars) 55, 165–175.

Lotharius, J., Brundin, P., 2002. Pathogenesis of Parkinson's disease: dopamine, vesicles and alpha-synuclein. Nat. Rev. Neurosci. 3, 932–942. https://doi.org/10.1038/nrn983.

Lozano, A.M., Fosdick, L., Chakravarty, M.M., Leoutsakos, J.-M., Munro, C., Oh, E., Drake, K.E., Lyman, C.H., Rosenberg, P.B., Anderson, W.S., Tang-Wai, D.F., Pendergrass, J.C., Salloway, S., Asaad, W.F., Ponce, F.A., Burke, A., Sabbagh, M., Wolk, D.A., Baltuch, G., Okun, M.S., Foote, K.D., McAndrews, M.P., Giacobbe, P., Targum, S.D., Lyketsos, C.G., Smith, G.S., 2016. A phase II study of fornix deep brain stimulation in mild Alzheimer's disease. J. Alzheimers Dis. 54, 777–787. https://doi.org/10.3233/JAD-160017.

McHaffie, J.G., Jiang, H., May, P.J., Coizet, V., Overton, P.G., Stein, B.E., Redgrave, P., 2006. A direct projection from superior colliculus to substantia nigra pars compacta in the cat. Neuroscience 138, 221–234. https://doi.org/10.1016/j.neuroscience.2005.11.015.

Mestre, T.A., Sidiropoulos, C., Hamani, C., Poon, Y.-Y., Lozano, A.M., Lang, A.E., Moro, E., 2016. Long-term double-blinded unilateral pedunculopontine area stimulation in Parkinson's disease. Mov. Disord. 31, 1570–1574. https://doi.org/10.1002/mds.26710.

Mesulam, M.M., Mufson, E.J., Levey, A.I., Wainer, B.H., 1983. Cholinergic innervation of cortex by the basal forebrain: cytochemistry and cortical connections of the septal area, diagonal band nuclei, nucleus basalis (substantia innominata), and hypothalamus in the rhesus monkey. J. Comp. Neurol. 214, 170–197. https://doi.org/10.1002/cne.902140206.

Miri, S., Glazman, S., Mylin, L., Bodis-Wollner, I., 2016. A combination of retinal morphology and visual electrophysiology testing increases diagnostic yield in Parkinson's disease. Parkinsonism Relat. Disord. 22 (Suppl. 1), S134–S137. https://doi.org/10.1016/j.parkreldis.2015.09.015.

Moro, E., Hamani, C., Poon, Y.-Y., Al-Khairallah, T., Dostrovsky, J.O., Hutchison, W.D., Lozano, A.M., 2010. Unilateral pedunculopontine stimulation improves falls in Parkinson's disease. Brain 133, 215–224. https://doi.org/10.1093/brain/awp261.

Moro, E., Bellot, E., Meoni, S., Pelissier, P., Hera, R., Dojat, M., Coizet, V., 2020. Superior Colliculus Study Group. Visual dysfunction of the superior colliculus in de novo Parkinsonian patients. Ann Neurol. PubMed PMID: 32030799. https://doi.org/10.1002/ana.25696. [Epub ahead of print].

Moschos, M.M., Tagaris, G., Markopoulos, I., Margetis, I., Tsapakis, S., Kanakis, M., Koutsandrea, C., 2011. Morphologic changes and functional retinal impairment in patients with Parkinson disease without visual loss. Eur. J. Ophthalmol. 21, 24–29. https://doi.org/10.5301/ejo.2010.1318.

Müller, T., Woitalla, D., Peters, S., Kohla, K., Przuntek, H., 2002. Progress of visual dysfunction in Parkinson's disease. Acta Neurol. Scand. 105, 256–260. https://doi.org/10.1034/j.1600-0404.2002.1o154.x.

Nandi, D., Liu, X., Winter, J.L., Aziz, T.Z., Stein, J.F., 2002. Deep brain stimulation of the pedunculopontine region in the normal non-human primate. J. Clin. Neurosci. 9, 170–174. https://doi.org/10.1054/jocn.2001.0943.

Nimura, T., Ando, T., Yamaguchi, K., Nakajima, T., Shirane, R., Itoh, M., Tominaga, T., 2004. The role of sigma-receptors in levodopa-induced dyskinesia in patients with advanced Parkinson disease: a positron emission tomography study. J. Neurosurg. 100, 606–610. https://doi.org/10.3171/jns.2004.100.4.0606.

Obeso, J.A., Rodríguez-Oroz, M.C., Rodríguez, M., Arbizu, J., Giménez-Amaya, J.M., 2002. The basal ganglia and disorders of movement: pathophysiological mechanisms. News Physiol. Sci. 17, 51–55. https://doi.org/10.1152/nips.01363.2001.

Oh, Y.-S., Kim, J.-S., Chung, S.-W., Song, I.-U., Kim, Y.-D., Kim, Y.-I., Lee, K.-S., 2011. Color vision in Parkinson's disease and essential tremor. Eur. J. Neurol. 18, 577–583. https://doi.org/10.1111/j.1468-1331.2010.03206.x.

Pahapill, P.A., Lozano, A.M., 2000. The pedunculopontine nucleus and Parkinson's disease. Brain 123 (Pt. 9), 1767–1783. https://doi.org/10.1093/brain/123.9.1767.

Palmiter, R.D., 2018. The parabrachial nucleus: CGRP neurons function as a general alarm. Trends Neurosci. 41, 280–293. https://doi.org/10.1016/j.tins.2018.03.007.

Pautrat, A., Rolland, M., Barthelemy, M., Baunez, C., Sinniger, V., Piallat, B., Savasta, M., Overton, P.G., David, O., Coizet, V., 2018. Revealing a novel nociceptive network that links the subthalamic nucleus to pain processing. Elife 7, e36607. https://doi.org/10.7554/eLife.36607.

Pelzer, E.A., Hintzen, A., Goldau, M., von Cramon, D.Y., Timmermann, L., Tittgemeyer, M., 2013. Cerebellar networks with basal ganglia: feasibility for tracking cerebello-pallidal and subthalamo-cerebellar projections in the human brain. Eur. J. Neurosci. 38, 3106–3114. https://doi.org/10.1111/ejn.12314.

Peppe, A., Stanzione, P., Pierelli, F., De Angelis, D., Pierantozzi, M., Bernardi, G., 1995. Visual alterations in de novo Parkinson's disease: pattern electroretinogram latencies are more delayed and more reversible by levodopa than are visual evoked potentials. Neurology 45, 1144–1148. https://doi.org/10.1212/wnl.45.6.1144.

Perera, T., Tan, J.L., Cole, M.H., Yohanandan, S.A.C., Silberstein, P., Cook, R., Peppard, R., Aziz, T., Coyne, T., Brown, P., Silburn, P.A., Thevathasan, W., 2018. Balance control systems in Parkinson's disease and the impact of pedunculopontine area stimulation. Brain 141, 3009–3022. https://doi.org/10.1093/brain/awy216.

Petersson, P., Halje, P., Cenci, M.A., 2019. Significance and translational value of high-frequency cortico-basal ganglia oscillations in Parkinson's disease. J. Parkinsons Dis. 9, 183–196.

Pieri, V., Diederich, N.J., Raman, R., Goetz, C.G., 2000. Decreased color discrimination and contrast sensitivity in Parkinson's disease. J. Neurol. Sci. 172, 7–11. https://doi.org/10.1016/s0022-510x(99)00204-x.

Pinto de Souza, C., Hamani, C., Oliveira Souza, C., Lopez Contreras, W.O., Dos Santos Ghilardi, M.G., Cury, R.G., Reis Barbosa, E., Jacobsen Teixeira, M., Talamoni Fonoff, E., 2017. Spinal cord stimulation improves gait in patients with Parkinson's disease previously treated with deep brain stimulation. Mov. Disord. 32, 278–282. https://doi.org/10.1002/mds.26850.

Piro, A., Tagarelli, A., Nicoletti, G., Fletcher, R., Quattrone, A., 2014. Color vision impairment in Parkinson's disease. J. Parkinsons Dis. 4, 317–319. https://doi.org/10.3233/JPD-140359.

Price, C.J., Wise, R.J., Watson, J.D., Patterson, K., Howard, D., Frackowiak, R.S., 1994. Brain activity during reading. The effects of exposure duration and task. Brain 117 (Pt. 6), 1255–1269. https://doi.org/10.1093/brain/117.6.1255.

Raudino, F., Leva, S., 2012. Involvement of the spinal cord in Parkinson's disease. Int. J. Neurosci. 122, 1–8. https://doi.org/10.3109/00207454.2011.613551.

Ray, N.J., Bradburn, S., Murgatroyd, C., Toseeb, U., Mir, P., Kountouriotis, G.K., Teipel, S.J., Grothe, M.J., 2018. In vivo cholinergic basal forebrain atrophy predicts cognitive decline in de novo Parkinson's disease. Brain 141, 165–176. https://doi.org/10.1093/brain/awx310.

Redgrave, P., Coizet, V., Comoli, E., McHaffie, J.G., Leriche, M., Vautrelle, N., Hayes, L.M., Overton, P., 2010. Interactions between the midbrain superior colliculus and the basal ganglia. Front. Neuroanat. 4, 132. https://doi.org/10.3389/fnana.2010.00132.

Regan, B.C., Freudenthaler, N., Kolle, R., Mollon, J.D., Paulus, W., 1998. Colour discrimination thresholds in Parkinson's disease: results obtained with a rapid computer-controlled colour vision test. Vision Res. 38, 3427–3431. https://doi.org/10.1016/s0042-6989(97)00402-1.

Ridder, A., Müller, M.L.T.M., Kotagal, V., Frey, K.A., Albin, R.L., Bohnen, N.I., 2017. Impaired contrast sensitivity is associated with more severe cognitive impairment in Parkinson disease. Parkinsonism Relat. Disord. 34, 15–19. https://doi.org/10.1016/j.parkreldis.2016.10.006.

Rietdijk, C.D., Perez-Pardo, P., Garssen, J., van Wezel, R.J.A., Kraneveld, A.D., 2017. Exploring Braak's hypothesis of Parkinson's disease. Front. Neurol. 8, 37. https://doi.org/10.3389/fneur.2017.00037.

Rosenthal, E., Brennan, L., Xie, S., Hurtig, H., Milber, J., Weintraub, D., Karlawish, J., Siderowf, A., 2010. Association between cognition and function in patients with Parkinson disease with and without dementia. Mov. Disord. 25, 1170–1176. https://doi.org/10.1002/mds.23073.

Rossignol, S., Dubuc, R., Gossard, J.-P., 2006. Dynamic sensorimotor interactions in locomotion. Physiol. Rev. 86, 89–154. https://doi.org/10.1152/physrev.00028.2005.

Saint-Cyr, J.A., Trépanier, L.L., Kumar, R., Lozano, A.M., Lang, A.E., 2000. Neuropsychological consequences of chronic bilateral stimulation of the subthalamic nucleus in Parkinson's disease. Brain 123, 2091–2108. https://doi.org/10.1093/brain/123.10.2091.

Samotus, O., Parrent, A., Jog, M., 2018. Spinal cord stimulation therapy for gait dysfunction in advanced Parkinson's disease patients. Mov. Disord. 33, 783–792. https://doi.org/10.1002/mds.27299.

Sanna, A., Follesa, P., Puligheddu, M., Cannas, A., Serra, M., Pisu, M.G., Dagostino, S., Solla, P., Tacconi, P., Marrosu, F., 2019. Cerebellar continuous theta burst stimulation reduces levodopa-induced dyskinesias and decreases serum BDNF levels. Neurosci. Lett. 716, 134653. https://doi.org/10.1016/j.neulet.2019.134653.

Santana, M.B., Halje, P., Simplício, H., Richter, U., Freire, M.A.M., Petersson, P., Fuentes, R., Nicolelis, M.A.L., 2014. Spinal cord stimulation alleviates motor deficits in a primate model of Parkinson disease. Neuron 84, 716–722. https://doi.org/10.1016/j.neuron.2014.08.061.

Sartucci, F., Porciatti, V., 2006. Visual-evoked potentials to onset of chromatic red-green and blue-yellow gratings in Parkinson's disease never treated with L-dopa. J. Clin. Neurophysiol. 23, 431–435. https://doi.org/10.1097/01.wnp.0000216127.53517.4d.

Schneider, J.S., 1986. Interactions between the basal ganglia, the pontine parabrachial region, and the trigeminal system in cat. Neuroscience 19, 411–425. https://doi.org/10.1016/0306-4522(86)90271-x.

Schroll, H., Hamker, F.H., 2013. Computational models of basal-ganglia pathway functions: focus on functional neuroanatomy. Front. Syst. Neurosci. 7, 122. https://doi.org/10.3389/fnsys.2013.00122.

Shih, L.C., Vanderhorst, V.G., Lozano, A.M., Hamani, C., Moro, E., 2013. Improvement of pisa syndrome with contralateral pedunculopontine stimulation. Mov. Disord. 28, 555–556. https://doi.org/10.1002/mds.25301.

Shimada, H., Hirano, S., Shinotoh, H., Aotsuka, A., Sato, K., Tanaka, N., Ota, T., Asahina, M., Fukushi, K., Kuwabara, S., Hattori, T., Suhara, T., Irie, T., 2009. Mapping of brain acetylcholinesterase alterations in Lewy body disease by PET. Neurology 73, 273–278. https://doi.org/10.1212/WNL.0b013e3181ab2b58.

Sil'kis, I.G., 2007. The contribution of synaptic plasticity in the basal ganglia to the processing of visual information. Neurosci. Behav. Physiol. 37, 779–790. https://doi.org/10.1007/s11055-007-0082-8.

Štenc Bradvica, I., Bradvica, M., Matić, S., Reisz-Majić, P., 2015. Visual dysfunction in patients with Parkinson's disease and essential tremor. Neurol. Sci. 36, 257–262. https://doi.org/10.1007/s10072-014-1930-2.

Sutton, A.C., Yu, W., Calos, M.E., Smith, A.B., Ramirez-Zamora, A., Molho, E.S., Pilitsis, J.G., Brotchie, J.M., Shin, D.S., 2013. Deep brain stimulation of the substantia nigra pars reticulata improves forelimb akinesia in the hemiparkinsonian rat. J. Neurophysiol. 109, 363–374. https://doi.org/10.1152/jn.00311.2012.

Takakusaki, K., 2017. Functional neuroanatomy for posture and gait control. J. Mov. Disord. 10, 1–17.https://doi.org/10.14802/jmd.16062.

Tewari, A., Fremont, R., Khodakhah, K., 2017. It's not just the basal ganglia: cerebellum as a target for dystonia therapeutics. Mov. Disord. 32, 1537–1545. https://doi.org/10.1002/mds.27123.

Thevathasan, W., Debu, B., Aziz, T., Bloem, B.R., Blahak, C., Butson, C., Czernecki, V., Foltynie, T., Fraix, V., Grabli, D., Joint, C., Lozano, A.M., Okun, M.S., Ostrem, J., Pavese, N., Schrader, C., Tai, C.-H., Krauss, J.K., Moro, E., Movement Disorders Society PPN DBS Working Groupin collaboration with the World Society for Stereotactic and Functional Neurosurgery, 2018. Pedunculopontine nucleus deep brain stimulation in Parkinson's disease: a clinical review. Mov. Disord. 33, 10–20. https://doi.org/10.1002/mds.27098.

Titova, N., Padmakumar, C., Lewis, S.J.G., Chaudhuri, K.R., 2017. Parkinson's: a syndrome rather than a disease? J. Neural Transm. (Vienna) 124, 907–914. https://doi.org/10.1007/s00702-016-1667-6.

Tokita, K., Inoue, T., Boughter, J.D., 2009. Afferent connections of the parabrachial nucleus in C57BL/6J mice. Neuroscience 161, 475–488. https://doi.org/10.1016/j.neuroscience.2009.03.046.

Valldeoriola, F., Muñoz, E., Rumià, J., Roldán, P., Cámara, A., Compta, Y., Martí, M.J., Tolosa, E., 2019. Simultaneous low-frequency deep brain stimulation of the substantia nigra pars reticulata and high-frequency stimulation of the subthalamic nucleus to treat levodopa unresponsive freezing of gait in Parkinson's disease: a pilot study. Parkinsonism Relat. Disord. 60, 153–157. https://doi.org/10.1016/j.parkreldis.2018.09.008.

Visanji, N.P., Brooks, P.L., Hazrati, L.-N., Lang, A.E., 2013. The prion hypothesis in Parkinson's disease: Braak to the future. Acta Neuropathol. Commun. 1, 2. https://doi.org/10.1186/2051-5960-1-2.

Weiss, D., Walach, M., Meisner, C., Fritz, M., Scholten, M., Breit, S., Plewnia, C., Bender, B., Gharabaghi, A., Wächter, T., Krüger, R., 2013. Nigral stimulation for resistant axial motor impairment in Parkinson's disease? A randomized controlled trial. Brain 136, 2098–2108. https://doi.org/10.1093/brain/awt122.

Weiss, D., Milosevic, L., Gharabaghi, A., 2019. Deep brain stimulation of the substantia nigra for freezing of gait in Parkinson's disease: is it about stimulation frequency? Parkinsonism Relat. Disord. 63, 229–230. https://doi.org/10.1016/j.parkreldis.2018.12.010.

Yadav, A.P., Nicolelis, M.A.L., 2017. Electrical stimulation of the dorsal columns of the spinal cord for Parkinson's disease. Mov. Disord. 32, 820–832. https://doi.org/10.1002/mds.27033.

Translational Therapeutics

Prodromal PD: A new nosological entity

10

Eva Schaeffer[a,*], **Ronald B. Postuma**[b], **Daniela Berg**[a]

[a]*Department of Neurology, Christian-Albrechts-University of Kiel, Kiel, Germany*
[b]*Department of Neurology, Montreal General Hospital, Montreal, QC, Canada*
Corresponding author: Tel.: +49-431-500-23983; Fax: +40-431-500 23994,
e-mail address: eva.schaeffer@uksh.de

Abstract

Recent years have brought a rapid growth in knowledge of the prodromal phase of Parkinson's disease (PD). It is now clear that the clinical phase of PD is preceded by a phase of progressing neurodegeneration lasting many years. This involves not only central nervous system structures outside the substantia nigra and neurotransmitter systems other than the dopaminergic system, but also the peripheral nervous systems. Different ways of alpha-synuclein spreading are presumed, corresponding to typical prodromal non-motor symptoms like constipation, REM sleep behavior disorder (RBD) and hyposmia. Moreover, many risk and prodromal markers have been identified and combined in the prodromal research criteria, which can be used to calculate an individual's probability of being in the prodromal phase of PD. Apart from specific genetic risk markers, including most importantly GBA- and LRRK2 mutations, RBD is currently the most important prodromal marker, predicting PD with a very high likelihood. This makes individuals with RBD a promising cohort for future clinical trials to detect and treat PD in its prodromal phase. New markers, especially those derived from tissue biopsies, quantitative motor assessment and imaging, appear very promising; these are paving the way for a better understanding of the prodromal phase and its potential clinicopathological subtypes, and a more precise probability calculation.

Keywords

Prodromal, Parkinson's disease, Biomarkers, REM sleep behavior disorder, Early detection, Risk factors

1 Introduction

For many decades Parkinson's disease (PD) was defined as a progressive degeneration of dopaminergic nerve cells in the substantia nigra (SN), leading to the cardinal motor symptoms of the disease. However, during the last years both clinical and

Progress in Brain Research, Volume 252, ISSN 0079-6123, https://doi.org/10.1016/bs.pbr.2020.01.003

pathological findings have led to the recognition that this concept covers only a small part of the whole picture of PD. On the one hand, it became evident that specific symptoms, like hyposmia or constipation, occur frequently long before the diagnosis (based on tremor, rigor and akinesia) is made. On the other hand, neuropathological findings showed that by the time cardinal symptoms occur the larger part of the do- paminergic nerve cells in the SN are already lost (Fearnley and Lees, 1991). Finally, the Braak stages, published 2003, confirmed that SN degeneration is just one aspect of a progressive neurodegenerative process, which comprises many other parts of the nervous system (Braak et al., 2003). Taken together the term "Parkinson's disease" is now the umbrella for three phases:

(1) the *preclinical phase*, in which the neurodegenerative process has already started, but does not cause any symptoms;
(2) the *prodromal phase*, in which a progressing neurodegeneration in different parts of the central nervous system (CNS) and peripheral nervous system (PNS) is leading to clinically evident symptoms and
(3) the *clinical phase*, in which cardinal motor symptoms become sufficiently evident to diagnose PD.

The definition of the *prodromal phase* can be considered an important milestone for PD research as it opens up a window to modify disease progression before the con- tinuing loss of dopaminergic neurons leads to burdensome and disabling motor symptoms. This chapter is compiled to give an overview on the current status of knowledge on pathophysiology, clinical markers and biomarkers as well as an out- look on promising approaches to improve diagnostic accuracy of the prodromal phase in the future.

2 Pathophysiology: Current concepts

An important contribution to the current understanding of the prodromal phase is the concept derived from the work of Braak et al. Thorough postmortem investigation of the whole brain led to the discovery of alpha-synuclein aggregates, formed into Lewy bodies and neurites, in vulnerable areas beyond the pars compacta of the substantia nigra (SN) (Fearnley and Lees, 1991). Comparison of many brains finally cumulated in the proposal of the so-called Braak stages, a staging of progressive alpha- synuclein pathology pathing its way from the medulla oblongata (including the ner- vus vagus and nervus glossopharyngeus or olfactory nucleus, named as stage one) followed by the pontine tegmentum (stage 2), only reaching the SN in the midbrain in stage 3 (Braak et al., 2003). Moreover, the typical alpha-synuclein pathology does not end in the SN but continues to spread over the brain, involving cortical structures in stages 4–6 of the disease. The findings and concept of Braak et al. have been val- idated in several neuropathological studies (Dickson et al., 2010; Kingsbury et al., 2010).

A further major contribution was the discovery that alpha-synuclein pathology in PD is not only restricted to the brain, but can also be found in the spinal cord and in many parts of the PNS, including the skin, the heart, and the enteral nervous system (Beach et al., 2010; Gelpi et al., 2014). Retrospective analyses of incidental Lewy Body Disease (iLBD) cases found that iLDB was associated with a higher rate of specific non-motor symptoms, including constipation and olfactory dysfunction, as seen, for example, in the Honolulu Asia Aging Study (Abbott et al., 2007; Driver-Dunckley et al., 2014; Ross et al., 2006). Braak et al. aggregated these findings in a dual-hit hypothesis, proposing two ways of entrance of a presumed pathogen: via the olfactory system or via the gut/stomach, which then induces an ascending alpha-synuclein pathology (Hawkes et al., 2007). Both potential paths are consistent with clinically findings preceding the diagnosis of PD, including early idiopathic hyposmia, corresponding to an olfactory entrance (Ross et al., 2008), and longstanding constipation, corresponding to an enteral entrance (Abbott et al., 2001). A potential propagation path from the enteral nervous system via the nervus vagus has gained recent attention, leading to the hypothesis that changes in the microbiome may play a role in PD etiology (see chapter "The gut microbiome in Parkinson's disease: A culprit or a bystander?" by Keshavarzian et al. for further information). Of particular interest was the finding that individuals receiving a vagotomy had a lower risk for developing PD in the future (Svensson et al., 2015), although it has to be noted that this finding has not been confirmed in all studies.

Irrespective of the starting point, how can alpha-synuclein pathology spread throughout the PNS and CNS in PD? An early answer came from postmortem findings in individuals who had received a transplantation of fetal dopaminergic nerve cells to improve symptoms of PD. On autopsy, the once healthy fetal dopaminergic nerve cells had developed Lewy body pathology after the transplantation, indicating a transmission of the alpha-synucleinopathy from the sick to the transplanted nerve cells (Kordower et al., 2008; Li et al., 2008). In vitro and animal models followed, showing a "seeding" mechanism of alpha-synuclein-fibrils, indicating a "prion-like" transmission route from one cell to the other (Desplats et al., 2009; Hansen et al., 2011; Luk et al., 2012; Volpicelli-Daley et al., 2011).

However, some clinical findings are still not answered sufficiently by the dual-hit hypothesis, including for instance individuals, in which cognitive decline precedes other symptoms (Marras and Chaudhuri, 2016). Other analogous staging systems have been proposed as well; some posit a possible "cortical first" route and so better account for the fact that dementia with Lewy bodies can often occur before development of Parkinson's disease (Beach et al., 2009; Borghammer and Van Den Berge, 2019).

3 Markers

The increasing awareness of a slowly-spreading premotor phase led to the search for early markers, to identify individuals for whom disease course modifying treatment might eventually be given before motor symptoms occur (Postuma and Berg, 2019). Two main types of markers were defined: (1) risk markers, which document increased PD risk, but do not reflect an already-ongoing neurodegenerative process

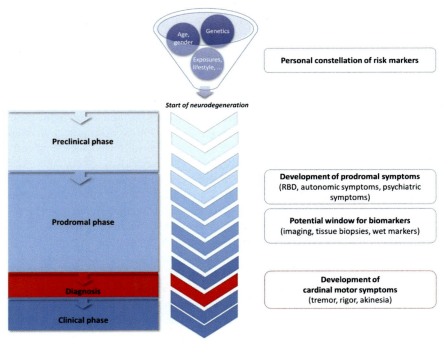

FIG. 1

Phases of Parkinson's disease displayed as currently applied. Once easily applicable biomarkers are at hand, these may already be used in individuals at risk or even as screening instruments before prodromal symptoms occur. RBD, REM sleep behavior disorder.

and (2) prodromal markers, which indicate that neurodegeneration has already begun, although clinical signs allowing the definite diagnosis are not yet fully manifest (see Fig. 1). As prodromal markers only occur after substantial neurodegeneration in the respective areas, a preclinical phase is postulated to precede the prodromal phase. The search for markers of the preclinical phase is ongoing, but has not yet been successful. Therefore, the following paragraphs will focus on risk and prodromal markers.

3.1 The research criteria for prodromal PD

Although many of the risk and prodromal markers occur frequently in PD patients, resulting in a high sensitivity, a main barrier for their use is their low specificity, as they also occur often in the general population. Therefore, a combination of markers is currently the most promising approach to better understand, whether an individual is in the prodromal phase (Ross et al., 2012). This principle was the basis for the development of the "Research Criteria on Prodromal Parkinson's Disease," compiled by a task force of the Movement Disorder Society (MDS) in 2015 (Berg et al., 2015). Based on likelihood rations derived from at least two prospective studies for each individual marker (see below) as well as the age dependent prior probability to develop PD, Bayesian probability statistics are applied to determine an individual's probability to be in the prodromal phase of PD. Since 2015 the criteria have been applied in several prospective and retrospective cohort studies and already showed good sensitivity and specificity (Mahlknecht et al., 2016, 2018; Mirelman et al., 2018). However, given rapid increase in knowledge, the research criteria have to be reevaluated regularly in order to further improve their diagnostic accuracy. A first update was thus already published in 2019 (Heinzel et al., 2019b). Currently, the criteria on prodromal PD are solely recommended for research purposes. If a likelihood of more than 80% is calculated, participation in disease-modification clinical studies can be envisioned, whereas individuals with lower likelihoods are rather recommended to be followed in prospective observational trials. Moreover, the knowledge of the most important risk and prodromal markers can be very helpful for early and differential diagnosis of PD. With the update of the criteria a freely available online calculator was provided to enable easy and proper application by research groups (Heinzel et al., 2019a).

The following section discusses the most important markers, which are part of the current MDS research criteria and can easily be filled in the dropdown menu of the web-based calculator, as well as new markers, that are still under evaluation (see Fig. 2 for an overview).

3.2 Risk markers

It is particularly noteworthy that although most risk factors are immutable, including age, gender or genetics, some of them (especially lifestyle habits) can potentially be influenced, making them particularly important for individuals with an already increased risk constellation. However, it should be noted that any observed correlation between risk markers and PD does not automatically imply causality (aside from genetics).

3.2.1 Genetic risk markers

Apart from the well-known monogenetic forms of PD which include dominant (like SNCA or LRRK2) or recessive genes (like Parkin or PINK1), about 90 risk loci for PD have been identified during the last years (Nalls et al., 2019). Of these, variants in the glucocerebrosidase (GBA) gene are of particular interest for the research community, as they have both a relatively high frequency in population and a high relative risk for PD (relative risk around five on average; Sidransky et al., 2009). Typical prodromal symptoms have been observed in cohorts of heterozygous GBA mutation carriers without clinical PD. Moreover, PD patients with a heterozygous GBA mutation seem to have a more rapid prodromal phase with a particularly high symptom burden (Beavan et al., 2015; Honeycutt et al., 2019; Zimmermann et al., 2019).

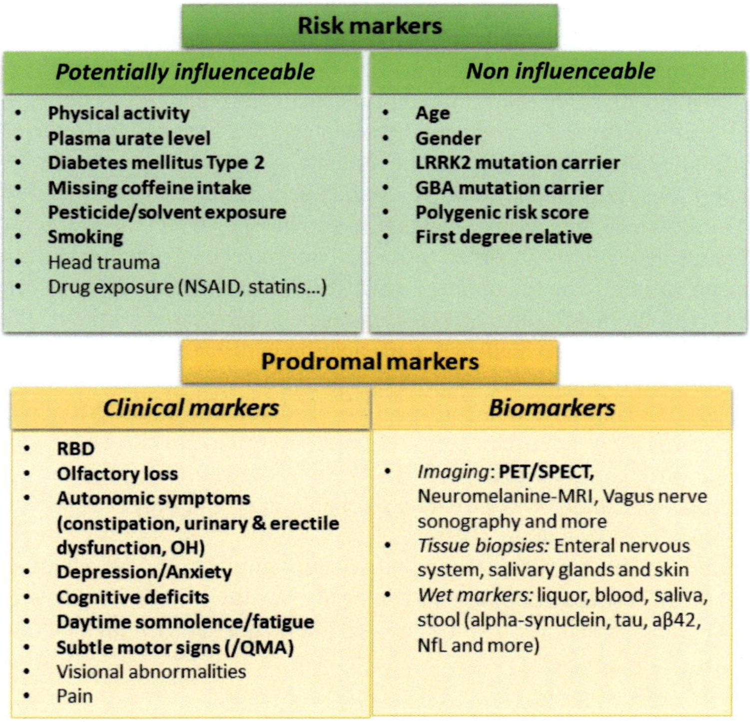

FIG. 2

Selection of risk and prodromal markers in Parkinson's disease. Markers listed in the official Movement Disorder Society research criteria for prodromal PD are printed in **bold**. GBA, β-glucocerebrosidase; LRRK2, leucine-rich repeat kinase 2; NSAID, non-steroidal anti-inflammatory drug; OH, orthostatic hypotension; PET, positron emission tomography; QMA, quantitative motor assessment; NfL, neurofilament light chain; RBD, REM sleep behavior disorder; SPECT, single photon emission computed tomography.

Mutations in LRRK2 contribute to rare highly penetrant dominant forms of PD, to medium risk variants with intermediate allele frequency (G2019S), and common low-risk variants, which may play a role in polygenic risk scores. Similar to GBA mutations, mutations in the LRRK2 gene, especially the G2019S mutation has a high prevalence in Ashkenazi Jews (Thaler et al., 2009), but also in North African Arabs (Lesage et al., 2006). Cohorts of G2019S mutation carriers are of great value for the understanding and identification of prodromal markers (Mirelman et al., 2016, 2018). Increasing understanding of the specific pathophysiological pathways in LRRK2 and GBA mutation carriers have led to the development of first causative therapeutic strategies, which are currently tested in first disease-modifying trials (see also chapter "Pharmacological targets for symptomatic and disease-modifying therapies—an update").

Apart from likelihood ratios of the GBA and LRRK2 G2019S variant associated with a moderate risk for PD, information on family history of PD (first-degree relatives) as well as data of polygenic risk scores (if available) are now included in the updated MDS prodromal criteria.

3.2.2 Non-genetic risk markers

Although PD can be observed in nearly all ages, higher age is the most important immutable risk factor for PD. The prodromal criteria appreciate this dependence of PD risk on age as "Prior Probability" in the risk calculation, raising to 4% in individuals older than 80 (Berg et al., 2015). Another commonly known risk factor is gender, with males having a 1.5 times higher probability for being in the prodromal phase of PD.

An important imaging risk marker, which has already found a way into clinical practice for many years, is SN hyperechogenicity, as assessed by transcranial sonography (Berg et al., 2011, 2013). The pathophysiology of this finding has not been completely clarified, although higher tissue iron content as well as microglia activation have been shown to contribute to the ultrasound signal (Berg et al., 2002). As this marker seems to occur already at very young age with little change over time, it has been attributed as a risk marker (Iova et al., 2004). However, resolution of images of new ultrasound machines seem to suggest slight changes of the marker over time, so in the future SN hyperechogenicity might eventually be assigned as a prodromal marker.

Many different lifestyle and nutritional habits have been associated with PD risk. One of the most discussed "protective" factors for PD is smoking. A negative association between smoking and the occurrence of PD was already recognized more than 50 years ago (Nefzger et al., 1968) and has been affirmed in many retrospective and prospective cohort studies thereafter (Gorell et al., 1999; Grandinetti et al., 1994). To date the question of causality has not been answered (i.e., studies suggest that reward mechanisms may differ in those with at-risk/prodromal PD, such that addiction to nicotine is less likely). Also, caffeine intake negatively correlates in a dose-dependent manner with the development of PD during lifetime (Ascherio et al., 2001; Fall et al., 1999). In the MDS prodromal criteria, those drinking less than three cups of coffee or six cups of tea per week have a 1.35-fold increased likelihood ratio (Berg et al., 2015). Further, neuroprotective effects of physical activity have been confirmed in many large epidemiological studies with up to 40% reduction in PD risk for those engaging in regular physical activity in middle age (Ahlskog, 2011; Hou et al., 2017; Xu et al., 2010; Yang et al., 2015). This makes physical inactivity an important influenceable risk factor for PD, and it has been added to the 2019 prodromal criteria update (Heinzel et al., 2019b). Finally, many other nutritional aspects have been suggested as having neuroprotective potential. For example, omega-3-fatty acids and polyphenols may positively influence underlying neuropathological mechanisms in PD; however, as sufficient prospective studies are lacking they have not entered the prodromal criteria yet (Bousquet et al., 2011; Kujawska and Jodynis-Liebert, 2018).

Another group of risk factors relate to the environment. Environmental factors have been seen to play an important role in risk assessment, with high evidence for pesticide and solvent exposure (Ascherio et al., 2006; Gorell et al., 1998), corresponding to likelihood ratios of 1.5.

Finally, newly validated risk markers include higher serum urate levels in men and diabetes mellitus type II (Hu et al., 2007; Simon et al., 2014). Still in discussion are traumatic head injuries as risk markers (Camacho-Soto et al., 2017; Gardner et al., 2015) and intake of specific medications as protective factors, including statins and anti-inflammatory drugs (Bai et al., 2016; Chen et al., 2003).

3.3 Prodromal markers

To date, the most important prodromal markers are clinical, as the search for preclinical markers, i.e., markers reflecting the neurodegenerative process without being associated to clinical symptoms has not yet led to validated conclusive results.

3.3.1 Clinical markers
3.3.1.1 REM sleep behavior disorder (RBD)

REM sleep behavior disorder (RBD) is caused by loss of the neural structures responsible for maintaining atonia (paralysis) during REM sleep (Schenck et al., 2013). Within the framework of spreading neurodegeneration in the prodromal phase of PD, it reflects a propagation of alpha-synucleinopathy through the brainstem (Boeve et al., 2007). Clinically, patients "act out" their dreams; for example, conversing, yelling, or thrashing in apparent response to dream content. RBD is extremely common in synucleinopathies (PD, DLB, and MSA), occurring in 35–80% of cases. It is uncommon in any other prevalent neurodegenerative condition, something that can help considerably in differential diagnosis of parkinsonian and dementia syndromes. Within neurodegenerative synucleinopathies, history is often sufficient to make empiric diagnoses, but the ultimate diagnosis of RBD requires a polysomnogram to document loss of REM atonia. Treatments primarily include clonazepam and melatonin, which in observational studies appear to help most patients (although randomized controlled trials have produced equivocal results). The overall prevalence of idiopathic/isolated RBD (i.e., RBD without another known degenerative disease) is approximately 1% after age 50 (Haba-Rubio et al., 2018).

RBD can occur many years before parkinsonism and dementia, and the risk of developing these conditions is extremely high. Large-scale multicenter studies have found that over 80% of patients with polysomnographic-proven idiopathic/isolated RBD will develop full neurodegenerative syndromes (almost always a synucleinopathy) (Postuma et al., 2019). This high-risk is unprecedented in its strength. To illustrate: in the revised MDS Prodromal criteria, the likelihood ratio of most clinical markers ranges from 1.5 to 6 (e.g., depression = 1.8, constipation = 2.5, olfaction = 6.4) (Heinzel et al., 2019b). The strongest known biomarker (dopaminergic functional imaging) has a likelihood ratio of 43. For RBD, this likelihood ratio is 130. Therefore, RBD is by far the strongest known predictive clinical or biomarker available. Any patient with polysomnographic-proven RBD without a clear alternate explanation should be considered as a likely prodromal PD patient. This indicates the need for regular follow-up to detect and treat symptoms of synucleinopathy. Moreover, it has substantial research implications, the most important of which include:

(1) Testing of other neurodegenerative predictors: It has been clearly documented that iRBD patients have numerous clinical prodromal markers (olfactory loss, autonomic symptoms and signs, mild motor deficits on quantitative testing, mild cognitive changes). Moreover, numerous biomarkers of early PD, including dopaminergic functional neuroimaging, resting state MRI changes, abnormal substantia nigra architecture, MIBG scintigraphy, etc., are also abnormal. This provides strong evidence that these markers can occur before full PD/DLB. Moreover, it can help stratify neurodegenerative risk: with some markers, rates of phenoconversion can rise from a baseline of 7% to 15–20% per year (Iranzo et al., 2017; Postuma et al., 2015a, 2019).

(2) Mapping the course of prodromal disease: If most patients develop disease, systematic follow-up of patients from their prodromal stages allows one to directly observe the evolution of disease manifestations. In a recent multivariate analysis of iRBD patients, the evolution of clinical prodromal markers closely resembled predictions of pathologic staging models, such that olfactory and autonomic manifestations developed first (estimated prodromal intervals = 15–20 years), followed by subtle motor features (5–8 years) and subtle cognitive changes (3–5 years) (Adler and Beach, 2016; Braak et al., 2003; Fereshtehnejad et al., 2019).

(3) Neuroprotective therapies: Patients with iRBD are ideal for neuroprotective trials because they are early in the degenerative process, have an extremely high-risk, and are still untreated (removing an important measurement confound). Recent sample size estimates suggest that depending on patient selection, outcome selection and treatment effectiveness, definitive neuroprotective trials in iRBD may require as few as 125–400 patients per group (Postuma et al., 2019). Therefore, using RBD patients to treat early stages may be the key to the first neuroprotective breakthrough in PD research.

3.3.1.2 Hyposmia

Hyposmia has been recognized as a highly frequent non-motor symptom in PD for many decades (Ansari and Johnson, 1975). Pathophysiologically, Lewy body pathology has been found in the olfactory bulb, which may precede neurodegeneration in the SN (stage I in the Braak staging scheme; Braak et al., 2003) and therefore might reflect a starting point of alpha-synuclein propagation (Hawkes et al., 2007). Moreover, an involvement of the olfactory cortex has been discussed

(Silveira-Moriyama et al., 2009). Correspondingly, prospective studies confirmed that idiopathic hyposmia is associated with an increased risk for the development of PD and therefore can be categorized as a prodromal symptom (Haehner et al., 2007; Mahlknecht et al., 2015; Ponsen et al., 2004). The Honolulu Asia Aging Study and Arizona Study of Aging and Neurodegenerative Disorders prospectively followed individuals with abnormality in olfactory testing until postmortem diagnosis, and found a relation between postmortem iLDB and pre-mortem hyposmia (Driver-Dunckley et al., 2014; Ross et al., 2006). Based on recent evidence the likelihood ratio of olfactory loss in the MDS prodromal criteria has been raised to 6.4 (Heinzel et al., 2019b). However, an important issue is the high overlap of hyposmia with other conditions, particularly dementia syndromes such as Alzheimer's disease (Dintica et al., 2019). Moreover, the sense of smell in general decreases with age, reducing specificity (Stevens et al., 1987). Taken together, hyposmia remains an important prodromal marker, but cannot be used as standalone prognostic factor.

3.3.1.3 Dysautonomia

Many autonomic symptoms that are frequently observed in the clinical phase of PD can already occur in the prodromal phase, indicating an early involvement of the autonomous nervous system (Gao et al., 2011). One of the most established symptoms is constipation, reflecting the clinical manifestation of a proposed start of alpha-synuclein pathology in the enteral nervous system as proposed in the dual-hit hypothesis of Braak (Hawkes et al., 2007). Constipation seems to be one of the earliest signs of ongoing neurodegeneration, occurring more than 15 years before motor symptoms of PD (Postuma et al., 2013). Severity of constipation seems to play a role (Lin et al., 2014). Moreover lower frequency of bowel movements has also been associated with iLDB (Abbott et al., 2007). Other impairments and diseases of the enteral nervous system, including inflammatory bowel disease and irritable bowel syndrome, are currently under discussion as potential risk/prodromal factors for PD (Lai et al., 2014; Villumsen et al., 2019).

Another autonomic symptom occurring very early before the clinical phase is orthostatic hypotension (Postuma et al., 2013; Schrag et al., 2015), which reflects the involvement of the sympathetic nervous system. This is also evident in cardiac sympathetic denervation, as visualized by MIBG (included in the MDS diagnostic criteria for Parkinson's disease) (Goldstein et al., 2002; Iwanaga et al., 1999; Postuma et al., 2015b). However, the correct diagnosis of orthostatic hypotension is often affected by confounders, including most importantly antihypertensives. The updated diagnostic criteria differentiate between symptomatic-only and confirmed neurogenic orthostatic hypotension, as diagnosed with quantitative tests (e.g., Schellong test) (Heinzel et al., 2019b).

Finally, urinary and erectile dysfunction have been observed to occur around 5–10 years before the conversion to clinical PD (Gao et al., 2007; Schrag et al., 2015) and have been associated with alpha-synuclein pathology in the spinal cord (Del Tredici and Braak, 2012; VanderHorst et al., 2015).

3.3.1.4 Psychiatric symptoms, daytime sleepiness and fatigue

Two of the most important psychiatric prodromal symptoms, which often have an important impact on quality of life, are the occurrence of depression and anxiety. Depression can occur as a very early prodromal symptom, potentially manifesting up to 20 years before motor PD (Gustafsson et al., 2015; Leentjens et al., 2003; Schuurman et al., 2002). Together with anxiety as another important prodromal marker (Bower et al., 2010; Weisskopf et al., 2003), depression is not only an expression of dopaminergic denervation, but has also be linked to an impairment of serotonergic and noradrenergic neuronal transmission (Mayeux et al., 1984; Remy et al., 2005). It thus stands to reason that pathology of the central nervous system in prodromal PD involves also other neurotransmitter circuits.

The same holds true for changes in cognitive function, which have been recently included in the updated version of the MDS criteria (Heinzel et al., 2019b). To date the question whether dementia with Lewy Bodies (DLB) should be defined a separate and mutually-exclusive disease entity is still under discussion. The Task Force of the Movement Disorder Society proposes that DLB and PD are both part of the same disease spectrum, meaning that early cognitive changes leading to DLB are not excluded from the prodromal diagnostic criteria (Berg et al., 2014). Moreover, alterations in cognitive function have also been observed in PD without dementia at onset or within the first year, in particular mild impairment of executive function and working memory (Fengler et al., 2017; Weintraub et al., 2018). Impairment of dual-tasking can also be attributed to deficits in executive function and attentional networks (Belghali et al., 2017; Mirelman et al., 2011). In clinical PD, higher rates of dementia and a more malign progression of the disease have been linked to RBD (Fereshtehnejad et al., 2015). Most interestingly, a faster deterioration of cognitive function has also been observed in the prodromal phase of individuals with RBD (Fantini et al., 2011; Gagnon et al., 2009). The occurrence of early cognitive changes expands the dual-hit hypothesis of Braak to include predominant, early neocortical involvement and affection of the cholinergic system, as has been described in other staging models (Beach et al., 2009; Marras and Chaudhuri, 2016).

Finally, another frequently described prodromal symptom is the occurrence of daytime sleepiness and fatigue (Ross et al., 2012). These symptoms have a huge overlap to other psychiatric symptoms and their pathophysiology is still not sufficiently understood (Abbott et al., 2005; Friedman et al., 2007). Perhaps as the field develops, a specifically-defined fatigue state could be included as a prodromal symptom.

Taken together, a predominance of psychiatric symptoms in the prodromal phase might reflect a different subtype of alpha-synuclein propagation than the predominant autonomic/RBD subtype, with different neurotransmitter systems of the CNS being primarily involved (Sauerbier et al., 2016).

3.3.1.5 Sensory symptoms

Although sensory deficits are frequently observed in clinical PD, their role in the prodromal phase has not been sufficiently investigated yet. Therefore, they are so far not included in the research criteria. While pain is an important early sign of the clinical phase, leading often to false diagnoses and treatments, its role in prodromal PD is still unclear, although first studies have shown an increased risk of PD in individuals with pain syndromes (Lin et al., 2013). Moreover, visual disturbances in the prodromal phase, including changes in color vision or pupil reaction, have been described (Armstrong, 2015; Postuma et al., 2011). Additionally, first results from optical coherence tomography have not only shown retinal thinning in PD, but also in non-manifest GBA mutation carriers (McNeill et al., 2013; Stemplewitz et al., 2015).

3.3.1.6 Subtle motor signs

Braak's staging scheme suggests that motor symptoms are an expression of advanced Lewy Body pathology. However, some subtle motor signs have been recorded years before the clinical phase of PD already decades ago. An impressive example are the video documentations of the football player Ray Kennedy, showing a reduced one-sided arm swing 14 years before diagnosis of PD as recognized in 1992 (Lees, 1992). Since then the detection of subtle motor alterations in prodromal PD has gained increasing attention. Apart from semiquantitative, clinician based rating tools (like the MDS Unified Parkinson's Disease Rating scale) and easy to assess quantitative techniques (like the pegboard or the timed-up-and-go-test), technological advances opened up new digital possibilities to detect slight abnormalities in movements by quantitative motor assessment tools. Therefore, the assessment of motor signs reflects a fluid transition from clinical to technical marker assessment. So far, the most promising prodromal motor signs include quantitative tests of motor speed in the hands (Postuma et al., 2019), sensitive measures of subtle gait dysfunction (especially under dual-task conditions, suggesting some overlap with cognitive prodromal changes), and reduced arm swing. These have been well-documented in examined high-risk LRRK2 and idiopathic RBD cohorts (Mirelman et al., 2011, 2016; Postuma et al., 2012). Furthermore, subtle voice changes, including variability of frequency and articulatory deficits (Harel et al., 2004a,b; Rusz et al., 2016) have been documented.

3.3.2 Non clinical biomarkers

Despite a large number of biomarker studies, so far only one imaging marker (abnormal dopaminergic PET/SPECT) has been included in the MDS prodromal criteria (Berg et al., 2015). However, it is to be expected that sensitivity and specificity of additional biomarkers will increase in the course of technical progress in the following years, which then can be used to improve diagnostic accuracy of prodromal PD.

3.3.2.1 Imaging markers

So far molecular imaging techniques are some of the most promising imaging tools for the detection of early functional changes in prodromal PD. With regard to CNS changes, dopaminergic SPECT and PET techniques have been validated as applicable tools to evaluate the probability for prodromal PD. Of particular interest are studies in high-risk cohorts, showing abnormalities of F-DOPA PET in PARK6,

Parkin and LRRK2 mutation carriers (Adams et al., 2005; Khan et al., 2002a,b, 2005), abnormalities in metabolic network patterns of RBD patients (Holtbernd et al., 2014; Meles et al., 2018; Wu et al., 2014) and a high correlation between dopaminergic imaging and other prodromal symptoms (Jennings et al., 2014; Noyce et al., 2018). In a prospective cohort of RBD individuals quantification of (Holtbernd et al., 2014) I-FP-CIT SPECT could detect progression as patients phenoconverted to clinical PD (Iranzo et al., 2017). The Parkinson Associated Risk Syndrome (PARS) study confirmed a very high predictive value of DAT-SPECT (in combination with hyposmia) (Jennings et al., 2014). Abnormal PET/SPECT is currently rated as having the second highest LR (43.3) after RBD in the MDS research criteria for prodromal PD (Heinzel et al., 2019b).

Apart from molecular imaging methods, some upcoming MRI techniques are promising candidates for the visualization of early CNS changes in prodromal PD. Reduced signal in locus coeruleus on neuromelanin-MRI sequences has been linked to RBD in individuals with and without PD, indicating that an impairment of the noradrenergic system might play an important role in the generation of RBD symptoms (García-Lorenzo et al., 2013; Knudsen et al., 2018; Sommerauer et al., 2018). Abnormal iron signal in the substantia nigra (e.g., the "swallow-tail" sign) is common in PD, and has recently been documented in patients with idiopathic RBD (Bae et al., 2018; Iranzo et al., 2017). Other techniques with potential include white matter abnormalities documented on diffusion MRI, volumetry of the olfactory tract (Rolheiser et al., 2011; Sobhani et al., 2017; Wang et al., 2011) and abnormalities of network function documented with resting state fMRI (Postuma, 2016).

While the focus of imaging has generally been placed on the CNS, there have been recent developments in imaging focused on the PNS, which in many cases seems to be affected earlier in the course of alpha-synuclein propagation. Changes in sympathetic innervation of the heart can be detected using MIBG-SPECT in most patients with PD as well as the large majority of patients with idiopathic RBD (Goldstein et al., 2012; Miyamoto et al., 2006; Sakakibara et al., 2014). Parasympathetic denervation in the enteral nervous system can be visualized with donepezil-PET (Fedorova et al., 2017; Gjerløff et al., 2015). Both techniques were combined in an imaging study in idiopathic RBD; this study used imaging techniques to visualize the Braak stages in vivo, confirming an ascending pathology from the enteral nervous system via the brainstem in this subgroup of prodromal PD patients (Knudsen et al., 2018).

Finally, a very easy to assess and cost-effective technique might be the sonographic measurement of the vagus nerve, for which an atrophy has been linked to PD (Pelz et al., 2018; Walter et al., 2018). Validation of diagnostic utility and prodromal interval is pending.

3.3.2.2 Tissue biopsies

An increasing amount of biomarker studies use easily accessible peripheral tissues for the detection of PD-related pathology, in particular alpha-synuclein deposits. However, all of these studies have still a huge variety in common with regard to sensitivity and specificity, limiting their use. One example is the use of enteral nerve biopsies. Using material from the esophagus, stomach, duodenum and colon, sensitivity and specificity has been documented over an extremely wide range (Ruffmann and Parkkinen, 2016). Several important issues have been discussed to explain this variance, including different techniques of probe extraction (for colonic biopsies it has to be assured that the submucosa is reached) and differing methods of detecting and defining pathological alpha-synuclein detection. A rostral to caudal gradient of alpha-synucleinopathy has been suggested, which may imply that proximal biopsies (e.g., submandibular gland) may have better sensitivity. Several small studies found alpha-synuclein in salivary glands, in particular the submandibular and minory glands (Adler et al., 2014; Beach et al., 2013; Cersósimo et al., 2011; Gao et al., 2015a). Finally, the proposed different subtypes of alpha-synuclein propagation have to be taken into account when validating enteral nerve biopsies. In fact, one enteral nerve biopsy study comparing alpha-synuclein deposits in the enteral nervous system in PD patients with and without RBD showed an occurrence of phosphorylated α-synuclein pathology in 64.3% of PD patients with RBD vs. 13.3% in PD patients without RBD (Leclair-Visonneau et al., 2017).

Skin biopsies, are easy to obtain with at low-risk, and have showed promising results. After phosphorylated alpha-synuclein in peripheral nerves of the skin was detected in postmortem studies (Ikemura et al., 2008), several blinded in vivo studies confirmed this finding, with a very high specificity up to 100%, but with varying sensitivity (Donadio et al., 2014; Michell et al., 2005). Studies have also clearly documented abnormal synuclein pathology in the majority of patients with idiopathic RBD, but almost never in controls (again using blinded evaluation) (Antelmi et al., 2017; Doppler et al., 2017).

Interestingly, alpha-synuclein deposits have also been found in postmortem studies of the retina (Beach et al., 2014; Bodis-Wollner et al., 2014), and in the cornea of PD patients after cataract surgery (Klettner et al., 2016), potentially helping to explain some of the visual disturbances seen in PD. However, obviously these findings will most likely not lead to potential in vivo biomarkers, due to the limited access to these tissues.

Reliable documentation of pathologic synuclein deposition during life would constitute a major advance for the field. In addition to providing confirmed diagnosis, it can help to select "on target" patients who are candidates for synuclein-based neuroprotective therapy. Given the ease of obtaining skin samples, these may have the most potential to revolutionize prodromal PD diagnosis.

3.3.2.3 Fluid-based bio markers

Many molecular markers in the blood and cerebrospinal fluid (CSF) are currently being investigated to determine their validity as prognostic and diagnostic markers in PD. However, most of them have only been examined in the clinical phase of PD and so far, none has shown sufficient prognostic reliability. A major focus is on the detection of pathological alpha-synuclein, with promising results in CSF studies, showing decreased alpha-synuclein levels in PD (Gao et al., 2015b; van Steenoven et al., 2018). Important technical developments are new amplification techniques, including RealTime Quaking-Induced Conversion (RT-QuIC) assays (Fairfoul et al., 2016; Garrido et al., 2019) and other prion-based assays. While it has been very difficult to detect alpha-synuclein in blood so far, due to high amounts of physiological alpha-synuclein in red blood cells (Barbour et al., 2008), the extraction of alpha-synuclein out of plasma exosomes might be a highly promising approach (Shi et al., 2014). Furthermore, plasma apolipoprotein levels (Qiang et al., 2013), tau, aβ42 and tau/aβ42 ratio (Kang et al., 2013; Parnetti et al., 2011), DJ-1 levels (Maita et al., 2008) and neurofilament light chain (Abdo et al., 2007; Bacioglu et al., 2016) are currently in discussion as suitable wet markers. Moreover, several studies suggested that saliva might be a useful biofluid, showing, for example, changes in alpha-synuclein and DJ-1 levels (Al-Nimer et al., 2014; Masters et al., 2015). Finally, changes in the microbiome, derived from stool samples, are part of several large studies (see chapter "The gut microbiome in Parkinson's disease: A culprit or a bystander?" by Keshavarzian et al. for further information).

4 Conclusion

Taken together, the potential importance of thorough scientific investigation of the prodromal phase is beyond doubt, as it leads to a better understanding of pathophysiological mechanisms in PD (Olanow and Obeso, 2012). Moreover reliable identification of prodromal stages will increase surveillance for troublesome non-motor symptoms, which are both common in prodromal PD and potentially treatable.

Most importantly, the identification of individuals in the prodromal phase sets the basis for future trials of disease-modifying treatment strategies. However, several challenges have to be faced until a clinical trial in the prodromal phase can be translated into reality. First, it is evident that a treatment in the prodromal phase can only be implemented if the prodromal criteria provide a sufficiently high diagnostic certainty. So far, there are only a few validation studies, showing a good, but not perfect diagnostic reliability, which depends also on interval between criteria assessment conversion to clinical manifest motor PD (Mahlknecht et al., 2018). A second important challenge is the identification of suitable outcome parameters, in particular, predicting the time of conversion. So far the MDS research criteria on prodromal PD provide a calculation for the probability of developing PD some time in the future, but do not estimate the actual time of conversion. If conversion to clinical PD/DLB is the most interesting outcome parameter for clinical trials, a clinical study will only be feasible if an individual is close to conversion, not 10 years away. Recently, it has been suggested that this goal is realistic for RBD patients. By further stratifying RBD patients using additional risk and prodromal markers, a neuroprotective trial with reasonable patient numbers and study duration seems achievable (Postuma et al., 2019). However, individuals with RBD are only a subgroup of prodromal PD patients, and to date the feasibility of neuroprotective trials for non-RBD individuals cannot be determined. Moving beyond conversion as an outcome, the search for objective, quantitative progression markers is therefore of highest importance. As non-motor symptoms seem to vary considerably during the prodromal phase, the search for reliable biomarkers of disease progression moves to the fore. Finally, it has to be noted that reliable differentiation between different synucleinopathies (multiple system atrophy, DLB and PD) in the prodromal phase is not yet feasible.

Future studies should aim for an interdisciplinary approach and close collaboration between clinical and basic scientists, to understand interactions and interrelationships between clinical and pathophysiological subtypes as well as biomarker findings. Finally, the increasing diagnostic accuracy for prodromal PD in the absence of yet available potential treatments will result in new ethical challenges that have to be faced and discussed.

References

Abbott, R.D., Petrovitch, H., White, L.R., Masaki, K.H., Tanner, C.M., Curb, J.D., et al., 2001. Frequency of bowel movements and the future risk of Parkinson's disease. Neurology 57 (3), 456–462.

Abbott, R.D., Ross, G.W., White, L.R., Tanner, C.M., Masaki, K.H., Nelson, J.S., et al., 2005. Excessive daytime sleepiness and subsequent development of Parkinson disease. Neurology 65 (9), 1442–1446.

Abbott, R.D., Ross, G.W., Petrovitch, H., Tanner, C.M., Davis, D.G., Masaki, K.H., et al., 2007. Bowel movement frequency in late-life and incidental Lewy bodies. Mov. Disord. 22 (11), 1581–1586.

Abdo, W.F., Bloem, B.R., Van Geel, W.J., Esselink, R.A.J., Verbeek, M.M., 2007. CSF neurofilament light chain and tau differentiate multiple system atrophy from Parkinson's disease. Neurobiol. Aging 28 (5), 742–747.

Adams, J.R., van Netten, H., Schulzer, M., Mak, E., Mckenzie, J., Strongosky, A., et al., 2005. PET in LRRK2 mutations: comparison to sporadic Parkinson's disease and evidence for presymptomatic compensation. Brain 128 (Pt. 12), 2777–2785.

Adler, C.H., Beach, T.G., 2016. Neuropathological basis of nonmotor manifestations of Parkinson's disease. Mov. Disord. 31 (8), 1114–1119.

Adler, C.H., Dugger, B.N., Hinni, M.L., Lott, D.G., Driver-Dunckley, E., Hidalgo, J., et al., 2014. Submandibular gland needle biopsy for the diagnosis of Parkinson disease. Neurology 82 (10), 858–864.

Ahlskog, J.E., 2011. Does vigorous exercise have a neuroprotective effect in Parkinson disease? Neurology 77 (3), 288–294.

Al-Nimer, M.S.M., Mshatat, S.F., Abdulla, H.I., 2014. Saliva α-synuclein and a high extinction coefficient protein: a novel approach in assessment biomarkers of Parkinson's disease. N. Am. J. Med. Sci. 6 (12), 633–637.

Ansari, K.A., Johnson, A., 1975. Olfactory function in patients with Parkinson's disease. J. Chronic Dis. 28 (9), 493–497.

Antelmi, E., Donadio, V., Incensi, A., Plazzi, G., Liguori, R., 2017. Skin nerve phosphorylated α-synuclein deposits in idiopathic REM sleep behavior disorder. Neurology 88 (22), 2128–2131.

Armstrong, R.A., 2015. Oculo-visual dysfunction in Parkinson's disease. J. Parkinsons Dis. 5 (4), 715–726.

Ascherio, A., Zhang, S.M., Hernán, M.A., Kawachi, I., Colditz, G.A., Speizer, F.E., et al., 2001. Prospective study of caffeine consumption and risk of Parkinson's disease in men and women. Ann. Neurol. 50 (1), 56–63.

Ascherio, A., Chen, H., Weisskopf, M.G., O'Reilly, E., McCullough, M.L., Calle, E.E., et al., 2006. Pesticide exposure and risk for Parkinson's disease. Ann. Neurol. 60 (2), 197–203.

Bacioglu, M., Maia, L.F., Preische, O., Schelle, J., Apel, A., Kaeser, S.A., et al., 2016. Neurofilament light chain in blood and CSF as marker of disease progression in mouse models and in neurodegenerative diseases. Neuron 91 (1), 56–66.

Bae, Y.J., Kim, J.-M., Kim, K.J., Kim, E., Park, H.S., Kang, S.Y., et al., 2018. Loss of substantia nigra hyperintensity at 3.0-T MR imaging in idiopathic REM sleep behavior disorder: comparison with 123I-FP-CIT SPECT. Radiology 287 (1), 285–293.

Bai, S., Song, Y., Huang, X., Peng, L., Jia, J., Liu, Y., et al., 2016. Statin use and the risk of Parkinson's disease: an updated meta-analysis. PLoS One 11 (3).

Barbour, R., Kling, K., Anderson, J.P., Banducci, K., Cole, T., Diep, L., et al., 2008. Red blood cells are the major source of alpha-synuclein in blood. Neurodegener. Dis. 5 (2), 55–59.

Beach, T.G., Adler, C.H., Lue, L., Sue, L.I., Bachalakuri, J., Henry-Watson, J., et al., 2009. Unified staging system for Lewy body disorders: correlation with nigrostriatal degeneration, cognitive impairment and motor dysfunction. Acta Neuropathol. 117 (6), 613–634.

Beach, T.G., Adler, C.H., Sue, L.I., Vedders, L., Lue, L., White Iii, C.L., et al., 2010. Multiorgan distribution of phosphorylated alpha-synuclein histopathology in subjects with Lewy body disorders. Acta Neuropathol. 119 (6), 689–702.

Beach, T.G., Adler, C.H., Dugger, B.N., Serrano, G., Hidalgo, J., Henry-Watson, J., et al., 2013. Submandibular gland biopsy for the diagnosis of Parkinson disease. J. Neuropathol. Exp. Neurol. 72 (2), 130–136.

Beach, T.G., Carew, J., Serrano, G., Adler, C.H., Shill, H.A., Sue, L.I., et al., 2014. Phosphorylated α-synuclein-immunoreactive retinal neuronal elements in Parkinson's disease subjects. Neurosci. Lett. 571, 34–38.

Beavan, M., McNeill, A., Proukakis, C., Hughes, D.A., Mehta, A., Schapira, A.H.V., 2015. Evolution of prodromal clinical markers of Parkinson disease in a glucocerebrosidase mutation positive cohort. JAMA Neurol. 72 (2), 201–208.

Belghali, M., Chastan, N., Cignetti, F., Davenne, D., Decker, L.M., 2017. Loss of gait control assessed by cognitive-motor dual-tasks: pros and cons in detecting people at risk of developing Alzheimer's and Parkinson's diseases. GeroScience 39 (3), 305–329.

Berg, D., Roggendorf, W., Schröder, U., Klein, R., Tatschner, T., Benz, P., et al., 2002. Echogenicity of the substantia nigra: association with increased iron content and marker for susceptibility to nigrostriatal injury. Arch. Neurol. 59 (6), 999–1005.

Berg, D., Seppi, K., Behnke, S., Liepelt, I., Schweitzer, K., Stockner, H., et al., 2011. Enlarged substantia nigra hyperechogenicity and risk for Parkinson disease: a 37-month 3-center study of 1847 older persons. Arch. Neurol. 68 (7), 932–937.

Berg, D., Behnke, S., Seppi, K., Godau, J., Lerche, S., Mahlknecht, P., et al., 2013. Enlarged hyperechogenic substantia nigra as a risk marker for Parkinson's disease. Mov. Disord. 28 (2), 216–219.

Berg, D., Postuma, R.B., Bloem, B., Chan, P., Dubois, B., Gasser, T., et al., 2014. Time to redefine PD? Introductory statement of the MDS Task Force on the definition of Parkinson's disease. Mov. Disord. 29 (4), 454–462.

Berg, D., Postuma, R.B., Adler, C.H., Bloem, B.R., Chan, P., Dubois, B., et al., 2015. MDS research criteria for prodromal Parkinson's disease. Mov. Disord. 30 (12), 1600–1611.

Bodis-Wollner, I., Kozlowski, P.B., Glazman, S., Miri, S., 2014. α-Synuclein in the inner retina in Parkinson disease. Ann. Neurol. 75 (6), 964–966.

Boeve, B.F., Silber, M.H., Saper, C.B., Ferman, T.J., Dickson, D.W., Parisi, J.E., et al., 2007. Pathophysiology of REM sleep behaviour disorder and relevance to neurodegenerative disease. Brain 130 (11), 2770–2788.

Borghammer, P., Van Den Berge, N., 2019. Brain-first versus gut-first Parkinson's disease: a hypothesis. J. Parkinsons Dis. 9 (s2), S281–S295.

Bousquet, M., Calon, F., Cicchetti, F., 2011. Impact of omega-3 fatty acids in Parkinson's disease. Ageing Res. Rev. 10 (4), 453–463.

Bower, J.H., Grossardt, B.R., Maraganore, D.M., Ahlskog, J.E., Colligan, R.C., Geda, Y.E., et al., 2010. Anxious personality predicts an increased risk of Parkinson's disease. Mov. Disord. 25 (13), 2105–2113.

Braak, H., Tredici, K.D., Rüb, U., de Vos, R.A.I., Jansen Steur, E.N.H., Braak, E., 2003. Staging of brain pathology related to sporadic Parkinson's disease. Neurobiol. Aging 24 (2), 197–211.

Camacho-Soto, A., Warden, M.N., Searles Nielsen, S., Salter, A., Brody, D.L., Prather, H., et al., 2017. Traumatic brain injury in the prodromal period of Parkinson's disease: a large epidemiological study using medicare data. Ann. Neurol. 82 (5), 744–754.

Cersósimo, M.G., Perandones, C., Micheli, F.E., Raina, G.B., Beron, A.M., Nasswetter, G., et al., 2011. Alpha-synuclein immunoreactivity in minor salivary gland biopsies of Parkinson's disease patients. Mov. Disord. 26 (1), 188–190.

Chen, H., Zhang, S.M., Hernán, M.A., Schwarzschild, M.A., Willett, W.C., Colditz, G.A., et al., 2003. Nonsteroidal anti-inflammatory drugs and the risk of Parkinson disease. Arch. Neurol. 60 (8), 1059–1064.

Del Tredici, K., Braak, H., 2012. Spinal cord lesions in sporadic Parkinson's disease. Acta Neuropathol. 124 (5), 643–664.

Desplats, P., Lee, H.-J., Bae, E.-J., Patrick, C., Rockenstein, E., Crews, L., et al., 2009. Inclusion formation and neuronal cell death through neuron-to-neuron transmission of alpha-synuclein. Proc. Natl. Acad. Sci. U. S. A. 106 (31), 13010–13015.

Dickson, D.W., Uchikado, H., Fujishiro, H., Tsuboi, Y., 2010. Evidence in favor of Braak staging of Parkinson's disease. Mov. Disord. 25 (S1), S78–S82.

Dintica, C.S., Marseglia, A., Rizzuto, D., Wang, R., Seubert, J., Arfanakis, K., et al., 2019. Impaired olfaction is associated with cognitive decline and neurodegeneration in the brain. Neurology 92 (7), e700–e709.

Donadio, V., Incensi, A., Leta, V., Giannoccaro, M.P., Scaglione, C., Martinelli, P., et al., 2014. Skin nerve α-synuclein deposits: a biomarker for idiopathic Parkinson disease. Neurology 82 (15), 1362–1369.

Doppler, K., Jentschke, H.-M., Schulmeyer, L., Vadasz, D., Janzen, A., Luster, M., et al., 2017. Dermal phospho-alpha-synuclein deposits confirm REM sleep behaviour disorder as prodromal Parkinson's disease. Acta Neuropathol. 133 (4), 535–545.

Driver-Dunckley, E., Adler, C.H., Hentz, J.G., Dugger, B.N., Shill, H.A., Caviness, J.N., et al., 2014. Olfactory dysfunction in incidental Lewy body disease and Parkinson's disease. Parkinsonism Relat. Disord. 20 (11), 1260–1262.

Fairfoul, G., McGuire, L.I., Pal, S., Ironside, J.W., Neumann, J., Christie, S., et al., 2016. Alpha-synuclein RT-QuIC in the CSF of patients with alpha-synucleinopathies. Ann. Clin. Transl. Neurol. 3 (10), 812–818.

Fall, P.A., Fredrikson, M., Axelson, O., Granérus, A.K., 1999. Nutritional and occupational factors influencing the risk of Parkinson's disease: a case-control study in southeastern Sweden. Mov. Disord. 14 (1), 28–37.

Fantini, M.L., Farini, E., Ortelli, P., Zucconi, M., Manconi, M., Cappa, S., et al., 2011. Longitudinal study of cognitive function in idiopathic REM sleep behavior disorder. Sleep 34 (5), 619–625.

Fearnley, J.M., Lees, A.J., 1991. Ageing and Parkinson's disease: substantia nigra regional selectivity. Brain 114 (Pt. 5), 2283–2301.

Fedorova, T.D., Seidelin, L.B., Knudsen, K., Schacht, A.C., Geday, J., Pavese, N., et al., 2017. Decreased intestinal acetylcholinesterase in early Parkinson disease: an 11C-donepezil PET study. Neurology 88 (8), 775–781.

Fengler, S., Liepelt-Scarfone, I., Brockmann, K., Schäffer, E., Berg, D., Kalbe, E., 2017. Cognitive changes in prodromal Parkinson's disease: a review. Mov. Disord. 32 (12), 1655–1666.

Fereshtehnejad, S.-M., Romenets, S.R., Anang, J.B.M., Latreille, V., Gagnon, J.-F., Postuma, R.B., 2015. New clinical subtypes of Parkinson disease and their longitudinal progression: a prospective cohort comparison with other phenotypes. JAMA Neurol. 72 (8), 863–873.

Fereshtehnejad, S.-M., Yao, C., Pelletier, A., Montplaisir, J.Y., Gagnon, J.-F., Postuma, R.B., 2019. Evolution of prodromal Parkinson's disease and dementia with Lewy bodies: a prospective study. Brain 142 (7), 2051–2067.

Friedman, J.H., Brown, R.G., Comella, C., Garber, C.E., Krupp, L.B., Lou, J.-S., et al., 2007. Fatigue in Parkinson's disease: a review. Mov. Disord. 22 (3), 297–308.

Gagnon, J.-F., Vendette, M., Postuma, R.B., Desjardins, C., Massicotte-Marquez, J., Panisset, M., et al., 2009. Mild cognitive impairment in rapid eye movement sleep behavior disorder and Parkinson's disease. Ann. Neurol. 66 (1), 39–47.

Gao, X., Chen, H., Schwarzschild, M.A., Glasser, D.B., Logroscino, G., Rimm, E.B., et al., 2007. Erectile function and risk of Parkinson's disease. Am. J. Epidemiol. 166 (12), 1446–1450.

Gao, X., Chen, H., Schwarzschild, M.A., Ascherio, A., 2011. A prospective study of bowel movement frequency and risk of Parkinson's disease. Am. J. Epidemiol. 174 (5), 546–551.

Gao, L., Chen, H., Li, X., Li, F., Ou-Yang, Q., Feng, T., 2015a. The diagnostic value of minor salivary gland biopsy in clinically diagnosed patients with Parkinson's disease: comparison with DAT PET scans. Neurol. Sci. 36 (9), 1575–1580.

Gao, L., Tang, H., Nie, K., Wang, L., Zhao, J., Gan, R., et al., 2015b. Cerebrospinal fluid alpha-synuclein as a biomarker for Parkinson's disease diagnosis: a systematic review and meta-analysis. Int. J. Neurosci. 125 (9), 645–654.

García-Lorenzo, D., Longo-Dos Santos, C., Ewenczyk, C., Leu-Semenescu, S., Gallea, C., Quattrocchi, G., et al., 2013. The coeruleus/subcoeruleus complex in rapid eye movement sleep behaviour disorders in Parkinson's disease. Brain 136 (Pt. 7), 2120–2129.

Gardner, R.C., Burke, J.F., Nettiksimmons, J., Goldman, S., Tanner, C.M., Yaffe, K., 2015. Traumatic brain injury in later life increases risk for Parkinson disease. Ann. Neurol. 77 (6), 987–995.

Garrido, A., Fairfoul, G., Tolosa, E.S., Martí, M.J., Green, A., Barcelona LRRK2 Study Group, 2019. α-Synuclein RT-QuIC in cerebrospinal fluid of LRRK2-linked Parkinson's disease. Ann. Clin. Transl. Neurol. 6 (6), 1024–1032.

Gelpi, E., Navarro-Otano, J., Tolosa, E., Gaig, C., Compta, Y., Rey, M.J., et al., 2014. Multiple organ involvement by alpha-synuclein pathology in Lewy body disorders. Mov. Disord. 29 (8), 1010–1018.

Gjerløff, T., Fedorova, T., Knudsen, K., Munk, O.L., Nahimi, A., Jacobsen, S., et al., 2015. Imaging acetylcholinesterase density in peripheral organs in Parkinson's disease with 11C-donepezil PET. Brain 138 (Pt. 3), 653–663.

Goldstein, D.S., Holmes, C.S., Dendi, R., Bruce, S.R., Li, S.-T., 2002. Orthostatic hypotension from sympathetic denervation in Parkinson's disease. Neurology 58 (8), 1247–1255.

Goldstein, D.S., Holmes, C., Sewell, L., Park, M.Y., Sharabi, Y., 2012. Sympathetic noradrenergic before striatal dopaminergic denervation: relevance to Braak staging of synucleinopathy. Clin. Auton. Res. 22 (1), 57–61.

Gorell, J.M., Johnson, C.C., Rybicki, B.A., Peterson, E.L., Richardson, R.J., 1998. The risk of Parkinson's disease with exposure to pesticides, farming, well water, and rural living. Neurology 50 (5), 1346–1350.

Gorell, J.M., Rybicki, B.A., Johnson, C.C., Peterson, E.L., 1999. Smoking and Parkinson's disease: a dose-response relationship. Neurology 52 (1), 115–119.

Grandinetti, A., Morens, D.M., Reed, D., MacEachern, D., 1994. Prospective study of cigarette smoking and the risk of developing idiopathic Parkinson's disease. Am. J. Epidemiol. 139 (12), 1129–1138.

Gustafsson, H., Nordström, A., Nordström, P., 2015. Depression and subsequent risk of Parkinson disease: a nationwide cohort study. Neurology 84 (24), 2422–2429.

Haba-Rubio, J., Frauscher, B., Marques-Vidal, P., Toriel, J., Tobback, N., Andries, D., et al., 2018. Prevalence and determinants of rapid eye movement sleep behavior disorder in the general population. Sleep 41 (2) .

Haehner, A., Hummel, T., Hummel, C., Sommer, U., Junghanns, S., Reichmann, H., 2007. Olfactory loss may be a first sign of idiopathic Parkinson's disease. Mov. Disord. 22 (6), 839–842.

Hansen, C., Angot, E., Bergström, A.-L., Steiner, J.A., Pieri, L., Paul, G., et al., 2011. α-Synuclein propagates from mouse brain to grafted dopaminergic neurons and seeds aggregation in cultured human cells. J. Clin. Invest. 121 (2), 715–725.

Harel, B., Cannizzaro, M., Snyder, P.J., 2004a. Variability in fundamental frequency during speech in prodromal and incipient Parkinson's disease: a longitudinal case study. Brain Cogn. 56 (1), 24–29.

Harel, B.T., Cannizzaro, M.S., Cohen, H., Reilly, N., Snyder, P.J., 2004b. Acoustic characteristics of Parkinsonian speech: a potential biomarker of early disease progression and treatment. J. Neurolinguistics 17 (6), 439–453.

Hawkes, C.H., Del Tredici, K., Braak, H., 2007. Parkinson's disease: a dual-hit hypothesis. Neuropathol. Appl. Neurobiol. 33 (6), 599–614.

Heinzel, S., Berg, D., Gasser, T., Chen, H., Yao, C., Postuma, R.B., 2019a. Prodromal Criteria Calculator. Version 2.1.3.html, Available from: www.movementdisorders.org/pdcalculator.

Heinzel, S., Berg, D., Gasser, T., Chen, H., Yao, C., Postuma, R.B., et al., 2019b. Update of the MDS research criteria for prodromal Parkinson's disease. Mov. Disord. 34, 1464–1470.

Holtbernd, F., Gagnon, J.-F., Postuma, R.B., Ma, Y., Tang, C.C., Feigin, A., et al., 2014. Abnormal metabolic network activity in REM sleep behavior disorder. Neurology 82 (7), 620–627.

Honeycutt, L., Montplaisir, J.Y., Gagnon, J.-F., Ruskey, J., Pelletier, A., Gan-Or, Z., et al., 2019. Glucocerebrosidase mutations and phenoconversion of REM sleep behavior disorder to parkinsonism and dementia. Parkinsonism Relat. Disord. 65, 230–233.

Hou, L., Chen, W., Liu, X., Qiao, D., Zhou, F.-M., 2017. Exercise-induced neuroprotection of the nigrostriatal dopamine system in Parkinson's disease. Front. Aging Neurosci. 9, 358.

Hu, G., Jousilahti, P., Bidel, S., Antikainen, R., Tuomilehto, J., 2007. Type 2 diabetes and the risk of Parkinson's disease. Diabetes Care 30 (4), 842–847.

Ikemura, M., Saito, Y., Sengoku, R., Sakiyama, Y., Hatsuta, H., Kanemaru, K., et al., 2008. Lewy body pathology involves cutaneous nerves. J. Neuropathol. Exp. Neurol. 67 (10), 945–953.

Iova, A., Garmashov, A., Androuchtchenko, N., Kehrer, M., Berg, D., Becker, G., et al., 2004. Postnatal decrease in substantia nigra echogenicity. Implications for the pathogenesis of Parkinson's disease. J. Neurol. 251 (12), 1451–1454.

Iranzo, A., Santamaría, J., Valldeoriola, F., Serradell, M., Salamero, M., Gaig, C., et al., 2017. Dopamine transporter imaging deficit predicts early transition to synucleinopathy in idiopathic rapid eye movement sleep behavior disorder. Ann. Neurol. 82 (3), 419–428.

Iwanaga, K., Wakabayashi, K., Yoshimoto, M., Tomita, I., Satoh, H., Takashima, H., et al., 1999. Lewy body-type degeneration in cardiac plexus in Parkinson's and incidental Lewy body diseases. Neurology 52 (6), 1269–1271.

Jennings, D., Siderowf, A., Stern, M., Seibyl, J., Eberly, S., Oakes, D., et al., 2014. Imaging prodromal Parkinson disease: the Parkinson Associated Risk Syndrome Study. Neurology 83 (19), 1739–1746.

Kang, J.-H., Irwin, D.J., Chen-Plotkin, A.S., Siderowf, A., Caspell, C., Coffey, C.S., et al., 2013. Association of cerebrospinal fluid β-amyloid 1-42, T-tau, P-tau181, and α-synuclein levels with clinical features of drug-naive patients with early Parkinson disease. JAMA Neurol. 70 (10), 1277–1287.

Khan, N.L., Valente, E.M., Bentivoglio, A.R., Wood, N.W., Albanese, A., Brooks, D.J., et al., 2002a. Clinical and subclinical dopaminergic dysfunction in PARK6-linked parkinsonism: an 18F-dopa PET study. Ann. Neurol. 52 (6), 849–853.

Khan, N.L., Brooks, D.J., Pavese, N., Sweeney, M.G., Wood, N.W., Lees, A.J., et al., 2002b. Progression of nigrostriatal dysfunction in a parkin kindred: an [18F] dopa PET and clinical study. Brain 125 (10), 2248–2256.

Khan, N.L., Scherfler, C., Graham, E., Bhatia, K.P., Quinn, N., Lees, A.J., et al., 2005. Dopaminergic dysfunction in unrelated, asymptomatic carriers of a single parkin mutation. Neurology 64 (1), 134–136.

Kingsbury, A.E., Bandopadhyay, R., Silveira-Moriyama, L., Ayling, H., Kallis, C., Sterlacci, W., et al., 2010. Brain stem pathology in Parkinson's disease: an evaluation of the Braak staging model. Mov. Disord. 25 (15), 2508–2515.

Klettner, A., Richert, E., Kuhlenbäumer, G., Nölle, B., Bhatia, K.P., Deuschl, G., et al., 2016. Alpha synuclein and crystallin expression in human lens in Parkinson's disease. Mov. Disord. 31 (4), 600–601.

Knudsen, K., Fedorova, T.D., Hansen, A.K., Sommerauer, M., Otto, M., Svendsen, K.B., et al., 2018. In-vivo staging of pathology in REM sleep behaviour disorder: a multimodality imaging case-control study. Lancet Neurol. 17 (7), 618–628.

Kordower, J.H., Chu, Y., Hauser, R.A., Freeman, T.B., Olanow, C.W., 2008. Lewy body-like pathology in long-term embryonic nigral transplants in Parkinson's disease. Nat. Med. 14 (5), 504–506.

Kujawska, M., Jodynis-Liebert, J., 2018. Polyphenols in Parkinson's disease: a systematic review of in vivo studies. Nutrients 10 (5), 642.

Lai, S.-W., Liao, K.-F., Lin, C.-L., Sung, F.-C., 2014. Irritable bowel syndrome correlates with increased risk of Parkinson's disease in Taiwan. Eur. J. Epidemiol. 29 (1), 57–62.

Leclair-Visonneau, L., Clairembault, T., Coron, E., Le Dily, S., Vavasseur, F., Dalichampt, M., et al., 2017. REM sleep behavior disorder is related to enteric neuropathology in Parkinson disease. Neurology 89 (15), 1612–1618.

Leentjens, A.F.G., den Akker, M.V., Metsemakers, J.F.M., Lousberg, R., Verhey, F.R.J., 2003. Higher incidence of depression preceding the onset of Parkinson's disease: a register study. Mov. Disord. 18 (4), 414–418.

Lees, A.J., 1992. When did Ray Kennedy's Parkinson's disease begin? Mov. Disord. 7 (2), 110–116.

Lesage, S., Dürr, A., Tazir, M., Lohmann, E., Leutenegger, A.-L., Janin, S., et al., 2006. LRRK2 G2019S as a cause of Parkinson's disease in North African Arabs. N. Engl. J. Med. 354 (4), 422–423.

Li, J.-Y., Englund, E., Holton, J.L., Soulet, D., Hagell, P., Lees, A.J., et al., 2008. Lewy bodies in grafted neurons in subjects with Parkinson's disease suggest host-to-graft disease propagation. Nat. Med. 14 (5), 501–503.

Lin, C.-H., Wu, R.-M., Chang, H.-Y., Chiang, Y.-T., Lin, H.-H., 2013. Preceding pain symptoms and Parkinson's disease: a nationwide population-based cohort study. Eur. J. Neurol. 20 (10), 1398–1404.

Lin, C.-H., Lin, J.-W., Liu, Y.-C., Chang, C.-H., Wu, R.-M., 2014. Risk of Parkinson's disease following severe constipation: a nationwide population-based cohort study. Parkinsonism Relat. Disord. 20 (12), 1371–1375.

Luk, K.C., Kehm, V., Carroll, J., Zhang, B., O'Brien, P., Trojanowski, J.Q., et al., 2012. Pathological α-synuclein transmission initiates Parkinson-like neurodegeneration in nontransgenic mice. Science 338 (6109), 949–953.

Mahlknecht, P., Iranzo, A., Högl, B., Frauscher, B., Müller, C., Santamaría, J., et al., 2015. Olfactory dysfunction predicts early transition to a Lewy body disease in idiopathic RBD. Neurology 84 (7), 654–658.

Mahlknecht, P., Gasperi, A., Willeit, P., Kiechl, S., Stockner, H., Willeit, J., et al., 2016. Prodromal Parkinson's disease as defined per MDS research criteria in the general elderly community: prodromal PD in the community. Mov. Disord. 31 (9), 1405–1408.

Mahlknecht, P., Gasperi, A., Djamshidian, A., Kiechl, S., Stockner, H., Willeit, P., et al., 2018. Performance of the Movement Disorders Society criteria for prodromal Parkinson's disease: a population-based 10-year study. Mov. Disord. 33 (3), 405–413.

Maita, C., Tsuji, S., Yabe, I., Hamada, S., Ogata, A., Maita, H., et al., 2008. Secretion of DJ-1 into the serum of patients with Parkinson's disease. Neurosci. Lett. 431 (1), 86–89.

Marras, C., Chaudhuri, K.R., 2016. Nonmotor features of Parkinson's disease subtypes. Mov. Disord. 31 (8), 1095–1102.

Masters, J.M., Noyce, A.J., Warner, T.T., Giovannoni, G., Proctor, G.B., 2015. Elevated salivary protein in Parkinson's disease and salivary DJ-1 as a potential marker of disease severity. Parkinsonism Relat. Disord. 21 (10), 1251–1255.

Mayeux, R., Stern, Y., Cote, L., Williams, J.B., 1984. Altered serotonin metabolism in depressed patients with Parkinson's disease. Neurology 34 (5), 642–646.

McNeill, A., Roberti, G., Lascaratos, G., Hughes, D., Mehta, A., Garway-Heath, D.F., et al., 2013. Retinal thinning in Gaucher disease patients and carriers: results of a pilot study. Mol. Genet. Metab. 109 (2), 221–223.

Meles, S.K., Renken, R.J., Janzen, A., Vadasz, D., Pagani, M., Arnaldi, D., et al., 2018. The metabolic pattern of idiopathic REM sleep behavior disorder reflects early-stage Parkinson's disease. J. Nucl. Med. 59, 1437–1444.

Michell, A.W., Luheshi, L.M., Barker, R.A., 2005. Skin and platelet α-synuclein as peripheral biomarkers of Parkinson's disease. Neurosci. Lett. 381 (3), 294–298.

Mirelman, A., Gurevich, T., Giladi, N., Bar-Shira, A., Orr-Utreger, A., Hausdorff, J.M., 2011. Gait alterations in healthy carriers of the LRRK2 G2019S mutation. Ann. Neurol. 69 (1), 193–197.

Mirelman, A., Bernad-Elazari, H., Thaler, A., Giladi-Yacobi, E., Gurevich, T., Gana-Weisz, M., et al., 2016. Arm swing as a potential new prodromal marker of Parkinson's disease. Mov. Disord. 31 (10), 1527–1534.

Mirelman, A., Saunders-Pullman, R., Alcalay, R.N., Shustak, S., Thaler, A., Gurevich, T., et al., 2018. Application of the movement disorder society prodromal criteria in healthy *G2019S-LRRK2* carriers: prodromal criteria nonmanifesting LRRK2 carriers. Mov. Disord. 33, 966–973 .

Miyamoto, T., Miyamoto, M., Inoue, Y., Usui, Y., Suzuki, K., Hirata, K., 2006. Reduced cardiac 123I-MIBG scintigraphy in idiopathic REM sleep behavior disorder. Neurology 67 (12), 2236–2238.

Nalls, M.A., Blauwendraat, C., Vallerga, C.L., Heilbron, K., Bandres-Ciga, S., Chang, D., Tan, M., et al., 2019. Identification of novel risk loci, causal insights, and heritable risk for Parkinson's disease: a meta-analysis of genome-wide association studies. The Lancet Neurology 18 (12), 1091–1102.

Nefzger, M.D., Quadfasel, F.A., Karl, V.C., 1968. A retrospective study of smoking in Parkinson's disease. Am. J. Epidemiol. 88 (2), 149–158.

Noyce, A.J., Dickson, J., Rees, R.N., Bestwick, J.P., Isaias, I.U., Politis, M., et al., 2018. Dopamine reuptake transporter-single-photon emission computed tomography and transcranial sonography as imaging markers of prediagnostic Parkinson's disease: Dat-Spect and TCS in subjects at risk of PD. Mov. Disord. 33 (3), 478–482.

Olanow, C.W., Obeso, J.A., 2012. The significance of defining preclinical or prodromal Parkinson's disease. Mov. Disord. 27 (5), 666–669.

Parnetti, L., Chiasserini, D., Bellomo, G., Giannandrea, D., De Carlo, C., Qureshi, M.M., et al., 2011. Cerebrospinal fluid Tau/α-synuclein ratio in Parkinson's disease and degenerative dementias. Mov. Disord. 26 (8), 1428–1435.

Pelz, J.O., Belau, E., Fricke, C., Classen, J., Weise, D., 2018. Axonal degeneration of the vagus nerve in Parkinson's disease—a high-resolution ultrasound study. Front. Neurol. 9, 951.

Ponsen, M.M., Stoffers, D., Booij, J., van Eck-Smit, B.L.F., Wolters, E.C., Berendse, H.W., 2004. Idiopathic hyposmia as a preclinical sign of Parkinson's disease. Ann. Neurol. 56 (2), 173–181.

Postuma, R.B., 2016. Resting state MRI: a new marker of prodromal neurodegeneration? Brain 139 (Pt. 8), 2106–2108.

Postuma, R.B., Berg, D., 2019. Prodromal Parkinson's disease: the decade past, the decade to come. Mov. Disord. 34 (5), 665–675.

Postuma, R.B., Gagnon, J.-F., Vendette, M., Desjardins, C., Montplaisir, J.Y., 2011. Olfaction and color vision identify impending neurodegeneration in rapid eye movement sleep behavior disorder. Ann. Neurol. 69 (5), 811–818.

Postuma, R.B., Lang, A.E., Gagnon, J.F., Pelletier, A., Montplaisir, J.Y., 2012. How does parkinsonism start? Prodromal parkinsonism motor changes in idiopathic REM sleep behaviour disorder. Brain 135 (6), 1860–1870.

Postuma, R.B., Gagnon, J.-F., Pelletier, A., Montplaisir, J., 2013. Prodromal autonomic symptoms and signs in Parkinson's disease and dementia with Lewy bodies. Mov. Disord. 28 (5), 597–604.

Postuma, R.B., Gagnon, J.-F., Bertrand, J.-A., Génier Marchand, D., Montplaisir, J.Y., 2015a. Parkinson risk in idiopathic REM sleep behavior disorder. Neurology 84 (11), 1104–1113.

Postuma, R.B., Berg, D., Stern, M., Poewe, W., Olanow, C.W., Oertel, W., et al., 2015b. MDS clinical diagnostic criteria for Parkinson's disease. Mov. Disord. 30 (12), 1591–1601.

Postuma, R.B., Iranzo, A., Hu, M., Högl, B., Boeve, B.F., Manni, R., et al., 2019. Risk and predictors of dementia and parkinsonism in idiopathic REM sleep behaviour disorder: a multicentre study. Brain 142 (3), 744–759.

Qiang, J.K., Wong, Y.C., Siderowf, A., Hurtig, H.I., Xie, S.X., VM-Y, L., et al., 2013. Plasma apolipoprotein A1 as a biomarker for Parkinson disease. Ann. Neurol. 74 (1), 119–127.

Remy, P., Doder, M., Lees, A., Turjanski, N., Brooks, D., 2005. Depression in Parkinson's disease: loss of dopamine and noradrenaline innervation in the limbic system. Brain 128 (Pt. 6), 1314–1322.

Rolheiser, T.M., Fulton, H.G., Good, K.P., Fisk, J.D., McKelvey, J.R., Scherfler, C., et al., 2011. Diffusion tensor imaging and olfactory identification testing in early-stage Parkinson's disease. J. Neurol. 258 (7), 1254–1260.

Ross, G.W., Abbott, R.D., Petrovitch, H., Tanner, C.M., Davis, D.G., Nelson, J., et al., 2006. Association of olfactory dysfunction with incidental Lewy bodies. Mov. Disord. 21 (12), 2062–2067.

Ross, G.W., Petrovitch, H., Abbott, R.D., Tanner, C.M., Popper, J., Masaki, K., et al., 2008. Association of olfactory dysfunction with risk for future Parkinson's disease. Ann. Neurol. 63 (2), 167–173.

Ross, G.W., Abbott, R.D., Petrovitch, H., Tanner, C.M., White, L.R., 2012. Pre-motor features of Parkinson's disease: the Honolulu-Asia Aging Study experience. Parkinsonism Relat. Disord. 18, S199–S202.

Ruffmann, C., Parkkinen, L., 2016. Gut feelings about α-synuclein in gastrointestinal biopsies: biomarker in the making? Mov. Disord. 31 (2), 193–202.

Rusz, J., Hlavnička, J., Tykalová, T., Bušková, J., Ulmanová, O., Růžička, E., et al., 2016. Quantitative assessment of motor speech abnormalities in idiopathic rapid eye movement sleep behaviour disorder. Sleep Med. 19, 141–147.

Sakakibara, R., Tateno, F., Kishi, M., Tsuyusaki, Y., Terada, H., Inaoka, T., 2014. MIBG myocardial scintigraphy in pre-motor Parkinson's disease: a review. Parkinsonism Relat. Disord. 20 (3), 267–273.

Sauerbier, A., Jenner, P., Todorova, A., Chaudhuri, K.R., 2016. Non motor subtypes and Parkinson's disease. Parkinsonism Relat. Disord. 22, S41–S46.

Schenck, C.H., Montplaisir, J.Y., Frauscher, B., Hogl, B., Gagnon, J.-F., Postuma, R., et al., 2013. Rapid eye movement sleep behavior disorder: devising controlled active treatment studies for symptomatic and neuroprotective therapy—a consensus statement from the International Rapid Eye Movement Sleep Behavior Disorder Study Group. Sleep Med. 14 (8), 795–806.

Schrag, A., Horsfall, L., Walters, K., Noyce, A., Petersen, I., 2015. Prediagnostic presentations of Parkinson's disease in primary care: a case-control study. Lancet Neurol. 14 (1), 57–64.

Schuurman, A.G., van den Akker, M., Ensinck, K.T.J.L., Metsemakers, J.F.M., Knottnerus, J.A., Leentjens, A.F.G., et al., 2002. Increased risk of Parkinson's disease after depression: a retrospective cohort study. Neurology 58 (10), 1501–1504.

Shi, M., Liu, C., Cook, T.J., Bullock, K.M., Zhao, Y., Ginghina, C., et al., 2014. Plasma exosomal α-synuclein is likely CNS-derived and increased in Parkinson's disease. Acta Neuropathol. 128 (5), 639–650.

Sidransky, E., Nalls, M.A., Aasly, J.O., Aharon-Peretz, J., Annesi, G., Barbosa, E.R., et al., 2009. Multicenter analysis of glucocerebrosidase mutations in Parkinson's disease. N. Engl. J. Med. 361 (17), 1651–1661.

Silveira-Moriyama, L., Holton, J.L., Kingsbury, A., Ayling, H., Petrie, A., Sterlacci, W., et al., 2009. Regional differences in the severity of Lewy body pathology across the olfactory cortex. Neurosci. Lett. 453 (2), 77–80.

Simon, K.C., Eberly, S., Gao, X., Oakes, D., Tanner, C.M., Shoulson, I., et al., 2014. Mendelian randomization of serum urate and parkinson disease progression. Ann. Neurol. 76 (6), 862–868.

Sobhani, S., Rahmani, F., Aarabi, M.H., Sadr, A.V., 2017. Exploring white matter microstructure and olfaction dysfunction in early parkinson disease: diffusion MRI reveals new insight. Brain Imaging Behav. 13, 210–219 .

Sommerauer, M., Fedorova, T.D., Hansen, A.K., Knudsen, K., Otto, M., Jeppesen, J., et al., 2018. Evaluation of the noradrenergic system in Parkinson's disease: an 11C-MeNER PET and neuromelanin MRI study. Brain 141 (2), 496–504.

Stemplewitz, B., Keserü, M., Bittersohl, D., Buhmann, C., Skevas, C., Richard, G., et al., 2015. Scanning laser polarimetry and spectral domain optical coherence tomography for the detection of retinal changes in Parkinson's disease. Acta Ophthalmol. 93 (8), e672–e677.

Stevens, J.C., Cain, W.S., Weinstein, D.E., Pierce, J.B., 1987. Aging impairs the ability to detect gas odor. Fire Technol. 23 (3), 198–204.

Svensson, E., Horváth-Puhó, E., Thomsen, R.W., Djurhuus, J.C., Pedersen, L., Borghammer, P., et al., 2015. Vagotomy and subsequent risk of Parkinson's disease. Ann. Neurol. 78 (4), 522–529.

Thaler, A., Ash, E., Gan-Or, Z., Orr-Utreger, A., Giladi, N., 2009. The LRRK2 G2019S mutation as the cause of Parkinson's disease in Ashkenazi Jews. J. Neural Transm. 116 (11), 1473.

van Steenoven, I., Majbour, N.K., Vaikath, N.N., Berendse, H.W., van der Flier, W.M., van de Berg, W.D.J., et al., 2018. α-Synuclein species as potential cerebrospinal fluid biomarkers for dementia with lewy bodies. Mov. Disord. 33 (11), 1724–1733.

VanderHorst, V.G., Samardzic, T., Saper, C.B., Anderson, M.P., Nag, S., Schneider, J.A., et al., 2015. α-Synuclein pathology accumulates in sacral spinal visceral sensory pathways. Ann. Neurol. 78 (1), 142–149.

Villumsen, M., Aznar, S., Pakkenberg, B., Jess, T., Brudek, T., 2019. Inflammatory bowel disease increases the risk of Parkinson's disease: a Danish nationwide cohort study 1977–2014. Gut 68 (1), 18–24.

Volpicelli-Daley, L.A., Luk, K.C., Patel, T.P., Tanik, S.A., Riddle, D.M., Stieber, A., et al., 2011. Exogenous α-synuclein fibrils induce Lewy body pathology leading to synaptic dysfunction and neuron death. Neuron 72 (1), 57–71.

Walter, U., Tsiberidou, P., Kersten, M., Storch, A., Löhle, M., 2018. Atrophy of the vagus nerve in Parkinson's disease revealed by high-resolution ultrasonography. Front. Neurol. 9, 805 .

Wang, J., You, H., Liu, J.-F., Ni, D.-F., Zhang, Z.-X., Guan, J., 2011. Association of olfactory bulb volume and olfactory sulcus depth with olfactory function in patients with Parkinson disease. AJNR Am. J. Neuroradiol. 32 (4), 677–681.

Weintraub, D., Tröster, A.I., Marras, C., Stebbins, G., 2018. Initial cognitive changes in Parkinson's disease. Mov. Disord. 33 (4), 511–519.

Weisskopf, M.G., Chen, H., Schwarzschild, M.A., Kawachi, I., Ascherio, A., 2003. Prospective study of phobic anxiety and risk of Parkinson's disease. Mov. Disord. 18 (6), 646–651.

Wu, P., Yu, H., Peng, S., Dauvilliers, Y., Wang, J., Ge, J., et al., 2014. Consistent abnormalities in metabolic network activity in idiopathic rapid eye movement sleep behaviour disorder. Brain 137 (Pt. 12), 3122–3128.

Xu, Q., Park, Y., Huang, X., Hollenbeck, A., Blair, A., Schatzkin, A., et al., 2010. Physical activities and future risk of Parkinson disease. Neurology 75 (4), 341–348.

Yang, F., Trolle Lagerros, Y., Bellocco, R., Adami, H.-O., Fang, F., Pedersen, N.L., et al., 2015. Physical activity and risk of Parkinson's disease in the Swedish National March Cohort. Brain 138 (Pt. 2), 269–275.

Zimmermann, M., Gaenslen, A., Prahl, K., Srulijes, K., Hauser, A.-K., Schulte, C., et al., 2019. Patient's perception: shorter and more severe prodromal phase in GBA-associated PD. Eur. J. Neurol. 26 (4), 694–698.

The gut microbiome in Parkinson's disease: A culprit or a bystander?

11

Ali Keshavarzian[a], Phillip Engen[a], Salvatore Bonvegna[b], Roberto Cilia[c],*

[a]*Department of Internal Medicine, Division of Digestive Disease and Nutrition, Rush University Medical Center, Chicago, IL, United States*
[b]*Parkinson Institute, ASST Gaetano Pini-CTO, Milan, Italy*
[c]*Fondazione IRCCS Istituto Neurologico Carlo Besta, Movement Disorders Unit, Milan, Italy*
Corresponding author: Tel.: +39-23942368; Fax: +39-23942539,
e-mail address: roberto.cilia@istituto-besta.it

Abstract

In recent years, large-scale metagenomics projects such as the Human Microbiome Project placed the gut microbiota under the spotlight of research on its role in health and in the pathogenesis several diseases, as it can be a target for novel therapeutical approaches. The emerging concept of a microbiota modulation of the gut-brain axis in the pathogenesis of neurodegenerative disorders has been explored in several studies in animal models, as well as in human subjects. Particularly, research on changes in the composition of gut microbiota as a potential trigger for alpha-synuclein (α-syn) pathology in Parkinson's disease (PD) has gained increasing interest. In the present review, we first provide the basis to the understanding of the role of gut microbiota in healthy subjects and the molecular basis of the gut-brain interaction, focusing on metabolic and neuroinflammatory factors that could trigger the alpha-synuclein conformational changes and aggregation. Then, we critically explored preclinical and clinical studies reporting on the changes in gut microbiota in PD, as compared to healthy subjects. Furthermore, we examined the relationship between the gut microbiota and PD clinical features, discussing data consistently reported across studies, as well as the potential sources of inconsistencies. As a further step toward understanding the effects of gut microbiota on PD, we discussed the relationship between dysbiosis and response to dopamine replacement therapy, focusing on Levodopa metabolism. We conclude that further studies are needed to determine whether the gut microbiota changes observed so far in PD patients is the cause or, instead, it is merely a consequence of lifestyle changes associated with the disease. Regardless, studies so far strongly suggest that changes in microbiota appears to be impactful in pathogenesis of neuroinflammation. Thus, dysbiotic microbiota in PD could influence the disease course and response to medication, especially Levodopa. Future research will assess the impact of microbiota-directed therapeutic intervention in PD patients.

Keywords

Parkinson's disease, Gut microbiota, Gut-brain axis, Neuroinflammation, Short-chain-fatty-acids, Levodopa

1 Gut microbiota in healthy subjects

1.1 Introductions and definitions

What is the **"Gut Microbiota"**? The gut microbiota is a complex ecological community composed by about 100 trillions of microbes inhabiting the human intestine, which influences both normal physiology and disease susceptibilities through its metabolic activities and host interactions (Lozupone et al., 2012). On the other hand, the term **"Microbiome"** refers to the collection of genetic material (genome)—bacteria, virus, archaea, fungi, and protozoa represented in the gut microbiota (Lozupone et al., 2012). The gut microbiome contains >3 million unique genes, outnumbering the number of human genes 150–1 (Proctor, 2011).

The composition of gut microbiota is established at birth (fetal gut is considered sterile, but this concepts require revisiting with additional studies using newly developed highly sensitive technique) by which the first colonization of bacteria is influenced by delivery (vaginal vs. cesarean section) and maternal feeding choice (breast-fed vs. formula-fed) and change throughout the lifetime (Ghaisas et al., 2016). It is greatly variable and the immense diversity of the microbiota, interpersonal variation and temporal fluctuations in composition, especially during disease and early development, make it an extremely complex and interesting topic. An adult-like microbiota is established by the first 3–5 years of life (Koenig et al., 2011; Palmer et al., 2007; Yatsunenko et al., 2012). Once established, the composition of the gut microbiota is relatively stable throughout adult life, but can be altered by antibiotics, stress, lifestyle choices, environmental exposures, bacterial infections, diet, geography and surgery (Ghaisas et al., 2016; Yatsunenko et al., 2012). Majority of gut microbes are harmless or beneficial to the host, as they protect against pathogens (Candela et al., 2008; Fukuda et al., 2011), contribute to normal immune function and influence the hypothalamic–pituitary–adrenal axis (Ghaisas et al., 2016; Olszak et al., 2012), regulate xenobiotics metabolism and energy production (Ghaisas et al., 2016; Sonnenburg et al., 2005; Yatsunenko et al., 2012).

The term **"Dysbiosis"** is generally used to define the disruption of the "normal" balance between the gut microbiota and the host (Lozupone et al., 2012). Gut dysbiosis has been associated with obesity (Ley et al., 2006; Turnbaugh et al., 2008), malnutrition (Kau et al., 2011), inflammatory bowel diseases (IBD) (Dicksved et al., 2008; Frank et al., 2007), cancer (Lupton, 2004), and neurological disorders (Gonzales et al., 2011). Understanding how the gut microbiota affects health and disease requires a shift in focus from individual pathogens, toward an ecological approach considering the community as a whole (Lozupone et al., 2012).

The aim of this first part of the chapter is to provide the tools to understand the symbiotic relationships of gut microbes with their hosts by characterizing the healthy gut microbiota, the measures used to define it and the methods used to analyze its composition. In the second part, we will apply these notions to highlight the changes in gut microbiota composition described in association with Parkinson's disease (PD), aiming to disentangle whether it may have a causative link with the pathogenesis of the disease or if further evidence is needed.

1.2 Microbial diversity: Definition and significance

1.2.1 Alpha- and beta-diversities

To understand how the gut microbiota interacts with the host shaping health status and affecting disease, we need to consider the community as a whole, instead of focusing on individual pathogens (Lozupone et al., 2012). Therefore, it is important to start our approach the interpretation of clinical studies by estimating microbiome diversity, as pathological changes in microbial composition are often associated with reduced diversity (Carding et al., 2015; Sarangi et al., 2019). Diversity is assessed using two separate types of measures: **alpha-** and **beta-diversities** (Fig. 1).

Alpha diversity (within-sample) is defined as the number of different species (richness) and their distribution (evenness) in a particular specimen. The Shannon index measures how evenly the microbes are distributed within a sample. Generally, high variability of alpha biodiversity of colonic microbiota is associated with a healthy status; whereas, low alpha diversity is commonly associated with pathological states, such as atopy, inflammatory bowel disease (IBD), autoimmunity, metabolic disease (e.g., type 2 diabetes mellitus), depression and colon-rectum carcinoma (Carding et al., 2015; Sarangi et al., 2019).

Beta diversity (between-sample) is defined as the difference in species composition of two groups of specimens (e.g., those from patients with PD vs. healthy individuals) and measures the number of species that differ between the two groups (Lozupone et al., 2012). Beta diversity calculates the difference in taxonomic abundance profiles between different microbial environments (Sarangi et al., 2019).

1.2.2 Taxonomic and functional diversity

1.2.2.1 Taxonomic diversity

Each human subject harbors over 1000 species-level phylotypes of bacteria, and most of them belong to just a few phyla (Claesson et al., 2009). In adults, *Bacteroidetes* and *Firmicutes* usually outnumber other phyla, whereas *Actinobacteria, Proteobacteria,* and *Verrucomicrobia* are frequent but generally minor components (Eckburg et al., 2005). Human gut microbiota also contains methanogenic archaea, eukarya (mainly yeasts), fungi and viruses (mainly bacteriophages) (Reyes et al., 2010). Nevertheless, culture-independent sequencing studies have repeatedly demonstrated a vast microbial diversity that is highly variable both over time and across human populations, challenging the initial notion that there is a "core" set of shared species in the human gut microbiota (Biagi et al., 2010; Yatsunenko et al., 2012).

1.2.2.2 Functional diversity

The composition of microbial community cannot be sufficient to understand community function. Functional information comes from (1) cultured isolates whose genome is well characterized and (2) ex-vivo phenotypes and from sequencing

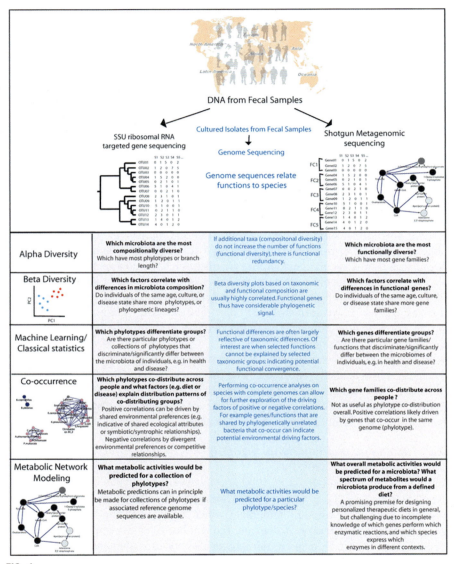

FIG. 1

Tools for understanding compositional and functional diversity of the microbiota.

From Lozupone, C.A., Stombaugh, J.I., Gordon, J.I., Jansson, J.K., Knight, R. 2012. Diversity, stability and resilience of the human gut microbiota. Nature 489(7415), 220–230.

community DNA (Lozupone et al., 2012). Identification of genes involved in specific metabolic pathways is limited by predicting functional effects, whereas it is only the analysis of mRNA, protein and metabolite profiling that is able to provide conclusive information. Functional gene profiles may be similar in different individuals (functional core microbiome), although gut microbiota compositions may be greatly different (in consistence with the notion that we are not sharing a "core microbiota" at taxonomic level) (Lozupone et al., 2012). Considering that many genes are expressed only under specific conditions, shotgun sequencing approaches that measure levels of mRNA (metatranscriptomics) or shotgun proteomics (metaproteomics) are needed to investigate functional variation caused by different factors (including disease, or environmental/lifestyle factors), that studies on DNA likely do not notice (Lozupone et al., 2012). In this scenario, it is important to disentangle *"functional redundancy"* by identifying which microbia have similar functional effects (Lozupone et al., 2012). The characterization of functional components associated with altered physiological states and pathology (such as in PD) relies on the identification of deviations from predictions about function from a microbiota composition (Lozupone et al., 2012).

1.2.3 Enterotypes

Studies of compositional and functional differences in the gut microbiota lay the foundation for relating these differences to human health, potentially explaining interpersonal variations in gut metabolic processes, including metabolism of xenobiotics, including medications (Clayton et al., 2009; Gonzales et al., 2011; Lozupone et al., 2012) such as Levodopa (Maini Rekdal et al., 2019). Considering that many of these metabolic pathways are outside the common functional core, they can underlie host-specific responses. Understanding how the microbiota varies across the human population, and correlating this variability with specific microbial functions, is emerging as a component of personalized medicine.

The human microbiome was proposed to form three distinct host-microbial symbiotic states conceptualized as "**enterotypes**," driven by groups of co-occurring species/genera, characterized by relatively high representation of the genera *Bacteroides, Prevotella,* or *Ruminococcus*, respectively (Fig. 2A–E). While individual host properties (e.g., body mass index, age, or gender) cannot explain the observed enterotypes, data-driven marker genes or functional modules have been identified for each enterotype (Arumugam et al., 2011). Recently, the *Bacteroides* enterotype has been subdivided into two parts (B1 and B2), as Butyrate-producing *Faecalibacterium* and *Coprococcus* bacteria have been linked with measures of quality of life and depression (Fig. 2F and G). Analysis of fecal metagenomes suggested that the bacterial synthesis of the dopamine metabolite 3,4-dihydroxyphenylacetic acid (DOPAC) positively correlated with higher mental quality of life and that the production of γ-aminobutyric acid was associated with depression (Valles-Colomer et al., 2019). It should be noted that the concept of distinct enterotypes has recently been challenged because most subjects living in western societies on mixed diet do not cleanly divided into these enterotypes.

1.3 Factors modulating gut microbiota

The gut microbiota is not equal in all populations (see above) and the significant intra- and inter-individual variability was discovered in the microbiota composition of healthy control subjects (Caporaso et al., 2011; Lozupone et al., 2012) and many diseases, where intestinal dysbiosis had been reported (Clemente et al., 2012; Gilbert et al., 2016). An explanation of this diversity comes from environmental factors,

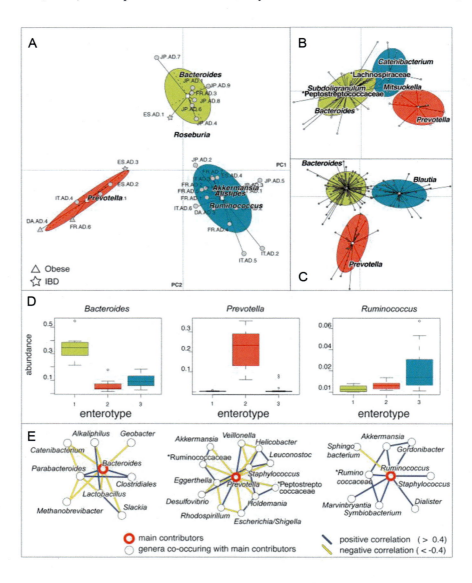

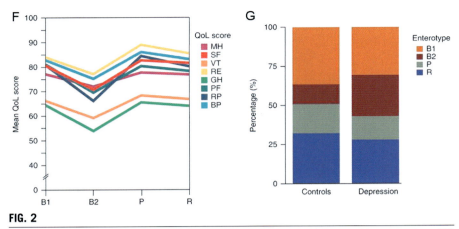

FIG. 2

(A–E) Phylogenetic differences between enterotypes; (F and G) Bacteroides enterotype B2 association with poor QoL and depression.

Panels (A-E): From Arumugam, M., Raes, J., Pelletier, E., Le Paslier, D., Yamada, T., Mende, D.R., et al.
Enterotypes of the human gut microbiome. 2011. Nature; 473(7346): 174–80.
Panels (F and G): From Valles-Colomer, M., Falony, G., Darzi, Y., Tigchelaar, E.F., Wang, J.,
Tito, R.Y., et al., 2019. The neuroactive potential of the human gut microbiota in quality of life and depression.
Nat. Microbiol. 4(4), 623–632.

which markedly affect microbiota community structure and composition. Therefore, it is expected that the intestinal microbiota in patients from the USA should be different from those living in Northern, Eastern or Western Europe or Asia. In fact, the intestinal microbiota is significantly different in individuals living in different communities in the city of Chicago, IL, USA (Miller et al., 2016). Understanding factors underlying compositional and functional changes will provide the basis in designing therapies that target the gut microbiota (Lozupone et al., 2012). The composition, diversity and function of gut microbiota can be modulated by many factors, briefly explained in this section and expanded later on this chapter.

1.3.1 Age

During the first years of life, the human microbiota undergo consecutive compositional and functional changes following initial colonization with an increase in diversity and stability (Koenig et al., 2011; Palmer et al., 2007; Yatsunenko et al., 2012), until a relatively stable community is established (Sharon et al., 2016; Fig. 3). The first colonization is influenced by delivery (vaginal vs. cesarean section) (Ghaisas et al., 2016), while the feeding status (breast-fed vs. formula-fed) and the use of antibiotics have large effects on the infant microbiota and may modulate individual susceptibility to immunologic diseases into adulthood (Kozyrskyj et al., 2011; Olszak et al., 2012). Although interindividual variability in microbial communities and

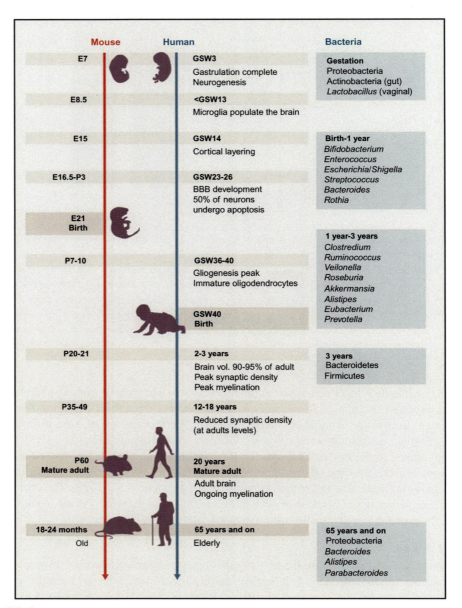

FIG. 3

Major events in mammalian brain development.

From Sharon, G., Sampson, T.R., Geschwind, D.H., Mazmanian, S.K. 2016. The central nervous system and the gut microbiome. Cell 167(4), 915–932.

functional gene repertoires is greater in infants than adults (Yatsunenko et al., 2012), infant microbiomes share characteristic properties across individuals and populations, including both composition (e.g., several *Bifidobacteria* and reduced species richness than adults) and function (e.g., genes encoding enzymes involved in folate biosynthesis are more represented) (Yatsunenko et al., 2012). Thus, age is an important confounder to consider in all studies on gut microbiota, including those on PD patients.

1.3.2 Genetics, environment and diet
The relative role of genetics and environment in modeling the human gut microbiota is still to be unraveled, in part due to the difficulties to disentangle the two from each other. A decade ago, human genetics has been suggested to influence the microbiota, because microbiota composition was shown to be similar among twins and mother-daughter (Dicksved et al., 2008; Turnbaugh et al., 2009; see also Section 3.1). However, shared environment rather than genes may drive familial similarities in monozygotic and dizygotic adult twins (Turnbaugh et al., 2009). Characteristic differences separate the gut microbiota in different populations (De Filippo et al., 2010; Yatsunenko et al., 2012) with genetic, environmental exposures, hygiene, diet and antibiotic use modulating these compositions (Arumugam et al., 2011; Wu et al., 2011). We will describe in detail the potential effects of all these confounders the second part of the chapter that focuses on the effects of gut microbiota in patients with PD.

1.4 *"Microbiota for dummies"*: Disentangling methods for studying gut microbiota
Whenever a neurologist approaches for the first time studies the gut microbiota in cohort of patients with PD, he/she is almost invariably tempted to skip the methodology section to read the results and their interpretation. However, it might happen that different studies provide contrasting results (as we are going to show in an ad-hoc designed table later on) so that he/she remains puzzled wondering which is the most reliable and why these discrepancies could occur.

Several methods have been used to study gut microbiota, and these have undergone a major change over time. Sarangi and colleagues recently described these methods in detail (Table 1). For example, in a recent study on PD, Aho et al. (2019) used three differential abundance comparison methods to analyze their sample with results not overlapping substantially to one another. Overall, it is important to examine not only the results, but note which methodology was implemented when interpreting data on PD microbiota.

Table 1 Techniques used to study microbiota in the gut as well as other body sites.

A. Culture-based methods
B. Molecular-based (nuclecic-acid based) methods
   a. Non-sequencing methods
      i. Fluorescence in situ hybridization flow cytometry
      ii. Pulsed field gel electrophoresis
      iii. Denaturing gradient gel electrophoresis
      iv. Temperature gradient gel electrophoresis
      v. Single-strand conformation polymorphism
   b. Sequence-based methods
      i. Sequencing of 16S rRNA genes or their hypervariable regions (targeted gene sequencing)
      ii. Whole bacterial genome DNA (metagenome) sequencing
      iii. Whole bacterial mRNA (meta-transcriptome) sequencing
C. Methods based on detection and quantification of small metabolites
   i. Gas chromatography mass spectrometry
   ii. Capillary electrophoresis coupled to mass spectrometry
   iii. Fourier-transform infrared spectroscopy
   iv. Nuclear and proton magnetic resonance spectroscopy

From Sarangi et al. (2019).

2 Gut microbiota in the pathogenesis of Parkinson's disease

Parkinson's disease is the second most common neurodegenerative disease in the United States affecting up to 1% of individuals older than age 60 (Tysnes and Storstein, 2017). Unfortunately, the incidence of Parkinson's disease is increasing in recent years, with expectations to affect over 3 million Americans with PD by 2035 (GBD 2016 Neurology Collaborators, 2018; Tysnes and Storstein, 2017). Symptoms and signs of PD are in large part due to loss of dopaminergic neurons in the substantia nigra that controls balance and movement (Kalia and Lang, 2015). PD is associated with significant morbidity and a huge burden to the caregivers and families of those suffering from PD and yet there are no available interventions to prevent PD or modify disease course. In fact, the current treatments of PD are only focused on symptomatic relief and are associated with significant side effects and typically lose effectiveness over time and are not universally effective in all patients (Kalia and Lang, 2015). Thus, there is a major unmet need to identify novel safer therapies for PD that could not only improve symptoms in most patients, but also be able to modify disease course and even prevent PD in high-risk individuals. This goal can only be achieved by better understanding of the pathogenesis of PD in order to identify the appropriate therapeutic targets.

In the following paragraphs, we will summarize scientific evidence supporting the hypothesis that the intestinal microbiota is the key mediator (thus lynchpin) of environmental factors that increase the risk of developing PD, suggesting that a disrupted microbiota community may be a critical player in the pathogenesis of PD that could act as a trigger and/or enabler for neuroinflammation and neurodegeneration.

2.1 **Gut microbiota and neuroinflammation**

2.1.1 *General remarks*

The hallmark pathology finding in PD of aggregated alpha-synuclein (a-syn) found as Lewy bodies and Lewy neurites in the SN appears to be one key for dopaminergic neurodegeneration (Obeso et al., 2017; Olanow and Brundin, 2013). Alpha-synuclein aggregation is a consequence of misfolding of the α-syn protein and this misfolding is thought to be a result of increased oxidative stress and neuroinflammation (Riederer et al., 2019). The source of this neuroinflammation is activated microglia and astro-cytes (Troncoso-Escudero et al., 2018). The critical question, which has yet to be decided, is why and how microglia are activated in PD (Borghammer, 2018; Johnson et al., 2019). Braak was the first to propose that the gastrointestinal (GI) tract is the site to initiate neuroinflammation and dopamine loss in PD patients via invasion of a pathogen or later via α-syn aggregation (Braak et al., 2003; Braak and Del Tredici, 2017; Lionnet et al., 2017). Through a series of elegant neuropathological studies in *post-mortem* PD patients, he proposed that toxins or pathogens enter in the host through the GI tract, primarily upper GI tract (or later nasal/olfactory bulb), and cause inflammation and misfolding/aggregation of α-syn in the ENS of the GI tract and these α-syn protein aggregates (Lewy bodies and Lewy neurites) move up to the CNS (prion-like) via the vagus nerve leading to DA loss in the SN and symptoms/signs of PD (Braak and Del Tredici, 2017; Hawkes et al., 2007). This proposed mechanism could explain why the PD patients develop gastrointestinal symptoms, loss of smell and REM sleep disorders (non-motor PD symptoms) prior to starting the cardinal motor and CNS symptoms and signs of PD (Braak and Del Tredici, 2017). Thus, he was the first to implicate the gut-brain axis in the pathogenesis of PD. Subsequent studies such as decreased frequency of PD in patients who underwent vagotomy (Liu et al., 2017; Svensson et al., 2015) and also detection of α-syn aggregates in the brain after injection of α-syn fibrils into the intestine or recently into the gastric wall in rodents or primates (Holmqvist et al., 2014; Kim et al., 2019) would support Braak's hypothesis. Nonetheless, in recent years the gut-to-brain propagation of α-syn pathology via the vagus nerve has been challenged, as we will review in the final paragraph of this chapter.

We propose that although toxins and pathogens could be triggers for neuroin-flammation in subsets of PD cases, there is no epidemiological evidence to support that such toxins/pathogens are the primary source of neuroinflammation in the majority of PD patients (Lionnet et al., 2017). However, the gastrointestinal tract could still be the site for environmental factors to trigger neuroinflammation because the GI tract is the largest surface area between the host and environment with over 300 square meters of surface area. Furthermore, the GI tract harbors the most com-plex microbiota community capable of producing proinflammatory factors (Blaser, 2014; Clemente et al., 2012; Gilbert et al., 2016). Indeed, humans carry over 10 Kg of microbiome mass with trillions of bacteria, archaea, fungi and viruses including over 1000 species of bacteria with many of them producing pro-inflammatory factors, like endotoxins (Blaser, 2014; Clemente et al., 2012; de Vos and de Vos, 2012). On the

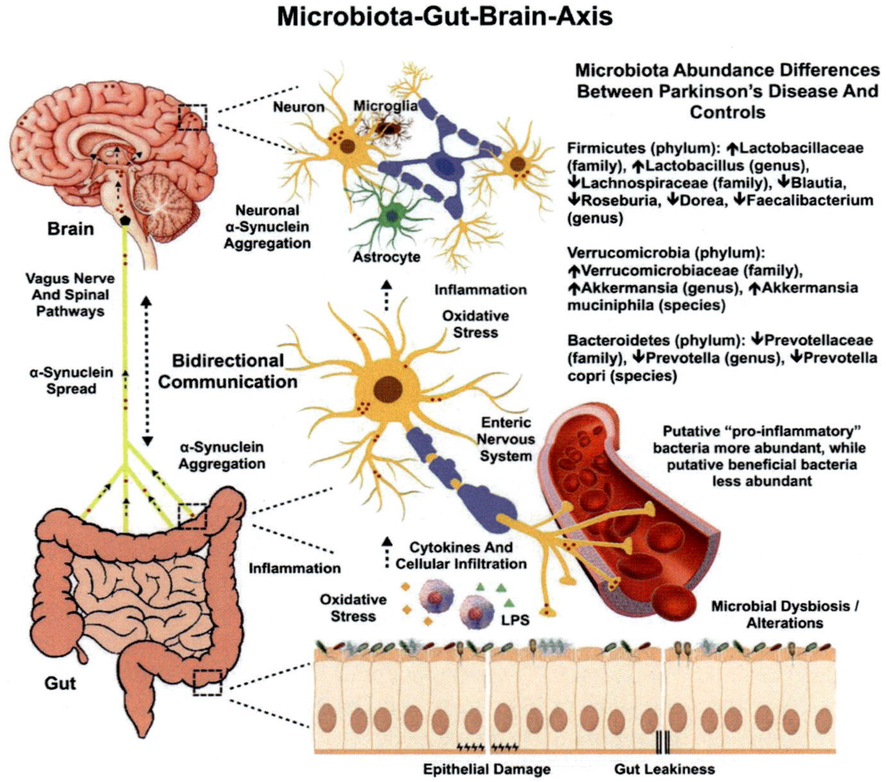

FIG. 4

The microbiota–gut–brian axis. Likely implicated pathways in the pathogenesis of Parkinson's disease.

From Lubomski, M., Davis, R.L., Sue, C.M. 2019a. The gut microbiota: a novel therapeutic target in Parkinson's disease? Parkinsonism Relat. Disord. 66, 265–266., Lim, S.Y., Holmes, A.J., Davis, R.L., Sue, C.M. 2019b. Parkinson's disease and the gastrointestinal microbiome. J Neurol. 1–17.

other hand, there are probiotic bacteria physiologically producing compounds with both local and systemic beneficial properties (e.g., enhancing enteric epithelial barrier and immune response) counteracting the effects of pathogenic bacteria (Fig. 4).

2.1.2 Molecular mediators: Toll-like receptors signaling

Bacterial metabolites from the gut microbiota are thought to potentially interact with the CNS principally via innate immune TLR (Toll-like receptors), NLR (NOD-like receptors) and RLR (RIG-1-like receptors) (Creagh and O'Neill, 2006;

Fiebich et al., 2018; Labzin et al., 2018). These innate immune receptors initiate biological processes (especially neuroinflammation) through engaging bacterial produced molecules called pathogen-associated molecular patterns (PAMPs) as well as endogenously produced (e.g., HMGB1) damage-associated molecular patterns (DAMPs). Thus far, 13 mammalian TLRs, 10 in humans and 13 in mice, have been identified. TLRs 1–9 are conserved among humans and mice, yet TLR10 is present only in humans and TLR11 is functional only in mice (West et al., 2006). Only microglia express all TLRs 1–9 in the CNS, and TLR4 is only expressed by microglia (neuron and astroglial TLR4 is debated) with neurons expressing TLR3,7,8,9 and astroglia expressing TLR2,3,9 (Hanke and Kielian, 2011). Activation of these TLRs triggers signaling cascades leading to activation of other key inflammatory molecules like NF-kB and eventually production of cytokines including proinflammatory cytokines like TNF alpha, IL-8 and interferon gamma (Hanke and Kielian, 2011; West et al., 2006). Intestinal epithelial cells, immune cells in the intestinal mucosa and more importantly cells in ENS and microglia in the brain have multiple TLRs including TLR4 and TLR2 and thus are capable of responding to bacterial products like Lipopolysaccharide (LPS) and initiating inflammatory cascades and a pro-inflammatory state in the intestine and brain. Much research has focused specifically on the role of TLR signaling in PD neuroinflammation and especially on microglial TLR4 and TLR2 signaling (Blaylock, 2017; Fiebich et al., 2018; Gerhard et al., 2006; Guillot-Sestier and Town, 2018; Labzin et al., 2018; Perez-Pardo et al., 2018b) (Fig. 5). Broadly established microbial ligands for TLR4 and TLR2 are the gram negative cell wall component LPS (TLR4) and gram positive cell wall component lipoteichoic acid (TLR2) (Taakeuchi et al., 2010). Significantly, both β-amyloid protein in AD and α-syn protein in PD can activate inflammatory microglial signaling via TLR4 and/or TLR2 (Codolo et al., 2013; Fellner et al., 2013; Halle et al., 2008). Also, LPS-TLR4 signaling promotes leakiness of the blood brain barrier (Banks et al., 2015; Jangula and Murphy, 2013).

Both Parkinson's and Alzheimer's disease (AD) neuroinflammation models currently propose initial microglial activation of inflammatory signaling to drive disease progression (Block and Hong, 2005; Fiebich et al., 2018; Liddelow et al., 2017; Yun et al., 2018). Current evidence supports a model in which chronic microglial TLR4 (and possibly TLR2) signaling drives NLRP3 inflammasome activation in PD (Anderson et al., 2018; Freeman et al., 2017; Gurung et al., 2015) with NLRP3 located predominantly in microglia in PD patients post mortem (Chakraborty et al., 2010; Gordon et al., 2018). The inflammasomes are multiprotein complex's that assemble in response to diverse inflammatory signals to promote caspase activation of IL-1β and IL-18 in the CNS and throughout the body, especially in immune cells and also in the intestine (Abderrazak et al., 2015; Chakraborty et al., 2010; Freeman and Ting, 2016). Recent studies support that inhibition of microglial NLRP3 is protective in many animal models of PD including α-syn preformed fibrils, MPTP, rotenone and at least one genetic PD model (Chakraborty et al., 2010; Gordon et al., 2018).

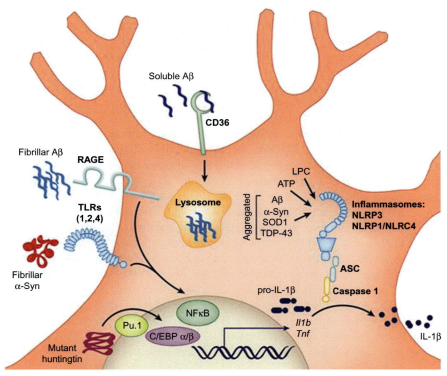

FIG. 5

The role of TLRs and inflammosomes in response to protein aggregates in PD and other neurodegenerative diseases.

From Labzin, L.I., Heneka, M.T., Latz, E. 2018. Innate immunity and neurodegeneration. Annu. Rev. Med. 69, 437–449.

2.1.3 Lipopolysaccharide and lipopolysaccharide-binding protein

Minato et al. (2017) found an increase of serum lipopolysaccharide-binding protein (LBP) over 2 years in PD patients with high intestinal counts of *Lactibacillus (L.) brevis* and *L. plantarum* subgroup. This finding support the hypothesis of protective properties of LBP against production of proinflammatory cytokines (such as tumor necrosis factor (TNF)-α; Ding and Jin, 2014), as also demonstrated by the relationship between reduced LBP levels and (1) increased LPS-induced expression of interleukin (IL)-1 and TNF-α in membrane-bound CD14 (mCD14)-positive cells, and (2) increased expression of IL-6 and IL-8 in mCD14-negative cells. *L. plantarum* subgroup has lipoteichoic acid, which blocks IL-8 production and thus exerts protective anti-inflammatory effects on human intestinal epithelial cells (Noh et al., 2015). Similarly, *L. brevis* is able to enhance intestinal barrier function in a murine model of colitis (Ueno et al., 2011). Accordingly, oral administration of L. *brevis* in aged mice was able to reduce (i) blood and colon fluid levels of LPS,

(ii) the levels of inflammatory markers (including TNF-α), and (iii) the activation of NF-κB (Jeong et al., 2016; Minato et al., 2017).

2.2 Metabolic mediators: The role of SCFAs in PD

2.2.1 Physiologic role in healthy subjects

Short-chain-fatty-acids (SCFA; acetate, propionate, butyrate—typically occurring in a 3:1:1 ratio) are bacterial products that are produced by a subset of intestinal bacteria, SCFA producers, by fermenting complex carbohydrate fibers (Koh et al., 2016; Makki et al., 2018). The gut microbiota composition influence the profile of the SCFAs produced. For example, while strains of *Lactobacillus* and *Bifidobacterium* spp. (e.g., *B. longum*, *B. bifidum*, *B. breve*, *B. adolescentis*, *B. dentium*, *B. pseudocatenulatum*, and *B. catenulatum*) are probiotics able to produce acetate, the majority of beneficial butyrate-producing bacteria are from the Firmicute phylum associated to both *Clostridium clusters IV and XIVa*) (Louis and Flint, 2009; Vital et al., 2014). Important *Clostridal cluster IV and XIVa* butyrate-producing bacterial species known to promote gut health are *Faecalibacterium prausnitizii*, *Eubacterium* spp. (e.g., *Eubacterium hallii*), *Roseburia* spp. (e.g., *Roseburia intestinalis*), *Anaerostipes* spp. (e.g., *Anaerostipes caccae*), *Coprococcus* spp. (e.g., *Coprococcus eutactus*, and *Coprococcus comes*) and *Butyricicoccus pullicaecorum*.

SCFA and specifically butyrate bacteria are the preferred energy substrate for epithelial cells in the colon and low levels of the SCFA butyrate are associated with disruption of intestinal barrier integrity and colonic inflammation (Hamer et al., 2008; Koh et al., 2016). Therefore, changes in the gut microbiota composition leading to a decrease in butyrate-producing microbiota (such as reduced *Firmicute* phylum) may be associated with increased mucosal permeability and mucosal inflammation. Furthermore, butyrate acts locally on the colonic mucosa but can also exert remote effects via the ENS, (e.g., butyrate alters the activity of enteric neurons by reversible hyperpolarization) (Kidd and Schneider, 2010). SCFA are also associated with other beneficial effects, such as promoting normal microglial development by other SCFA-responsive genes such as the histone deacetylates that modulate gene expression (Bourassa et al., 2016; Erny et al., 2015), may cross the blood-brain-barrier and impact the physiology of cells in the CNS (Mitchell et al., 2011), increase colonic contractility (Soret et al., 2010), suppressing inflammation (Koh et al., 2016) and stimulating vagal signaling and production of glucagon-like peptide-1 (GLP-1) by intestinal L-cells (De Vadder et al., 2014; Everard and Cani, 2014). GLP-1 and related agonists have recently received much attention for their beneficial effects in AD and PD patients as well as animal models of PD (Athauda and Foltynie, 2018; Athauda et al., 2017; Zhang et al., 2018).

SCFAs may influence CNS functions by several mechanisms, including immune, endocrine, vagal and other humoral pathways (Fig. 6). Relevant for PD, the nourishing properties of SCFAs for colonocytes may explain the association between reduced levels of SCFAs and constipation as well as increased intestinal barrier

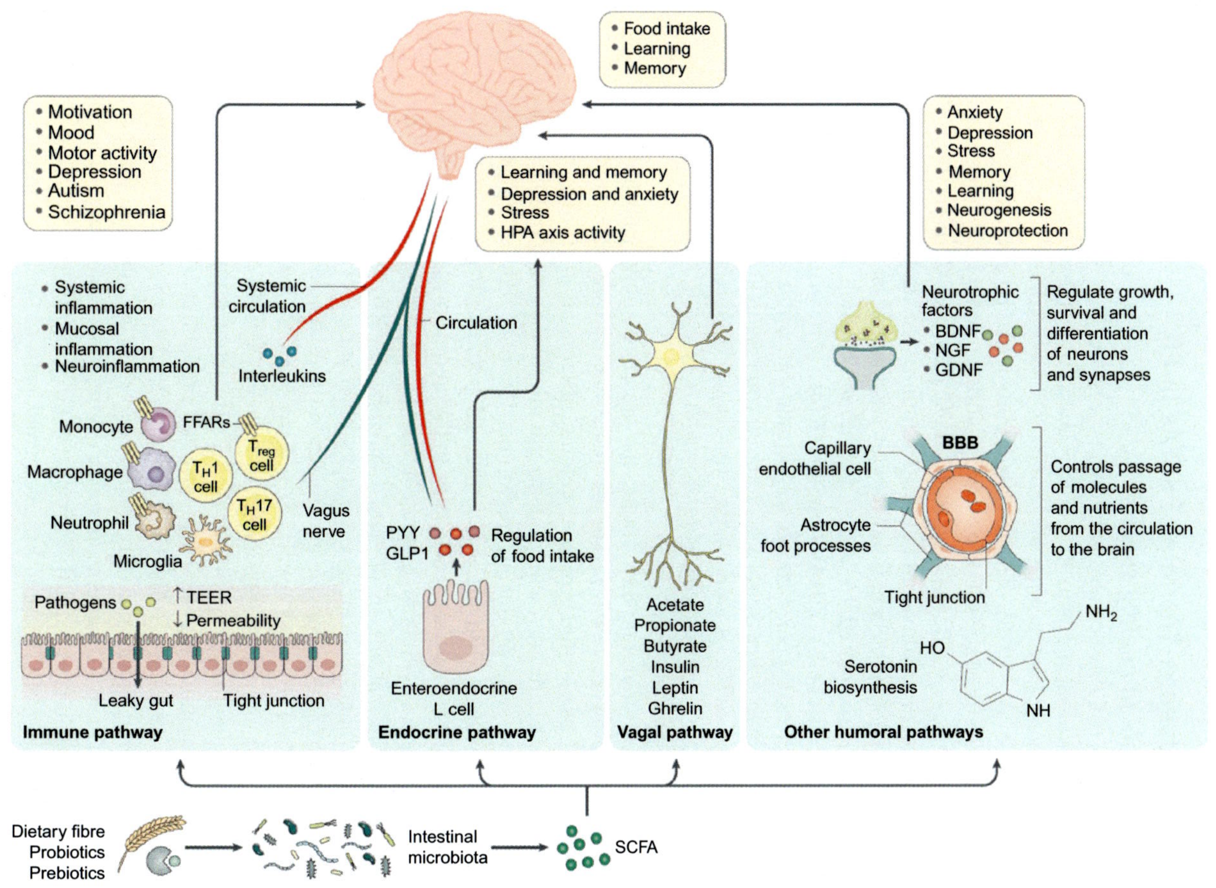

FIG. 6

Potential gut-brain pathways through which Short Chain Fatty Acids (SCFAs) might modulate brain function.

From Dalile, B., Van Oudenhove, L., Vervliet, B., Verbeke, K., 2019. The role of short-chain fatty acids in microbiota-gut-brain communication. Nat. Rev. Gastroenterol. Hepatol. 16(8), 461–478.

leakiness (Bedarf et al., 2017; Ganapathy et al., 2013; Soret et al., 2010). Furthermore, SCFA and specifically butyrate have mucosal and systemic anti-inflamamotry properties in multiple pathological conditions including IBD, alcohol associated pathology and metabolic syndrome (Dalile et al., 2019). Additionally, butyrate is a stong HDAC inhibitor that could also contribute to its anti-inflammatory and immune modulatory activites (Kidd and Schneider, 2010; Wu et al., 2008). Nonetheless, some harmful effect of SCFAs have been described, primarily in germ free mice (Sampson et al., 2016). Thus, the net effects of SCFAs on neuronal development and function and nuero-inflamamtion and neurodegeneration in PD patients require additional studies.

Another potential molecular mechanism for a dysbiotic microbiota-mediated gut leakiness to endotoxin in PD is the low level of beneficial bacterial products required for maintaining intestinal barrier integrity due to the loss of "good" bacteria (Fig. 6). The effects of SCFAs in the pathogenesis of PD have been explored in preclinical and clinical studies.

2.2.2 Studies in vitro and in animal models of PD

The histone deacetylase inhibitor sodium-butyrate showed protective properties on dopaminergic neurons in vitro (Kidd and Schneider, 2010) and prevented motor impairment in a toxin-induced drosophila model of PD (St Laurent et al., 2013). A germ-free transgenic mouse model over-expressing α-syn receiving human gut microbiota from PD donor displayed a significantly altered SCFA profile (lower concentration of acetate and higher relative abundance of propionate and butyrate) and enhanced motor dysfunction compared with animals colonized with microbes from healthy controls (Sampson et al., 2016). In this model, depletion of gut bacteria (Germ-Free) reduced microglia activation and α-syn pathology, supporting the hypothesis of a key role of altered gut microbiota in triggering microglia activation and α-syn pathology. Interestingly, however, SCFA administration contra-intuitively promoted motor dysfunction and α-syn reactive microglia α-syn pathology without microbiota colonization (Sampson et al., 2016). Since oral administration of heat-killed bacteria had no effect on motor performance, we should conclude that PD pathogenesis needs metabolically active microbiota (Sampson et al., 2016).

2.2.3 Patients with PD

Changes in the gut microbiota composition and reduced SCFA production has been consistently reported in PD patients as compared to healthy controls. Specifically, PD-derived feces were shown to contain less SCFA butyrate-producing bacteria, such as from the taxonomic family Lachnospiraceae (Keshavarzian et al., 2015; Hill-Burns et al., 2017; Petrov et al., 2017; Barichella et al., 2019), as well as species *Faecalibacterium prausnitzii* (Keshavarzian et al., 2015; Unger et al., 2016), which were previously attributed to exert putative anti-inflammatory effects. These changes were consistently reported both in sigmoid mucosal biopsies (Keshavarzian et al., 2015) and in fecal samples (Barichella et al., 2019; Hill-Burns et al., 2017, Petrov et al., 2017, Unger et al., 2016). *It is intriguing and probably significant that one of the hallmarks of dysbiotic microbiota in PD is low abundance of SCFA producers, which could clearly be the mechanism of leaky gut and neuroinflammation in PD.* Interestingly, a reduction in butyrate-producing genera *Faecalibacterium* and *Coprococcus* bacteria have been associated with a decrease in the dopamine metabolite DOPAC and a relative reduction in measures of quality of life and depression (Valles-Colomer et al., 2019). Nevertheless, as suggested by Hill-Burns et al. (2017), it may be speculated that a shortage of SCFA-producing bacteria may influence PD pathogenesis and that replenishing the microbiome with SCFA-producing bacteria may prevent PD and reverse the disease in those who are affected. On the other hand, the evidence of depletion of SCFA-producing bacteria in several non-PD disorders (Guo et al., 2016; Pozuelo et al., 2015; Qin et al., 2012, 2014; Valles-Colomer et al., 2019; Yamada et al., 2015), may underlie the notion that a decrease in SCFA can also be a consequence, rather than a specific cause of disease. Therefore, reduced SCFA-producing bacteria in general (and butyrate-producing in particular) needs further investigation, before being considered a biomarker. While inconclusive at the moment, prospective research on SCFA gene expression and metabolomics profiles of microbiota in health and disease will shed further light on this aspect.

2.3 Is PD associated with a "leaky gut"?

2.3.1 Mucosal microbiota vs. luminal microbiota

It had been hypothesized that gut microbiota composition changes along the longitudinal axis of the gut. However, more recent evidence suggests that the gut microbiota can also be differentiated radially, correlating with radial oxygen gradient and the distribution of its nourishment such as tissue-associated mucus (Albenberg et al., 2014). Oxygenation in the intestinal lumen increases upon increase in host oxygenation, causing a diffusion of oxygen from the host tissue into the intestinal lumen and suggesting the presence of a spatial oxygen gradient (Albenberg et al., 2014). This would lead to an increase in oxygen-tolerant microbia adherent to the epithelium (Albenberg et al., 2014; Tang et al., 2015). Thus, any change in host oxygenation is able to modulate the composition of the gut microbiota (Albenberg et al., 2014). Increased tissue oxygenation could directly influence microbes (e.g., decreasing genus *Anaerostipes*) but could also have an effect on host immune system, which in turn could indirectly influence the gut microbiota composition (Albenberg et al., 2014; Thom, 2011). Furthemore, changes in the host oxygenation influence mucosally adherent bacterial populations differently than those in the lumen (Albenberg et al., 2014).

The intestinal mucosa and lumen are two distinct microbial habitats and bacteria that reside on the surface of the mucosa encounter an environment specialized to prevent their translocation across the intestinal epithelial barrier, which is different from the luminal habitat (Tang et al., 2015; Vinogradov and Wilson, 2012). As a consequence, the mucosal and luminal microbiota have distinct profiles of biodiversity and bacterial taxa, with a substantially lower diversity (reduced number of operational taxonomic units (OTUs), species richness and species evenness) for the mucosal microbiota as compared to the luminal microbiota (Tang et al., 2015; Vinogradov and Wilson, 2012). Beta-diversity of the mucosal and luminal microbiota suggests that the differences in microbial composition could be clustered by sample type: the main phyla driving these differences are *Firmicutes, Bacteroidetes and Proteobacteria* (Tang et al., 2015). *Proteobacteria* is the most abundant phylum in the mucosa, whereas Firmicutes are most abundant in fecal samples, followed by Bacteroidetes (Tang et al., 2015). Functional pathways enriched in the mucosal microbiota are similar to those enriched under inflammatory conditions: the functional differences between luminal samples and biopsies include a switch toward lipid and amino acid metabolism in the mucosal microbiota and from carbohydrate and nucleotide metabolism in the luminal microbiota (Tang et al., 2015). Genetic information processing, replication and repair are also enriched in the luminal microbiota, suggesting greater bacterial turnover and replication then the mucosa (Tang et al., 2015). Concerning studies on gut microbiota in PD, these critical differences are to be taken into account when comparing data obtained from fecal samples as compared to those obtained from colonic biopsies.

It has been recently hypothesized that the intestinal mucosa of most healthy individuals in a steady state is colonized with mucosal bacteria that are controlled by the luminal communities, as well as epithelial barrier function and immune response (Fig. 7A) (Tang et al., 2015). During pathological conditions increasing mucosal permeability, this relationship may break down and specific components of the mucosal microbiota may increase, spilling over into the lumen and/or translocating across the intestinal epithelial barrier and thus trigger a strong inflammatory response (Fig. 7B) (Tang et al., 2015). The overlap of bacterial taxa and microbial functional pathways between bioptic vs. fecal samples, and between active inflammation vs. normal mucosa supports the proposed oxygen-gradient model (Albenberg et al., 2014; Tang et al., 2015). Overall, these data would support the hypothesis that the mucosal microbiota may be a reservoir for critical species that may contribute to the inflammatory response and, potentially, to the pathogenesis of several diseases when the mucosal permeability is increased.

2.3.2 The "leaky gut" in PD
One of the consequences of intestinal microbiota dysbiosis and an associated inflammatory state in the intestinal epithelial layer is disruption of the intestinal barrier integrity (gut leakiness). Intestinal hyperpermeability can then allow bacterial inflammatory products like LPS to leak into the mucosa and systemic circulation causing mucosal, systemic and also a neuroinflammatory state in the brain (Buford, 2017; Fasano, 2012; Köhler et al., 2016; Maes et al., 2008; Obrenovich, 2018). Recent studies on *post-mortem* brains of Alzheimer's disease patients show dramatically increased LPS staining in the hippocampus and endotoxemia (Zhang et al., 2009a, 2009b; Zhao et al., 2017a, 2017b). Thus, one possible mechanism by which dysbiotic microbiota

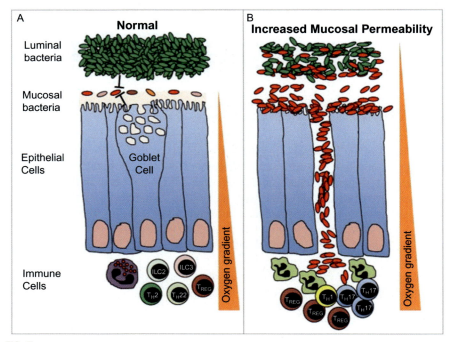

FIG. 7

(A) Oxygen gradient in a steady-state; (B) During inflammation the intestinal barrier breaks down and oxygen gradient disappeared.

Adapted from Tang, M.S., Poles, J., Leung, J.M., Wolff, M.J., Davenport, M., Lee, S.C., Lim, Y.A., Chua, K.H., Loke, P., Cho, I. 2015. Inferred metagenomic comparison of mucosal and fecal microbiota from individuals undergoing routine screening colonoscopy reveals similar differences observed during active inflammation. Gut Microbes 6(1), 48–56.

promote neuroinflammation in PD is by disruption of intestinal barrier integrity and endotoxin leak (Lubomski et al., 2019a, 2019b; Perez-Pardo et al., 2017a, 2017b; Sampson, 2019). Indeed, Forsyth et al. (2011) showed that newly diagnosed and untreated PD patients have increased intestinal permeability, endotoxin leak and evidence of endotoxemia. Similarly, Clairembault et al. (2015) demonstrated that morphological changes in the intestinal epithelial barrier occur in PD patients, thus strengthening the possible role of the "leaky gut" in the pathophysiology of PD.

The site of intestinal leak in PD patients appears to be primarily in the colon (site of most of the microbiota) because 24 h urinary sucralose (marker of total gut permeability) (Arrieta et al., 2006; Shaikh et al., 2015) was increased while urinary mannitol or urinary lactulose (both markers of small intestinal permeability) (Arrieta et al., 2006) were normal (Forsyth et al., 2011; Shaikh et al., 2015). A subsequent study (Perez-Pardo et al., 2018a, 2018b) also showed that increased intestinal permeability in PD patients, defined by increase urinary sucralose after ingestion of sugar cocktail containing sucralose, was significantly correlated with intestinal tissue pro-inflammatory cytokines and increased TLR4 and TLR2 expression, strongly suggesting that increased intestinal permeability to endotoxin in PD patients is a consequence of microbiota-initiated intestinal inflammation (Perez-Pardo et al., 2018a, 2018b). More direct evidence for this view comes from an associated study in mice that showed rotenone-induced disruption of intestinal barrier integrity, intestinal inflammation and more importantly, PD like pathology in the brain and PD like behavior were significantly mitigated in TLR-4 knock out mice (Perez-Pardo et al., 2018a, 2018b).

2.4 The role of virus (phageome)

2.4.1 Healthy subjects

The gut microbiome is shaped by bacteriophages, which are viruses able to infect and replicate within bacterial hosts. The healthy gut phageome (defined "core phageome") includes core and common bacteriophages and plays a pivotal role in keeping in check the healthy gut microbiota (Manrique et al., 2016). The healthy gut phageome has been found to be reduced in individuals with gastrointestinal disease, such as IBD. Although its role it is not fully understood yet, the existence of a healthy gut phageome suggests that the net influence of bacteria–bacteriophages interactions in the human gut is not deleterious, but rather beneficial by influencing the stability and maintenance of a healthy ecosystem through active lytic interactions in a "prey–predator" dynamic (Manrique et al., 2016; Rodriguez-Valera et al., 2009). Future studies are needed to address the role of the gut phageome in health and disease (including PD), and to assess the potential use of core bacteriophages for clinical therapeutics and controlled manipulation of the human gut ecosystem.

2.4.2 Parkinson's disease

There is limited evidence from studies of the human gut phageome in patients with PD. Bedarf et al. (2017) showed that total virus abundance (reflecting bacterial and archaeal phages) is lowered in PD patients as compared to healthy control subjects, whereas the abundances for prophages and plasmids was similar. Noteworthy, the assessment of the relative load of virus and phage was limited as it was entirely dependent on the corresponding protein families in the ACLAME database. In this study, the authors found no correlation between the viral load abundance and bacterial family (Bedarf et al., 2017). Although preliminary data on PD is very limited, the interaction between viruses, gut microbiota composition, host cells and immune response warrants further research.

3 Genetic factors

3.1 Twin studies

Several microbial taxa and functional modules in the gut microbiome have been shown to be heritable (Xie et al., 2016). There is a widespread concordance in the composition, Single Nucleotide Polymorphisms (SNPs), and functional capacity of the gut microbiome between twins (specifically an increased concordance among monozygotic twins over dizygotic twins), supporting the hypothesis of a substantial host genetic influence (Xie et al., 2016). Immune traits are also highly heritable and human genetic associations have been found for all major immune cell types (Roederer et al., 2015). To date, only a few human genes (such as Fut2 and Nod2) and the major histocompatibility complex locus have been suggested to influence the relationship between gut microbiota and host immune response and metabolism (Xie et al., 2016).

It has been suggested that these co-occurrence patterns could derive from different scenarios: (1) multiple taxa may be heritable and co-occur while each taxon is affected by host genetics independently, or alternatively (2) a few taxa may be heritable and other taxa correlate with host genetics due to their co-occurrence with these key heritable taxa (Goodrich et al., 2016).

Among microbial taxa whose abundances were influenced by host genetics, one of the most highly heritable taxa was the family Christensenellaceae (Goodrich et al., 2016; Hansen et al., 2011). On the other hand, according to the Twins UK and Missouri twin's datasets, the *Ruminococcaceae* and *Lachnospiraceae* families had the highest heritability estimates. Considering the consistent association between these bacteria families and PD (see below Section 6.2), we do suggest future research to include also information about family history and known genetic factors in the equation on the causality link between gut microbiota and the risk for PD. Nevertheless, as abovementioned, we should keep in mind how it is challenging to disentangle the effects on gut microbiota played by shared environment (especially in early life) from the role of genome in studies on adult twins (Turnbaugh et al., 2009). Therefore, multiple environmental confounders, including those related to early-life events, are always to be present along with genetic ones.

3.2 PD-causing genes and neuroinflammation

Genetics has been proposed as one underlying cause of neuro-inflammation and dopaminergic neuron loss in Parkinson's. However, most cases of Parkinson's disease occur in people with no apparent family history of the disease (Tysnes and Storstein, 2017) These so called sporadic cases may have genetic susceptibility or even an inheritance pattern that is unknown. Several genetic variants have been reported to result in increased risk for Parkinson's disease. Only an estimated 5–10% of PD patients suffer from monogenic forms of PD due to Mendelian inheritance (Deng et al., 2018). Nonetheless, this observation highlights the potential importance of genetics in neuroinflammation and DA loss. Indeed, first degree family history of PD is a major risk factor for PD (Ferreira and Massano, 2016) and GWAS studies have identified mutations in several genes such as LRRK2, and SNCA (autosomal dominant) and PARK7, PINK1, PRKN (autosomal recessive) that were found to be present more commonly in patients with Parkinson's disease (Deng et al., 2018). Glucocerebrosidase-1 (GBA1) gene mutations confer a 20- to 30-fold increased risk for the development of PD. About 7–10% of PD patients have a GBA1 mutation (Migdalska-Richards and Schapira, 2016).

Nevertheless, despite the increased risk for PD, mutations on genes involved in autosomal-dominant forms of PD—such as LRRK2 and GBA—present with low-penetrance, as <30% of carriers actually develop PD features (Blandini et al., 2019). In this scenario, environmental factors may play a substantial role in the modulation of the risk for PD and one of these factor is likely to involve the microbiota and the gut-brain axis.

Inflammation is clearly associated with PD pathogenesis, mechanisms whereby it contributes to the disease process are unclear. Increasing evidence implicates these PD-associated genes in immune and inflammatory pathways (Dzamko et al., 2015). Evidence emerging from genetic studies, suggests that defects in inflammation or innate immune homeostasis are likely to contribute significantly to the risk for PD (Holmans et al., 2013; Raj et al., 2014). At least 23 loci and 19 disease-causing genes and various genetic risk factors for Parkinsonism have been identified so far (Deng et al., 2018). Several of these genes are involved in signaling pathways such as inflammatory signaling in microglia. One example is heme-oxygenase that can prime microglia and augment its response to pro-inflammatory factors like endotoxin (LPS) that could lead to neuroinflammation and α-syn aggregation (Mateo et al., 2010). It is noteworthy that knockout, mutation or overexpression of the majority of the robust familial PD-causing genes seemingly affects production of pro-inflammatory cytokines, particularly TNFα, IL-6 and IL-1b, when macrophage/microglial cells are treated with agonists of TLR signaling, such as LPS. When released by activated microglia, these cytokines can have detrimental effects on neurons. Moreover, dying neurons release DAMPS, such as α-syn that can activate microglia thereby exacerbating the inflammatory process.

3.2.1 Genes associated with autosomal-dominant PD
3.2.1.1 Alpha-Synuclein
The relationship between a-syn and inflammation involves TLRs (see Section 2.1.2), particularly the ability of a-syn to act as a DAMPs (Dzamko et al., 2015). In contrast to the innate immune activating PAMPs, DAMPs are self-originating and can induce "sterile inflammation" through activation and/or modulation of TLRs (Dzamko et al., 2015; Piccinini and Midwood, 2010). In particular, a-syn-induced microglial activation seems to be mediated by TLR4 (Fellner et al., 2013) and potentiated by peripheral LPS (Couch et al., 2011; Dzamko et al., 2015). Therefore, both native or pathogenic forms of a-syn are able to promote microglial activation, a process resulting in production of inflammatory cytokines and reactive oxygen species (ROS) that contribute to progressive nigral degeneration in PD (Dzamko et al., 2015).

3.2.1.2 Leucine-rich repeat kinase 2 (LRRK2)
Besides being one of the greatest contributor to autosomal-dominant PD, mutations in the LRRK2 gene are also involved in the regulation of intestinal microbiota and associated with other inflammatory diseases, including IBD (Liu et al., 2011; Umeno et al., 2011; Van Limbergen et al., 2009) and increased susceptibility to

Micobacteria infection (Cardoso et al., 2011; Marcinek et al., 2013; Zhang et al., 2009a, 2009b). In particular, the LRRK2 N2081D risk allele and the N551K/R1398H protective alleles, as well as numerous other variants within the LRRK2 locus, revealed shared genetic effects between Crohn's Disease and PD risk, providing a potential biological basis for clinical co-occurrence (Hui et al., 2018). Human peripheral blood mononuclear cells express LRRK2 (Dzamko et al., 2013; Gardet et al., 2010; Hakimi et al., 2011; Thévenet et al., 2011), and siRNA knockdown of LRRK2 impaired clearance of foreign pathogens (Gardet et al., 2010). LRRK2 can also directly modulate inflammatory pathways downstream of TLRs. Transcriptional profiling of PBMCs from control and human PD LRRK2 mutation carriers suggested dysregulation of inflammatory pathways in PD only and not in controls (Mutez et al., 2011). LRRK2 may modulate the magnitude of cytokine production following TLR-activation, although the mechanisms underlying the abnormal inflammatory response related to PD-causing LRRK2 mutations is still to be elucidated (Dzamko et al., 2015).

3.2.1.3 Glucocerebrosidase (GBA1)

Loss-of-function mutations in GBA1, which cause the autosomal recessive lysosomal storage disease, Gaucher disease (GD), are also a key genetic risk factor for the α-synucleinopathies, including PD and dementia with Lewy bodies (Blandini et al., 2019). In a zebrafish model of Gaucher's disease (gba1 −/−), it has been recently found early microglial activation with marked neuroinflammation in association to a marked increase of transcript levels of a master regulator of inflammation (i.e., miR-155), that has been implicated in a wide range of different neurodegenerative disorders (Watson et al., 2019). Further study is needed to investigate the potential relationship between GBA1 mutations and neuroinflammation in the pathogenesis of GBA-related PD.

3.2.2 Genes associated with autosomal-recessive PD

3.2.2.1 Parkin

Parkin is required for clearance of intracellular pathogens via ubiquitin-mediated autophagy (Dzamko et al., 2015; Manzanillo et al., 2013). Parkin-deficient drosophila have increased susceptibility to a range of intracellular pathogens and dysregulated innate immune signaling pathways (Dzamko et al., 2015; Greene et al., 2005; Manzanillo et al., 2013). Parkin-null mice have increased susceptibility to LPS-induced neuronal loss (Dzamko et al., 2015; Frank-Cannon et al., 2008). However, exact mechanisms by which Parkin increases pro-inflammatory cytokines are still unclear.

3.2.2.2 PTEN-induced putative kinase 1 (PINK1)

The pathophysiologies of PINK1- and Parkin-associated PD are similar, which is expected as both of them regulate mitophagy along the same pathway. Similar to Parkin, PINK1 seems to modulate the production of inflammatory cytokines, as demonstrated by the higher levels of IL-1b, IL-12, and TNFa in response to systemic LPS shown in brain homogenates from PINK1-null mice compared to wild type (Akundi et al., 2011).

3.2.2.3 DJ-1

DJ-1 expression is markedly upregulated in murine macrophages in response to LPS (Mitsumoto and Nakagawa, 2001). Astrocytes from DJ-1-null mice have heightened responses to LPS and produce more inflammatory cytokines, possibly via hyperactivation of TLR4/MAPK signaling and subsequent inflammatory responses (Cornejo Castro et al., 2010; Dzamko et al., 2015; Waak et al., 2009).

3.3 Genes involved in the modulation of immune response to microbiota

Host genome may interact with microbiota by modulating immune response to structural components of the bacterial cell wall, such as peptidoglycan. Interestingly, an increased risk for PD has been variants in genes that encode peptidoglycan recognition proteins, which regulate the immune response to both commensal and harmful bacteria and are this involved in the maintenance of a healthy gut microbiota (Goldman et al., 2014).

In conclusion, further studies are needed to link PD susceptibility genes with regulation of gut microbiota.

4 Environmental factors/lifestyle habits modulating gut microbiota

4.1 Environmental factors

Although genetics is involved in shaping our microbiota community (Gilbert et al., 2018), environmental factors have far more impact in determining the type of microbiota community we have in our GI tract (Gilbert et al., 2018; Hirschberg et al., 2019). For example, although the intestinal microbiota community is more similar between twins compared to unrelated subjects, the microbiota community is significantly different even between identical twins (Faith et al., 2013; Turnbaugh et al., 2009). Thus, environmental factors like diet, place of living (rural vs. urban living; western vs. eastern societies) dictates the type of microbiota community resides in the intestine (Faith et al., 2013; Lozupone et al., 2012).

Several environmental toxins have been associated with PD. For example, exposure to narcotic by-product MPTP causes severe Parkinsonism (Burns et al., 1985; Kohutnicka et al., 1998). The pesticide rotenone was associated with increased risk of Parkinson in farmers who were exposed to this mitochondrial toxin and it is now a model for PD in rodents (Greenamyre et al., 2003; Liu et al., 2003). In PD, there is evidence for increased xenobiotics degradation in the gut, including pesticides and herbicides, which are known to increase the risk for PD by causing dopaminergic neuronal degeneration (Hill-Burns et al., 2017; Pezzoli and Cereda, 2013). Future studies may investigate on the role of xenobiotics in triggering gut dysbiosis along with individual predisposition. Rodents models exposed to Manganese showed increased gut transit time and an altered gut metabolic profile (Ghaisas et al., 2016), suggesting that exposure to environmental metals can influence the gut microbiome profile with potentially harmful effects. However, the overwhelming majority of patients with PD do not have clear exposure to neurotoxins. Nonetheless, several epidemiological studies support the role of environmental factors in PD (Chen and Ritz, 2018; Marras et al., 2019). For example, PD is more common in Western societies and the incidence of PD is rising in developing countries where Western a lifestyle has been adopted (Chen and Ritz, 2018; Marras et al., 2019). One example of environmental factors associated with Weston lifestyle is stress (Metz, 2007; Smith et al., 2002), which has been suggested to be a risk factor for PD (Marras et al., 2019). Animal studies have also shown that stress can exacerbate neuroinflammation, dopamine loss and Parkinson-like behavior (Dodiya et al., 2020; Smith et al., 2008). The question that remains to be answered is how environmental factors modulate neuroinflammation and neurodegeneration in genetically susceptible individuals that leads to symptomatic PD (Chen and Ritz, 2018; Marras et al., 2019).

It may be hypothesized that environmental factors trigger neuroinflammatory cascade through disrupting the gut microbiota-brain axis leading to microglial activation and the neuroinflammation that drives PD (Dinan and Cryan, 2017; Lubomski et al., 2019a, 2019b; Mulak and Bonaz, 2015; Sampson, 2019). Indeed, several epidemiological, clinical and experimental studies have clearly showed that these environmental factors also affect intestinal microbiota community providing compelling evidence that environmental factors promote PD through their impacts on microbiota. Diet is the most potent factor that affect microbiota community and function (David et al., 2014; Muegge et al., 2011). For example, high sugar/refined carbohydrate and low fiber typical for western diet (Cassani et al., 2017; Chen et al., 2003; Maraki et al., 2019), promote dysbiotic microbiota community characterized by increase abundance of putative pro-inflammatory bacteria like LPS producers, while decreasing abundance of putative anti-inflammatory bacteria like short chain fatty acids producers (Noble et al., 2017; Singh et al., 2017; Zinocker and Lindseth, 2018). Similar disruption of microbiota community is associated with stress and poor sleep and disrupted circadian rhythms, common features of modern societies and risk factors for PD (Paschos and FitzGerald, 2017; Videnovic and Golombek, 2013; Voigt et al., 2016). These PD associated environmental factors can create a pro-inflammatory microenvironment in the gut capable of producing a pro-inflammatory state in the intestinal mucosa, systemically and in the CNS leading to activation of enteric glia and brain microglia that could promote α-syn aggregation and eventually DA loss and symptoms and signs of PD. Thus, intestinal microbiota can be a linchpin between environmental factors and PD risk.

Microbiota not only can be the mediator of the environmental factors to promote PD, it may also be the mechanism for several diseases that are risk factors for PD. For example, several studies have shown that metabolic syndrome and diabetes are risk factors for PD (Hu et al., 2007; Xu et al., 2011). It is well known that metabolic syndrome and diabetes are associated with abnormal microbiota community (so called dysbiosis) which is characterized by increased abundance of pro-inflammatory bacteria (Qin et al., 2012; Upadhyaya and Banerjee, 2015). Recent studies have also reported that inflammatory bowel disease (IBD) is a risk factor for PD

(Villumsen et al., 2019; Weimers et al., 2018) and IBD is also associated with pro-inflammatory dysbiotic microbiota community in patients (Sartor, 2010; Vindigni et al., 2016) as well as animal models of IBD (Hartog et al., 2015). One recent study also showed that DSS colitis worsens PD pathology and behavior in mouse model of PD and this exacerbation correlate significantly with severity of intestinal inflammation and gut-derived inflammatory state (Kishimoto et al., 2019).

Collectively these epidemiological, clinical and experimental studies strongly support the importance of the gut-brain axis in pathogenesis of PD and also provide compelling evidence that intestinal microbiota is the critical element of the gut-brain axis associated with PD pathogenesis. Critical questions are whether PD patients indeed have an abnormal intestinal microbiota community and whether the intestinal microbiota changes in PD are characterized by pro-inflammatory features.

4.2 Lifestyle habits

There is compelling evidence that the risk of PD is reduced by approximately 33% and 36–50% in individuals with a history of coffee consumption and smoking, respectively, with an inverse dose-response relationship for the latter (Noyce et al., 2012; Scheperjans et al., 2015a). Although caffeine and nicotine are the most evident candidates because of their neuroprotective properties in experimental settings (Chen et al., 2001; Maggio et al., 1998), the exact mechanisms behind these associations are largely unknown. Evidences support the hypothesis that the reduced risk for PD in association with smoking and coffee drinking may be due to a modulation of gut microbiota (Derkinderen et al., 2014).

4.2.1 Smoking
It has been hypothesized that cigarette smoking may reduce the risk for PD by acting on the gut with two mechanisms, partly overlapping. First, smoking has been shown to improve gut barrier function in mice (Wang et al., 2012) and in humans (Prytz et al., 1989). In a second instance, smoking seems to have an effect on gut microbiome composition (Scheperjans et al., 2015a, 2015b). Indeed, smokers have higher abundance of *Bacteroides/Prevotella* in their feces, and this abundance decreases together with that of *Proteobacteria* after smoking cessations, while levels of *Firmicutes* and *Actinobacteria* increase (Biedermann et al., 2013). Finally, an improvement in smoking-related colonic barrier function and inflammation were associated with a decrease of the family *Enterobacteriaceae* (Scheperjans et al., 2015a, 2015b) and species *Ruminococcus albus* (Wang et al., 2012). Nevertheless, it remains to be established whether these simultaneous changes are causally related to each other and eventually to PD.

4.2.2 Coffee
Coffee drinking has marked effects on gut microbiota, as it up-regulates activity and proportion of genus *Bifidobacterium*, that exert anti-inflammatory properties (Jaquet et al., 2009) and it is rapidly metabolized into SCFAs and causes a marked expansion of Bacteroides/Prevotella bacteria (Gniechwitz et al., 2007). Consumption of coffee in both mice and humans induced a significant increase in the *Bifidobacterium* population

without major impact on the dominant microbes (Jaquet et al., 2009; Nakayama and Oishi, 2013). It is important to consider that alterations in gut motility, as found in PD, and gut microbiome composition could be independently related to each other.

In conclusion, data has thus been suggested that the changes in gut bacteria observed after coffee consumption and cigarette smoking affect the risk of PD being triggered in the gastrointestinal tract and, specifically, that in the absence of coffee drinking and cigarette smoking, the microbiota would shift toward a pro-inflammatory state (Derkinderen et al., 2014; Scheperjans et al., 2015a, 2015b). This would promote chronic gastrointestinal inflammation and an enteric glial reaction, which actually occur in the early stage of PD (Devos et al., 2013) (Fig. 8). As an alternative explanation, smoking and coffee consumption may reduce central nervous system neurodegeneration, by decreasing the release of proinflammatory cytokines from the gut to the bloodstream (Derkinderen et al., 2014).

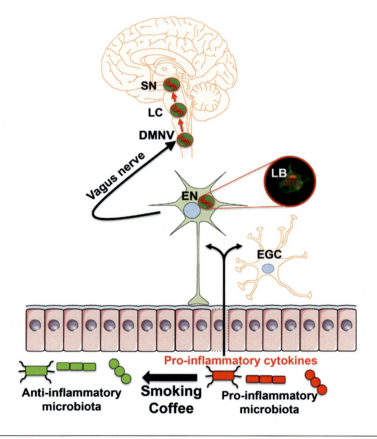

FIG. 8

Possible role of smoking and coffee consumption on microbiota-gut-brain-axis and the development of Parkinson's disease.

From Derkinderen, P., Shannon, K.M., Brundin, P. 2014. Gut feelings about smoking and coffee in Parkinson's disease. Mov. Disord. 29(8): 976-9.

5 Upper gut microbiota (nasal, stomach and small intestine) and PD

5.1 Nasal and oral microbiota

The intestinal microbiota does not appear to be the only microbiota that is disrupted in PD patients. To date, there are two studies that interrogated the nasal and oral microbiota community structure and composition in PD patients. Pereira et al. interrogated both nasal and oral microbiota profiles (using oro-nasal wash sample) between PD patients and healthy controls (Pereira et al., 2017). The oral microbiota composition was significantly altered in PD patients, compared to healthy controls, predominantly by higher relative abundance of opportunistic pathogens. The nasal microbiota lacked strong significant individual taxa differences, but trended toward an overall difference in the microbial composition between groups.

Heintz-Buschart and colleagues found no differences in nasal microbiota between PD patients and healthy control subjects, indicating clear differences between gut and nasal microbiota in PD (Heintz-Buschart et al., 2018). Several studies consistently evidenced greater variation in nasal microbiota over the different individuals (Bassis et al., 2014; Faust et al., 2012; Heintz-Buschart et al., 2018; Pereira et al., 2017) with gender being the strongest grouping factor (Heintz-Buschart et al., 2018). Critically, more bacterial families were found to be potentially affected by PD medication than to be different in relation to PD.

In contrast, Mihaila et al. (2019) interrogated the oral microbiota using saliva samples through shotgun metatranscriptomic profiling and found significant changes in the microbiota community structure, composition and function in PD patients. They found several similarities between dysbiotic oral microbiota and dysbiotic fecal microbiota in PD patients, when they compared their findings with previously published human PD studies. Dysbiotic oral microbiota once again was characterized by higher relative abundance of putative pro-inflammatory bacteria (Mihaila et al., 2019). This finding is potentially important in PD pathogenesis because one proposed site of initial injury in PD is the olfactory bulb, which is in close proximity to oro-nasal space, as proposed by Braak et al. (2003) and Hawkes et al. (2007).

5.2 Stomach: *Helicobacter pylori* and response to Levodopa

Increasing evidence supports the causative link between Helicobacter pylori (HP) and motor fluctuations in PD, caused by a clinical relevant interference in the absorption of levodopa, leading to an increased prevalence of delayed on-time and decreased daily on-time (Fasano et al., 2015; Lee et al., 2008; Narożańska et al., 2014; Pierantozzi et al., 2006). The effect of HP eradication on motor fluctuations has been confirmed in a double-blind study (Pierantozzi et al., 2006) with a larger improvement in motor fluctuations at 3-month follow-up, suggesting a time-dependent decrease in HP-related inflammatory changes in gastrointestinal mucosa (Fasano et al., 2015; Lee et al., 2008).

5.3 Small intestine: The role of bacterial overgrowth (SIBO)

In healthy subjects, the OTUs and composition of small intestine microbiota is kept under control by gastric acid and biliary and pancreatic secretions (limiting the growth of the majority of bacteria entering the stomach from the food and oral resident flora), the intestinal mucus (trapping bacteria), and the ileocecal valve (inhibiting retrograde migration of colonic bacteria) (Fasano et al., 2015; Grace et al., 2013). A prospective study provided evidence for an increased prevalence of SIBO in PD patients that was associated with motor disability (Gabrielli et al., 2011). Fasano et al. (2013) showed an increased prevalence of SIBO in fluctuating PD (55% vs. 20% in healthy controls) and more severe motor fluctuations (OFF-time, delayed-ON time and No-ON time). These findings are similar to patients taking proton-pump inhibitors, which increase the risk of SIBO due to reduced the gastric acid-related ability to kill bacteria (Fasano et al., 2015; Tan et al., 2014).

Besides the abnormalities in Levodopa absorption and motor response, SIBO may play a role in PD pathogenesis. The peripheral inflammatory state induced by SIBO might contribute to increased intestinal permeability, thus promoting the translocation of bacteria and endotoxins across the intestinal epithelium, triggering microglial activation and exacerbating the neurodegenerative process (Tan et al., 2014). Therefore, SIBO might create a proinflammatory environment (Quigley and Quera, 2006), supported by a study that quantified local inflammation and enteric glial reaction in gastrointestinal biopsies of patients with PD (Devos et al., 2013; Fasano et al., 2015).

In conclusion, understanding the relationship between HP infection and SIBO in the response to Levodopa and, potentially, their role in PD colonic gut microbiota changes and inflammation is needed, as it might allow new treatment approaches.

6 Lower gut microbiota (colonic) and PD
6.1 Preclinical studies

Regardless of whether microbiota is the initial trigger or not, it could still be a major player in the pathogenesis of PD because bacterial associated pro-inflammatory factors may trigger or sustain neuroinflammation that is required for DA loss in PD. Several in vitro and rodent studies support this notion.

6.1.1 In vitro studies

Co-culture of microglia with LPS results in microglia activation (so called "priming") (Perry and Holmes, 2014) and production of a series of pro-inflammatory cytokines and neurotoxins (Block and Hong, 2005; Lull and Block, 2010; Norden and Godbout, 2012).

6.1.1.1 Rodent models

Inflammation resulting from infection with H5N1 influenza virus may trigger the development of pathogenic forms of α-syn, first by infecting the enteric nervous system and then disseminating (Jang et al., 2009). Microglial activation has been associated with increased aggregation and phosphorylation of α-syn at Ser129, which in turn may accumulate and cause neurodegeneration (Jang et al., 2009).

Changes in the intestinal microbiota in PD rodent models further support associated evidence from PD human studies. Several rodent studies support the importance of pro-inflammatory dysbiotic microbiota in the pathogenesis of PD (Table 2). Changes in the microbiota composition were discovered in pharmacologically-induced PD rodent models via neurotoxins to promote PD-like neurological symptoms including rotenone (Dodiya et al., 2020; Johnson et al., 2018; Perez-Pardo et al., 2018a, 2018b, 2019; Yang et al., 2017) and 1-methyl-4-phenyl-1,2,3,6-tetrahydropyridine (MPTP) (Choi et al., 2018; Lai et al., 2018), as well as with restraint stress (Dodiya et al., 2020) and using fecal microbiota transplantation from human PD patients (Sampson, 2019). Additionally, genotyping H67D HFE variant (the mouse homologue of the H63D variant) and toll-like-receptor 4 (TLR4) +/− knock out were used in a rotenone neurotoxicity model to show the importance of microbiota and bacterial products like LPS in pathogenesis of PD (Perez-Pardo et al., 2018a, 2018b; Nixon et al., 2018).

Changes in fecal microbiota composition in multiple models of PD were similar to PD patients and showed increased abundance of putative pro-inflammatory bacteria like of LPS-producing bacteria and decreased abundance of putative anti-inflammatory bacteria like SCFA producers. However, the only direct evidence for the hypothesis that abnormal microbiota is required for PD pathology and triggers symptomatic PD comes from a combined human PD-animal study. Notably, Sampson et al. showed that unlike conventional mice, germ free α-syn overexpressing (ASO) transgenic PD model mice have reduced microglia activation, α-syn inclusions, no DA loss and exhibit no PD phenotype compared to animals with a complex microbiota (Sampson, 2019). More importantly, the study showed that germ free ASO mice who received fecal transplant from PD patients developed greater PD pathology and phenotype than with fecal microbiota transplantation from healthy donors, including activated microglia in the substantia nigra and typical PD behavior. Treatment with microbial produced SCFAs restores all major features of disease in germ-free mice, identifying potential molecular mediators involved in gut-brain signaling. This study showed that dysbiotic stool microbiota, transplanted from human PD patients, can trigger PD pathology and behavior in a genetically susceptible host.

Systemic injection of LPS to mice causes neurodegeneration in the substantia nigra (Qin et al., 2007) and worsens PD pathology and PD behaviors in rodent models of PD (Dutta et al., 2008; Henry et al., 2009; Hoogland et al., 2015). Of course, we propose the primary source of LPS in human PD is increased gram negative bacteria characteristic of the dysbiotic gut microbiota in most PD studies.

Table 2 Changes in the intestinal microbiota in Parkinson's disease rodent models.

Reference	Study design	Methodology	Major taxa altered in Parkinson's disease animal model	Predictive KEGG functional pathways	Clinical features in Parkinson's disease animal model	Major finding
Sampson et al. (2016)	Germ Free WT and Germ Free Alpha-Synuclein-Overexpressing (ASO) mice Fecal microbiota from PD naïve treatment patients or healthy controls were transplanted into individual groups of GF recipient mice via oral gavage	16S rRNA V4 gene amplicon sequencing Fecal samples	**Altered in Mice with PD Fecal Transplant** ↓*Lachnospiraceae Unclassified* genus, ↓*Rikenellaceae Unclassified* genus, ↑*Peptostreptococcaceae Unclassified* genus, ↑*Proteus* genus, ↑*Bilophila* genus, ↑*Roseburia* genus, ↑*Butyricicoccus* genus **Altered only in GF-ASO Mice** ↓*Lachnospiraceae Unclassified* genus, ↑*Proteus* genus, ↑*Bilophila* genus, ↓*Veillonellaceae Unclassified* genus, ↑*Bacteroides* genus, ↑*Clostridium* genus, ↓*Butyricicoccus* genus, ↓*Clostridiaceae Unclassified* genus, ↓*Peptostreptococcaceae Unclassified* genus **Changes independent of Mouse Genotype** ↓*Rikenellaceae Unclassified* genus, ↑*Roseburia* genus, ↓*Enterococcus* genus, ↓*RC4–4* genus, ↑*Pseudoramibacter Eubacterium* genus	The abundance of three SCFA-producing KEGG families were increased in mice that received fecal microbes derived from PD donors: K00929, butyrate kinase; K01034 and K01035, acetate CoA/acetoacetate CoA transferase alpha and beta	Animals receiving PD derived microbiota demonstrated a significantly altered SCFA profile, with a higher relative abundances of propionate and butyrate and lower concentration of acetate compared to animals colonized with microbes from HC Gut Microbes promote the hallmark motor and intestinal dysfunction for preclinical PD: Decrease in gross and fine motor functions 1. Time to traverse beam apparatus 2. Time to descend pole 3. Time to remove adhesive from nasal bridge 4. Hind-limb clasping reflex score 5. Time course of fecal output in a novel environment over 15min 6. Total fecal pellets produced in 15min	Differences in fecal microbiota between control and PD patients can be maintained following transfer into mice These findings reveal that gut bacteria regulate movement disorders in mice, and suggest that alterations in the human microbiome represent a risk factor for PD The disease status of the donor had a strong effect on the microbiota within recipient mice. Genotype also effected the microbial community configuration Alpha-synuclein overexpression causes distinct alterations to the gut microbiome profile after fecal transplant Gut microbes promote α-synuclein-mediated motor deficits and brain pathology Depletion of gut bacteria reduces microglia activation SCFAs modulate microglia and enhance PD pathophysiology

Continued

| Perez-Pardo et al. (2018a, 2018b) | 7 week old C57BL/6J male mice, housed under a 12 h light/ dark cycle 28 days of treatment: Vehicle-treated and Rotenone-treated mice | 16S rRNA gene amplicon sequencing Caecum Mucosa and Caecum Luminal Content samples | **Caecum Mucosa**
Rotenone Effect:
↑**Bacteroidetes phylum:**
↑Rikenellaceae family, ↑S24-7 family ↑*Rikenellaceae Unclassified* genus, ↑S24-7 *Unclassified* genus.
↑**Firmicutes phylum:**
↑Clostridiales Unclassified family,
↑Erysipelotrichaceae family,
↑Ruminococcaceae family.
↑*Clostridiales Unclassified Unclassified* genus, ↑*Allobaculum* genus
↓**Actinobacteria phylum:**
↓Bifidobacteriaceae family,
↓*Bifidobacterium genus*
Firmicutes-to-Bacteroidetes ratio significantly lower in Rotenone Caecum Mucosa
Caecum Content
Rotenone Effect:
↑**Bacteroidetes phylum,**
↑Rikenellaceae family,
↑*Rikenellaceae Unclassified* genus
↑**Firmicutes phylum,**
↑Erysipelotrichaceae family,
↑*Allobaculum genus*
↓**Actinobacteria phylum,**
↓Bifidobacteriaceae family,
↓*Bifidobacterium genus* | **Caecum Mucosa**
KEGG pathways upregulated in Rotenone:
Glycosphingolipid biosynthesis – ganglio series, glycosaminoglycan degradation, biosynthesis of siderophore group nonribosomal peptides, beta-alanine metabolism, aminobenzoate degradation, butirosin and neomycin biosynthesis, lysine degradation, lipid metabolism, biosynthesis and biodegradation of secondary metabolites, amino acid metabolism, bisphenol degradation, glycerolipid metabolism
KEGG pathways downregulated in Rotenone:
Chlorocyclohexane and chlorobenzene degradation, retinal metabolism, metabolism of xenobiotics by cytochrome P450
Caecum Content
KEGG pathways downregulated in Rotenone:
Chlorocyclohexane and chlorobenzene degradation, retinol metabolism, metabolism of xenobiotics by cytochrome P450, xylene degradation, dioxin degradation, naphthalene degradation, lipoic acid metabolism, chloroalkane and chloroalkene degradation | Microbial alteration was correlated with ZO-1 expression in brain, CD3+ T-cells in colon, and alpha-synuclein in the colonic plexi. There was no correlation between microglial activation in substantia nigra with any bacterial family in caecal mucosa and content. The Dopaminergic cell number in the substantia nigra was positively correlated with *Bifidobacteriaceae and* significantly inversely correlated with Ruminococcaceae, Rikenellaceae, Erysipelotrichaceae, and S24-7 (only in mucosal associated bacteria) Alpha-synuclein in the colonic plexi correlated with mentioned microbiota above in caecal content and mucosa except for Ruminococcaceae in the caecal content | Significant changes were found in the composition of caecum mucosal associated and luminal content microbiota and the associated metabolic pathways in a rotenone-induced mouse model for PD. The mouse model for PD, induced by the pesticide rotenone, was associated with an imbalance in the gut microbiota, characterized by a significant decrease in the relative abundance of the beneficial commensal bacteria genus *Bifidobacterium* and enhanced abundances of putative proinflammatory bacteria *Rikenellaceae Unclassified and Allobaculum* Dysbiotic microbiota is correlated with PD-like pathological and functional changes in intestine and brain of rotenone-treated mice. Microbial alteration was correlated with ZO-1 expression in brain, CD3+ T-cells in colon, and alpha-synuclein in the colonic plexi |

Table 2 Changes in the intestinal microbiota in Parkinson's disease rodent models.—cont'd

Reference	Study design	Methodology	Major taxa altered in Parkinson's disease animal model	Predictive KEGG functional pathways	Clinical features in Parkinson's disease animal model	Major finding
Dodiya et al. (2020)	6–8 week male C57BL/6 mice Total (n = 28): Vehicle (n = 7), Rotenone (n = 7), Restraint Stress (n = 7), Restraint Stress +Rotenone (n = 7) Treatment started at week 3 12 weeks of treatment (6 weeks Restraint Stress treatment followed by 6 weeks Restraint Stress +/−rotenone) Scarified at week 15	16S rRNA V4 gene amplicon sequencing Fecal samples	**Baseline** **Restraint Stress vs. Control:** No taxa differences **6 Weeks of ↑ Restraint Stress** **Restraint Stress:** **Firmicutes phylum:** ↓Lactobacillaceae family, ↓*Lactobacillus* genus, ↑*Lachnospiraceae Unclassified* genus **12 Weeks of Restraint Stress +/− Rotenone** **Rotenone:** ↓**Actinobacteria phylum** **Restraint stress:** **Firmicutes:** ↑ Clostridiales Unclassified Family, ↑*Clostridiales Unclassified* genus **Restraint Stress + Rotenone:** ↓**Actinobacteria phylum:** ↓Coriobacteriaceae family ↓**Firmicutes phylum:** ↓Clostridiales Unclassified Family, ↓*Clostridiales Unclassified* genus ↑**Verrucomicrobia phylum:** ↑Verrucomicrobiaceae family, ↑*Akkermansia* genus	**6 Weeks** **KEGG pathways upregulated in Restraint Stress:** Alpha-Linolenic acid metabolism **KEGG pathways downregulated in Restraint Stress:** Nucleotide metabolism, Amino sugar and nucleotide sugar metabolism, Fructose and mannose metabolism **12 Weeks** **KEGG pathways upregulated in Restraint Stress:** Electron transfer carriers **KEGG pathways downregulated in Restraint Stress:** 1,1,1-Trichloro-2,2-bis (4-chlorophenyl)ethane (DDT) degradation, Flavonoid biosynthesis **KEGG pathways upregulated in Restraint Stress + Rotenone:** Carotenoid biosynthesis, Drug metabolism—cytochrome P450, Ubiquinone biosynthesis, Bacterial secretion system **KEGG pathways downregulated in Restraint Stress + Rotenone:** D-Arginine and D-ornithine metabolism, Electron transfer carriers, Transporters, Flavone and flavonol biosynthesis, Citrate cycle (TCA cycle)	The initial 6 weeks of Restraint Stress caused: significantly higher urinary cortisol, intestinal hyperpermeability, and decreased abundance of putative *Lactobacillus* genus compared to non-stressed mice. Rotenone alone (without Restraint Stress) disrupted the expression of the ZO-1, increased oxidative stress, increased myenteric plexus enteric glial cell GFAP expression and increased α-synuclein (α-syn) protein levels in the colon compared to controls Restraint Stress exacerbated these rotenone-induced changes. Restraint stress potentiated rotenone-induced effects in the colon: intestinal hyper-permeability, disruption of ZO-1, Occludin, Claudin1, oxidative stress (N-tyrosine), inflammation in glial cells (GFAP + enteric glia cells), α-syn, increased fecal *Akkermansia* genus (mucin-degrading Gram-negative bacteria), and endotoxemia. Restraint Stress also promoted a number of rotenone-induced effects in the brain: reduced number of resting	The chronic stress induced gut-derived, pro-inflammatory milieu exacerbates in the PD phenotype via a dysfunctional microbiota-gut-brain Axis. Chronic stress decreased putative anti-inflammatory bacteria and increased in putative pro-inflammatory bacteria. These changes may contribute to neuroinflammation and neurodegeneration in Restraint Stress + rotenone mice *Akkermansia* genus was significantly positively correlated with dystrophic/phagocytic microglial counts in the substantia nigra. *Akkermansia* genus was significantly inversely correlated with hanging grip test behavior data Fecal butyrate levels were significantly positively correlated with striatal DA concentrations

| Sun et al. (2018) | 8 week-old male C57BL/6 mice Control +/− treatment (n = 7); MPTP + PBS (n = 7); MPTP + FMT (n = 8) FMT is fecal from healthy control mice | 16S rRNA V4 gene amplicon sequencing Fecal samples | ***Control vs. MPTP ± PBS*** **MPTP + PBS:** ↑**Proteobacteria phylum:** ↑Enterobacteriales order **Bacteroidetes phylum:** ↑Bacteroidales order ↓**Firmicutes phylum:** ↓Clostridiales order, ↑Turicibacterales order ***MPTP ± PBS vs. MPTP ± FMT*** **MPTP + FMT:** ↓**Proteobacteria phylum:** ↓Enterobacteriales order **Firmicutes phylum:** ↓Clostridiales order, ↓Turicibacterales order ***Control vs. MPTP ± FMT*** **MPTP + FMT:** ↑**Actinobacteria phylum:** ↑Bifidobacteriales order ↑**Bacteroidetes phylum:** ↑Bacteroidales order | Did Not Analyze | Gut microbiota from PD mice induced motor impairment and striatal neurotransmitter decrease on normal mice FMT alleviates neuroinflammation and gut inflammation by the signaling pathway TLR4/TBK1/NF-jB/TNF-a **Short Chain Fatty Acids** **Control vs. MPTP + PBS** **MPTP + PBS:** ↑Acetic acid, ↑Propionic acid, ↑Butyric acid, ↑n-valeric acid **MPTP + PBS vs. MPTP + FMT** **MPTP + FMT:** ↓Acetic acid, ↓Propionic acid, ↓Butyric acid, ↓n-valeric acid **Control vs. MPTP + FMT** No Significance | Fecal samples from PD mice exhibit manifestations of gut microbial dysbiosis: decreases in the phylum Firmicutes & order Clostridiales, & increases in the phylum Proteobacteria, order Turicibacterales & Enterobacterales. It also demonstrated increased SCFAs Fecal microbiota transplantation (FMT) reduced gut microbial dysbiosis, decreased fecal SCFAs, alleviated physical impairment, and increased striatal DA and 5-HT content of PD mice. Further, FMT reduced the activation of microglia and astrocytes in the substantia nigra, and reduced expression of TLR4/TNF-a signaling pathway components in gut and brain. Gut microbial dysbiosis is involved in PD pathogenesis, and FMT can protect PD mice by suppressing neuroinflammation and reducing TLR4/TNF-a signaling |

Continued

Table 2 Changes in the intestinal microbiota in Parkinson's disease rodent models.—cont'd

Reference	Study design	Methodology	Major taxa altered in Parkinson's disease animal model	Predictive KEGG functional pathways	Clinical features in Parkinson's disease animal model	Major finding
Johnson et al. (2018)	8-week old male Sprague-Dawley rats 26 rats Total: Control ($n = 12$) Rotenone ($n = 14$) Treatment for 5 days a week (4 weeks long)	16S rRNA V3–V4 gene amplicon sequencing RT-PCR Colon, small intestine samples	**Rotenone Effect on Small Intestine & Colon** ↑**Actinobacteria phylum:** ↑Bifidobacteriales order, ↑Bifidobacteriaceae family, ↑*Bifidobacterium* genus ↑**Proteobacteria phylum:** ↑Burkholderiales order, ↑Alcaligenaceae family, ↑*Sutterella* genus ↓**Bacteroidetes phylum:** ↓Bacteroidales order, ↓S24–7 family, ↓Prevotellaceae family, ↓Paraprevotellaceae family, ↓*Prevotella* genus, ↓S24-7 *Unclassified* genus ↓**Cyanobacteria phylum:** ↓YS2 family ↑**Firmicutes phylum:** ↓Clostridiales order, ↓Lactobacillales order, ↑Turicibacterales order, ↓Ruminococcaceae family, ↓Lachnospiraceae family, ↑Clostridiaceae family, ↑*Lactobacillus* genus, ↑*Turicibacter* genus, ↓*Oscillospira* genus	Did Not Analyze	Despite RT-PCR only finding one statistically significant change in the microbiota of the rotenone group (*Bifidobacterium* genus higher relative abundance), there was an overall trend for rotenone rats to have a greater amount of bacteria per stool weight for all bacteria tested, in both the small intestine and colon, suggesting an overgrowth of bacteria may be occurring. The increase in *Lactobacillus* and *Bifidobacterium* genera aligned with the increases seen for these bacteria via RT-PCR, and as reported in human PD patients	Rats exposed to rotenone had more days with evidence of diarrhea and significantly delayed gastric emptying, reproducing the clinical symptom of gastroparesis. In comparison to control, PD Rotenone model exhibited alteration in the small intestine and colon which is consistent with changes in PD patients. Rotenone-treated rats displayed increased Clostridiaceae and reduced Lachnospiraceae and Prevotellaceae, as reported in PD patients PD rotenone model demonstrated mucosal thickening and goblet cell hyperplasia in the colon of rotenone rats which could be a possible adaptive response to the toxin or changes in GI microbiota

Lai et al. (2018)	Male C57BL/6 mice specific-pathogen-free (SPF) chronic low-dose MPTP model Control and MPTP groups (20 per group) The MPTP group was treated with MPTP (18mg/kg) twice a week in the afternoon (Tuesday and Saturday) for 5 weeks, whereas the control group was injected with a standard suspension vehicle (0.9% NaCl, w/v)	16S rRNA V3–V4 gene amplicon sequencing Fecal Samples	**MPTP 2 Days** **Bacteroidetes phylum:** ↑Prevotellaceae family ↓**Proteobacteria phylum** **MPTP 3 Weeks** **Firmicutes phylum:** ↓Clostridiales order, ↓Lachnospiraceae family, ↑Erysipelotrichales order, ↑Erysipelotrichaceae family ↓**Proteobacteria phylum**	Did Not Analyze	Chronic low-dose MPTP administration leads to: 1. Gut microbiota dysbiosis 2. Reduces striatal and ileal DA content 3. Decreased TH and Increased α-Syn Expression in the Striatum & Ileum of mice 4. General conditions, such as the decreased motor ability, body weight, and defecation status of mice MPTP causes inflammation in substantia nigra and ileum as measured by Microglia activation (High Iba-1 expression, IL-17, IL-1β and TNF-α) and increased iNOS expression, plus expression levels of IL-17, IL-1β and TNF-α in the ileum	MPTP caused GI dysfunction and intestinal pathology prior to motor dysfunction The composition of the gut microbiota was changed; in particular, the change in the abundance of phyla Proteobacteria, orders Clostridiales, Erysipelotrichales, and families Lachnospiraceae, Erysipelotrichaceae, Prevotellaceae Data indicates that a chronic low-dose MPTP model can be used to evaluate the progression of intestinal pathology and gut microbiota dysbiosis in the early stage of PD, which may provide new insights into the pathogenesis of PD

Continued

Table 2 Changes in the intestinal microbiota in Parkinson's disease rodent models.—cont'd

Reference	Study design	Methodology	Major taxa altered in Parkinson's disease animal model	Predictive KEGG functional pathways	Clinical features in Parkinson's disease animal model	Major finding
Yang et al. (2017)	Male C57BL/6 mice Aged between 8 and 9 weeks Rotenone administered once daily by gavage 30 mg/kg for 4 weeks One group ($n = 5$) was for in vivo tests (fecal pellet collection, open field test, and pole test) weekly during the period The other five groups ($n = 6$–8 per group) were sacrificed at the indicated time point, colon, and brain tissues were collected for biochemical examinations	16S rRNA V4–V5 gene amplicon sequencing Fecal Samples	***Longitudinal Rotenone Effect*** **1 week:** **Firmicutes phylum:** ↓*Clostridium* genus, ↓*Lactococcus* genus **Proteobacteria phylum:** ↓*Sutterella* genus **Bacteroidetes phylum:** ↓*Paraprevotella* genus **2 week:** **Firmicutes phylum:** ↓*Clostridium* genus, ↓*Lactococcus* genus **Proteobacteria phylum:** ↓*Sutterella* genus, ↓*Desulfovibrio* genus **Actinobacteria phylum:** ↓*Adlercreutzia* genus **3 week:** ↑**Firmicutes phylum:** ↓*Clostridium* genus, ↓*Lactococcus* genus, ↑*Lactobacillus* genus **Proteobacteria phylum:** ↓*Sutterella* genus, ↓*Desulfovibrio* genus ↓**Bacteroidetes phylum:** ↓*Paraprevotella* genus **4 week:** ↑**Firmicutes phylum:** ↓*Clostridium* genus, ↑*Lactobacillus* genus **Proteobacteria phylum:** ↓*Sutterella* genus, ↓*Desulfovibrio* genus ↓**Bacteroidetes phylum** **At both 3 and 4 weeks, the Firmicutes-to-Bacteroidetes ratio was higher**	**KEGG pathways upregulated in Rotenone:** 11 pathways predicted as enriched in the fecal microbiome of rotenone treated mice at 3 weeks: Membrane Transport, Immune System Diseases, Signaling Molecules and Interaction, Neurodegenerative Diseases, Nucleotide Metabolism, Translation, Replication and Repair, Transcription, Xenobiotics Biodegradation and Metabolism, Infectious Diseases and Cell Growth and Death pathways **KEGG pathways downregulated in Rotenone:** Fourteen pathways were lower at 4 weeks: Endocrine System, Metabolism of Cofactors and Vitamins, Amino Acid Metabolism, Immune System, Cellular Processes and Signaling, Transport and Catabolism, Glycan Biosynthesis and Metabolism, Folding Sorting and Degradation, Energy Metabolism, Biosynthesis of Other Secondary Metabolites, Metabolism of Other Amino Acids, Digestive System, Metabolism, Metabolism of Terpenoids, and Polyketides pathways	The genus *Desulfovibrio* was positively associated with stool weight, stool water content, locomotor activity, and inversely correlated with climbing time. The genus *Lactobacillus* was inversely correlated with stool weight, stool water content, locomotor activity, and positively associated with climbing time. The genera *Clostridium* and *Adlercreutzia* were inversely associated with stool weight. The genus *Sutterella* had no correlation with stool weight, but was positively associated with stool water content and motor functions Chronic oral treatment by rotenone for 4 weeks induced decrease in stool frequency and replicates neurodegeneration of the substantia nigra, induced ENS dysfunction and delayed gastric emptying	Alteration of fecal microbiota preceded the major motor dysfunction and CNS pathology of PD Gut microbiome perturbation may contribute to rotenone toxicity in the initiation of PD

Reference	Subjects	Method	Findings		Conclusions

| Nixon et al. (2018) | Male C57BL/6 J mice (*n* = 129) mice Aged 3-month old Wild-type (WT) or Knock-in of H67D alleles | 16S rRNA V2 gene amplicon sequencing Fecal Samples | ***H67D Effect***
 ↑**Acidobacteria phylum**
 Firmicutes phylum:
 ↓*Erysipelotrichi class*
 Proteobacteria phylum:
 ↑*Deltaproteobacteria class,*
 ↓*Betaproteobacteria class,*
 ↓*Alphaproteobacteria class*
 ↑**Tenericutes phylum**
 ↓**Bacteroidetes phylum**
 ↓**Actinobacteria phylum**
 ↓**Verrucomicrobia phylum**
 A high number (over 45) differentiating taxa in the microbial community composition between WT and H67D mice | Did Not Analyze | Bacteria from the butyrate-producing Firmicutes phyla are increased in H67D HFE, with the Lachnospiraceae family and *Blautia* genus. This data is consistent with results in the human population where neurologically normal controls have increased levels of these bacteria, as compared to Parkinson's disease patients

 The Bacteroidetes phyla are decreased in H67D HFE mice, predominantly the *Akkermansia* genus in the Verrucomicrobiaceae family, while they are elevated in Parkinson's patients |

| Perez-Pardo et al. (2018a, 2018b) | 7 weeks old C57BL/6 J mice Wild-type (WT) & TLR4-KO mice +/− rotenone were housed under a 12-h light/dark cycle (*N* = 9–10) per group | 16S rRNA V4 gene amplicon sequencing Caecum Mucosa and Caecum Luminal Content samples | ***Significant effects of Rotenone or Genotype (TLR4KO) on Cecum Mucosa***
 WT Rotenone vs. WT Vehicle
 WT Rotenone: ↑*Rikenellaceae Unclassified genus,* ↑*S24–7 Unclassified genus,* ↑*Clostridiales Unclassified genus,* ↑*Allobaculum genus,* ↓*Bifidobacterium genus*
 TLR4-KO Rotenone vs. TLR4-KO-Vehicle
 TLR4-KO Rotenone: ↑*Clostridiales Unclassified Unclassified genus,* ↑*Lachnospiraceae Unclassified genus,* ↑*Lactobacillus genus,* ↓*Bifidobacterium genus*
 TLR4-KO Vehicle vs. WT Vehicle
 TLR4-KO Vehicle: ↑*Rikenellaceae Unclassified genus,* ↑*S24–7 Unclassified genus,* ↑*Bacteroides genus,* ↓*Bifidobacterium genus,* ↓*Lactobacillus genus*
 TLR4-KO-Rotenone vs. WT-Rotenone
 TLR4-KO-Rotenone: ↓*Lactobacillus genus*
 Firmicutes-to-Bacteroidetes | Did Not Analyze | Altered microbiota profiles was observed in the TLR4-KO mice and rotenone treatment resulted in similar microbiota profiles irrespective of the genotype. The caecum mucosa taxa relative abundance differences were greater than the luminal content taxa

 The TLR 4-mediated inflammation plays an important role in intestinal and/or brain inflammation, which may be one of the key factors leading to neurodegeneration in PD. Rotenone treatment in TLR 4-KO mice revealed less intestinal inflammation, intestinal and motor dysfunction, neuroinflammation and neurodegeneration, relative to rotenone-treated wild-type animals despite the presence of dysbiotic microbiota in TLR 4-KO mice |

TLR4-KO mice were protected against PD-like consequences of rotenone-induced pathology. Rotenone-treated mice had increased number of CD3+ T cells and TLR4+ cells in the colonic mucosa and this pro-inflammatory state was associated with increased number of GFAP + enteric glial cells and α-syn pathology in the colonic myenteric plexuses. The loss of TLR4 significantly mitigated the effect of rotenone on motor function impairment, intestinal barrier integrity, myenteric plexus GFAP expression, colonic α-syn, SN microglial activation & dopaminergic cell loss. It shows that knocking out TLR4 mitigates neuroinflammation in the brain This findings in both human samples and a mouse model strongly suggest a possible role for TLR4-mediated signaling in gut leakiness, gut-derived inflammation, neuroinflammation and neurodegeneration in PD. The observed TLR4-mediated

HFE genotype impacts the expression of tyrosine hydroxylase in the substantia nigra, the gut microbiome and the response to paraquat providing additional support that the HFE genotype is a disease modifier for Parkinson's disease

Continued

Table 2 Changes in the intestinal microbiota in Parkinson's disease rodent models.—cont'd

Reference	Study design	Methodology	Major taxa altered in Parkinson's disease animal model	Predictive KEGG functional pathways	Clinical features in Parkinson's disease animal model	Major finding
			ratio in the cecal mucosa showed a significant rotenone treatment effect, genotype effect, and interaction between rotenone and genotype effect *Significant effects of Rotenone or Genotype (TLR4KO) on Cecum Content* Similar individual taxa differential abundances were noted in Cecum Content, as shown in Cecum Mucosa The Firmicutes-to-Bacteroidetes ratio in the luminal content showed a significant effect of interaction between rotenone and genotype effect		immune activation in the colon and the brain in PD could be secondary to changes in the intestinal microbiota. Rotenone treatment and loss of key bacterial recognition receptor, TLR4, resulted in a 'pro-inflammatory' dysbiotic microbiota compositions Increased relative abundances of putative pro-inflammatory intestinal bacterial genera *Rickenellaceae unclassified, Allobaculum* and *Bacteroides,* decreased F/B ratios, reduction of the relative abundances of putative anti-inflammatory bacteria genera *Bifidobacterium* and/or *Lactobacillus,* and in WT-rotenone and/or TLR4-KO-vehicle mice	
Choi et al. (2018)	Male C57BL/6 mice 12-week-old MPTP/probenecid (p) induced PD mice 7-week-old mice MPTP-induced PD mice 7-week old mice administrated with MPTP + Proteus mirabilis 7-week old mice with intra-rectal injection of LPS Proteus mirabilis 7-week old mice with administrated Proteus mirabilis alone Male ICR 6-week-old mice (6-OHDA-induced PD mice)	RT-PCR 16S rRNA gene amplicon sequencing Fecal Samples	*MPTP/p-induced PD model* **Proteobacteria phylum:** ↑Enterobacteriaceae family, ↑Proteus genus→Proteus mirabilis *MPTP-induced PD model* **Proteobacteria phylum:** ↑ Enterobacteriaceae family, ↑Proteus species *6-OHDA-induced PD model* **Proteobacteria phylum:** ↑ Enterobacteriaceae family	Did Not Analyze	Administration of Proteus mirabilis isolated from PD mice significantly induced motor deficits, selectively caused dopaminergic neuronal damage and inflammation in the substantia nigra and striatum, and stimulated α-synuclein aggregation in the brain as well as in the colon	*Proteus mirabilis directly induced PD symptoms and dopaminergic neuronal damage in the mouse brain* Data suggested that the number of Enterobacteriaceae, particularly *Proteus mirabilis,* markedly and commonly increased in PD mouse models Lipopolysaccharides, a virulence factor of *Proteus mirabilis,* may be associated in the pathological changes via gut leakage and inflammatory actions

| Zhou et al. (2019) (Abstract) | Mice Groups: NS+AL, NS+FMD, MPTP+AL, MPTP+FMD Fasting 3 days followed by 4 days of refeeding for three 1-week cycles | 16S and 18S rRNA gene amplicon sequencing Fecal Samples | ***MPTP ± FMD*** **↑Firmicutes phylum** **↑Tenericutes phylum** **↑Opisthokonta phylum** **↓Proteobacteria phylum** | Not Available (Abstract only) | Fasting for 3 days, followed by 4 days of refeeding for three 1-week cycles, accelerated the retention of motor function and attenuated the loss of dopaminergic neurons in the substantia nigra in MPTP-induced PD mice BDNF, known to promote the survival of dopaminergic neurons, were increased in PD mice after FMD, suggesting an involvement of BDNF in FMD-mediated neuroprotection FMD decreased the number of glial cells as well as the release of TNF-α and IL-1β in PD mice, showing that FMD also inhibited neuro-inflammation Gas chromatography-mass spectrometry and liquid chromatography-mass spectrometry revealed that FMD modulated the MPTP-induced lower propionic acid and isobutyric acid, and higher butyric acid and valeric acid and other metabolites | Transplantation of fecal microbiota, from normal mice with fasting mimicking diet (FMD) treatment to antibiotic-pretreated PD mice increased dopamine levels in the recipient PD mice, suggesting that gut microbiota contributed to the neuroprotection of FMD for PD Data demonstrates that FMD can be a new means of preventing and treating PD through promoting a favorable gut microbiota composition and metabolites |

Thus, it is essential to know how microbiota and microbiota-products trigger neuroinflammation.

Longitudinal and interventional studies are needed to establish the causal link between dysbiotic microbiota and development of PD (Scheperjans et al., 2018).

6.2 Clinical studies

6.2.1 Case-control studies

Since Scheperjans et al. (2015a, 2015b) and Keshavarzian et al. (2015) first reported abnormal intestinal microbiota community structure and composition in PD patients, there has been 15 additional studies from USA, Northern, Western, Eastern Europe and Asia that confirm the initial reports that PD patients have "dysbiotic" intestinal microbiota community (Table 3) (Lubomski et al., 2019a, 2019b; Sampson, 2019). Parkinson's disease subjects demonstrated significantly altered intestinal microbial compositions in comparison to healthy controls with some trends worthy of comment. Overall, current studies reported PD subjects to have:

- Increased relative abundance of genera *Akkermansia* (seven studies) (Barichella et al., 2019; Bedarf et al., 2017; Heintz-Buschart et al., 2018; Hill-Burns et al., 2017; Keshavarzian et al., 2015; Li et al., 2019; Unger et al., 2016), *Bifidobacterium* (five studies) (Aho et al., 2019; Hill-Burns et al., 2017; Lin et al., 2018; Petrov et al., 2017; Unger et al., 2016), and *Lactobacillus* (seven studies) (Aho et al., 2019; Barichella et al., 2019; Hasegawa et al., 2015; Hill-Burns et al., 2017; Hopfner et al., 2017; Petrov et al., 2017; Scheperjans et al, 2015b).
- Decreased abundance of genus *Prevotella* (seven studies) (Aho et al., 2019; Bedarf et al., 2017; Li et al., 2017; Minato et al., 2017; Petrov et al., 2017; Scheperjans, 2018; Unger et al., 2016) and the family *Lachnospiraceae* (six studies) (Aho et al., 2019; Barichella et al., 2019; Hill-Burns et al., 2017; Keshavarzian et al., 2015; Li et al., 2017; Lin et al., 2018) along with its lower taxonomic hierarchal putative SCFA-producing genera *Faecalibacterium* (five studies) (Hill-Burns et al., 2017; Keshavarzian et al., 2015; Li et al., 2017; Petrov et al., 2017; Unger et al., 2016), *Roseburia* (four studies) (Aho et al., 2019; Barichella et al., 2019; Hill-Burns et al., 2017; Keshavarzian et al., 2015), *Blautia* (five studies) (Aho et al., 2019; Hill-Burns et al., 2017; Keshavarzian et al., 2015; Li et al., 2017; Petrov et al., 2017), *Coprococcus* and *Dorea* (two studies) (Keshavarzian et al., 2015; Petrov et al., 2017) (Table 3).

Furthermore, a few of the studies evaluated predicted functional gene content profiling (PICRUSt) (Langille et al., 2013) to infer changes in microbiota function (Table 3). Keshavarzian et al. (2015) discovered PD subject's fecal samples had significantly higher abundant genes involved in lipopolysaccharide (LPS) biosynthesis, ubiquinone and other terpenoid-quinone biosynthesis and type III bacterial secretions systems, whereas a large number of genes in metabolism were significantly lower abundant.

Reference	Study design	Methodology	Major taxa altered in Parkinson's disease	Predictive KEGG functional pathways	Parkinson's disease characteristics	Major findings
Hasegawa et al. (2015)	Case-Control study PD (N = 52), Random Healthy Controls (N = 36)	Real-time Quantitative PCR 16s rRNA or 23S rRNA Fecal samples	**Firmicutes phylum:** ↑*Lactobacillus* genus, ↓*Clostridium coccoides* species, ↓*Clostridium leptum* species **Bacteroidetes phylum:** ↓*Bacteroides fragilis* species	Did Not Analyze	Disease duration: 9.5 years Hoehn-Yahr score: 2.7	In comparison to the control cohort, patients with PD exhibited an altered gut microbial composition, specifically an increase in abundance of *Lactobacillus* genus Disease duration had a positive correlation with *Lactobacillus gasseri* species and a negative correlation with *Clostridium coccoides*
Keshavarzian et al. (2015)	Case-Control study PD (N = 38): untreated naïve (N = 12), treated (N = 26) Random Healthy Controls (N = 34)	16S rRNA V4 gene amplicon sequencing Fecal and Colonic Sigmoid Mucosa samples	***Mucosa:*** **Firmicutes phylum:** ↓*Faecalibacterium* genus, ↓*Dorea* genus **Proteobacteria phylum:** ↑*Oxalobacteraceae* family, ↑*Ralstonia* genus **Actinobacteria phylum:** ↑*Coriobacteriaceae* family ***Feces:*** Ratio of Firmicutes-to-Bacteroidetes was significantly higher in PD patients **Bacteroidetes phylum:** ↑*Bacteroidaceae* family, ↑*Bacteroides* genus **Proteobacteria phylum** **Verrucomicrobia phylum:** ↑*Verrucomicrobiaceae* family, ↑*Akkermansia* genus **Firmicutes phylum:** ↑*Clostridiaceae* family, ↑*Lachnospiraceae* family, ↑*Coprobacillaceae* family, ↑*Oscillospira* genus, ↓*Blautia* genus, ↓*Coprococcus* genus, ↓ *Roseburia* genus, ↓*Dorea* genus	1248 KO's had significantly different abundances between PD and healthy control fecal samples 14 KO's within LPS biosynthesis pathway and 8 KO's within ubiquinone and other terpenoid-quinone biosynthesis were more abundant in PD compared to healthy control **Upregulated KEGG pathways in PD:** Two-component system, oxidative phosphorylation, bacterial secretion system, folate biosynthesis, citrate cycle, LPS biosynthesis pathway, ubiquinone and other terpenoid-quinone biosynthesis, genes from type III secretion systems **Downregulated pathways in PD:** metabolic pathways, biosynthesis of secondary metabolites, microbial metabolism in diverse environments, biosynthesis of amino acids, carbon metabolism ABC transporters, ribosome, purine metabolism, pyrimidine metabolism, methane metabolism, amino sugar nucleotide metabolism, pyruvate metabolism, amino-acyl-tRNA biosynthesis, porphyrin and chlorophyll metabolism, carbon fixation pathways in prokaryotes, cysteine and methionine metabolism, homologous recombination, arginine and proline metabolism, glyoxylate and dicarboxylate metabolism	Disease duration: 6.4 years Hoehn-Yahr score: 2.1	PD patients demonstrated significantly altered GI microbial composition in fecal and mucosal communities in comparison to healthy controls PD duration had positive correlations with Bacteroidetes and Proteobacterium phylum and negative correlations with Firmicutes phylum, Lachnospiraceae family, and *Blautia* genus. This further suggests the presence of pro-inflammatory dysbiosis in PD patients, potentially triggering the misfolding of α-synuclein protein Predictive functional analysis demonstrated a higher abundance of pathways including LPS biosynthesis, ubiquinone and other terpenoid-quinone biosynthesis, and type III secretion systems within fecal samples of PD patients

Continued

Table 3 Changes in the human intestinal microbiota in Parkinson's disease compared to healthy controls.—cont'd

Reference	Study design	Methodology	Major taxa altered in Parkinson's disease	Predictive KEGG functional pathways	Parkinson's disease characteristics	Major findings
Scheperjans et al. (2015a, 2015b)	Case-Control study PD ($N = 72$), Random Healthy Controls ($N = 72$)	16S rRNA V1-V3 gene amplicon sequencing Fecal samples	**Proteobacteria phylum:** ↑Bradyrhizobiaceae family **Firmicutes phylum:** ↑Lactobacillaceae family, ↑Clostridiales Incertae Sedis IV family **Verrucomicrobia phylum:** ↑Verrucomicrobiaceae family **Bacteroidetes phylum:** ↓Prevotellaceae family	Did Not Analyze	**PD Drug effects:** Levodopa ($N = 39$) COMT inhibitor ($N = 11$) Dopamine agonist ($N = 56$) MAO inhibitor ($N = 51$) Anticholinergic ($N = 4$) Warfarin ($N = 1$) Statin ($N = 15$) Disease duration: 6.5 years Hoehn-Yahr score: Not reported	In comparison to healthy controls, patients with PD had significantly altered GI microbial compositions, specifically demonstrating a reduction in Prevotellaceae family, thus indicating increased gut permeability PD patients with severe postural instability and gait difficulty displayed a higher abundance of Enterobacteriaceae family
Unger et al. (2016)	Case-Control study PD ($N = 34$), Random Healthy Controls ($N = 34$)	Real time-Quantitative PCR Fecal samples	**_PD:_** **Firmicutes phylum:** ↓Lactobacillaceae family, ↓Enterococcaceae family ↓_Faecalibacterium prausnitzii_ species **Actinobacteria phylum:** ↑_Bifidobacterium_ genus **Bacteroidetes phylum:** ↓Prevotellaceae family **Verrucomicrobia phylum:** ↑_Akkermansia muciniphila_ species **Proteobacteria phylum:** ↑Enterobacteriaceae family **_PD with COMT inhibitor:_** ↓**Firmicutes phylum:** ↓_Faecalibacterium prausnitzii_ species	Did Not Analyze	**PD Drug effects:** COMT inhibitor Levodopa Disease duration: 6.8 years Hoehn-Yahr score: 2.5	PD patients exhibited significantly altered GI microbial compositions when compared to healthy controls Metabolomics of SCFA in PD patients demonstrated a reduction in absolute concentrations of butyrate, acetate and propionate, and a reduction in the relative concentrations of butyrate Metabolomics of SCFA of PD patients taking COMT inhibitor medication demonstrated a reduction in absolute and relative concentrations of butyrate
Bedarf et al. (2017)	Case-Control study PD ($N = 31$), Random Healthy Controls ($N = 28$)	Metagenomics shotgun sequencing Fecal samples	**L-Dopa naïve PD patients:** **Firmicutes phylum:** ↓Erysipelotrichaceae family, ↓_Eubacterium_ genus, ↓_Clostridium_ genus, ↓_Eubacterium biforme_ species, ↓_Clostridium saccarolyticum_ species **Verrucomicrobia phylum:** ↑Verrucomicrobiaceae family, ↑_Akkermansia_ genus, ↑_Akkermansia muciniphila_ species,	**Bacterial metabolic pathways:** 436 KEGG pathways upregulated, 389 KEGG pathways downregulated **Significant KEGG pathways upregulated in PD:** Tryptophan metabolism, tryptophan degradation **Significant KEGG pathways downregulated in PD:** D- Glucoronate degradation, Beta-D-glucuronide degradation	**PD Drug effects:** Amantadine ($N = 26$) Dopamine agonist ($N = 11$) MAO-inhibitor ($N = 28$) Statin ($N = 1$) Metformin ($N = 1$) Acetyl salicylic acid ($N = 2$) Disease Duration: Not reported Hoehn-Yahr score: Not reported	L-Dopa naïve PD patients had significant altered GI microbial composition in comparison to healthy controls Predictive functional analysis of L-Dopa naïve PD patients indicated differences in regulation of B-glucoronate and Tryptophan degradation pathways

Continued

| Heintz-Buschart et al. (2018) | Case-Control study PD (N = 76), Random Healthy Controls (N = 78) | 16S and 18S rRNA V4 gene amplicon sequencing, metagenomic shotgun sequencing 16S rRNA: Nasal Wash (N = 147) and Fecal (N = 84) samples 18S rRNA: 61 Fecal samples | **Bacteroidetes phylum:** ↓Prevotellaceae family, ↓Prevotella genus, ↑Alistipes genus, ↑Alistipes shahii species, ↓Prevotella copri species

Fecal PD:
Firmicutes phylum:
↑Acidaminococcaceae class, ↑Erysipelotrichaceae order, ↑Ruminococcaceae family, ↑Clostridiales family, ↑Lachnospiraceae family, ↑Flavonifractor genus, ↑Ruminococcus genus, ↑Megasphaera genus, ↑Mitsuokella genus, ↑Clostridium XVIa genus, ↑Anaerotruncus genus
Verrucomicrobia phylum:
↑Verrucomicrobiales order, ↑Verrucomicrobiae class, ↑Verrucomicrobiaceae family, ↑Akkermansia genus,
Actinobacteria phylum:
↑Olsenella genus
Proteobacteria phylum:
↑Sutterella genus
Bacteroidetes phylum:
↑Prevotellaceae family, ↑Prevotella genus, ↑Bacteroides genus, ↑Butyricimonas genus
Nasal fluid of PD (taxa trending toward significance):
Firmicutes phylum:
↑Listeriaceae family, ↑Bacillaceae family, ↑Ruminococcaceae family, ↑Carnobacteriaceae family
Actinobacteria phylum:
↑Micrococcaceae family
Proteobacteria phylum:
↑Hydrogenophilaceae family, ↑Hyphomicrobiaceae family, ↑Desulfovibrionaceae family | Did Not Analyze | **PD Drug effects:**
Levodopa (N = 66)
Dopamine agonist (N = 52)
COMT inhibitor (N = 4)
MAO-B inhibitors (N = 52)
Metformin (N = 7)
Disease duration: 6 years
Hoehn-Yahr score: 2.14 | In comparison to healthy controls, PD patients had significant altered GI microbial composition, but lacked a strong significant alteration in nasal microbiome
Within the PD patients, Anaerotruncus genus, Clostridium genus, and Lachnospiraceae family were found to be related to motor symptoms. Anaerotruncus genus and Akkermansia genus were found to be related to nonmotor symptoms |

Table 3 Changes in the human intestinal microbiota in Parkinson's disease compared to healthy controls.—cont'd

Reference	Study design	Methodology	Major taxa altered in Parkinson's disease	Predictive KEGG functional pathways	Parkinson's disease characteristics	Major findings
Hill-Burns et al. (2017)	Case-Control study **Total PD** (N = 197): Random PD (N = 143) Spousal PD (N = 54) **Total HC** (N = 130): Random Healthy Controls (N = 76), Spousal Healthy Controls (N = 54)	16S rRNA gene amplicon sequencing Fecal samples	***Non-medicated PD:*** **Actinobacteria phylum:** ↑Bifidobacteriaceae family ↑*Bifidobacterium* OTU, ↑*Bifidobacterium* genus **Bacteroidetes phylum:** ↑*Parabacteroides* OTU, ↑*Prevotella* OTU **Firmicutes phylum:** ↑Lactobacillaceae family, ↑Tissierellaceae family, ↑Christensenellaceae family, ↓Lachnospiraceae family, ↓*Blautia* OTU, ↓*Coprococcus* OTU, ↓*Roseburia* OTU, ↓*Unclassified* OTU, ↓*Faecalibacterium* OTU, ↓*Lactobacillus* genus, ↑*Unclassified* genus, ↓*Blautia* genus, ↓*Roseburia* genus, ↓*Faecalibacterium* genus **Verrucomicrobia phylum:** ↑Verrucomicrobiaceae family, ↑*Akkermansia* genus **Proteobacteria phylum:** ↑Pasteurellaceae family ***Medicated PD:*** **Actinobacteria phylum:** ↑Bifidobacteriaceae family ↑*Bifidobacterium* OTU, ↑*Bifidobacterium* genus, **Bacteroidetes phylum:** ↑*Prevotella* OTU **Firmicutes phylum:** ↑Lactobacillaceae family, ↑Tissierellaceae family, ↑Christensenellaceae family, ↓Lachnospiraceae family ↓*Blautia* OTU ↓*Coprococcus* OTU, ↓*Roseburia* OTU, ↓*Unclassified* OTU, ↓*Faecalibacterium* OTU, ↓*Lactobacillus* genus, ↑*Unclassified* genus, ↓*Blautia* genus, ↓*Roseburia* genus, ↓*Faecalibacterium* genus	**Bacterial metabolic pathways:** 17 KEGG upregulated, 9 KEGG downregulated **17 KEGG pathways upregulated in PD:** Citrate cycle, carbon fixation pathways in prokaryotes, biosynthesis of unsaturated fatty acids, pyruvate metabolism, folate biosynthesis, prenyltransferases, naphthalene degradation, fatty acid biosynthesis, chloroalkane and chloroalkene degradation, atrazine degradation, glycine, serine and threonine metabolism, steroid hormone synthesis, steroid biosynthesis, N-glycan biosynthesis, carotenoid biosynthesis, flavonoid biosynthesis, caffeine metabolism **9 KEGG pathways downregulated in PD:** starch and sucrose metabolism, porphrin and chlorophyll metabolism, galactose metabolism, photosynthesis, photosynthesis proteins, cyanoamino acid metabolism, phenylpropanoid biosynthesis, nicotinate and nicotinamide metabolism, flavone and flavonol biosynthesis	**PD Drug effects:** COMT inhibitor (N = 37) Anticholinergic (N = 7) Carbidopa/levodopa (N = 168) Amantadine (N = 49) Dopamine agonist (N = 99) MAO-B inhibitor (N = 71) Disease Duration: 13.7 years Hoehn-Yahr score: Not reported	In comparison to healthy controls, PD patients were found to have significant altered GI microbial community Ruminococcaceae family were found to be associated with disease duration (>10 years) Predictive functional analysis indicated 17 upregulated KEGG pathways and 9 downregulated pathways, including xenobiotics degradation and metabolism of plant-derived compounds

Reference	Study design	Method	Microbial findings		Clinical data	Summary
Li et al. (2017)	Case-Control study PD (N = 24), Random Healthy Controls (N = 14)	16S rRNA V3–V5 gene amplicon sequencing Fecal samples	↑Verrucomicrobiaceae family, ↑*Akkermansia* genus **Proteobacteria phylum:** ↓Pasteurellaceae family **Actinobacteria phylum:** ↑Coriobacteriaceae family **Firmicutes phylum:** ↑Bacilli class, ↑Negativicutes class, ↑Veillonellaceae family, ↑Erysipelotrichaceae family, ↑Enterococcaceae family, ↑*Acidaminococcus* genus, ↑*Enterococcus* genus, ↑*Megamonas* genus, ↑*Megasphaera* genus, ↑*Streptococcus* genus, ↓Ruminococcus family, ↓Lachnospiraceae family, ↓*Blautia* genus, ↓*Faecalibacterium* genus, ↓*Ruminococcus* genus **Proteobacteria phylum:** ↑Enterobacteriaceae family, ↑Moraxellaceae family, ↑*Acinetobacter* genus, ↑*Escherichia-shigella* genus, ↑*Proteus* genus **Bacteroidetes phylum:** ↓Prevotellaceae family	Did Not Analyze	**PD Drug effects:** Anti-parkinsonian medication (N = 22) Disease duration: 5 years Hoehn-Yahr score: Not reported	GI microbial composition was significantly altered in PD participants compared to healthy controls *Faecalibacterium* genus was found to be significantly lower, while *Megasphaera* genus was significantly higher in severe PD cohort Greater PD duration was associated with an increase in *Proteus* genus, *Enterococcus* genus, *Escherichia-Shigella* genus, *Megasphaera* genus, all of which are putative pathobionts. Greater PD duration was associated with a decrease in *Blautia* genus, *Ruminococcus* genus, *Sporobacter* genus and *Haemophilus* genus, all of which are cellulose degraders
Minato et al. (2017)	2-year Longitudinal study PD (N = 36): deteriorated PD (N = 18), stable PD (N = 18) No controls	Quantitative PCR of 16S and 23S rRNA Fecal samples	***Deteriorated versus stable PD—year 1:*** Deteriorated PD: **Actinobacteria phylum:** ↓*Bifidobacterium* genus ***Deteriorated versus stable PD—year 2:*** Deteriorated PD **Firmicutes phylum:** ↓*Lactobacillus gasseri* species **Bacteroidetes phylum:** ↓*Prevotella* genus Stable PD: **Actinobacteria phylum:** ↓*Bifidobacterium* genus, ↓*Aptobium* cluster **Firmicutes phylum:** ↓*Enterococcus* genus, ↓*Lactobacillus gasseri* species, ↓*Lactobacillus reuteri* species, ↓*Clostridium leptum* species	Did Not Analyze	Disease duration: Year 0: 9.2 years Year 2: 9.8 years Hoehn-Yahr score: Deteriorated PD: 2.5 Stable PD: 2.6 Total UPDRS score: Deteriorated PD: 51.4 Stable PD: 36.3	Fecal samples were obtained from year 1 and year 2 from deteriorated and stable PD cohorts In comparison to year 1, deteriorated and stable PD groups exhibited a significant alteration in gut microbial communities in comparison to healthy controls When compared to stable PD group, deteriorated PD cohort demonstrated a significantly differentiated bacterial community. The change in bacterial communities between both cohorts, across 2 years, further correlates with a progressive PD pathology

Continued

Table 3 Changes in the human intestinal microbiota in Parkinson's disease compared to healthy controls.—cont'd

Reference	Study design	Methodology	Major taxa altered in Parkinson's disease	Predictive KEGG functional pathways	Parkinson's disease characteristics	Major findings
Pereira et al. (2017)	Case-Control study Oral: PD (N = 72), Random Healthy Controls (N = 76) Nasal: PD (N = 69), Random Healthy Controls (N = 67)	16S rRNA V3-V4 gene amplicon sequencing Nasal Swab and Oral Swab samples	**Bacteroidetes phylum:** ↓*Prevotella* genus, ↓*Bacteroides fragilis* species **Oral swab microbiota in PD:** **Bacteroidetes phylum:** ↑Prevotellaceae family, ↑*Prevotella* genus, ↓*Capnocytophaga* genus, ↓*unclassified Flavobacteriaceae* OTU **Firmicutes phylum:** ↑Veillonellaceae family, ↑Lactobacillaceae family, ↑Erysipelotrichaceae family, ↑Carnobacteriaceae family, ↑*Veillonella* genus, *Solobacterium* genus, ↑*Moryella* genus, ↓*Gemella* genus, **Fusobacteria phylum:** ↓Leptotrichiaceae family, ↓*Leptotrichia* genus, **Proteobacteria phylum:** ↓Nisseriaceae family, ↓Pasteurellaceae family, ↓Micrococcaceae family, ↓Corynebacteriaceae family, ↓*Kingella* genus, ↓*Haemophilus* genus, ↓*Neisseria* genus, ↓*Granulicatella* genus **Actinobacteria phylum:** ↑Coriobacteriaceae family, ↓*Rothia* genus, ↓*Actinomyces* genus, ↓*Corynebacterium* genus, **Nasal swab microbiota in PD:** **Firmicutes phylum:** ↑*Staphylococcus* OTU **Actinobacteria phylum:** ↑*Marmoricola* genus **Bacteroidetes phylum:** ↑Flavobacteriaceae family	Did Not Analyze	Disease duration: Not reported Total UPDRS score: Oral: 45.9 Nasal: 46.0	Oral swab microbiota composition was significantly altered in PD patients in comparison to healthy controls Within the oral microbiota of PD patients, there were a high abundance of opportunistic pathogens Nasal swab microbiota in PD patients demonstrated differences in abundances of bacterial taxa, however lacked strong significance

| Petrov et al. (2017) | Case-Control study PD (N = 89), Random Healthy Controls (N = 66) | 16S rRNA V3-V4 gene amplicon sequencing Fecal samples | **Firmicutes phylum:** ↑Christensenella genus, ↑Lactobacillus genus, ↑Oscillospira genus, ↑Catabacter genus, ↓Dorea genus, ↓Faecalibacterium genus, ↑Christensenella minuta species, ↑Catabacter hongkongenesis species, ↑Lactobacillus mucosae species, ↑Ruminococcus bromii species, ↑Papillibacter cinnamivorans species, ↓Stoquefichus massiliensis species, ↓Blautia glucerasea species, ↓Dorea longicatena species, ↓Coprococcus eutactus species, ↓Rumino coccuscallidus species **Actinobacteria phylum:** ↑Bifidobacterium genus **Bacteroidetes phylum:** ↓Bacteroides genus, ↓Prevotella genus, ↓Bacteroides massiliensis species, ↓Bacteroides coprocola species, ↓Bacteroides dorei species, ↓Bacteroides plebeus species, ↓Prevotella copri species | Did Not Analyze | Not Reported | Patients with PD demonstrated significantly altered GI microbial composition when compared to healthy controls, causing changes in 9 genera and 15 species |

Continued

Table 3 Changes in the human intestinal microbiota in Parkinson's disease compared to healthy controls.—cont'd

Reference	Study design	Methodology	Major taxa altered in Parkinson's disease	Predictive KEGG functional pathways	Parkinson's disease characteristics	Major findings
Barichella et al. (2019)	Case-Control study PD (N = 193), Random Healthy Controls (N = 113)	16S rRNA V3–V4 gene amplicon sequencing Fecal samples	**PD:** **Verrucomicrobia phylum:** ↑Verrucomicrobiaceae family, ↑*Akkermansia* genus **Proteobacteria phylum:** ↑Enterobacteriaceae family **Firmicutes phylum:** ↑Christensenellaceae family, ↑Lactobacillaceae family, ↓Lachnospiraceae family, ↑*Ruminococcus* genus, ↓*Oscillospira* genus, ↓*Roseburia* genus **Actinobacteria phylum:** ↑Coriobacteriaceae family, ↑Bifidobacteriaceae family **Bacteroidetes phylum:** ↑*Parabacteroides* genus **PD with COMT inhibitors:** **Bacteroidetes phylum:** ↑Porphyromonadaceae family **Proteobacteria phylum** **Firmicutes phylum:** ↑Lactobacillaceae family, ↓Lachnospiraceae family, ↑Ruminococcaceae family ↑**Actinobacteria phylum** **PD with Proton Pump Inhibitors:** **Firmicutes phylum:** ↑Christensenellaceae family, ↑Lactobacillaceae family	**Bacterial metabolic pathways:** 11 upregulated, 15 downregulated **11 KEGG pathways upregulated in PD:** Signal transduction, signaling molecules and interaction, folding, sorting and degradation, infectious diseases, lipid metabolism, xenobiotics biodegradation and metabolism, cellular processing and signaling, genetic information processing, and metabolism **15 KEGG pathways downregulated in PD:** Cell growth and death, replication and repair, translation, infectious disease, amino acid metabolism, energy metabolism, glycan biosynthesis and metabolism, metabolism of cofactors and vitamins	**PD Drug effects:** COMT inhibitors (PD = 29) proton pump inhibitors (HC = 22, PD = 22) Disease duration: early-stage (≤5 years, N = 57), mid-stage (6–10 years, N = 53), advanced (≥11 years, N = 44) Hoehn-Yahr score: 2.0	In comparison to healthy controls, PD patients demonstrated significantly altered GI microbial community, specifically a reduction in Lachnospiraceae and an increase in Lactobacillaecea and Christensenellaceae. Such alterations are associated with increased severity in PD Predictive functional analysis revealed 11 upregulated KEGG pathways and 15 downregulated KEGG pathways in de novo PD patients, when compared to healthy controls. Many of the altered pathways are involved in cellular processes, human diseases, metabolism, and more

Study	Study type	Method	Microbiota findings	Functional analysis	PD Drug effects	Conclusions
Lin et al. (2018)	Case-Control study PD (N = 75), Spousal Healthy Controls (N = 45)	16s rRNA V4 gene amplicon sequencing Fecal samples	**Firmicutes phylum:** ↑Eubacteriaceae family, ↑Aerococcaceae family, ↓Lachnospiraceae family, ↓Streptococcaceae family, ↓Gemellaceae family **Actinobacteria phylum:** ↑Bifidobacteriaceae family, ↑Actinomycetaceae family, ↓Micrococcaceae family, ↑Intrasporangeaceae family, ↑Brevibacteriaceae family **Proteobacteria phylum:** ↑Desulfovibrioaceae family, ↑Pasteurellaceae family, ↑Methylobacteriaceae family, ↓Comamonadaceae family, ↓Halomonadaceae family, ↓Hyphomonadaceae family, ↓Brucellaceae family, ↓Xanthomonadaceae family, ↓Sphingomonadaceae family, ↓Idiomarinaceae family **Euryarchaeota phylum:** ↓Methanobacteriaceae family	Did Not Analyze	**PD Drug effects:** COMT inhibitors (PD, N = 4), Carbidopa/levodopa (PD, N = 55) Disease duration: PD < 5 years (N = 44) PD ≥ 5 years (N = 30) early-onset PD (age < 50 years, N = 23) late-onset PD (age ≥ 50 years, N = 51) Hoehn-Yahr score: Not reported	Patients with PD exhibited significant altered abundance of GI microbial community in comparison to healthy controls
Qian et al. (2018)	Case-Control study PD (N = 45), Spousal Healthy Controls (N = 45)	16s rRNA V3-V4 gene amplicon sequencing Fecal Samples	**PD:** **Proteobacteria phylum:** ↑Sphingomonas genus, ↑Aquabacterium genus **Firmicutes phylum:** ↑Clostridium IV genus, ↑Clostridium XVIII genus, ↑Butyricicoccus genus, ↑Holdemania genus, ↑Anaerotruncus genus, ↓Lactobacillus genus **Bacteroidetes phylum:** ↓Sediminibacterium genus **PD with Levodopa Equivalent Doses:** **Firmicutes phylum:** ↓Dorea genus, ↓Phascolarctobacterium genus	27KO's were significantly different in abundances between PD patients and healthy controls **Bacterial metabolic pathways:** 4 upregulated, 3 downregulated **4 KEGG pathways upregulated in PD:** energy metabolism, fatty acid biosynthesis, flavone and flavonal biosynthesis, and apoptotic pathways **3 KEGG pathways downregulated in PD:** Porphyrin and chlorophyll metabolism, metabolism of cofactors and vitamins, and biotin metabolism	**PD Drug effects:** Dopamine agonists, COMT inhibitors, MAO-B inhibitors, Anticholinergic, Amantadine, LED Disease Duration: 5.7 years Hoehn-Yahr score: 2.2	Patients with PD exhibited significant altered GI microbial composition when compared to healthy controls Further disease duration was associated with a reduction in Esherichia/Shigella genus Predictive functional analysis demonstrated pathways involving metabolism of cofactors and vitamins, porphyrin and chlorophyll metabolism, and biotin metabolism were less abundant in PD fecal samples. However, pathways involving energy metabolism, flavone and flavonal biosynthesis, fatty acid biosynthesis and apoptosis were more abundant in PD fecal samples

Continued

Table 3 Changes in the human intestinal microbiota in Parkinson's disease compared to healthy controls.—cont'd

Reference	Study design	Methodology	Major taxa altered in Parkinson's disease	Predictive KEGG functional pathways	Parkinson's disease characteristics	Major findings
Aho et al. (2019)	Longitudinal study (follow up (2–2.5 year) Scheperjans et al., 2015a, 2015b) PD (N =64), Random Healthy Controls (N =64)	16S rRNA V3–V4 gene amplicon sequencing Fecal samples	**Baseline PD:** **Bacteroidetes phylum:** ↑Rikenellaceae family, ↑OTU 0300 *Alistipes*, ↑OTU 0098 *Bacteroides* **Firmicutes phylum:** ↑OTU 0513 *Anaerotruncus*, ↓Lachnospiraceae family, ↓*Roseburia* genus, ↓*Blautia* genus, ↓*Clostridium XlVa* genus, ↓Roseburia OTU, ↓Blautia OTU **Verrucomicrobia phylum:** ↓Puniceicoccaceae family **Second Year Follow up in PD:** **Actinobacteria phylum:** ↑Bifidobacteriaceae family, ↑*Bifidobacterium* genus **Firmicutes phylum:** ↑Lactobacillaceae family, ↑*Lactobacillus* genus, ↓*Roseburia* genus, ↓*Clostridium XlVa* genus, ↓*Roseburia* OTU, ↓*Blautia* OTU, ↓Ruminococcus OTU **Bacteroidetes phylum:** ↓Prevotellaceae family, ↑OTU 0379 *Alstipes*, ↑ OTU 0464 *Lactobacillus*, ↓*Prevotella* genus, ↓Bacteroides OTU **Verrucomicrobia phylum:** ↓Puniceicoccaceae family	Did Not Analyze	Disease duration: Not reported Hoehn-Yahr score: Baseline: 2.5 Follow-up 2.5	Gut microbial compositions exhibited significant alterations in PD patients when compared to healthy controls A two-year follow-up demonstrated a significant alteration in microbial communities in PD patients in comparison to healthy controls

Reference	Study design	Method	Altered taxa	Analysis	Disease/drug data	Main findings
Li et al. (2019)	Case-Control study PD (N=51) Healthy Controls (N=48) Spousal Healthy Controls (N=39) Random Healthy Controls (N=9)	16S rRNA V4 gene amplicon sequencing Fecal samples	**Verrucomicrobia phylum**, ↑Verrucomicrobiaceae family, ↑Verrucomicrobiales order, ↑Akkermansia genus **Firmicutes phylum**: ↑Clostridia class, ↑Negativicutes order, ↑Clostridiales order, ↑Selenomonadales order, ↑Ruminococcaceae family, ↑Veillonellaceae family, ↑Acidaminococcaceae family, ↑Ruminococcus genus, ↑Eubacterium coprostanoligenes group, ↑Coprococcus genus, ↑Phascolarctobacterium genus, ↑Roseburia genus, ↑Lachnospiraceae family, ↑Ruminococcus callidus species, ↑Roseburia inulinivorans species, ↑Ruminococcus torques species, ↓Lactobacillus genus, ↓Bacilli class, ↓Lactobacillales order, ↓Lactobacillaceae family, ↓Streptococcus genus, ↓Eubacterium hallii species **Eurarchaeota phylum**: ↑Methanobrevibacter genus, ↑Methanobrevibacter smithii species **Bacteroidetes phylum**: ↑Porphyromonadaceae family, ↑Rikenellaceae family, ↑Parabacteroides genus, ↑Alistipes genus, ↑Parabacteroides merdae species, ↑Prevotella copri species	Did Not Analyze	Disease duration: 4.55 years Hoehn-Yahr score: 1.70	PD patients showed significant altered gut microbial taxa in comparison to healthy controls PD patients also demonstrated a decrease in species richness and phylogenetic diversity in comparison to healthy controls
Hopfner et al. (2017)	Case-Control study PD (N=29), Random Healthy Controls (N=29)	16S rRNA V1-V2 gene amplicon sequencing Fecal samples	**Firmicutes phylum**: ↑Lactobacillaceae family, ↑Enterococcaceae family **Bacteroidetes phylum**: ↑Barnesiellaceae family	Did Not Analyze	**PD Drug effects**: MAO-B inhibitor (N=4) Dopamine agonist (N=16) Levodopa (N=20) COMT inhibitor (N=4) Disease duration: 11.2 years Hoehn-Yahr score: Not reported	In comparison to healthy controls, PD participants were found to have significantly altered GI microbial community, in specifically a high abundance of Lactobacillaceae family, Enterococcaceae family, and Barnesiellaceae family

Hill-Burns et al., 2017 reported 17 upregulated pathways and 9 downregulated pathways, including xenobiotics degradation and metabolism of plant-derived compounds in PD subjects. Barichella et al. (2019) found 11 upregulated pathways and 15 downregulated pathways in de novo PD subjects, compared to healthy controls. The majority of these pathways are involved in cellular processes, human diseases and metabolism. Qian et al. (2018) predictive functional analysis indicated four metabolic pathways upregulated and three pathways downregulated. Finally, Bedarf et al. (2017) used the detailed metagenomics shotgun analysis and showed differences in microbiota metabolism in PD subjects involving the β-glucuronate and tryptophan metabolism.

Nevertheless, it is worth highlighting that several studies comparing PD vs. healthy control subjects reported conflicting results that we summarized in Table 4. These discrepancies may be due to either methodological differences, subject inclusion/exclusion criteria, PD medications, disease duration, constipation, geography, diet, or to the lack of adjustment of the multiple confounders that may greatly influence gut microbiota composition.

6.2.2 Longitudinal studies

Longitudinal studies assessing the progression of PD motor and nonmotor features according to baseline microbiota composition are essential to investigate the causative link between gut dysbiosis and the neurodegenerative process as well as any putative prognostic effect on the evolution of clinical features. The striking finding is that patients with PD have abnormal intestinal microbiota communities (dysbiotic microbiota) regardless of where they live and the microbiota community remains abnormal even after 2 years of follow up (Aho et al., 2019; Minato et al., 2017). Baseline and follow-up features of the two longitudinal studies have been summarized in Table 3.

Minato et al. (2017) subdivided 36 PD patients in two groups according to stability vs. progression of clinical features at follow-up and found that the deteriorated group had lower genus *Bifidobacterium* counts than the stable group at baseline, suggesting that lower counts of *Bifidobacterium* at baseline may be predictive of progression of PD in 2 years. In addition, the count of *Bifidobacterium*, the species *Bacteroides fragilis* group negatively correlated with changes of nonmotor PD features as assessed by the UPDRS part I scores. According to this study, it could be speculated that a low count of *Bifidobacterium* may have a negative prognostic effects on the progression of PD symptoms and, on the other hand, *Bifidobacterium* may be protective against progression of PD, including cognitive functions (Minato et al., 2017).

Recently, Aho et al. (2019) reported a 2-year follow-up study of their previous study (Scheperjans et al., 2015a, 2015b) and used three different tools to look for taxa that differ between PD patients and controls. Regarding disease progression, PD patients who progressed were more likely to have a Firmicutes-dominated enterotype than stable patients or control subjects. In addition, *Prevotella* was less

Table 4 Consistent versus inconsistent findings in studies on gut microbiota in PD populations compared to healthy control subjects.

Phylum	Family	Genus	PD versus HC	
			Increased	Reduced
Actinobacteria	Bifidobacteriaceae	Bifidobacterium	Lin et al. (2018) Unger et al. (2016), Hill-Burns et al. (2017), Barichella et al. (2019), Petrov et al. (2017)	
	Coriobacteriaceae	Olsenella	Barichella et al. (2019), Li et al. (2017)	
	Coriobacteriaceae		Heintz-Buschart et al. (2018)	Keshavarzian et al. (2015)
	Micrococcaceae			Lin et al. (2018)
	Actinomycetaceae			Lin et al. (2018)
	Intrasporangeaceae			Lin et al. (2018)
	Brevibacteriaceae			Lin et al. (2018)
Bacteroidetes	Prevotellaceae	Prevotella	Heintz-Buschart et al. (2018)	Unger et al. (2016)
		Prevotella (copri)	Heintz-Buschart et al. (2018), Hill-Burns et al. (2017), Li et al. (2019)	Unger et al. (2016), Li et al. (2017), Scheperjans et al. (2015b) Petrov et al. (2017)
	Rikenellaceae	Alistipes (shahii)	Li et al. (2019)	
	Parabacteroides	Parabacteroides	Bedarf et al. (2017), Li et al. (2019)	Bedarf et al. (2017), Petrov et al. (2017)
		Parabacteroides (merdae)	Barichella et al. (2019), Li et al. (2019)	
	Bacteroidaceae	Bacteroides	Li et al. (2019)	
		Bacteroides (massiliensi, coprocola, dorei, plebus)	Keshavarzian et al. (2015), Heintz-Buschart et al. (2018), Aho et al. (2019)	Hasegawa et al. (2015), Petrov et al. (2017) Petrov et al. (2017)
	Odoribacteraceae	Butyricimonas	Heintz-Buschart et al. (2018)	
	Chitinophagaceae	Sediminibacterium	Hopfner et al. (2017)	
	Barnesiellaceae			Qian et al. (2018)

Continued

Table 4 Consistent versus inconsistent findings in studies on gut microbiota in PD populations compared to healthy control subjects.—cont'd

Phylum	Family	Genus	PD versus HC Increased	PD versus HC Reduced
Firmicutes	Porphyromonadaceae		Li et al. (2019)	Keshavarzian et al. (2015)
	Gemellaceae			Lin et al. (2018)
	Enterococcaceae	Flavonifractor	Heintz-Buschart et al. (2018)	Unger et al. (2016)
		Enterococcus	Li et al. (2017), Hopfner et al. (2017)	
	Lactobacillaceae		Li et al. (2017)	Unger et al. (2016), Li et al. (2019)
		Lactobacillus	Scheperjans et al. (2015b), Hill-Burns et al. (2017), Hopfner et al. (2017), Barichella et al. (2019)	Qian et al. (2018), Li et al. (2019)
		Lactobacillus (mucosae)	Hasegawa et al. (2015), Hill – Burns et al. (2017)	
	Lachnospiraceae		Petrov et al. (2017)	Hill-Burns et al. (2017), Keshavarzian et al. (2015), Li et al. (2017), Barichella et al. (2019), Lin et al. (2018), Aho et al. (2019)
		Roseburia	Heintz-Buschart et al. (2018), Li et al. (2019)	Barichella et al. (2019),
		Roseburia (inulinivorans)	Li et al., 2019	Keshavarzian et al. (2015), Hill-Burns et al. (2017), Aho et al. (2019)
		Dorea	Li et al. (2019)	Keshavarzian et al. (2015), Petrov et al. (2017), Qian et al. (2018)
		Dorea (longicatena)		Petrov et al. (2017)
		Blautia		Hill-Burns et al. (2017), Li et al. (2017), Aho et al. (2019)
		Blautia (glucerasea)		Petrov et al. (2017)
	Ruminococcaceae	Faecalibacterium	Scheperjans et al., 2015a, 2015b Heintz-Buschart et al. (2018), Li et al. (2019)	Barichella et al. (2019)
		Faecalibacterium (prausnitzii)		Keshavarzian et al. (2015), Hill-Burns et al. (2017), Li et al. (2017), Petrov et al. (2017), Unger et al. (2016)

Family	Genus (species)	References	References
	Ruminococcus	Barichella et al. (2019), Li et al. (2019)	Li et al. (2017) Petrov et al. (2017)
	Ruminococcus (callidus)		
	Ruminococcus (torques)	Li et al. (2019)	
	Ruminococcus (callidus)	Li et al. (2019)	
	Oscillospira	Keshavarzian et al. (2015), Petrov et al. (2017), Barichella et al. (2019)	
Erysipelotrichaceae	Eubacterium (biforme)	Li et al. (2017), Pereira et al. (2017)	Bedarf et al. (2017) Petrov et al. (2017)
	Stoquefichus (massiliensis)		
	Holdemania	Qian et al. (2018)	
Clostridiaceae	Clostridium (saccharolyticum)	Keshavarzian et al. (2015), Heintz-Buschart et al. (2018)	Bedarf et al. (2017)
	Clostridium (coccoides)		Hasegawa et al. (2015)
	Anaerotruncus	Heintz-Buschart et al. (2018), Qian et al. (2018), Aho et al. (2019)	
	Clostridium (IV)	Qian et al. (2018)	
	Clostridium (XVIII)	Qian et al. (2018)	
	Clostridium (XIV)	Qian et al. (2018)	Aho et al. (2019)
	Butyricicoccus	Scheperjans et al. (2015b)	
Clostridiales incertae sedis IV			
Christensenellaceae	Christensenella (minuta)	Barichella et al. (2019), Hill-Burns et al. (2017), Petrov et al. (2017)	
Coprobacillaceae	Coprococcus	Li et al. (2019)	Keshavarzian et al. (2015)
	Coprococcus (eutactus)		Keshavarzian et al. (2015), Hill-Burns et al. (2017), Petrov et al. (2017)
Veillonellaceae	Megasphaera	Li et al. (2017), Li et al. (2019)	
	Megamonas	Heintz-Buschart et al. (2018), Li et al. (2017)	Petrov et al. (2017)
		Li et al. (2017)	

Continued

Table 4 Consistent versus inconsistent findings in studies on gut microbiota in PD populations compared to healthy control subjects.—cont'd

Phylum	Family	Genus	PD versus HC	
			Increased	Reduced
	Acidamminococcaceae		Li et al. (2019)	
		Mitsuokella	Heintz-Buschart et al. (2018)	
		Acidaminococcus	Li et al. (2017)	
		Papillibacter (cinnamivorans)	Petrov et al. (2017)	
		Phascolarctobacterium	Li et al. (2019)	Qian et al. (2018)
	Tissirellaceae		Hill-Burns et al. (2017)	Lin et al. (2018)
	Streptococcaceae	Streptococcus	Li et al. (2017)	Li et al. (2019)
	Catabacteriaceae	Catabacter (hongkongenesis)	Petrov et al. (2017)	
	Eubacteriaceae	Eubacterium (alii)	Lin et al. (2018)	Li et al. (2019)
	Aerococcaceae		Lin et al. (2018) Keshavarzian et al. (2015)	
Proteobacteria	Enterobacteriaceae		Unger et al. (2016), Li et al. (2017), Barichella et al. (2019)	
		Escherichia Coli	Forsyth et al. (2011), Li et al. (2017)	
		Preteus	Li et al. (2017)	
	Ralstoniaceae	Ralstonia	Keshavarzian et al. (2015)	
	Bradyrhizobiaceae		Scheperjans et al. (2015b)	
	Sutterellaceae	Sutterella	Heintz-Buschart et al. (2018)	
	Pasteurellaceae		Li et al. (2017)	
	Moraxellaceae	Acinetobacter	Li et al. (2017)	Hill-Burns et al. (2017)
	Pasteurellaceae			
	Desulfovibrioaceae		Lin et al. (2018)	Lin et al. (2018)
	Methylobacteriaceae			Lin et al. (2018)
	Comamonadaceae			Lin et al. (2018)

Phylum	Family	Genus	Reference	Reference
	Halomonadaceae	Aquabacterium	Qian et al. (2018)	Lin et al. (2018)
	Hyphomonadaceae			Lin et al. (2018)
	Brucellaceae			Lin et al. (2018)
	Xanthomonadaceae			Lin et al. (2018)
	Sphingomonadaceae			Lin et al. (2018)
	Idiomarinaceae	Sphingomonas	Qian et al. (2018)	Lin et al. (2018)
Euryarchaeota	Methanobacteriaceae	Methanobrevibacter	Li et al. (2019)	Lin et al., 2018
		Methanobrevibacter (smithii)	Li et al. (2019)	
Verrucomicrobia	Puniceicoccaceae		Scheperjans et al. (2015a, 2015b), Heintz-Buschart et al. (2018), Barichella et al. (2019), Li et al. (2019)	
	Verrucomicrobiaceae	Akkermansia (muciniphila)	Keshavarzian et al. (2015), Unger et al. (2016), Bedarf et al. (2017), Hill-Burns et al. (2017), Heintz-Buschart et al. (2018), Barichella et al. (2019), Li et al. (2019)	Aho et al. (2019)

abundant in patients with faster disease progression. Aho et al. (2019) did not replicate the findings reported by Minato et al. (2017), especially those concerning the prognostic role of *Bifidobacterium* and *Bacteroides fragilis*. These discrepancies might be due to either methodological differences or the lack of a strong microbial signal associated with PD progression. Further studies in larger PD cohort and longer follow-up are needed to ascertain the effects of gut microbiota on PD progression.

6.2.3 Microbiota in subjects at high risk for PD (iRBD) and de novo PD patients

6.2.3.1 Idiopathic rapid eye movement sleep behavior disorder (iRBD)

The question whether the observed microbiota changes in PD are specific in only PD and no other neurodegenerative disease was further answered by four other investigators who evaluated atypical Parkinsonism or iRBD in prodromal PD (Table 5). Heintz-Buschart et al. (2018) compared iRBD PD patients to healthy controls testing both nasal wash and fecal samples. Nasal wash microbial profiles did not differ between study groups. Fecal samples of iRBD PD patients showed 41 OTUs that were differentially abundant, of which over 75% indicated a similar microbiota change as found in their corresponding PD cohort compared to healthy controls. Specifically, common patterns in PD and iRBD were observed for *Anaerotruncus spp.*, *Clostridium XIVb* and several *Bacteroidetes* (Heintz-Buschart et al., 2018). In this study, the authors reported neither significant changes in butyrate producing bacteria nor differences in any butyrate-producing enzymes (Heintz-Buschart et al., 2018). Interestingly, this study showed that the decrease in the relative abundance of genus *Prevotella* preceded the motor phase in PD patients suffering from premotor iRBD. Thus, dysbiosis is not limited in PD, but appears to be present in other neurodegenerative disorders.

6.2.3.2 De novo PD patients

Considering the effects of medications and the heterogeneity of disease progression, the most suitable population of PD patients to assess whether the changes in gut microbiota are associated to the disease or to disease- or drug-related confounders is de novo untreated PD subjects.

Barichella et al. (2019) compared gut microbiota of 39 drug naïve PD patients with a group of 113 healthy control subjects, adjusting for multiple potential confounders, including type of delivery and type of lactation, dietary habits, lifestyle habits (smoking and coffee), and constipation. They found lower abundance of the Lachnospiraceae family (and its genus *Roseburia*). Compared to PD patients on medication at different disease stages, the lower abundance in Lachnospiraceae remained significant, along with a significant trend effect for disease duration on increasing levels of Lactobacillaceae family (and *Lactobacillus* genus) and the coabundant genus of *Akkermansia* (as well as of the phylum Verrucomicrobia and the family Verrucomicrobiaceae). Considering that Lachnospiraceae are involved in the production of SCFAs (butyrate, propionate, and acetate), which are an important energy source for intestinal epithelial cells, their reduction in de novo PD may

Table 5 Changes in the human intestinal microbiota in atypical Parkinsonism or prodromal Parkinson's disease compared to healthy controls.

Reference	Study design	Methodology	Major taxa altered in Parkinsonism	Predictive KEGG functional pathways	Parkinsonism characteristics	Major finding
Engen et al. (2017)	Case-Control study MSA (N = 6); Healthy Controls (N = 11)	16S rRNA V4 gene amplicon sequencing Fecal and Colonic Sigmoid Mucosa samples	**Multiple System Atrophy (MSA) Feces:** ↑**Bacteroidetes phylum**, ↑Rikenellaceae family ↓**Firmicutes phylum**, ↓Coprobacillaceae family, ↓Clostridiaceae family, ↓Blautia genus, ↓Dorea genus *↑**Firmicutes-to Bacteroidetes Ratio significantly lower in MSA Feces** *SCFA butyrate-producing taxa trended lower in MSA Feces **Multiple System Atrophy (MSA) Sigmoid Mucosa:** **Firmicutes phylum:** ↓Coprobacillaceae family **Proteobacteria phylum:** ↑Oxalobacteraceae family, ↑Ralstonia genus **Bacteroidetes phylum:** ↑Porphyromonadaceae family	**Bacterial metabolic pathways** Feces: 3 upregulated, 16 downregulated Mucosa: 1 upregulated, 2 downregulated **KEGG pathways upregulated in MSA:** Lipopolysaccharide biosynthesis (feces & mucosa), Lipopolysaccharide biosynthesis proteins (feces), Ubiquinone and other terpenoid-quinone biosynthesis (feces) 33 genes were annotated to the LPS-Biosynthesis pathways, of which the relative abundance of 12 KOs significantly increased in MSA feces; 9 KOs significantly increased in MSA sigmoid mucosa **KEGG pathways downregulated in MSA:** Phosphotransferase system (feces & mucosa), sporulation (feces & mucosa), ABC transporters, transporters, cytoskeleton proteins, lysine biosynthesis, novobiocin biosynthesis, nicotinate and nicotinamide metabolism, benzoate degradation, DNA replication proteins, pyrimidine metabolism, fatty acid metabolism, homologous recombination, translation factors, drug metabolism-other enzymes, and energy metabolism	Disease Duration MSA: 5.5 years Hoehn-Yahr MSA: 3.4	MSA subjects suggested high relative abundance of Gram-negative, putative proinflammatory bacteria from the phylum Bacteroidetes and Proteobacteria, in both feces and sigmoid mucosa. The relative abundance of putative anti-inflammatory SCFA butyrate-producing bacteria, from the families Lachnospiraceae and Ruminococcaceae, were less abundant in MSA feces Predictive functional analysis indicated that the relative abundance of a number of genes involved in metabolism were lower in MSA feces, whereas the relative abundance of genes involved in lipopolysaccharide biosynthesis were higher in both MSA feces and mucosa compared to healthy controls
Tan et al. (2018)	Case-Control study MSA (N = 17); Age matched Spouse/Sibling (N = 17)	16S rRNA V3-V4 gene amplicon sequencing Fecal samples	**Multiple System Atrophy (MSA)** **Bacteroidetes phylum:** ↓Paraprevotella clara OTU, ↓Paraprevotella genus, ↓Bacteroides genus ↑Bacteroides genus	No Significant Differences	Disease Duration MSA: 3.7 years Hoehn-Yahr MSA: 4.5	Significantly altered fecal microbiome and metabolome in MSA compared to household controls

Continued

Table 5 Changes in the human intestinal microbiota in atypical Parkinsonism or prodromal Parkinson's disease compared to healthy controls.—cont'd

Reference	Study design	Methodology	Major taxa altered in Parkinsonism	Predictive KEGG functional pathways	Parkinsonism characteristics	Major finding
Barichella et al. (2019)	Case-Control study MSA (N = 22); PSP (N = 22); Random Healthy Controls (N = 113)	16S rRNA V3-V4 gene amplicon sequencing Fecal samples	*Multiple System Atrophy (MSA):* **↑Verrucomicrobia phylum:** ↑Verrucomicrobiaceae family, ↑*Akkermansia* genus **Firmicutes phylum:** ↑Lactobacillaceae family, ↑*Christensenellaceae Unclassified* genus, ↑*Ruminococcaceae Other* genus, ↓*Faecalibacterium* genus **Bacteroidetes phylum:** ↓ Prevotellaceae family, ↑*Parabacteroides* genus *Progressive Supranuclear Palsy (PSP):* **↑Verrucomicrobia phylum,** ↑Verrucomicrobiaceae family, ↑*Akkermansia* genus **Firmicutes phylum:** ↑Christensenellaceae family, ↑*Christensenellaceae Unclassified* genus, ↑*Oscillospira* genus, ↓*Ruminococcaceae Other* genus, ↓*Roseburia* genus, ↓*Lachnospiraceae Unclassified* genus, ↓Lachnospiraceae family, ↓*Streptococcus* genus **Bacteroidetes phylum:** ↑*Parabacteroides* genus	Did Not Analyze	Disease Duration MSA: 5.8 years Hoehn-Yahr MSA: 3.1 Disease Duration PSP: 6.7 years Hoehn-Yahr PSP: 3.2	When compared with HC subjects, both MSA and PSP subjects had similar bacterial taxa changes to PD subjects, with a few exceptions: MSA family Lachnospiraceae and genus *Roseburia* relative abundances were not lower, and family Prevotellaceae was significantly reduced compared to HC PSP family Lactobacillaceae was similar, and family Streptococcaceae and genus *Streptococcus* relative abundances were reduced, compared to HC
Heintz-Buschart et al. (2018)	Case-Control study iRBD (N = 21); Healthy Controls (N = 78)	16S and 18s rRNA V4 gene amplicon sequencing V3 whole metagenome shotgun sequencing Nasal Wash and Fecal samples	*Idiopathic Rapid Eye Movement Sleep Behavior Disorder (iRBD)* **Bacteroidetes phylum:** ↓Prevotellaceae OTU, ↓*Prevotella* genus, ↓*Bacteroides* OTU **Actinobacteria phylum:** ↓*Olsenella* OTU **Firmicutes phylum:** ↓*Ruminococcaceae* OTU, ↑*Flavonifactor* OTU, ↑*Clostridium XlVb* genus, ↑*Anaerotruncus* genus **Proteobacteria phylum:** ↓*Sutterella* OTU, ↓*α-proteobacterium* OTU	Whole Metagenome Sequencing of Selected Samples (N = 2; HC and iRBD) found OTUs 171 & 469 to be highly abundant OTU469 was classified as *α-proteobacterium*. Function indicated fermentative lifestyle, auxotrophy for most vitamins, and motility OTU171 was classified as cyanobacterium *Melainabacterium MelB1*. Function indicated fermentative metabolism, flagellation, and B vitamin synthesis	Disease Duration PD: 6.0 years Hoehn-Yahr PD: 2.14 RBD by PSG Number PD: 40 RBD by PSG Number iRBD: 21	Compared to HC, 41 operational taxonomic units (OTUs) in feces were differentially abundant in iRBD of which over 75% indicated a similar change, as was discovered in their PD cohort compared to HC Data suggests that a decreased abundance of *Prevotella* genus precedes the motor phase in PD subjects suffering from premotor RBD No nasal wash OTUs differed significantly between study groups

support the hypothesis of a defective integrity of the local epithelial barrier and immune activation (Furusawa et al., 2013; Sampson, 2019; Segain et al., 2000). The findings by Barichella et al. (2019) are in contrast with the study by Bedarf et al. (2017) in early-stage levodopa-naïve patients, reporting an increase in Verrucomicrobiaceae (genus *Akkermansia*), and a reduction in *Prevotellaceae* family. The inconsistency could be explained either by different methodological approaches or by a weak effect. Therefore, further study in larger cohort of de novo PD patients are needed.

6.2.3.3 Functional pathways involved in early de novo PD

Bedarf et al. (2017) observed a putative reduction in microbiota β-glucuronidase activity and a trend toward an increased tryptophan degradation in early stage PD participants. On the one hand, decreased β-glucuronidation has been hypothesized to influence drug availability (still to be determined experimentally though). The trend toward an increased tryptophan degradation gene copy number in PD might be associated with a decrease of L-tryptophan (the precursor for serotonin) within PD patients' brains. According to the Kyoto Encyclopedia of Genes and Genomes (KEGG), Barichella et al. (2019) found that de novo PD compared to healthy controls could present differences in pathways related to metabolism regarded amino acids, energy, glycans, lipids, cofactors and vitamins, and xenobiotics. The functional changes derived from the gut microbiome profile need to be directly assessed with specific studies assessing metabolomics.

6.2.4 Atypical primary parkinsonism (MSA and PSP)

Engen et al. (2017) were the first to show that patients with multiple system atrophy (MSA) compared to healthy controls, had increased relative abundance of gram-negative, putative pro-inflammatory bacteria associated to the phyla *Proteobacteria* and *Bacteroidetes*, in both fecal and sigmoid mucosa samples (Table 5). They further showed the relative abundance of putative anti-inflammatory SCFA butyrate-producing bacteria was diminished in feces of MSA patients. Interestingly, the predictive functional gene content profiles of MSA patients indicated that the relative abundance of a number of genes involved in metabolism were lower in MSA feces microbiota, whereas the relative abundance of microbiota genes involved in LPS biosynthesis were higher in both MSA feces and mucosal microbiota samples. This finding was similar to the inferred functional outcomes found in PD subjects by Keshavarzian et al. (2015).

Tan et al. (2018) also found significantly altered pro-inflammatory dysbiotic fecal microbiota in MSA patients compared to household controls. Barichella et al. (2019) analyzed both MSA and Progressive Supranuclear Palsy (PSP) patients, compared to healthy controls and found that both MSA and PSP patients had similar bacterial profile alterations compared to PD subjects. Compared to healthy controls, the MSA patients had significantly reduced relative abundance of family *Prevotellaceae*, whereas the PSP patients had lower relative abundance of both family *Streptococcaceae* and genus *Streptococcus*. They also noted increased and comparable levels of Lactobacillaceae in MSA and PSP patients, respectively. On the other hand, considering the analysis of PD patients and the important role of disease duration and severity on *Lactobacillaceae* (Barichella et al., 2019), it could not be excluded that some of the observed differences in MSA and/or PSP might be a consequence rather than a cause of the diseases.

7 Are gut microbiota changes a cause or a consequence of Parkinson's disease?

The causal link between dysbiotic microbiota and the development of PD is yet to be established. The debate is whether these changes in microbiota community structure and composition in PD is the trigger for PD, or are a consequence of PD. Indeed, several studies have shown a correlation between changes in microbiota and duration of the disease and dysbiosis is more pronounced in those with longer duration of PD (Hasegawa et al., 2015; Keshavarzian et al., 2015; Li et al., 2017; Minato et al., 2017). This is not surprising because PD patients commonly change their life habit to better cope with their symptoms and this life style change can impact microbiota composition. For example, GI symptoms are common in PD patients (Fasano et al., 2015; Pfeiffer, 2011) and thus they typically change diet that could affect their microbiota. Although several studies did not find a correlation between diet and dysbiosis in PD patients, multiple studies have shown the impact of diet on microbiota and dietary habit of PD patients like increased consumption of dairy products and low fiber/high-refined carbohydrate diet that can affect microbiota community (Cassani et al., 2017; Hughes et al., 2017; Perez-Pardo et al., 2017a, 2017b; Zhou et al., 2019).

Constipation is very common in PD patients and typically occurs years before onset of CNS symptoms (Abbott et al., 2001; Fasano et al., 2015; Knudsen et al., 2017) and constipation can impact the microbiota community (Khalif et al., 2005). However, dysbiosis was also found to occur in those PD patients who did not suffer from constipation (Keshavarzian et al., 2015).

Patients with PD have poor sleep and reversal of sleep/wake cycles that can cause disruption of circadian rhythms (Videnovic and Golombek, 2013; Videnovic and Willis, 2016) and both disrupted sleep and circadian disruption can cause dysbiotic microbiota in both humans and rodents (Paschos and FitzGerald, 2017; Voigt et al., 2016). Additionally, PD medication correlates with dysbiosis (Heintz-Buschart et al., 2018; Hill-Burns et al., 2017; Qian et al., 2018). However, dysbiosis was still present in early onset and de novo PD patients never exposed to any PD medication (Barichella et al., 2019; Bedarf et al., 2017; Keshavarzian et al., 2015). More importantly, dysbiosis has been reported in patients with iRBD (i.e., prodromal PD) (Heintz-Buschart et al., 2018). Thus, even though life style changes from PD symptoms and PD medication contribute to changes in microbiota composition, it does not appear to explain the observed dysbiosis in PD patients. These findings support that abnormal microbiota composition plays a critical role in the pathogenesis of PD (as a trigger or enabler) and thus a contributor for development of symptomatic PD.

7.1 Relationship between gut microbiota and PD clinical features

7.1.1 Motor features

To date there are inconsistent findings (Table 4). On the one hand, Scheperjans et al. (2015a, 2015b) reported a greater abundance of *Enterobacteriaceae* in PD with a postural instability and gait difficulty (PIGD) phenotype compared to tremor-dominant PD patients. However, Unger et al. (2016) did not confirm this finding, as he found no significant difference in the abundance of *Enterobacteriaceae* according to PD phenotype. More recently, Lin et al. (2018) have found that the genus *Roseburia* (*Lachnospiraceae* family) was significantly more abundant in patients with a non-tremor dominant phenotype. On the other hand, Bedarf et al. (2017) found no significant taxonomic associations between relative microbiota abundance and clinical data, although some interesting trends were reported between the severity of motor symptoms PD (assessed by the UPDRS III) and three different Eubacteria strains (*E. eligens*, *E. rectale*, and *E. hallii*). A positive association between total UPDRS score and abundance of *Enterococcus* and *Escherichia-Shigella* which potentially reduce the production of SCFAs and produce more endotoxins and neurotoxins– and a negative correlation with putative cellulose degraders (e.g., *Faecalibacterium* and *Ruminococcus*) have been reported in a Chinese population (Li et al., 2017). A significant correlation between motor symptoms (assessed by MDS-UPDRS part III) and the relative abundance of *Anaerotruncus* spp., *Clostridium XIVa*, and Lachnospiraceae was reported by Heintz-Buschart et al. (2018). Barichella et al. (2019) found a significant relationship between increased *Lactobacillaceae* and UPDRS part III score as a whole and at a separate analysis of nondopaminergic features and postural instability, while reduced *Lachnospiraceae* negatively correlated with gait disturbances (Barichella et al., 2019). The relationship between *Lachnospiraceae* and PD motor features is consistent with a previous study (Heintz-Buschart et al., 2018).

7.1.2 Non-motor features

As abovementioned, specific microbiota enterotypes have been associated with depression and reduced quality of life in non-PD populations (Valles-Colomer et al., 2019). In consistence with this putative relationship between gut microbiota and nonmotor symptoms (including depression), Heintz-Buschart et al. (2018) found a negative association between *Akkermansia* and nonmotor symptom as assessed by the MDS-UPDRS part I. Recently, a worse nonmotor symptom burden was found in association with higher abundance in *Christensenellaceae* in a cohort a 193 PD patients, whose cognitive dysfunctions was correlated with increased levels of *Lactobacillaceae* (Barichella et al., 2019).

7.2 Interaction between gut microbiota and PD medications

Given the important influence of pharmaceuticals on the gut microbiota (Forslund et al., 2015), several cross-sectional studies tested the effects of PD medications on gut microbiota profile.

7.2.1 Levodopa

Increased levels of DA in the brain by administration of levodopa is a cornerstone of current PD treatment (Lewitt, 2008). The current approach is to combine levodopa with a tyrosine decarboxylase inhibitor to protect Levodopa from breakdown by the host decarboxylases. However, the tyrosine decarboxylase inhibitors in the current PD Levodopa regimen (carbidopa and benserazide) are not effective to inhibit bacterial decarboxylases that are also capable of metabolizing Levodopa (van Kessel et al., 2019). Two recent studies have shown that bacteria that are part of the intestinal microbiota community (such as *Enterobacter faecalis*) express tyrosine decarboxylases, which are capable of metabolizing levodopa (Maini Rekdal et al., 2019; van Kessel et al., 2019). The study by van Kessel et al. (2019) also showed that stools from PD patients who require a high dose of levodopa have significantly higher bacterial tyrosine decarboxylase content, that primarily belong to the bacilli family, than those PD patients who required lower doses of levodopa to control their symptoms. Thus, loss of efficiency of Levodopa in subsets of patients could be due to increased abundance of these bacteria leading to breakdown of levodopa in the intestine compromising its absorption and then release of dopamine in the brain. In addition, Maini Rekdal et al. (2019) showed that species *Enterobacter lenta* used another specific reaction (i.e., catechol dihydroxylation) to metabolize the dopamine resulting from Levodopa decarboxylation (Fig. 9). These studies provide compelling evidence of a substantial effect of gut microbiota on the metabolism of endogenous dopamine, which is likely to underlie different features of PD, ranging from gut dismotility to response to levodopa.

Finally, Heintz-Buschart et al. (2018) compared PD patients receiving levodopa, dopamine agonists, catechol-*O*-methyl transferase (COMT) inhibitor, and/or mono-amine oxidase type B (MAO-B) inhibitors versus patients who were treatment-naïve,

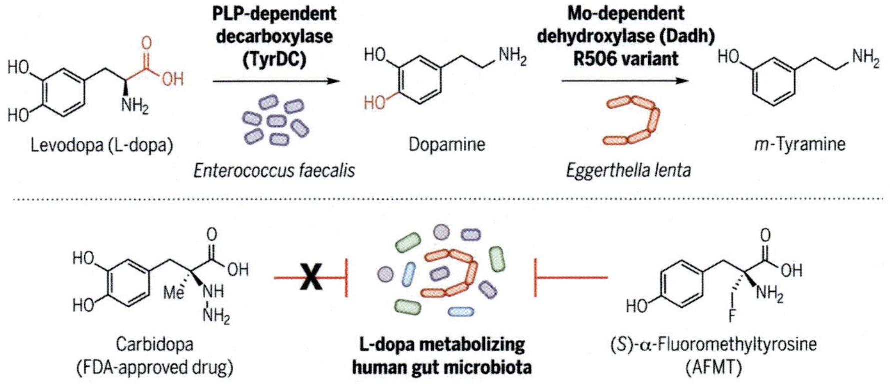

FIG. 9

Gut bacteria metabolize Levodopa.

From Maini Rekdal, V., Bess, E.N., Bisanz, J.E., Turnbaugh, P.J., Balskus, E.P. 2019. Discovery and inhibition of an interspecies gut bacterial pathway for Levodopa metabolism. Science 364(6445), eaau6323.

found that the relative abundance of the family *Bacillaceae* was influenced by the treatment with Levodopa rather than PD. On the other hand, Bedarf et al. (2017) found no significant differences in taxa abundances according to the various combinations of anti-PD medication and acknowledged that this negative finding was likely to be due to the relatively small samples size.

7.2.2 COMT inhibitors

Considering that one of the most common side effects on COMT inhibitors (iCOMT) entacapone and tolcapone is diarrhea, it is not surprising that this class of anti-PD medications is likely to influence gut microbiota composition and thus represent a confounder to be ruled out in the statistical analyses on treated PD patients. Unger et al. (2016) found a strong negative correlation between the use of entacapone and the abundances of *Firmicutes* and *Faecalibacterium prausnitzii*. Entacapone also showed a negative correlation with fecal butyrate concentrations, which is a putative metabolic product of *Faecalibacterium prausnitzii* (Unger et al., 2016). Scheperjans et al. (2015a, 2015b) found a significant association between iCOMT and *Enterobacteriaceae*. Hill-Burns et al. (2017) found significant signals for iCOMT, anticholinergics, and a borderline signal for carbidopa/levodopa. After excluding patients who were on iCOMT or anticholinergics to confirm whether these medications were driving the associations, they found a reduction in the association signal for family *Lachnospiraceae* and genera *Bifidobacterium* and *Blautia* (Hill-Burns et al., 2017). In the study by Barichella et al. (2019), the use of iCOMT influenced the level of several taxa, including *Firmicutes*, *Proteobacteria*, *Lachnospiraceae* and *Ruminococcaceae*, *Actinobacteria*, *Porphyromonadaceae*, and *Lactobacillaceae*. Overall, the analysis of the relationship between the use of iCOMT and gut microbiota yielded discordant findings, except for an overlap on *Firmicutes* (between Unger et al., 2016 and Barichella et al., 2019) and *Lachnospiraceae* (between Hill-Burns et al., 2017 and Barichella et al., 2019).

8 Interventions on gut microbiota as potential disease modifying strategies?

Taken as a whole, the data present in this book chapter supports the notion that gut microbiota-directed interventions could be the novel therapy that may not only improve symptoms and augment the effects of current therapy, but could also modify disease course and even prevent the onset of disease by reducing neuroinflammation.

Taken together, as discussed above, multiple cross-sectional clinical studies have now shown that PD patients exhibit a dysbiotic, "pro-inflammatory" microbiota and that in some cases these changes correlate with disease markers (Lubomski et al., 2019a, 2019b; Scheperjans et al., 2018). Consistent with these data, animal models of PD also support a role for microbiota dysbiosis playing a role in central neuroinflammation and dopamine loss in PD.

Importantly, in some animal models dietary intervention was beneficial (Perez-Pardo et al., 2017a, 2017b) providing a potential rationale for considering microbiota directed interventions for treatment of Parkinson's disease. Multiple studies have shown that diet, probiotics and prebiotics can normalize dysbiotic microbiota (Bailey and Holscher, 2018; Bourassa et al., 2016; Dinan and Cryan, 2017; Noble et al., 2017; Perez-Pardo et al., 2017a, 2017b; Xu and Knight, 2015) and thus could potentially be beneficial in treatment of PD. Dietary modification is an attractive intervention to favor a more beneficial microbiota community. For example, prebiotics (fiber) are capable of favoring growth of SCFA-producing bacteria, increasing SCFA and in particular the butyrate/SCFA ratio in the colon (Koh et al., 2016; Makki et al., 2018). If prebiotics are given to patients with PD, increased SCFA could modify disease course by preventing and/or treating intestinal leakiness to endotoxin and colonic inflammation and promoting GLP-1 and vagal stimulation of brain BDNF resulting in decreased neuroinflammation and less DA neuron loss (Everard and Cani, 2014; Kim et al., 2017; Matt et al., 2018; Park et al., 2011; Yadav et al., 2013; Zhao et al., 2018). However, well designed clinical trials in patients with Parkinson's disease are required to determine whether microbiota-directed interventions can be beneficial in patients with PD.

Finally, another potential use of microbiota directed intervention is in management of patients with PD to optimize efficacy of current PD medications (Lubomski et al., 2019a, 2019b). As above mentioned (see Section 7.2), increased abundance of a few bacteria within the gut microbiota are capable of metabolizing levodopa and may account for impaired Levodopa bioavailability in subsets of PD patients (Maini Rekdal et al., 2019; van Kessel et al., 2019). Furthermore, infection by HP and/or SIBO has been shown to interfere with levodopa pharmacokinetics and increase delayed-on and no-on periods (Fasano et al., 2015).

These data provide a compelling rationale for considering microbiota-directed intervention to optimize Levodopa therapy in those who require high and frequent doses of levodopa and suffer from levodopa's life altering side effects (Fasano et al., 2015; van Kessel et al., 2019). However, once again, well-designed clinical trials in these PD patients are required to first identify the best approach to selectively decrease the effects of the bacterial decarboxylases (by using specific inhibitors) or interventions to decrease abundance of these bacteria in the intestine (such as selective antibiotic, prebiotic or even fecal microbiota transplant). A well-designed randomized clinical trial is required to determine the beneficial effects of this novel approach.

9 Does PD start in the gut?

In the present chapter, we have described the evidences suggesting a potential causative role of gut dysbiosis (triggered by environmental factors in genetically susceptible host) in the pathogenesis of local (gut and CNS) and systemic inflammation and its relationship with microglia activation in animal models and in human subjects. According to the Braak's hypothesis, the initial site of aggregation of misfolded α-syn may be either the gut or the olfactory bulb. Concerning the gut-to-brain propagation of α-syn pathology, Braak's hypothesis would implicate a chain of events: (a) a pathogen passes the GI mucosal; (b) α-Syn misfolding is triggered in post-ganglionic ENS terminals and aggregates in Lewy bodies within GI walls; (c) α-syn aggregates propagate along the vagus nerve toward the DMV; (d) misfolded α-syn reaches the DMV and aggregates in Lewy bodies; (e) α-Syn spreads and aggregates in Lewy bodies throughout the CNS, reaching and (f) triggering neurodegeneration in the locus coeruleus and the substantia nigra pars compacta. However, evidences supporting each of these steps have not been conclusively demonstrated due to some conflicting findings, as we described in the following paragraphs.

9.1 A pathogen passes the gastrointestinal mucosa

Pathogens (or pathobionts) need to reach the ENS by passing the intestinal epithelial barrier due to an increased epithelial permeability (Clairembault et al., 2015; Forsyth et al., 2011). Indeed, disrupted intestinal barrier integrity, i.e., intestinal hyperpermeability, has been reported in PD patients (Forsyth et al., 2011). Furthermore, diseases associated with increased intestinal permeability like IBD seem to be a risk factor for PD, although epidemiological data on the relationship between IBD and PD provided conflicting results so far (Fujioka et al., 2017; Lin et al., 2016).

9.2 α-Syn aggregates in post-ganglionic ENS terminals

α-Syn aggregates have been found in post-ganglionic ENS terminals. The large majority of neurons receiving vagal fibers are located in the myenteric Auerbach's plexus and the submucosal Meissner's plexus, whereas α-staining in PD biopsies has been mostly confined to mucosal nerve fibers or to the submucosal ganglionic cells where the fibers originate. Furthermore, α-Syn aggregates has been found in the sigmoid colon where there is no vagal innervation. Also, not all PD patients had α-Syn aggregates in their GI tract. Indeed, immunohistochemical methods to detect α-Syn aggregates in GI mucosal biopsy is not adequate to be used as biomarkers to predict PD due to the lack of sensibility and specificity (Corbillé et al., 2016; Ruffmann and Parkkinen, 2016).

9.3 Does α-syn pathology spread from gut to brain?

It has been reported that the risk to develop PD is reduced in subjects previously undergoing truncal vagotomy as compared to selective vagotomy (Svensson et al., 2015). Nevertheless, according to the follow-up 1977–1995 within the Danish registry there is a nonsignificant lower PD risk at 5 years follow-up (adjusted HR 0.85; 95% CI 0.56–1.27) with only a marginal significance only > 20-year follow-up (HR 0.53, 95% CI 0.53–0.99) due to wide associated confidence intervals. In a subsequent study on the same Danish population (Tysnes et al., 2015), an extended follow-up 1977–2011 reported that truncal vagotomy was associated with a modest (12%) and o a nonsignificant lower PD risk (HR 0.88, 95% CI 0.55–1.21) as well as a modest (14%) and nonsignificant elevated PD risk >20 years after the surgery (HR 1.14, 95% CI 0.23–2.05). Similar finding was noted in the Swedish registry including 9430 vagotomized patients (3445 truncal and 5978 selective) with a 30-year follow-up (1970–2010), showing that truncal vagotomy was associated with a modest and a nonsignificant lower PD risk (HR 0.78, 95% CI 0.55–1.09; Liu et al., 2017). Therefore, a history of truncal vagotomy does not provide a significant risk reduction for the development of PD even at 20-year follow-up.

The hypothesis that α-Syn (monomeric, oligomeric and aggregated fibrillary forms) can be retrogradely transported to the brain via the vagus nerve reaching the DMV has been demonstrated in animal models (Holmqvist et al., 2014). Interestingly, no α-syn deposits were detected in the brainstem of controls after injection of bovine serum albumin, suggesting that α-syn may have specific properties allowing the retrograde transportation along the vagues nerve fibers. However, despite transportation of α-syn in DMV, there was no evidence of α-syn pathology and DMV neuronal death. The lack of DMV pathology has been further evidenced in humans in a study including whole-body autopsies of 417 elderly subjects without any neurodegenerative disorder that found involvement of dorsal medulla limited to 5 of the 55 subjects (9%) showing incidental Lewy bodies (Adler and Beach, 2016). This is a key point, because the involvement of DMV is mandatory according to the Braak's hypothesis.

Several studies challenge Braak's hypothesis of a gut-to-brain spread of α-syn pathology. For example, Manfredsson reported that persistent α-syn pathology in the ENS does not result in sustained spread to the CNS in two distinct animal models of α-syn pathology (rodents and nonhuman primates (Manfredsson et al., 2018). The evidence of a substantial lack Lewy Body pathology in the ENS without concomitant involvement of the CNS, further challenge the Braaks' hypothesis of α-syn pathology starting in the gut and spreading to the brain. It should also be noted that there is still no evidence of retrograde trans-synaptic spread in humans.

It is worthwhile emphasizing that α-SYN can be transported anterogradely and retrogradely with similar efficiency, as overexpressed human α-syn in rodent models of Parkinsonism can travel from the ventral mesencephalon into the DMV. Then, via vagal motor fibers, α-syn reaches preganglionic vagal terminals into the gastric wall, possibly allowed via a novel nigro-vagal pathway (Anselmi et al., 2017; Ulusoy et al., 2017). Thus, it is tempting to consider both *"Gut-to-Brain"* hypothesis of propagation of α-syn pathology as wells as *"Brain-to-Gut"* propagation of α-syn pathology.

Finally, study by Adler and Beach (2016) suggest "olfactory bulb—brain axis" rather than "gut-brain axis" for propagation of α-syn pathology in PD. In this postmortem pathological study of 417 elderly individuals, they found that olfactory bulb was the only site for α-syn pathology in 52 subjects out of the 55 found with incidental Lewy bodies (Adler and Beach, 2016). The authors proposed a novel unified staging system for Lewy body disorders with the involvement of the olfactory bulb and of the brainstem as stage I and stage IIa, respectively. The evidence of a substantial lack Lewy Body pathology in the ENS without concomitant involvement of the CNS, further challenge the Braaks' hypothesis of α-syn pathology starting in the gut and then spreading to the brain via the vagus nerve.

9.4 How does α-syn spread throughout the CNS?

Although there is now compelling evidence supporting the hypothesis that α-syn pathology spreads throughout the CNS, this seems to be limited to a subset of neurons whose phenotype renders them susceptible to α-syn spreading (Brundin and Melki, 2017; Surmeier et al., 2017). Recent study suggested that pathological transmission of α-syn aggregates is favored by receptors uptaking aggregation-prone species of α-syn, such as the lymphocyte-activation gene 3 (Mao et al., 2016).

9.5 Do Lewy bodies induce neuronal death?

Another critical aspect of Braak's hypothesis is the assumption that the presence of Lewy bodies would be followed by neuronal death. Indeed, there is still an unclear relationship between α-syn (severity of Lewy Body pathology) and neuronal loss (Lang and Espay, 2018; Surmeier et al., 2017). First despite the presence of α-syn aggregates in Meissner's plexus in colonic biopsy samples of PD patients, there is no difference between PD and healthy subjects in the expression levels, phosphorylation or aggregation status of α-Syn in colonic biopsies, consistent with the notion that myenteric neuronal loss is not a required feature of PD and GI symptoms are not necessarily associated with ENS neuronal death (Annerino et al., 2012; Corbillé et al., 2017). Second, this assumption is also true in the CNS, where the extent and severity of α-Syn and Lewy bodies does not match the severity of neuronal loss (Surmeier et al., 2017). A final question that still needs to be conclusively answered is *"What's the role of α-syn in neurodegeneration?"*

10 Conclusions

In this review, we propose that: (1) A disrupted microbiota gut-brain axis plays a critical role in the pathogenesis of PD and dysbiotic microbiota could be the trigger and/or enabler for neuro-inflammation that is required for neurodegeneration and DA loss in PD; (2) Dysbiotic microbiota could be the mechanism by which risk factors like environmental factors (stress, diet, disrupted circadian/sleep) and/or diseases (inflammatory bowel disease, metabolic syndrome/diabetes) promote PD and thus microbiota is the lynchpin between environment and PD; (3) Intestinal microbiota-directed interventions are a promising novel therapy for PD that have a potential to augment/improve pharmacokinetics of current medications like levodopa, modify disease course and even prevent/delay the onset of PD in genetically susceptible individuals. We have tried to provide a strong rationale to encourage clinician scientists, pharma companies, PD focused foundation/societies and other stakeholders to design and start high quality clinical trials to test the tolerability and efficacy of gut microbiota-directed interventions like diet, prebiotics/probiotics and even fecal microbiota transplant in PD patients.

Acknowledgments

A.K. and P.E. would like to thank Drs. Christopher B. Forsyth, Robin M. Voigt, Geethika Earthineni and Ms. Vivian Ramirez, and Shohreh Raeisi for their contribution to this book chapter. A.K. would also like to acknowledge philanthropy funding from Mrs. Barbara and Mr. Larry Field, Mrs. Ellen and Mr. Philip Glass, and Mrs. Marcia and Mr. Silas Keehn. R.C. would like to thank Dr. Emanuele Cereda for brainstorming on the relationship between gut microbiota and PD pathophysiology as well as his helpful comments during the manuscript revision. S.B. and R.C. are thankful to the "Fondazione Grigioni per il Morbo di Parkinson" (Milano, Italy) for supporting the clinical studies on gut microbiota in patients with Parkinsonism.

References

Abbott, R.D., Petrovitch, H., White, L.R., Masaki, K.H., Tanner, C.M., Curb, J.D., Grandinetti, A., Blanchette, P.L., Popper, J.S., Ross, G.W., 2001. Frequency of bowel movements and the future risk of Parkinson's disease. Neurology 57, 456–462.

Abderrazak, A., Syrovets, T., Couchie, D., El Hadri, K., Friguet, B., Simmet, T., Rouis, M., 2015. NLRP3 inflammasome: from a danger signal sensor to a regulatory node of oxidative stress and inflammatory diseases. Redox Biol. 4, 296–307.

Adler, C.H., Beach, T.G., 2016. Neuropathological basis of nonmotor manifestations of Parkinson's disease. Mov. Disord. 31 (8), 1114–1119.

Aho, V.T.E., Pereira, P.A.B., Voutilainen, S., Paulin, L., Pekkonen, E., Auvinen, P., Scheperjans, F., 2019. Gut microbiota in Parkinson's disease: temporal stability and relations to disease progression. EBioMedicine 44, 691–707.

Akundi, R.S., Huang, Z., Eason, J., Pandya, J.D., Zhi, L., Cass, W.A., Sullivan, P.G., Büeler, H., 2011. Increased mitochondrial calcium sensitivity and abnormal expression of innate immunity genes precede dopaminergic defects in Pink1-deficient mice. PLoS One 6, e16038.

Albenberg, L., Esipova, T.V., Judge, C.P., Bittinger, K., Chen, J., Laughlin, A., Grunberg, S., Baldassano, R.N., Lewis, J.D., Li, H., Thom, S.R., Bushman, F.D., Vinograd, S.A., Wu, G.D., 2014. Correlation between intraluminal oxygen gradient and radial partitioning of intestinal microbiota. Gastroenterology 147 (5), 1055–1063.

Anderson, F.L., Coffey, M.M., Berwin, B.L., Havrda, M.C., 2018. Inflammasomes: an emerging mechanism translating environmental toxicant exposure into neuroinflammation in Parkinson's disease. Toxicol. Sci. 166, 3–15.

Annerino, D.M., Arshad, S., Taylor, G.M., Adler, C.H., Beach, T.G., Greene, J.G., 2012. Parkinson's disease is not associated with gastrointestinal myenteric ganglion neuron loss. Acta Neuropathol. 124 (5), 665–680.

Anselmi, L., Toti, L., Bove, C., Hampton, J., Travagli, R.A., 2017. A Nigro-Vagal pathway controls gastric motility and is affected in a rat model of Parkinsonism. Gastroenterology 153 (6), 1581–1593.

Arrieta, M.C., Bistritz, L., Meddings, J.B., 2006. Alterations in intestinal permeability. Gut 55, 1512–1520.

Arumugam, M., Raes, J., Pelletier, E., Le Paslier, D., Yamada, T., Mende, D.R., Fernandes, G.R., Tap, J., Bruls, T., Batto, J.M., Bertalan, M., Borruel, N., Casellas, F., Fernandez, L., Gautier, L., Hansen, T., Hattori, M., Hayashi, T., Kleerebezem, M., Kurokawa, K., Leclerc, M., Levenez, F., Manichanh, C., Nielsen, H.B., Nielsen, T., Pons, N., Poulain, J., Qin, J., Sicheritz-Ponten, T., Tims, S., Torrents, D., Ugarte, E., Zoetendal, E.G., Wang, J., Guarner, F., Pedersen, O., de Vos, W.M., Brunak, S., Doré, J., MetaHIT Consortium, Antolín, M., Artiguenave, F., Blottiere, H.M., Almeida, M., Brechot, C., Cara, C., Chervaux, C., Cultrone, A., Delorme, C., Denariaz, G., Dervyn, R., Foerstner, K.U., Friss, C., van de Guchte, M., Guedon, E., Haimet, F., Huber, W., van Hylckama-Vlieg, J., Jamet, A., Juste, C., Kaci, G., Knol, J., Lakhdari, O., Layec, S., Le Roux, K., Maguin, E., Mérieux, A., Melo Minardi, R., M'rini, C., Muller, J., Oozeer, R., Parkhill, J., Renault, P., Rescigno, M., Sanchez, N., Sunagawa, S., Torrejon, A., Turner, K., Vandemeulebrouck, G., Varela, E., Winogradsky, Y., Zeller, G., Weissenbach, J., Ehrlich, S.D., Bork, P., 2011. Enterotypes of the human gut microbiome. Nature 473 (7346), 174–180.

Athauda, D., Foltynie, T., 2018. Protective effects of the GLP-1 mimetic exendin-4 in Parkinson's disease. Neuropharmacology 136, 260–270.

Athauda, D., Maclagan, K., Skene, S.S., Bajwa-Joseph, M., Letchford, D., Chowdhury, K., Hibbert, S., Budnik, N., Zampedri, L., Dickson, J., Li, Y., Aviles-Olmos, I., Warner, T.T., Limousin, P., Lees, A.J., Greig, N.H., Tebbs, S., Foltynie, T., 2017. Exenatide once weekly versus placebo in Parkinson's disease: a randomised, double-blind, placebo-controlled trial. Lancet 390, 1664–1675.

Bailey, M.A., Holscher, H.D., 2018. Microbiome-mediated effects of the Mediterranean diet on inflammation. Adv. Nutr. 9, 193–206.

Banks, W.A., Gray, A.M., Erickson, M.A., Salameh, T.S., Damodarasamy, M., Sheibani, N., Meabon, J.S., Wing, E.E., Morofuji, Y., Cook, D.G., Reed, M.J., 2015. Lipopolysaccharide-induced blood-brain barrier disruption: roles of cyclooxygenase, oxidative stress, neuroinflammation, and elements of the neurovascular unit. J. Neuroinflammation 12, 223.

Barichella, M., Severgnini, M., Cilia, R., Cassani, E., Bolliri, C., Caronni, S., Ferri, V., Cancello, R., Ceccarani, C., Faierman, S., Pinelli, G., De Bellis, G., Zecca, L., Cereda, E., Consolandi, C., Pezzoli, G., 2019. Unraveling gut microbiota in Parkinson's disease and atypical Parkinsonism. Mov. Disord. 34 (3), 396–405.

Bassis, C.M., Tang, A.L., Young, V.B., Pynnonen, M.A., 2014. The nasal cavity microbiota of healthy adults. Microbiome 2, 27.

Bedarf, J.R., Hildebrand, F., Coelho, L.P., Sunagawa, S., Bahram, M., Goeser, F., Bork, P., Wüllner, U., 2017. Functional implications of microbial and viral gut metagenome changes in early stage L-DOPA-naïve Parkinson's disease patients. Genome Med. 9 (1), 39.

Biagi, E., Nylund, L., Candela, M., Ostan, R., Bucci, L., Pini, E., Nikkïla, J., Monti, D., Satokari, R., Franceschi, C., Brigidi, P., De Vos, W., 2010. Through ageing, and beyond: gut microbiota and inflammatory status in seniors and centenarians. PLoS One 5, e10667.

Biedermann, L., Zeitz, J., Mwinyi, J., Sutter-Minder, E., Rehman, A., Ott, S.J., Steurer-Stey, C., Frei, A., Frei, P., Scharl, M., Loessner, M.J., Vavricka, S.R., Fried, M., Schreiber, S., Schuppler, M., Rogler, G., 2013. Smoking cessation induces profound changes in the composition of the intestinal microbiota in humans. PLoS One 8 (3), e59260.

Blandini, F., Cilia, R., Cerri, S., Pezzoli, G., Schapira, A.H.V., Mullin, S., Lanciego, J.L., 2019. Glucocerebrosidase mutations and synucleinopathies: toward a model of precision medicine. Mov. Disord. 34 (1), 9–21.

Blaser, M.J., 2014. The microbiome revolution. J. Clin. Invest. 124, 4162–4165.

Blaylock, R.L., 2017. Parkinson's disease: microglial/macrophage-induced immunoexcito-toxicity as a central mechanism of neurodegeneration. Surg. Neurol. Int. 8, 65.

Block, M.L., Hong, J.S., 2005. Microglia and inflammation-mediated neurodegeneration: multiple triggers with a common mechanism. Prog. Neurobiol. 76, 77–98.

Borghammer, P., 2018. How does Parkinson's disease begin? Perspectives on neuroanatomical pathways, prions, and histology. Mov. Disord. 33, 48–57.

Bourassa, M.W., Alim, I., Bultman, S.J., Ratan, R.R., 2016. Butyrate, neuroepigenetics and the gut microbiome: can a high fiber diet improve brain health? Neurosci. Lett. 625, 56–63.

Braak, H., Del Tredici, K., 2017. Neuropathological staging of brain pathology in sporadic Parkinson's disease: separating the wheat from the chaff. J. Parkinsons Dis. 7, S73–S87.

Braak, H., Rub, U., Gai, W.P., Del Tredici, K., 2003. Idiopathic Parkinson's disease: possible routes by which vulnerable neuronal types may be subject to neuroinvasion by an unknown pathogen. J. Neural Transm. 110, 517–536.

Brundin, P., Melki, R., 2017. Prying into the prion hypothesis for Parkinson's disease. J. Neurosci. 37 (41), 9808–9818.

Buford, T.W., 2017. (Dis)Trust your gut: the gut microbiome in age-related inflammation, health, and disease. Microbiome 5, 80.

Burns, R.S., LeWitt, P.A., Ebert, M.H., Pakkenberg, H., Kopin, I.J., 1985. The clinical syndrome of striatal dopamine deficiency. Parkinsonism induced by 1-methyl-4-phenyl-1,2,3,6-tetrahydropyridine (MPTP). N. Engl. J. Med. 312, 1418–1421.

Candela, M., Perna, F., Carnevali, P., Vitali, B., Ciati, R., Gionchetti, P., Rizzello, F., Campieri, M., Brigidi, P., 2008. Interaction of probiotic Lactobacillus and Bifidobacterium strains with human intestinal epithelial cells: adhesion properties, competition against enteropathogens and modulation of IL-8 production. Int. J. Food Microbiol. 125, 286–292.

Caporaso, J.G., Lauber, C.L., Costello, E.K., Berg-Lyons, D., Gonzalez, A., Stombaugh, J., Knights, D., Gajer, P., Ravel, J., Fierer, N., Gordon, J.I., Knight, R., 2011. Moving pictures of the human microbiome. Genome Biol. 12, R50.

Carding, S., Verbeke, K., Vipond, D.T., Corfe, B.M., Owen, L.J., 2015. Dysbiosis of the gut microbiota in disease. Microb. Ecol. Health Dis. 26, 26191.

Cardoso, C.C., Pereira, A.C., de Sales Marques, C., Moraes, M.O., 2011. Leprosy susceptibility: genetic variations regulate innate and adaptive immunity, and disease outcome. Future Microbiol. 6, 533–549.

Cassani, E., Barichella, M., Ferri, V., Pinelli, G., Iorio, L., Bolliri, C., Caronni, S., Faierman, S.A., Mottolese, A., Pusani, C., Monajemi, F., Pasqua, M., Lubisco, A., Cereda, E., Frazzitta, G., Petroni, M.L., Pezzoli, G., 2017. Dietary habits in Parkinson's disease: adherence to Mediterranean diet. Parkinsonism Relat. Disord. 42, 40–46.

Chakraborty, S., Kaushik, D.K., Gupta, M., Basu, A., 2010. Inflammasome signaling at the heart of central nervous system pathology. J. Neurosci. Res. 88, 1615–1631.

Chen, H., Ritz, B., 2018. The search for environmental causes of Parkinson's disease: moving forward. J. Parkinsons Dis. 8, S9–S17.

Chen, J.F., Xu, K., Petzer, J.P., Staal, R., Xu, Y.H., Beilstein, M., Sonsalla, P.K., Castagnoli, K., Castagnoli, N.J., Schwarzschild, M.A., 2001. Neuroprotection by caffeine and A(2A) adenosine receptor inactivation in a model of Parkinson's disease. J. Neurosci. 21 (10), RC143.

Chen, H., Zhang, S.M., Hernán, M.A., Willett, W.C., Ascherio, A., 2003. Dietary intakes of fat and risk of Parkinson's disease. Am. J. Epidemiol. 157, 1007–1014.

Choi, J.G., Kim, N., Ju, I.G., Eo, H., Lim, S.M., Jang, S.E., Kim, D.H., Oh, M.S., 2018. Oral administration of *Proteus mirabilis* damages dopaminergic neurons and motor functions in mice. Sci. Rep. 8, 1275.

Claesson, M.J., O'Sullivan, O., Wang, Q., Nikkilä, J., Marchesi, J.R., Smidt, H., de Vos, W.M., Ross, R.P., O'Toole, P.W., 2009. Comparative analysis of pyrosequencing and a phylogenetic microarray for exploring microbial community structures in the human distal intestine. PLoS One 4, e6669.

Clairembault, T., Leclair-Visonneau, L., Coron, E., Bourreille, A., Le Dily, S., Vavasseur, F., Heymann, M.F., Neunlist, M., Derkinderen, P., 2015. Structural alterations of the intestinal epithelial barrier in Parkinson's disease. Acta Neuropathol. Commun. 3, 12.

Clayton, T.A., Baker, D., Lindon, J.C., Everett, J.R., Nicholson, J.K., 2009. Pharmacometabonomic identification of a significant host-microbiome metabolic interaction affecting human drug metabolism. Proc. Natl. Acad. Sci. U. S. A. 106, 14728–14733.

Clemente, J.C., Ursell, L.K., Parfrey, L.W., Knight, R., 2012. The impact of the gut microbiota on human health: an integrative view. Cell 148, 1258–1270.

Codolo, G., Plotegher, N., Pozzobon, T., Brucale, M., Tessari, I., Bubacco, L., de Bernard, M., 2013. Triggering of inflammasome by aggregated alpha-synuclein, an inflammatory response in synucleinopathies. PLoS One 8, e55375.

Corbillé, A.G., Letournel, F., Kordower, J.H., Lee, J., Shanes, E., Neunlist, M., Munoz, D.G., Derkinderen, P., Beach, T.G., 2016. Evaluation of alpha-synuclein immunohistochemical methods for the detection of Lewy-type synucleinopathy in gastrointestinal biopsies. Acta Neuropathol. Commun. 4, 35.

Corbillé, A.G., Preterre, C., Rolli-Derkinderen, M., Coron, E., Neunlist, M., Lebouvier, T., Derkinderen, P., 2017. Biochemical analysis of α-synuclein extracted from control and Parkinson's disease colonic biopsies. Neurosci. Lett. 641, 81–86.

Cornejo Castro, E.M., Waak, J., Weber, S.S., Fiesel, F.C., Oberhettinger, P., Schütz, M., Autenrieth, I.B., Springer, W., Kahle, P.J., 2010. Parkinson's disease-associated DJ-1 modulates innate immunity signaling in *Caenorhabditis elegans*. J. Neural Transm. 117, 599–604.

Couch, Y., Alvarez-Erviti, L., Sibson, N.R., Wood, M.J., Anthony, D.C., 2011. The acute inflammatory response to intranigral alpha-synuclein differs significantly from intranigral lipopolysaccharide and is exacerbated by peripheral inflammation. J. Neuroinflammation 8, 166.

Creagh, E.M., O'Neill, L.A., 2006. TLRs, NLRs and RLRs: a trinity of pathogen sensors that co-operate in innate immunity. Trends Immunol. 27, 352–357.

Dalile, B., Van Oudenhove, L., Vervliet, B., Verbeke, K., 2019. The role of short-chain fatty acids in microbiota-gut-brain communication. Nat. Rev. Gastroenterol. Hepatol. 16 (8), 461–478.

David, L.A., Maurice, C.F., Carmody, R.N., Gootenberg, D.B., Button, J.E., Wolfe, B.E., Ling, A.V., Devlin, A.S., Varma, Y., Fischbach, M.A., Biddinger, S.B., Dutton, R.J., Turnbaugh, P.J., 2014. Diet rapidly and reproducibly alters the human gut microbiome. Nature 505, 559–563.

De Filippo, C., Cavalieri, D., Di Paola, M., Ramazzotti, M., Poullet, J.B., Massart, S., Collini, S., Pieraccini, G., Lionetti, P., 2010. Impact of diet in shaping gut microbiota revealed by a comparative study in children from Europe and rural Africa. Proc. Natl. Acad. Sci. U. S. A. 107, 14691–14696.

De Vadder, F., Kovatcheva-Datchary, P., Goncalves, D., Vinera, J., Zitoun, C., Duchampt, A., Bäckhed, F., Mithieux, G., 2014. Microbiota-generated metabolites promote metabolic benefits via gut-brain neural circuits. Cell 156, 84–96.

de Vos, W.M., de Vos, E.A., 2012. Role of the intestinal microbiome in health and disease: from correlation to causation. Nutr. Rev. 70 (Suppl. 1), S45–S56.

Deng, H., Wang, P., Jankovic, J., 2018. The genetics of Parkinson disease. Ageing Res. Rev. 42, 72–85.

Derkinderen, P., Shannon, K.M., Brundin, P., 2014. Gut feelings about smoking and coffee in Parkinson's disease. Mov. Disord. 29 (8), 976–979.

Devos, D., Lebouvier, T., Lardeux, B., Biraud, M., Rouaud, T., Pouclet, H., Coron, E., Bruley, d., Varannes, S., Naveilhan, P., Nguyen, J.M., Neunlist, M., Derkinderen, P., 2013. Colonic inflammation in Parkinson's disease. Neurobiol. Dis. 50, 42–48.

Dicksved, J., Halfvarson, J., Rosenquist, M., Järnerot, G., Tysk, C., Apajalahti, J., Engstrand, L., Jansson, J.K., 2008. Molecular analysis of the gut microbiota of identical twins with Crohn's disease. ISME J. 2, 716–727.

Dinan, T.G., Cryan, J.F., 2017. Gut instincts: microbiota as a key regulator of brain development, ageing and neurodegeneration. J. Physiol. 595 (2), 489–503.

Ding, P.H., Jin, L.J., 2014. The role of lipopolysaccharide-binding protein in innate immunity: a revisit and its relevance to oral/periodontal health. J. Periodontal Res. 49 (1), 1–9.

Dodiya, H.B., Forsyth, C.B., Voigt, R.M., Engen, P.A., Patel, J., Shaikh, M., Green, S.J., Naqib, A., Roy, A., Kordower, J.H., Pahan, K., Shannon, K.M., Keshavarzian, A., 2020. Chronic stress-induced gut dysfunction exacerbates Parkinson's disease phenotype and pathology in a rotenone-induced mouse model of Parkinson's disease. Neurobiol. Dis. 135, 104352.

Dutta, G., Zhang, P., Liu, B., 2008. The lipopolysaccharide Parkinson's disease animal model: mechanistic studies and drug discovery. Fundam. Clin. Pharmacol. 22, 453–464.

Dzamko, N., Chua, G., Ranola, M., Rowe, D.B., Halliday, G.M., 2013. Measurement of LRRK2 and Ser910/935 phosphorylated LRRK2 in peripheral blood mononuclear cells from idiopathic Parkinson's disease patients. J. Parkinsons Dis. 3, 145–152.

Dzamko, N., Geczy, C.L., Halliday, G.M., 2015. Inflammation is genetically implicated in Parkinson's disease. Neuroscience 302, 89–102.

Eckburg, P.B., Bik, E.M., Bernstein, C.N., Purdom, E., Dethlefsen, L., Sargent, M., Gill, S.R., Nelson, K.E., Relman, D.A., 2005. Diversity of the human intestinal microbial flora. Science 308, 1635–1638.

Engen, P.A., Dodiya, H.B., Naqib, A., Forsyth, C.B., Green, S.J., Voigt, R.M., Kordower, J.H., Mutlu, E.A., Shannon, K.M., Keshavarzian, A., 2017. The potential role of gut-derived inflammation in multiple system atrophy. J. Parkinsons Dis. 7, 331–346.

Erny, D., Hrabě de Angelis, A.L., Jaitin, D., Wieghofer, P., Staszewski, O., David, E., Keren-Shaul, H., Mahlakoiv, T., Jakobshagen, K., Buch, T., Schwierzeck, V., Utermöhlen, O., Chun, E., Garrett, W.S., McCoy, K.D., Diefenbach, A., Staeheli, P., Stecher, B., Amit, I., Prinz, M., 2015. Host microbiota constantly control maturation and function of microglia in the CNS. Nat. Neurosci. 18, 965–977.

Everard, A., Cani, P.D., 2014. Gut microbiota and GLP-1. Rev. Endocr. Metab. Disord. 15, 189–196.

Faith, J.J., Guruge, J.L., Charbonneau, M., et al., 2013. The long-term stability of the human gut microbiota. Science 341, 1237439.

Fasano, A., 2012. Leaky gut and autoimmune diseases. Clin. Rev. Allergy Immunol. 42, 71–78.

Fasano, A., Bove, F., Gabrielli, M., Petracca, M., Zocco, M.A., Ragazzoni, E., Barbaro, F., Piano, C., Fortuna, S., Tortora, A., Di Giacopo, R., Campanale, M., Gigante, G., Lauritano, E.C., Navarra, P., Marconi, S., Gasbarrini, A., Bentivoglio, A.R., 2013. The role of small intestinal bacterial overgrowth in Parkinson's disease. Mov. Disord. 28 (9), 1241–1249.

Fasano, A., Visanji, N.P., Liu, L.W., Lang, A.E., Pfeiffer, R.F., 2015. Gastrointestinal dysfunction in Parkinson's disease. Lancet Neurol. 14, 625–639.

Faust, K., Sathirapongsasuti, J.F., Izard, J., Segata, N., Gevers, D., Raes, J., Huttenhower, C., 2012. Microbial co-occurrence relationships in the human microbiome. PLoS Comput. Biol. 8 (7), e1002606.

Fellner, L., Irschick, R., Schanda, K., Reindl, M., Klimaschewski, L., Poewe, W., Wenning, G.K., Stefanova, N., 2013. Toll-like receptor 4 is required for alpha-synuclein dependent activation of microglia and astroglia. Glia 61, 349–360.

Ferreira, M., Massano, J., 2016. An updated review of Parkinson's disease genetics and clinicopathological correlations. Acta Neurol. Scand. 135, 273–284.

Fiebich, B.L., Batista, C.R.A., Saliba, S.W., Yousif, N.M., de Oliveira, A.C.P., 2018. Role of microglia TLRs in neurodegeneration. Front. Cell. Neurosci. 12, 329.

Forslund, K., Hildebrand, F., Nielsen, T., Falony, G., Le Chatelier, E., Sunagawa, S., Prifti, E., Vieira-Silva, S., Gudmundsdottir, V., Pedersen, H.K., Arumugam, M., Kristiansen, K., Voigt, A.Y., Vestergaard, H., Hercog, R., Costea, P.I., Kultima, J.R., Li, J., Jørgensen, T., Levenez, F., Dore, J., MetaHIT Consortium, Nielsen, H.B., Brunak, S., Raes, J., Hansen, T., Wang, J., Ehrlich, S.D., Bork, P., Pedersen, O., 2015. Disentangling type 2 diabetes and metformin treatment signatures in the human gut microbiota. Nature 528 (7581), 262–266.

Forsyth, C.B., Shannon, K.M., Kordower, J.H., Voigt, R.M., Shaikh, M., Jaglin, J.A., Estes, J.D., Dodiya, H.B., Keshavarzian, A., 2011. Increased intestinal permeability correlates with sigmoid mucosa alpha-synuclein staining and endotoxin exposure markers in early Parkinson's disease. PLoS One 6, e28032.

Frank, D.N., St Amand, A.L., Feldman, R.A., Boedeker, E.C., Harpaz, N., Pace, N.R., 2007. Molecular-phylogenetic characterization of microbial community imbalances in human inflammatory bowel diseases. Proc. Natl. Acad. Sci. U. S. A. 104, 13780–13785.

Frank-Cannon, T.C., Tran, T., Ruhn, K.A., Martinez, T.N., Hong, J., Marvin, M., Hartley, M., Treviño, I., O'Brien, D.E., Casey, B., Goldberg, M.S., Tansey, M.G., 2008. Parkin deficiency increases vulnerability to inflammation-related nigral degeneration. J. Neurosci. 28, 10825–10834.

Freeman, L.C., Ting, J.P., 2016. The pathogenic role of the inflammasome in neurodegenerative diseases. J. Neurochem. 136 (Suppl. 1), 29–38.

Freeman, L., Guo, H., David, C.N., Brickey, W.J., Jha, S., Ting, J.P., 2017. NLR members NLRC4 and NLRP3 mediate sterile inflammasome activation in microglia and astrocytes. J. Exp. Med. 214, 1351–1370.

Fujioka, S., Curry, S.E., Kennelly, K.D., Tacik, P., Heckman, M.G., Tsuboi, Y., Strongosky, A.J., van Gerpen, J.A., Uitti, R.J., Ross, O.A., Ikezu, T., Wszolek, Z.K., 2017. Occurrence of Crohn's disease with Parkinson's disease. Parkinsonism Relat. Disord. 37, 116–117.

Fukuda, S., Toh, H., Hase, K., Oshima, K., Nakanishi, Y., Yoshimura, K., Tobe, T., Clarke, J.M., Topping, D.L., Suzuki, T., Taylor, T.D., Itoh, K., Kikuchi, J., Morita, H., Hattori, M., Ohno, H., 2011. Bifidobacteria can protect from enteropathogenic infection through production of acetate. Nature 469, 543–547.

Furusawa, Y., Obata, Y., Fukuda, S., Endo, T.A., Nakato, G., Takahashi, D., Nakanishi, Y., Uetake, C., Kato, K., Kato, T., Takahashi, M., Fukuda, N.N., Murakami, S., Miyauchi, E., Hino, S., Atarashi, K., Onawa, S., Fujimura, Y., Lockett, T., Clarke, J.M., Topping, D.L., Tomita, M., Hori, S., Ohara, O., Morita, T., Koseki, H., Kikuchi, J., Honda, K., Hase, K., Ohno, H., 2013. Commensal microbe-derived butyrate induces the differentiation of colonic regulatory T cells. Nature 504 (7480), 446–450.

Gabrielli, M., Bonazzi, P., Scarpellini, E., Bendia, E., Lauritano, E.C., Fasano, A., Ceravolo, M.G., Capecci, M., Rita Bentivoglio, A., Provinciali, L., Tonali, P.A., Gasbarrini, A., 2011. Prevalence of small intestinal bacterial overgrowth in Parkinson's disease. Mov. Disord. 26 (5), 889–892.

Ganapathy, V., Thangaraju, M., Prasad, P.D., Martin, P.M., Singh, N., 2013. Transporters and receptors for short-chain fatty acids as the molecular link between colonic bacteria and the host. Curr. Opin. Pharmacol. 13 (6), 869–874.

Gardet, A., Benita, Y., Li, C., Sands, B.E., Ballester, I., Stevens, C., Korzenik, J.R., Rioux, J.D., Daly, M.J., Xavier, R.J., Podolsky, D.K., 2010. LRRK2 is involved in the IFN-gamma response and host response to pathogens. J. Immunol. 185, 5577–5585.

GBD 2016 Neurology Collaborators, 2018. Global, regional, and national burden of Parkinson's disease, 1990–2016: a systematic analysis for the Global Burden of Disease Study 2016. Lancet Neurol. 17, 939–953.

Gerhard, A., Pavese, N., Hotton, G., Turkheimer, F., Es, M., Hammers, A., Eggert, K., Oertel, W., Banati, R.B., Brooks, D.J., 2006. In vivo imaging of microglial activation with [11C](R)-PK11195 PET in idiopathic Parkinson's disease. Neurobiol. Dis. 21, 404–412.

Ghaisas, S., Maher, J., Kanthasamy, A., 2016. Gut microbiome in health and disease: linking the microbiome-gut-brain axis and environmental factors in the pathogenesis of systemic and neurodegenerative diseases. Pharmacol. Ther. 158, 52–62.

Gilbert, J.A., Quinn, R.A., Debelius, J., Xu, Z.Z., Morton, J., Garg, N., Jansson, J.K., Dorrestein, P.C., Knight, R., 2016. Microbiome-wide association studies link dynamic microbial consortia to disease. Nature 535, 94–103.

Gilbert, J.A., Blaser, M.J., Caporaso, J.G., Jansson, J.K., Lynch, S.V., Knight, R., 2018. Current understanding of the human microbiome. Nat. Med. 24, 392–400.

Gniechwitz, D., Reichardt, N., Blaut, M., Steinhart, H., Bunzel, M., 2007. Dietary fiber from coffee beverage: degradation by human fecal microbiota. J. Agric. Food Chem. 55 (17), 6989–6996.

Goldman, S.M., Kamel, F., Ross, G.W., Jewell, S.A., Marras, C., Hoppin, J.A., Umbach, D.M., Bhudhikanok, G.S., Meng, C., Korell, M., Comyns, K., Hauser, R.A., Jankovic, J., Factor, S.A., Bressman, S., Lyons, K.E., Sandler, D.P., Langston, J.W., Tanner, CM., 2014. Peptidoglycan recognition protein genes and risk of Parkinson's disease. Mov. Disord. 29 (9), 1171–1180.

Gonzales, A., Decourt, B., Walker, A., Condjella, R., Nural, H., Sabbagh, M.N., 2011. Development of a specific ELISA to measure BACE1 levels in human tissues. J. Neurosci. Methods 202 (1), 70–76.

Goodrich, J.K., Davenport, E.R., Beaumont, M., Jackson, M.A., Knight, R., Ober, C., Spector, T.D., Bell, J.T., Clark, A.G., Ley, R.E., 2016. Genetic determinants of the gut microbiome in UK twins. Cell Host Microbe 19, 731–743.

Gordon, R., Albornoz, E.A., Christie, D.C., Langley, M.R., Kumar, V., Mantovani, S., Robertson, A.A.B., Butler, M.S., Rowe, D.B., O'Neill, L.A., Kanthasamy, A.G., Schroder, K., Cooper, M.A., Woodruff, T.M., 2018. Inflammasome inhibition preventsα-synuclein pathology and dopaminergic neurodegeneration in mice. Sci. Transl. Med. 10 (465), eaah4066

Grace, E., Shaw, C., Whelan, K., Andreyev, H.J., 2013. Review article: small intestinal bacterial overgrowth—prevalence, clinical features, current and developing diagnostic tests, and treatment. Aliment. Pharmacol. Ther. 38 (7), 674–688.

Greenamyre, J.T., Betarbet, R., Sherer, T.B., 2003. The rotenone model of Parkinson's disease: genes, environment and mitochondria. Parkinsonism Relat. Disord. 9 (Suppl. 2), S59–S64.

Greene, J.C., Whitworth, A.J., Andrews, L.A., Parker, T.J., Pallanck, L.J., 2005. Genetic and genomic studies of Drosophila parkin mutants implicate oxidative stress and innate immune responses in pathogenesis. Hum. Mol. Genet. 14, 799–811.

Guillot-Sestier, M.V., Town, T., 2018. Let's make microglia great again in neurodegenerative disorders. J. Neural Transm. (Vienna) 125, 751–770.

Guo, Z., Zhang, J., Wang, Z., Ang, K.Y., Huang, S., Hou, Q., Su, X., Qiao, J., Zheng, Y., Wang, L., Koh, E., Danliang, H., Xu, J., Lee, Y.K., Zhang, H., 2016. Intestinal microbiota distinguish gout patients from healthy humans. Sci. Rep. 6, 20602.

Gurung, P., Li, B., Subbarao Malireddi, R.K., Lamkanfi, M., Geiger, T.L., Kanneganti, T.D., 2015. Chronic TLR stimulation controls NLRP3 inflammasome activation through IL-10 mediated regulation of NLRP3 expression and caspase-8 activation. Sci. Rep. 5, 14488.

Hakimi, M., Selvanantham, T., Swinton, E., Padmore, R.F., Tong, Y., Kabbach, G., Venderova, K., Girardin, S.E., Bulman, D.E., Scherzer, C.R., LaVoie, M.J., Gris, D., Park, D.S., Angel, J.B., Shen, J., Philpott, D.J., Schlossmacher, M.G., 2011. Parkinson's disease-linked LRRK2 is expressed in circulating and tissue immune cells and upregulated following recognition of microbial structures. J. Neural Transm. 118, 795–808.

Halle, A., Hornung, V., Petzold, G.C., Stewart, C.R., Monks, B.G., Reinheckel, T., Fitzgerald, K.A., Latz, E., Moore, K.J., Golenbock, D.T., 2008. The NALP3 inflammasome is involved in the innate immune response to amyloid-beta. Nat. Immunol. 9, 857–865.

Hamer, H.M., Jonkers, D., Venema, K., Vanhoutvin, S., Troost, F.J., Brummer, R.J., 2008. Review article: the role of butyrate on colonic function. Aliment. Pharmacol. Ther. 27, 104–119.

Hanke, M.L., Kielian, T., 2011. Toll-like receptors in health and disease in the brain: mechanisms and therapeutic potential. Clin. Sci. (Lond.) 121, 367–387.

Hansen, E.E., Lozupone, C.A., Rey, F.E., Wu, M., Guruge, J.L., Narra, A., Goodfellow, J., Zaneveld, J.R., McDonald, D.T., Goodrich, J.A., Heath, A.C., Knight, R., Gordon, J.I., 2011. Pan-genome of the dominant human gut-associated archaeon, *Methanobrevibacter smithii*, studied in twins. Proc. Natl. Acad. Sci. U. S. A. 108 (Suppl. 1), 4599–4606.

Hartog, A., Belle, F.N., Bastiaans, J., de Graaff, P., Garssen, J., Harthoorn, L.F., Vos, A.P., 2015. A potential role for regulatory T-cells in the amelioration of DSS induced colitis by dietary non-digestible polysaccharides. J. Nutr. Biochem. 26, 227–233.

Hasegawa, S., Goto, S., Tsuji, H., Okuno, T., Asahara, T., Nomoto, K., Shibata, A., Fujisawa, Y., Minato, T., Okamoto, A., Ohno, K., Hirayama, M., 2015. Intestinal dysbiosis and lowered serum lipopolysaccharide-binding protein in Parkinson's disease. PLoS One 10, e0142164.

Hawkes, C.H., Del Tredici, K., Braak, H., 2007. Parkinson's disease: a dual-hit hypothesis. Neuropathol. Appl. Neurobiol. 33, 599–614.

Heintz-Buschart, A., Pandey, U., Wicke, T., Sixel-Döring, F., Janzen, A., Sittig-Wiegand, E., Trenkwalder, C., Oertel, W.H., Mollenhauer, B., Wilmes, P., 2018. The nasal and gut microbiome in Parkinson's disease and idiopathic rapid eye movement sleep behavior disorder. Mov. Disord. 33, 88–98.

Henry, C.J., Huang, Y., Wynne, A.M., Godbout, J.P., 2009. Peripheral lipopolysaccharide (LPS) challenge promotes microglial hyperactivity in aged mice that is associated with exaggerated induction of both pro-inflammatory IL-1beta and anti-inflammatory IL-10 cytokines. Brain Behav. Immun. 23, 309–317.

Hill-Burns, E.M., Debelius, J.W., Morton, J.T., Wissemann, W.T., Lewis, M.R., Wallen, Z.D., Peddada, S.D., Factor, S.A., Molho, E., Zabetian, C.P., Knight, R., Payami, H., 2017. Parkinson's disease and Parkinson's disease medications have distinct signatures of the gut microbiome. Mov. Disord. 32, 739–749.

Hirschberg, S., Gisevius, B., Duscha, A., Haghikia, A., 2019. Implications of diet and the gut microbiome in neuroinflammatory and neurodegenerative diseases. Int. J. Mol. Sci. 20 (12), E3109.

Holmans, P., Moskvina, V., Jones, L., Sharma, M., International Parkinson's Disease Genomics Consortium, Vedernikov, A., Buchel, F., Saad, M., Bras, J.M., Bettella, F., Nicolaou, N., Simón-Sánchez, J., Mittag, F., Gibbs, J.R., Schulte, C., Durr, A., Guerreiro, R., Hernandez, D., Brice, A., Stefánsson, H., Majamaa, K., Gasser, T., Heutink, P., Wood, N.W., Martinez, M., Singleton, A.B., Nalls, M.A., Hardy, J., Morris, H.R., Williams, N.M., 2013. A pathway-based analysis provides additional support for an immune-related genetic susceptibility to Parkinson's disease. Hum. Mol. Genet. 22, 1039–1049.

Holmqvist, S., Chutna, O., Bousset, L., Aldrin-Kirk, P., Li, W., Björklund, T., Wang, Z.Y., Roybon, L., Melki, R., Li, J.Y., 2014. Direct evidence of Parkinson pathology spread from the gastrointestinal tract to the brain in rats. Acta Neuropathol. 128, 805–820.

Hoogland, I.C., Houbolt, C., van Westerloo, D.J., van Gool, W.A., van de Beek, D., 2015. Systemic inflammation and microglial activation: systematic review of animal experiments. J. Neuroinflammation 12, 114.

Hopfner, F., Künstner, A., Müller, S.H., Künzel, S., Zeuner, K.E., Margraf, N.G., Deuschl, G., Baines, J.F., Kuhlenbäumer, G., 2017. Gut microbiota in Parkinson disease in a northern German cohort. Brain Res. 1667, 41–45.

Hu, G., Jousilahti, P., Bidel, S., Antikainen, R., Tuomilehto, J., 2007. Type 2 diabetes and the risk of Parkinson's disease. Diabetes Care 30, 842–847.

Hughes, K.C., Gao, X., Kim, I.Y., Wang, M., Weisskopf, M.G., Schwarzschild, M.A., Ascherio, A., 2017. Intake of dairy foods and risk of Parkinson disease. Neurology 89, 46–52.

Hui, K.Y., Fernandez-Hernandez, H., Hu, J., Schaffner, A., Pankratz, N., Hsu, N.Y., Chuang, L.S., Carmi, S., Villaverde, N., Li, X., Rivas, M., Levine, A.P., Bao, X., Labrias, P.R., Haritunians, T., Ruane, D., Gettler, K., Chen, E., Li, D., Schiff, E.R., Pontikos, N., Barzilai, N., Brant, S.R., Bressman, S., Cheifetz, A.S., Clark, L.N., Daly, M.J., Desnick, R.J., Duerr, R.H., Katz, S., Lencz, T., Myers, R.H., Ostrer, H., Ozelius, L., Payami, H., Peter, Y., Rioux, J.D., Segal, A.W., Scott, W.K., Silverberg, M.S., Vance, J.M., Ubarretxena-Belandia, I., Foroud, T., Atzmon, G., Pe'er, I., Ioannou, Y., McGovern, D.P.B., Yue, Z., Schadt, E.E., Cho, J.H., Peter, I., 2018. Functional variants in the LRRK2 gene confer shared effects on risk for Crohn's disease and Parkinson's disease. Sci. Transl. Med. 10 (423), eaai7795.

Jang, H., Boltz, D., Sturm-Ramirez, K., Shepherd, K.R., Jiao, Y., Webster, R., Smeyne, R.J., 2009. Highly pathogenic H5N1 influenza virus can enter the central nervous system and induce neuroinflammation and neurodegeneration. Proc. Natl. Acad. Sci. U. S. A. 106 (33), 14063–14068.

Jangula, A., Murphy, E.J., 2013. Lipopolysaccharide-induced blood brain barrier permeability is enhanced by alpha-synuclein expression. Neurosci. Lett. 551, 23–27.

Jaquet, M., Rochat, I., Moulin, J., Cavin, C., Bibiloni, R., 2009. Impact of coffee consumption on the gut microbiota: a human volunteer study. Int. J. Food Microbiol. 130 (2), 117–121.

Jeong, J.J., Kim, K.A., Hwang, Y.J., Han, M.J., Kim, D.H., 2016. Anti-inflammaging effects of *Lactobacillus brevis* OW38 in aged mice. Benefic. Microbes 7 (5), 707–718.

Johnson, M.E., Stringer, A., Bobrovskaya, L., 2018. Rotenone induces gastrointestinal pathology and microbiota alterations in a rat model of Parkinson's disease. Neurotoxicology 65, 174–185.

Johnson, M.E., Stecher, B., Labrie, V., Brundin, L., Brundin, P., 2019. Triggers, facilitators, and aggravators: redefining Parkinson's disease pathogenesis. Trends Neurosci. 42, 4–13.

Kalia, L.V., Lang, A.E., 2015. Parkinson's disease. Lancet 386, 896–912.

Kau, A.L., Ahern, P.P., Griffin, N.W., Goodman, A.L., Gordon, J.I., 2011. Human nutrition, the gut microbiome and the immune system. Nature 474, 327–336.

Keshavarzian, A., Green, S.J., Engen, P.A., Voigt, R.M., Naqib, A., Forsyth, C.B., Mutlu, E., Shannon, K.M., 2015. Colonic bacterial composition in Parkinson's disease. Mov. Disord. 30, 1351–1360.

Khalif, I.L., Quigley, E.M., Konovitch, E.A., Maximova, I.D., 2005. Alterations in the colonic flora and intestinal permeability and evidence of immune activation in chronic constipation. Dig. Liver Dis. 37, 838–849.

Kidd, S.K., Schneider, J.S., 2010. Protection of dopaminergic cells from MPP+-mediated toxicity by histone deacetylase inhibition. Brain Res. 1354, 172–178.

Kim, D.S., Choi, H.I., Wang, Y., Luo, Y., Hoffer, B.J., Greig, N.H., 2017. A new treatment strategy for Parkinson's disease through the gut-brain axis: the glucagon-like peptide-1 receptor pathway. Cell Transplant. 26, 1560–1571.

Kim, S., Kwon, S.H., Kam, T.I., Panicker, N., Karuppagounder, S.S., Lee, S., Lee, J.H., Kim, W.R., Kook, M., Foss, C.A., Shen, C., Lee, H., Kulkarni, S., Pasricha, P.J., Lee, G., Pomper, M.G., Dawson, V.L., Dawson, T.M., Ko, H.S., 2019. Transneuronal propagation of pathologic alpha-synuclein from the gut to the brain models Parkinson's Disease. Neuron 103 (4), 627–641.

Kishimoto, Y., Zhu, W., Hosoda, W., Sen, J.M., Mattson, M.P., 2019. Chronic mild gut inflammation accelerates brain neuropathology and motor dysfunction in alpha-synuclein mutant mice. Neuromolecular Med. 21 (3), 239–249.

Knudsen, K., Krogh, K., Østergaard, K., Borghammer, P., 2017. Constipation in parkinson's disease: subjective symptoms, objective markers, and new perspectives. Mov. Disord. 32, 94–105.

Koenig, J.E., Spor, A., Scalfone, N., Fricker, A.D., Stombaugh, J., Knight, R., Angenent, L.T., Ley, R.E., 2011. Succession of microbial consortia in the developing infant gut microbiome. Proc. Natl. Acad. Sci. U. S. A. 108 (Suppl. 1), 4578–4585.

Koh, A., De Vadder, F., Kovatcheva-Datchary, P., Bäckhed, F., 2016. From dietary fiber to host physiology: short-chain fatty acids as key bacterial metabolites. Cell 165, 1332–1345.

Köhler, C.A., Maes, M., Slyepchenko, A., Berk, M., Solmi, M., Lanctôt, K.L., Carvalho, A.F., 2016. The gut-brain axis, including the microbiome, leaky gut and bacterial translocation: mechanisms and pathophysiological role in Alzheimer's disease. Curr. Pharm. Des. 22, 6152–6166.

Kohutnicka, M., Lewandowska, E., Kurkowska-Jastrzebska, I., Członkowski, A., Członkowska, A., 1998. Microglial and astrocytic involvement in a murine model of Parkinson's disease induced by 1-methyl-4-phenyl-1,2,3,6-tetrahydropyridine (MPTP). Immunopharmacology 39, 167–180.

Kozyrskyj, A.L., Bahreinian, S., Azad, M.B., 2011. Early life exposures: impact on asthma and allergic disease. Curr. Opin. Allergy Clin. Immunol. 11, 400–406.

Labzin, L.I., Heneka, M.T., Latz, E., 2018. Innate immunity and neurodegeneration. Annu. Rev. Med. 69, 437–449.

Lai, F., Jiang, R., Xie, W., Liu, X., Tang, Y., Xiao, H., Gao, J., Jia, Y., Bai, Q., 2018. Intestinal pathology and gut microbiota alterations in a methyl-4-phenyl-1,2,3,6-tetrahydropyridine (MPTP) mouse model of Parkinson's disease. Neurochem. Res. 43, 1986–1999.

Lang, A.E., Espay, A.J., 2018. Disease Modification in Parkinson's Disease: Current Approaches, Challenges, and Future Considerations. Mov Disord. 33 (5), 660–677.

Langille, M.G., Zaneveld, J., Caporaso, J.G., McDonald, D., Knights, D., Reyes, J.A., Clemente, J.C., Burkepile, D.E., Vega Thurber, R.L., Knight, R., Beiko, R.G., Huttenhower, C., 2013. Predictive functional profiling of microbial communities using 16S rRNA marker gene sequences. Nat. Biotechnol. 31, 814–821.

Lee, W.Y., Yoon, W.T., Shin, H.Y., Jeon, S.H., Rhee, P.L., 2008. Helicobacter pylori infection and motor fluctuations in patients with Parkinson's disease. Mov. Disord. 23 (12), 1696–1700.

Lewitt, P.A., 2008. Levodopa for the treatment of Parkinson's disease. N. Engl. J. Med. 359, 2468–2476.

Ley, R.E., Turnbaugh, P.J., Klein, S., Gordon, J.I., 2006. Microbial ecology: human gut microbes associated with obesity. Nature 444, 1022–1023.

Li, W., Wu, X., Hu, X., Wang, T., Liang, S., Duan, Y., Jin, F., Qin, B., 2017. Structural changes of gut microbiota in Parkinson's disease and its correlation with clinical features. Sci. China Life Sci. 60 (11), 1223.

Li, C., Cui, L., Yang, Y., Miao, J., Zhao, X., Zhang, J., Cui, G., Zhang, Y., 2019. Gut micro-biota differs between Parkinson's disease patients and healthy controls in northeast China. Front. Mol. Neurosci. 12, 171.

Liddelow, S.A., Guttenplan, K.A., Clarke, L.E., Bennett, F.C., Bohlen, C.J., Schirmer, L., Bennett, M.L., Münch, A.E., Chung, W.S., Peterson, T.C., Wilton, D.K., Frouin, A., Napier, B.A., Panicker, N., Kumar, M., Buckwalter, M.S., Rowitch, D.H., Dawson, V.L., Dawson, T.M., Stevens, B., Barres, B.A., 2017. Neurotoxic reactive astro-cytes are induced by activated microglia. Nature 541, 481–487.

Lin, J.C., Lin, C.S., Hsu, C.W., Lin, C.L., Kao, C.H., 2016. Association between Parkinson's disease and inflammatory bowel disease: a nationwide Taiwanese retrospective cohort study. Inflamm. Bowel Dis. 22 (5), 1049–1055.

Lin, A., Zheng, W., He, Y., Tang, W., Wei, X., He, R., Huang, W., Su, Y., Huang, Y., Zhou, H., Xie, H., 2018. Gut microbiota in patients with Parkinson's disease in southern China. Parkinsonism Relat. Disord. 53, 82–88.

Lionnet, A., Leclair-Visonneau, L., Neunlist, M., Murayama, S., Takao, M., Adler, C.H., Derkinderen, P., Beach, T.G., 2017. Does Parkinson's disease start in the gut? Acta Neu-ropathol. 135, 1–12.

Liu, B., Gao, H.M., Hong, J.S., 2003. Parkinson's disease and exposure to infectious agents and pesticides and the occurrence of brain injuries: role of neuroinflammation. Environ. Health Perspect. 111, 1065–1073.

Liu, Z., Lee, J., Krummey, S., Lu, W., Cai, H., Lenardo, M.J., 2011. The kinase LRRK2 is a regulator of the transcription factor NFAT that modulates the severity of inflammatory bowel disease. Nat. Immunol. 12, 1063–1070.

Liu, B., Fang, F., Pedersen, N.L., Tillander, A., Ludvigsson, J.F., Ekbom, A., Svenningsson, P., Chen, H., Wirdefeldt, K., 2017. Vagotomy and Parkinson disease: a Swedish register-based matched-cohort study. Neurology 88, 1996–2002.

Louis, P., Flint, H.J., 2009. Diversity, metabolism and microbial ecology of butyrate-producing bacteria from the human large intestine. FEMS Microbiol. Lett. 294, 1–8.

Lozupone, C.A., Stombaugh, J.I., Gordon, J.I., Jansson, J.K., Knight, R., 2012. Diversity, sta-bility and resilience of the human gut microbiota. Nature 489 (7415), 220–230.

Lubomski, M., Davis, R.L., Sue, C.M., 2019a. The gut microbiota: a novel therapeutic target in Parkinson's disease? Parkinsonism Relat. Disord. 66, 265–266.

Lubomski, M., Tan, A.H., Lim, S.Y., Holmes, A.J., Davis, R.L., Sue, C.M., 2019b. Parkinson's disease and the gastrointestinal microbiome. J. Neurol. 1–17. [Epub ahead of print].

Lull, M.E., Block, M.L., 2010. Microglial activation and chronic neurodegeneration. Neurotherapeutics 7, 354–365.

Lupton, J.R., 2004. Microbial degradation products influence colon cancer risk: the butyrate controversy. J. Nutr. 134, 479–482.

Maes, M., Kubera, M., Leunis, J.C., 2008. The gut-brain barrier in major depression: intestinal mucosal dysfunction with an increased translocation of LPS from gram negative enterobacteria (leaky gut) plays a role in the inflammatory pathophysiology of depression. Neuro Endocrinol. Lett. 29, 117–124.

Maggio, R., Riva, M., Vaglini, F., Fornai, F., Molteni, R., Armogida, M., Racagni, G., Corsini, G.U., 1998. Nicotine prevents experimental Parkinsonism in rodents and induces striatal increase of neurotrophic factors. J. Neurochem. 71 (6), 2439–2446.

Maini Rekdal, V., Bess, E.N., Bisanz, J.E., Turnbaugh, P.J., Balskus, E.P., 2019. Discovery and inhibition of an interspecies gut bacterial pathway for Levodopa metabolism. Science 364 (6445), eaau6323.

Makki, K., Deehan, E.C., Walter, J., Bäckhed, F., 2018. The impact of dietary fiber on gut microbiota in host health and disease. Cell Host Microbe 23, 705–715.

Manfredsson, F.P., Luk, K.C., Benskey, M.J., Gezer, A., Garcia, J., Kuhn, N.C., Sandoval, I.M., Patterson, J.R., O'Mara, A., Yonkers, R., Kordower, J.H., 2018. Induction of alpha-synuclein pathology in the enteric nervous system of the rat and non-human primate results in gastrointestinal dysmotility and transient CNS pathology. Neurobiol. Dis. 112, 106–118.

Manrique, P., Bolduc, B., Walk, S.T., van der Oost, J., de Vos, W.M., Young, M.J., 2016. Healthy human gut phageome. Proc. Natl. Acad. Sci. U. S. A. 113 (37), 10400–10405.

Manzanillo, P.S., Ayres, J.S., Watson, R.O., Collins, A.C., Souza, G., Rae, C.S., Schneider, D.S., Nakamura, K., Shiloh, M.U., Cox, J.S., 2013. The ubiquitin ligase parkin mediates resistance to intracellular pathogens. Nature 501, 512–516.

Mao, X., Ou, M.T., Karuppagounder, S.S., Kam, T.I., Yin, X., Xiong, Y., Ge, P., Umanah, G.E., Brahmachari, S., Shin, J.H., Kang, H.C., Zhang, J., Xu, J., Chen, R., Park, H., Andrabi, S.A., Kang, S.U., Gonçalves, R.A., Liang, Y., Zhang, S., Qi, C., Lam, S., Keiler, J.A., Tyson, J., Kim, D., Panicker, N., Yun, S.P., Workman, C.J., Vignali, D.A., Dawson, V.L., Ko, H.S., Dawson, T.M., 2016. Pathological α-synuclein transmission initiated by binding lymphocyte-activation gene 3. Science 353 (6307), aah3374.

Maraki, M.I., Yannakoulia, M., Stamelou, M., Stefanis, L., Xiromerisiou, G., Kosmidis, M.H., Dardiotis, E., Hadjigeorgiou, G.M., Sakka, P., Anastasiou, C.A., Simopoulou, E., Scarmeas, N., 2019. Mediterranean diet adherence is related to reduced probability of prodromal Parkinson's disease. Mov. Disord. 34, 48–57.

Marcinek, P., Jha, A.N., Shinde, V., Sundaramoorthy, A., Rajkumar, R., Suryadevara, N.C., Neela, S.K., van Tong, H., Balachander, V., Valluri, V.L., Thangaraj, K., Velavan, T.P., 2013. LRRK2 and RIPK2 variants in the NOD 2-mediated signaling pathway are associated with susceptibility to *Mycobacterium leprae* in Indian populations. PLoS One 8, e73103.

Marras, C., Canning, C.G., Goldman, S.M., 2019. Environment, lifestyle, and Parkinson's disease: implications for prevention in the next decade. Mov. Disord. 34, 801–811.

Mateo, I., Infante, J., Sánchez-Juan, P., García-Gorostiaga, I., Rodríguez-Rodríguez, E., Vázquez-Higuera, J.L., Berciano, J., Combarros, O., 2010. Serum heme oxygenase-1 levels are increased in Parkinson's disease but not in Alzheimer's disease. Acta Neurol. Scand. 121, 136–138.

Matt, S.M., Allen, J.M., Lawson, M.A., Mailing, L.J., Woods, J.A., Johnson, R.W., 2018. Butyrate and dietary soluble fiber improve neuroinflammation associated with aging in mice. Front. Immunol. 9, 1832.

Metz, G.A., 2007. Stress as a modulator of motor system function and pathology. Rev. Neurosci. 18, 209–222.

Migdalska-Richards, A., Schapira, A.H., 2016. The relationship between glucocerebrosidase mutations and Parkinson disease. J. Neurochem. 139 (Suppl. 1), 77–90.

Mihaila, D., Donegan, J., Barns, S., LaRocca, D., Du, Q., Zheng, D., Vidal, M., Neville, C., Uhlig, R., Middleton, F.A., 2019. The oral microbiome of early stage Parkinson's disease and its relationship with functional measures of motor and non-motor function. PLoS One 14, e0218252.

Miller, G.E., Engen, P.A., Gillevet, P.M., Shaikh, M., Sikaroodi, M., Forsyth, C.B., Mutlu, E., Keshavarzian, A., 2016. Lower neighborhood socioeconomic status associated with reduced diversity of the colonic microbiota in healthy adults. PLoS One 11, e0148952.

Minato, T., Maeda, T., Fujisawa, Y., Tsuji, H., Nomoto, K., Ohno, K., Hirayama, M., 2017. Progression of Parkinson's disease is associated with gut dysbiosis: two-year follow-up study. PLoS One 12, e0187307.

Mitchell, R.W., On, N.H., Del Bigio, M.R., Miller, D.W., Hatch, GM., 2011. Fatty acid transport protein expression in human brain and potential role in fatty acid transport across human brain microvessel endothelial cells. J. Neurochem. 117 (4), 735–746.

Mitsumoto, A., Nakagawa, Y., 2001. DJ-1 is an indicator for endogenous reactive oxygen species elicited by endotoxin. Free Radic. Res. 35, 885–893.

Muegge, B.D., Kuczynski, J., Knights, D., Clemente, J.C., González, A., Fontana, L., Henrissat, B., Knight, R., Gordon, J.I., 2011. Diet drives convergence in gut microbiome functions across mammalian phylogeny and within humans. Science 332, 970–974.

Mulak, A., Bonaz, B., 2015. Brain-gut-microbiota axis in Parkinson's disease. World J. Gastroenterol. 21, 10609–10620.

Mutez, E., Larvor, L., Leprêtre, F., Mouroux, V., Hamalek, D., Kerckaert, J.P., Pérez-Tur, J., Waucquier, N., Vanbesien-Mailliot, C., Duflot, A., Devos, D., Defebvre, L., Kreisler, A., Frigard, B., Destée, A., Chartier-Harlin, M.C., 2011. Transcriptional profile of Parkinson blood mononuclear cells with LRRK2 mutation. Neurobiol. Aging 32, 1839–1848.

Nakayama, T., Oishi, K., 2013. Influence of coffee (*Coffea arabica*) and galacto-oligosaccharide consumption on intestinal microbiota and the host responses. FEMS Microbiol. Lett. 343 (2), 161–168.

Narożańska, E., Białecka, M., Adamiak-Giera, U., Gawrońska-Szklarz, B., Sołtan, W., Schinwelski, M., Robowski, P., Madaliński, M.H., Sławek, J., 2014. Pharmacokinetics of levodopa in patients with Parkinson disease and motor fluctuations depending on the presence of *Helicobacter pylori* infection. Clin. Neuropharmacol. 37 (4), 96–99.

Nixon, A.M., Meadowcroft, M.D., Neely, E.B., Snyder, A.M., Purnell, C.J., Wright, J., Lamendella, R., Nandar, W., Huang, X., Connor, J.R., 2018. HFE genotype restricts the response to paraquat in a mouse model of neurotoxicity. J. Neurochem. 145, 299–311.

Noble, E.E., Hsu, T.M., Kanoski, S.E., 2017. Gut to brain dysbiosis: mechanisms linking western diet consumption, the microbiome, and cognitive impairment. Front. Behav. Neurosci. 11, 9.

Noh, S.Y., Kang, S.S., Yun, C.H., Han, S.H., 2015. Lipoteichoic acid from *Lactobacillus plantarum* inhibits Pam2CSK4-induced IL-8 production in human intestinal epithelial cells. Mol. Immunol. 64 (1), 183–189.

Norden, D.M., Godbout, J.P., 2012. Review: microglia of the aged brain: primed to be activated and resistant to regulation. Neuropathol. Appl. Neurobiol. 39, 19–34.

Noyce, A.J., Bestwick, J.P., Silveira-Moriyama, L., Hawkes, C.H., Giovannoni, G., Lees, A.J., Schrag, A., 2012. Meta-analysis of early nonmotor features and risk factors for Parkinson disease. Ann. Neurol. 72, 893–901.

Obeso, J.A., Stamelou, M., Goetz, C.G., Poewe, W., Lang, A.E., Weintraub, D., Burn, D., Halliday, G.M., Bezard, E., Przedborski, S., Lehericy, S., Brooks, D.J., Rothwell, J.C., Hallett, M., DeLong, M.R., Marras, C., Tanner, C.M., Ross, G.W., Langston, J.W., Klein, C., Bonifati, V., Jankovic, J., Lozano, A.M., Deuschl, G., Bergman, H., Tolosa, E., Rodriguez-Violante, M., Fahn, S., Postuma, R.B., Berg, D., Marek, K., Standaert, D.G., Surmeier, D.J., Olanow, C.W., Kordower, J.H., Calabresi, P., Schapira, A.H.V., Stoessl, A.J., 2017. Past, present, and future of Parkinson's disease: a special essay on the 200th Anniversary of the Shaking Palsy. Mov. Disord. 32, 1264–1310.

Obrenovich, M.E.M., 2018. Leaky gut, leaky brain? Microorganisms 6 (4), E107.

Olanow, C.W., Brundin, P., 2013. Parkinson's disease and alpha synuclein: is Parkinson's disease a prion-like disorder? Mov. Disord. 28, 31–40.

Olszak, T., An, D., Zeissig, S., Vera, M.P., Richter, J., Franke, A., Glickman, J.N., Siebert, R., Baron, R.M., Kasper, D.L., Blumberg, R.S., 2012. Microbial exposure during early life has persistent effects on natural killer T cell function. Science 336 (6080), 489–493.

Palmer, C., Bik, E.M., DiGiulio, D.B., Relman, D.A., Brown, P.O., 2007. Development of the human infant intestinal microbiota. PLoS Biol. 5 (7), e177.

Park, Y., Subar, A.F., Hollenbeck, A., Schatzkin, A., 2011. Dietary fiber intake and mortality in the NIH-AARP diet and health study. Arch. Intern. Med. 171, 1061–1068.

Paschos, G.K., FitzGerald, G.A., 2017. Circadian clocks and metabolism: implications for microbiome and aging. Trends Genet. 33, 760–769.

Pereira, P.A.B., Aho, V.T.E., Paulin, L., Pekkonen, E., Auvinen, P., Scheperjans, F., 2017. Oral and nasal microbiota in Parkinson's disease. Parkinsonism Relat. Disord. 38, 61–67.

Perez-Pardo, P., Dodiya, H.B., Broersen, L.M., Douna, H., van Wijk, N., Lopes da Silva, S., Garssen, J., Keshavarzian, A., Kraneveld, A.D., 2017a. Gut-brain and brain-gut axis in Parkinson's disease models: effects of a uridine and fish oil diet. Nutr. Neurosci. 21, 391–402.

Perez-Pardo, P., Kliest, T., Dodiya, H.B., Broersen, L.M., Garssen, J., Keshavarzian, A., Kraneveld, A.D., 2017b. The gut-brain axis in Parkinson's disease: possibilities for food-based therapies. Eur. J. Pharmacol. 817, 86–95.

Perez-Pardo, P., Dodiya, H.B., Engen, P.A., Naqib, A., Forsyth, C.B., Green, S.J., Garssen, J., Keshavarzian, A., Kraneveld, A.D., 2018a. Gut bacterial composition in a mouse model of Parkinson's disease. Benefic. Microbes 9, 799–814.

Perez-Pardo, P., Dodiya, H.B., Engen, P.A., Forsyth, C.B., Huschens, A.M., Shaikh, M., Voigt, R.M., Naqib, A., Green, S.J., Kordower, J.H., Shannon, K.M., Garssen, J., Kraneveld, A.D., Keshavarzian, A., 2018b. Role of TLR4 in the gut-brain axis in Parkinson's disease: a translational study from men to mice. Gut 68, 829–843.

Perry, V.H., Holmes, C., 2014. Microglial priming in neurodegenerative disease. Nat. Rev. Neurol. 10, 217–224.

Petrov, V.A., Saltykova, I.V., Zhukova, I.A., Alifirova, V.M., Zhukova, N.G., Dorofeeva, Y.B., Tyakht, A.V., Kovarsky, B.A., Alekseev, D.G., Kostryukova, E.S., Mironova, Y.S., Izhboldina, O.P., Nikitina, M.A., Perevozchikova, T.V., Fait, E.A., Babenko, V.V., Vakhitova, M.T., Govorun, V.M., Sazonov, A.E., 2017. Analysis of gut microbiota in patients with Parkinson's disease. Bull. Exp. Biol. Med. 162, 734–737.

Pezzoli, G., Cereda, E., 2013. Exposure to pesticides or solvents and risk of Parkinson disease. Neurology 80 (22), 2035–2041.

Pfeiffer, R.F., 2011. Gastrointestinal dysfunction in Parkinson's disease. Parkinsonism Relat. Disord. 17, 10–15.

Piccinini, A.M., Midwood, K.S., 2010. DAMPening inflammation by modulating TLR signalling. Mediat. Inflamm. 2010, 672395.

Pierantozzi, M., Pietroiusti, A., Brusa, L., Galati, S., Stefani, A., Lunardi, G., Fedele, E., Sancesario, G., Bernardi, G., Bergamaschi, A., Magrini, A., Stanzione, P., Galante, A., 2006. Helicobacter pylori eradication and l-dopa absorption in patients with PD and motor fluctuations. Neurology 66 (12), 1824–1829.

Pozuelo, M., Panda, S., Santiago, A., Mendez, S., Accarino, A., Santos, J., Guarner, F., Azpiroz, F., Manichanh, C., 2015. Reduction of butyrate- and methane-producing microorganisms in patients with Irritable Bowel Syndrome. Sci. Rep. 5, 12693.

Proctor, L.M., 2011. The Human Microbiome Project in 2011 and beyond. Cell Host Microbe 10 (4), 287–291.

Prytz, H., Benoni, C., Tagesson, C., 1989. Does smoking tighten the gut? Scand. J. Gastroenterol. 24, 1084–1088.

Qian, Y., Yang, X., Xu, S., Wu, C., Song, Y., Qin, N., Chen, S.D., Xiao, Q., 2018. Alteration of the fecal microbiota in Chinese patients with Parkinson's disease. Brain Behav. Immun. 70, 194–202.

Qin, L., Wu, X., Block, M.L., Liu, Y., Breese, G.R., Hong, J.S., Knapp, D.J., Crews, F.T., 2007. Systemic LPS causes chronic neuroinflammation and progressive neurodegeneration. Glia 55, 453–462.

Qin, J., Li, Y., Cai, Z., Li, S., Zhu, J., Zhang, F., Liang, S., Zhang, W., Guan, Y., Shen, D., Peng, Y., Zhang, D., Jie, Z., Wu, W., Qin, Y., Xue, W., Li, J., Han, L., Lu, D., Wu, P., Dai, Y., Sun, X., Li, Z., Tang, A., Zhong, S., Li, X., Chen, W., Xu, R., Wang, M., Feng, Q., Gong, M., Yu, J., Zhang, Y., Zhang, M., Hansen, T., Sanchez, G., Raes, J., Falony, G., Okuda, S., Almeida, M., LeChatelier, E., Renault, P., Pons, N., Batto, J.M., Zhang, Z., Chen, H., Yang, R., Zheng, W., Li, S., Yang, H., Wang, J., Ehrlich, S.D., Nielsen, R., Pedersen, O., Kristiansen, K., Wang, J., 2012. A metagenome-wide association study of gut microbiota in type 2 diabetes. Nature 490, 55–60.

Qin, N., Yang, F., Li, A., Prifti, E., Chen, Y., Shao, L., Guo, J., Le Chatelier, E., Yao, J., Wu, L., Zhou, J., Ni, S., Liu, L., Pons, N., Batto, J.M., Kennedy, S.P., Leonard, P., Yuan, C., Ding, W., Chen, Y., Hu, X., Zheng, B., Qian, G., Xu, W., Ehrlich, S.D., Zheng, S., Li, L., 2014. Alterations of the human gut microbiome in liver cirrhosis. Nature 513 (7516), 59–64.

Quigley, E.M., Quera, R., 2006. Small intestinal bacterial overgrowth: roles of antibiotics, prebiotics, and probiotics. Gastroenterology 130 (2 Suppl. 1), S78–S90, Review.

Raj, T., Rothamel, K., Mostafavi, S., Ye, C., Lee, M.N., Replogle, J.M., Feng, T., Lee, M., Asinovski, N., Frohlich, I., Imboywa, S., Von Korff, A., Okada, Y., Patsopoulos, N.A., Davis, S., McCabe, C., Paik, H.I., Srivastava, G.P., Raychaudhuri, S., Hafler, D.A., Koller, D., Regev, A., Hacohen, N., Mathis, D., Benoist, C., Stranger, B.E., De Jager, P.L., 2014. Polarization of the effects of autoimmune and neurodegenerative risk alleles in leukocytes. Science 344, 519–523.

Reyes, A., Haynes, M., Hanson, N., Angly, F.E., Heath, A.C., Rohwer, F., Gordon, J.I., 2010. Viruses in the faecal microbiota of monozygotic twins and their mothers. Nature 466, 334–338.

Riederer, P., Berg, D., Casadei, N., Cheng, F., Classen, J., Dresel, C., Jost, W., Krüger, R., Müller, T., Reichmann, H., Rieß, O., Storch, A., Strobel, S., van Eimeren, T.,

Völker, H.U., Winkler, J., Winklhofer, K.F., Wüllner, U., Zunke, F., Monoranu, C.M., 2019. Alpha-synuclein in Parkinson's disease: causal or bystander? J. Neural Transm. (Vienna) 126 (7), 815–840.

Rodriguez-Valera, F., Martin-Cuadrado, A.B., Rodriguez-Brito, B., Pasić, L., Thingstad, T.F., Rohwer, F., Mira, A., 2009. Explaining microbial population genomics through phage predation. Nat. Rev. Microbiol. 7 (11), 828–836.

Roederer, M., Quaye, L., Mangino, M., Beddall, M.H., Mahnke, Y., Chattopadhyay, P., Tosi, I., Napolitano, L., Terranova Barberio, M., Menni, C., Villanova, F., Di Meglio, P., Spector, T.D., Nestle, F.O., 2015. The genetic architecture of the human immune system: a bioresource for autoimmunity and disease pathogenesis. Cell 161 (2), 387–403.

Ruffmann, C., Parkkinen, L., 2016. Gut feelings about α-synuclein in gastrointestinal biopsies: biomarker in the making? Mov. Disord. 31 (2), 193–202.

Sampson, T., 2019. The impact of indigenous microbes on Parkinson's disease. Neurobiol. Dis. 135, 104426.

Sampson, T.R., Debelius, J.W., Thron, T., Janssen, S., Shastri, G.G., Ilhan, Z.E., Challis, C., Schretter, C.E., Rocha, S., Gradinaru, V., Chesselet, M.F., Keshavarzian, A., Shannon, K.M., Krajmalnik-Brown, R., Wittung-Stafshede, P., Knight, R., Mazmanian, S.K., 2016. Gut microbiota regulate motor deficits and neuroinflammation in a model of Parkinson's disease. Cell 167, 1469–1480.e12.

Sarangi, A.N., Goel, A., Aggarwal, R., 2019. Methods for studying gut microbiota: a primer for physicians. J. Clin. Exp. Hepatol. 9 (1), 62–73.

Sartor, R.B., 2010. Genetics and environmental interactions shape the intestinal microbiome to promote inflammatory bowel disease versus mucosal homeostasis. Gastroenterology 139, 1816–1819.

Scheperjans, F., 2018. The prodromal microbiome. Mov. Disord. 33, 5–7.

Scheperjans, F., Pekkonen, E., Kaakkola, S., Auvinen, P., 2015a. Linking smoking, coffee, urate, and Parkinson's disease—A role for gut microbiota? J. Parkinsons Dis. 5, 255–262.

Scheperjans, F., Aho, V., Pereira, P.A., Koskinen, K., Paulin, L., Pekkonen, E., Haapaniemi, E., Kaakkola, S., Eerola-Rautio, J., Pohja, M., Kinnunen, E., Murros, K., Auvinen, P., 2015b. Gut microbiota are related to Parkinson's disease and clinical phenotype. Mov. Disord. 30 (3), 350–358.

Scheperjans, F., Derkinderen, P., Borghammer, P., 2018. The gut and Parkinson's disease: hype or hope? J. Parkinsons Dis. 8, S31–S39.

Segain, J.P., Raingeard de la Blétière, D., Bourreille, A., Leray, V., Gervois, N., Rosales, C., Ferrier, L., Bonnet, C., Blottière, H.M., Galmiche, J.P., 2000. Butyrate inhibits inflammatory responses through NFkappaB inhibition: implications for Crohn's disease. Gut 47 (3), 397–403.

Shaikh, M., Rajan, K., Forsyth, C.B., Voigt, R.M., Keshavarzian, A., 2015. Simultaneous gas-chromatographic urinary measurement of sugar probes to assess intestinal permeability: use of time course analysis to optimize its use to assess regional gut permeability. Clin. Chim. Acta 442C, 24–32.

Sharon, G., Sampson, T.R., Geschwind, D.H., Mazmanian, S.K., 2016. The central nervous system and the gut microbiome. Cell 167 (4), 915–932.

Singh, R.K., Chang, H.W., Yan, D., Lee, K.M., Ucmak, D., Wong, K., Abrouk, M., Farahnik, B., Nakamura, M., Zhu, T.H., Bhutani, T., Liao, W., 2017. Influence of diet on the gut microbiome and implications for human health. J. Transl. Med. 15, 73.

Smith, A.D., Castro, S.L., Zigmond, M.J., 2002. Stress-induced Parkinson's disease: a working hypothesis. Physiol. Behav. 77, 527–531.

Smith, L.K., Jadavji, N.M., Colwell, K.L., Katrina Perehudoff, S., Metz, G.A., 2008. Stress accelerates neural degeneration and exaggerates motor symptoms in a rat model of Parkinson's disease. Eur. J. Neurosci. 27, 2133–2146.

Sonnenburg, J.L., Xu, J., Leip, D.D., Chen, C.H., Westover, B.P., Weatherford, J., Buhler, J.D., Gordon, J.I., 2005. Glycan foraging in vivo by an intestine-adapted bacterial symbiont. Science 307, 1955–1959.

Soret, R., Chevalier, J., De Coppet, P., Poupeau, G., Derkinderen, P., Segain, J.P., Neunlist, M., 2010. Short-chain fatty acids regulate the enteric neurons and control gastrointestinal motility in rats. Gastroenterology 138, 1772–1782.

St Laurent, R., O'Brien, L.M., Ahmad, S.T., 2013. Sodium butyrate improves locomotor impairment and early mortality in a rotenone-induced Drosophila model of Parkinson's disease. Neuroscience 246, 382–390.

Sun, M.F., Zhu, Y.L., Zhou, Z.L., Jia, X.B., Xu, Y.D., Yang, Q., Cui, C., Shen, Y.Q., 2018. Neuroprotective effects of fecal microbiota transplantation on MPTP-induced Parkinson's disease mice: gut microbiota, glial reaction and TLR4/TNF-alpha signaling pathway. Brain Behav. Immun. 70, 48–60.

Surmeier, D.J., Obeso, J.A., Halliday, G.M., 2017. Parkinson's disease is not simply a prion disorder. J. Neurosci. 37 (41), 9799–9807.

Svensson, E., Horváth-Puhó, E., Thomsen, R.W., Djurhuus, J.C., Pedersen, L., Borghammer, P., Sørensen, H.T., 2015. Vagotomy and subsequent risk of Parkinson's disease. Ann. Neurol. 78, 522–529.

Tan, A.H., Mahadeva, S., Thalha, A.M., Gibson, P.R., Kiew, C.K., Yeat, C.M., Ng, S.W., Ang, S.P., Chow, S.K., Tan, C.T., Yong, H.S., Marras, C., Fox, S.H., Lim, S.Y., 2014. Small intestinal bacterial overgrowth in Parkinson's disease. Parkinsonism Relat. Disord. 20 (5), 535–540.

Tan, A.H., Chong, C.W., Song, S.L., Teh, C.S.J., Yap, I.K.S., Loke, M.F., Tan, Y.Q., Yong, H.S., Mahadeva, S., Lang, A.E., Lim, S.Y., 2018. Altered gut microbiome and metabolome in patients with multiple system atrophy. Mov. Disord. 33, 174–176.

Tang, M.S., Poles, J., Leung, J.M., Wolff, M.J., Davenport, M., Lee, S.C., Lim, Y.A., Chua, K.H., Loke, P., Cho, I., 2015. Inferred metagenomic comparison of mucosal and fecal microbiota from individuals undergoing routine screening colonoscopy reveals similar differences observed during active inflammation. Gut Microbes 6 (1), 48–56.

Thévenet, J., Pescini Gobert, R., Hooft van Huijsduijnen, R., Wiessner, C., Sagot, Y.J., 2011. Regulation of LRRK2 expression points to a functional role in human monocyte maturation. PLoS One 6, e21519.

Thom, S.R., 2011. Hyperbaric oxygen: its mechanisms and efficacy. Plast. Reconstr. Surg. 127 (Suppl. 1), 131S–141S.

Troncoso-Escudero, P., Parra, A., Nassif, M., Vidal, R.L., 2018. Outside in: unraveling the role of neuroinflammation in the progression of Parkinson's disease. Front. Neurol. 9, 860.

Turnbaugh, P.J., Bäckhed, F., Fulton, L., Gordon, J.I., 2008. Diet-induced obesity is linked to marked but reversible alterations in the mouse distal gut microbiome. Cell Host Microbe 3, 213–223.

Turnbaugh, P.J., Hamady, M., Yatsunenko, T., Cantarel, B.L., Duncan, A., Ley, R.E., Sogin, M.L., Jones, W.J., Roe, B.A., Affourtit, J.P., Egholm, M., Henrissat, B., Heath, A.C., Knight, R., Gordon, J.I., 2009. A core gut microbiome in obese and lean twins. Nature 457, 480–484.

Tysnes, O.B., Storstein, A., 2017. Epidemiology of Parkinson's disease. J. Neural Transm. (Vienna) 124, 901–905.

Tysnes, O.B., Kenborg, L., Herlofson, K., Steding-Jessen, M., Horn, A., Olsen, J.H., Reichmann, H., 2015. Does vagotomy reduce the risk of Parkinson's disease? Ann. Neurol. 78 (6), 1011–1012.

Ueno, N., Fujiya, M., Segawa, S., Nata, T., Moriichi, K., Tanabe, H., Mizukami, Y., Kobayashi, N., Ito, K., Kohgo, Y., 2011. Heat-killed body of *Lactobacillus brevis* SBC8803 ameliorates intestinal injury in a murine model of colitis by enhancing the intestinal barrier function. Inflamm. Bowel Dis. 17 (11), 2235–2250.

Ulusoy, A., Phillips, R.J., Helwig, M., Klinkenberg, M., Powley, T.L., Di Monte, D.A., 2017. Brain-to-stomach transfer of α-synuclein via vagal preganglionic projections. Acta Neuropathol. 133 (3), 381–393.

Umeno, J., Asano, K., Matsushita, T., Matsumoto, T., Kiyohara, Y., Iida, M., Nakamura, Y., Kamatani, N., Kubo, M., 2011. Meta-analysis of published studies identified eight additional common susceptibility loci for Crohn's disease and ulcerative colitis. Inflamm. Bowel Dis. 17, 2407–2415.

Unger, M.M., Spiegel, J., Dillmann, K.U., Grundmann, D., Philippeit, H., Bürmann, J., Faßbender, K., Schwiertz, A., Schäfer, K.H., 2016. Short chain fatty acids and gut microbiota differ between patients with Parkinson's disease and age-matched controls. Parkinsonism Relat. Disord. 32, 66–72.

Upadhyaya, S., Banerjee, G., 2015. Type 2 diabetes and gut microbiome: at the intersection of known and unknown. Gut Microbes 6, 85–92.

Valles-Colomer, M., Falony, G., Darzi, Y., Tigchelaar, E.F., Wang, J., Tito, R.Y., Schiweck, C., Kurilshikov, A., Joossens, M., Wijmenga, C., Claes, S., Van Oudenhove, L., Zhernakova, A., Vieira-Silva, S., Raes, J., 2019. The neuroactive potential of the human gut microbiota in quality of life and depression. Nat. Microbiol. 4 (4), 623–632.

van Kessel, S.P., Frye, A.K., El-Gendy, A.O., Castejon, M., Keshavarzian, A., van Dijk, G., El Aidy, S., 2019. Gut bacterial tyrosine decarboxylases restrict levels of levodopa in the treatment of Parkinson's disease. Nat. Commun. 10, 310.

Van Limbergen, J., Wilson, D.C., Satsangi, J., 2009. The genetics of Crohn's disease. Annu. Rev. Genomics Hum. Genet. 10, 89–116.

Videnovic, A., Golombek, D., 2013. Circadian and sleep disorders in Parkinson's disease. Exp. Neurol. 243, 45–56.

Videnovic, A., Willis, G.L., 2016. Circadian system—a novel diagnostic and therapeutic target in Parkinson's disease? Mov. Disord. 31, 260–269.

Villumsen, M., Aznar, S., Pakkenberg, B., Jess, T., Brudek, T., 2019. Inflammatory bowel disease increases the risk of Parkinson's disease: a Danish nationwide cohort study 1977–2014. Gut 68 (1), 18–24.

Vindigni, S.M., Zisman, T.L., Suskind, D.L., Damman, C.J., 2016. The intestinal microbiome, barrier function, and immune system in inflammatory bowel disease: a tripartite pathophysiological circuit with implications for new therapeutic directions. Ther. Adv. Gastroenterol. 9, 606–625.

Vinogradov, S.A., Wilson, D.F., 2012. Porphyrin-dendrimers as biological oxygen sensors. In: Capagna, S., Ceroni, P. (Eds.), Designing Dendrimers. Wiley, New York.

Vital, M., Howe, A.C., Tiedje, J.M., 2014. Revealing the bacterial butyrate synthesis pathways by analyzing (meta)genomic data. MBio 5, e00889–14.

Voigt, R.M., Forsyth, C.B., Green, S.J., Engen, P.A., Keshavarzian, A., 2016. Circadian rhythm and the gut microbiome. Int. Rev. Neurobiol. 131, 193–205.

Waak, J., Weber, S.S., Waldenmaier, A., Görner, K., Alunni-Fabbroni, M., Schell, H., Vogt-Weisenhorn, D., Pham, T.T., Reumers, V., Baekelandt, V., Wurst, W., Kahle, P.J., 2009. Regulation of astrocyte inflammatory responses by the Parkinson's disease-associated gene DJ-1. FASEB J. 23, 2478–2489.

Wang, H., Zhao, J.X., Hu, N., Ren, J., Du, M., Zhu, M.J., 2012. Sidestream smoking reduces intestinal inflammation and increases expression of tight junction proteins. World J. Gastroenterol. 18, 2180–2187.

Watson, L., Keatinge, M., Gegg, M., Bai, Q., Sandulescu, M.C., Vardi, A., Futerman, A.H., Schapira, A.H.V., Burton, E.A., Bandmann, O., 2019. Ablation of the pro-inflammatory master regulator miR-155 does not mitigate neuroinflammation or neurodegeneration in a vertebrate model of Gaucher's disease. Neurobiol. Dis. 127, 563–569.

Weimers, P., Halfvarson, J., Sachs, M.C., Saunders-Pullman, R., Ludvigsson, J.F., Peter, I., Burisch, J., Olén, O., 2018. Inflammatory bowel disease and Parkinson's disease: a nation-wide Swedish cohort study. Inflamm. Bowel Dis. 25, 111–123.

West, A.P., Koblansky, A.A., Ghosh, S., 2006. Recognition and signaling by toll-like receptors. Annu. Rev. Cell Dev. Biol. 22, 409–437.

Wu, X., Chen, P.S., Dallas, S., Wilson, B., Block, M.L., Wang, C.C., Kinyamu, H., Lu, N., Gao, X., Leng, Y., Chuang, D.M., Zhang, W., Lu, R.B., Hong, J.S., 2008. Histone deacetylase inhibitors up-regulate astrocyte GDNF and BDNF gene transcription and protect dopaminergic neurons. Int. J. Neuropsychopharmacol. 11 (8), 1123–1134.

Wu, G.D., Chen, J., Hoffmann, C., Bittinger, K., Chen, Y.Y., Keilbaugh, S.A., Bewtra, M., Knights, D., Walters, W.A., Knight, R., Sinha, R., Gilroy, E., Gupta, K., Baldassano, R., Nessel, L., Li, H., Bushman, F.D., Lewis, J.D., 2011. Linking long-term dietary patterns with gut microbial enterotypes. Science 334 (6052), 105–108.

Xie, H., Guo, R., Zhong, H., Feng, Q., Lan, Z., Qin, B., Ward, K.J., Jackson, M.A., Xia, Y., Chen, X., Chen, B., Xia, H., Xu, C., Li, F., Xu, X., Al-Aama, J.Y., Yang, H., Wang, J., Kristiansen, K., Wang, J., Steves, C.J., Bell, J.T., Li, J., Spector, T.D., Jia, H., 2016. Shotgun metagenomics of 250 adult twins reveals genetic and environmental impacts on the gut microbiome. Cell Syst. 3 (6), 572–584. e3.

Xu, Z., Knight, R., 2015. Dietary effects on human gut microbiome diversity. Br. J. Nutr. 113 (Suppl), S1–S5.

Xu, Q., Park, Y., Huang, X., Hollenbeck, A., Blair, A., Schatzkin, A., Chen, H., 2011. Diabetes and risk of Parkinson's disease. Diabetes Care 34, 910–915.

Yadav, H., Lee, J.H., Lloyd, J., Walter, P., Rane, S.G., 2013. Beneficial metabolic effects of a probiotic via butyrate-induced GLP-1 hormone secretion. J. Biol. Chem. 288, 25088–25097.

Yamada, T., Shimizu, K., Ogura, H., Asahara, T., Nomoto, K., Yamakawa, K., Hamasaki, T., Nakahori, Y., Ohnishi, M., Kuwagata, Y., Shimazu, T., 2015. Rapid and sustained long-term decrease of fecal short-chain fatty acids in critically Ill patients with systemic inflammatory response syndrome. JPEN J. Parenter. Enteral Nutr. 39 (5), 569–577.

Yang, X., Qian, Y., Xu, S., Song, Y., Xiao, Q., 2017. Longitudinal analysis of fecal microbiome and pathologic processes in a rotenone induced mice model of Parkinson's disease. Front. Aging Neurosci. 9, 441.

Yatsunenko, T., Rey, F.E., Manary, M.J., Trehan, I., Dominguez-Bello, M.G., Contreras, M., Magris, M., Hidalgo, G., Baldassano, R.N., Anokhin, A.P., Heath, A.C., Warner, B., Reeder, J., Kuczynski, J., Caporaso, J.G., Lozupone, C.A., Lauber, C., Clemente, J.C.,

Knights, D., Knight, R., Gordon, J.I., 2012. Human gut microbiome viewed across age and geography. Nature 486, 222–227.

Yun, S.P., Kam, T.I., Panicker, N., Kim, S., Oh, Y., Park, J.S., Kwon, S.H., Park, Y.J., Karuppagounder, S.S., Park, H., Kim, S., Oh, N., Kim, N.A., Lee, S., Brahmachari, S., Mao, X., Lee, J.H., Kumar, M., An, D., Kang, S.U., Lee, Y., Lee, K.C., Na, D.H., Kim, D., Lee, S.H., Roschke, V.V., Liddelow, S.A., Mari, Z., Barres, B.A., Dawson, V.L., Lee, S., Dawson, T.M., Ko, H.S., 2018. Block of A1 astrocyte conversion by microglia is neuroprotective in models of Parkinson's disease. Nat. Med. 24, 931–938.

Zhang, F.R., Huang, W., Chen, S.M., Sun, L.D., Liu, H., Li, Y., Cui, Y., Yan, X.X., Yang, H.T., Yang, R.D., Chu, T.S., Zhang, C., Zhang, L., Han, J.W., Yu, G.Q., Quan, C., Yu, Y.X., Zhang, Z., Shi, B.Q., Zhang, L.H., Cheng, H., Wang, C.Y., Lin, Y., Zheng, H.F., Fu, X.A., Zuo, X.B., Wang, Q., Long, H., Sun, Y.P., Cheng, Y.L., Tian, H.Q., Zhou, F.S., Liu, H.X., Lu, W.S., He, S.M., Du, W.L., Shen, M., Jin, Q.Y., Wang, Y., Low, H.Q., Erwin, T., Yang, N.H., Li, J.Y., Zhao, X., Jiao, Y.L., Mao, L.G., Yin, G., Jiang, Z.X., Wang, X.D., Yu, J.P., Hu, Z.H., Gong, C.H., Liu, Y.Q., Liu, R.Y., Wang, D.M., Wei, D., Liu, J.X., Cao, W.K., Cao, H.Z., Li, Y.P., Yan, W.G., Wei, S.Y., Wang, K.J., Hibberd, M.L., Yang, S., Zhang, X.J., Liu, J.J., 2009a. Genomewide association study of leprosy. N. Engl. J. Med. 361, 2609–2618.

Zhang, R., Miller, R.G., Gascon, R., Champion, S., Katz, J., Lancero, M., Narvaez, A., Honrada, R., Ruvalcaba, D., McGrath, M.S., 2009b. Circulating endotoxin and systemic immune activation in sporadic amyotrophic lateral sclerosis (sALS). J. Neuroimmunol. 206, 121–124.

Zhang, L., Li, L., Holscher, C., 2018. Neuroprotective effects of the novel GLP-1 long acting analogue semaglutide in the MPTP Parkinson's disease mouse model. Neuropeptides 71, 70–80.

Zhao, Y., Cong, L., Jaber, V., Lukiw, W.J., 2017a. Microbiome-derived lipopolysaccharide enriched in the perinuclear region of Alzheimer's disease brain. Front. Immunol. 8, 1064.

Zhao, Y., Jaber, V., Lukiw, W.J., 2017b. Secretory products of the human GI tract microbiome and their potential impact on Alzheimer's disease (AD): detection of lipopolysaccharide (LPS) in AD hippocampus. Front. Cell. Infect. Microbiol. 7, 318.

Zhao, L., Zhang, F., Ding, X., Wu, G., Lam, Y.Y., Wang, X., Fu, H., Xue, X., Lu, C., Ma, J., Yu, L., Xu, C., Ren, Z., Xu, Y., Xu, S., Shen, H., Zhu, X., Shi, Y., Shen, Q., Dong, W., Liu, R., Ling, Y., Zeng, Y., Wang, X., Zhang, Q., Wang, J., Wang, L., Wu, Y., Zeng, B., Wei, H., Zhang, M., Peng, Y., Zhang, C., 2018. Gut bacteria selectively promoted by dietary fibers alleviate type 2 diabetes. Science 359, 1151–1156.

Zhou, Z.L., Jia, X.B., Sun, M.F., Zhu, Y.L., Qiao, C.M., Zhang, B.P., Zhao, L.P., Yang, Q., Cui, C., Chen, X., Shen, Y.Q., 2019. Neuroprotection of fasting mimicking diet on MPTP-induced Parkinson's disease mice via gut microbiota and metabolites. Neurotherapeutics 16 (3), 741–760.

Zinöcker, M.K., Lindseth, I.A., 2018. The Western diet-microbiome-host interaction and its role in metabolic disease. Nutrients 10 (3), 365, pii: E365.

Further reading

Austin, K.W., Ameringer, S.W., Cloud, L.J., 2016. An integrated review of psychological stress in Parkinson's disease: biological mechanisms and symptom and health outcomes. Parkinsons Dis. 2016, 9869712.

Berg, D., Postuma, R.B., Bloem, B., Chan, P., Dubois, B., Gasser, T., Goetz, C.G., Halliday, G.M., Hardy, J., Lang, A.E., Litvan, I., Marek, K., Obeso, J., Oertel, W., Olanow, C.W., Poewe, W., Stern, M., Deuschl, G., 2014. Time to redefine PD? Introductory statement of the MDS Task Force on the definition of Parkinson's disease. Mov. Disord. 29, 454–462.

Biedermann, L., Brulisauer, K., Zeitz, J., Frei, P., Scharl, M., Vavricka, S.R., Fried, M., Loessner, M.J., Rogler, G., Schuppler, M., 2014. Smoking cessation alters intestinal microbiota: insights from quantitative investigations on human fecal samples using FISH. Inflamm. Bowel Dis. 20, 1496–1501.

Blesa, J., Phani, S., Jackson-Lewis, V., Przedborski, S., 2012. Classic and new animal models of Parkinson's disease. J. Biomed. Biotechnol. 2012, 845618.

Borghammer, P., Hamani, C., 2017. Preventing Parkinson disease by vagotomy: fact or fiction? Neurology 88 (21), 1982–1983.

Chen, H., O'Reilly, E., McCullough, M.L., Rodriguez, C., Schwarzschild, M.A., Calle, E.E., Thun, M.J., Ascherio, A., 2007. Consumption of dairy products and risk of Parkinson's disease. Am. J. Epidemiol. 165, 998–1006.

Cui, S.S., Du, J.J., Liu, S.H., Meng, J., Lin, Y.Q., Li, G., He, Y.X., Zhang, P.C., Chen, S., Wang, G., 2019. Serum soluble lymphocyte activation gene-3 as a diagnostic biomarker in Parkinson's disease: a pilot multicenter study. Mov. Disord. 34 (1), 138–141.

Freire, C., Koifman, S., 2012. Pesticide exposure and Parkinson's disease: epidemiological evidence of association. Neurotoxicology 33 (5), 947–971.

Fujiwara, H., Hasegawa, M., Dohmae, N., Kawashima, A., Masliah, E., Goldberg, M.S., Shen, J., Takio, K., Iwatsubo, T., 2002. Alpha-synuclein is phosphorylated in synucleinopathy lesions. Nat. Cell Biol. 4 (2), 160–164.

Gatto, N.M., Cockburn, M., Bronstein, J., Manthripragada, A.D., Ritz, B., 2009. Well-water consumption and Parkinson's disease in rural California. Environ. Health Perspect. 117 (12), 1912–1918.

Gonzalez, A., Stombaugh, J., Lozupone, C., Turnbaugh, P.J., Gordon, J.I., Knight, R., 2011. The mind-body-microbial continuum. Dialogues Clin. Neurosci. 13, 55–62.

Grider, J.R., Piland, B.E., 2007. The peristaltic reflex induced by short-chain fatty acids is mediated by sequential release of 5-HT and neuronal CGRP but not BDNF. Am. J. Physiol. Gastrointest. Liver Physiol. 292, G429–G437.

Harrison, I.F., Dexter, D.T., 2013. Epigenetic targeting of histone deacetylase: therapeutic potential in Parkinson's disease? Pharmacol. Ther. 140, 34–52.

Human Microbiome Project Consortium, 2012a. Structure, function and diversity of the healthy human microbiome. Nature 486, 207–214.

Human Microbiome Project Consortium, 2012b. Structure, function and diversity of the healthy human microbiome. Nature 486, 207–214.

Jansson, J., Willing, B., Lucio, M., Fekete, A., Dicksved, J., Halfvarson, J., Tysk, C., Schmitt-Kopplin, P., 2009. Metabolomics reveals metabolic biomarkers of Crohn's disease. PLoS One 4, e6386.

Jellinger, K.A., 2009. Formation and development of Lewy pathology: a critical update. J. Neurol. 256, 270–279.

Kalaitzakis, M.E., Graeber, M.B., Gentleman, S.M., Pearce, R.K., 2008. The dorsal motor nucleus of the vagus is not an obligatory trigger site of Parkinson's disease: a critical analysis of alpha-synuclein staging. Neuropathol. Appl. Neurobiol. 34 (3), 284–295.

Lacassagne, O., Kessler, J.P., 2000. Cellular and subcellular distribution of the amino-3-hydroxy-5-methyl-4-isoxazole propionate receptor subunit GluR2 in the rat dorsal vagal complex. Neuroscience 99, 557–563.

Lal, S., Kirkup, A.J., Brunsden, A.M., Thompson, D.G., Grundy, D., 2001. Vagal afferent responses to fatty acids of different chain length in the rat. Am. J. Physiol. Gastrointest. Liver Physiol. 281, G907–G915.

Lauretti, E., Di Meco, A., Merali, S., Praticò, D., 2016. Circadian rhythm dysfunction: a novel environmental risk factor for Parkinson's disease. Mol. Psychiatry 22, 280–286.

Lees, A.J., Hardy, J., Revesz, T., 2009. Parkinson's disease. Lancet 373, 2055–2066.

Lei, E., Vacy, K., Boon, W.C., 2016. Fatty acids and their therapeutic potential in neurological disorders. Neurochem. Int. 95, 75–84.

Liu, Z., Lenardo, M.J., 2012. The role of LRRK2 in inflammatory bowel disease. Cell Res. 22, 1092–1094.

Lopez de Maturana, R., Aguila, J.C., Sousa, A., Vazquez, N., Del Rio, P., Aiastui, A., Gorostidi, A., Lopez de Munain, A., Sanchez-Pernaute, R., 2014. Leucine-rich repeat kinase 2 modulates cyclooxygenase 2 and the inflammatory response in idiopathic and genetic Parkinson's disease. Neurobiol. Aging 35, 1116–1124.

MacFabe, D.F., 2015. Enteric short-chain fatty acids: microbial messengers of metabolism, mitochondria, and mind: implications in autism spectrum disorders. Microb. Ecol. Health Dis. 26, 28177.

Maes, M., Leunis, J.C., 2008. Normalization of leaky gut in chronic fatigue syndrome (CFS) is accompanied by a clinical improvement: effects of age, duration of illness and the translocation of LPS from gram-negative bacteria. Neuro Endocrinol. Lett. 29, 902–910.

Nankova, B.B., Agarwal, R., MacFabe, D.F., La Gamma, E.F., 2014. Enteric bacterial metabolites propionic and butyric acid modulate gene expression, including CREB-dependent catecholaminergic neurotransmission, in PC12 cells—possible relevance to autism spectrum disorders. PLoS One 9, e103740.

Nelson, K.E., Weinstock, G.M., Highlander, S.K., Worley, K.C., Creasy, H.H., Wortman, J.R., Rusch, D.B., Mitreva, M., Sodergren, E., Chinwalla, A.T., Feldgarden, M., Gevers, D., Haas, B.J., Madupu, R., Ward, D.V., Birren, B.W., Gibbs, R.A., Methe, B., Petrosino, J.F., Strausberg, R.L., Sutton, G.G., White, O.R., Wilson, R.K., Durkin, S., Giglio, M.G., Gujja, S., Howarth, C., Kodira, C.D., Kyrpides, N., Mehta, T., Muzny, D.M., Pearson, M., Pepin, K., Pati, A., Qin, X., Yandava, C., Zeng, Q., Zhang, L., Berlin, A.M., Chen, L., Hepburn, T.A., Johnson, J., McCorrison, J., Miller, J., Minx, P., Nusbaum, C., Russ, C., Sykes, S.M., Tomlinson, C.M., Young, S., Warren, W.C., Badger, J., Crabtree, J., Markowitz, V.M., Orvis, J., Cree, A., Ferriera, S., Fulton, L.L., Fulton, R.S., Gillis, M., Hemphill, L.D., Joshi, V., Kovar, C., Torralba, M., Wetterstrand, K.A., Abouelleil, A., Wollam, A.M., Buhay, C.J., Ding, Y., Dugan, S., FitzGerald, M.G., Holder, M., Hostetler, J., Clifton, S.W., Allen-Vercoe, E., Earl, A.M., Farmer, C.N., Liolios, K., Surette, M.G., Xu, Q., Pohl, C., Wilczek-Boney, K., Zhu, D., 2010. A catalog of reference genomes from the human microbiome. Science 328, 994–999.

Sharma, M., 2018. Brain and gut: partners in crime. Mov. Disord. 33 (7), 1098.

Takeuchi, O., Akira, S., 2010. Pattern recognition receptors and inflammation. Cell 140, 805–820.

Tran, T.A., Nguyen, A.D., Chang, J., Goldberg, M.S., Lee, J.K., Tansey, M.G., 2011. Lipopolysaccharide and tumor necrosis factor regulate Parkin expression via nuclear factor-kappa B. PLoS One 6, e23660.

Ventura, M., Turroni, F., Lugli, G.A., van Sinderen, D., 2014. Bifidobacteria and humans: our special friends, from ecological to genomics perspectives. J. Sci. Food Agric. 94 (2), 163–168.

Verberkmoes, N.C., Russell, A.L., Shah, M., Godzik, A., Rosenquist, M., Halfvarson, J., Lefsrud, M.G., Apajalahti, J., Tysk, C., Hettich, R.L., Jansson, J.K., 2009. Shotgun meta-proteomics of the human distal gut microbiota. ISME J. 3, 179–189.

Weimers, P., Halfvarson, J., Sachs, M.C., Ludvigsson, J.F., Peter, I., Olén, O., Burisch, J., 2019. Association between inflammatory bowel disease and Parkinson's disease: seek and you shall find? Gut 68 (1), 175–176.

Westfall, S., Lomis, N., Kahouli, I., Dia, S.Y., Singh, S.P., Prakash, S., 2017. Microbiome, probiotics and neurodegenerative diseases: deciphering the gut brain axis. Cell. Mol. Life Sci. 74, 3769–3787.

Zhu, F., Li, C., Gong, J., Zhu, W., Gu, L., Li, N., 2019. The risk of Parkinson's disease in inflammatory bowel disease: a systematic review and meta-analysis. Dig. Liver Dis. 51 (1), 38–42.

Novel approaches to counter protein aggregation pathology in Parkinson's disease

Simon R.W. Stott[a], Richard K. Wyse[a], Patrik Brundin[b],*

[a]*The Cure Parkinson's Trust, London, United Kingdom*
[b]*Center for Neurodegenerative Science, Van Andel Institute, Grand Rapids, MI, United States*
Corresponding author: Tel.: +1-616-234-5312, e-mail address: patrik.brundin@vai.org

Abstract

The primary neuropathological characteristics of the Parkinsonian brain are the loss of nigral dopamine neurons and the aggregation of alpha synuclein protein. Efforts to development potentially disease-modifying treatments have largely focused on correcting these aspects of the condition. In the last decade treatments targeting protein aggregation have entered the clinical pipeline. In this chapter we provide an overview of ongoing clinical trial programs for different therapies attempting to reduce protein aggregation pathology in Parkinson's disease. We will also briefly consider various novel approaches being proposed—and being developed preclinically—to inhibit/reduce aggregated protein pathology in Parkinson's.

Keywords

Aggregation, Alpha synuclein, Immunotherapy, Lewy body, Neurodegeneration

1 Introduction

Parkinson's disease (PD) is the second most common neurodegenerative disease after Alzheimer's disease (AD), and a neurological condition with an incidence that is growing faster than the rate of population growth (Dorsey and Bloem, 2018; Dorsey et al., 2018). Given that age is the primary factor associated with neurodegeneration, conditions like PD will inflict a greater burden on society as the elderly populations continue to grow in Western countries. Clinically, PD is characterized by a combination of motor features, including bradykinesia, rigidity, and a resting tremor, in addition to a collection of non-motor issues (such as gastrointestinal complaints, hyposmia, depression, cognitive decline, and sleep disturbance).

Progress in Brain Research, Volume 252, ISSN 0079-6123, https://doi.org/10.1016/bs.pbr.2019.10.007

The motor features of the condition are associated with the loss of dopamine neurons in the substantia nigra. Disease-modifying therapies that can halt the course of PD and improve quality of life are urgently needed.

The last 20 years have been marked by a tremendous increase in the number of clinical trials for PD. This trend has resulted both from progress where large genome analysis studies provided information on rare inherited forms of PD, which are also possibly relevant to the common idiopathic form, as well as the emergence of animal models with a high degree of construct validity that reflect the slow progressive nature of PD. Both have provided much more understanding of the biology underlying the condition than was available 20 years ago. As a consequence, many recent clinical trials have targeted actual disease modification for PD—attempting to "slow, stop or reverse" the progress of the condition.

The exact cause of the neurodegeneration observed in PD is unknown, but in many cases it is believed to be linked with the build-up of the protein alpha synuclein (α-syn). The strong focus on α-syn in the context of PD has stemmed from two primary sources. First, point mutations in the *SNCA* gene (which encodes α-syn) are associated with autosomal dominant PD (Polymeropoulos et al., 1997), indicating that a gain-of-function toxicity may be involved in these particular cases of familial PD (Cookson et al., 2008). This genetic connection is further supported by the observation that individuals with an *SNCA* duplication and triplication develop neurological conditions with PD-like features, and those with triplications exhibit increased pathology compared to *SNCA* duplication cases (Fuchs et al., 2007). In addition, polymorphisms close to the *SNCA* have been associated with an altered risk of developing PD, and it has been suggested, although not proven, that this is due to small changes in the levels of α-syn (Hadjigeorgiou et al., 2006).

Second, α-syn is a major component of the intracellular inclusions—Lewy bodies and Lewy neurites—which are one of the classical pathological features of PD (Spillantini et al., 1998, 1997). The discovery of the gradual development of Lewy pathology in transplanted cells (Kordower et al., 2008; Li et al., 2008, 2016), however, has led to the proposal that a "prion-like" propagation may influence the progression of PD (Steiner et al., 2018). Preclinical data have supported this idea, demonstrating that monomeric α-syn and certain α-syn assemblies can be passed from cell-to-cell, and can seed α-syn aggregation as well as triggering cellular dysfunction and death in the recipient cell (Froula et al., 2019; Grassi et al., 2018; Volpicelli-Daley and Brundin, 2018). As a result of the attention on α-syn in PD, numerous clinical trials now seek to limit the aggregation and intracellular transfer of this protein (see Table 1). The path toward clinical trials for novel therapies is long and arduous for any approach aimed at slowing neurodegenerative disease, and targeting α-syn in PD has its own set of hurdles (Merchant et al., 2019). The different steps required to take an experimental therapeutic targeting α-syn in PD from the laboratory into the clinic first requires extensive preclinical testing in cell and animal models of PD (Merchant et al., 2019). Some features of the roadmap detailing how to translate an experimental α-syn therapy into proof-of-concept clinical trials are particularly challenging, for example, there is an absence of a validated biofluid biomarker or imaging modality that can be used to assess target engagement in the brain.

Table 1 Clinical trial programs for counter protein aggregation pathology in Parkinson's.

Immunotherapy—Passive

Roche/Prothena—PASADENA study	PRX002	Phase I	https://clinicaltrials.gov/ct2/show/NCT02095171
	PRX002	Phase I	https://clinicaltrials.gov/ct2/show/NCT02157714
	Prasinezumab	Phase II	https://clinicaltrials.gov/ct2/show/NCT03100149
Biogen/Neurimmune—SPARK study	BIIB054	Phase I	https://clinicaltrials.gov/ct2/show/NCT02459886
	BIIB054	Phase I	https://clinicaltrials.gov/ct2/show/NCT03716570
	BIIB054	Phase II	https://clinicaltrials.gov/ct2/show/NCT03318523
Lundbeck/Genmab	Lu AF82422	Phase I	https://clinicaltrials.gov/ct2/show/NCT03611569
Astra Zeneca/Takeda	MEDI1341	Phase I	https://clinicaltrials.gov/ct2/show/NCT03272165
BioArctic/AbbVie	ABBV-0805	Phase I	https://clinicaltrials.gov/ct2/show/NCT04127695

Immunotherapy—Active

AFFiRiS	PD01A	Phase I	https://clinicaltrials.gov/ct2/show/NCT01568099
	PD01A	Phase I	https://clinicaltrials.gov/ct2/show/NCT02216188
	PD01A	Phase I	https://clinicaltrials.gov/ct2/show/NCT01885494
	PD03A	Phase I	https://clinicaltrials.gov/ct2/show/NCT02267434
United Neuroscience	UB-312	Phase I	https://clinicaltrials.gov/ct2/show/NCT04075318

Small molecule inhibitors of aggregation

Neuropore Therapies/UCB	NPT200-11	Phase I	https://clinicaltrials.gov/ct2/show/NCT02606682
	NPT520-34	Phase I	https://clinicaltrials.gov/ct2/show/NCT03954600
Enterin—RASMET study	ENT-01	Phase I	https://clinicaltrials.gov/ct2/show/NCT03047629
Enterin—KARMET study	ENT-01	Phase II	https://clinicaltrials.gov/ct2/show/NCT03781791
	ENT-01	Phase I	https://clinicaltrials.gov/ct2/show/NCT03938922
Proclara Biosciences	NPT 088	Phase I	https://clinicaltrials.gov/ct2/show/NCT03008161
Hadassah Medical Org	Mannitol	Phase II	https://clinicaltrials.gov/ct2/show/NCT03823638
MODAG	Anle138b	Phase I	Not yet registered
Annovis	ANVS-401	Phase II	Not yet registered (Q1 2020)

Lysosomal-base enhancers

AiM-PD study	Ambroxol	Phase II	https://clinicaltrials.gov/ct2/show/NCT02941822
	Ambroxol	Phase II	https://clinicaltrials.gov/ct2/show/NCT02914366
Lysosomal Therapeutics	LTI-291	Phase I	https://www.trialregister.nl/trial/6516
	LTI-291	Phase I	https://www.trialregister.nl/trial/7061
Prevail Therapeutics—PROPEL study	PR001	Phase I	https://clinicaltrials.gov/ct2/show/NCT04127578
Sanofi Genzyme—MOVES-PD study	Venglustat	Phase II	https://clinicaltrials.gov/ct2/show/NCT02906020

Continued

Table 1 Clinical trial programs for counter protein aggregation pathology in Parkinson's.—Cont'd

Autophagy inducers

PD Nilotinib	Phase I	Nilotinib	https://clinicaltrials.gov/ct2/show/NCT02281474
	Phase II	Nilotinib	https://clinicaltrials.gov/ct2/show/NCT02954978
Nilo PD	Phase II	Nilotinib	https://clinicaltrials.gov/ct2/show/NCT03205488
Sun SPARC—PROSEEK study	Phase I	K0706	https://clinicaltrials.gov/ct2/show/NCT02629692
	Phase II	K0706	https://clinicaltrials.gov/ct2/show/NCT03655236
resTORbio	Phase I	RTB101	ACTRN12619000372189, https://www.anzctr.org.au/Trial/Registration/TrialReview.aspx?id=376782

LRRK2 inhibition

Denali Therapeutics	Phase I	DNL201	https://www.trialregister.nl/trial/7350
	Phase I	DNL201	https://clinicaltrials.gov/ct2/show/NCT03710707
			https://www.trialregister.nl/trial/6778
Biogen/Ionis—REASON study	Phase I	DNL151	https://clinicaltrials.gov/ct2/show/NCT04056689
	Phase I	BIIB094	https://clinicaltrials.gov/ct2/show/NCT03976349

Iron chelation

ApoPharma—SKY study	Phase II	Deferiprone	https://clinicaltrials.gov/ct2/show/NCT02728843
FAIRPARKII study	Phase II	Deferiprone	https://clinicaltrials.gov/ct2/show/NCT02655315
Alterity (Prana) Therapeutics/Tekada	Phase I	PBT434	ACTRN12618000541202

GLP-1 agonists

CPT/Van Andel/MJFF	Phase II	Exenatide	https://clinicaltrials.gov/ct2/show/NCT01174810
	Phase II	Exenatide	https://clinicaltrials.gov/ct2/show/NCT01971242
	Phase III	Exenatide	http://www.isrctn.com/ISRCTN14552789
University of Florida	Phase I	Exenatide	https://clinicaltrials.gov/ct2/show/NCT03456687
Peptron	Phase I	PT320	https://clinicaltrials.gov/ct2/show/NCT00964262
	Phase II	PT320	Not yet registered
CPT/Van Andel/Novo Nordisk A/S	Phase II	Liraglutide	https://clinicaltrials.gov/ct2/show/NCT02953665
CPT/Van Andel/Sanofi	Phase II	Lixisenatide	https://clinicaltrials.gov/ct2/show/NCT03439943
Oslo University Hospital – GIPD study	Phase II	Semaglutide	https://clinicaltrials.gov/ct2/show/NCT03659682
Neuraly	Phase I	NLY01	https://clinicaltrials.gov/ct2/show/NCT03672604

Despite the great interest in α-syn as a therapeutic target, it is notable that it has also been debated whether α-syn oligomerization and aggregation is really an upstream pathogenic event worthy of targeting or whether it is just an epiphenomenon or even a neuroprotective response to other pathogenic events. It is beyond the scope of this review to discuss in detail all the theoretical arguments that speak against targeting α-syn in PD, and they have been described elsewhere (Espay et al., 2019). One of the arguments is that anti-amyloid therapies have repeatedly failed in clinical trials in AD, and by analogy strategies to reduce α-syn aggregation in PD might be destined for failure. In this context, it is worth noting that none of the therapies designed to target α-syn pathology which are described in this review have actually gone through full clinical testing so presently it is fair to conclude that it is too early to dispel Lewy pathology as a valid therapeutic target. That said, Lewy bodies contain >300 different proteins, of which 90 have been confirmed by immunohistochemistry (Wakabayashi et al., 2013). These additional proteins include SOD1 (Nishiyama et al., 1995; Trist et al., 2017) and TAU (Arima et al., 1999; Pollanen et al., 1992), as well as PD-linked gene products, such as PINK1 (Gandhi et al., 2006), LRRK2 (Zhu et al., 2006), and PARKIN (Schlossmacher et al., 2002). Thus, in addition to the α-syn targeted approaches, there are also clinical trials attempting to enhance autophagy and lysosomal dysfunction, and therefore to better support neurons which are stressed and over-burdened by protein deposits.

In our overview of the current state of clinical efforts to target protein aggregation in PD, we have divided the ongoing clinical trials into seven subgroups:

- Immunotherapy
- Small molecule inhibitors of aggregation
- Lysosome-based enhancers
- Autophagy inducers
- LRRK2 inhibition
- Iron chelation
- GLP-1 agonists

In the following sections, we describe and evaluate the clinical research being conducting in each of these areas. We also provide a brief overview over therapeutic strategies that target α-syn aggregation which are still at an experimental stage but which are approaching clinical trials.

2 Immunotherapy

Determining exactly how to address α-syn as a target has proven to be a challenge. The finding that α-syn pathology appears in healthy cells transplanted into PD patients raised the possibility that cell-to-cell transfer of α-syn could be involved in PD progression. This has led many to speculate about the utility of immunotherapy as a treatment approach for PD (Fig. 1). Antibodies designed to bind and aid in the removal of monomeric α-syn or different forms of α-syn assemblies have exhibited

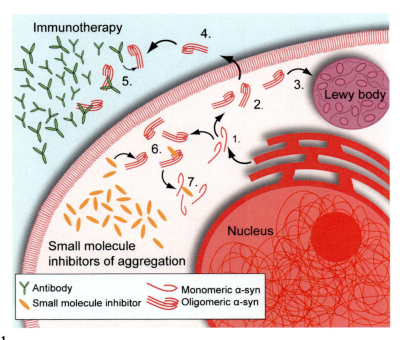

FIG. 1

A cartoon representation illustrating how immunotherapy and small molecule inhibitors are being tested in PD. Monomeric α-syn is produced (1), but in PD the protein aggregates for unknown reasons (2). This oligomerization is believed to lead to pathology, such as the formation of Lewy bodies (3). Aggregated α-syn is also released extraneuronally (4). Immunotherapies (such as Prasinezumab, BIIB054, and PD01A) targeting extracellular α-syn oligomers (5) are being clinically evaluated to determine if they can slow the spread of the condition. In addition, small molecule inhibitors of α-syn aggregation (such as Anle138b and ENT01)—which are not necessarily limited to the extracellular space—are able to bind to α-syn oligomers (6) and disaggregate them (7), reducing levels of both extra- and intracellular oligomeric α-syn.

positive results in preclinical models of PD (Chen et al., 2013; Ghochikyan et al., 2014; Masliah et al., 2005; Sanchez-Guajardo et al., 2013; Ugen et al., 2015), and these findings have resulted to the initiation of clinical trials using passive and active immunization.

2.1 Immunotherapy—Passive

Several pharmaceutical companies are exploring passive immunotherapy for PD. These therapies entail repeated intravenous infusion of antibodies that are intended to target α-syn present in the brain, and direct this α-syn for degradation, presumably via microglia. The first program to enter clinical trials was conducted by Roche in partnership with the biotech company Prothena (https://www.roche.com/media/re leases/med-cor-2013-12-11p.htm). This study is focused on a monoclonal antibody called Prasinezumab (previously called PRX002/RG7935). Prasinezumab is a humanized IgG1 monoclonal antibody, which is directed against the C-terminus of α-syn (Games et al., 2013). Testing in mouse models of α-syn pathology demonstrated that treatment reduced pathology and improved motor function (Games et al., 2014; Masliah et al., 2011).

The results of a Phase I clinical trial of Prasinezumab demonstrated that the treatment is safe and well tolerated, with a half-life of 18.2 days, and a dose-dependent lowering of "free" α-syn (Schenk et al., 2017), although the cerebrospinal fluid (CSF) antibody concentrations were only 0.3% of those recorded in plasma (Jankovic et al., 2018), which is in line with previous immunotherapy treatments targeting the brain (Freskgård and Urich, 2017).

In June 2017, Roche and Prothena initiated the PASADENA study—a 2-year Phase II clinical trial comparing the efficacy of two doses of intravenous Prasinezumab to placebo in 300 people with early PD (https://clinicaltrials.gov/ct2/show/ NCT03100149). This study involves a 1-year placebo-controlled double-blind phase, to be followed by another year in which all participants receive the antibody. During this second phase of the study, the participants will be blinded to the actual dose that they will be receiving. The primary outcome of this study will be the change in the Movement Disorder Society-sponsored revision of the Unified Parkinson's Disease Rating Scale (MDS-UPDRS), over baseline scores, while the secondary outcomes include DaTscan changes, as well as a range of clinical, safety, and pharmacokinetic measures. This trial is set to report results in 2021/22.

A second major passive immunotherapy program for PD is being conducted by the pharmaceutical company Biogen. They licensed their monoclonal antibody—called BIIB054 (also known as NI-202)—from the biotech firm Neurimmune (http://www. neurimmune.com/newsartikel/20122010-biogen-idec-and-neurimmune-announce-agreement-on-three-neurodegenerative-disease-programs-.html). BIIB054 is a fully human IgG1 monoclonal antibody targeting the N-terminus of α-syn, which demonstrates a high affinity for aggregated forms of α-syn (800-fold higher affinity for fibrillar vs monomeric recombinant α-syn). It has also been reported to reduce levels of aggregated α-syn in three different mouse models of PD (Weihofen et al., 2019).

The clinical trial program for BIIB054—called the SPARK study—has been through Phase I testing (https://clinicaltrials.gov/ct2/show/NCT03716570; https:// n.neurology.org/content/90/15_Supplement/S26.001) and, in December 2017, Biogen started a 2-year Phase II trial involving over 300 people with PD. It compares monthly infusions of three doses of BIIB054 to placebo in order to evaluate safety, pharmacodynamic effects of this antibody on preservation of nigrostriatal dopaminergic nerve terminals, pharmacokinetics and immunogenicity of BIIB054. The trial, is conducted at 17 US sites, and is expected to report results in 2022 (https:// clinicaltrials.gov/ct2/show/NCT03318523).

In addition to these two more advanced programs, there are a number of additional immunotherapy clinical programs under development. Pharmaceutical company Lundbeck has a monoclonal called Lu AF82422 currently in Phase I clinical testing (https://clinicaltrials.gov/ct2/show/NCT03611569). Little is known about Lu AF82422, which was developed by Genmab (https://ir.genmab.com/news-releases/news-release-details/genmab-announces-antibody-development-collaboration-lundbeck). In addition, Astra Zeneca and Takeda also have a α-syn-targeting monoclonal antibody—MEDI1341. Preclinical data demonstrates that MEDI1341 significantly reduces α-syn accumulation and propagation by binding to both monomeric and aggregated forms of α-syn (Schofield et al., 2019), and Phase I clinical testing of this antibody has started in healthy volunteers (https://clinicaltrials.gov/ct2/show/NCT03272165). BioArctic and AbbVie also have a monoclonal antibody—ABBV-0805 (also known as BAN0805) initiating clinical evaluation (https://clinicaltrials.gov/ct2/show/NCT04127695). We can expect to see considerable data coming from the passive immunotherapy space for PD over the next few years.

2.2 Immunotherapy—Active

Active vaccine-based immunotherapies are also currently being developed for PD. In this paradigm, the patient is immunized with an antigen present in aggregated α-syn and which is supposed to trigger an endogenous immune response and antibody production intended to reduce Lewy pathology. One example of this is being developed by the Austrian biotech firm, AFFiRiS (Schneeberger et al., 2016). The company has two vaccines (PD01A and PD03A) which have been designed to target the aggregated forms of α-syn. Preclinical screening of peptides, and then testing in models of synucleinopathies, resulted in reduced accumulation of α-syn oligomers in neurons, less neurodegeneration, and improvements in motor deficits (Mandler et al., 2015, 2014). AFFiRiS has subsequently undertaken multiple Phase I clinical trials of these vaccines (https://clinicaltrials.gov/ct2/show/NCT01568099; https://clinicaltrials.gov/ct2/show/NCT02216188; https://clinicaltrials.gov/ct2/show/NCT01885494), which has resulted in more than 90 PD and multiple system atrophy (MSA) patients having been injected with either PD01A or PD03A, with safety data and clinical observations now spanning out to more than 4 years. The company is continuing to follow up these treated individuals, while exploring new Phase II testing of these vaccines.

A second active vaccine approach is being developed is a synthetic peptide-based "endobody" called UB-312, which is owned by the biotech company United Neuroscience Ltd. UB-312 has been designed based on the company's proprietary UBITh® platform (Wang et al., 2007). United Neuroscience also has an endobody for β-amyloid called UB-311 (Wang et al., 2017a), which is now in a clinical trial (https://clinicaltrials.gov/ct2/show/NCT02551809; https://clinicaltrials.gov/ct2/show/NCT03531710). No preclinical data regarding UB-312 has yet been published, but the ongoing Phase I clinical trial (https://clinicaltrials.gov/ct2/show/NCT04075318) should report in 2021.

3 Small molecule anti-aggregates

There are a growing number of biotech companies developing small molecules that target protein aggregation in PD. These are brain penetrant small molecules designed to directly inhibit α-syn aggregation, and circumventing the problem with poor blood brain barrier penetrance of the antibodies involved in immunotherapies (Fig. 1).

The San Diego-based company, Neuropore Therapies, and their pharma collaborator UCB are developing such small molecules which interfere with the interaction between α-syn and membranes, resulting in reduced α-syn oligomerization. Orally administered NPT200-11 has been shown to inhibit the development of α-syn pathology in a transgenic mouse model of α-syn pathology (Price et al., 2018). NPT200-11 is based on a previous compound (NPT100-18A) which lacked the pharmacokinetic properties suitable for clinical testing. The Phase I safety testing of NPT200-11 in 55 healthy volunteers was completed in early 2016 (https://clinicaltrials.gov/ct2/show/NCT02606682; https://www.businesswire.com/news/home/20160322005337/en/Neuropore-Announces-Successful-Completion-Phase-Lead-Compound). The results of that study have not been published and it is not clear whether Neuropore plans to develop this compound further. The company has, however, initiated a Phase I clinical trial of a new misfolded protein targeting compound called NPT520-34 (https://www.businesswire.com/news/home/20190508005227/en/Neuropore-Initiates-Phase-1-Clinical-Trial-Healthy; https://clinicaltrials.gov/ct2/show/NCT03954600).

Another set of small, aggregation inhibiting molecules being developed are based on the naturally occurring compound, squalamine. Derived from dogfish sharks, this molecule has previously exhibited anticancer and antiviral properties (Márquez-Garbán et al., 2019; Zasloff et al., 2011). It has also recently been reported to affect α-syn aggregation, both in vitro and in vivo. Squalamine was found to displace α-syn from the surfaces of lipid vesicles, which blocks its aggregation process. When administered to models of α-syn aggregation in vivo, squalamine reduced levels of protein aggregation (Perni et al., 2017).

ENT-01 is a synthetic version of squalamine being developed by a biotech firm called Enterin. Unfortunately, neither squalamine nor ENT-01 access the central nervous system (CNS) in sufficient quantities, so the company has instead focused its clinical development of ENT-01 on reducing the levels of aggregated α-syn in the enteric nerves surrounding the gastrointestinal system. This compound has been Phase I testing (the RASMET study; https://clinicaltrials.gov/ct2/show/NCT03047629; https://www.sciencedirect.com/science/article/pii/S2590112519300015). The results of this open-label study suggest that the drug was well tolerated and resulted in an increase in bowel movements in 80% of the participants. Systemic absorption was only <0.3%; however, suggesting that the effect of the drug is localized to the gastrointestinal system. A Phase II clinical trial of ENT-01 is currently being conducted (the KARMET study; https://clinicaltrials.gov/ct2/show/NCT03781791). This study is a multi-center, double-blind, placebo-controlled trial involving 72 patients, randomized (3:1) to treatment or placebo. The primary endpoints are safety and measures of constipation.

Given the issues with CNS-penetrance of ENT-01, Enterin has also developed a derivative of squalamine called Trodusquemine, which has been tested in models of PD and found to access the brain (Perni et al., 2018). Notably, Enterin has recently initiated a Phase Ib clinical trial of ENT-01 for PD dementia (https://clinicaltrials.gov/ct2/show/NCT03938922), with the primary outcome being a cognitive score.

A third example of a small molecule approach is an experimental treatment called NPT 088. This drug is particularly interesting as it incorporates a general amyloid interaction motif, which means that it binds all amyloid proteins. The filamentous bacteriophage M13 has previously been reported to improve cognition and reduce β-amyloid plaque load in transgenic AD mouse models following chronic administration. Further characterization of this anti-amyloid activity revealed that M13 has the ability to bind and remodel different types of misfolded protein aggregates, including β-amyloid, phosphorylated tau, and α-syn. Critically, this activity occurs without affecting the monomeric forms of these proteins (Krishnan et al., 2014). M13 achieves this via a two-domain fragment of the phage capsid protein g3p.

NPT 088 is a fusion protein that has been developed by Proclara Biosciences (formerly NeuroPhage Pharmaceuticals Inc.), which was engineered to contain of the active fragment of g3p and human-IgG1-Fc. In preclinical testing, NPT 088 was found to significantly reduce β-amyloid plaques and improve cognitive performance of aged Tg2576 mice. In addition, NPT 088 also reduced phosphorylated-tau pathology, improved cognition, and reduced brain atrophy in rTg4510 mice (Levenson et al., 2016). Currently, a clinical trial is ongoing on NPT 088 in individuals with probable AD (https://clinicaltrials.gov/ct2/show/NCT03008161), and the company has indicated on their website this putative therapeutic would likely also be applicable to PD (http://www.proclarabio.com/our-programs).

Another small molecule that is being clinically tested in PD with the aim of reducing α-syn aggregation is an FDA-approved osmotic diuretic agent, Mannitol. In vitro analyses indicated that low concentrations of mannitol inhibit the formation of fibrils, while higher concentrations significantly decrease oligomerization. These results were supported by the reversal or prevention of pathological changes in vivo in drosophila and mouse models of PD (Shaltiel-Karyo et al., 2013). More recently, a conjugation of Mannitol and naphthoquinone-tryptophan was found to be more effective than the two parent molecules (Paul et al., 2019). A Phase II clinical trial is underway assessing the safety and tolerability of 36 weeks Mannitol treatment in a dose finding study of 60 individuals with PD (https://clinicaltrials.gov/ct2/show/NCT03823638), with the goal of subsequently conducting a larger assessment of efficacy.

A novel small protein aggregation inhibitor is called Anle138b. Treatment with this diphenyl-pyrazole-based compound not only inhibits α-syn protein aggregation, but also disassembles mature fibrils. Of particular interest for this compound is that delayed onset of the treatment in preclinical models of PD was also effective (Levin et al., 2014; Wagner et al., 2013; Wegrzynowicz et al., 2019). A German biotech firm, called MODAG, recently secured funding to begin clinical testing of Anle138b (https://www.businesswire.com/news/home/20190627005315/en/MODAG-Launches-Stealth-Mode-Series-Financing-EUR). After Phase I safety/tolerability trials are completed, the company will be focusing their attention on MSA—an α-synucleinopathy related to PD. Anle138b has beneficial effects in MSA models (Fellner et al., 2016; Heras-Garvin et al., 2019).

Another potential method of limiting α-syn levels is preventing translation of *SNCA* mRNA. Posiphen is a translational inhibitor of α-syn that targets the 5′ untranslated region (UTR) of *SNCA* mRNA which contains an Iron-Response Element (IRE). Posiphen increases the affinity of Iron Regulatory Protein-1 to the IRE of *SNCA* mRNA, which in turn prevents ribosomes from associating with the mRNA and thereby represses translation (Mikkilineni et al., 2012; Rogers et al., 2011). Interestingly, Posiphen also suppresses the translation of AD-associated amyloid precursor protein (APP). The mRNA of *APP* contains an IRE homologous to that of *SNCA* mRNA (Cahill et al., 2009; Shaw et al., 2001; Venti et al., 2004).

The translational inhibition of *APP* by Posiphen has been well documented in models of AD (Lahiri et al., 2007; Lilja et al., 2013; Teich et al., 2018). In addition, in Phase I clinical trial of Posiphen—which is being developed by a biotech firm called Annovis (formerly QR Pharma)—the treatment was reported to be safe and well tolerated in 120 healthy volunteers. Following on from that result, Posiphen reduced the level of soluble APP, Aβ42 and tau in the CSF of four individuals with mild cognitively impaired (https://clinicaltrials.gov/ct2/show/NCT01072812; Maccecchini et al., 2012). Posiphen is now being tested a multi-dose, Phase I/II clinical study in 24 patients with early AD (https://clinicaltrials.gov/ct2/show/NCT02925650).

Preclinical research in models of PD have demonstrated that Posiphen also reduces levels of α-syn in rodent primary neurons and human cell lines (Mikkilineni et al., 2012; Rogers et al., 2011; Yu et al., 2013). Furthermore, a recent report suggests that 21 weeks of Posiphen treatment (10 mg/kg) reduces α-syn levels in the gut of h*SNCA-A53T* mice, while higher doses (50–65 mg/kg) also reduces levels in the CNS (Kuo et al., 2019). If the ongoing Posiphen clinical trial in early AD reveals beneficial effects on disease progression, it would be a relatively straightforward task to launch clinical trials of this experimental therapy in PD patients.

Whether inhibition of α-syn translation, aggregation or cell-to-cell transfer can halt the clinical progression of PD will be determined in the next few years. As these clinical immunotherapy and small molecule aggregate inhibitor trials are being conducted, additional studies are attempting to tackle the issue from a different angle by enhancing the degradation of protein aggregates.

4 Lysosomal-based enhancers

Lysosomal and autophagy dysfunction is an acknowledged feature of PD (Nguyen et al., 2019), and many of the genetic variants associated with the condition are in genes that have specific roles regulating these processes (Robak et al., 2017).

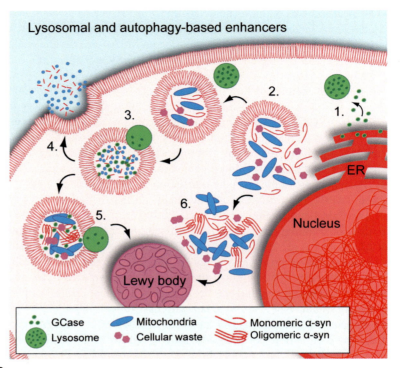

FIG. 2

Lysosomal and autophagic clearance of waste is a necessary cellular function. It begins with the production and lysosomal uptake of digestive enzymes (1), such as Glucocerebrosidase (GCase). In parallel, waste is collected in autophagosome vesicles (2), which then dock and fuse with lysosomes (3). After digestion, the contents of the vesicle can be exocytosed (4) or recycled. The lysosomal dysfunction observed in PD may involve faulty or reduced digestive enzyme activity (for example GCase), which might remain attached to the endoplasmic reticulum (ER) or fail to correctly break down proteins in the lysosome, resulting in incomplete digestion of vesicular contents (5). This situation may cause cellular stress, leading to pathology. Intracellular stress may also be induced by the accumulation of un-recycled waste (6), which could also lead to PD pathology. Lysosomal therapies being clinically tested for PD are primarily focused on enhancing GCase activity (such as Ambroxol, LTI-291, and PR001), while autophagy inducers (including Nilotinib, K0706, and RTB101) are being evaluated for their ability to increase waste disposal and reduce cellular stress.

It has long been recognized; however, that correcting dysfunction of the lysosomal autophagy system, or enhancing cellular waste disposal, might represent a therapeutic approach to apply to PD, and many efforts have been made toward identifying lysosomal/autophagy inducers (Fig. 2). In this section, we focus on strategies to promote lysosomal function, and in the subsequent section we will discuss approaches aimed at inducing autophagy.

Ambroxol hydrochloride (Ambroxol) is an expectorant which is used in the treatment of respiratory diseases that are associated with mucus hypersecretion. In 2009, it was identified in a screening of US Food and Drug Administration-approved drugs as a chaperone of the lysosomal enzyme β-glucocerebrosidase (GCase—Maegawa et al., 2009). Genetic variations in the *GBA* gene, which encodes GCase, are associated with Gaucher disease and PD (Sidransky et al., 2009). *GBA*-associated PD has a similar phenotype to idiopathic PD, although carriers generally have earlier onset of symptoms and higher risk of rapid decline and cognitive impairment (Beavan and Schapira, 2013).

Normally, GCase is synthesized, folded and translocated to the lysosome, but mutant forms of GCase fail to be correctly folded and become arrested in the endoplasmic reticulum. They are subsequently redirected to undergo proteasome degradation, resulting in reduced levels of GCase in the lysosomes (Maor et al., 2013). The decrease in GCase activity can lead to altered glycosphingolipid homeostasis and membrane composition, further resulting in lysosomal/endosomal dysfunction, compromised vesicular transport, and α-syn aggregation (Do et al., 2019; Hallett et al., 2018). Ambroxol has been shown to improve the translocation of mutant GCase to the lysosome, increasing GCase activity in cells carrying *GBA* mutations (Maor et al., 2016).

Ambroxol treatment has been shown to raise brain GCase levels in transgenic mouse models of PD (Migdalska-Richards et al., 2016), in fibroblasts from PD patients with *GBA* variants (McNeill et al., 2014), and in nonhuman primates (Migdalska-Richards et al., 2017). These and other findings have led to a clinical trial of Ambroxol in people with PD—called AiM-PD (or Ambroxol in Disease Modification in Parkinson Disease—https://clinicaltrials.gov/ct2/show/NCT02941822). There is also a second clinical trial of Ambroxol in PD dementia (https://clinicaltrials.gov/ct2/show/NCT02914366; Silveira et al., 2019). This study is being conducted in Canada and involves 75 participants who are receiving daily administration of Ambroxol (or placebo) for 52 weeks. Both of these studies should be reporting their results within the next 12 months.

Another GCase enzyme-based approach for PD is a small molecule called LTI-291, which is being developed by Lysosomal Therapeutics. This drug is an activator of GCase and while very little information about this drug has been published, the company has conducted safety and imaging studies in Europe (https://www.trialregister.nl/trial/6516; https://www.trialregister.nl/trial/7061). Lysosomal Therapeutics is now seeking to evaluate the compound in Phase II testing.

Additional GCase activators/chaperones are also being developed preclinically. For example, the NIH Chemical Genomics Center has been screening and testing GCase activators (Aflaki et al., 2016, 2014; Mazzulli et al., 2016; Patnaik et al., 2012). One ambitious approach toward correcting GCase activity in *GBA*-associated PD is a gene therapy treatment currently being proposed by the biotech firm Prevail Therapeutics. Their primary candidate for clinical development is PR001, an AAV9 viral vector encoding the *GBA* gene. Gene therapy approaches for *GBA*-associated PD have demonstrated efficacy in animal models of the condition (Sardi et al., 2013, 2011), and it has been reported to rescue α-syn pathology in additional models of PD (Rocha et al., 2015). The US FDA accepted Prevail's Investigational New Drug (IND) application for PR001, and the company has now initiated Phase I safety/tolerability testing of the treatment (https://clinicaltrials.gov/ct2/show/NCT04127578). The proposed route of administration is a single injection into the intra cisterna magna which is intended to lead to widespread AAV9 viral vector-mediated expression of GCase in the CNS.

An alternative means of correcting altered glycosphingolipid homeostasis is substrate reduction therapy—inhibiting the production of glycolipid that accumulates in the absence of normal GCase activity. An example of this is a compound called Venglustat (also known as Ibiglustat and GZ/SAR402671), which is being developed by Sanofi Genzyme for lysosomal dysfunction disorders, such as Fabry, Gaucher disease, and PD. It is an oral inhibitor of the enzyme glucosylceramide synthase (GCS), which converts ceramide into glucosylceramide—one of the substrates of GCase that accumulates in *GBA*-associated PD.

GCS inhibitors have been shown to reduce α-syn pathological in models of PD (with and without *GBA* mutations) (Kim et al., 2018; Zunke et al., 2018), and a novel brain-penetrant GCS inhibitor similar to Venglustat (called GZ667161) was reported to rescue mouse models of GBA-related synucleinopathy and α-syn overexpression (Sardi et al., 2017). These results led to the initiation of a large (240+ participant), two part Phase II clinical trial of Venglustat, which will assess safety/tolerability (part 1) and efficacy of the drug (part 2) in individuals with *GBA*-associated PD (https://www.clinicaltrials.gov/ct2/show/NCT02906020).

5 Autophagy inducers

Autophagy is a critical component of cellular waste disposal. Therapeutic manipulation of this process is seen by many molecular biology researchers as an important goal, the main aim of which is to improve how the body removes various proteins that implicated in human diseases (Boland et al., 2018). In PD, activation of autophagy has been found to have beneficial effects in many different models of the condition (Fowler and Moussa, 2018), and these results have led to numerous clinical trials evaluating a number of molecules for their ability to clear aggregated/accumulated proteins (Fig. 2).

An example of this is the effort to repurpose the cancer treatment, Nilotinib, for PD. This is a brain-penetrant, non-receptor tyrosine kinase Abelson (c-Abl) inhibitor, which is used for the treatment of chronic myeloid leukemia. c-Abl is a kinase which is present in most cells and has a variety of physiological functions, including autophagy, DNA repair, cell survival, regulation of cell growth and motility (Hantschel and Superti-Furga, 2004). In cells, c-Abl is normally present in an inactive form. It becomes activated by DNA damage and cellular stress. Preclinical data has indicated that Nilotinib is effective in inducing autophagy and clearing α-syn accumulation in models of PD (Hebron et al., 2013; Karuppagounder et al., 2015; Mahul-Mellier et al., 2014). A small unblinded pilot clinical study aimed at assessing safety tentatively suggested potential clinical benefits in individuals with advanced PD dementia and dementia with Lewy bodies (Pagan et al., 2016). This small, open-label study prompted the launch of two large, placebo-controlled Phase II clinical trials of Nilotinib in PD: "PD Nilotinib" (https://clinicaltrials.gov/ct2/show/NCT02954978), which has already released some results (Pagan et al., 2019), and "Nilo PD" (https://clinicaltrials.gov/ct2/show/NCT03205488). There are concerns regarding the use of this c-Abl inhibitor as it carries a black box warning due to the chance of QT interval prolongation in some patients. In addition, myelosuppression and sudden cardiac death have been reported in some patients taking Nilotinib (http://www.bloodjournal.org/content/118/21/2761). The doses currently being tested in PD, however, are lower than those used in oncology, and it is thought that this will hopefully reduce the risk of these undesirable side effects.

There are further efforts being made to develop additional CNS penetrant c-Abl inhibitors for use in PD. For example, the pharmaceutical company Sun/SPARC Pharma Advanced Research Company Ltd. has a c-Abl inhibitor (called K-0706) which has been evaluated in two Phase I studies of healthy individuals (https://clinicaltrials.gov/ct2/show/NCT03316820), and is being clinically investigated for the treatment of chronic myelogenous Leukemia (https://clinicaltrials.gov/ct2/show/NCT02629692). The company has also conducted a Phase I clinical study of K-0706 in healthy individuals to assess CSF penetrance (https://clinicaltrials.gov/ct2/show/NCT03445338) and safety/tolerability in 32 participants with PD (https://clinicaltrials.gov/ct2/show/NCT02970019). A Phase II clinical evaluation of K-0706 in 504 people with PD has recently been commenced (the PROSEEK study; https://clinicaltrials.gov/ct2/show/NCT03655236), and a Phase II study assessing K-0706 has also been initiated in Lewy body dementia (https://clinicaltrials.gov/ct2/show/NCT03996460). In addition to the aforementioned c-Abl inhibitors, there are further c-Abl being developed for testing in PD. Biotech firms, such as Neuraly (Lee et al., 2018) and Inhibikase Therapeutics, have specifically designed c-Abl inhibitors for use in CNS conditions, (https://www.inhibikase.com/news/press-releases/inhibikase-therapeutics-files-two-investigational-new-drug-applications-for-ikt-148009-a-disease-modifying-therapy-for-parkinsons-disease/).

Another autophagy-inducing drug that is being considered for repurposing for PD is the mitochondrial pyruvate carrier inhibitor, MSDC-0160. Originally developed for type II diabetes (Colca et al., 2013), this compound was recently reported to be an effective inhibitor of the mammalian target of rapamycin (mTOR) pathway (Quansah et al., 2018). MSDC-0160 rescued human neurons in vitro and mice against the toxic effects of MPTP and the loss of one allele of the transcription factor *Engrailed 1*, as well as being protective in an α-syn *C. elegans* model (Ghosh et al., 2016). Given the good safety profile of MSDC-0160—and its ability to access the CNS—it could be a candidate for clinical evaluation in PD.

A robust approach to increasing autophagy is the use of TORC1 inhibition. The serine/threonine-specific protein kinase, mammalian target of rapamycin (mTOR) is a major regulator of cellular metabolism (Saxton and Sabatini, 2017). It acts via two protein complexes, mTORC1 and mTORC2. mTORC1 negatively regulates autophagy (Kim et al., 2011) and its inhibition is associated with increased autophagy activity (Noda, 2017). Given this particular property, researchers have focused on mTORC1 as a novel therapeutic target for PD (Lan et al., 2017) and significant efforts have been made to identify CNS penetrant mTORC1 specific inhibitors. One of these research programs has been conducted by the biotech firm resTORbio which is developing a TORC1 inhibitor called RTB101. The drug has recently entered a Phase I clinical trial in New Zealand (https://www.anzctr.org.au/Trial/Registra tion/TrialReview.aspx?id=376782&isReview=true). The company plans to evaluate RTB101 in both idiopathic and *GBA*-associated PD (with and without the mTOR inhibitor Sirolimus (Rapamycin)—which is believed to enhance the potency of RTB101).

One final autophagy enhancing treatment that is being evaluated in PD is the urea cycle disorder treatment, Sodium Phenylbutyrate. This prodrug of phenylacetate has been reported to have beneficial properties in multiple models of PD (Gardian et al., 2004; Inden et al., 2007; Ono et al., 2009; Roy et al., 2012; Zhou et al., 2011) as well as other neurodegenerative conditions (Gardian et al., 2005; Minamiyama et al., 2004; Sturm et al., 2016; Ying et al., 2006). As a result of these findings, in 2014 the University of Colorado initiated a Phase I clinical trial of the FDA-approved small molecule glycerol phenylbutyrate, which is approved for the chronic management of urea cycle disorders. The open label, non-randomized trial involved 40 participants (20 with idiopathic PD and 20 age/sex matched controls), who would be taking 20 g/day of phenylbutyrate in the liquid phenylbutyrate-triglyceride form, three times per day. Changes in plasma α-syn on day 1, 7, 14, and 21 days of the study were the primary end point (https://clinicaltrials.gov/ct2/show/ NCT02046434). Although the results of the study have not been published, conference presentations indicate the treatment is tolerable and it nearly doubled plasma α-syn. The researchers implied that sodium phenylbutyrate promoted α-syn clearance from the brain (https://www.alzforum.org/news/conference-coverage/poten tial-parkinsons-treatments-target-synuclein-cell-replacement).

6 LRRK2 inhibition

Leucine-rich repeat kinase 2 (LRRK2; also known as PARK8) is a multidomain protein, containing two enzymatic domains—a GTPase and a kinase—as well as several protein/protein interaction domains. It has been implicated in a number of cellular processes, and autosomal dominant mutations in the LRRK2 locus have been found to cause a familial form of PD with a penetrance estimated at 40–80%. In addition, polymorphisms close to the LRRK2 gene have also been identified as risk factors for the sporadic form of PD (International Parkinson Disease Genomics Consortium et al., 2011; Ross et al., 2011). One of the most common PD risk variants— G2019S—results in an increase in LRRK2 kinase activity. This enhanced kinase activity has been shown to disrupt autophagy (Manzoni and Lewis, 2017; Ramonet et al., 2011). Inhibitors of LRRK2 have been developed and preclinical data indicates that they not only prevent neurodegeneration in models of PD (Lee et al., 2010), but also promote autophagy (Ho et al., 2018; Manzoni et al., 2013; Saez-Atienzar et al., 2014; Schapansky et al., 2018). On the back of this supportive data, clinical evaluation of LRRK2 inhibitors has commenced.

Denali has two LRRK2 inhibitors (DNL201 and DNL151) in clinical testing. Phase Ia testing of DNL201 in healthy volunteers demonstrated more than 90% inhibition of LRRK2 kinase activity at peak concentrations, and more than 50% inhibition at trough concentrations (http://investors.denalitherapeutics.com/news-releases/news-release-details/denali-therapeutics-announces-advancement-and-expansion-its#ir-pages). Phase Ia testing of DNL151 has also taken place (https://www.trialregister.nl/trial/7350). Denali is currently conducting a Phase Ib multicenter study of multiple oral doses of DNL201 in 30 PD patients with and without LRRK2 mutations in the United States (https://denalitherapeutics. com/investors/press-release/denali-therapeutics-announces-first-patient-dosed-in-phase-1b-study-of-dnl201-for-parkinsons-disease-1; https://clinicaltrials.gov/ct2/show/study/NCT03710707), and a Phase 1b safety and tolerability study of DNL151 in 24 subjects with Parkinson's in the Netherlands (https://clinicaltrials.gov/ct2/show/NCT04056689).

A second approach for LRRK2 inhibition appears to be very close to clinical testing. The pharmaceutical company Biogen, and the antisense oligonucleotide firm Ionis Pharmaceutics, have recently registered a clinical trial to "evaluate the safety, tolerability, and pharmacokinetics of BIIB094 in adults with PD" (https://clinicaltrials.gov/ct2/show/NCT03976349). BIIB094 is a LRRK2 targeting antisense oligonucleotide approach. G2019S-LRRK2 expression has been reported to augment α-syn aggregation (Volpicelli-Daley et al., 2016), and preclinical data has demonstrated that LRRK2 antisense oligonucleotides ameliorate this pathology in vivo (Zhao et al., 2017).

Both the Denali and Biogen clinical trial programs are testing LRRK2 inhibition in both mutant LRRK2-associated PD and idiopathic PD, as there is preclinical data suggesting that LRRK2 inhibitors may be useful for some cases of idiopathic PD (Di Maio et al., 2018). However, concerns have been raised regarding LRRK2 inhibition approaches. Phenotypic analysis of LRRK2 knockout rodents has reported morphologic changes in lungs and kidneys (Baptista et al., 2013; Fuji et al., 2015), thus careful monitoring will be required during these trials.

7 Iron chelation

Various metals have been found to accumulate in certain tissues in PD, and this can stimulate protein aggregation in the brain. For example, iron is known to accelerate α-syn aggregation (Ostrerova-Golts et al., 2000), inducing oxidative stress and neuronal cell death (Deas et al., 2016). α-syn also acts on iron as a ferrireductase, reducing Fe^{3+} to Fe^{2+} and influencing iron homeostasis (Davies et al., 2011). Given these processes, iron chelation has been proposed as a therapeutic option for PD, and preclinical data has been supportive (Carboni et al., 2017; Das et al., 2017; Finkelstein et al., 2016). There are currently two large Phase II clinical trials of the iron chelator Deferiprone in PD (https://clinicaltrials.gov/ct2/show/NCT02728843; https://clinicaltrials.gov/ct2/show/NCT02655315).

A second iron chelator that is about to enter clinical testing in synucleinopathies is a compound called PBT424, which is being developed by Alterity (formerly Prana) Therapeutics (in collaboration with Takeda). PBT434 has demonstrated impressive anti-protein aggregation properties, rescuing the motor performance of multiple mouse models, including α-syn transgenic models (Finkelstein et al., 2017). A Phase I clinical trial to assess the safety, tolerability and pharmacokinetics of PBT434 in healthy volunteers has recently been completed (https://alteritytherapeutics.com/investor-centre/news/2019/07/29/alterity-therapeutics-announces-successful-completion-of-phase-1-clinical-trial/; https://www.anzctr.org.au/Trial/Registration/TrialReview.aspx?id=374741), suggesting that the drug is safe and well tolerated. Alterity Therapeutics is now planning Phase II studies in the synucleinopathy MSA.

8 GLP-1 agonists

The results of a Phase II clinical trial exploring the use of a Glucagon-like peptide-1 (GLP-1) receptor agonist in PD generated a lot of excitement about this class of drugs (Athauda et al., 2017; https://clinicaltrials.gov/ct2/show/NCT01971242). GLP-1 receptor agonists are a frontline therapy for individuals with type 2 diabetes, stimulating glucose level–dependent insulin release, β islet cell proliferation, reduction of β islet cell apoptosis, and weight loss (Buse et al., 2004). GLP-1 receptors have also been found throughout the brain, however, and evidence supports that Exenatide and other GLP-1 receptor agonists are beneficial in PD models (Chen et al., 2018; Harkavyi et al., 2008; Li et al., 2009; Liu et al., 2015). Most of these studies have involved classical neurotoxin PD models, but recently, GLP-1 receptor agonists been also tested in the context of synucleinopathies and found to reduce the accumulation of α-syn (Yun et al., 2018; Zhang et al., 2019). In a first open label Phase II study of exenatide in PD (https://clinicaltrials.gov/ct2/show/NCT01174810), exenatide was found to be well tolerated, and resulted in an average improvement on the MDS-UPDRS of 2.7 points, compared to a decline of 2.2 points in control patients at 12 months of treatment (Aviles-Olmos et al., 2013). While the initial study suffered from the shortcoming that there was no double-blind placebo control, of particular interest was the finding that follow up assessments indicated beneficial effects persisted 12 months after cessation of exenatide (Aviles-Olmos et al., 2015).

Given the open-label nature of that first study, a randomized, double-blind Phase II trial was initiated. After 48 weeks on exenatide, the treated group was found to have a statistically significant reduction in the progression of their motor features—as determined by the MDS-UPDRS III—when compared to the placebo control group (Athauda et al., 2017; https://clinicaltrials.gov/ct2/show/NCT01971242). Following up on those results, a Phase III clinical trial has recently been registered (http://www.isrctn.com/ISRCTN14552789). The study is recruiting 200 people with PD, who will be randomized and blindly treated with either exenatide or placebo for 2 years. The results are expected in 2024.

Since the encouraging Phase II trial results were published, another research group at the University of Florida has started an open label Phase I clinical trial of exenatide in PD (https://clinicaltrials.gov/ct2/show/NCT03456687). Additional GLP-1 receptor agonists are also being clinically evaluated in PD. A Phase II study of Liraglutide in 60 participants with PD and insulin resistance is being conducted in California (https://clinicaltrials.gov/ct2/show/NCT02953665), and a Phase II trial of Lixisenatide in 158 patients with early stage PD is being run across 21 hospitals in France (https://clinicaltrials.gov/ct2/show/NCT03439943).

Based on the preclinical and clinical success of GLP-1 receptor agonists in both diabetes and PD, several biotech and pharmaceutical companies have begun developing drugs of this nature. Neuraly's NLY01 is a pegylated long-acting GLP-1 receptor agonist with an extended half-life and favorable pharmacodynamic profiles in non-human primates. NLY01 reduced the behavioral deficits and pathology in both the human A53T α-syn transgenic mouse model and the α-syn preformed fibril (PFF) models of PD (Yun et al., 2018). The results of this study suggest that NLY01 not only reduced levels of phosphorylated α-syn in the brain, but also provided neuroprotection via the prevention of microglia-mediated conversion of astrocytes to a neurotoxic A1 phenotype. A Phase I trial has now been registered (https://clinicaltrials.gov/ct2/show/NCT03672604), to test NLY01 in healthy individuals, with the ultimate goal of testing in PD cohorts. In addition, Novo Nordisk's longer acting GLP-1 receptor agonist Semaglutide is being tested in a large Phase II trial of 120 people with PD by the Oslo University Hospital—GIPD study (https://clinicaltrials.gov/ct2/show/NCT03659682). The Korean company, Peptron, are planning a Phase II study of their GLP-1 receptor agonist, PT320 (which is a sustained-release version of exenatide). Phase I testing has indicated that PT320 is safe and well tolerated (https://clinicaltrials.gov/ct2/show/NCT00964262).

9 Preclinical developments

In addition to these numerous ongoing clinical trial programs focused, some in part, on reducing α-syn aggregation in PD, there are various novel approaches being preclinically developed. In this final section of the chapter, we provide a brief overview of some of these efforts.

9.1 β2-adrenergic receptor agonists

β2 adrenergic receptors (β2AR) are coupled to a stimulatory G protein of adenylyl cyclase. Agonists of this receptor increase cyclic adenosine monophosphate (cAMP) levels. In the lung, cAMP decreases calcium concentrations within cells and activates protein kinase A. β2AR agonism causes bronchodilation, providing relief for individuals affected by asthma and chronic obstructive pulmonary disease (COPD).

β2AR agonists have previously been shown to have neuroprotective properties (Xu et al., 2018), and the effects of both short-acting (salbutamol) and the longer-acting β2AR agonists (bambuterol, clenbuterol, formoterol, and salmeterol) have been evaluated on different models of PD. Results suggest that low doses of all of the β2AR agonists were neuroprotective against LPS-induced neurotoxicity (Qian et al., 2011). A recent screen of 1126 FDA-approved compounds identified 35 compounds which lowered endogenous *SNCA* expression by more than 35%. When they validated these compounds, they found that of the top four compounds, three were β2AR agonists (Mittal et al., 2017). In addition, salbutamol was associated with reduced risk of developing PD, and the investigators found that Propranolol (a β2AR antagonist) was correlated with an increased risk of PD.

Efforts to replicate the association reported in this initial paper have been mixed. A population-based, case-control study of United States Medicare beneficiaries failed to find any association (Searles Nielsen et al., 2018), but two subsequent analyses of two large databases in Israel did report a reduced risk of PD with β2AR agonists use (Gronich et al., 2018; Koren et al., 2019). There have been three small clinical trials in PD (Alexander et al., 1994; Hishida et al., 1992; Uc et al., 2003) which have demonstrated improved response to levodopa, but the duration of these trials has not been long enough to evaluate any changes in disease progression.

9.2 Intrabodies

Given the brain penetrance issues associated with current immunotherapy approaches, efforts have been made to develop smaller intrabody or "nanobody" alternatives. Intrabodies are small antibody fragments that target antigens intracellularly. At only 140–250 amino acids in length, intrabodies can be delivered as proteins or genes. These single chain antibody fragments have been evaluated in reducing the aggregation of α-syn (Emadi et al., 2004; Iljina et al., 2017; Zhou et al., 2004).

Two particular intrabodies have been characterized in vitro: VH14 (as known as NAC14) and NbSyn87 (El-Turk et al., 2016; Lynch et al., 2008), and they have subsequently been tested in vivo (Chatterjee et al., 2018). Although these experimental therapeutics show some promise, there are significant challenges in taking them forward to the clinic, not least when it comes to delivery. Maintaining high levels of the intrabodies in the brain for prolonged periods of time has required the employment of gene therapy approaches and direct delivery.

9.3 PARP inhibition

Several publications suggest that poly(adenosine 5′-diphosphate-ribose) (PAR) polymerase-1 (PARP-1) is activated in both neurotoxin-induced models of PD (Mandir et al., 2002, 1999) and in brains of PD patients (Kam et al., 2018). PARP1 inhibition has also been found to be neuroprotective in neurotoxin-induced models of PD (Kim et al., 2013; Lee et al., 2013; Yokoyama et al., 2009) and in genetic models of PD (Delgado-Camprubi et al., 2017). Recently it was reported that preformed fibrils of α-syn also activate PARP-1, and that the accumulation of PAR further accelerates the formation of aggregated α-syn (Kam et al., 2018). This feed forward loop results in cell death via parthanatos, a form of cell death which is defined as being dependent upon PARP-1 activity. This same study demonstrated that treatment with PARP1 inhibitors reduced the aggregation of α-syn, suggesting that this class of drugs, which primarily have been developed for oncology, could potentially be repurposed for therapeutic use in PD.

9.4 Stearoyl-CoA desaturase inhibition

Lipodomic analysis has indicated that preventing lipid droplet formation increases α-syn toxicity in yeast, rodent and human cells. Specifically, augmenting oleic acid appears to enhance the detrimental effects of accumulating α-syn (Fanning et al., 2019). In agreement with this finding is the discovery that suppressing the oleic acid-generating enzyme stearoyl-CoA-desaturase (SCD) is protective in various models of α-syn toxicity (including overexpression of α-syn, patient-derived *SNCA* triplication cells, patient-derived E46K genetic variant cells, and in transgenic E46K mice—Fanning et al., 2019; Vincent et al., 2018). The biotech firm Yumanity has developed a novel, brain penetrant stearoyl-CoA desaturase inhibitor called YTX-7739, which they have recently initiated a Phase I clinical trial of in 40 healthy volunteers (https://www.businesswire.com/news/home/20191007005107/en/Yumanity-Therapeutics-Initiates-Phase-1-Clinical-Trial).

9.5 Inflammasome inhibition

Inflammation has long been associated with PD, but only recently have very specific forms been identified as potentially playing key roles in the condition. Of particular interest is the activation of the microglial NLR family pyrin domain containing 3 (NLRP3) inflammasome. Inhibition or deficiency of NLRP3 has been shown to rescue models of PD (Fan et al., 2017; Lee et al., 2019; Qiao et al., 2018). Aggregated α-syn has been reported to activate NLRP3 (Codolo et al., 2013), and oral administration of a potent NLRP3 inhibitor in multiple models of PD not only inhibited inflammasome activation, but also rescued motor deficits, prevented neurodegeneration, and also reduced the accumulation of α-syn aggregates (Gordon et al., 2018). The Irish biotech firm Inflazome is now developing several small molecule NLRP3 inhibitors which they are hoping to start clinically testing soon (http://inflazome.com/news-articles/2019/3/24/inflazome-receives-funding-from-the-michael-j-fox-foundation-for-parkinsons-research). Inflazome has recently initiated Phase I testing of their lead compound, an NLRP3 inhibitor called Inzomelid (https://clinicaltrials.gov/ct2/show/NCT04015076).

9.6 Proteasome activation

The ubiquitin-proteasome system is the primary cellular mechanism responsible for the degradation of intracellular proteins. Disrupting the proteasome has been suggested to result in the induction of α-syn aggregation (Ebrahimi-Fakhari et al., 2011), and oligomeric α-syn may actually inhibit proteasomal function (Snyder et al., 2003). Numerous small molecule activators of proteasomal activity have been found to reduce levels of aggregated protein in preclinical PD models (Chatterjee et al., 2018; Njomen et al., 2018; Yuan et al., 2019; Zhou et al., 2019). Researchers have also devised powerful methods to manipulate proteasome degradation for therapeutic purposes. Proteolysis targeting chimera (PROTAC) is one such approach. It involves a two-headed chimeric molecule capable of inducing selective intracellular proteolysis (Sakamoto et al., 2001). A PROTAC contains two linked protein-binding molecules, the first is capable of binding an E3 ubiquitin ligase, while the second attaches to the protein targeted for degradation. Such a system could be utilized in the targeting of aggregated proteins in PD. Oncology has led the way with this approach and the first clinical trial of PROTAC is underway, assessing this methodology in prostate cancer (https://clinicaltrials.gov/ct2/show/NCT03888612).

9.7 Deubiquitinating enzymes

In addition to augmenters of proteasomal activation, manipulation of deubiquitinating enzymes is also being explored. For example, inhibition of the deubiquitinase USP9X promotes the accumulation of α-syn and enhances the formation of inclusions (Rott et al., 2011). Likewise, knockdown of Usp8 in Drosophila and human cells increased the lysosomal degradation of α-syn and reduced cell death (Alexopoulou et al., 2016). Deubiquitination is intimately associated with PD, as genetic variants in two genes encoding deubiquitinating enzymes are associated with an increased risk of developing the condition (Ubiquitin C-terminal hydrolase L1 or UCL1/PARK5 and Ubiquitin specific peptidase 24/PARK10). Multiple biotech firms are attempting to exploit deubiquitinating enzymes, but delivery and toxicity issues have slowed the first clinical trials of deubiquitinating enzyme inhibitors (for example, VLX1570 for multiple myeloma—https://clinicaltrials.gov/ct2/show/NCT02372240).

10 Concluding remarks

Despite tremendous progress, significant challenges remain for experimental therapies targeting protein aggregation pathology. For example, it is still not entirely clear that the formation of α-syn aggregates is a pathological event in the human brain. Transgenic models of PD—in which large amounts of α-syn aggregates are generated—suggest that they can impair neuronal function and cause neurodegeneration. Post-mortem analysis of human brains suggests protein aggregation is a common feature of the aged brain (Markesbery et al., 2009) and around 10% of aged individuals exhibit α-syn aggregates at death, without displaying evident neurological disease. These individuals are classified as having incidental Lewy body disorder, despite not clearly having a "disorder." On the other hand, it is possible that they would have developed PD or another related synucleinopathy if they had lived longer (Delle Donne et al., 2008; Dickson et al., 2009; Frigerio et al., 2011). Why then might some "normal" people have α-syn aggregates in the brain? Accumulating evidence suggests that some protein aggregation may be anti-microbial/viral, forming the basis of a cellular defense mechanism (Beatman et al., 2016) and this might be the case for α-syn (Tulisiak et al., 2019). In addition, immunotherapy efforts to target aggregated β-amyloid in AD have thus far failed to slow the neurodegenerative process of the condition, despite substantial clearance of the protein (Wang et al., 2017b).

Another challenge facing the clinical evaluation of these experimental therapies is the determination of target engagement. Efforts to image aggregation-prone proteins, such as α-syn, have struggled as the levels of aggregated protein constitute just a tiny fraction of the total protein in the brain and it has consequently been difficult to devise methods to specifically detect the aggregates, either using imaging or biochemistry. Recent studies have introduced methods which amplify and quantify small quantities of oligomeric α-syn from CSF, or even plasma, and have reported that the levels are elevated in people suffering from synucleinopathies like PD (Concha-Marambio et al., 2019; Groveman et al., 2018; Paciotti et al., 2018; Yuan et al., 2019). Whether changes in the levels of α-syn oligomers in the CSF are also a good measure of target engagement remains to be determined. In the future, the issue of target engagement could be addressed with the use of brain-derived exosomes extracted from serum samples (Athauda et al., 2019), which would represent an enrichment of the α-syn oligomers found in the circulation, but the validation of this method is still being tested and there are no publications yet where this has been applied to α-syn oligomers.

While the path to clinical success of α-syn based therapies in PD is not simple, and there is definitely no guarantee that any of these proposed therapies will be effective, we believe our review gives reasons for hope. First of all, it is remarkable to consider the wide range of approaches that are currently being used to target the likely pathogenicity of α-syn aggregates. Second, attacking the same problem from different angles means that there might also be scope for combination therapies that could be synergistic or additive, even when the effects of the individual therapies alone are insufficient to provide clinical benefit. Third, it is clear that the biotech and pharmaceutical industries are showing great interest in α-syn as a therapeutic target. In light of the great costs involved in developing new CNS drugs, significant financial investments are required to take any α-syn based therapy to market and the current engagement of the industry is therefore very good news for the PD patient community.

Acknowledgments

P.B. acknowledges the Van Andel Institute and many individuals and corporations that financially support research into neurodegenerative disease at the Institute. P.B. is supported by grants from the National Institutes of Health (1R01DC016519-01, 5R21NS093993-02, 1R21NS106078-01A1), the Office of the Assistant Secretary of Defense for Health Affairs (Parkinson's Research Program, Award No. W81XWH-17-1-0534), The Michael J Fox Foundation and Cure Parkinson's Trust, which are relevant to this review.

Conflicts of interest

P.B. has received commercial support as a consultant from Axial Biotherapeutics, CuraSen, Fujifilm-Cellular Dynamics International, IOS Press Partners, LifeSci Capital LLC, Lundbeck A/S and Living Cell Technologies LTD. He has received commercial support for grants/research from Lundbeck A/S and Roche. He has ownership interests in Acousort AB and Axial Biotherapeutics and is on the steering committee of the NILO-PD trial. R.K.W. is also a member of the steering committee of the NILO-PD trial. Both S.R.W.S and R.K.W declare no other conflicts, but they are employed by an international PD grant-giving charity.

References

Aflaki, E., Stubblefield, B.K., Maniwang, E., Lopez, G., Moaven, N., Goldin, E., Marugan, J., Patnaik, S., Dutra, A., Southall, N., Zheng, W., Tayebi, N., Sidransky, E., 2014. Macrophage models of Gaucher disease for evaluating disease pathogenesis and candidate drugs. Sci. Transl. Med. 6, 240ra73. https://doi.org/10.1126/scitranslmed.3008659.

Aflaki, E., Borger, D.K., Moaven, N., Stubblefield, B.K., Rogers, S.A., Patnaik, S., Schoenen, F.J., Westbroek, W., Zheng, W., Sullivan, P., Fujiwara, H., Sidhu, R., Khaliq, Z.M., Lopez, G.J., Goldstein, D.S., Ory, D.S., Marugan, J., Sidransky, E., 2016. A new glucocerebrosidase chaperone reduces α-synuclein and glycolipid levels in iPSC-derived dopaminergic neurons from patients with Gaucher disease and parkinsonism. J. Neurosci. 36, 7441–7452. https://doi.org/10.1523/JNEUROSCI.0636-16.2016.

Alexander, G.M., Schwartzman, R.J., Nukes, T.A., Grothusen, J.R., Hooker, M.D., 1994. Beta 2-adrenergic agonist as adjunct therapy to levodopa in Parkinson's disease. Neurology 44, 1511–1513. https://doi.org/10.1212/wnl.44.8.1511.

Alexopoulou, Z., Lang, J., Perrett, R.M., Elschami, M., Hurry, M.E.D., Kim, H.T., Mazaraki, D., Szabo, A., Kessler, B.M., Goldberg, A.L., Ansorge, O., Fulga, T.A., Tofaris, G.K., 2016. Deubiquitinase Usp8 regulates α-synuclein clearance and modifies its toxicity in Lewy body disease. Proc. Natl. Acad. Sci. U. S. A. 113, E4688–E4697. https://doi.org/10.1073/pnas.1523597113.

Arima, K., Hirai, S., Sunohara, N., Aoto, K., Izumiyama, Y., Uéda, K., Ikeda, K., Kawai, M., 1999. Cellular co-localization of phosphorylated tau- and NACP/alpha-synuclein-epitopes in Lewy bodies in sporadic Parkinson's disease and in dementia with Lewy bodies. Brain Res. 843, 53–61. https://doi.org/10.1016/s0006-8993(99)01848-x.

Athauda, D., Maclagan, K., Skene, S.S., Bajwa-Joseph, M., Letchford, D., Chowdhury, K., Hibbert, S., Budnik, N., Zampedri, L., Dickson, J., Li, Y., Aviles-Olmos, I.,

Warner, T.T., Limousin, P., Lees, A.J., Greig, N.H., Tebbs, S., Foltynie, T., 2017. Exenatide once weekly versus placebo in Parkinson's disease: a randomised, double-blind, placebo-controlled trial. Lancet 390, 1664–1675. https://doi.org/10.1016/S0140-6736(17)31585-4.

Athauda, D., Gulyani, S., Karnati, H.K., Li, Y., Tweedie, D., Mustapic, M., Chawla, S., Chowdhury, K., Skene, S.S., Greig, N.H., Kapogiannis, D., Foltynie, T., 2019. Utility of neuronal-derived exosomes to examine molecular mechanisms that affect motor function in patients with Parkinson disease. JAMA Neurol. 76, 420. https://doi.org/10.1001/jamaneurol.2018.4304.

Aviles-Olmos, I., Dickson, J., Kefalopoulou, Z., Djamshidian, A., Ell, P., Soderlund, T., Whitton, P., Wyse, R., Isaacs, T., Lees, A., Limousin, P., Foltynie, T., 2013. Exenatide and the treatment of patients with Parkinson's disease. J. Clin. Invest. 123, 2730–2736. https://doi.org/10.1172/JCI68295.

Aviles-Olmos, I., Dickson, J., Kefalopoulou, Z., Djamshidian, A., Kahan, J., Ell, P., Whitton, P., Wyse, R., Isaacs, T., Lees, A., Limousin, P., Foltynie, T., 2015. Motor and cognitive advantages persist 12 months after exenatide exposure in Parkinson's disease. J. Park. Dis. 4, 337–344. https://doi.org/10.3233/JPD-140364.

Baptista, M.A.S., Dave, K.D., Frasier, M.A., Sherer, T.B., Greeley, M., Beck, M.J., Varsho, J.S., Parker, G.A., Moore, C., Churchill, M.J., Meshul, C.K., Fiske, B.K., 2013. Loss of leucine-rich repeat kinase 2 (LRRK2) in rats leads to progressive abnormal phenotypes in peripheral organs. PLoS One 8, e80705. https://doi.org/10.1371/journal.pone.0080705.

Beatman, E.L., Massey, A., Shives, K.D., Burrack, K.S., Chamanian, M., Morrison, T.E., Beckham, J.D., 2016. Alpha-synuclein expression restricts RNA viral infections in the brain. J. Virol. 90, 2767–2782. https://doi.org/10.1128/JVI.02949-15.

Beavan, M.S., Schapira, A.H.V., 2013. Glucocerebrosidase mutations and the pathogenesis of Parkinson disease. Ann. Med. 45, 511–521. https://doi.org/10.3109/07853890.2013.849003.

Boland, B., Yu, W.H., Corti, O., Mollereau, B., Henriques, A., Bezard, E., Pastores, G.M., Rubinsztein, D.C., Nixon, R.A., Duchen, M.R., Mallucci, G.R., Kroemer, G., Levine, B., Eskelinen, E.-L., Mochel, F., Spedding, M., Louis, C., Martin, O.R., Millan, M.J., 2018. Promoting the clearance of neurotoxic proteins in neurodegenerative disorders of ageing. Nat. Rev. Drug Discov. 17, 660–688. https://doi.org/10.1038/nrd.2018.109.

Buse, J.B., Henry, R.R., Han, J., Kim, D.D., Fineman, M.S., Baron, A.D., Exenatide-113 Clinical Study Group, 2004. Effects of exenatide (exendin-4) on glycemic control over 30 weeks in sulfonylurea-treated patients with type 2 diabetes. Diabetes Care 27, 2628–2635. https://doi.org/10.2337/diacare.27.11.2628.

Cahill, C.M., Lahiri, D.K., Huang, X., Rogers, J.T., 2009. Amyloid precursor protein and alpha synuclein translation, implications for iron and inflammation in neurodegenerative diseases. Biochim. Biophys. Acta 1790, 615–628. https://doi.org/10.1016/j.bbagen.2008.12.001.

Carboni, E., Tatenhorst, L., Tönges, L., Barski, E., Dambeck, V., Bähr, M., Lingor, P., 2017. Deferiprone rescues behavioral deficits induced by mild iron exposure in a mouse model of alpha-synuclein aggregation. Neuromolecular Med. 19, 309–321. https://doi.org/10.1007/s12017-017-8447-9.

Chatterjee, D., Bhatt, M., Butler, D., De Genst, E., Dobson, C.M., Messer, A., Kordower, J.H., 2018. Proteasome-targeted nanobodies alleviate pathology and functional decline in an α-synuclein-based Parkinson's disease model. npj Park. Dis. 4, 25. https://doi.org/10.1038/s41531-018-0062-4.

Chen, Z., Yang, Y., Yang, X., Zhou, C., Li, F., Lei, P., Zhong, L., Jin, X., Peng, G., 2013. Immune effects of optimized DNA vaccine and protective effects in a MPTP model of Parkinson's disease. Neurol. Sci. 34, 1559–1570. https://doi.org/10.1007/s10072-012-1284-6.

Chen, S., Yu, S.-J., Li, Y., Lecca, D., Glotfelty, E., Kim, H.K., Choi, H.-I., Hoffer, B.J., Greig, N.H., Kim, D.S., Wang, Y., 2018. Post-treatment with PT302, a long-acting Exendin-4 sustained release formulation, reduces dopaminergic neurodegeneration in a 6-hydroxydopamine rat model of Parkinson's disease. Sci. Rep. 8, 10722. https://doi.org/10.1038/s41598-018-28449-z.

Codolo, G., Plotegher, N., Pozzobon, T., Brucale, M., Tessari, I., Bubacco, L., de Bernard, M., 2013. Triggering of inflammasome by aggregated α-synuclein, an inflammatory response in synucleinopathies. PLoS One 8, e55375. https://doi.org/10.1371/journal.pone.0055375.

Colca, J.R., Vander Lugt, J.T., Adams, W.J., Shashlo, A., McDonald, W.G., Liang, J., Zhou, R., Orloff, D.G., 2013. Clinical proof-of-concept study with MSDC-0160, a prototype mTOT-modulating insulin sensitizer. Clin. Pharmacol. Ther. 93, 352–359. https://doi.org/10.1038/clpt.2013.10.

Concha-Marambio, L., Shahnawaz, M., Soto, C., 2019. Detection of misfolded α-synuclein aggregates in cerebrospinal fluid by the protein misfolding cyclic amplification platform. Methods Mol. Biol. 1948, 35–44. https://doi.org/10.1007/978-1-4939-9124-2_4.

Cookson, M.R., Hardy, J., Lewis, P.A., 2008. Genetic neuropathology of Parkinson's disease. Int. J. Clin. Exp. Pathol. 1, 217–231.

Das, B., Rajagopalan, S., Joshi, G.S., Xu, L., Luo, D., Andersen, J.K., Todi, S.V., Dutta, A.K., 2017. A novel iron (II) preferring dopamine agonist chelator D-607 significantly suppresses α-syn- and MPTP-induced toxicities in vivo. Neuropharmacology 123, 88–99. https://doi.org/10.1016/j.neuropharm.2017.05.019.

Davies, P., Moualla, D., Brown, D.R., 2011. Alpha-synuclein is a cellular ferrireductase. PLoS One 6, e15814. https://doi.org/10.1371/journal.pone.0015814.

Deas, E., Cremades, N., Angelova, P.R., Ludtmann, M.H.R., Yao, Z., Chen, S., Horrocks, M.H., Banushi, B., Little, D., Devine, M.J., Gissen, P., Klenerman, D., Dobson, C.M., Wood, N.W., Gandhi, S., Abramov, A.Y., 2016. Alpha-synuclein oligomers interact with metal ions to induce oxidative stress and neuronal death in Parkinson's disease. Antioxid. Redox Signal. 24, 376–391. https://doi.org/10.1089/ars.2015.6343.

Delgado-Camprubi, M., Esteras, N., Soutar, M.P., Plun-Favreau, H., Abramov, A.Y., 2017. Deficiency of Parkinson's disease-related gene Fbxo7 is associated with impaired mitochondrial metabolism by PARP activation. Cell Death Differ. 24, 120–131. https://doi.org/10.1038/cdd.2016.104.

Delle Donne, A., Klos, K.J., Fujishiro, H., Ahmed, Z., Parisi, J.E., Josephs, K.A., Frigerio, R., Burnett, M., Wszolek, Z.K., Uitti, R.J., Ahlskog, J.E., Dickson, D.W., 2008. Incidental Lewy body disease and preclinical Parkinson disease. Arch. Neurol. 65, 1074–1080. https://doi.org/10.1001/archneur.65.8.1074.

Di Maio, R., Hoffman, E.K., Rocha, E.M., Keeney, M.T., Sanders, L.H., De Miranda, B.R., Zharikov, A., Van Laar, A., Stepan, A.F., Lanz, T.A., Kofler, J.K., Burton, E.A., Alessi, D.R., Hastings, T.G., Greenamyre, J.T., 2018. LRRK2 activation in idiopathic Parkinson's disease. Sci. Transl. Med. 10, eaar5429. https://doi.org/10.1126/scitranslmed.aar5429.

Dickson, D.W., Braak, H., Duda, J.E., Duyckaerts, C., Gasser, T., Halliday, G.M., Hardy, J., Leverenz, J.B., Del Tredici, K., Wszolek, Z.K., Litvan, I., 2009. Neuropathological assessment of Parkinson's disease: refining the diagnostic criteria. Lancet Neurol. 8, 1150–1157. https://doi.org/10.1016/S1474-4422(09)70238-8.

Do, J., McKinney, C., Sharma, P., Sidransky, E., 2019. Glucocerebrosidase and its relevance to Parkinson disease. Mol. Neurodegener. 14, 36. https://doi.org/10.1186/s13024-019-0336-2.

Dorsey, E.R., Bloem, B.R., 2018. The Parkinson pandemic—a call to action. JAMA Neurol. 75, 9. https://doi.org/10.1001/jamaneurol.2017.3299.

Dorsey, E.R., Elbaz, A., Nichols, E., Abd-Allah, F., Abdelalim, A., Adsuar, J.C., Ansha, M.G., Brayne, C., Choi, J.-Y.J., Collado-Mateo, D., Dahodwala, N., Do, H.P., Edessa, D., Endres, M., Fereshtehnejad, S.-M., Foreman, K.J., Gankpe, F.G., Gupta, R., Hankey, G.J., Hay, S.I., Hegazy, M.I., Hibstu, D.T., Kasaeian, A., Khader, Y., Khalil, I., Khang, Y.-H., Kim, Y.J., Kokubo, Y., Logroscino, G., Massano, J., Mohamed Ibrahim, N., Mohammed, M.A., Mohammadi, A., Moradi-Lakeh, M., Naghavi, M., Nguyen, B.T., Nirayo, Y.L., Ogbo, F.A., Owolabi, M.O., Pereira, D.M., Postma, M.J., Qorbani, M., Rahman, M.A., Roba, K.T., Safari, H., Safiri, S., Satpathy, M., Sawhney, M., Shafieesabet, A., Shiferaw, M.S., Smith, M., Szoeke, C.E. I., Tabarés-Seisdedos, R., Truong, N.T., Ukwaja, K.N., Venketasubramanian, N., Villafaina, S., Gidey Weldegwergs, K., Westerman, R., Wijeratne, T., Winkler, A.S., Xuan, B.T., Yonemoto, N., Feigin, V.L., Vos, T., Murray, C.J.L., 2018. Global, regional, and national burden of Parkinson's disease, 1990–2016: a systematic analysis for the Global Burden of Disease Study 2016. Lancet Neurol. 17, 939–953. https://doi.org/10.1016/S1474-4422(18)30295-3.

Ebrahimi-Fakhari, D., Cantuti-Castelvetri, I., Fan, Z., Rockenstein, E., Masliah, E., Hyman, B.T., McLean, P.J., Unni, V.K., 2011. Distinct roles in vivo for the ubiquitin-proteasome system and the autophagy-lysosomal pathway in the degradation of α-synuclein. J. Neurosci. 31, 14508–14520. https://doi.org/10.1523/JNEUROSCI.1560-11.2011.

El-Turk, F., Newby, F.N., De Genst, E., Guilliams, T., Sprules, T., Mittermaier, A., Dobson, C.M., Vendruscolo, M., 2016. Structural effects of two camelid nanobodies directed to distinct C-terminal epitopes on α-synuclein. Biochemistry 55, 3116–3122. https://doi.org/10.1021/acs.biochem.6b00149.

Emadi, S., Liu, R., Yuan, B., Schulz, P., McAllister, C., Lyubchenko, Y., Messer, A., Sierks, M.R., 2004. Inhibiting aggregation of α-synuclein with human single chain antibody fragments. Biochemistry 43, 2871–2878. https://doi.org/10.1021/bi036281f.

Espay, A.J., Vizcarra, J.A., Marsili, L., Lang, A.E., Simon, D.K., Merola, A., Josephs, K.A., Fasano, A., Morgante, F., Savica, R., Greenamyre, J.T., Cambi, F., Yamasaki, T.R., Tanner, C.M., Gan-Or, Z., Litvan, I., Mata, I.F., Zabetian, C.P., Brundin, P., Fernandez, H.H., Standaert, D.G., Kauffman, M.A., Schwarzschild, M.A., Sardi, S.P., Sherer, T., Perry, G., Leverenz, J.B., 2019. Revisiting protein aggregation as pathogenic in sporadic Parkinson and Alzheimer diseases. Neurology 92, 329–337. https://doi.org/10.1212/WNL.0000000000006926.

Fan, Z., Liang, Z., Yang, H., Pan, Y., Zheng, Y., Wang, X., 2017. Tenuigenin protects dopaminergic neurons from inflammation via suppressing NLRP3 inflammasome activation in microglia. J. Neuroinflammation 14, 256. https://doi.org/10.1186/s12974-017-1036-x.

Fanning, S., Haque, A., Imberdis, T., Baru, V., Barrasa, M.I., Nuber, S., Termine, D., Ramalingam, N., Ho, G.P.H., Noble, T., Sandoe, J., Lou, Y., Landgraf, D., Freyzon, Y., Newby, G., Soldner, F., Terry-Kantor, E., Kim, T.-E., Hofbauer, H.F., Becuwe, M., Jaenisch, R., Pincus, D., Clish, C.B., Walther, T.C., Farese, R.V., Srinivasan, S., Welte, M.A., Kohlwein, S.D., Dettmer, U., Lindquist, S., Selkoe, D., 2019. Lipidomic analysis of α-synuclein neurotoxicity identifies stearoyl CoA desaturase as a target for Parkinson treatment. Mol. Cell 73, 1001–1014.e8. https://doi.org/10.1016/j.molcel.2018.11.028.

Fellner, L., Kuzdas-Wood, D., Levin, J., Ryazanov, S., Leonov, A., Griesinger, C., Giese, A., Wenning, G.K., Stefanova, N., 2016. Anle138b partly ameliorates motor deficits despite failure of neuroprotection in a model of advanced multiple system atrophy. Front. Neurosci. 10, 99. https://doi.org/10.3389/fnins.2016.00099.

Finkelstein, D.I., Hare, D.J., Billings, J.L., Sedjahtera, A., Nurjono, M., Arthofer, E., George, S., Culvenor, J.G., Bush, A.I., Adlard, P.A., 2016. Clioquinol improves cognitive, motor function, and microanatomy of the alpha-synuclein hA53T transgenic mice. ACS Chem. Nerosci. 7, 119–129. https://doi.org/10.1021/acschemneuro.5b00253.

Finkelstein, D.I., Billings, J.L., Adlard, P.A., Ayton, S., Sedjahtera, A., Masters, C.L., Wilkins, S., Shackleford, D.M., Charman, S.A., Bal, W., Zawisza, I.A., Kurowska, E., Gundlach, A.L., Ma, S., Bush, A.I., Hare, D.J., Doble, P.A., Crawford, S., Gautier, E.C., Parsons, J., Huggins, P., Barnham, K.J., Cherny, R.A., 2017. The novel compound PBT434 prevents iron mediated neurodegeneration and alpha-synuclein toxicity in multiple models of Parkinson's disease. Acta Neuropathol. Commun. 5, 53. https://doi.org/10.1186/s40478-017-0456-2.

Fowler, A.J., Moussa, C.E.-H., 2018. Activating autophagy as a therapeutic strategy for Parkinson's disease. CNS Drugs 32, 1–11. https://doi.org/10.1007/s40263-018-0497-5.

Freskgård, P.-O., Urich, E., 2017. Antibody therapies in CNS diseases. Neuropharmacology 120, 38–55. https://doi.org/10.1016/j.neuropharm.2016.03.014.

Frigerio, R., Fujishiro, H., Ahn, T.-B., Josephs, K.A., Maraganore, D.M., DelleDonne, A., Parisi, J.E., Klos, K.J., Boeve, B.F., Dickson, D.W., Ahlskog, J.E., 2011. Incidental Lewy body disease: do some cases represent a preclinical stage of dementia with Lewy bodies? Neurobiol. Aging 32, 857–863. https://doi.org/10.1016/j.neurobiolaging.2009.05.019.

Froula, J.M., Castellana-Cruz, M., Anabtawi, N.M., Camino, J.D., Chen, S.W., Thrasher, D.R., Freire, J., Yazdi, A.A., Fleming, S., Dobson, C.M., Kumita, J.R., Cremades, N., Volpicelli-Daley, L.A., 2019. Defining α-synuclein species responsible for Parkinson's disease phenotypes in mice. J. Biol. Chem. 294, 10392–10406. https://doi.org/10.1074/jbc.RA119.007743.

Fuchs, J., Nilsson, C., Kachergus, J., Munz, M., Larsson, E.-M., Schüle, B., Langston, J.W., Middleton, F.A., Ross, O.A., Hulihan, M., Gasser, T., Farrer, M.J., 2007. Phenotypic variation in a large Swedish pedigree due to *SNCA* duplication and triplication. Neurology 68, 916–922. https://doi.org/10.1212/01.wnl.0000254458.17630.c5.

Fuji, R.N., Flagella, M., Baca, M., Baptista, M.A., Brodbeck, J., Chan, B.K., Fiske, B.K., Honigberg, L., Jubb, A.M., Katavolos, P., Lee, D.W., Lewin-Koh, S.-C., Lin, T., Liu, X., Liu, S., Lyssikatos, J.P., O'Mahony, J., Reichelt, M., Roose-Girma, M., Sheng, Z., Sherer, T., Smith, A., Solon, M., Sweeney, Z.K., Tarrant, J., Urkowitz, A., Warming, S., Yaylaoglu, M., Zhang, S., Zhu, H., Estrada, A.A., Watts, R.J., 2015. Effect of selective LRRK2 kinase inhibition on nonhuman primate lung. Sci. Transl. Med. 7, 273ra15. https://doi.org/10.1126/scitranslmed.aaa3634.

Games, D., Seubert, P., Rockenstein, E., Patrick, C., Trejo, M., Ubhi, K., Ettle, B., Ghassemiam, M., Barbour, R., Schenk, D., Nuber, S., Masliah, E., 2013. Axonopathy in an α-synuclein transgenic model of Lewy body disease is associated with extensive accumulation of C-terminal-truncated α-synuclein. Am. J. Pathol. 182, 940–953. https://doi.org/10.1016/j.ajpath.2012.11.018.

Games, D., Valera, E., Spencer, B., Rockenstein, E., Mante, M., Adame, A., Patrick, C., Ubhi, K., Nuber, S., Sacayon, P., Zago, W., Seubert, P., Barbour, R., Schenk, D., Masliah, E., 2014. Reducing C-terminal-truncated alpha-synuclein by immunotherapy attenuates neurodegeneration and propagation in Parkinson's disease-like models. J. Neurosci. 34, 9441–9454. https://doi.org/10.1523/JNEUROSCI.5314-13.2014.

Gandhi, S., Muqit, M.M.K., Stanyer, L., Healy, D.G., Abou-Sleiman, P.M., Hargreaves, I., Heales, S., Ganguly, M., Parsons, L., Lees, A.J., Latchman, D.S., Holton, J.L., Wood, N.W., Revesz, T., 2006. PINK1 protein in normal human brain and Parkinson's disease. Brain 129, 1720–1731. https://doi.org/10.1093/brain/awl114.

Gardian, G., Yang, L., Cleren, C., Calingasan, N.Y., Klivenyi, P., Beal, M.F., 2004. Neuroprotective effects of phenylbutyrate against MPTP neurotoxicity. Neuromolecular Med. 5, 235–242. https://doi.org/10.1385/NMM:5:3:235.

Gardian, G., Browne, S.E., Choi, D.-K., Klivenyi, P., Gregorio, J., Kubilus, J.K., Ryu, H., Langley, B., Ratan, R.R., Ferrante, R.J., Beal, M.F., 2005. Neuroprotective effects of phenylbutyrate in the N171-82Q transgenic mouse model of Huntington's disease. J. Biol. Chem. 280, 556–563. https://doi.org/10.1074/jbc.M410210200.

Ghochikyan, A., Petrushina, I., Davtyan, H., Hovakimyan, A., Saing, T., Davtyan, A., Cribbs, D.H., Agadjanyan, M.G., 2014. Immunogenicity of epitope vaccines targeting different B cell antigenic determinants of human α-synuclein: feasibility study. Neurosci. Lett. 560, 86–91. https://doi.org/10.1016/j.neulet.2013.12.028.

Ghosh, A., Tyson, T., George, S., Hildebrandt, E.N., Steiner, J.A., Madaj, Z., Schulz, E., Machiela, E., McDonald, W.G., Escobar Galvis, M.L., Kordower, J.H., Van Raamsdonk, J.M., Colca, J.R., Brundin, P., 2016. Mitochondrial pyruvate carrier regulates autophagy, inflammation, and neurodegeneration in experimental models of Parkinsons disease. Sci. Transl. Med. 8, 368ra174. https://doi.org/10.1126/scitranslmed.aag2210.

Gordon, R., Albornoz, E.A., Christie, D.C., Langley, M.R., Kumar, V., Mantovani, S., Robertson, A.A.B., Butler, M.S., Rowe, D.B., O'Neill, L.A., Kanthasamy, A.G., Schroder, K., Cooper, M.A., Woodruff, T.M., 2018. Inflammasome inhibition prevents α-synuclein pathology and dopaminergic neurodegeneration in mice. Sci. Transl. Med. 10, eaah4066. https://doi.org/10.1126/scitranslmed.aah4066.

Grassi, D., Howard, S., Zhou, M., Diaz-Perez, N., Urban, N.T., Guerrero-Given, D., Kamasawa, N., Volpicelli-Daley, L.A., LoGrasso, P., Lasmézas, C.I., 2018. Identification of a highly neurotoxic α-synuclein species inducing mitochondrial damage and mitophagy in Parkinson's disease. Proc. Natl. Acad. Sci. U. S. A. 115, E2634–E2643. https://doi.org/10.1073/pnas.1713849115.

Gronich, N., Abernethy, D.R., Auriel, E., Lavi, I., Rennert, G., Saliba, W., 2018. β2-adrenoceptor agonists and antagonists and risk of Parkinson's disease. Mov. Disord. 33, 1465–1471. https://doi.org/10.1002/mds.108.

Groveman, B.R., Orrù, C.D., Hughson, A.G., Raymond, L.D., Zanusso, G., Ghetti, B., Campbell, K.J., Safar, J., Galasko, D., Caughey, B., 2018. Rapid and ultra-sensitive quantitation of disease-associated α-synuclein seeds in brain and cerebrospinal fluid by αSyn RT-QuIC. Acta Neuropathol. Commun. 6, 7. https://doi.org/10.1186/s40478-018-0508-2.

Hadjigeorgiou, G.M., Xiromerisiou, G., Gourbali, V., Aggelakis, K., Scarmeas, N., Papadimitriou, A., Singleton, A., 2006. Association of α-synuclein Rep1 polymorphism and Parkinson's disease: influence of Rep1 on age at onset. Mov. Disord. 21, 534–539. https://doi.org/10.1002/mds.20752.

Hallett, P.J., Huebecker, M., Brekk, O.R., Moloney, E.B., Rocha, E.M., Priestman, D.A., Platt, F.M., Isacson, O., 2018. Glycosphingolipid levels and glucocerebrosidase activity are altered in normal aging of the mouse brain. Neurobiol. Aging 67, 189–200. https://doi.org/10.1016/j.neurobiolaging.2018.02.028.

Hantschel, O., Superti-Furga, G., 2004. Regulation of the c-Abl and Bcr–Abl tyrosine kinases. Nat. Rev. Mol. Cell Biol. 5, 33–44. https://doi.org/10.1038/nrm1280.

Harkavyi, A., Abuirmeileh, A., Lever, R., Kingsbury, A.E., Biggs, C.S., Whitton, P.S., 2008. Glucagon-like peptide 1 receptor stimulation by exendin-4 reverses key deficits in distinct rodent models of Parkinson's disease. J. Neuroinflammation 5, 19. https://doi.org/10.1186/1742-2094-5-19.

Hebron, M.L., Lonskaya, I., Moussa, C.E.-H., 2013. Nilotinib reverses loss of dopamine neurons and improves motor behavior via autophagic degradation of α-synuclein in Parkinson's disease models. Hum. Mol. Genet. 22, 3315–3328. https://doi.org/10.1093/hmg/ddt192.

Heras-Garvin, A., Weckbecker, D., Ryazanov, S., Leonov, A., Griesinger, C., Giese, A., Wenning, G.K., Stefanova, N., 2019. Anle138b modulates α-synuclein oligomerization and prevents motor decline and neurodegeneration in a mouse model of multiple system atrophy. Mov. Disord. 34, 255–263. https://doi.org/10.1002/mds.27562.

Hishida, R., Kurahashi, K., Narita, S., Baba, T., Matsunaga, M., 1992. "Wearing-off" and beta 2-adrenoceptor agonist in Parkinson's disease. Lancet 339, 870. https://doi.org/10.1016/0140-6736(92)90313-r.

Ho, D.H., Kim, H., Nam, D., Sim, H., Kim, J., Kim, H.G., Son, I., Seol, W., 2018. LRRK2 impairs autophagy by mediating phosphorylation of leucyl-tRNA synthetase. Cell Biochem. Funct. 36, 431–442. https://doi.org/10.1002/cbf.3364.

Iljina, M., Hong, L., Horrocks, M.H., Ludtmann, M.H., Choi, M.L., Hughes, C.D., Ruggeri, F.S., Guilliams, T., Buell, A.K., Lee, J.-E., Gandhi, S., Lee, S.F., Bryant, C.E., Vendruscolo, M., Knowles, T.P.J., Dobson, C.M., De Genst, E., Klenerman, D., 2017. Nanobodies raised against monomeric α-synuclein inhibit fibril formation and destabilize toxic oligomeric species. BMC Biol. 15, 57. https://doi.org/10.1186/s12915-017-0390-6.

Inden, M., Kitamura, Y., Takeuchi, H., Yanagida, T., Takata, K., Kobayashi, Y., Taniguchi, T., Yoshimoto, K., Kaneko, M., Okuma, Y., Taira, T., Ariga, H., Shimohama, S., 2007. Neurodegeneration of mouse nigrostriatal dopaminergic system induced by repeated oral administration of rotenone is prevented by 4-phenylbutyrate, a chemical chaperone. J. Neurochem. 101, 1491–1504. https://doi.org/10.1111/j.1471-4159.2006.04440.x.

International Parkinson Disease Genomics Consortium, Nalls, M.A., Plagnol, V., Hernandez, D.G., Sharma, M., Sheerin, U.-M., Saad, M., Simón-Sánchez, J., Schulte, C., Lesage, S., Sveinbjörnsdóttir, S., Stefánsson, K., Martinez, M., Hardy, J., Heutink, P., Brice, A., Gasser, T., Singleton, A.B., Wood, N.W., 2011. Imputation of sequence variants for identification of genetic risks for Parkinson's disease: a meta-analysis of genome-wide association studies. Lancet 377, 641–649. https://doi.org/10.1016/S0140-6736(10)62345-8.

Jankovic, J., Goodman, I., Safirstein, B., Marmon, T.K., Schenk, D.B., Koller, M., Zago, W., Ness, D.K., Griffith, S.G., Grundman, M., Soto, J., Ostrowitzki, S., Boess, F.G., Martin-Facklam, M., Quinn, J.F., Isaacson, S.H., Omidvar, O., Ellenbogen, A., Kinney, G.G., 2018. Safety and tolerability of multiple ascending doses of PRX002/RG7935, an anti-α-synuclein monoclonal antibody, in patients with Parkinson disease. JAMA Neurol. 75, 1206. https://doi.org/10.1001/jamaneurol.2018.1487.

Kam, T.-I., Mao, X., Park, H., Chou, S.-C., Karuppagounder, S.S., Umanah, G.E., Yun, S.P., Brahmachari, S., Panicker, N., Chen, R., Andrabi, S.A., Qi, C., Poirier, G.G., Pletnikova, O., Troncoso, J.C., Bekris, L.M., Leverenz, J.B., Pantelyat, A., Ko, H.S., Rosenthal, L.S., Dawson, T.M., Dawson, V.L., 2018. Poly(ADP-ribose) drives pathologic α-synuclein neurodegeneration in Parkinson's disease. Science 362, eaat8407. https://doi.org/10.1126/science.aat8407.

Karuppagounder, S.S., Brahmachari, S., Lee, Y., Dawson, V.L., Dawson, T.M., Ko, H.S., 2015. The c-Abl inhibitor, Nilotinib, protects dopaminergic neurons in a preclinical animal model of Parkinson's disease. Sci. Rep. 4, 4874. https://doi.org/10.1038/srep04874.

Kim, J., Kundu, M., Viollet, B., Guan, K.-L., 2011. AMPK and mTOR regulate autophagy through direct phosphorylation of Ulk1. Nat. Cell Biol. 13, 132–141. https://doi.org/10.1038/ncb2152.

Kim, T.W., Cho, H.M., Choi, S.Y., Suguira, Y., Hayasaka, T., Setou, M., Koh, H.C., Mi Hwang, E., Park, J.Y., Kang, S.J., Kim, H.S., Kim, H., Sun, W., 2013. (ADP-ribose) polymerase 1 and AMP-activated protein kinase mediate progressive dopaminergic neuronal degeneration in a mouse model of Parkinson's disease. Cell Death Dis. 4, e919. https://doi.org/10.1038/cddis.2013.447.

Kim, S., Yun, S.P., Lee, S., Umanah, G.E., Bandaru, V.V.R., Yin, X., Rhee, P., Karuppagounder, S.S., Kwon, S.-H., Lee, H., Mao, X., Kim, D., Pandey, A., Lee, G., Dawson, V.L., Dawson, T.M., Ko, H.S., 2018. GBA1 deficiency negatively affects physiological α-synuclein tetramers and related multimers. Proc. Natl. Acad. Sci. U. S. A. 115, 798–803. https://doi.org/10.1073/pnas.1700465115.

Kordower, J.H., Chu, Y., Hauser, R.A., Freeman, T.B., Olanow, C.W., 2008. Lewy body–like pathology in long-term embryonic nigral transplants in Parkinson's disease. Nat. Med. 14, 504–506. https://doi.org/10.1038/nm1747.

Koren, G., Norton, G., Radinsky, K., Shalev, V., 2019. Chronic use of β-blockers and the risk of Parkinson's disease. Clin. Drug Investig. 39, 463–468. https://doi.org/10.1007/s40261-019-00771-y.

Krishnan, R., Tsubery, H., Proschitsky, M.Y., Asp, E., Lulu, M., Gilead, S., Gartner, M., Waltho, J.P., Davis, P.J., Hounslow, A.M., Kirschner, D.A., Inouye, H., Myszka, D.G., Wright, J., Solomon, B., Fisher, R.A., 2014. A bacteriophage capsid protein provides a general amyloid interaction motif (GAIM) that binds and remodels misfolded protein assemblies. J. Mol. Biol. 426, 2500–2519. https://doi.org/10.1016/j.jmb.2014.04.015.

Kuo, Y.-M., Nwankwo, E.I., Nussbaum, R.L., Rogers, J., Maccecchini, M.L., 2019. Translational inhibition of α-synuclein by Posiphen normalizes distal colon motility in transgenic Parkinson mice. Am. J. Neurodegener. Dis. 8, 1–15.

Lahiri, D.K., Chen, D., Maloney, B., Holloway, H.W., Yu, Q., Utsuki, T., Giordano, T., Sambamurti, K., Greig, N.H., 2007. The experimental Alzheimer's disease drug posiphen [(+)-phenserine] lowers amyloid-β peptide levels in cell culture and mice. J. Pharmacol. Exp. Ther. 320, 386–396. https://doi.org/10.1124/jpet.106.112102.

Lan, A., Chen, J., Zhao, Y., Chai, Z., Hu, Y., 2017. mTOR signaling in Parkinson's disease. Neuromolecular Med. 19, 1–10. https://doi.org/10.1007/s12017-016-8417-7.

Lee, B.D., Shin, J.-H., Van Kampen, J., Petrucelli, L., West, A.B., Ko, H.S., Lee, Y.-I., Maguire-Zeiss, K.A., Bowers, W.J., Federoff, H.J., Dawson, V.L., Dawson, T.M., 2010. Inhibitors of leucine-rich repeat kinase-2 protect against models of Parkinson's disease. Nat. Med. 16, 998–1000. https://doi.org/10.1038/nm.2199.

Lee, Y., Karuppagounder, S.S., Shin, J.-H., Lee, Y.-I., Ko, H.S., Swing, D., Jiang, H., Kang, S.-U., Lee, B.D., Kang, H.C., Kim, D., Tessarollo, L., Dawson, V.L., Dawson, T.M., 2013. Parthanatos mediates AIMP2-activated age-dependent dopaminergic neuronal loss. Nat. Neurosci. 16, 1392–1400. https://doi.org/10.1038/nn.3500.

Lee, S., Kim, S., Park, Y.J., Yun, S.P., Kwon, S.-H., Kim, D., Kim, D.Y., Shin, J.S., Cho, D.J., Lee, G.Y., Ju, H.S., Yun, H.J., Park, J.H., Kim, W.R., Jung, E.A., Lee, S., Ko, H.S., 2018. The c-Abl inhibitor, radotinib HCl, is neuroprotective in a preclinical Parkinson's disease mouse model. Hum. Mol. Genet. 27, 2344–2356. https://doi.org/10.1093/hmg/ddy143.

Lee, E., Hwang, I., Park, S., Hong, S., Hwang, B., Cho, Y., Son, J., Yu, J.-W., 2019. MPTP-driven NLRP3 inflammasome activation in microglia plays a central role in dopaminergic neurodegeneration. Cell Death Differ. 26, 213–228. https://doi.org/10.1038/s41418-018-0124-5.

Levenson, J.M., Schroeter, S., Carroll, J.C., Cullen, V., Asp, E., Proschitsky, M., Chung, C.H.-Y., Gilead, S., Nadeem, M., Dodiya, H.B., Shoaga, S., Mufson, E.J., Tsubery, H., Krishnan, R., Wright, J., Solomon, B., Fisher, R., Gannon, K.S., 2016. NPT088 reduces both amyloid-β and tau pathologies in transgenic mice. Alzheimers Dement 2, 141–155. https://doi.org/10.1016/j.trci.2016.06.004.

Levin, J., Schmidt, F., Boehm, C., Prix, C., Bötzel, K., Ryazanov, S., Leonov, A., Griesinger, C., Giese, A., 2014. The oligomer modulator anle138b inhibits disease progression in a Parkinson mouse model even with treatment started after disease onset. Acta Neuropathol. 127, 779–780. https://doi.org/10.1007/s00401-014-1265-3.

Li, J.-Y., Englund, E., Holton, J.L., Soulet, D., Hagell, P., Lees, A.J., Lashley, T., Quinn, N.P., Rehncrona, S., Björklund, A., Widner, H., Revesz, T., Lindvall, O., Brundin, P., 2008. Lewy bodies in grafted neurons in subjects with Parkinson's disease suggest host-to-graft disease propagation. Nat. Med. 14, 501–503. https://doi.org/10.1038/nm1746.

Li, Y., Perry, T., Kindy, M.S., Harvey, B.K., Tweedie, D., Holloway, H.W., Powers, K., Shen, H., Egan, J.M., Sambamurti, K., Brossi, A., Lahiri, D.K., Mattson, M.P., Hoffer, B.J., Wang, Y., Greig, N.H., 2009. GLP-1 receptor stimulation preserves primary cortical and dopaminergic neurons in cellular and rodent models of stroke and Parkinsonism. Proc. Natl. Acad. Sci. U. S. A. 106, 1285–1290. https://doi.org/10.1073/pnas.0806720106.

Li, W., Englund, E., Widner, H., Mattsson, B., van Westen, D., Lätt, J., Rehncrona, S., Brundin, P., Björklund, A., Lindvall, O., Li, J.-Y., 2016. Extensive graft-derived dopaminergic innervation is maintained 24 years after transplantation in the degenerating parkinsonian brain. Proc. Natl. Acad. Sci. U. S. A. 113, 6544–6549. https://doi.org/10.1073/pnas.1605245113.

Lilja, A.M., Luo, Y., Yu, Q., Röjdner, J., Li, Y., Marini, A.M., Marutle, A., Nordberg, A., Greig, N.H., 2013. Neurotrophic and neuroprotective actions of (−)- and (+)-phenserine, candidate drugs for Alzheimer's disease. PLoS One 8, e54887. https://doi.org/10.1371/journal.pone.0054887.

Liu, W., Jalewa, J., Sharma, M., Li, G., Li, L., Hölscher, C., 2015. Neuroprotective effects of lixisenatide and liraglutide in the 1-methyl-4-phenyl-1,2,3,6-tetrahydropyridine mouse model of Parkinson's disease. Neuroscience 303, 42–50. https://doi.org/10.1016/j.neuroscience.2015.06.054.

Lynch, S.M., Zhou, C., Messer, A., 2008. An scFv intrabody against the nonamyloid component of α-synuclein reduces intracellular aggregation and toxicity. J. Mol. Biol. 377, 136–147. https://doi.org/10.1016/j.jmb.2007.11.096.

Maccecchini, M.L., Chang, M.Y., Pan, C., John, V., Zetterberg, H., Greig, N.H., 2012. Posiphen as a candidate drug to lower CSF amyloid precursor protein, amyloid-β peptide and τ levels: target engagement, tolerability and pharmacokinetics in humans. J. Neurol. Neurosurg. Psychiatry 83, 894–902. https://doi.org/10.1136/jnnp-2012-302589.

Maegawa, G.H.B., Tropak, M.B., Buttner, J.D., Rigat, B.A., Fuller, M., Pandit, D., Tang, L., Kornhaber, G.J., Hamuro, Y., Clarke, J.T.R., Mahuran, D.J., 2009. Identification and characterization of ambroxol as an enzyme enhancement agent for gaucher disease. J. Biol. Chem. 284, 23502–23516. https://doi.org/10.1074/jbc.M109.012393.

Mahul-Mellier, A.-L., Fauvet, B., Gysbers, A., Dikiy, I., Oueslati, A., Georgeon, S., Lamontanara, A.J., Bisquertt, A., Eliezer, D., Masliah, E., Halliday, G., Hantschel, O., Lashuel, H.A., 2014. c-Abl phosphorylates α-synuclein and regulates its degradation: implication for α-synuclein clearance and contribution to the pathogenesis of Parkinson's disease. Hum. Mol. Genet. 23, 2858–2879. https://doi.org/10.1093/hmg/ddt674.

Mandir, A.S., Przedborski, S., Jackson-Lewis, V., Wang, Z.-Q., Simbulan-Rosenthal, C.M., Smulson, M.E., Hoffman, B.E., Guastella, D.B., Dawson, V.L., Dawson, T.M., 1999. Poly(ADP-ribose) polymerase activation mediates 1-methyl-4-phenyl-1,2,3,6-tetrahydropyridine (MPTP)-induced parkinsonism. Proc. Natl. Acad. Sci. U. S. A. 96, 5774–5779. https://doi.org/10.1073/pnas.96.10.5774.

Mandir, A.S., Simbulan-Rosenthal, C.M., Poitras, M.F., Lumpkin, J.R., Dawson, V.L., Smulson, M.E., Dawson, T.M., 2002. A novel in vivo post-translational modification of p53 by PARP-1 in MPTP-induced parkinsonism. J. Neurochem. 83, 186–192. https://doi.org/10.1046/j.1471-4159.2002.01144.x.

Mandler, M., Valera, E., Rockenstein, E., Weninger, H., Patrick, C., Adame, A., Santic, R., Meindl, S., Vigl, B., Smrzka, O., Schneeberger, A., Mattner, F., Masliah, E., 2014. Next-generation active immunization approach for synucleinopathies: implications for Parkinson's disease clinical trials. Acta Neuropathol. 127, 861–879. https://doi.org/10.1007/s00401-014-1256-4.

Mandler, M., Valera, E., Rockenstein, E., Mante, M., Weninger, H., Patrick, C., Adame, A., Schmidhuber, S., Santic, R., Schneeberger, A., Schmidt, W., Mattner, F., Masliah, E., 2015. Active immunization against alpha-synuclein ameliorates the degenerative pathology and prevents demyelination in a model of multiple system atrophy. Mol. Neurodegener. 10, 10. https://doi.org/10.1186/s13024-015-0008-9.

Manzoni, C., Lewis, P.A., 2017. LRRK2 and autophagy. In: Rideout, H.J. (Ed.), Advances in Neurobiology. Springer Nature, pp. 89–105. https://doi.org/10.1007/978-3-319-49969-7_5.

Manzoni, C., Mamais, A., Dihanich, S., Abeti, R., Soutar, M.P.M., Plun-Favreau, H., Giunti, P., Tooze, S.A., Bandopadhyay, R., Lewis, P.A., 2013. Inhibition of LRRK2 kinase activity stimulates macroautophagy. Biochim. Biophys. Acta 1833, 2900–2910. https://doi.org/10.1016/j.bbamcr.2013.07.020.

Maor, G., Rencus-Lazar, S., Filocamo, M., Steller, H., Segal, D., Horowitz, M., 2013. Unfolded protein response in Gaucher disease: from human to Drosophila. Orphanet J. Rare Dis. 8, 140. https://doi.org/10.1186/1750-1172-8-140.

Maor, G., Cabasso, O., Krivoruk, O., Rodriguez, J., Steller, H., Segal, D., Horowitz, M., 2016. The contribution of mutant *GBA* to the development of Parkinson disease in *Drosophila*. Hum. Mol. Genet. 25, ddw129. https://doi.org/10.1093/hmg/ddw129.

Markesbery, W.R., Jicha, G.A., Liu, H., Schmitt, F.A., 2009. Lewy body pathology in normal elderly subjects. J. Neuropathol. Exp. Neurol. 68, 816–822. https://doi.org/10.1097/NEN.0b013e3181ac10a7.

Márquez-Garbán, D.C., Gorrín-Rivas, M., Chen, H.-W., Sterling, C., Elashoff, D., Hamilton, N., Pietras, R.J., 2019. Squalamine blocks tumor-associated angiogenesis and growth of human breast cancer cells with or without HER-2/neu overexpression. Cancer Lett. 449, 66–75. https://doi.org/10.1016/j.canlet.2019.02.009.

Masliah, E., Rockenstein, E., Adame, A., Alford, M., Crews, L., Hashimoto, M., Seubert, P., Lee, M., Goldstein, J., Chilcote, T., Games, D., Schenk, D., 2005. Effects of α-synuclein immunization in a mouse model of Parkinson's disease. Neuron 46, 857–868. https://doi.org/10.1016/j.neuron.2005.05.010.

Masliah, E., Rockenstein, E., Mante, M., Crews, L., Spencer, B., Adame, A., Patrick, C., Trejo, M., Ubhi, K., Rohn, T.T., Mueller-Steiner, S., Seubert, P., Barbour, R., McConlogue, L., Buttini, M., Games, D., Schenk, D., 2011. Passive immunization reduces behavioral and neuropathological deficits in an alpha-synuclein transgenic model of Lewy body disease. PLoS One 6, e19338. https://doi.org/10.1371/journal.pone.0019338.

Mazzulli, J.R., Zunke, F., Tsunemi, T., Toker, N.J., Jeon, S., Burbulla, L.F., Patnaik, S., Sidransky, E., Marugan, J.J., Sue, C.M., Krainc, D., 2016. Activation of β-glucocerebrosidase reduces pathological α-synuclein and restores lysosomal function in Parkinson's patient midbrain neurons. J. Neurosci. 36, 7693–7706. https://doi.org/10.1523/JNEUROSCI.0628-16.2016.

McNeill, A., Magalhaes, J., Shen, C., Chau, K.-Y., Hughes, D., Mehta, A., Foltynie, T., Cooper, J.M., Abramov, A.Y., Gegg, M., Schapira, A.H.V., 2014. Ambroxol improves lysosomal biochemistry in glucocerebrosidase mutation-linked Parkinson disease cells. Brain 137, 1481–1495. https://doi.org/10.1093/brain/awu020.

Merchant, K.M., Cedarbaum, J.M., Brundin, P., Dave, K.D., Eberling, J., Espay, A.J., Hutten, S.J., Javidnia, M., Luthman, J., Maetzler, W., Menalled, L., Reimer, A.N., Stoessl, A.J., Weiner, D.M., The Michael J. Fox Foundation Alpha Synuclein Clinical Path Working Group, 2019. A proposed roadmap for Parkinson's disease proof of concept clinical trials investigating compounds targeting alpha-synuclein. J. Park. Dis. 9, 31–61. https://doi.org/10.3233/JPD-181471.

Migdalska-Richards, A., Daly, L., Bezard, E., Schapira, A.H.V., 2016. Ambroxol effects in glucocerebrosidase and α-synuclein transgenic mice. Ann. Neurol. 80, 766–775. https://doi.org/10.1002/ana.24790.

Migdalska-Richards, A., Ko, W.K.D., Li, Q., Bezard, E., Schapira, A.H.V., 2017. Oral ambroxol increases brain glucocerebrosidase activity in a nonhuman primate. Synapse 71, e21967. https://doi.org/10.1002/syn.21967.

Mikkilineni, S., Cantuti-Castelvetri, I., Cahill, C.M., Balliedier, A., Greig, N.H., Rogers, J.T., 2012. The anticholinesterase phenserine and its enantiomer posiphen as 5′ untranslated-region-directed translation blockers of the Parkinson's alpha synuclein expression. Parkinsons Dis. 2012, 1–13. https://doi.org/10.1155/2012/142372.

Minamiyama, M., Katsuno, M., Adachi, H., Waza, M., Sang, C., Kobayashi, Y., Tanaka, F., Doyu, M., Inukai, A., Sobue, G., 2004. Sodium butyrate ameliorates phenotypic expression in a transgenic mouse model of spinal and bulbar muscular atrophy. Hum. Mol. Genet. 13, 1183–1192. https://doi.org/10.1093/hmg/ddh131.

Mittal, S., Bjørnevik, K., Im, D.S., Flierl, A., Dong, X., Locascio, J.J., Abo, K.M., Long, E., Jin, M., Xu, B., Xiang, Y.K., Rochet, J.-C., Engeland, A., Rizzu, P., Heutink, P., Bartels, T., Selkoe, D.J., Caldarone, B.J., Glicksman, M.A., Khurana, V., Schüle, B., Park, D.S., Riise, T., Scherzer, C.R., 2017. β2-Adrenoreceptor is a regulator of the α-synuclein gene driving risk of Parkinson's disease. Science 357, 891–898. https://doi.org/10.1126/science.aaf3934.

Nguyen, M., Wong, Y.C., Ysselstein, D., Severino, A., Krainc, D., 2019. Synaptic, mitochondrial, and lysosomal dysfunction in Parkinson's disease. Trends Neurosci. 42, 140–149. https://doi.org/10.1016/j.tins.2018.11.001.

Nishiyama, K., Murayama, S., Shimizu, J., Ohya, Y., Kwak, S., Asayama, K., Kanazawa, I., 1995. Cu/Zn superoxide dismutase-like immunoreactivity is present in Lewy bodies from Parkinson disease: a light and electron microscopic immunocytochemical study. Acta Neuropathol. 89, 471–474. https://doi.org/10.1007/bf00571500.

Njomen, E., Osmulski, P.A., Jones, C.L., Gaczynska, M., Tepe, J.J., 2018. Small molecule modulation of proteasome assembly. Biochemistry 57, 4214–4224. https://doi.org/10.1021/acs.biochem.8b00579.

Noda, T., 2017. Regulation of autophagy through TORC1 and mTORC1. Biomolecules 7, 52. https://doi.org/10.3390/biom7030052.

Ono, K., Ikemoto, M., Kawarabayashi, T., Ikeda, M., Nishinakagawa, T., Hosokawa, M., Shoji, M., Takahashi, M., Nakashima, M., 2009. A chemical chaperone, sodium 4-phenylbutyric acid, attenuates the pathogenic potency in human α-synuclein A30P +A53T transgenic mice. Parkinsonism Relat. Disord. 15, 649–654. https://doi.org/10.1016/j.parkreldis.2009.03.002.

Ostrerova-Golts, N., Petrucelli, L., Hardy, J., Lee, J.M., Farer, M., Wolozin, B., 2000. The A53T alpha-synuclein mutation increases iron-dependent aggregation and toxicity. J. Neurosci. 20, 6048–6054.

Paciotti, S., Bellomo, G., Gatticchi, L., Parnetti, L., 2018. Are we ready for detecting α-synuclein prone to aggregation in patients? The case of "protein-misfolding cyclic amplification" and "real-time quaking-induced conversion" as Diagnostic Tools. Front. Neurol. 9, 415. https://doi.org/10.3389/fneur.2018.00415.

Pagan, F., Hebron, M., Valadez, E.H., Torres-Yaghi, Y., Huang, X., Mills, R.R., Wilmarth, B.M., Howard, H., Dunn, C., Carlson, A., Lawler, A., Rogers, S.L., Falconer, R.A., Ahn, J., Li, Z., Moussa, C., 2016. Nilotinib effects in Parkinson's disease and dementia with Lewy bodies. J. Park. Dis. 6, 503–517. https://doi.org/10.3233/JPD-160867.

Pagan, F.L., Hebron, M.L., Wilmarth, B., Torres-Yaghi, Y., Lawler, A., Mundel, E.E., Yusuf, N., Starr, N.J., Arellano, J., Howard, H.H., Peyton, M., Matar, S., Liu, X., Fowler, A.J., Schwartz, S.L., Ahn, J., Moussa, C., 2019. Pharmacokinetics and pharmacodynamics of a single dose Nilotinib in individuals with Parkinson's disease. Pharmacol. Res. Perspect. 7, e00470. https://doi.org/10.1002/prp2.470.

Patnaik, S., Zheng, W., Choi, J.H., Motabar, O., Southall, N., Westbroek, W., Lea, W.A., Velayati, A., Goldin, E., Sidransky, E., Leister, W., Marugan, J.J., 2012. Discovery, structure–activity relationship, and biological evaluation of noninhibitory small molecule chaperones of glucocerebrosidase. J. Med. Chem. 55, 5734–5748. https://doi.org/10.1021/jm300063b.

Paul, A., Zhang, B.-D., Mohapatra, S., Li, G., Li, Y.-M., Gazit, E., Segal, D., 2019. Novel mannitol-based small molecules for inhibiting aggregation of α-synuclein amyloids in Parkinson's disease. Front. Mol. Biosci. 6, 16. https://doi.org/10.3389/fmolb.2019.00016.

Perni, M., Galvagnion, C., Maltsev, A., Meisl, G., Müller, M.B.D., Challa, P.K., Kirkegaard, J.B., Flagmeier, P., Cohen, S.I.A., Cascella, R., Chen, S.W., Limboker, R., Sormanni, P., Heller, G.T., Aprile, F.A., Cremades, N., Cecchi, C., Chiti, F., Nollen, E.A.A., Knowles, T.P.J., Vendruscolo, M., Bax, A., Zasloff, M., Dobson, C.M., 2017. A natural product inhibits the initiation of α-synuclein aggregation and suppresses its toxicity. Proc. Natl. Acad. Sci. U. S. A. 114, E1009–E1017. https://doi.org/10.1073/pnas.1610586114.

Perni, M., Flagmeier, P., Limbocker, R., Cascella, R., Aprile, F.A., Galvagnion, C., Heller, G.T., Meisl, G., Chen, S.W., Kumita, J.R., Challa, P.K., Kirkegaard, J.B., Cohen, S.I.A., Mannini, B., Barbut, D., Nollen, E.A.A., Cecchi, C., Cremades, N., Knowles, T.P.J., Chiti, F., Zasloff, M., Vendruscolo, M., Dobson, C.M., 2018. Multistep inhibition of α-synuclein aggregation and toxicity in vitro and in vivo by trodusquemine. ACS Chem. Biol. 13, 2308–2319. https://doi.org/10.1021/acschembio.8b00466.

Pollanen, M.S., Bergeron, C., Weyer, L., 1992. Detergent-insoluble cortical lewy body fibrils share epitopes with neurofilament and tau. J. Neurochem. 58, 1953–1956. https://doi.org/10.1111/j.1471-4159.1992.tb10074.x.

Polymeropoulos, M.H., Lavedan, C., Leroy, E., Ide, S.E., Dehejia, A., Dutra, A., Pike, B., Root, H., Rubenstein, J., Boyer, R., Stenroos, E.S., Chandrasekharappa, S., Athanassiadou, A., Papapetropoulos, T., Johnson, W.G., Lazzarini, A.M., Duvoisin, R.C., Di Iorio, G., Golbe, L.I., Nussbaum, R.L., 1997. Mutation in the α-synuclein gene identified in families with Parkinson's disease. Science 276, 2045–2047. https://doi.org/10.1126/science.276.5321.2045.

Price, D.L., Koike, M.A., Khan, A., Wrasidlo, W., Rockenstein, E., Masliah, E., Bonhaus, D., 2018. The small molecule alpha-synuclein misfolding inhibitor, NPT200-11, produces multiple benefits in an animal model of Parkinson's disease. Sci. Rep. 8, 16165. https://doi.org/10.1038/s41598-018-34490-9.

Qian, L., Wu, H., Chen, S.-H., Zhang, D., Ali, S.F., Peterson, L., Wilson, B., Lu, R.-B., Hong, J.-S., Flood, P.M., 2011. β2-adrenergic receptor activation prevents rodent dopaminergic neurotoxicity by inhibiting microglia via a novel signaling pathway. J. Immunol. 186, 4443–4454. https://doi.org/10.4049/jimmunol.1002449.

Qiao, C., Zhang, Q., Jiang, Q., Zhang, T., Chen, M., Fan, Y., Ding, J., Lu, M., Hu, G., 2018. Inhibition of the hepatic Nlrp3 protects dopaminergic neurons via attenuating systemic inflammation in a MPTP/p mouse model of Parkinson's disease. J. Neuroinflammation 15, 193. https://doi.org/10.1186/s12974-018-1236-z.

Quansah, E., Peelaerts, W., Langston, J.W., Simon, D.K., Colca, J., Brundin, P., 2018. Targeting energy metabolism via the mitochondrial pyruvate carrier as a novel approach to attenuate neurodegeneration. Mol. Neurodegener. 13, 28. https://doi.org/10.1186/s13024-018-0260-x.

Ramonet, D., Daher, J.P.L., Lin, B.M., Stafa, K., Kim, J., Banerjee, R., Westerlund, M., Pletnikova, O., Glauser, L., Yang, L., Liu, Y., Swing, D.A., Beal, M.F., Troncoso, J.C., McCaffery, J.M., Jenkins, N.A., Copeland, N.G., Galter, D., Thomas, B., Lee, M.K., Dawson, T.M., Dawson, V.L., Moore, D.J., 2011. Dopaminergic neuronal loss, reduced neurite complexity and autophagic abnormalities in transgenic mice expressing G2019S mutant LRRK2. PLoS One 6, e18568. https://doi.org/10.1371/journal.pone.0018568.

Robak, L.A., Jansen, I.E., van Rooij, J., Uitterlinden, A.G., Kraaij, R., Jankovic, J., Heutink, P., Shulman, J.M., Nalls, M.A., Plagnol, V., Hernandez, D.G., Sharma, M., Sheerin, U.-M., Saad, M., Simón-Sánchez, J., Schulte, C., Lesage, S., Sveinbjörnsdóttir, S., Arepalli, S., Barker, R., Ben, Y., Berendse, H.W., Berg, D., Bhatia, K., de Bie, R.M.A., Biffi, A., Bloem, B., Bochdanovits, Z., Bonin, M., Bras, J.M., Brockmann, K., Brooks, J., Burn, D.J., Majounie, E., Charlesworth, G., Lungu, C., Chen, H., Chinnery, P.F., Chong, S., Clarke, C.E., Cookson, M.R., Mark Cooper, J., Corvol, J.C., Counsell, C., Damier, P., Dartigues, J.-F., Deloukas, P., Deuschl, G., Dexter, D.T., van Dijk, K.D., Dillman, A., Durif, F., Dürr, A., Edkins, S., Evans, J.R., Foltynie, T., Dong, J., Gardner, M., Raphael Gibbs, J., Goate, A., Gray, E., Guerreiro, R., Harris, C., van Hilten, J.J., Hofman, A., Hollenbeck, A., Holton, J., Hu, M., Huang, X., Wurster, I., Mätzler, W., Hudson, G., Hunt, S.E., Huttenlocher, J., Illig, T., Jónsson, P.V., Lambert, J.-C., Langford, C., Lees, A., Lichtner, P., Limousin, P., Lopez, G., Lorenz, D., Lungu, C., McNeill, A., Moorby, C., Moore, M., Morris, H.R., Morrison, K.E., Escott-Price, V., Mudanohwo, E., O'Sullivan, S.S., Pearson, J., Perlmutter, J.S., Pétursson, H., Pollak, P., Post, B., Potter, S., Ravina, B.,

Revesz, T., Riess, O., Rivadeneira, F., Rizzu, P., Ryten, M., Sawcer, S., Schapira, A., Scheffer, H., Shaw, K., Shoulson, I., Shulman, J., Sidransky, E., Smith, C., Spencer, C.C. A., Stefánsson, H., Bettella, F., Stockton, J.D., Strange, A., Talbot, K., Tanner, C.M., Tashakkori-Ghanbaria, A., Tison, F., Trabzuni, D., Traynor, B.J., Uitterlinden, A.G., Velseboer, D., Vidailhet, M., Walker, R., van de Warrenburg, B., Wickremaratchi, M., Williams, N., Williams-Gray, C.H., Winder-Rhodes, S., Stefánsson, K., Martinez, M., Wood, N.W., Hardy, J., Heutink, P., Brice, A., Gasser, T., Singleton, A.B., 2017. Excessive burden of lysosomal storage disorder gene variants in Parkinson's disease. Brain 140, 3191–3203. https://doi.org/10.1093/brain/awx285.

Rocha, E.M., Smith, G.A., Park, E., Cao, H., Brown, E., Hayes, M.A., Beagan, J., McLean, J.R., Izen, S.C., Perez-Torres, E., Hallett, P.J., Isacson, O., 2015. Glucocerebrosidase gene therapy prevents α-synucleinopathy of midbrain dopamine neurons. Neurobiol. Dis. 82, 495–503. https://doi.org/10.1016/j.nbd.2015.09.009.

Rogers, J.T., Mikkilineni, S., Cantuti-Castelvetri, I., Smith, D.H., Huang, X., Bandyopadhyay, S., Cahill, C.M., Maccecchini, M.L., Lahiri, D.K., Greig, N.H., 2011. The alpha-synuclein 5′untranslated region targeted translation blockers: anti-alpha synuclein efficacy of cardiac glycosides and Posiphen. J. Neural Transm. 118, 493–507. https://doi.org/10.1007/s00702-010-0513-5.

Ross, O.A., Soto-Ortolaza, A.I., Heckman, M.G., Aasly, J.O., Abahuni, N., Annesi, G., Bacon, J.A., Bardien, S., Bozi, M., Brice, A., Brighina, L., Van Broeckhoven, C., Carr, J., Chartier-Harlin, M.-C., Dardiotis, E., Dickson, D.W., Diehl, N.N., Elbaz, A., Ferrarese, C., Ferraris, A., Fiske, B., Gibson, J.M., Gibson, R., Hadjigeorgiou, G.M., Hattori, N., Ioannidis, J.P., Jasinska-Myga, B., Jeon, B.S., Kim, Y.J., Klein, C., Kruger, R., Kyratzi, E., Lesage, S., Lin, C.-H., Lynch, T., Maraganore, D.M., Mellick, G.D., Mutez, E., Nilsson, C., Opala, G., Park, S.S., Puschmann, A., Quattrone, A., Sharma, M., Silburn, P.A., Sohn, Y.H., Stefanis, L., Tadic, V., Theuns, J., Tomiyama, H., Uitti, R.J., Valente, E.M., van de Loo, S., Vassilatis, D.K., Vilariño-Güell, C., White, L.R., Wirdefeldt, K., Wszolek, Z.K., Wu, R.-M., Farrer, M.J., Genetic Epidemiology Of Parkinson's Disease (GEO-PD) Consortium, 2011. Association of LRRK2 exonic variants with susceptibility to Parkinson's disease: a case–control study. Lancet Neurol. 10, 898–908. https://doi.org/10.1016/S1474-4422(11)70175-2.

Rott, R., Szargel, R., Haskin, J., Bandopadhyay, R., Lees, A.J., Shani, V., Engelender, S., 2011. α-synuclein fate is determined by USP9X-regulated monoubiquitination. Proc. Natl. Acad. Sci. U. S. A. 108, 18666–18671. https://doi.org/10.1073/pnas.1105725108.

Roy, A., Ghosh, A., Jana, A., Liu, X., Brahmachari, S., Gendelman, H.E., Pahan, K., 2012. Sodium phenylbutyrate controls neuroinflammatory and antioxidant activities and protects dopaminergic neurons in mouse models of Parkinson's disease. PLoS One 7, e38113. https://doi.org/10.1371/journal.pone.0038113.

Saez-Atienzar, S., Bonet-Ponce, L., Blesa, J.R., Romero, F.J., Murphy, M.P., Jordan, J., Galindo, M.F., 2014. The LRRK2 inhibitor GSK2578215A induces protective autophagy in SH-SY5Y cells: involvement of Drp-1-mediated mitochondrial fission and mitochondrial-derived ROS signaling. Cell Death Dis. 5, e1368. https://doi.org/10.1038/cddis.2014.320.

Sakamoto, K.M., Kim, K.B., Kumagai, A., Mercurio, F., Crews, C.M., Deshaies, R.J., 2001. Protacs: chimeric molecules that target proteins to the Skp1-Cullin-F box complex for ubiquitination and degradation. Proc. Natl. Acad. Sci. U. S. A. 98, 8554–8559. https://doi.org/10.1073/pnas.141230798.

Sanchez-Guajardo, V., Annibali, A., Jensen, P.H., Romero-Ramos, M., 2013. α-synuclein vaccination prevents the accumulation of Parkinson disease-like pathologic inclusions in striatum in association with regulatory T cell recruitment in a rat model. J. Neuropathol. Exp. Neurol. 72, 624–645. https://doi.org/10.1097/NEN.0b013e31829768d2.

Sardi, S.P., Clarke, J., Kinnecom, C., Tamsett, T.J., Li, L., Stanek, L.M., Passini, M.A., Grabowski, G.A., Schlossmacher, M.G., Sidman, R.L., Cheng, S.H., Shihabuddin, L.S., 2011. CNS expression of glucocerebrosidase corrects α-synuclein pathology and memory in a mouse model of Gaucher-related synucleinopathy. Proc. Natl. Acad. Sci. U. S. A. 108, 12101–12106. https://doi.org/10.1073/pnas.1108197108.

Sardi, S.P., Clarke, J., Viel, C., Chan, M., Tamsett, T.J., Treleaven, C.M., Bu, J., Sweet, L., Passini, M.A., Dodge, J.C., Yu, W.H., Sidman, R.L., Cheng, S.H., Shihabuddin, L.S., 2013. Augmenting CNS glucocerebrosidase activity as a therapeutic strategy for parkinsonism and other Gaucher-related synucleinopathies. Proc. Natl. Acad. Sci. U. S. A. 110, 3537–3542. https://doi.org/10.1073/pnas.1220464110.

Sardi, S.P., Viel, C., Clarke, J., Treleaven, C.M., Richards, A.M., Park, H., Olszewski, M.A., Dodge, J.C., Marshall, J., Makino, E., Wang, B., Sidman, R.L., Cheng, S.H., Shihabuddin, L.S., 2017. Glucosylceramide synthase inhibition alleviates aberrations in synucleinopathy models. Proc. Natl. Acad. Sci. U. S. A. 114, 2699–2704. https://doi.org/10.1073/pnas.1616152114.

Saxton, R.A., Sabatini, D.M., 2017. mTOR signaling in growth, metabolism, and disease. Cell 168, 960–976. https://doi.org/10.1016/j.cell.2017.02.004.

Schapansky, J., Khasnavis, S., DeAndrade, M.P., Nardozzi, J.D., Falkson, S.R., Boyd, J.D., Sanderson, J.B., Bartels, T., Melrose, H.L., LaVoie, M.J., 2018. Familial knockin mutation of LRRK2 causes lysosomal dysfunction and accumulation of endogenous insoluble α-synuclein in neurons. Neurobiol. Dis. 111, 26–35. https://doi.org/10.1016/j.nbd.2017.12.005.

Schenk, D.B., Koller, M., Ness, D.K., Griffith, S.G., Grundman, M., Zago, W., Soto, J., Atiee, G., Ostrowitzki, S., Kinney, G.G., 2017. First-in-human assessment of PRX002, an anti-α-synuclein monoclonal antibody, in healthy volunteers. Mov. Disord. 32, 211–218. https://doi.org/10.1002/mds.26878.

Schlossmacher, M.G., Frosch, M.P., Gai, W.P., Medina, M., Sharma, N., Forno, L., Ochiishi, T., Shimura, H., Sharon, R., Hattori, N., Langston, J.W., Mizuno, Y., Hyman, B.T., Selkoe, D.J., Kosik, K.S., 2002. Parkin localizes to the Lewy bodies of Parkinson disease and dementia with Lewy bodies. Am. J. Pathol. 160, 1655–1667. https://doi.org/10.1016/S0002-9440(10)61113-3.

Schneeberger, A., Tierney, L., Mandler, M., 2016. Active immunization therapies for Parkinson's disease and multiple system atrophy. Mov. Disord. 31, 214–224. https://doi.org/10.1002/mds.26377.

Schofield, D.J., Irving, L., Calo, L., Bogstedt, A., Rees, G., Nuccitelli, A., Narwal, R., Petrone, M., Roberts, J., Brown, L., Cusdin, F., Dosanjh, B., Lloyd, C., Dobson, C., Gurrell, I., Fraser, G., McFarlane, M., Rockenstein, E., Spencer, B., Masliah, E., Spillantini, M.G., Tan, K., Billinton, A., Vaughan, T., Chessell, I., Perkinton, M.S., 2019. Preclinical development of a high affinity α-synuclein antibody, MEDI1341, that can enter the brain, sequester extracellular α-synuclein and attenuate α-synuclein spreading in vivo. Neurobiol. Dis. 132, 104582. https://doi.org/10.1016/j.nbd.2019.104582.

Searles Nielsen, S., Gross, A., Camacho-Soto, A., Willis, A.W., Racette, B.A., 2018. β2-adrenoreceptor medications and risk of Parkinson disease. Ann. Neurol. 84, 683–693. https://doi.org/10.1002/ana.25341.

Shaltiel-Karyo, R., Frenkel-Pinter, M., Rockenstein, E., Patrick, C., Levy-Sakin, M., Schiller, A., Egoz-Matia, N., Masliah, E., Segal, D., Gazit, E., 2013. A blood-brain barrier (BBB) disrupter is also a potent α-synuclein (α-syn) aggregation inhibitor. J. Biol. Chem. 288, 17579–17588. https://doi.org/10.1074/jbc.M112.434787.

Shaw, K.T.Y., Utsuki, T., Rogers, J., Yu, Q.-S., Sambamurti, K., Brossi, A., Ge, Y.-W., Lahiri, D.K., Greig, N.H., 2001. Phenserine regulates translation of -amyloid precursor protein mRNA by a putative interleukin-1 responsive element, a target for drug development. Proc. Natl. Acad. Sci. U. S. A. 98, 7605–7610. https://doi.org/10.1073/pnas.131152998.

Sidransky, E., Nalls, M.A., Aasly, J.O., Aharon-Peretz, J., Annesi, G., Barbosa, E.R., Bar-Shira, A., Berg, D., Bras, J., Brice, A., Chen, C.-M., Clark, L.N., Condroyer, C., De Marco, E.V., Dürr, A., Eblan, M.J., Fahn, S., Farrer, M.J., Fung, H.-C., Gan-Or, Z., Gasser, T., Gershoni-Baruch, R., Giladi, N., Griffith, A., Gurevich, T., Januario, C., Kropp, P., Lang, A.E., Lee-Chen, G.-J., Lesage, S., Marder, K., Mata, I.F., Mirelman, A., Mitsui, J., Mizuta, I., Nicoletti, G., Oliveira, C., Ottman, R., Orr-Urtreger, A., Pereira, L.V., Quattrone, A., Rogaeva, E., Rolfs, A., Rosenbaum, H., Rozenberg, R., Samii, A., Samaddar, T., Schulte, C., Sharma, M., Singleton, A., Spitz, M., Tan, E.-K., Tayebi, N., Toda, T., Troiano, A.R., Tsuji, S., Wittstock, M., Wolfsberg, T.G., Wu, Y.-R., Zabetian, C.P., Zhao, Y., Ziegler, S.G., 2009. Multicenter analysis of glucocerebrosidase mutations in Parkinson's disease. N. Engl. J. Med. 361, 1651–1661. https://doi.org/10.1056/NEJMoa0901281.

Silveira, C.R.A., Mac Kinley, J., Coleman, K., Li, Z., Finger, E., Bartha, R., Morrow, S.A., Wells, J., Borrie, M., Tirona, R.G., Rupar, C.A., Zou, G., Hegele, R.A., Mahuran, D., Mac Donald, P., Jenkins, M.E., Jog, M., Pasternak, S.H., 2019. Ambroxol as a novel disease-modifying treatment for Parkinson's disease dementia: protocol for a single-centre, randomized, double-blind, placebo-controlled trial. BMC Neurol. 19, 20. https://doi.org/10.1186/s12883-019-1252-3.

Snyder, H., Mensah, K., Theisler, C., Lee, J., Matouschek, A., Wolozin, B., 2003. Aggregated and monomeric α-synuclein bind to the S6′ proteasomal protein and inhibit proteasomal function. J. Biol. Chem. 278, 11753–11759. https://doi.org/10.1074/jbc.M208641200.

Spillantini, M.G., Schmidt, M.L., Lee, V.M.-Y., Trojanowski, J.Q., Jakes, R., Goedert, M., 1997. α-synuclein in Lewy bodies. Nature 388, 839–840. https://doi.org/10.1038/42166.

Spillantini, M.G., Crowther, R.A., Jakes, R., Hasegawa, M., Goedert, M., 1998. α-synuclein in filamentous inclusions of Lewy bodies from Parkinson's disease and dementia with Lewy bodies. Proc. Natl. Acad. Sci. U. S. A. 95, 6469–6473. https://doi.org/10.1073/pnas.95.11.6469.

Steiner, J.A., Quansah, E., Brundin, P., 2018. The concept of alpha-synuclein as a prion-like protein: ten years after. Cell Tissue Res. 373, 161–173. https://doi.org/10.1007/s00441-018-2814-1.

Sturm, E., Fellner, L., Krismer, F., Poewe, W., Wenning, G.K., Stefanova, N., 2016. Neuroprotection by epigenetic modulation in a transgenic model of multiple system atrophy. Neurotherapeutics 13, 871–879. https://doi.org/10.1007/s13311-016-0447-1.

Teich, A.F., Sharma, E., Barnwell, E., Zhang, H., Staniszewski, A., Utsuki, T., Padmaraju, V., Mazell, C., Tzekou, A., Sambamurti, K., Arancio, O., Maccecchini, M.L., 2018. Translational inhibition of APP by Posiphen: efficacy, pharmacodynamics, and pharmacokinetics in the APP/PS1 mouse. Alzheimers Dement. 4, 37–45. https://doi.org/10.1016/j.trci.2017.12.001.

Trist, B.G., Davies, K.M., Cottam, V., Genoud, S., Ortega, R., Roudeau, S., Carmona, A., De Silva, K., Wasinger, V., Lewis, S.J.G., Sachdev, P., Smith, B., Troakes, C., Vance, C., Shaw, C., Al-Sarraj, S., Ball, H.J., Halliday, G.M., Hare, D.J., Double, K.L., 2017. Amyotrophic lateral sclerosis-like superoxide dismutase 1 proteinopathy is associated with neuronal loss in Parkinson's disease brain. Acta Neuropathol. 134, 113–127. https://doi.org/10.1007/s00401-017-1726-6.

Tulisiak, C.T., Mercado, G., Peelaerts, W., Brundin, L., Brundin, P., 2019. Can infections trigger alpha-synucleinopathies? In: Teplow, D.B. (Ed.), Molecular Biology of Neurodegenerative Diseases: Visions for the Future. Elsevier Inc., pp. 1–24. https://doi.org/10.1016/bs.pmbts.2019.06.002.

Uc, E.Y., Lambert, C.P., Harik, S.I., Rodnitzky, R.L., Evans, W.J., 2003. Albuterol improves response to levodopa and increases skeletal muscle mass in patients with fluctuating Parkinson disease. Clin. Neuropharmacol. 26, 207–212. https://doi.org/10.1097/00002826-200307000-00011.

Ugen, K.E., Lin, X., Bai, G., Liang, Z., Cai, J., Li, K., Song, S., Cao, C., Sanchez-Ramos, J., 2015. Evaluation of an α synuclein sensitized dendritic cell based vaccine in a transgenic mouse model of Parkinson disease. Hum. Vaccin. Immunother. 11, 922–930. https://doi.org/10.1080/21645515.2015.1012033.

Venti, A., Giordano, T., Eder, P., Bush, A.I., Lahiri, D.K., Greig, N.H., Rogers, J.T., 2004. The integrated role of desferrioxamine and phenserine targeted to an iron-responsive element in the APP-mRNA 5′-untranslated region. Ann. N. Y. Acad. Sci. 1035, 34–48. https://doi.org/10.1196/annals.1332.003.

Vincent, B.M., Tardiff, D.F., Piotrowski, J.S., Aron, R., Lucas, M.C., Chung, C.Y., Bacherman, H., Chen, Y., Pires, M., Subramaniam, R., Doshi, D.B., Sadlish, H., Raja, W.K., Solís, E.J., Khurana, V., Le Bourdonnec, B., Scannevin, R.H., Rhodes, K.J., 2018. Inhibiting stearoyl-CoA desaturase ameliorates α-synuclein cytotoxicity. Cell Rep. 25, 2742–2754.e31. https://doi.org/10.1016/j.celrep.2018.11.028.

Volpicelli-Daley, L., Brundin, P., 2018. Prion-like propagation of pathology in Parkinson disease. In: Pocchiari, M., Manson, J. (Eds.), Handbook of Clinical Neurology, pp. 321–335. https://doi.org/10.1016/B978-0-444-63945-5.00017-9.

Volpicelli-Daley, L.A., Abdelmotilib, H., Liu, Z., Stoyka, L., Daher, J.P.L., Milnerwood, A.J., Unni, V.K., Hirst, W.D., Yue, Z., Zhao, H.T., Fraser, K., Kennedy, R.E., West, A.B., 2016. G2019S-LRRK2 expression augments-synuclein sequestration into inclusions in neurons. J. Neurosci. 36, 7415–7427. https://doi.org/10.1523/JNEUROSCI.3642-15.2016.

Wagner, J., Ryazanov, S., Leonov, A., Levin, J., Shi, S., Schmidt, F., Prix, C., Pan-Montojo, F., Bertsch, U., Mitteregger-Kretzschmar, G., Geissen, M., Eiden, M., Leidel, F., Hirschberger, T., Deeg, A.A., Krauth, J.J., Zinth, W., Tavan, P., Pilger, J., Zweckstetter, M., Frank, T., Bähr, M., Weishaupt, J.H., Uhr, M., Urlaub, H., Teichmann, U., Samwer, M., Bötzel, K., Groschup, M., Kretzschmar, H., Griesinger, C., Giese, A., 2013. Anle138b: a novel oligomer modulator for disease-modifying therapy of neurodegenerative diseases such as prion and Parkinson's disease. Acta Neuropathol. 125, 795–813. https://doi.org/10.1007/s00401-013-1114-9.

Wakabayashi, K., Tanji, K., Odagiri, S., Miki, Y., Mori, F., Takahashi, H., 2013. The Lewy body in Parkinson's disease and related neurodegenerative disorders. Mol. Neurobiol. 47, 495–508. https://doi.org/10.1007/s12035-012-8280-y.

Wang, C.Y., Finstad, C.L., Walfield, A.M., Sia, C., Sokoll, K.K., Chang, T.-Y., De Fang, X., Hung, C.H., Hutter-Paier, B., Windisch, M., 2007. Site-specific UBITh® amyloid-β vaccine for immunotherapy of Alzheimer's disease. Vaccine 25, 3041–3052. https://doi.org/10.1016/j.vaccine.2007.01.031.

Wang, C.Y., Wang, P.-N., Chiu, M.-J., Finstad, C.L., Lin, F., Lynn, S., Tai, Y.-H., De Fang, X., Zhao, K., Hung, C.-H., Tseng, Y., Peng, W.-J., Wang, J., Yu, C.-C., Kuo, B.-S., Frohna, P.A., 2017a. UB-311, a novel UBITh® amyloid β peptide vaccine for mild Alzheimer's disease. Alzheimers Dement. 3, 262–272. https://doi.org/10.1016/j.trci.2017.03.005.

Wang, Y., Yan, T., Lu, H., Yin, W., Lin, B., Fan, W., Zhang, X., Fernandez-Funez, P., 2017b. Lessons from anti-amyloid-β immunotherapies in Alzheimer disease: aiming at a moving target. Neurodegener. Dis. 17, 242–250. https://doi.org/10.1159/000478741.

Wegrzynowicz, M., Bar-On, D., Calo', L., Anichtchik, O., Iovino, M., Xia, J., Ryazanov, S., Leonov, A., Giese, A., Dalley, J.W., Griesinger, C., Ashery, U., Spillantini, M.G., 2019. Depopulation of dense α-synuclein aggregates is associated with rescue of dopamine neuron dysfunction and death in a new Parkinson's disease model. Acta Neuropathol. 138, 575–595. https://doi.org/10.1007/s00401-019-02023-x.

Weihofen, A., Liu, Y., Arndt, J.W., Huy, C., Quan, C., Smith, B.A., Baeriswyl, J.-L., Cavegn, N., Senn, L., Su, L., Marsh, G., Auluck, P.K., Montrasio, F., Nitsch, R.M., Hirst, W.D., Cedarbaum, J.M., Pepinsky, R.B., Grimm, J., Weinreb, P.H., 2019. Development of an aggregate-selective, human-derived α-synuclein antibody BIIB054 that ameliorates disease phenotypes in Parkinson's disease models. Neurobiol. Dis. 124, 276–288. https://doi.org/10.1016/j.nbd.2018.10.016.

Xu, H., Rajsombath, M.M., Weikop, P., Selkoe, D.J., 2018. Enriched environment enhances β-adrenergic signaling to prevent microglia inflammation by amyloid-β. EMBO Mol. Med. 10, e8931. https://doi.org/10.15252/emmm.201808931.

Ying, M., Xu, R., Wu, X., Zhu, H., Zhuang, Y., Han, M., Xu, T., 2006. Sodium butyrate ameliorates histone hypoacetylation and neurodegenerative phenotypes in a mouse model for DRPLA. J. Biol. Chem. 281, 12580–12586. https://doi.org/10.1074/jbc.M511677200.

Yokoyama, H., Kuroiwa, H., Tsukada, T., Uchida, H., Kato, H., Araki, T., 2009. Poly(ADP-ribose)polymerase inhibitor can attenuate the neuronal death after 1-methyl-4-phenyl-1,2,3,6-tetrahydropyridine-induced neurotoxicity in mice. J. Neurosci. Res. 88, 1522–1536. https://doi.org/10.1002/jnr.22310.

Yu, Q.-S., Reale, M., Kamal, M.A., Holloway, H.W., Luo, W., Sambamurti, K., Ray, B., Lahiri, D.K., Rogers, J.T., Greig, N.H., 2013. Synthesis of the Alzheimer drug Posiphen into its primary metabolic products (+)-N1-norPosiphen, (+)-N8-norPosiphen and (+)-N1, N8-bisnorPosiphen, their inhibition of amyloid precursor protein, α-Synuclein synthesis, interleukin-1β release, and cholinergic action. Antiinflamm. Antiallergy Agents Med. Chem. 12, 117–128.

Yuan, N.-N., Cai, C.-Z., Wu, M.-Y., Zhu, Q., Su, H., Li, M., Ren, J., Tan, J.-Q., Lu, J.-H., 2019. Canthin-6-one accelerates alpha-synuclein degradation by enhancing UPS activity: drug target identification by CRISPR-Cas 9 whole genome-wide screening technology. Front. Pharmacol. 10, 16. https://doi.org/10.3389/fphar.2019.00016.

Yun, S.P., Kam, T.-I., Panicker, N., Kim, S., Oh, Y., Park, J.-S., Kwon, S.-H., Park, Y.J., Karuppagounder, S.S., Park, H., Kim, S., Oh, N., Kim, N.A., Lee, S., Brahmachari, S., Mao, X., Lee, J.H., Kumar, M., An, D., Kang, S.-U., Lee, Y., Lee, K.C., Na, D.H., Kim, D., Lee, S.H., Roschke, V.V., Liddelow, S.A., Mari, Z., Barres, B.A., Dawson, V.L., Lee, S., Dawson, T.M., Ko, H.S., 2018. Block of A1 astrocyte conversion by microglia is neuroprotective in models of Parkinson's disease. Nat. Med. 24, 931–938. https://doi.org/10.1038/s41591-018-0051-5.

Zasloff, M., Adams, A.P., Beckerman, B., Campbell, A., Han, Z., Luijten, E., Meza, I., Julander, J., Mishra, A., Qu, W., Taylor, J.M., Weaver, S.C., Wong, G.C.L., 2011. Squalamine as a broad-spectrum systemic antiviral agent with therapeutic potential. Proc. Natl. Acad. Sci. U. S. A. 108, 15978–15983. https://doi.org/10.1073/pnas.1108558108.

Zhang, L., Zhang, L.Y., Li, L., Hölscher, C., 2019. Semaglutide is neuroprotective and reduces α-synuclein levels in the chronic MPTP mouse model of Parkinson's disease. J. Park. Dis. 9, 157–171. https://doi.org/10.3233/JPD-181503.

Zhao, H.T., John, N., Delic, V., Ikeda-Lee, K., Kim, A., Weihofen, A., Swayze, E.E., Kordasiewicz, H.B., West, A.B., Volpicelli-Daley, L.A., 2017. LRRK2 antisense oligonucleotides ameliorate α-synuclein inclusion formation in a Parkinson's disease mouse model. Mol. Ther. Nucleic Acids 8, 508–519. https://doi.org/10.1016/j.omtn.2017.08.002.

Zhou, C., Emadi, S., Sierks, M.R., Messer, A., 2004. A human single-chain Fv intrabody blocks aberrant cellular effects of overexpressed α-synuclein. Mol. Ther. 10, 1023–1031. https://doi.org/10.1016/j.ymthe.2004.08.019.

Zhou, W., Bercury, K., Cummiskey, J., Luong, N., Lebin, J., Freed, C.R., 2011. Phenylbutyrate Up-regulates the DJ-1 protein and protects neurons in cell culture and in animal models of Parkinson disease. J. Biol. Chem. 286, 14941–14951. https://doi.org/10.1074/jbc. M110.211029.

Zhou, H., Shao, M., Guo, B., Li, C., Lu, Y., Yang, X., Shengnan, L., Li, H., Zhu, Q., Zhong, H., Wang, Y., Zhang, Z., Lu, J., Lee, S.M., 2019. Tetramethylpyrazine analogue T-006 promotes the clearance of alpha-synuclein by enhancing proteasome activity in Parkinson's disease models. Neurotherapeutics https://doi.org/10.1007/s13311-019-00759-8.

Zhu, X., Siedlak, S.L., Smith, M.A., Perry, G., Chen, S.G., 2006. LRRK2 protein is a component of Lewy bodies. Ann. Neurol. 60, 617–618. https://doi.org/10.1002/ana.20928.

Zunke, F., Moise, A.C., Belur, N.R., Gelyana, E., Stojkovska, I., Dzaferbegovic, H., Toker, N.J., Jeon, S., Fredriksen, K., Mazzulli, J.R., 2018. Reversible conformational conversion of α-synuclein into toxic assemblies by glucosylceramide. Neuron 97, 92–107.e10. https://doi.org/10.1016/j.neuron.2017.12.012.

Repurposing anti-diabetic drugs for the treatment of Parkinson's disease: Rationale and clinical experience

13

Tom Foltynie[*], **Dilan Athauda**

Department of Clinical and Movement Neurosciences, UCL Institute of Neurology, London, United Kingdom

[*]*Corresponding author: Tel.: +44-0203-4488-726, e-mail address: t.foltynie@ucl.ac.uk*

Abstract

The most pressing need in Parkinson's disease (PD) clinical practice is to identify agents that might slow down, stop or reverse the neurodegenerative process of Parkinson's disease and therefore avoid the onset of the most disabling, dopa-refractory symptoms of the disease. These include dementia, speech and swallowing problems, poor balance and falling.

To date, there have been no agents which have yet had robust trial data to confirm positive effects at slowing down the neurodegenerative disease process of PD. In this chapter we will review the reasons why there is growing interest in drugs currently licensed for the treatment of diabetes as agents which may slow down disease progression in PD, including a review of the published trials regarding exenatide, a GLP-1 receptor agonist licensed to treat type 2 diabetes, and recently shown to be associated with reduced severity of PD in a randomized, placebo controlled washout design trial of 60 patients treated for 48 weeks.

This subject is now a major area of interest for multiple pharmaceutical companies hoping to bring GLP-1 receptor agonists forward as treatment options in PD.

Keywords

Parkinson's disease, Diabetes, Glucagon-like peptide 1

While there are a number of symptomatic treatment options for PD including dopaminergic replacement therapies and deep brain stimulation surgery, none of the conventional treatment options have any impact on the relentless progression of the neurodegenerative process. This means that over time, patients will develop dopamine refractory problems including speech and swallowing, gait and balance,

Progress in Brain Research, Volume 252, ISSN 0079-6123, https://doi.org/10.1016/bs.pbr.2019.10.008

and cognitive/psychiatric issues that can greatly impact on quality of life. There is therefore an urgent need for the identification of therapies that may slow down, stop or reverse this condition. In this review, we will discuss the reasons why there is interest in trying to re-purpose drugs, licensed for the treatment of type 2 diabetes mellitus (T2DM), as potential disease modifying treatment options for patients with Parkinson's disease (PD).

As with many areas of science, basic laboratory findings documenting possible relationships between pathophysiological processes related to these two diseases, have been complemented by careful epidemiological observations, which have been further informed by biological specimens collected from patients. Of greatest interest are the initial results already emerging from clinical trials in which patients with PD have been exposed to agents licensed for the treatment of T2DM, during which the severity of their Parkinson's disease has been carefully monitored.

1 Links between diabetes and Parkinson's disease

As a starting point regarding whether there might be a role for repurposing anti-diabetic agents in PD patients, it is reasonable to query whether there are any obvious links between these two diseases. Epidemiological methods allow us to critically evaluate whether there are any clear relationships between the risk of developing T2DM and the risk of developing PD. In studies using case-control methodology, "cases," i.e., patients with Parkinson's disease are questioned regarding whether they have a history of diabetes mellitus, and their response rates compared with "controls," i.e., from people without a diagnosis of PD. From these studies, there is inconsistency regarding the frequency with which patients with PD report a concurrent diagnosis of T2DM than people without PD (Cereda et al., 2011). Many of these studies are based on very small numbers of participants, which likely explains the inconsistency. Furthermore, the general criticism of case-control methods relates to the uncertainty regarding misclassification of diabetes (patients self-report rather than undergoing formal testing), and PD (patients may subsequently develop PD at some later timepoint), as well as potential bias in the selection of control populations.

In contrast to the findings of the case-control studies, cohort studies on the other hand tend to find a positive association between the diagnosis of T2DM and the diagnosis of PD (Yue et al., 2016). Cohort studies tend to follow larger numbers of individuals over time and are therefore less prone to the biases associated with case-control studies. Whereas self-reporting inaccuracies may still occur, the longitudinal nature of cohort studies and the consequent reduction in the selection biases may explain the discrepancy between these and some of the case-control data. The largest study, based on UK hospital records data evaluated >2 million T2DM patients and >6 million non-diabetic individuals and found a hazard ratio of 1.32 of developing PD among the diabetic group which rose to 3.81 in individuals who developed T2DM at a younger age (De Pablo-Fernandez et al., 2018). The use of such a very large database allows a very high level of confidence in the precision of these results. As such it can be concluded that, in the UK population at least, there is a definite increased risk of PD among patients with T2DM and that this risk increases either with the number of years of the disease, or is highest among those individuals in whom diabetes presents at a younger age.

The potential mechanism underlying this association will be further discussed later on in this chapter, but an immediate question that emerges relates to the potential misdiagnosis of vascular parkinsonism instead of the neurodegenerative form of PD. Since T2DM is a well-known risk factor for both large vessel and small vessel atherosclerosis, it might be argued that patients with a long history of T2DM have a much higher risk of developing vascular parkinsonism (cerebrovascular disease mimicking the clinical phenotype of neurodegenerative PD). Furthermore patients with poorly controlled T2DM are vulnerable to other neurological complication such as peripheral neuropathy, which may lead to balance issues (although this is less likely to lead to a misdiagnosis of PD).

Related to this, and as a next step to help explore further this epidemiological association between T2DM and PD, other teams have explored whether the concurrent presence of T2DM, influences the rate of progression of PD symptoms and signs. In one series, PD patients with comorbid T2DM had worse levels of postural instability even after adjusting for disease duration, while (in relation to the question regarding misdiagnosis of vascular parkinsonism) there was no difference in the extent of leucoaraiosis (nor peripheral sensory loss) between the two groups (Kotagal et al., 2013). Co-morbid diabetes in PD has also been shown to be associated with more severe motor symptoms, higher equivalent doses of levodopa, and lower striatal dopamine transporter binding (Cereda et al., 2012; Pagano et al., 2018).

In addition the degree of cognitive impairment has been found to be greater among PD patients with T2DM than those without T2DM although this was not related to the degree of dopaminergic or cholinergic deficits identified using PET imaging indicating that other processes are likely to be contributory (Bohnen et al., 2014; Cereda et al., 2012). Further compelling data however comes from the study of non-diabetic PD patients with and without dementia, using oral glucose tolerance tests to objectively measure the extent of insulin resistance in these patients (Bosco et al., 2012). Peripheral insulin resistance is measured using the HOMA (homeostasis model assessment) formula which requires simultaneous measurement of peripheral insulin and glucose levels. This study found that 62% of patients with PD dementia met criteria for insulin resistance compared to 35% of patients with PD in the absence of dementia, despite not having received any diagnosis of insulin resistance/impaired glucose tolerance or T2DM on recruitment. Five percent of the patients in each group met criteria for T2DM.

The most useful data exploring the relationship between PD, T2DM and insulin resistance can be gained from longitudinal cohorts of PD patients in whom measures of insulin resistance or diabetes have been measured at baseline. This has been done in the DeNoPa cohort, in which de novo PD patients with elevated HBA1c, even in the pre-diabetic range, had faster rates of decline in cognitive performance than non-diabetic/normoglycemic patients with PD (Simuni et al., 2018).

2 Causation or shared patho-etiology?

The cause of this intriguing association between T2DM and PD has been further considered. Possible shared genetic risks for the two diseases have been explored, finding a major degree of overlap between genes known to increase the risk of T2DM and genes known to increase the risk of PD (Santiago and Potashkin, 2013, 2014). This might suggest that for some patients with T2DM there is an inevitability of developing PD based on inherited genetic risks that lead to both pathologies. If this were the case this might open novel pathways to intervention. Intriguingly, one of the shared genes is Akt1 which encodes for the protein Akt, critically involved in cellular survival pathways, and which will be discussed in further detail later in this chapter.

Alternatively, it has been proposed that it is simply the higher circulating levels of glucose that occur in patients with T2DM, or even with the modestly raised glucose levels seen in pre-diabetes, that may in itself be a risk factor for neurodegeneration, mediated through elevated levels of advanced glycation end products (AGE's) and the impact of these on alpha synuclein aggregation and pro-inflammatory pathways (Vicente Miranda et al., 2016; Videira and Castro-Caldas, 2018). A critical player may be the level of methylglyoxal, a major glycation agent produced as a by-product of glucose metabolism, and which has been shown to increase the oligomerization and aggregation of alpha synuclein in Drosophila and mice (Vicente Miranda et al., 2017). Conversely aggregations of alpha synuclein have been shown to be reduced using methylglyoxal inhibitors.

A further hypothesis that is being explored relates to the potential overlap between peripheral insulin resistance in T2DM and the concept of central insulin resistance in PD. This is a relatively novel concept in PD, but it has been well established in patients with Alzheimer's disease that central insulin resistance occurs and is associated with a loss of neuronal insulin signaling (Talbot et al., 2012).

Insulin receptors are expressed in all cell types throughput the brain, with particular high densities in the olfactory bulb, hippocampus, striatum and cerebellum (Werther et al., 1987). The source of insulin within the brain remains under debate, with some small studies suggesting de novo synthesis from hippocampal, pyramidal, hippocampal and olfactory bulb neurons (Devaskar et al., 1994; Kuwabara et al., 2011), however the majority of insulin in the CNS thought to derive from circulating pancreatic insulin (Banks, 2004; Havrankova et al., 1978; Sankar et al., 2002).

While central insulin does play a role in controlling behaviors related to feeding/satiety through glucose sensing neurons in the hippocampus, it is increasingly evident that insulin signaling also plays an important role in neuronal survival pathways. Understanding the phenomenon of insulin sensitivity vs insulin resistance requires an understanding of a tightly controlled system that follows binding of insulin to the insulin receptor.

Insulin receptor stimulation leads to phosphorylation of the insulin receptor substrate 1 (IRS-1) on tyrosine residues which then phosphorylate downstream effectors that activate secondary messenger pathways (Hotamisligil et al., 1996). Intact insulin signaling relies on the stability of IRS-1 proteins, which have numerous phosphorylation sites at serine and tyrosine residues and act as a critical node for transmitting the insulin signal to successive downstream intracellular mediators (Gual et al., 2005). While phosphorylation of IRS-1 at tyrosine residues is needed for maintaining insulin signaling and sensitivity, phosphorylation of IRS-1 at serine residues leads to dissociation of IRS-1 from the insulin receptor and promotes its degradation by the proteasome (Herschkovitz et al., 2007). Thus the net effect of IRS-1 serine phosphorylation is to increase insulin resistance whereas IRS-1 tyrosine phosphorylation tends to enhance insulin sensitivity.

The major downstream effectors in the insulin signaling are the Akt pathway, which modulates activity of intracellular proteins including glycogen synthase kinase 3 (GSK3) and mechanistic target of rapamycin (mTOR), among others; and the MAP kinase pathway, involved in controlling transcription factors such as cAMP-responsive element-binding protein (CREB). These in turn regulate a variety of processes including apoptosis, autophagy, inflammation, nerve cell metabolism, protein synthesis and synaptic plasticity (Fig. 1) (Akintola and van Heemst, 2015; Bassil et al., 2014).

The presence of insulin resistance can be detected indirectly in human post mortem brain tissue by quantifying the degree and type of phosphorylation of IRS-1 (Talbot et al., 2012). This has been confirmed to occur in PD patients (Bassil et al., 2017), together with evidence of reduced expression of downstream components of the insulin signaling pathway (Sekar and Taghibiglou, 2018). More circumstantially, insulin receptors are also reduced in the substantia nigra of PD patients (Moroo et al., 1994; Takahashi et al., 1996). In contrast to the effects of insulin resistance peripherally, insulin resistance centrally leads to an imbalance in intra-neuronal processes related to neuronal survival, known to be relevant to PD, i.e., autophagy, mitochondrial function and neuroinflammation (Athauda and Foltynie, 2016a, b).

A further intriguing theory relates to abnormal protein mis-folding in both T2DM and PD. It is generally accepted that abnormal spreading of pathological alpha synuclein, in a prion-like manner, causes disease propagation in PD, and similarly, aberrant folding of islet amyloid polypeptide (IAPP) protein in pancreatic beta islets is thought to contribute to the development of pancreatic β-cell dysfunction, cell death, and development of T2DM (Mukherjee et al., 2015). In addition, there appears to be the potential for interaction between these two proteins. In a small study, phosphorylated alpha synuclein was found in pancreatic tissue in 90% of patients with a synucleinopathy and almost 70% of patients with T2DM, with evidence of co-localization between these two proteins (Martinez-Valbuena et al., 2018). A single study also demonstrated that IAPP can accelerate alpha synuclein aggregation in vitro (though the converse is not true), providing a simple theoretical justification for why T2DM is a risk factor for PD, whereas patients with PD do have an increased risk of developing T2DM (Horvath and Wittung-Stafshede, 2016). Further questions remain regarding the nature of this interaction and to date, there are no studies demonstrating co-localization of these proteins in PD brains, but aberrant heterologous cross seeding of these proteins remains a novel theory on how these diseases may be related.

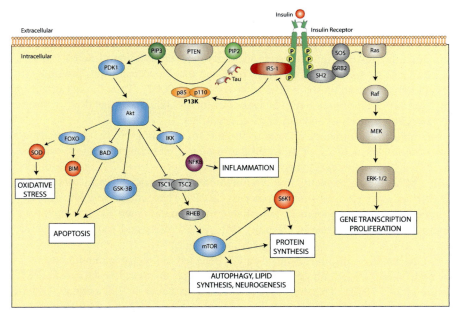

FIG. 1

Insulin binds to extracellular subunits of the insulin receptor, triggering autophosphorylation at intracellular tyrosine residues, leading to activation of IRS-1. Activated IRS-1 then phosphorylates PI3K. PI3K then in turn phosphorylates PIP2 to form PIP3. Formation of PIP3 (over PIP2) is dependent on the phosphatase activity of PTEN, of which a fraction is dependent on physiological tau protein binding. PIP3 is then able to activate PDK1, which then phosphorylates Akt, a major node of the insulin signaling pathway. Akt phosphorylates (and inhibits activity of) GSK-3B involved in regulating autophagy, cellular proliferation, apoptosis. In addition AKT-mediated activation of mTOR modulates protein synthesis and other aspects of cell metabolism including growth, survival and autophagy. mTOR also provides important negative feedback of IRS-1, through promotion of serine phosphorylation. AKT also activates proteins such as BAD, FOXO and IKK, which regulate apoptosis, cytokine production and cell survival. In parallel, insulin receptor activation also activates the Ras-MAPK pathway involved in synaptic plasticity, cell proliferation, and differentiation. Forkhead box protein (FOX); inhibitor of nuclear factor-κB kinase (IKK); glycogen synthase kinase-3B (GSK-3B); GRB2, growth factor receptor-bound protein 2; MEK, MAPK/ERK kinase, 90kDa ribosomal protein S6 kinase 1 (S6K1); PDK1, 3-phophoinositide-dependent protein kinase 1; PIP2, phosphatidylinositol 4,5-bisphosphate; PIP3, phosphatidylinositol (3,4,5)-trisphosphate; SHC, SHC-transforming protein.

Given the growing links between PD and T2DM it is perhaps not surprising that drugs used in the treatment of T2DM are among the most promising treatments currently being prioritized for repositioning as possible novel treatments for PD (Brundin et al., 2013) in an attempt to restore this signaling pathway.

3 Anti-diabetic agents

An individual presenting with elevated blood glucose leading to a diagnosis of type 2 diabetes should be given advice regarding diet, exercise and lifestyle. Where drug treatment becomes necessary, the initial recommended choice is metformin. If this is ineffective at controlling blood glucose levels, then the recommended options are to add a second line agent which may be a DPP-4 inhibitor, pioglitazone, a sulfon-lyurea or a sodium-glucose transport protein 2 (SGLT2) inhibitor. If necessary, triple combinations of these therapies may be instituted, and if still not helpful then insulin or a GLP-1 receptor agonist are then recommended (https://www.nice.org.uk/guid ance/ng28/chapter/1-Recommendations#drug-treatment-2).

Of the anti-diabetic agents, there are a number that have been proposed as potentially possessing direct/indirect actions that may be relevant to people with PD. The current exceptions are the sulfonylureas which to date have not been proposed to be of any relevance to PD, and the SGLT2 inhibitors of which there is as yet limited but encouraging data which suggests they may play a role in improving brain insulin signaling in rodents (Sa-Nguanmoo et al., 2017).

4 Metformin and PD

Metformin is an orally active biguanide currently used as a first-line treatment for T2DM and is also classed as an insulin sensitizer. The potential role of metformin as a neuroprotective agent in PD has been supported by demonstrating protective effects in the toxin based MPTP rodent models leading to improvements in motor functioning (Katila et al., 2017; Lu et al., 2016; Patil et al., 2014), though some studies reported metformin exerted no protective effects (Tayara et al., 2018), or even accelerated dopaminergic neuronal loss (Ismaiel et al., 2016). However, any of these beneficial effects have had to been considered in the context that metformin may simply interfere with the toxic action of MPTP on mitochondrial function in neurons, which may have no relevance to the mechanisms related to human PD. Beyond this however, metformin has also been shown, via inhibition of downstream mTOR and enhanced PP2A activity, to reduce alpha synuclein phosphorylation in the brain of healthy mice (Pérez-Revuelta et al., 2014), which is far more relevant as a potential disease modifying mechanism in the pathogenesis of PD.

A detailed review of the merits of metformin in neurodegenerative diseases exploring its potential mechanisms of action has been published (Rotermund et al., 2018) but are thought to broadly involve regulation of cell death and survival mechanisms. The divergent actions of metformin are thought to involve activation of AMP kinase (AMPK); reduction of oxidative phosphorylation via inhibition of complex I in mitochondria; modulation of insulin signaling via enhancement of peripheral GLP-1 expression; and reduction of inflammatory pathways via NF-κB inhibition, though it is not clear if all, or any are needed for its protective effects.

Clinical studies to date have only evaluated metformin compared to, or in combination with other anti-glycemic agents. Epidemiological studies have suggested that metformin might rescue the elevated risk of PD among Taiwanese patients with T2DM compared with T2DM patients using sulfonylureas, (Wahlqvist et al., 2012), and more than 4 years treatment with metformin was associated with a lower incidence of PD in a longitudinal study involving elderly patients with T2DM in a US population (Shi et al., 2019); however a separate study suggested patients with T2DM taking glitazones had a significantly lower incidence of PD compared to patients on metformin alone (Brakedal et al., 2017).

There have not as yet been any prospective trials evaluating whether metformin may have any beneficial effects in PD, although there is considerable interest in exploring this further, perhaps depending on results of trials of other anti-diabetic medications already being set-up/in progress.

5 Thiozolidinediones and PD

Pioglitazone belongs to the class of drugs known as thiozolidinediones. These drugs work as agonists for the peroxisome proliferator activated receptor gamma (PPAR-gamma). These receptors are not only expressed in peripheral insulin sensitive tissues but also expressed in nigral and putaminal nuclei (Swanson and Emborg, 2014). Pioglitazone and rosiglitazone (since suspended from clinical use due to an increased risk of heart attack and stroke) have demonstrated neuroprotective effects across a range of animal toxin models of PD, including the MPTP (Dehmer et al., 2004; Laloux et al., 2012; Quinn et al., 2008; Schintu et al., 2009; Swanson et al., 2011), LPS (Hunter et al., 2008), 6-OHDA (Lee et al., 2012; Machado et al., 2019) and rotenone models (Corona et al., 2014), resulting in improvements in behavioral and motor responses. These effects are thought to be due to inhibition of pro-inflammatory pathways, and modulation of mitochondrial function and oxidative stress responses (Corona and Duchen, 2015).

Human studies regarding glitazones and PD have been conflicting. Two large retrospective cohort studies evaluating the use of glitazones in a cohort of Norwegian patients and a subset of a UK population with T2DM found a 28% reduced risk of PD compared with the frequency of PD seen among patients prescribed other antiglycemics (including metformin) (Brakedal et al., 2017; Brauer et al., 2015). However there was no significant association between the incidence of glitazone use and PD incidence in a US Medicare population (Connolly et al., 2015) and a large Taiwanese population of diabetics (Wu et al., 2018).

However based on the combination of the epidemiological data and encouraging preclinical work, pioglitazone has been formally tested in a clinical trial in PD patients as a potential disease modifying agent.

A trial of 210 patients compared outcomes among patients randomized to 1 of 2 doses of Pioglitazone over 44 weeks against Placebo using the total UPDRS score as the outcome measure (NINDS Exploratory Trials in Parkinson Disease (NET-PD) FS-ZONE Investigators, 2015). There was a modest deterioration seen in all three subgroups with only a 1.1 point advantage in the 45 mg dose, and 1.8 point advantage in the 15 mg dose compared to placebo which was not significantly different and did not meet the a priori threshold of three points difference, leading to the authors concluding that this agent was futile. The authors comment that the duration of follow up may have been insufficient to allow any advantage to emerge in the active treatment groups and this may account for the discrepancy between the trial findings and the epidemiological and preclinical data. The lack of short term effect however highlights the issue that short term efficacy in the toxin based animal models of PD is insufficient evidence that an intervention will have equivalent efficacy in human PD. At present, there are no further trials planned of pioglitazone in PD.

Despite the failure of recent clinical trials and the potential risks associated with this class of drugs, promising in vivo data highlights the need for an improved understanding of the factors influencing pioglitazone's potential for treating PD. Following the discovery that the insulin sensitizing effects of pioglitazone may occur independently of PPARγ (Grahame Hardie, 2014; LeBrasseur et al., 2006), and its mitochondrial enhancing effects may involve binding directly to complexes on the inner mitochondrial membrane (identified as mTOT—mitochondrial target of thiozolidinediones) (Colca et al., 2013a, b, 2004), novel compounds that activate similar pathways but are "PPAR-sparing" are in early development in testing against animal models of PD, which may offer similar benefits of neuroprotection with limited adverse effects (Colca et al., 2013a, b).

6 Insulin as a therapy for PD

The relationship between insulin resistance and neurodegeneration has led to teams considering whether administration of exogenous insulin itself may ameliorate the pathological processes of PD. The obvious limitation to this idea is that peripheral administration of insulin is inevitably limited by the potential to induce hypoglycemia. Instead, interest has focused predominantly on the intranasal route of administration, which allows insulin to access the central nervous system without the peripherally mediated hypoglycemic effects. In patients with Alzheimer's disease, acute exposure to intranasal insulin improved cognitive performance among ApoE4 negative individuals (Reger et al., 2008a, b). This has been followed by a small trial of intranasal insulin in 16 patients with PD or a related neurodegenerative condition called multiple system atrophy (MSA) (Novak et al., 2019). Patients self-administered 40 IU of intranasal insulin once daily or placebo for 4 weeks, finding that the insulin treated patients had an improvement in both verbal fluency and motor severity at the end of the exposure period compared with baseline scores, and importantly there was no report of changes in serum glucose or hypoglycemia.

7 GLP-1 receptor agonists, DPP-4 inhibitors and PD

Glucagon-like peptide 1 (GLP-1) is a hormone released by the cells in the small intestine and colon in response to ingestion of a meal (Baggio and Drucker, 2007). GLP-1 levels rise within minutes of eating, suggesting that there is a neural control over its release, not simply a mechanical trigger caused by food transit. GLP-1 is highly conserved across species suggesting it has fundamental importance in metabolic pathways in mammals. It is an "incretin" hormones, named for its role in mediating the incretin effect, i.e., the observation of a greater level of insulin release in response to an enteral glucose load compared to an equivalent intravenous glucose load.

GLP-1 circulates in the bloodstream and binds to GLP-1 receptors in multiple tissue types. GLP-1 receptors were originally identified on the beta islet cells of the pancreas and it was discovered that on binding to these receptors, there followed an increase in the release of insulin and suppression of the release of glucagon. This hormonal system therefore clearly has a role in blood glucose control. Importantly insulin release resulting from GLP-1 receptor stimulation occurs in a glucose level dependent manner and therefore high circulating levels of blood glucose are lowered but normal circulating levels of blood glucose are not lowered to hypoglycemic levels.

Circulating GLP-1 is rapidly broken down in the blood stream by an enzyme called dipeptidyl peptidase 4 (DPP-4). There is therefore tight control of the duration of action of GLP-1 following its release in response to a meal. In addition to the effect of GLP-1 receptor stimulation on blood insulin levels, it has been identified that GLP-1 receptor stimulation has trophic effects on beta islet cells, enhancing islet beta cell proliferation differentiation, inhibiting apoptosis, and enhancing cell survival (Drucker et al., 2010; Lovshin and Drucker, 2009).

In patients with T2DM, the incretin effect is substantially reduced or lost, and lower levels of GLP-1 are detectable following a meal. The routine use of DPP-4 inhibitors as a means of potentiation of GLP-1 stimulation is now established as a treatment for T2DM.

7.1 DPP4 inhibitors and PD

Dipeptidyl peptidase-4 (DPP-4) inhibitors were originally developed to minimize the rapid cleavage of GLP-1 and thus enhancing its anti-glycemic effects (Andersen et al., 2018). In studies, DPP-4 inhibitors have been shown to enhance insulin secretion, suppress glucagon secretion as well as induce beta islet cell proliferation mimicking the actions of GLP-1 receptor agonists (Mousa and Ayoub, 2019). Although currently licensed DPP-IV inhibitors do not cross the blood-brain-barrier (Srinivas, 2015), their neuroprotective effects are thought to mediated by an increase in circulating GLP-1. Inhibitors of the DPP-4 enzyme have been commercially produced, are active following oral administration and result in stabilization and increase of circulating GLP-1 by about 2–4 fold (He et al., 2007). DPP-4 inhibitors also stabilize the levels of other substrates including oxyntomodulin and glucose-dependent insulinotropic polypeptide (GIP—see later) which may contribute to their effects (Thornberry and Gallwitz, 2009). DPP-4 inhibitors have been found to have improved efficacy over time, due to the trophic effects of GLP-1 on the beta islet cells of the pancreas.

Vildagliptin was found to restore the impairment of neuronal insulin receptor function in rats fed a high fat diet, which was associated with an increase in plasma and brain GLP-1 levels (Pintana et al., 2015; Pipatpiboon et al., 2013). More recently, a novel intranasal formulation of omarigliptin has been found to be the first gliptin to cross the blood brain barrier successfully which was accompanied by a 2.6-fold increase in brain GLP-1 concentration (Mousa and Ayoub, 2019).

The potential use of DPP-4 inhibition in neurodegeneration is also supported by a number of studies in rodent models of PD. Sitagliptin, saxagliptin and vildagliptin have been tested in the rotenone animal toxin model of PD, leading to improvements in motor and cognitive performance, alongside preservation of dopaminergic cells, with evidence that this was mediated through anti-inflammatory and anti-apoptotic mechanisms (Abdelsalam and Safar, 2015; Badawi et al., 2017; Li et al., 2018; Nassar et al., 2015). However, when utilizing other animal toxin models, other groups have found saxagliptin was not able to restore dopaminergic function in the 6-OHDA toxin rodent model (Turnes et al., 2018), and similarly, rats acutely or chronically pretreated with supramaximal doses of sitagliptin were not protected against MPTP-induced striatal dopaminergic degeneration (Ribeiro et al., 2012).

Human studies involving DPP-IV inhibitors are limited. A single case-control study of 1000 diabetic patients with PD from a Swedish national database indicated exposure to DDP-4 inhibitors compared to other oral anti-glycemics was associated with a reduced risk of PD (OR 0.23, CI 0.07–0.73) (Svenningsson et al., 2016). In addition the addition of sitaglipitin to an anti-glycemic regimen has been studied for its effects on cognitive function in 253 elderly patients with T2DM (Isik et al., 2017). Results indicated that in comparison to metformin, 6 months treatment with sitagliptin was associated with an increase in Mini-Mental State Examination (MMSE) scores in patients without cognitive impairment and with known AD. None of the DPP-4 inhibitors have yet been tested in a trial for potential effects in PD. The most potent DPP4 inhibitor with the longest half-life is Alogliptin (Andukuri et al., 2009). This drug is the subject of a clinical trial in PD patients to gage its potential as a disease modifying agent, currently being set up in Australia.

7.2 Exenatide

An alternative means of improving GLP-1 receptor stimulation is via the use of direct GLP-1 receptor agonist drugs. The first GLP-1 receptor agonist identified was Exendin-4 (Eng et al., 1992). This was identified in the saliva of the Glia monster, a lizard resident in the Arizona desert. This animal is of interest as it eats very infrequently (only on a few occasions per year), and therefore needs to have a very tight control on its own metabolism. Exendin-4 is highly resistant to degradation by DPP-4 and thus enables prolonged GLP-1 receptor stimulation. It has to be given by subcutaneous injection to avoid being metabolized in the stomach. A synthetic version of Exendin-4 has been manufactured (Exenatide), a single injection of which has been shown to stimulate GLP-1 receptors for 6–8 h (Nielsen et al., 2004). Exenatide has been the subject of a number of phase 3 trials in T2DM, demonstrating good long term safety and tolerability, and beneficial effects on glycemic control, accompanied by 2–5 kg weight loss (Best et al., 2009; Drucker et al., 2008). The most common adverse effect is nausea, which tends to lessen with prolonged use. Exenatide was approved for use in T2DM patients in the USA in 2005 and in Europe in 2006.

7.3 The rationale for using GLP-1 receptor agonists in PD

As well as their presence on beta islet cells, GLP-1 receptors have also been identified in adipose tissue, hepatocytes, cardiac myocytes, lung, kidney and central nervous system tissue (Seufert and Gallwitz, 2014). As such there has been interest in the manipulation of the GLP-1 receptor system in illnesses affecting all of these organs.

Interest in the function of GLP-1 receptors in nervous tissue started in 2002 by a team lead by Nigel Greig at NIH. This team demonstrated that both GLP-1 and Exendin-4 were able to induce neurite outgrowth to a similar extent to nerve growth factor (NGF), promoting neuronal differentiation and rescuing degenerating neurons (Perry et al., 2002b). Following this, the same team showed that exendin-4 could protect against excitotoxic damage caused by either glutamate or ibotenic acid (Perry et al., 2002a). Following these observations, there has been a plethora of preclinical studies exploring the potential role of GLP-1 receptor agonists in either Alzheimer's disease or PD models.

7.4 GLP-1 in vitro

Exenatide has been shown to increase transcription of tyrosine hydroxylase (TH) (the rate limiting enzyme in dopamine synthesis) in brainstem catecholaminergic neurons (Yamamoto et al., 2002). These effects are blocked by GLP-1 receptor antagonists confirming that these actions are mediated through the GLP-1 receptor.

Shiraishi et al. evaluated the effects of exenatide on macrophages (Shiraishi et al., 2012). GLP-1 receptors are expressed on macrophages and in the presence of exenatide, human monocyte derived macrophages develop an M2 phenotype, through activation of STAT3 leading to up-regulation of anti-inflammatory molecules such as interleukin-10 and TGF-beta. Whether the effects of exenatide are mediated through an increase in anti-inflammatory molecules or a decrease in pro-inflammatory molecules is likely to be indistinguishable.

The impact of exenatide on mitochondrial number and function has been evaluated by Fan et al. researching its mechanisms of action in type 2 diabetes (Fan et al., 2010). In vitro work performed using human amyloid polypeptide as a toxin for insulinoma cells showed that exenatide increased cell survival through a reduction in apoptosis. This was then shown to be mediated through activation of the Akt pathway known to be a critical step in normal mitochondrial function. Furthermore, it was shown that Exenatide induced mitochondrial gene expression and led to recovery of mitochondrial enzyme activity and mitochondrial number.

A modified version of exenatide has been created called NLY01. This modification has involved the addition of a polyethylene glycol (Pegylation) which acts to increase its ability to penetrate the central nervous system as well as prolong its circulating half-life. Its biologic activity remains comparable to exenatide. In an elegant series of experiments, preformed fibrils of alpha synuclein added to cultures of microglia led to their activation and production of inflammatory mediators. In the absence of NLY01, the microglia become active and secrete the inflammatory mediators—TNF-alpha, interleukin 1-alpha and C1q. The media containing these inflammatory mediators when added to astrocyte cultures causes activation to toxic A1 astrocytes, which in turn secrete inflammatory mediators, which ultimately leads to neuronal toxicity and death. The addition of NLY01 to the microglial cultures prevents this inflammatory cascade of events (Yun et al., 2018).

7.5 GLP-1 in animal models

An increasing number of groups have independently investigated and confirmed beneficial effects of exenatide administration in multiple rodent models of PD.

In MPTP animal toxin models, pre-treatment with exenatide provided complete protection against dopaminergic cell loss, suppressed MPTP-induced activation of microglia and attenuated expression of pro-inflammatory molecules leading to improved performance on motor assessments that were essentially no different to controls (Kim et al., 2009; Li et al., 2009). Similarly, rats unilaterally administered 6-OHDA or LPS into the median forebrain bundle led to significant reductions in striatal TH+activity and dopamine concentration (Harkavyi et al., 2008), with the animals exhibiting marked apomorphine and amphetamine-induced rotational behavior (indicative of the severity of nigrostriatal lesion) (Bertilsson et al., 2008). In these studies, exenatide was administered well after the nigrostriatal lesion had been allowed to establish and still resulted in normalization of apomorphine- and amphetamine-induced circling in a dose dependent manner. Furthermore, immunostaining of the striatum demonstrated exendin-4 significantly increased the number of both TH- and VMAT-2-positive neurons in the substantia nigra above the control values, suggesting exenatide may be able to halt and reverse established nigrostriatal lesions. In a rotenone rodent model, exenatide reduced the loss of dopaminergic neurons in the striatum alongside a reduction in the abnormal behavioral consequences, and reduced levels of tumor necrosis factor alpha (Aksoy et al., 2017).

In study utilizing models more representative of human PD, i.e., the alpha synuclein preformed fibril mouse model, and the A53T alpha synuclein transgenic mouse model the pegylated form of exenatide (NLY01) protected against the loss of dopaminergic neurons (Yun et al., 2018).

7.6 Results of exenatide in clinical trials

7.6.1 Exenatide PD-1

There have been two randomized clinical trials, which have evaluated the use of exenatide in PD patients. The first of these was an open label trial, by necessity given that the investigating team, despite their best efforts, were unable to gain access to placebo versions of the exenatide injection pen devices (Aviles-Olmos et al., 2013). This open label trial was therefore very much designed as a proof of concept, to assess safety, tolerability and gather preliminary measures of efficacy of exenatide in PD patients. The trial was funded in its entirety by the Cure Parkinson's Trust, a UK charity focused on developing treatments for PD that might slow, stop or reverse the relentless progression of the disease.

Forty-four patients were recruited from a single center. Inclusion criteria were quite broad. All participants were already on treatment with L-dopa and reported wearing off phenomena to allow a window into PD severity by performing assessments in the early morning before a participant had taken their first dose of dopaminergic medication. By recruiting patients already using L-dopa, this provided greater certainty regarding the diagnosis of PD, and minimized the risk of patient dropout due to symptom progression necessitating increases in dopaminergic treatment, and prevented the major changes in PD severity that are seen when dopaminergic treatment is introduced for the first time. Patients were between 45 and 70 years old and were independently mobile when on medication with at least 33% improvement in response to L-dopa.

The trial was a parallel group design with patients allocated at random to two groups, either to (a) continue their best medical treatment only or (b) to continue best medical treatment in addition to self-administration of subcutaneous injections of exenatide in the form of Byetta 5 mcg bd for 1 month followed by 10 mcg bd for the subsequent 11 months. They were followed up and assessed every 3 months for the whole treatment period, i.e., for 12 months, then seen again after a 2 month washout period, i.e., at 14 months. This was chosen to assess whether any effects were persistent at a timepoint when exenatide should have disappeared and thus distinguish any effects relating to "symptomatic" benefits from effects potentially attributable to disease modification.

All the motor assessments at every timepoint were video recorded. This allowed for the possibility of independent rating of the severity of disability by people trained on the use of the MDS UPDRS, but without knowledge of the randomization allocation of the patients. It is however impossible to judge limb rigidity using video recordings of assessments and therefore the single blinded ratings excluded the rigidity components of the MDS UPDRS part 3 scale.

The recruited cohort comprised patients who had a mean age of 60 years, had a mean PD duration of about 10 years and were using ~975 mg of L-dopa equivalent dose. Analysis of the data at the end of the trial revealed an advantage in the group randomized to self-administer exenatide. Given the well-known placebo effects observed in PD trials, this was interpreted cautiously. What was more compelling was that the advantage of exenatide was seen not only in terms of MDS UPDRS part 3 scores, (4.9 UPDRS points at 48 weeks and 4.4 UPDRS points at 14 months), but also there were clear advantages seen in cognitive performance as measured by the Mattis Dementia rating scale. While these data were presented as the major outcomes of the trial, further exploration of the data, e.g., that added in the unblinded rigidity scores, or that summed the data from MDS UPDRS 1, 2 and 3, all strongly favored the exenatide treated patients.

While exenatide was generally well tolerated, three patients dropped out form the exenatide arm due to poor absorption of L-dopa, weight loss and dysgeusia (loss of taste) respectively. As a group, the exenatide treated patients lost a mean of 3.2 kg compared to the control group who lost a mean of 0.8 kg.

Given the open label design of the trial, and prior to the publication of results indicating whether exenatide had conferred any advantage or not, it was possible to perform a long term follow up assessment, 12 months after all patients had ceased administering exenatide injections. The purpose of this assessment was to collect further information that might indicate whether the observed advantages at the 14 month timepoint were still present. To minimize the effect of observer bias, all scores were rated using video assessments, again necessarily excluding items relating to limb rigidity. At this timepoint, patients who had been allocated to the exenatide arm still had a clear advantage in the severity of the motor and cognitive features of the PD. The MDS UPDRS part 3 scores were 5.6 points better than the group in the control arm, while the Mattis Dementia rating scale scores were 5.3 points better (Aviles-Olmos et al., 2014).

7.6.2 Exenatide PD-2

The results of the first trial served to provide a degree of reassurance that the positive data reported by multiple laboratories regarding the neuroprotective properties of exenatide might indeed have clinical relevance in patients with PD. As such the second trial could be planned with more confidence and the investigators were able to convince Bristol Myers Squibb who, at the time were manufacturing exenatide, to provide placebo versions of the exenatide injection devices, to enable a double blind placebo controlled trial to be designed. The observation that the Byetta version of exenatide could lead to slowing down of gastric emptying and the risk of L-dopa dose failures, further influenced the choice of intervention. By this point, patients with diabetes were generally using the once weekly formulation of exenatide (known as Bydureon). This preparation was known to have much less impact on gastric emptying and therefore was anticipated to be even better tolerated by patients with PD, dependent on timely absorption of L-dopa doses. While the dose (2 mg) sounds much higher than the dose provided as Byetta (10 mcg), Bydureon contains exenatide encapsulated in 0.06 mm diameter microspheres of medical grade poly-D,L-lactide-co-glycolide. Only 1% of the exenatide on the surface of the microspheres is released in the first few hours, with the fully encapsulated exenatide being gradually released over the subsequent 7 weeks (Fineman et al., 2011).

The funding for the remaining trial infrastructure; costs for patient travel, patient assessments, DATSCAN imaging, randomization, project management, pharmacy dispensing, monitoring and specimen collection, storage and analysis, alongside data entry and analysis was provided by the Michael J Fox Foundation.

This trial was designed as a parallel group, double blind, randomized controlled trial with a sample size of 60 patients allocated into two groups, either to receive best medical treatment and self-administer exenatide 2 mg once weekly, or to receive best medical treatment and self-administer placebo injections once weekly. The exenatide and placebo injections were prepared each week by the patient by reconstituting a powder into solution, drawing the solution up into a syringe and then injecting subcutaneously.

The primary outcome of the trial was again chosen to be the MDS UPDRS part 3 assessed in the practically defined Off dopaminergic medication state, performed first thing in the morning before a participant had taken their usual dopaminergic medications (Athauda et al., 2017a, b). Patients were instructed to stop any long acting dopaminergic agents at least 36 h before the assessments. By using this off medication assessment, rather than restricting recruitment to untreated (L-dopa naive) patients only, it allowed greater numbers of patients to be eligible for recruitment, which enabled more rapid recruitment through a single recruiting center, and also meant that the results would have relevance for the broader, prevalent population of PD patients.

The trial was again designed as a washout trial, with a 48 week period of treatment exposure followed by a 12 week washout. Assessments were performed every 12 weeks and comprised standard clinical assessments; MDS UPDRS parts 1, 2, 3 and 4; Mattis Dementia rating scale; Montgomery-Asberg; Depression Rating scale; non-motor symptoms severity scale; Unified Dyskinesia Rating Scale; Hauser diaries; PDQ39; EQ-5D; Timed motor tests; L-dopa equivalent dose.

Patients had collection of blood and urine samples at each timepoint and had CSF collection at the 12 week and 48 week timepoints to measure exenatide levels in the CSF. The mean age of participants in this trial was again about 60 years, but with a shorter mean disease duration (6.4 years), and on a lower dose of Levodopa equivalent ($\sim$800 mg/day) than those recruited in the first trial.

The primary outcome was the comparison of the motor severity of PD using the MDS UPDRS part three scores in the off medication state at the 60 week timepoint. The analysis was defined a priori to incorporate an adjustment for any differences in the baseline severity of MDS UPDRS scores. This subsequently turned out to be quite important given that, by chance there was a difference in the baseline severity between the two groups with exenatide group being approximately five points worse at baseline despite random allocation. Even after adjustment for the baseline differences, the analysis of the data at the end of the trial showed a statistically significant difference between the two groups favoring exenatide of 3.5 points (95% CI −6.7 to −0.3, $P = 0.0318$) (Athauda et al., 2017a, b).

Understandably, these results attracted a great deal of comment from peer reviewers during the publication process. Other explanations for the difference observed in MDS UPDRS part three scores were sought. This included repeated analyses: (1) making further adjustments for differences in the amount of L-dopa equivalent used throughout the trial follow up, (2) comparing the subclasses of concomitant medication used between the two groups at baseline and during the follow up period, (3) comparing the frequency of adverse effects experienced between the two groups lest this might have led to unblinding of the participants and to a greater likelihood of placebo effects. None of these issues could explain the difference observed between the two groups (Athauda et al., 2017a, b).

The most objective tool we have to compare disease severity between groups is in theory the use of DATscan imaging, which allows labeling and quantification of the number of dopamine transporters on presynaptic dopaminergic terminals. While quantification of DATscan uptake can be performed on a longitudinal basis, this tool has still not been validated as an objective biomarker of disease progression in PD. In a comparison between the two groups, DATscan uptake declined in both groups but there was a slower rate of decline in three striatal regions in the exenatide treated group compared with the control group.

A detailed exploration of other measures collected during the trial was also performed. While the overall scores on the non-motor symptoms severity scale were not different between the two groups, a post hoc analysis in which non-motor symptoms were compared in more detail revealed that exenatide might also have favorable impact on mood/apathy scores during the period of treatment, which disappeared on treatment cessation (Athauda et al., 2018).

7.6.3 Exenatide PD-3

The positive data reported in the double blind trial led to a further funding application for an investigator initiated trial to seek to confirm whether exenatide has efficacy as a disease modifying drug in PD. Formal phase 3 efficacy trials need to be multicenter to ensure that the data emerging are relevant to the broad population of patients that are anticipated to benefit from the intervention.

Various trial designs were considered. The delayed start design has been criticized based on the limited time period during which patients are allocated to placebo treatment, during which any disease modifying effect must emerge. The long term simple design is preferred but is more expensive and concerns exist regarding the adherence to randomized allocation over long term follow up periods. Any long term designs chosen to explore interventions with potential disease modifying properties have to carefully consider the impact that disease progression will have on the need for conventional symptomatic medication, and the consequences that this will have on patient disease severity scores and the ability to judge the impact of the intervention under study.

A focus group meeting with patients was helpful in the decision making process for the exenatide PD-3 trial. There was clear feedback from the patient focus group that a placebo controlled arm that required once weekly subcutaneous injections for a 3 year period would be unacceptable, whereas a 2 year period would be far more acceptable. It was therefore decided, that a 2 year parallel group design would be adopted, with a 1:1 randomization allocation between active treatment in the form of Bydureon 2 mg subcutaneous injection once weekly and matched placebo and while the primary outcome would be the difference between the 2 arms at 2 years, that a planned comparison between any effect size at 1 year would be compared to any effect size at 2 years, to help distinguish between any static (i.e., likely symptomatic effect) from any cumulative (i.e., disease modifying) effect.

The motor subsection of the MDS UPDRS was again chosen to be used as the primary outcome. This aligns with the primary outcome measure for the previous trials and while it has its weaknesses, it is the most widely adopted outcome measure in PD trials with general acceptance with its validity and reliability. Given the long term design, the participants will be scored in the early morning in the practically defined "off medication" state, i.e., having had an overnight period without L-dopa, and at least 36 h without any long acting dopaminergic medications, e.g., Ropinirole, Pramipexole, Rotigotine. This will allow the definitive identification of any effects of exenatide on the motor severity of PD, while minimizing the inevitable effect of conventional dopaminergic medication replacement.

A distinction between a symptomatic effect and a cumulative effect is of major importance in the judgment of what might be the minimally clinically relevant effect size. This has been discussed in the context of interventions with a symptomatic effect, but has not yet been discussed in the context of disease modifying approaches, given none exist thus far. Any indication that exenatide has a cumulative advantage according to duration of exposure will therefore be taken as evidence in support of a disease modifying effect.

Other teams have combined the interest in exenatide as a neuroprotective agent with the interest in developing a structural imaging biomarker of disease progression in PD (Burciu et al., 2017). The team in Gainsville, Florida are recruiting 15 patients to each receive exenatide injections once weekly for 1 year, accompanied by free-water MRI scans at baseline and end of the 1 year exposure period (ClinicalTrials.gov Identifier: NCT03456687).

7.6.4 Exenatide-NLY01

The manufacturers of NLY01 (Neuraly) are keen to move their modified version of exenatide into clinical trials in PD. As a first step, they have set up a phase 1, first in human, single ascending, followed by multiple ascending dose trials in healthy volunteers to gage/confirm the safety and tolerability of NLY01 (ClinicalTrials. gov Identifier: NCT03672604).

7.6.5 Exenatide-PT302

A further modified formulation of exenatide called PT302 has been created by a South Korean company called Peptron. This is also a slow release formulation using the D,L-lactide-*co*-glycolide (PLGA) release control agent. Microparticles of exenatide are coated with L-lysine which suppresses the initial burst of exenatide after injection, and allows for a smaller needle size for the administration of the injection. Following the standard pathway for drug development, the company has completed a phase 1, first in human single dose, dose escalating study to explore the pharmaco-kinetics of exenatide following injection of PT302, with the intention to explore the potential disease modifying effects in PD patients.

7.6.6 Liraglutide

Although exenatide was first-in-class, there is evidence that other GLP-1 agonists may achieve tighter glycemic control. A head to head comparison between liraglu-tide (administered once daily) and exenatide (in the form of Byetta administered twice daily) confirmed an advantage in terms of blood glucose and HbA1c with liraglutide (Buse et al., 2009). Thus there is growing interest in repurposing other GLP-1 receptor agonists as a result of the positive data emerging from the first exenatide trials alongside laboratory data.

In neuroblastoma cells exposed to methylglyoxal, liraglutide restored cell viabil-ity and reduced apoptosis (Sharma et al., 2014). In animal toxin models of PD, liraglutide has demonstrated protective effects in both the rotenone (Badawi et al., 2017) and MPTP models (Feng et al., 2018; Liu et al., 2015; Yuan et al., 2017; Zhang et al., 2018) though did not rescue dopaminergic neuronal loss post lesioning in the 6-OHDA rodent model (Hansen et al., 2016).

There is also emerging limited data comparing efficacy of GLP-1 agonists in models of neurodegeneration. Although liraglutide has been consistently found to be more potent than exenatide in comparisons in laboratory models, comparisons have so far not been based on experiments with equivalent molar concentrations, or using dose response demonstrations of maximal potency (Liu et al., 2015).

A team in California led by Dr. Tagliati are following up a cohort of 57 patients with PD recruited to a trial of Liraglutide. The design of this trial is very similar to the Exenatide PD-2 trial with the exception that the primary outcome will include UPDRS part 3, non-motor symptoms scale and cognitive performance, using the Mattis Dementia rating scale (ClinicalTrials.gov Identifier: NCT02953665). It is estimated that this study will be completed in December 2020.

7.6.7 Lixisenatide

Following the observations that Lixisenatide has greater CNS penetration than either liraglutide or exenatide and greater neuroprotective properties at equivalent doses in in vitro models of neurodegeneration (Liu et al., 2015; McClean and Hölscher, 2014) another large GLP-1 receptor agonist trial in PD has also been set up and is recruiting patients in France. This has been jointly funded through both charity and the commercial manufacturer of lixisenatide, Sanofi.

The trial will recruit 158 early stage PD patients (<3 years since diagnosis) from multiple centers across France and randomize them to receive Lixisenatide injections 20 µg once daily or placebo for 12 months followed by a 2 month washout period. The primary outcome will be a comparison of MDS UPDRS part 3 scores at the end of the 12 month treatment period.

7.6.8 Semaglutide

There is also interest in the potential for the newest GLP-1R agonist, semaglutide as a potential disease modifying treatment in PD. Semaglutide has also been shown to have neuroprotective properties in a rodent model of PD. In comparison to the effects of liraglutide, semaglutide administered on alternate days had greater potency at reducing the loss of tyrosine hydroxylase positive neurons induced by MPTP, reducing inflammation and restoring behavior (Zhang et al., 2018). Of interest is that there is now an orally active formulation of semaglutide which has been shown to have equivalent beneficial effects to subcutaneous semaglutide in T2DM patients (Davies et al., 2017).

Novo Nordisk are setting up a 270 patient trial in Scandinavia. This is the most ambitious trial to date and will involve a 2 year period in which patients are randomly allocated to self-administer Semaglutide or placebo once weekly for 2 years using double blind methodology. This will be followed by a further 2 year open label period in which all participants will self-administer the active drug. Patients will be compared on the basis of their motor disability using MDS UPDRS part 3, their cognitive performance as well as non-motor tasks and DATscan imaging.

7.6.9 What might be the mechanism of action of GLP-1 receptor agonists in PD?

While there have been a wide range of mechanisms of action of GLP-1 RAs seen in animal models of neurodegeneration, none of these can adequately recapitulate the processes involved in PD neurodegeneration in humans. Whether any of the beneficial effects seen in the early exenatide trials relate to effects on apoptosis, mitochondrial function, insulin resistance, neuroinflammation or neurogenesis remains to be fully elucidated (Athauda and Foltynie, 2016a, b).

One way of trying to assess the potential impact of interventions on neuronal cellular processes has been through the examination of the contents of serum extracellular vesicles. In the exenatide PD-2 trial, participants had blood samples collected at each timepoint, and serum stored. From these serum samples, it was possible to extract and precipitate extracellular vesicles according to their size. These vesicles are nanosized membranous particles secreted by virtually all cells, and contain cargo representative of the physiological state of their tissue of origin. From the mixed population of vesicles it was then further possible to use immune capture techniques to separate out vesicles bearing the L1CAM surface molecule and thus greatly enrich for vesicles of neuronal origin.

This purified population of neuronal vesicles was then lysed and the proteins they contained were quantified using electrochemiluminescence. In this experiment it was found that patients who had been treated with exenatide had an increase in the tyrosine phosphorylation of IRS-1, an increase in total Akt, phosphorylated Akt and phosphorylated mTOR compared to patients treated with placebo. These increases largely disappeared by the end of the 12-week washout period (Athauda et al., 2019a).

These changes, while in need of replication, provide human in vivo evidence to suggest that exenatide engages with the insulin signaling pathway as hypothesized, and has resulting downstream effects on survival processes mediated by Akt and mTOR. Whether these effects are the sole explanations for the clinical effects seen in association with exenatide treatment in PD patients requires further study.

8 Which PD patients might do best on GLP-1 agonist medications?

Intuitively, the PD patients who would gain the most from using GLP-1 agonist medications would be those with concurrent Type 2 diabetes with unsatisfactory control despite their usual hypoglycemic regimes. It might be expected that any beneficial effects on PD will also be compounded by additional benefits mediated through better diabetes control. This may include a reduction in neuropathy, retinopathy and cerebrovascular disease, all of which might complicate the disability experienced by PD patients, as well as reduce the rate of neurodegeneration due to any contributions from elevated alpha synuclein glycation, neuroinflammation, mitochondrial dysfunction and central insulin resistance.

In the two exenatide PD trials conducted thus far, none of the patients had type 2 diabetes given this was an a priori exclusion criterion, and the trials were seeking to identify effects of exenatide not mediated through an anti-diabetic action. In a post hoc analysis of the exenatide PD trial, patients that had the greatest magnitude of improvement had; higher levels of tremor, and lower overall motor severity on the MDS UPDRS part 3 scale. In contrast, while some improvements were still noted, older patients, and those with longer duration of disease responded less well (Athauda et al., 2019b).

It is likely however that patients with type 2 diabetes or impaired glucose tolerance would be most likely to gain an advantage for both their diabetes control and concurrent PD, irrespective of the mechanism of action of the GLP-1 agonist group. Whether this emerges to be the case might depend on the extent/severity of insulin resistance within an individual patient. It could be counter-hypothesized that patients with the highest level of peripheral insulin resistance or the least well controlled serum glucose, might have the highest level of central insulin resistance, alpha synuclein glycation or neuroinflammation and therefore be the most challenging patients to treat despite the putative mechanistic effects of exenatide on PD progression.

8.1 Dual incretin agonists

In a similar way to GLP-1, another hormone known as glucose-dependent insulinotropic polypeptide (GIP) is secreted by the cells of the small intestine and acts as an incretin hormone to stimulate insulin release in a glucose level dependent manner. GIP, like GLP-1 is also metabolized by DPP-4. Activation of the GIP receptors (present in pancreas, brain, bone, cardiovascular system and gastrointestinal tract) triggers a similar cascade of trophic and anti-apoptotic effects mediated by Akt and MAPkinase (Baggio and Drucker, 2007).

In view of the parallel effects of GLP-1 and GIP, dual agonists for the GLP-1 and GIP receptor have been developed. These dual agonists have been found to be superior to liraglutide in reversing the motor impairment triggered by MPTP (Cao et al., 2016; Feng et al., 2018; Yuan et al., 2017). Not surprisingly therefore there is interest in developing dual incretin agonists as potential treatment options for PD.

9 Concluding remarks

Interest in the links between T2DM and PD and the potential for anti-diabetes drugs as treatment options in PD are only relatively recently established. There is however now major momentum behind the evaluation of the anti-diabetic drugs, particularly the GLP-1 receptor agonists as disease modifying drugs in PD. While exenatide has already accumulated some clinical trial data to indicate potential efficacy, further data are required before this drug can be recommended for routine use in PD. The relative safety, tolerability and efficacy of each of the GLP-1 receptor agonists will have to be further explored, and head to head studies between these drugs and the dual incretin agonists may also be necessary. Inevitably, major commercial investment in this field will accelerate the development of repurposing of this drug class far more effectively than is possible from investigator initiated trials alone.

References

Abdelsalam, R.M., Safar, M.M., 2015. Neuroprotective effects of vildagliptin in rat rotenone Parkinson's disease model: role of RAGE-NFκB and Nrf2-antioxidant signaling pathways. J. Neurochem. 133 (5), 700–707. https://doi.org/10.1111/jnc.13087.

Akintola, A.A., van Heemst, D., 2015. Insulin, aging, and the brain: mechanisms and implications. Front. Endocrinol. 6 (January), Frontiers: 13. https://doi.org/10.3389/fendo.2015.00013.

Aksoy, D., Solmaz, V., Çavuşoğlu, T., Meral, A., Ateş, U., Erbaş, O., 2017. Neuroprotective effects of exenatide in a rotenone-induced rat model of Parkinson's disease. Am. J. Med. Sci. 354 (3), 319–324. https://doi.org/10.1016/j.amjms.2017.05.002.

Andersen, E.S., Deacon, C.F., Holst, J.J., 2018. Do we know the true mechanism of action of the DPP-4 inhibitors? Diabetes Obes. Metab. 20 (1), 34–41. https://doi.org/10.1111/dom.13018.

Andukuri, R., Drincic, A., Rendell, M., 2009. Alogliptin: a new addition to the class of DPP-4 inhibitors. Diabetes Metab. Syndr. Obes. 2 (July), 117–126. Dove Press.

Athauda, D., Foltynie, T., 2016a. Insulin resistance and Parkinson's disease: a new target for disease modification? Prog. Neurobiol. 145–146, 98–120. https://doi.org/10.1016/j.pneurobio.2016.10.001.

Athauda, D., Foltynie, T., 2016b. The glucagon-like peptide 1 (GLP) receptor as a therapeutic target in Parkinson's disease: mechanisms of action. Drug Discov. Today 21 (5), 802–818. https://doi.org/10.1016/j.drudis.2016.01.013.

Athauda, D., Wyse, R., Brundin, P., Foltynie, T., 2017a. Is exenatide a treatment for Parkinson's disease? J. Parkinsons Dis. 7 (3), 451–458. https://doi.org/10.3233/JPD-171192.

Athauda, D., Maclagan, K., Skene, S.S., Bajwa-Joseph, M., Letchford, D., Chowdhury, K., Hibbert, S., et al., 2017b. Exenatide once weekly versus placebo in Parkinson's disease: a randomised, double-blind, placebo-controlled trial. Lancet 390, 1664–1675. https://doi.org/10.1016/S0140-6736(17)31585-4.

Athauda, D., Maclagan, K., Budnik, N., Zampedri, L., Hibbert, S., Skene, S.S., Chowdhury, K., Aviles-Olmos, I., Limousin, P., Foltynie, T., 2018. What effects might exenatide have on non-motor symptoms in Parkinson's disease: a post hoc analysis. J. Parkinsons Dis. 8 (2), 247–258. https://doi.org/10.3233/JPD-181329.

Athauda, D., Gulyani, S., Karnati, H., Li, Y., Tweedie, D., Mustapic, M., Chawla, S., et al., 2019a. Utility of neuronal-derived exosomes to examine molecular mechanisms that affect motor function in patients with Parkinson disease: a secondary analysis of the exenatide-PD trial. JAMA Neurol. 76, 420–429. https://doi.org/10.1001/jamaneurol.2018.4304.

Athauda, D., Maclagan, K., Budnik, N., Zampedri, L., Hibbert, S., Aviles-Olmos, I., Chowdhury, K., Skene, S.S., Limousin, P., Foltynie, T., 2019b. Post hoc analysis of the exenatide-PD trial-factors that predict response. Eur. J. Neurosci. 49 (3), 410–421. https://doi.org/10.1111/ejn.14096.

Aviles-Olmos, I., Dickson, J., Kefalopoulou, Z., Djamshidian, A., Ell, P., Soderlund, P., Whitton, P., et al., 2013. Exenatide and the treatment of patients with Parkinson's disease. J. Clin. Invest. 123 (6), 2370–2736. https://doi.org/10.1172/JCI68295.2730.

Aviles-Olmos, I., Dickson, J., Kefalopoulou, Z., Djamshidian, A., Kahan, J., Fmedsci, P.E., Whitton, P., et al., 2014. Motor and cognitive advantages persist 12 months after exenatide exposure in Parkinson's disease. J. Parkinsons Dis. 4, 337–344. https://doi.org/10.3233/JPD-140364.

Badawi, G.A., Abd El Fattah, M.A., Zaki, H.F., El Sayed, M.I., 2017. Sitagliptin and liraglutide reversed nigrostriatal degeneration of rodent brain in rotenone-induced Parkinson's disease. Inflammopharmacology 25 (3), 369–382. https://doi.org/10.1007/s10787-017-0331-6.

Baggio, L.L., Drucker, D.J., 2007. Biology of incretins: GLP-1 and GIP. Gastroenterology 132 (6), 2131–2157. https://doi.org/10.1053/j.gastro.2007.03.054.

Banks, W.A., 2004. The source of cerebral insulin. Eur. J. Pharmacol. 490 (1–3), 5–12. https://doi.org/10.1016/j.ejphar.2004.02.040.

Bassil, F., Fernagut, P.-O., Bezard, E., Meissner, W.G., 2014. Insulin, IGF-1 and GLP-1 signaling in neurodegenerative disorders: targets for disease modification. Prog. Neurobiol. 118, 1–18. https://doi.org/10.1016/j.pneurobio.2014.02.005.

Bassil, F., Canron, M.-H., Dutheil, N., Vital, A., Bezard, E., Fernagut, P.-O., Meissner, W., 2017. Brain insulin resistance in Parkinson's disease [abstract]. Mov Disord. 32 (Suppl. 2). https://www.mdsabstracts.org/abstract/brain-insulin-resistance-in-parkinsons-disease/.

Bertilsson, G., Patrone, C., Zachrisson, O., Andersson, A., Dannaeus, K., Heidrich, J., Kortesmaa, J., et al., 2008. Peptide hormone exendin-4 stimulates subventricular zone neurogenesis in the adult rodent brain and induces recovery in an animal model of Parkinson's disease. J. Neurosci. Res. 86 (2), 326–338. https://doi.org/10.1002/jnr.21483.

Best, J.H., Boye, K.S., Rubin, R.R., Cao, D., Kim, T.H., Peyrot, M., 2009. Improved treatment satisfaction and weight-related quality of life with exenatide once weekly or twice daily. Diabet. Med. 26 (7), 722–728. https://doi.org/10.1111/j.1464-5491.2009.02752.x.

Bohnen, N.I., Kotagal, V., Müller, M.L.T.M., Koeppe, R.A., Scott, P.J.H., Albin, R.L., Frey, K.A., Petrou, M., 2014. Diabetes mellitus is independently associated with more severe cognitive impairment in Parkinson disease. Parkinsonism Relat. Disord. 20 (12), 1394–1398. https://doi.org/10.1016/j.parkreldis.2014.10.008.

Bosco, D., Plastino, M., Cristiano, D., Colica, C., Ermio, C., De Bartolo, M., Mungari, P., et al., 2012. Dementia is associated with insulin resistance in patients with Parkinson's disease. J. Neurol. Sci. 315 (1–2), 39–43. https://doi.org/10.1016/j.jns.2011.12.008.

Brakedal, B., Flønes, I., Reiter, S.F., Torkildsen, Ø., Dölle, C., Assmus, J., Haugarvoll, K., Tzoulis, C., 2017. Glitazone use associated with reduced risk of Parkinson's disease. Mov. Disord. 32 (11), 1594–1599. https://doi.org/10.1002/mds.27128.

Brauer, R., Bhaskaran, K., Chaturvedi, N., Dexter, D.T., Smeeth, L., Douglas, I., 2015. Glitazone treatment and incidence of Parkinson's disease among people with diabetes: a retrospective Cohort study. PLoS Med. 12 (7), e1001854. https://doi.org/10.1371/journal.pmed.1001854.

Brundin, P., Barker, R.A., Conn, P.J., Dawson, T.M., Kieburtz, K., Lees, A.J., Schwarzschild, M.A., et al., 2013. Linked clinical trials—the development of new clinical learning studies in Parkinson's disease using screening of multiple prospective new treatments. J. Parkinsons Dis. 3 (3), 231–239. https://doi.org/10.3233/JPD-139000.

Burciu, R.G., Ofori, E., Archer, D.B., Wu, S.S., Pasternak, O., McFarland, N.R., Okun, M.S., Vaillancourt, D.E., 2017. Progression marker of Parkinson's disease: a 4-year multi-site imaging study. Brain 140 (8), 2183–2192. https://doi.org/10.1093/brain/awx146.

Buse, J.B., Rosenstock, J., Sesti, G., Schmidt, W.E., Montanya, E., Brett, J.H., Zychma, M., Blonde, L., LEAD-6 Study Group, 2009. Liraglutide once a day versus Exenatide twice a day for type 2 diabetes: a 26-week randomised, parallel-group, multinational, open-label trial (LEAD-6). Lancet 374 (9683), 39–47. https://doi.org/10.1016/S0140-6736(09)60659-0.

Cao, L., Li, D., Feng, P., Li, L., Xue, G.-F., Li, G., Hölscher, C., 2016. A novel dual GLP-1 and GIP incretin receptor agonist is neuroprotective in a mouse model of Parkinson's disease by reducing chronic inflammation in the brain. Neuroreport 27 (6), 384–391. https://doi.org/10.1097/WNR.0000000000000548.

Cereda, E., Barichella, M., Pedrolli, C., Klersy, C., Cassani, E., Caccialanza, R., Pezzoli, G., 2011. Diabetes and risk of Parkinson's disease: a systematic review and meta-analysis. Diabetes Care 34 (12), 2614–2623. https://doi.org/10.2337/dc11-1584.

Cereda, E., Barichella, M., Cassani, E., Caccialanza, R., Pezzoli, G., 2012. Clinical features of Parkinson disease when onset of diabetes came first: a case-control study. Neurology 78 (19), 1507–1511. https://doi.org/10.1212/WNL.0b013e3182553cc9.

Colca, J.R., McDonald, W.G., Waldon, D.J., Leone, J.W., Lull, J.M., Bannow, C.A., Lund, E.T., Mathews, W.R., 2004. Identification of a novel mitochondrial protein ('mitoNEET') cross-linked specifically by a thiazolidinedione photoprobe. Am. J. Physiol. Endocrinol. Metab. 286 (2), E252–E260. https://doi.org/10.1152/ajpendo.00424.2003.

Colca, J.R., Vander Lugt, J.T., Adams, W.J., Shashlo, A., McDonald, W.G., Liang, J., Zhou, R., Orloff, D.G., 2013a. Clinical proof-of-concept study with MSDC-0160, a prototype mTOT-modulating insulin sensitizer. Clin. Pharmacol. Ther. 93 (4), 352–359. https://doi.org/10.1038/clpt.2013.10.

Colca, J.R., McDonald, W.G., Cavey, G.S., Cole, S.L., Holewa, D.D., Brightwell-Conrad, A.S., Wolfe, C.L., et al., 2013b. Identification of a mitochondrial target of thiazolidine-dione insulin sensitizers (mTOT)—relationship to newly identified mitochondrial pyruvate carrier proteins. PLoS One 8 (5). e61551. https://doi.org/10.1371/journal.pone.0061551.

Connolly, J.G., Bykov, K., Gagne, J.J., 2015. Thiazolidinediones and Parkinson disease: a cohort study. Am. J. Epidemiol. 182 (11), 936–944. https://doi.org/10.1093/aje/kwv109.

Corona, J.C., Duchen, M.R., 2015. PPARγ and PGC-1α as therapeutic targets in Parkinson's. Neurochem. Res. 40 (2), 308–316. https://doi.org/10.1007/s11064-014-1377-0.

Corona, J.C., de Souza, S.C., Duchen, M.R., 2014. PPARγ activation rescues mitochondrial function from inhibition of complex I and loss of PINK1. Exp. Neurol. 253 (March), 16–27. https://doi.org/10.1016/j.expneurol.2013.12.012.

Davies, M., Pieber, T.R., Hartoft-Nielsen, M.-L., Hansen, O.K.H., Jabbour, S., Rosenstock, J., 2017. Effect of oral semaglutide compared with placebo and subcutaneous semaglutide on glycemic control in patients with type 2 diabetes: a randomized clinical trial. JAMA 318 (15), 1460. https://doi.org/10.1001/jama.2017.14752.

De Pablo-Fernandez, E., Goldacre, R., Pakpoor, J., Noyce, A.J., Warner, T.T., 2018. Association between diabetes and subsequent Parkinson disease: a record-linkage cohort study. Neurology 91 (2), e139–e142. https://doi.org/10.1212/WNL.0000000000005771.

Dehmer, T., Heneka, M.T., Sastre, M., Dichgans, J., Schulz, J.B., 2004. Protection by pioglitazone in the MPTP model of Parkinson's disease correlates with I kappa B alpha induction and block of NF kappa B and iNOS activation. J. Neurochem. 88 (2), 494–501.

Devaskar, S.U., Giddings, S.J., Rajakumar, P.A., Carnaghi, L.R., Menon, R.K., Zahm, D.S., 1994. Insulin gene expression and insulin synthesis in mammalian neuronal cells. J. Biol. Chem. 269 (11), 8445–8454.

Drucker, D.J., Buse, J.B., Taylor, K., Kendall, D.M., Trautmann, M., Zhuang, D., Porter, L., DURATION-1 Study Group, 2008. Exenatide once weekly versus twice daily for the treatment of type 2 diabetes: a randomised, open-label, non-inferiority study. Lancet 372 (9645), 1240–1250. https://doi.org/10.1016/S0140-6736(08)61206-4.

Drucker, D.J., Sherman, S.I., Gorelick, F.S., Bergenstal, R.M., Sherwin, R.S., Buse, J.B., 2010. Incretin-based therapies for the treatment of type 2 diabetes: evaluation of the risks and benefits. Diabetes Care 33 (2), 428–433. https://doi.org/10.2337/dc09-1499.

Eng, J., Kleinman, W.A., Singh, L., Singh, G., Raufman, J.P., 1992. Isolation and characterization of exendin-4, an exendin-3 analogue, from *Heloderma suspectum* venom. Further evidence for an exendin receptor on dispersed acini from Guinea pig pancreas. J. Biol. Chem. 267 (11), 7402–7405.

Fan, R., Li, X., X, G., Chan, J.C.N., G, X., 2010. Exendin-4 protects pancreatic beta cells from human islet amyloid polypeptide-induced cell damage: potential involvement of AKT and mitochondria biogenesis. Diabetes Obes. Metab. 12, 815–824.

Feng, P., Zhang, X., Li, D., Ji, C., Yuan, Z., Wang, R., Xue, G., Li, G., Hölscher, C., 2018. Two novel dual GLP-1/GIP receptor agonists are neuroprotective in the MPTP mouse model of Parkinson's disease. Neuropharmacology 133 (May), 385–394. https://doi.org/10.1016/j.neuropharm.2018.02.012.

Fineman, M., Flanagan, S., Taylor, K., Aisporna, M., Shen, L.Z., Mace, K.F., Walsh, B., et al., 2011. Pharmacokinetics and pharmacodynamics of exenatide extended-release after single and multiple dosing. Clin. Pharmacokinet. 50 (1), 65–74. https://doi.org/10.2165/11585880-000000000-00000.

Grahame Hardie, D., 2014. AMP-activated protein kinase: a key regulator of energy balance with many roles in human disease. J. Intern. Med. 276 (6), 543–559. https://doi.org/10.1111/joim.12268.

Gual, P., Le Marchand-Brustel, Y., Tanti, J.-F., 2005. Positive and negative regulation of insulin signaling through IRS-1 phosphorylation. Biochimie 87 (1), 99–109. https://doi.org/10.1016/j.biochi.2004.10.019.

Hansen, H.H., Fabricius, K., Barkholt, P., Mikkelsen, J.D., Jelsing, J., Pyke, C., Knudsen, L.B., Vrang, N., 2016. Characterization of liraglutide, a glucagon-like peptide-1 (GLP-1) receptor agonist, in rat partial and full nigral 6-hydroxydopamine lesion models of Parkinson's disease. Brain Res. 1646 (September), 354–365. https://doi.org/10.1016/j.brainres.2016.05.038.

Harkavyi, A., Abuirmeileh, A., Lever, R., Kingsbury, A.E., Biggs, C.S., Whitton, P.S., 2008. Glucagon-like peptide 1 receptor stimulation reverses key deficits in distinct rodent models of Parkinson's disease. J. Neuroinflammation 5 (January), 19. https://doi.org/10.1186/1742-2094-5-19.

Havrankova, J., Schmechel, D., Roth, J., Brownstein, M., 1978. Identification of insulin in rat brain. Proc. Natl. Acad. Sci. U. S. A. 75 (11), 5737–5741.

He, Y.-L., Wang, Y., Bullock, J.M., Deacon, C.F., Holst, J.J., Dunning, B.E., Ligueros-Saylan, M., Foley, J.E., 2007. Pharmacodynamics of vildagliptin in patients with type 2 diabetes during OGTT. J. Clin. Pharmacol. 47 (5), 633–641. https://doi.org/10.1177/0091270006299137.

Herschkovitz, A., Liu, Y.-F., Ilan, E., Ronen, D., Boura-Halfon, S., Zick, Y., 2007. Common inhibitory serine sites phosphorylated by IRS-1 kinases, triggered by insulin and inducers of insulin resistance. J. Biol. Chem. 282 (25), 18018–18027. https://doi.org/10.1074/jbc.M610949200.

Horvath, I., Wittung-Stafshede, P., 2016. Cross-talk between amyloidogenic proteins in type-2 diabetes and Parkinson's disease. Proc. Natl. Acad. Sci. U. S. A. 113 (44), 12473–12477. https://doi.org/10.1073/pnas.1610371113.

Hotamisligil, G.S., Peraldi, P., Budavari, A., Ellis, R., White, M.F., Spiegelman, B.M., 1996. IRS-1-mediated inhibition of insulin receptor tyrosine kinase activity in TNF-alpha- and obesity-induced insulin resistance. Science 271 (5249), 665–668.

Hunter, R.L., Choi, D.-Y., Ross, S.A., Bing, G., 2008. Protective properties afforded by pioglitazone against Intrastriatal LPS in Sprague–Dawley rats. Neurosci. Lett. 432 (3), 198–201. https://doi.org/10.1016/j.neulet.2007.12.019.

Isik, A.T., Soysal, P., Yay, A., Usarel, C., 2017. The effects of sitagliptin, a DPP-4 inhibitor, on cognitive functions in elderly diabetic patients with or without Alzheimer's disease. Diabetes Res. Clin. Pract. 123 (January), 192–198. https://doi.org/10.1016/J.DIABRES.2016.12.010.

Ismaiel, A.A.K., Espinosa-Oliva, A.M., Santiago, M., García-Quintanilla, A., Oliva-Martín, M.J., Herrera, A.J., Venero, J.L., de Pablos, R.M., 2016. Metformin, besides exhibiting strong in vivo anti-inflammatory properties, increases mptp-induced damage to the nigrostriatal dopaminergic system. Toxicol. Appl. Pharmacol. 298 (May), 19–30. https://doi.org/10.1016/j.taap.2016.03.004.

Katila, N., Bhurtel, S., Shadfar, S., Srivastav, S., Neupane, S., Ojha, U., Jeong, G.-S., Choi, D.-Y., 2017. Metformin lowers α-synuclein phosphorylation and upregulates neurotrophic factor in the MPTP mouse model of Parkinson's disease. Neuropharmacology 125 (October), 396–407. https://doi.org/10.1016/j.neuropharm.2017.08.015.

Kim, S., Moon, M., Park, S., 2009. Exendin-4 protects dopaminergic neurons by inhibition of microglial activation and matrix metalloproteinase-3 expression in an animal model of Parkinson's disease. J. Endocrinol. 202 (3), 431–439. https://doi.org/10.1677/JOE-09-0132.

Kotagal, V., Albin, R.L., Müller, M.L.T.M., Koeppe, R.A., Frey, K.A., Bohnen, N.I., 2013. Diabetes is associated with postural instability and gait difficulty in Parkinson disease. Parkinsonism Relat. Disord. 19 (5), 522–526. https://doi.org/10.1016/j.parkreldis.2013.01.016.

Kuwabara, T., Kagalwala, M.N., Onuma, Y., Ito, Y., Warashina, M., Terashima, K., Sanosaka, T., Nakashima, K., Gage, F.H., Asashima, M., 2011. Insulin biosynthesis in neuronal progenitors derived from adult hippocampus and the olfactory bulb. EMBO Mol. Med. 3 (12), 742–754. https://doi.org/10.1002/emmm.201100177.

Laloux, C., Petrault, M., Lecointe, C., Devos, D., Bordet, R., 2012. Differential susceptibility to the PPAR-γ agonist pioglitazone in 1-methyl-4-phenyl-1, 2,3,6-tetrahydropyridine and 6-hydroxydopamine rodent models of Parkinson's disease. Pharmacol. Res. 65 (5), 514–522. https://doi.org/10.1016/j.phrs.2012.02.008.

LeBrasseur, N.K., Kelly, M., Tsao, T.-S., Farmer, S.R., Saha, A.K., Ruderman, N.B., Tomas, E., 2006. Thiazolidinediones can rapidly activate AMP-activated protein kinase in mammalian tissues. Am. J. Physiol. Endocrinol. Metab. 291 (1), E175–E181. https://doi.org/10.1152/ajpendo.00453.2005.

Lee, E.Y., Lee, J.E., Park, J.H., Shin, I.C., Koh, H.C., 2012. Rosiglitazone, a PPAR-γ agonist, protects against striatal dopaminergic neurodegeneration induced by 6-OHDA lesions in the substantia nigra of rats. Toxicol. Lett. 213 (3), 332–344. https://doi.org/10.1016/j.toxlet.2012.07.016.

Li, Y., Perry, T., Kindy, M.S., Harvey, B.K., Tweedie, D., Holloway, H.W., Powers, K., et al., 2009. GLP-1 receptor stimulation preserves primary cortical and dopaminergic neurons in cellular and rodent models of stroke and parkinsonism. PNAS 106 (4), 1285–1290.

Li, J., Zhang, S., Li, C., Li, M., Ma, L., 2018. Sitagliptin rescues memory deficits in parkinsonian rats via upregulating BDNF to prevent neuron and dendritic spine loss. Neurol. Res. 40 (9), 736–743. https://doi.org/10.1080/01616412.2018.1474840.

Liu, W., Jalewa, J., Sharma, M., Li, G., Li, L., Hölscher, C., 2015. Neuroprotective effects of lixisenatide and liraglutide in the 1-methyl-4-phenyl-1,2,3,6-tetrahydropyridine mouse model of Parkinson's disease. Neuroscience 303 (September), 42–50. https://doi.org/10.1016/j.neuroscience.2015.06.054.

Lovshin, J.A., Drucker, D.J., 2009. Incretin-based therapies for type 2 diabetes mellitus. Nat. Rev. Endocrinol. 5 (5), 262–269. https://doi.org/10.1038/nrendo.2009.48.

Lu, M., Su, C., Qiao, C., Bian, Y., Ding, J., Hu, G., 2016. Metformin prevents dopaminergic neuron death in MPTP/P-induced mouse model of Parkinson's disease via autophagy and mitochondrial ROS clearance. Int. J. Neuropsychopharmacol. 19 (9). pyw047. https://doi.org/10.1093/ijnp/pyw047.

Machado, M.M.F., Bassani, T.B., Cóppola-Segovia, V., Moura, E.L.R., Zanata, S.M., Andreatini, R., Vital, M.A.B.F., 2019. PPAR-γ agonist pioglitazone reduces microglial proliferation and NF-κB activation in the substantia nigra in the 6-hydroxydopamine model of Parkinson's disease. Pharmacol. Rep. 71 (4), 556–564. https://doi.org/10.1016/j.pharep.2018.11.005.

Martinez-Valbuena, I., Amat-Villegas, I., Valenti-Azcarate, R., Carmona-Abellan, M.d.M., Marcilla, I., Tuñon, M.-T., Luquin, M.-R., 2018. Interaction of amyloidogenic proteins in pancreatic β cells from subjects with synucleinopathies. Acta Neuropathol. 135 (6), 877–886. https://doi.org/10.1007/s00401-018-1832-0.

McClean, P.L., Hölscher, C., 2014. Lixisenatide, a drug developed to treat type 2 diabetes, shows neuroprotective effects in a mouse model of Alzheimer's disease. Neuropharmacology 86 (November), 241–258. https://doi.org/10.1016/j.neuropharm.2014.07.015.

Moroo, I., Yamada, T., Makino, H., Tooyama, I., McGeer, P.L., McGeer, E.G., Hirayama, K., 1994. Loss of insulin receptor immunoreactivity from the substantia Nigra pars compacta neurons in Parkinson's disease. Acta Neuropathol. 87 (4), 343–348.

Mousa, S.A., Ayoub, B.M., 2019. Repositioning of dipeptidyl peptidase-4 inhibitors and glucagon like peptide-1 agonists as potential neuroprotective agents. Neural Regen. Res. 14 (5), 745. https://doi.org/10.4103/1673-5374.249217.

Mukherjee, A., Morales-Scheihing, D., Butler, P.C., Soto, C., 2015. Type 2 diabetes as a protein misfolding disease. Trends Mol. Med. 21 (7), 439–449. https://doi.org/10.1016/j.molmed.2015.04.005.

Nassar, N.N., Al-Shorbagy, M.Y., Arab, H.H., Abdallah, D.M., 2015. Saxagliptin: a novel antiparkinsonian approach. Neuropharmacology 89 (February), 308–317. https://doi.org/10.1016/j.neuropharm.2014.10.007.

Nielsen, L.L., Young, A.A., Parkes, D.G., 2004. Pharmacology of exenatide (synthetic exendin-4): a potential therapeutic for improved glycemic control of type 2 diabetes. Regul. Pept. 117 (2), 77–88.

NINDS Exploratory Trials in Parkinson Disease (NET-PD) FS-ZONE Investigators, 2015. Pioglitazone in early Parkinson's disease: a phase 2, multicentre, double-blind, randomised trial. Lancet Neurol. 14 (8), 795–803. https://doi.org/10.1016/S1474-4422(15)00144-1.

Novak, P., Maldonado, D.A.P., Novak, V., 2019. Safety and preliminary efficacy of intranasal insulin for cognitive impairment in parkinson disease and multiple system atrophy: a double-blinded placebo-controlled pilot study. PLoS One 14 (4), e0214364. https://doi.org/10.1371/journal.pone.0214364.

Pagano, G., Polychronis, S., Wilson, H., Giordano, B., Ferrara, N., Niccolini, F., Politis, M., 2018. Diabetes mellitus and Parkinson disease. Neurology 90 (19), e1654–e1662. https://doi.org/10.1212/WNL.0000000000005475.

Patil, S.P., Jain, P.D., Ghumatkar, P.J., Tambe, R., Sathaye, S., 2014. Neuroprotective effect of metformin in MPTP-induced Parkinson's disease in mice. Neuroscience 277 (September), 747–754. https://doi.org/10.1016/j.neuroscience.2014.07.046.

Pérez-Revuelta, B.I., Hettich, M., Ciociaro, A., Rotermund, C., Kahle, P.J., Krauss, S., Di Monte, D.A., 2014. Metformin lowers ser-129 phosphorylated α-synuclein levels via mTOR-dependent protein phosphatase 2A activation. Cell Death Dis. 5 (January), e1209. https://doi.org/10.1038/cddis.2014.175.

Perry, T.A., Haughey, N.J., Mattson, M.P., Egan, J.M., Greig, N.H., 2002a. Protection and reversal of excitotoxic neuronal damage by glucagon-like peptide-1 and exendin-4. J. Pharmacol. Exp. Ther. 302 (3), 881–888. https://doi.org/10.1124/jpet.102.037481.

Perry, T.A., Lahiri, D.K., Chen, D., Zhou, J., Shaw, K.T.Y., Egan, J.M., Greig, N.H., 2002b. A novel neurotrophic property of glucagon-like peptide 1: a promoter of nerve growth factor-mediated differentiation in PC12 cells. J. Pharmacol. Exp. Ther. 300 (3), 958–966.

Pintana, H., Pongkan, W., Pratchayasakul, W., Chattipakorn, N., Chattipakorn, S.C., 2015. Dipeptidyl peptidase 4 inhibitor improves brain insulin sensitivity, but fails to prevent cognitive impairment in orchiectomy obese rats. J. Endocrinol. 226 (2), M1–11. https://doi.org/10.1530/JOE-15-0099.

Pipatpiboon, N., Pintana, H., Pratchayasakul, W., Chattipakorn, N., Chattipakorn, S.C., 2013. DPP4-inhibitor improves neuronal insulin receptor function, brain mitochondrial function and cognitive function in rats with insulin resistance induced by high-fat diet consumption. Eur. J. Neurosci. 37 (5), 839–849. https://doi.org/10.1111/ejn.12088.

Quinn, L.P., Crook, B., Hows, M.E., Vidgeon-Hart, M., Chapman, H., Upton, N., Medhurst, A.D., Virley, D.J., 2008. The PPARgamma agonist pioglitazone is effective in the MPTP mouse model of Parkinson's disease through inhibition of monoamine oxidase B. Br. J. Pharmacol. 154 (1), 226–233. https://doi.org/10.1038/bjp.2008.78.

Reger, M.A., Watson, G.S., Green, P.S., Wilkinson, C.W., Baker, L.D., Cholerton, B., Fishel, M.A., et al., 2008a. Intranasal insulin improves cognition and modulates beta-amyloid in early AD. Neurology 70 (6), 440–448. https://doi.org/10.1212/01.WNL. 0000265401.62434.36.

Reger, M.A., Watson, G.S., Green, P.S., Baker, L.D., Cholerton, B., Fishel, M.A., Plymate, S.R., et al., 2008b. Intranasal insulin administration dose-dependently modulates verbal memory and plasma amyloid-beta in memory-impaired older adults. J. Alzheimers Dis. 13 (3), 323–331.

Ribeiro, C.A., Silva, A.M., Viana, S.D., Pereira, F.C., 2012. Sitagliptin does not protect against MPTP-induced dopaminergic striatal toxicity. In: 6th European Congress of Pharmacology (EPHAR 2012).

Rotermund, C., Machetanz, G., Fitzgerald, J.C., 2018. The therapeutic potential of metformin in neurodegenerative diseases. Front. Endocrinol. 9, 400. https://doi.org/10.3389/fendo. 2018.00400.

Sa-Nguanmoo, P., Tanajak, P., Kerdphoo, S., Jaiwongkam, T., Pratchayasakul, W., Chattipakorn, N., Chattipakorn, S.C., 2017. SGLT2-inhibitor and DPP-4 inhibitor improve brain function via attenuating mitochondrial dysfunction, insulin resistance, inflammation, and apoptosis in HFD-induced obese rats. Toxicol. Appl. Pharmacol. 333 (October), 43–50. https://doi.org/10.1016/J.TAAP.2017.08.005.

Sankar, R., Thamotharan, S., Shin, D., Moley, K.H., Devaskar, S.U., 2002. Insulin-responsive glucose transporters-GLUT8 and GLUT4 are expressed in the developing mammalian brain. Brain Res. Mol. Brain Res. 107 (2), 157–165.

Santiago, J.A., Potashkin, J.A., 2013. Integrative network analysis unveils convergent molecular pathways in Parkinson's disease and diabetes. PLoS One 8 (12), e83940. https://doi.org/10.1371/journal.pone.0083940.

Santiago, J.A., Potashkin, J.A., 2014. System-based approaches to decode the molecular links in Parkinson's disease and diabetes. Neurobiol. Dis. 72 (Pt. A), 84–91. https://doi.org/ 10.1016/j.nbd.2014.03.019.

Schintu, N., Frau, L., Ibba, M., Caboni, P., Garau, A., Carboni, E., Carta, A.R., 2009. PPAR-gamma-mediated neuroprotection in a chronic mouse model of Parkinson's disease. Eur. J. Neurosci. 29 (5), 954–963. https://doi.org/10.1111/j.1460-9568.2009. 06657.

Sekar, S., Taghibiglou, C., 2018. Elevated nuclear phosphatase and tensin homolog (PTEN) and altered insulin Signaling in substantia nigral region of patients with Parkinson's disease. Neurosci. Lett. 666 (February), 139–143. https://doi.org/10.1016/j.neulet. 2017.12.049.

Seufert, J., Gallwitz, B., 2014. The extra-pancreatic effects of GLP-1 receptor agonists: a focus on the cardiovascular, gastrointestinal and central nervous systems. Diabetes Obes. Metab. 16 (8), 673–688. https://doi.org/10.1111/dom.12251.

Sharma, M.K., Jalewa, J., Hölscher, C., 2014. Neuroprotective and anti-apoptotic effects of liraglutide on SH-SY5Y cells exposed to methylglyoxal stress. J. Neurochem. 128 (3), 459–471. https://doi.org/10.1111/jnc.12469.

Shi, Q., Liu, S., Fonseca, V.A., Thethi, T.K., Shi, L., 2019. Effect of metformin on neurodegenerative disease among elderly adult US veterans with type 2 diabetes mellitus. BMJ Open 9 (7), e024954. https://doi.org/10.1136/BMJOPEN-2018-024954.

Shiraishi, D., Fujiwara, Y., Komohara, Y., Mizuta, H., Takeya, M., 2012. Glucagon-like peptide-1 (GLP-1) induces M2 polarization of human macrophages via STAT3 activation. Biochem. Biophys. Res. Commun. 425 (2), 304–308. https://doi.org/10.1016/j.bbrc.2012.07.086.

Simuni, T., Caspell-Garcia, C., Coffey, C.S., Weintraub, D., Mollenhauer, B., Lasch, S., Tanner, C.M., et al., 2018. baseline prevalence and longitudinal evolution of non-motor symptoms in early Parkinson's disease: the PPMI cohort. J. Neurol. Neurosurg. Psychiatry 89 (1), 78–88. https://doi.org/10.1136/jnnp-2017–316213.

Srinivas, N.R., 2015. Linagliptin-role in the reversal of Aβ-mediated impairment of insulin signaling and reduced neurotoxicity in AD pathogenesis: some considerations. CNS Neurosci. Ther. 21 (12), 962–963. https://doi.org/10.1111/cns.12475.

Svenningsson, P., Wirdefeldt, K., Yin, L., Fang, F., Markaki, I., Efendic, S., Ludvigsson, J.F., 2016. Reduced incidence of Parkinson's disease after dipeptidyl peptidase-4 inhibitors-a nationwide case-control study. Mov. Disord. 31, 1422–1423. https://doi.org/10.1002/mds.26734.

Swanson, C., Emborg, M., 2014. Expression of peroxisome proliferator-activated receptor-gamma in the substantia nigra of hemiparkinsonian nonhuman primates. Neurol. Res. 36 (7), 634–646. https://doi.org/10.1179/1743132813Y.0000000305.

Swanson, C.R., Joers, V., Bondarenko, V., Brunner, K., Simmons, H.A., Ziegler, T.E., Kemnitz, J.W., Johnson, J.A., Emborg, M.E., 2011. The PPAR-γ agonist pioglitazone modulates inflammation and induces neuroprotection in Parkinsonian monkeys. J. Neuroinflammation 8 (January), 91. https://doi.org/10.1186/1742-2094-8-91.

Takahashi, M., Yamada, T., Tooyama, I., Moroo, I., Kimura, H., Yamamoto, T., Okada, H., 1996. Insulin receptor mRNA in the substantia nigra in Parkinson's disease. Neurosci. Lett. 204 (3), 201–204. https://doi.org/10.1016/0304-3940(96)12357-0.

Talbot, K., Wang, H.-Y., Kazi, H., Han, L.-Y., Bakshi, K.P., Stucky, A., Fuino, R.L., et al., 2012. Demonstrated brain insulin resistance in Alzheimer's disease patients is associated with IGF-1 resistance, IRS-1 dysregulation, and cognitive decline. J. Clin. Investig. 122 (4), 1316–1338. https://doi.org/10.1172/JCI59903.

Tayara, K., Espinosa-Oliva, A.M., García-Domínguez, I., Ismaiel, A.A., Boza-Serrano, A., Deierborg, T., Machado, A., Herrera, A.J., Venero, J.L., de Pablos, R.M., 2018. Divergent effects of metformin on an inflammatory model of Parkinson's disease. Front. Cell. Neurosci. 12, 440. https://doi.org/10.3389/FNCEL.2018.00440.

Thornberry, N.A., Gallwitz, B., 2009. Mechanism of action of inhibitors of dipeptidyl-peptidase-4 (DPP-4). Best Pract. Res. Clin. Endocrinol. Metab. 23 (4), 479–486. https://doi.org/10.1016/j.beem.2009.03.004.

Turnes, J.d.M., Bassani, T.B., Souza, L.C., Vital, M.A.B.F., 2018. Ineffectiveness of saxagliptin as a neuroprotective drug in 6-OHDA-lesioned rats. J. Pharm. Pharmacol. 70 (8), 1059–1068. https://doi.org/10.1111/jphp.12936.

Vicente Miranda, H., El-Agnaf, O.M.A., Outeiro, T.F., 2016. Glycation in Parkinson's disease and Alzheimer's disease. Mov. Disord. 31, 782–790. https://doi.org/10.1002/mds.26566.

Vicente Miranda, H., Szegő, É.M., Oliveira, L.M.A., Breda, C., Darendelioglu, E., de Oliveira, R.M., Ferreira, D.G., et al., 2017. Glycation potentiates α-synuclein-associated neurodegeneration in synucleinopathies. Brain 140 (5), 1399–1419. https://doi.org/10.1093/brain/awx056.

Videira, P.A.Q., Castro-Caldas, M., 2018. Linking glycation and glycosylation with inflammation and mitochondrial dysfunction in Parkinson's disease. Front. Neurosci. 12 (June), 381. https://doi.org/10.3389/fnins.2018.00381.

Wahlqvist, M.L., Lee, M.-S., Hsu, C.-C., Chuang, S.-Y., Lee, J.-T., Tsai, H.-N., 2012. Metformin-inclusive sulfonylurea therapy reduces the risk of Parkinson's disease occurring with type 2 diabetes in a Taiwanese population cohort. Parkinsonism Relat. Disord. 18 (6), 753–758. https://doi.org/10.1016/j.parkreldis.2012.03.010.

Werther, G.A., Hogg, A., Oldfield, B.J., Mckinley, M.J., Figdor, R., Allen, A.M., Mendelsohn, F.A.O., 1987. Localization and characterization of insulin receptors in rat brain and pituitary gland using *in vitro* autoradiography and computerized densitometry. Endocrinology 121 (4), 1562–1570. https://doi.org/10.1210/endo-121-4-1562.

Wu, H.-F., Kao, L.-T., Shih, J.-H., Kao, H.-H., Chou, Y.-C., Li, I.-H., Kao, S., 2018. Pioglitazone use and Parkinson's disease: a retrospective cohort study in Taiwan. BMJ Open 8 (8), e023302. https://doi.org/10.1136/bmjopen-2018-023302.

Yamamoto, H., Lee, C.E., Marcus, J.N., Williams, T.D., Overton, J.M., Lopez, M.E., Hollenberg, A.N., et al., 2002. Glucagon-like peptide-1 receptor stimulation increases blood pressure and heart rate and activates autonomic regulatory neurons. J. Clin. Invest. 110 (1), 43–52. https://doi.org/10.1172/JCI15595.

Yuan, Z., Li, D., Feng, P., Xue, G., Ji, C., Li, G., Hölscher, C., 2017. A novel GLP-1/GIP dual agonist is more effective than liraglutide in reducing inflammation and enhancing GDNF release in the MPTP mouse model of Parkinson's disease. Eur. J. Pharmacol. 812 (October), 82–90. https://doi.org/10.1016/j.ejphar.2017.06.029.

Yue, X., Li, H., Yan, H., Zhang, P., Chang, L., Li, T., 2016. Risk of Parkinson disease in diabetes mellitus. Medicine 95 (18). e3549. https://doi.org/10.1097/MD.0000000000003549.

Yun, S.P., Kam, T.-I., Panicker, N., Kim, S.M., Yumin, O., Park, J.-S., Kwon, S.-H., et al., 2018. Block of A1 astrocyte conversion by microglia is neuroprotective in models of Parkinson's disease. Nat. Med. 24, 931–938. https://doi.org/10.1038/s41591-018-0051-5.

Zhang, L., Zhang, L., Li, L., Hölscher, C., 2018. Neuroprotective effects of the novel GLP-1 long acting analogue semaglutide in the MPTP Parkinson's disease mouse model. Neuropeptides 71 (October), 70–80. https://doi.org/10.1016/j.npep.2018.07.003.

Basal ganglia oscillations as biomarkers for targeting circuit dysfunction in Parkinson's disease

14

Per Petersson[a,b,*], Andrea A. Kühn[c], Wolf-Julian Neumann[c], Romulo Fuentes[d]

[a]*Umeå University, Umeå, Sweden*
[b]*Lund University, Lund, Sweden*
[c]*Charité—Universitätsmedizin Berlin, Berlin, Germany*
[d]*Universidad de Chile, Santiago, Chile*
*Corresponding author: Tel.: +46-72-2526072, e-mail address: per.petersson@umu.se

Abstract

Oscillations are a naturally occurring phenomenon in highly interconnected dynamical systems. However, it is thought that excessive synchronized oscillations in brain circuits can be detrimental for many brain functions by disrupting neuronal information processing. Because synchronized basal ganglia oscillations are a hallmark of Parkinson's disease (PD), it has been suggested that aberrant rhythmic activity associated with symptoms of the disease could be used as a physiological biomarker to guide pharmacological and electrical neuromodulatory interventions. We here briefly review the various manifestations of basal ganglia oscillations observed in human subjects and in animal models of PD. In this context, we also review the evidence supporting a pathophysiological role of different oscillations for the suppression of voluntary movements as well as for the induction of excessive motor activity. In light of these findings, it is discussed how oscillations could be used to guide a more precise targeting of dysfunctional circuits to obtain improved symptomatic treatment of PD.

Keywords

Hypokinesia, Hyperkinesia, Levodopa, Non-motor symptoms, LFP, Neuronal circuits, Neurophysiology

Progress in Brain Research, Volume 252, ISSN 0079-6123, https://doi.org/10.1016/bs.pbr.2020.02.002

1 Introduction

1.1 Why circuit level pathophysiological processes are important to consider in PD

Many of the symptoms of PD can be primarily attributed to the extensive neurodegeneration of the nigrostriatal pathways, as observed for the first time already in 1919 by Tretiakoff (reviewed, e.g., by Fahn, 2003). However, while striatum is considered to be the main neuronal target of dopaminergic projections, the complex projection patterns arising from midbrain dopaminergic cell groups (A8–A10; Björklund and Dunnett, 2007; German et al., 1989; Prensa and Parent, 2001; Smith and Kieval, 2000), suggest that several other structures are also affected. This includes, for example, thalamus (Freeman et al., 2001; Sanchez-Gonzalez et al., 2005), pallidum (Lindvall and Björklund, 1979), subthalamic nucleus (STN; Meibach and Katzman, 1979), substantia nigra (Cheramy et al., 1981) and cerebral cortex (Berger et al., 1986; Brown et al., 1979; Gaspar et al., 1989, see also Smith and Villalba, 2008). Moreover, the abnormal neural activity patterns that comprise the pathophysiological features of the disease may arise partly as a consequence of maladaptive circuit plasticity in the remaining neuronal circuitry, developing secondary to a decreased dopaminergic tone (Picconi et al., 2012; Villalba and Smith, 2018). Thus, to get a deeper understanding of the core pathophysiology of the disease, the various adaptations taking place in different brain structures, and the mechanisms of action of the existing therapies—including both electrical neuromodulation and pharmacotherapy—it is crucial to investigate not only the direct consequences of midbrain dopaminergic cell-loss, but also the secondary circuit-level mechanisms causing the abnormal activity that gives rise to symptoms.

A complicating factor for our understanding of PD pathophysiology is that several conceptual models exist, each emphasizing different physiological features out of a multifaceted pattern of altered neuronal activity in different parts of the cortico-basal ganglia-thalamic circuits. For example, while experimental evidence exists for the specific pattern of changes in firing rates of neurons in the indirect and direct basal ganglia pathways, as would be predicted by the classical model presented by Albin et al. (1989) and DeLong (1990), it is clear that this model cannot fully explain other experimental findings (Nambu et al., 2015). Similarly, evidence has also been presented for an alternative pathophysiological model emphasizing neuronal synchrony and excessive oscillatory activity (Hammond et al., 2007). But also in this case, it remains unclear if the observed neuronal oscillations are in fact a key pathophysiological mechanistic component or merely an epiphenomenon that arises as a consequence of maladaptive network changes. In the current review, we primarily focus on basal ganglia oscillations as potential biomarkers to guide the targeting of circuit dysfunctions (Fig. 1; for a more extensive discussion on the role of basal ganglia oscillations in PD and other neurologic and psychiatric conditions, see e.g., Halje et al., 2019). Finally, from a methodological perspective, it should be noted that the signal constituting the oscillatory phenomena discussed herein principally is

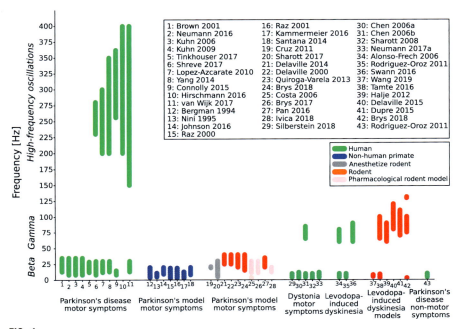

FIG. 1

Summary of articles cited in this review demonstrating oscillatory activity in specific frequency bands associated with PD signs that might serve as biomarkers to guide therapeutic interventions. The bars represent the frequency range described in each reference (numbers on x-axis indicate citation as can be found in the inset). The leftmost group corresponds to data collected from patients with Parkinson's disease while in off-medication (in cases where two frequency ranges are given, high-frequency oscillations display amplitude coupling to the phase of the slower beta oscillation). The second and third groups correspond to non-human primate and rodent PD models, respectively, where increased beta oscillations have been associated with the parkinsonian condition, as indicated on a behavioral level by brady- and hypokinesia. Note that the frequency ranges associated with dystonia are below beta range, but may also include gamma oscillations. Levodopa-induced dyskinesia humans and in PD models include two components (typically in theta range and in a narrow part of the high-gamma band, respectively). Finally, one report is included associating low-frequency oscillations with impulse control deficiency in PD patients.

derived from the low-frequency component (<400 Hz) of extracellular recordings obtained from electrodes located inside the brain parenchyma—that is, the local field potential (LFP). The LFP is thought to mainly reflect the local synchronization of dendritic currents induced by excitatory and inhibitory synaptic inputs to the population of cells surrounding each recording electrode (Buzsáki et al., 2012; Lindén et al., 2010).

1.2 The emergence and spreading of synchronized oscillatory activity in cortico-basal ganglia-thalamic structures in PD

Oscillations at different scales are known to emerge under a range of different circumstances in highly interconnected networks, for example, in association with state transitions or external perturbations. Depending on the relative dampening of different frequencies, each network will have a propensity to sustain oscillations at different frequencies (Jenkins, 2013). In neuronal circuits, physiological features of the cells making up the network will together determine the oscillatory properties (Singer, 2018; Wang, 2010). Hence, it is expected that the long-term dopaminergic denervation associated with PD will cause neurochemical, physiological and structural changes that in turn affect the intrinsic oscillation properties and the frequency response of neuronal networks in different parts of the cortico-basal ganglia-thalamic circuits (Cenci et al., 2018; Richter et al., 2013). The factors that contribute to the emergence of oscillatory activity are therefore likely to involve mechanisms at several different levels. For example, changes in the excitability of certain groups of neurons due to altered expression of voltage sensitive or shunting ion channels, or other similar changes that affect intrinsic membrane properties (Bevan et al., 2007), changes in the relative strength of synaptic connectivity (Chu et al., 2015; Fieblinger et al., 2014) or changes in electrical coupling mediated via gap junctions (Hjorth et al., 2009; Phookan et al., 2015; Schwab et al., 2014). Similarly, intrinsically in the network, the interaction between principal cells and interneurons or the balance between excitatory and inhibitory activity could lead to changes in network frequency tuning properties (Gittis and Kreitzer, 2012; Gittis et al., 2011; Lindenbach et al., 2016).

While several factors may contribute to the generation of oscillatory activity, changes in connection properties between different brain structures can, in turn, facilitate the further propagation of this emerging oscillatory activity across widely distributed brain circuits, such as within the cortico-basal ganglia-thalamic loop (Belić et al., 2016; Herz et al., 2014; Petersson et al., 2019; Santana et al., 2014). In the classical view of information flow through the cortico-basal ganglia brain circuits, patterned activity in one structure will directly influence the next downstream brain structure via synaptic excitation or inhibition (Albin et al., 1989; DeLong, 1990). In this way, oscillatory activity can be passed on from one structure to the next in a stepwise chain of events. This box-and-arrow flowchart model of the cortico-basal ganglia circuits is however complicated by, among other things, the more extensive anatomical network containing multiple reciprocal connections that has been revealed in later studies (Bar-Gad and Bergman, 2001). An example of reciprocal connections of potential physiological relevance is the recurrently connected network of excitatory and inhibitory neurons in the STN and GPe, which has been suggested to serve as a core element in the generation of basal ganglia oscillations (Bevan et al., 2002; Plenz and Kital, 1999). While interaction between these nuclei alone may not be sufficient to generate basal ganglia oscillations in PD (Loucif et al., 2005; Wilson et al., 2006), it is possible that synchronous cortical input to these two nuclei, can induce oscillatory activity via mechanisms involving recip-rocal interaction between the nuclei (Baufreton et al., 2005; Holgado et al., 2010; Richter et al., 2013; Terman et al., 2002; Tseng et al., 2001).

Taken together, altered physiological coupling may arise by different mecha-nisms and can have a synchronizing role on a systems level, by coupling of indepen-dently oscillating cell groups in different structures (Cumin and Unsworth, 2006; Restrepo et al., 2006).

Finally, it should be noted that both cortex and the basal ganglia nuclei are under the strong influence of thalamus. Therefore, rhythmic thalamic activity, or any other type of activity pattern that facilitate downstream oscillatory activity, has the poten-tial to induce oscillations in several parts of the circuit via first-order synaptic connections (Cole et al., 2017; Reis et al., 2019; Sherman et al., 2016).

1.3 The value of translational neurophysiological biomarkers for improved treatment of PD

Deep brain stimulation (DBS) is an effective treatment alternative for patients with medically intractable movement disorders, such as PD (Deuschl et al., 2006; Schuepbache et al., 2013; Weaver et al., 2009), dystonia (Kupsch et al., 2006; Vidailhet et al., 2005, 2007; Volkmann et al., 2014) and essential tremor (Benabid et al., 1991; Flora et al., 2010; Kumar et al., 2003; Schuurman et al., 2000). In addition to its clinical efficacy, functional neurosurgery for DBS-electrode implantation provides unprecedented access to neurophysiological characterization of human neural population activity from basal ganglia nuclei residing deep in the brain (Kühn and Volkmann, 2017; Neumann et al., 2019). This has allowed the interrogation of oscillatory network dynamics of basal ganglia circuit activity in behaving human patients that has led to the characterization of pathological and nor-mal physiological patterns of basal ganglia activation. The resulting findings have in turn significantly contributed to our current understanding of the complex mecha-nisms of action of high-frequency neuromodulation in the basal ganglia cortical cir-cuit. It is now clear that DBS related modulation of neural populations leads to neural activity changes from the synapse to the systems levels (Kühn and Volkmann, 2017). For example, locally DBS suppresses cell firing through synaptic depletion and changes in short-term plasticity through facilitation and inhibition differentially depending on neurotransmitter systems (Milosevic et al., 2018a,b; Steiner et al., 2019). On the network level, synchronization patterns in oscillatory circuits are mod-ulated with a net desynchronizing effect on the major circuit hallmark of motor sys-tem physiology: the beta rhythm (Cagnan et al., 2019b). As will be discussed in more detail below, the withdrawal of therapeutic dopaminergic medication in PD patients leads to rhythmic discharges of neural ensembles in the subthalamic nucleus and internal pallidum in the beta frequency band (13–35 Hz) (Brown et al., 2001). Administration of a fast acting dopaminergic agent reverses this discharge pattern, which is distinct from the activity recorded in patients with idiopathic dystonia (Silberstein et al., 2003). Basal ganglia nuclei in patients with dystonia showed stronger low-frequency activity (3–12 Hz) when compared to PD patients (Silberstein et al., 2003). These early findings led to the definition of disease specific oscillatory patterns and the rise of an entire research field dedicated to the analysis of the significance of basal ganglia oscillations. Today, the analysis of neural population activity in relation to neuromodulatory interventions makes it possible to: (a) describe physiological circuit mechanisms of neural communication in health and neurological and psychiatric disease, and (b) identify therapeutic avenues for the development of next-generation neurotechnological interventions (Neumann et al., 2019). This latter aspects will be discussed in further detail below, in the context of adaptive DBS (Little et al., 2013; Rosin et al., 2011).

Translational physiological biomarkers may become useful in a similar way for the characterization of pharmacological effects in novel treatment approaches. Linking pharmacological effects related to the stimulation or depression of different modulatory neurotransmitter systems to distinct changes in neurophysiological activity patterns at the network level is, however, a daunting task. Not the least because the induced effects are normally highly dependent on the current state of the network, which typically changes substantially over time. Nevertheless, it is increasingly being recognized that a deeper mechanistic understanding of both the pathophysiology causing symptoms and physiological action of different pharmacological interventions to treat CNS disease could provide essential clues for the development of improved therapies. In human subjects, physiological features associated with disease are generally limited to data obtained using non-invasive techniques (such as functional magnetic resonance imaging [fMRI] or electroencephalogram [EEG]). However, in the case of PD, recordings performed in association with DBS treatment gives access also to intracranial neuronal recordings. Hence, it should be possible to identify specific translational physiological biomarkers associated with both the therapeutic effects of pharmacotherapy as well as various side effects. So far this possibility has not been extensively explored, but it is highly likely that physiological biomarkers will have an important role to guide the targeting of circuit dysfunctions in PD via pharmacotherapy in the future (Petersson et al., 2019).

2 Oscillations associated with hypokinesia

2.1 Recordings in PD patients

Neurophysiological studies on the beta rhythm in Parkinson's disease are still carving out the pathological and functional role of this rhythmic activity in relation to specific symptoms. Beta activity was shown to correlate directly with parkinsonian motor signs as measured clinically with the unified Parkinson's disease rating scale, part III; UPDRS-III (Neumann et al., 2016a). Moreover, suppression of beta activity through dopaminergic medication (Kühn et al., 2006, 2009) and DBS (Kühn et al., 2008; Oswal et al., 2016) was shown to correlate with the amount of symptom alleviation through the respective therapeutic intervention. Importantly, these correlations rely on the presence of hypokinetic/rigid symptoms as beta activity does not reflect tremor severity (Neumann and Kühn, 2017), which in itself can suppress beta oscillations (Hirschmann et al., 2016). Similarly, voluntary movement is known to suppress beta activity in the basal ganglia. Nevertheless, even during movement, higher amplitudes of beta activity could be demonstrated to reflect the concurrent presence of bradykinetic motor signs (Steiner et al., 2017). The beta activity associated with bradykinesia exhibits an inverse relation to higher frequency activity in the gamma frequency range (60–80 Hz) that was shown to correlate directly with movement velocity and parallels the decline thereof in the dopaminergic OFF state (Lofredi et al., 2018).

Recent studies that have focused on the temporal dynamics of pathological oscillations in PD demonstrate that beta activity occurs in brief bursts of 100–1000 ms lengths (Tinkhauser et al., 2017b). These analyses propose that burst duration rather than amplitude reflects the hypokinetic state (Tinkhauser et al., 2018), and found that beta burst durations are higher in PD when compared to dystonic patients (Lofredi et al., 2019a) and also present during motor execution (Lofredi et al., 2019b).

Toward the other edge of the time scale, beta activity has also been characterized in terms of its longevity, as all the studies mentioned above only recorded LFP in the acute DBS implantation stage, without long-term follow-up. Here, technical advances in the design of implantable pulse generators for therapeutic DBS have now enabled the recording of beta activity in the long-term, and the first studies have demonstrated that elevated beta activity can be reliably detected even months after implantation (Neumann et al., 2016b, 2017b; Quinn et al., 2015; Trager et al., 2016).

In conclusion, clinical and experimental research from invasive neural recordings in PD patients have established rhythmic basal ganglia activity characteristics associated with bradykinesia which may be utilized to improve neuromodulatory therapy through technical innovations in DBS methodology (Kühn and Volkmann, 2017).

2.2 Recordings in primate models of PD

One of the earlier pieces of evidence presented for the oscillatory nature of neuronal activity changes associated with the parkinsonian state came from non-human primate models. These studies were originally aimed to test the firing rate changes during the parkinsonian state, as predicted by the Albin-Young hypothesis, for example, that the tonic firing rate of STN neurons would be increased. Yet, while firing rates indeed changed after induction of a parkinsonian condition, another perhaps more striking change was observed: an increase in oscillatory activity, both at unitary and population level. In the first such experiments, performed in grivet monkeys treated with MPTP (1-methyl-4-phenyl-1,2,3,6-tetrahydropyridine) to induce parkinsonism, 16% of the recorded STN neurons displayed rhythmic bursts at 4–8 Hz, while 25% and 13% of the globus pallidus pars interna (GPi) neurons display rhythmic bursts at 4–8 Hz and 8–20 Hz, respectively (Bergman et al., 1994; Wichmann et al., 1994). Similarly, rhesus monkeys, treated with MPTP until they displayed parkinsonian motor signs, had 35% of their pallidal neurons oscillating with burst at 5–11 Hz, and in the same frequency range, 19% of them displayed phase-locked oscillations with other neurons of the same area (Nini et al., 1995). In the same rhesus model, LFPs also presents powerful oscillations with a peak around 14 Hz (Johnson et al., 2016). Along the same lines, later studies in MPTP-treated grivet monkeys showed an increase in the proportion of globus pallidus pars externa (GPe) and GPi neurons with significant oscillatory activity in the range 3–19 Hz (GPe: from 11% to 39%; GPi: from 3% to 43%, Raz et al., 2000). MPTP treatment has been reported to cause an increase from 9% to 59% of the pairs of putamen-pallidal neurons firing together in a periodic oscillatory way in the frequency range 3–19 Hz (Raz et al., 2001). Beside pallidum, STN, and striatum, other structures of the cortico-basal ganglia-thalamic circuit have been found to present strong oscillatory low-frequency activity in the parkinsonian state. In the ventral thalamus receiving input from basal ganglia, MPTP-treated rhesus monkeys show increased oscillatory power at 3–13 Hz in the spiking activity (Kammermeier et al., 2016).

In the much smaller primate, the common marmoset, rendered parkinsonian by injections of 6-hydroxydopamine (6-OHDA) into the medial forebrain bundle, chronic recordings from several structures in cortex, the basal ganglia and thalamus, in parallel, have revealed synchronous distinct low-frequency oscillations with peak frequencies ranging between 10 and 20 Hz associated with hypokinesia (Santana et al., 2014). These oscillations were found in the LFPs recorded in the different structures but were also distinguishable in rhythmic firing patterns of single units (Santana et al., 2014).

2.3 Recordings in rodent models of PD

Non-human primates have undoubtedly provided vital information on basal ganglia pathophysiology in PD (Wichmann et al., 2018), but, for the vast majority of preclinical research studies, rodent PD models have been the preferred choice—in particular, the 6-OHDA lesioned rat. However, while rodent models have been important for the elucidation of pathophysiological mechanisms, certain differences appear to exist between rodents and primates, including the detailed oscillation patterns observed in parkinsonian animals. Thus, to allow for comparisons between monophyletic groups of species, the corresponding phenomena needs to be identified. An additional complication is that oscillatory patterns are highly dependent on brain state. Consequently, specific information on the alertness and activity state of the experimental animal is crucial to any interpretation of oscillatory phenomena and for direct comparisons between different experiments or between species. For example, while studies in anesthetized rats typically have reported oscillatory activity in the lower beta band and an associated phase-locking of action potential firing (Cagnan et al., 2019a; Cruz et al., 2011; Magill et al., 2000; Sharott et al., 2017), the most robust PD-associated oscillation in awake, behaving rats is normally found in the higher end of the beta band (25–40 Hz; Delaville et al., 2014, 2015). It remains to be clarified if these two oscillations should both be considered rodent counterparts of the human beta, which in the awake state typically is found in the lower beta range but which varies greatly between individuals (with peak frequencies 8–35 Hz; Kühn et al., 2009).

Finally, it can be noted that, in rodent PD models, pronounced beta oscillations are generally only discernible during states of severe dopaminergic depletion (Costa et al., 2006), and not during milder or moderate depletion (Brys et al., 2017; Quiroga-Varela et al., 2013). Moreover, akinesia/bradykinesia can be pharmacologically induced without simultaneously increased beta oscillations (see, for example, Ivica et al., 2018; Pan et al., 2016), suggesting that parkinsonian beta oscillations in rodents are at least not directly causing these motor impairments.

2.4 Beta oscillations in neuronal networks and the role abnormal connectivity

The exact mechanisms underlying beta oscillations are unknown, and to identify specific cellular components generating oscillatory patterns at a network level is notoriously difficult. In the case of beta oscillations, not the least since practically all the structures making up the cortico-basal ganglia-thalamic loop are to some extent affected by the substantial loss of midbrain dopaminergic innervation in PD. This could result in several different types of alterations of connections strengths and resonance properties both within and between different brain structures (Rommelfanger and Wichmann, 2010). Changes in functional connectivity may therefore develop both within and between structures. However, given the severely reduced dopaminergic innervation density of the striatum associated with PD, striatal mechanisms leading to altered connectivity have in particular been implicated. For example, in mice exposed to chronic dopamine depletion, an enhanced connectivity between striatal fast-spiking interneurons and D2 type medium spiny neurons (MSNs) has been found, tending to increase the synchrony between MSNs of the indirect pathway (Gittis et al., 2011). Dopamine is also known to affect cholinergic intrastriatal signaling, which can have secondary effects; for example, acute pharmacological manipulations of the cholinergic striatal interneuronal network has been shown to induce striatal beta oscillations on a population level (McCarthy et al., 2011).

The role of cells in the indirect pathway has also been studied in relation to cortical patterning of striatal activity that in turn affect basal ganglia signal processing on a circuit level. In specific, cells in the indirect pathway of the basal ganglia have been shown to be particularly easily entrained to oscillatory activity (Sharott et al., 2017). Such synchronized oscillatory activity involving the indirect pathway could help facilitate oscillations downstream of the striatum. Indeed, both STN and GPe have been shown to fire action potentials coherently with cortical slow oscillations in dopamine depleted states (Magill et al., 2000). In addition, reciprocal connections between GPe and STN, which have been found to be functionally and structurally potentiated after depletion of DA (Cruz et al., 2011; Fan et al., 2012), could help maintain and even amplify such rhythmic network activity (Bevan et al., 2002; Plenz and Kital, 1999).

Further evidence for a disturbed network connectivity in PD stems from computer modeling. For instance, dynamic causal modeling of the propagation of low-frequency oscillations have shown that the STN could be a major network component that helps promoting beta synchrony arising from cortical patterning of basal ganglia activity (Marreiros et al., 2013), which is also in agreement with simulations (Holgado et al., 2010; Pavlides et al., 2015; Terman et al., 2002). This kind of synchrony-driven coupling has also been observed in PD patients, specifically between parietal-frontal cortices and the subthalamic region, and is enhanced in the beta band during the resting state (with indications from partial directed coherence analyses that cortex is primarily leading STN; Litvak et al., 2011).

Finally, as previously discussed, it is possible that beta oscillations do not have a single origin within the cortico-basal ganglia-thalamic circuit (Brittain and Brown, 2014). This notion would be in line with some recent computational modeling studies (Liu et al., 2017), suggesting that abnormal beta band activity in PD emerge in the lower part of the band (12–20 Hz) due to dysfunctional coupling in the striatum-GPe-STN network, and that these oscillations under certain circumstances can be driven to higher frequencies (21–35 Hz) via cortical afferents (Liu et al., 2017).

Importantly, direct evidence for an association between signs of hypokinesia and aberrant functional connectivity on the circuit level has been established in a primate model of PD (Santana et al., 2014). By recording neuronal activity from a large number of structures throughout the cortico-basal ganglia-thalamic circuit in parallel, a significantly increased functional connectivity linked to oscillatory activity in the beta range was revealed. This aberrant network connectivity was effectively reversed by levodopa and with neuromodulatory anti-parkinsonian treatment (Santana et al., 2014). This suggests, that the pathological effect of increased beta oscillations might arise not from its mere occurrence in local brain structures, but because the complete circuit, comprised by motor cortices, basal ganglia, and thalamus, is engaged in the same rhythmic activity, thus preventing local neuronal operations to initiate and maintain motor activity. Elaborating further, in terms of therapeutic intervention, any perturbation to this global synchronized activity, either by pharmacologically altering the properties of local networks (like replacing dopamine in the caudate-putamen), or through the input of neuromodulating currents that are not coherent with the dominant rhythm (like deep brain or spinal cord stimulation), will have the effect of interrupting the global synchrony, and thus allowing the circuit to functionally improve.

2.5 Utilizing beta activity for target selection and programming in DBS

Improvement in DBS-electrode localization methodology in human patients, using combined analysis of preoperative and postoperative CT in standardized MNI space, as has been semi-automatically implemented in the toolbox "Lead-DBS," has significantly extended our understanding of DBS circuit effects (de Almeida Marcelino et al., 2019; Horn et al., 2019; Neumann et al., 2018b). In combination with neurophysiological recordings, it allows the characterization of neural population activity in 3D atlas space. By mapping beta oscillatory activity in PD patients to a 3D representation of the subthalamic nucleus, it was demonstrated that beta activity spatially peaks in the dorsolateral STN, which is also known to be the optimal therapeutic target (Horn et al., 2017). This spatial overlap, together with the finding that peak beta activity can predict optimal stimulation contact (Chen et al., 2006c; Pogosyan et al., 2010), allows use of the physiological information for refinement of intraoperative targeting and postoperative contact selection. The latter will become increasingly relevant with the addition of multiple spatial degrees of freedom through growing numbers of stimulation contacts in modern electrode designs. For example, using directional leads, it has now be shown that beta activity can aid the correct identification of therapeutically efficacious stimulation contacts through beta activity amplitude measurements (Tinkhauser et al., 2018). This could counteract the time and financial burden of the standardized monopolar reprogramming with multi-contact electrodes in the clinical routine, but implementation ultimately depends on DBS manufacturers who have not included this functionality to current implantable devices.

3 Oscillations related to hyperkinesia

3.1 Oscillations associated with dystonia

Early LFP studies have utilized the comparison of basal ganglia activity from PD patients against patients with idiopathic dystonia to demonstrate disease specificity. Compared to PD, patients with dystonia exhibited elevated levels of activity in the theta-alpha/low-frequency range (Silberstein et al., 2003) that was shown to correlate and drive EMG discharge (Chen and Brown, 2007; Chen et al., 2006a,b; Sharott et al., 2008). Recently, in a large cohort of patients with cervical dystonia it could be shown that oscillatory activity in the low-frequency range correlates directly with dystonic symptom severity (measured with Toronto Western Spasmodic Torticollis Rating Scale, TWSTRS) (Neumann et al., 2017a). Analysis of recording locations again showed a clear association of low-frequency amplitude with proximity to optimal DBS target location, and the same study showed a significant correlation between low-frequency activity and clinical outcome after long-term DBS. This hints toward a similar clinical utility for DBS target selection and programming as described above for beta activity. Using a combination of LFP-MEG another study has analyzed the circuit connectivity of pallidal oscillations with cortex. Here, it was found that the basal ganglia exhibit theta (4–8 Hz) coupling with temporal areas and beta (13–30 Hz) coupling with motor cortex. Moreover, a distinct oscillatory network with central cerebellar nuclei was identified for the alpha range (7–13 Hz) (Neumann et al., 2015). Importantly, the amount of coupling was negatively correlated with dystonic symptom severity in these patients, which extended previous findings from diffusion MRI based structural connectivity measures to the functional oscillatory domain (Kühn and Volkmann, 2017).

While most of the previously referenced studies focused on across-cohort comparisons and correlations, for the first time, a recent study has utilized the opportunity to record from sensing enabled implantable pulse generators in dystonia. Here, recordings were performed repeatedly after cessation of long-term DBS (up to 5 h) with parallel documentation of clinical symptom severity (Scheller et al., 2019). The study shows that dystonic symptom severity is correlated with low-frequency activity within subjects across different recording time points, and highlights the utility of low-frequency activity as a real-time biomarker for dystonia. The symptom specificity of this biomarker can, however, be questioned by the resemblance of pallidal and thalamic activity in other hyperkinetic movement disorders (Priori et al., 2013). Here, a systematic account on pallidal and thalamic oscillations in patients with Tourette's syndrome reported presence of low-frequency activity in the internal pallidum and centromedian/parafascicular nuclei of the intralaminar thalamus (Neumann et al., 2018a). Thus, low-frequency activity may be a hallmark of the hyperkinetic state that may facilitate voluntary and involuntary movement in PD (Alonso-Frech et al., 2006), dystonia (Neumann et al., 2017a) and Tourette's syndrome (Neumann et al., 2018a) alike. Similarly, PD and essential tremor shows robust activity in the tremor frequency range (3–6 Hz) that can be recorded from the basal ganglia, thalamus and cortex and can be used for the classification and detection of the presence of tremulous movements (Hirschmann et al., 2013, 2017; Pedrosa et al., 2014, 2018).

3.2 Oscillations associated with dyskinesia

When it comes to hyperkinesia induced by dopaminergic pharmacotherapy or by excessive neuromodulatory electrical stimulation, neuronal recordings obtained from cortico-basal ganglia structures in PD patients following DBS-electrode implantation have indicated two separate frequency bands. First, the wide theta band (4–10 Hz) was implicated in a seminal study by Alonso-Frech and co-workers, where STN LFP-activity was recorded OFF and ON levodopa in 11 dyskinetic patients (Alonso-Frech et al., 2006). These findings were corroborated in a later study, pointing to a specific involvement of the dorsal aspect of the STN in hyperkinesia reflected by exaggerated oscillatory activity in this region, mainly in the higher end of this frequency band (Rodriguez-Oroz et al., 2011).

Rodent models of levodopa-induced dyskinesia (LID) have subsequently confirmed the presence of theta oscillations in LFPs recorded throughout the cortico-basal ganglia-thalamic loop in the dyskinetic state (Tamtè et al., 2016; Wang et al., 2019). It should be noted, however, that in the unilateral 6-OHDA rat model of LID, a relative increase in LFP theta power has been observed concomitantly in both hemispheres even though abnormal involuntary movements in this model are principally limited to muscle groups contralateral to the lesioned hemisphere

(Tamtè et al., 2016). It is therefore possible that the relative increase in theta power is partly related to a general increase in motor activity rather than to dyskinetic motor signs, per se.

The second frequency range that has been associated with hyperkinesia is within the gamma band, in a relatively narrow frequency window around 70–90 Hz. Also for these oscillations, electrophysiological recordings obtained during the early postoperative period in patients being implanted with DBS electrodes provided the first pieces of evidence. However, initial studies suggested that this activity, which was recorded with deep electrodes in STN and GPi (and with parallel EEG-recordings in some patients), was primarily related to the therapeutic pro-kinetic effect of levodopa rather than to dyskinetic symptoms (Alonso-Frech et al., 2006; Brown et al., 2001; Cassidy et al., 2002; Fogelson et al., 2005; Williams et al., 2002). This notion was further strengthened by the observation that gamma band oscillations increased in power in conjunction with voluntary movements (Alegre et al., 2005; Cassidy et al., 2002; Lalo et al., 2008; Litvak et al., 2012). On the other hand, it was also noted that these high-frequency oscillations were often most evident during dyskinesia (Fogelson et al., 2005). Thus, while a role of narrowband gamma oscillations in the generation of LID could not be ruled out based on these findings, it was proposed that a relative increase in high-gamma power in these deep basal ganglia nuclei is primarily associated with increased motor activity and/or a state of arousal that may help enabling motor activity (reviewed by Jenkinson et al., 2013).

However, subsequent experiments in animal models of LID presented complementary, and to some extent conflicting, evidence on the association between narrowband gamma oscillations and dyskinetic motor signs. Based on experiments using microwire electrodes to obtain long-term recordings from motor cortex and the dorsal striatum in unilaterally 6-hydroxydopamine (6-OHDA) lesioned rats, a direct pathological role of 60–90 Hz oscillations in the motor cortex was instead proposed (Halje et al., 2012). These authors found distinct 80 Hz oscillations that were only present in the lesioned hemisphere and during LID. They could also demonstrate that the topical application of a dopamine type 1 receptor (D1R) antagonist onto the cortical surface was sufficient to break the oscillation and concomitantly suppress signs of dyskinesia. Hence, in this PD model, persistent narrowband gamma oscillations in cortical structures are strongly linked to dyskinesia.

Importantly, the findings by Halje et al. were later confirmed in PD patients, as a result of the first long-term (~1 year) recordings performed in dyskinetic patients using a combined deep brain stimulation and electrocorticogram (DBS-ECoG) device (Swann et al., 2016). The use of chronically implantable bidirectional electrodes may have been critical for this discovery, as this type of long-term recordings can help circumvent experimental limitations associated with the early postoperative phase following DBS-electrode implantation. Especially, during a period following electrode implantation, symptoms are often significantly reduced (i.e., even when no current is passed through the stimulation electrode)—indicating that symptomatic relief in this early phase primarily arises as a consequence of the lesion inflicted by the electrode (Groiss et al., 2009). In the study by Swann et al. neuronal activity was recorded over motor cortical areas and in the STN for 12 months. Through this experimental design the authors could demonstrate that dyskinesia in PD patients is indeed tightly linked to the same type of cortical oscillations that are observed in the rat model of PD/LID. Furthermore, a more detailed analysis of the high-gamma oscillations in the STN clearly suggested that this narrowband activity is in fact predominantly pathological rather than pro-kinetic, as the oscillations were found to be minimally affected by voluntary movements while their presence proved to be a very reliable biomarker of dyskinesia.

Since the original finding by Halje et al., the role of narrowband gamma oscillations in rodent models of LID has been further explored (Belić et al., 2016; Delaville et al., 2014; Dupre et al., 2015; Tamtè et al., 2016). In the study by Dupre and colleagues, the presence and relative power of gamma oscillations was assessed together with dyskinetic symptoms induced by daily levodopa administration during a 1-week priming period (Dupre et al., 2015), and oscillations were found to become more distinct as LID became gradually more severe. The authors also showed that the oscillation can be induced in L-DOPA-primed rats by independently activating either D1 or D2 dopamine receptors with more selective dopamine agonists (Dupre et al., 2015). In addition, that rat model has been used to investigate which brain structures display simultaneous narrowband gamma oscillations. These studies have demonstrated that oscillations are particularly powerful in the LFPs recorded in corticostriatal circuits, but are also found in both the rodent globus pallidus (corresponding to the external pallidal segment in primates) and in motor nuclei of the thalamus, but appear to be less pronounced in the STN compared to in patients (Brys et al., 2018; Swann et al., 2016; Tamtè et al., 2016; although, in this context it is important to acknowledge that precise targeting of the STN is experimentally challenging in the rat brain—leaving the possibility that this apparent species difference could be partly due to experimental limitations of the rodent model).

Taken together, neurophysiological data from animal models and PD patients appear to be very well aligned and jointly point to oscillations in the two frequency bands: 4–10 Hz and 70–90 Hz, as robust electrophysiological biomarkers that could be used to classify the dyskinetic state, for example, for the purpose of functionally adapting characteristics of electrical neuromodulation. Indeed, to directly test the usefulness of different electrophysiological biomarkers for classification of PD and LID states, Tamtè and co-authors quantitatively compared different spectral components in eight different brain structures in parallel in rats. As expected, classification performance was found to improve steadily with inclusion of a broader spectral range and the addition of several brain structures. But, in particular, oscillations around 80 Hz in the rostral forelimb area (a premotor/supplementary motor area in rodents) were found to be a useful physiological marker of LID (Tamtè et al., 2016). This finding is in agreement with ECoG recordings carried out by Swann et al. in PD patients (Swann et al., 2016). This type of reliable translational physiological biomarkers opens up for more rapid development of new therapies. Indeed, in the long-term study by Swann and colleagues, the authors present data that point to how problems with stimulation-induced dyskinesia could potentially be overcome by feedback control of the stimulator based on ECoG/STN on-line recordings (Swann et al., 2016). The usefulness of this approach was subsequently confirmed in two patients, where significant energy savings could be achieved using adaptive DBS based on these oscillations (Swann et al., 2018; see below).

Narrowband gamma oscillations consequently serves as a good example of a biomarker used for targeting circuit dysfunction in Parkinson's disease. Importantly, however, it remains to be explored if these aberrant activity patterns are also causally linked to dyskinesia (for a review, see Petersson et al., 2019).

4 Oscillations at very high frequencies (>100 Hz)

Oscillations in the recorded field potentials at very high frequencies (100–500 Hz) have been observed in the basal ganglia of PD patients and in animal models of PD; often referred to as high-frequency oscillations (HFOs; Connolly et al., 2015; Foffani et al., 2003; Hirschmann et al., 2016; van Wijk et al., 2017; Wang et al., 2014; Yang et al., 2014). On a technical note, it should be cautioned that separating LFPs from spiking activity can be technically challenging for very high frequencies but methods have been developed that can help preventing the leakage of action potential into the field potential signal (see, e.g., Wang et al., 2014).

However, while beta on the one hand, and theta and narrowband gamma oscillations on the other have been directly associated with signs of hypo- and hyperkinesia, respectively, HFOs have been more diffusely linked to motor signs in PD and seem to be modulated by behavioral state.

In PD patients, intraoperative STN LFPs recordings have suggested that HFOs (200–400 Hz) may occur in a large fraction of patients (e.g., HFOs were found in 41 out of 47 recorded STNs in the study by Shreve et al. (2017)). Moreover, phase-amplitude coupling between concurrent HFOs and beta oscillations was detected in >98% of cases and was stronger in the more affected hemisphere (Shreve et al., 2017). Other authors have reported HFOs in the subthalamic nucleus of PD patients both OFF and ON levodopa. In the ON state, this high-frequency activity is modulated by movement, and in the OFF state the beta (12–30 Hz) phase is coupled to the amplitude of HFOs (Lopez-Azcarate et al., 2010). Finally, HFOs have also been associated with resting tremor (with modulation within the tremor cycle; Hirschmann et al., 2016).

These reports, together with findings that beta-gamma phase-amplitude coupling in motor cortex correlates with motor symptom severity and is reduced by DBS (de Hemptinne et al., 2015; Shimamoto et al., 2013), provide support for HFOs and beta-HFO phase-amplitude coupling (as well as the underlying spiking activity) as biomarkers that could potentially be used to target circuit dysfunctions (Meidahl et al., 2019).

Surprisingly, even though several studies have reported HFOs in recordings from PD patients, so far these oscillations have not been systematically investigated in animal models of the disease.

However, a recent rodent study reported fast gamma oscillations with a peak around 130 Hz in cortico-basal ganglia-thalamic circuits of 6-OHDA lesioned rats (Brys et al., 2018). These oscillations were phase-amplitude coupled to slow delta waves (2–3 Hz), in a similar way as the human beta-HFO coupling has previously been described. Levodopa treatment in these animals, resulting in levodopa-induced dyskinesia, lowered the oscillation peak-frequency while increasing the power and cross-frequency coupling to slow oscillations. Consequently, while the role of HFOs and the phase-amplitude coupling to lower oscillations remains largely unknown, this phenomenon could potentially serve as a translational biomarker and deserves to be further explored.

5 Oscillations associated with non-motor symptoms

It is increasingly being recognized that non-motor symptoms constitute a significant complication in PD. Disturbed sleep patterns and mood disorders including anxiety and depression are common complications affecting about half of all PD patients, which may develop even before the manifestation of the cardinal motor symptoms (Faivre et al., 2019; Pontone et al., 2009; Reijnders et al., 2008). Even more troublesome non-motor complications relating to cognitive decline are unfortunately also very common, in particular during later phases of the disease (Hely et al., 2008). Moreover, persistent psychotic symptoms are known to develop in a large fraction of patients, in particular in older PD patients and often in association with cognitive deterioration (Cenci and Odin, 2009). Symptoms may include, for example, hallucinations, delusion and excitement, and are thought to be induced or exacerbated by L-DOPA and other dopaminergic treatments for PD (Fénelon and Alves, 2010). Other non-motor complications, which have also been chiefly attributed to dopaminergic pharmacotherapy, are impulse control disorders—that is, the failure to resist an impulse or temptation to control an act or specific behavior, which is ultimately harmful to oneself or others and interferes in major areas of the patient's life (Gatto and Aldinio, 2019).

Interestingly, in the same way that specific brain activity states have been associated with motor symptoms in PD/LID, certain electrophysiological features have been linked to various non-motor symptoms. For example, while 4–10 Hz oscillations have been found in the more dorsal part of the STN in recordings from patients with dyskinesia, PD patients that display impulse control deficiencies have instead been reported to show excessive theta oscillations in the ventral, more limbic part of the STN, and preferentially in the lower part of the 4–10 Hz frequency band (Rodriguez-Oroz et al., 2011). Thus, it is possible that very similar oscillatory phenomena, which involve closely situated neuronal populations in different subregions of the STN, can give rise to vastly different types of symptoms, being manifested either in the motor domain or in the cognitive/limbic domain.

For obvious reasons, this category of symptoms are challenging to investigate in animal models of PD. For PD-psychosis, non-human primates have offered the best possibly to model the disease and a neuropsychiatric rating scale has been developed which includes behavior abnormalities in the domains of hyperkinesia, repetitive grooming, responses to non-apparent stimuli and stereotypies (Fox et al., 2010). It remains to be established what neurophysiological features are associated with these aberrant behaviors.

On the other hand, outside the context of PD, HFOs have been observed in several species and has been linked to cognitive and limbic processes in rodents (Tort et al., 2013). Moreover, a growing number of studies have investigated the link between psychosis-like states induced in rats by psychotomimetic drugs (typically NMDA antagonists or $5HT_{2A}$ agonists). Interestingly, although acting via different receptors, both NMDA antagonists (e.g., PCP, MK-801 or ketamine), and $5HT_{2A}$ receptor agonists (e.g., LSD or DOI) dramatically increase HFOs at 130–150 Hz in cognitive/limbic circuits (Goda et al., 2013; Hunt and Kasicki, 2013) but also in basal ganglia structures (Cordon et al., 2015).

Thus, while the physiological underpinning of non-motor symptoms in PD largely remains to be explored, and reliable translational biomarkers are currently lacking, it can nevertheless be speculated that the 130 Hz oscillations coupled to slow delta waves (2–3 Hz) reported by Brys et al. (2018), represents a pathophysiological phenomenon related primarily to non-motor rather than motor symptoms, given the great similarity to the activity patterns observed after treatment with psychotomimetic compounds.

6 Oscillations as biomarkers for targeting circuit dysfunction
6.1 Oscillations as biomarkers in neuromodulation

The symptom and disease specificity has inspired the idea of closed-loop adaptive deep brain stimulation (aDBS) that adapts stimulation parameters to the concurrent clinical demand, as estimated based on a neurophysiological feedback signal. This was first tested in a non-human primate model of PD, using motor cortex and globus pallidus (GPi) recordings to adapt stimulation of GPi (Rosin et al., 2011). This approach is now widely investigated by international research teams with the aim to improve therapeutic outcome in DBS patients (Arlotti et al., 2018; Hebb et al., 2014; Herron et al., 2017; Hoang et al., 2017; Malekmohammadi et al., 2016; Meidahl et al., 2017; Molina et al., 2018; Piña-Fuentes et al., 2017; Rosa et al., 2017; Ryapolova-Webb et al., 2014). In a seminal paper by Little et al. (2013), beta activity recorded in the STN was used to activate stimulation whenever a threshold of activity as an index for therapeutic demand was crossed (Little et al., 2013). The results were promising with improved efficacy and a better side-effect profile when compared to conventional chronic and random burst stimulation (Little et al., 2013, 2016a,b). The beneficial effect of this stimulation has later been attributed to a truncation of particularly long beta bursts (Tinkhauser et al., 2017a). In parallel, the first results from portable devices were published, allowing the first interrogation of effects in freely moving patients (Rosa et al., 2015, 2017). This has led to a recent clinical study reporting long-term adaptive stimulation over 8 h (Arlotti et al., 2018), with therapeutic success and considerably lower energy delivered to the patients. Unfortunately, the study did not compare results against chronic non-adaptive DBS. Also in PD, a first study using electrocorticography has reported an aDBS paradigm in two parkinsonian patients (Swann et al., 2018). Interestingly, contrary to the other studies mentioned above, the main ECoG signature was instead gamma activity, which was shown to reflect a dyskinetic state, thus increases in cortical gamma was used to indicate excessive stimulation. This has led to significant energy savings of 38–45% with stable clinical response (Swann et al., 2018). The optimal adaptive stimulation strategies are still under debate and first studies are exploring more complex algorithms, e.g., balancing the delivered current between two defined amplitudes using upper and lower biomarker thresholds (Velisar et al., 2019). Unfortunately, few studies compare their effects against optimal chronic stimulation to benchmark the true additional benefit of adaptive DBS. This may reflect a narrow window for optimization of an already highly efficacious therapy. Further innovations based on artificial intelligence and concurrent behavioral decoding using multichannel recordings in freely moving patients may in the future elevate the therapeutic potential for neuromodulation by allowing specific and comprehensive DBS parameter adaptations to the patients' needs during walking, talking or resting (Neumann et al., 2019).

A recently introduced prospective complement to DBS is electrical stimulation of the spinal cord (SCS). Currently used SCS devices consist of cylindrical or flat electrodes with several contacts distributed over the epidural membrane facing the dorsal surface of the spinal cord. The electrodes typically deliver chronic biphasic square current pulses at fixed frequency. While this mode of stimulation likely affects local spinal circuits to some extent, it stands clear that ascending somatosensory pathways of the dorsal columns-medial lemniscal system are effectively activated and thereby modulate neuronal activity in, among other structures, the ventral posterior lateral thalamic nucleus and the primary somatosensory cortex (Holsheimer and Buitenweg, 2015; Molnar and Barolat, 2014; Santana et al., 2014). Thus, from a mechanistic standpoint, SCS could potentially interrupt excessive synchronization and oscillatory activity in sensorimotor circuits in the PD brain and thereby restore neuronal activity to a state closer to normal physiology (Fuentes et al., 2010). The first demonstrations of the efficacy of spinal cord stimulation for treating motor symptoms of PD came from studies in rodent and non-human primate models of the disease (Fuentes et al., 2009; Santana et al., 2014). Subjects treated with SCS presented substantial improvements in akinesia and bradykinesia and a concomitant change in LFP oscillatory patterns of cortico-basal ganglia structures (Fuentes et al., 2009; Santana et al., 2014). It was noted that the SCS-induced shift from low- to high-frequency LFP power closely resembled the state-shift taking place prior to spontaneous initiation of locomotion. Thus, it appears that SCS interferes with the hypokinetic state, characterized by synchronized low-frequency oscillations, and thereby creates the conditions needed for initiation of motor actions. Indeed, in the primate PD study, it was shown that the rapid motor improvements induced by SCS were associated with a disruption of coherent beta oscillations on a global scale (Santana et al., 2014).

Following these promising results in animal models of PD, single case studies and clinical trials have reported that SCS applied to PD patients effectively alleviates a range of different symptoms, including tremor, rigidity, body bradykinesia, campto-cormia and freezing, as well as inducing improvements of gait and posture, quality of life, mood and locomotion speed (Agari and Date, 2012; Akiyama et al., 2017; de Souza et al., 2018; Fénelon et al., 2011; Hassan et al., 2013; Hubsch et al., 2019; Kobayashi et al., 2018; Landi et al., 2012; Nishioka and Nakajima, 2015; Samotus et al., 2018). While these findings need to be followed up in larger random-ized double-blinded studies, it clearly indicates a potential use for SCS as a novel therapy in PD. To date, we are however still lacking information on the effects of SCS on neuronal oscillatory patterns in humans. Interestingly, SCS appears to be par-ticularly effective for the treatment of axial symptoms in PD (de Souza et al., 2018; Samotus et al., 2018), which are difficult to treat with DBS and conventional phar-macotherapy. It is therefore an interesting possibility to combine these two neuro-modulatory interventions. Under such conditions, intracranial neurophysiological measurements could permit intelligent adaptive designs to be applied to both types of stimulation in parallel.

6.2 Oscillations as biomarkers in drug development

To clarify the pathophysiological mechanisms that give rise to various symptoms in PD is for obvious reasons a highly prioritized research aim. However, even without a full understanding of the disease causing mechanisms, the oscillatory phenomena reviewed herein could potentially be used as biomarkers of specific pathological brain states for which different novel pharmacological treatments can be evaluated and benchmarked against other compounds. In particular, because several of the discussed phenomena are very similar in PD patients and in animal models of disease, the effects produced by an experimental therapeutic intervention is more likely to be translationally relevant.

Electrophysiological biomarkers have so far not been extensively used in drug development but it stands clear that this type of information can complement data obtained from neurochemical, histological and behavioral experiments in impor-tant ways. This has, for example, been illustrated in studies aimed at evaluating putative antidyskinetic compounds in hemiparkinsonian 6-OHDA lesioned rats (Brys et al., 2018; Tamtè et al., 2016). Perhaps more importantly, with respect to non-motor symptoms, a neurophysiological readout of the effects induced by experimental pharmacological treatments offers a unique opportunity to evaluate new neuropsychiatric drug candidates in animals. Indeed, a similar approach has been used in the development of new treatments for schizophrenia, where, for example, genetically modified mice carrying human mutations associated with increased risk of schizophrenia are assessed in translationally relevant testing paradigms applying auditory stimuli known to produce deviating EEG response patterns in schizophrenic patients (Kwon et al., 1999; Thelin et al., 2016). It is not yet known if similar experimental paradigms can be applied in animal models of PD-psychosis, but if successful, the combination of spontaneously emerging oscillatory phenomena and specific patterns of evoked activity based on translationally relevant testing paradigms could, no doubt, become an invaluable tool for the targeting of circuit dysfunction in PD.

7 Conclusions

Taken together, while the pathophysiological role of the different oscillatory phenomena that we have here reviewed remains to be clarified, accumulating evidence clearly indicates that basal ganglia oscillations are very informative biomarkers for more precise targeting of circuit dysfunction in PD. To date, this realization has primarily made an impact in the context of improving DBS therapy but, as discussed herein, a growing body of data point to a much wider applicability in the treatment of PD—including both the design of alternative strategies for electrical neuromodulation and the development and subsequent fine-tuning of pharmacological treatments.

References

Agari, T., Date, I., 2012. Spinal cord stimulation for the treatment of abnormal posture and gait disorder in patients with Parkinson's disease. Neurol. Med. Chir. (Tokyo) 52, 470–474.

Akiyama, H., Nukui, S., Akamatu, M., Hasegawa, Y., Nishikido, O., Inoue, S., 2017. Effectiveness of spinal cord stimulation for painful camptocormia with Pisa syndrome in Parkinson's disease: a case report. BMC Neurol. 17, 148.

Albin, R.L., Young, A.B., Penney, J.B., 1989. The functional anatomy of basal ganglia disorders. Trends Neurosci. 12, 366–375.

Alegre, M., Alonso-Frech, F., Rodríguez-Oroz, M.C., Guridi, J., Zamarbide, I., Valencia, M., Manrique, M., Obeso, J.A., Artieda, J., 2005. Movement-related changes in oscillatory activity in the human subthalamic nucleus: ipsilateral vs. contralateral movements. Eur. J. Neurosci. 22, 2315–2324.

Alonso-Frech, F., Zamarbide, I., Alegre, M., Rodríguez-Oroz, M.C., Guridi, J., Manrique, M., Valencia, M., Artieda, J., Obeso, J.A., 2006. Slow oscillatory activity and levodopa-induced dyskinesias in Parkinson's disease. Brain 129, 1748–1757.

Arlotti, M., Marceglia, S., Foffani, G., Volkmann, J., Lozano, A.M., Moro, E., Cogiamanian, F., Prenassi, M., Bocci, T., Cortese, F., Rampini, P., Barbieri, S., Priori, A., 2018. Eight-hours adaptive deep brain stimulation in patients with Parkinson disease. Neurology 90, e971–e976.

Bar-Gad, I., Bergman, H., 2001. Stepping out of the box: information processing in the neural networks of the basal ganglia. Curr. Opin. Neurobiol. 11, 689–695.

Baufreton, J., Atherton, J.F., Surmeier, D.J., Bevan, M.D., 2005. Enhancement of excitatory synaptic integration by GABAergic inhibition in the subthalamic nucleus. J. Neurosci. 25, 8505–8517.

Belić, J.J., Halje, P., Richter, U., Petersson, P., Hellgren, K.J., 2016. Untangling cortico-striatal connectivity and cross-frequency coupling in L-DOPA-induced dyskinesia. Front. Syst. Neurosci. 10, 26.

Benabid, A.L., Pollak, P., Hoffmann, D., Gervason, C., Hommel, M., Perret, J.E., de Rougemont, J., Gao, D.M., 1991. Long-term suppression of tremor by chronic stimulation of the ventral intermediate thalamic nucleus. Lancet 337, 403–406.

Berger, B., Trottier, S., Gaspar, P., Verney, C., Alvarez, C., 1986. Major dopamine innervation of the cortical motor areas in the cynomolgus monkey. A radioautographic study with comparative assessment of serotonergic afferents. Neurosci. Lett. 72, 121–127.

Bergman, H., Wichmann, T., Karmon, B., DeLong, M.R., 1994. The primate subthalamic nucleus. II. Neuronal activity in the MPTP model of parkinsonism. J. Neurophysiol. 72, 507–520.

Bevan, M.D., Magill, P.J., Terman, D., Bolam, J.P., Wilson, C.J., 2002. Move to the rhythm: oscillations in the subthalamic nucleus-external globus pallidus network. Trends Neurosci. 25, 525–531.

Bevan, M.D., Hallworth, N.E., Baufreton, J., 2007. GABAergic control of the subthalamic nucleus. Prog. Brain Res. 160, 173–188.

Björklund, A., Dunnett, S.B., 2007. Dopamine neuron systems in the brain: an update. Trends Neurosci. 30, 194–202.

Brittain, J.S., Brown, P., 2014. Oscillations and the basal ganglia: motor control and beyond. Neuroimage 85, 637–647.

Brown, R.M., Crane, A.M., Goldman, P.S., 1979. Regional distribution of monoamines in the cerebral cortex and subcortical structures of the rhesus monkey: concentrations and in vivo synthesis rates. Brain Res. 168, 133–150.

Brown, P., Oliviero, A., Mazzone, P., Insola, A., Tonali, P., Di Lazzaro, V., 2001. Dopamine dependency of oscillations between subthalamic nucleus and pallidum in Parkinson's disease. J. Neurosci. 21, 1033–1038.

Brys, I., Nunes, J., Fuentes, R., 2017. Motor deficits and beta oscillations are dissociable in an alpha-synuclein model of Parkinson's disease. Eur. J. Neurosci. 46, 1906–1917.

Brys, I., Halje, P., Scheffer-Teixeira, R., Varney, M., Newman-Tancredi, A., Petersson, P., 2018. Neurophysiological effects in cortico-basal ganglia-thalamic circuits of antidyskinetic treatment with 5-HT 1A receptor biased agonists. Exp. Neurol. 302, 155–168.

Buzsáki, G., Anastassiou, C.A., Koch, C., 2012. The origin of extracellular fields and currents—EEG, ECoG, LFP and spikes. Nat. Rev. Neurosci. 13, 407–420.

Cagnan, H., Denison, T., McIntyre, C., Brown, P., 2019a. Emerging technologies for improved deep brain stimulation. Nat. Biotechnol. 37, 1024–1033.

Cagnan, H., Mallet, N., Moll, C.K.E., Gulberti, A., Holt, A.B., Westphal, M., Gerloff, C., Engel, A.K., Hamel, W., Magill, P.J., Brown, P., Sharott, A., 2019b. Temporal evolution of beta bursts in the parkinsonian cortical and basal ganglia network. Proc. Natl. Acad. Sci. U. S. A. 116, 16095–16104.

Cassidy, M., Mazzone, P., Oliviero, A., Insola, A., Tonali, P., Di Lazzaro, V., Brown, P., 2002. Movement-related changes in synchronization in the human basal ganglia. Brain 125, 1235–1246.

Cenci, M.A., Odin, P., 2009. Dopamine replacement therapy in Parkinson's disease: past, present and future. In: Tseng, K.-Y. (Ed.), Cortico-Subcortical Dynamics in Parkinsons Disease. Humana Press, Springer, pp. 309–334.

Cenci, M.A., Jörntell, H., Petersson, P., 2018. On the neuronal circuitry mediating L-DOPA-induced dyskinesia. J. Neural Transm. 125, 1157–1169.

Chen, C.-C., Brown, P., 2007. The role of synchronised low frequency activity in globus pallidus interna in dystonia. Acta Neurol. Taiwan. 16, 1–6.

Chen, C., Kuhn, A., Trottenberg, T., Kupsch, A., Schneider, G., Brown, P., 2006a. Neuronal activity in globus pallidus interna can be synchronized to local field potential activity over 3–12 Hz in patients with dystonia. Exp. Neurol. 202, 480–486.

Chen, C.C., Kühn, A.A., Hoffmann, K.-T., Kupsch, A., Schneider, G.-H., Trottenberg, T., Krauss, J.K., Wöhrle, J.C., Bardinet, E., Yelnik, J., Brown, P., 2006b. Oscillatory pallidal local field potential activity correlates with involuntary EMG in dystonia. Neurology 66, 418–420.

Chen, C.C., Pogosyan, A., Zrinzo, L.U., Tisch, S., Limousin, P., Ashkan, K., Yousry, T., Hariz, M.I., Brown, P., 2006c. Intra-operative recordings of local field potentials can help localize the subthalamic nucleus in Parkinson's disease surgery. Exp. Neurol. 198, 214–221.

Cheramy, A., Leviel, V., Glowinski, J., 1981. Dendritic release of dopamine in the substantia nigra. Nature 289, 537–542.

Chu, H.-Y., Atherton, J.F., Wokosin, D., Surmeier, D.J., Bevan, M.D., 2015. Heterosynaptic regulation of external globus pallidus inputs to the subthalamic nucleus by the motor cortex. Neuron 85, 364–376.

Cole, S.R., van der Meij, R., Peterson, E.J., de Hemptinne, C., Starr, P.A., Voytek, B., 2017. Nonsinusoidal beta oscillations reflect cortical pathophysiology in Parkinson's disease. J. Neurosci. 37, 4830–4840.

Connolly, A.T., Jensen, A.L., Bello, E.M., Netoff, T.I., Baker, K.B., Johnson, M.D., Vitek, J.L., 2015. Modulations in oscillatory frequency and coupling in globus pallidus with increasing parkinsonian severity. J. Neurosci. 35, 6231–6240.

Cordon, I., Nicolás, M.J., Arrieta, S., Lopetegui, E., López-Azcárate, J., Alegre, M., Artieda, J., Valencia, M., 2015. Coupling in the cortico-basal ganglia circuit is aberrant in the ketamine model of schizophrenia. Eur. Neuropsychopharmacol. 25, 1375–1387.

Costa, R.M., Lin, S.C., Sotnikova, T.D., Cyr, M., Gainetdinov, R.R., Caron, M.G., Nicolelis, M.A.L., 2006. Rapid alterations in corticostriatal ensemble coordination during acute dopamine-dependent motor dysfunction. Neuron 52, 359–369.

Cruz, A.V., Mallet, N., Magill, P.J., Brown, P., Averbeck, B.B., 2011. Effects of dopamine depletion on information flow between the subthalamic nucleus and external globus pallidus. J. Neurophysiol. 106, 2012–2023.

Cumin, D., Unsworth, C.P., 2006. Generalising the Kuramoto model for the study of neuronal synchronisation in the brain. Phys. D Nonlinear Phenom. 226, 181–196.

de Almeida Marcelino, A.L., Horn, A., Krause, P., Kühn, A.A., Neumann, W.-J., 2019. Subthalamic neuromodulation improves short-term motor learning in Parkinson's disease. Brain 142, 2198–2206.

de Hemptinne, C., Swann, N.C., Ostrem, J.L., Ryapolova-Webb, E.S., San Luciano, M., Galifianakis, N.B., Starr, P.A., 2015. Therapeutic deep brain stimulation reduces cortical phase-amplitude coupling in Parkinson's disease. Nat. Neurosci. 18, 779–786.

de Souza, C.P., dos Santos, M.G.G., Hamani, C., Fonoff, E.T., 2018. Spinal cord stimulation for gait dysfunction in Parkinson's disease: essential questions to discuss. Mov. Disord. 33, 1828–1829.

Delaville, C., Cruz, A.V., McCoy, A.J., Brazhnik, E., Avila, I., Novikov, N., Walters, J.R., 2014. Oscillatory activity in basal ganglia and motor cortex in an awake behaving rodent model of Parkinson's disease. Basal Ganglia 3, 221–227.

Delaville, C., McCoy, A.J., Gerber, C.M., Cruz, A.V., Walters, J.R., 2015. Subthalamic nucleus activity in the awake hemiparkinsonian rat: relationships with motor and cognitive networks. J. Neurosci. 35, 6918–6930.

DeLong, M.R., 1990. Primate models of movement disorders of basal ganglia origin. Trends Neurosci. 13, 281–285.

Deuschl, G., Schade-Brittinger, C., Krack, P., Volkmann, J., Schäfer, H., Bötzel, K., Daniels, C., Deutschländer, A., Dillmann, U., Eisner, W., Gruber, D., Hamel, W., Herzog, J., Hilker, R., Klebe, S., Kloß, M., Koy, J., Krause, M., Kupsch, A., Lorenz, D., Lorenzl, S., Mehdorn, H.M., Moringlane, J.R., Oertel, W., Pinsker, M.O., Reichmann, H., Reuß, A., Schneider, G.-H., Schnitzler, A., Steude, U., Sturm, V., Timmermann, L., Tronnier, V., Trottenberg, T., Wojtecki, L., Wolf, E., Poewe, W., Voges, J., German Parkinson Study Group, Neurostimulation Section, 2006. A randomized trial of deep-brain stimulation for Parkinson's disease. N. Engl. J. Med. 355, 896–908.

Dupre, K.B., Cruz, A.V., McCoy, A.J., Delaville, C., Gerber, C.M., Eyring, K.W., Walters, J.R., 2015. Effects of L-dopa priming on cortical high beta and high gamma oscillatory activity in a rodent model of Parkinson's disease. Neurobiol. Dis. 86, 1–15.

Fahn, S., 2003. Description of Parkinson's disease as a clinical syndrome. Ann. N. Y. Acad. Sci. 991, 1–14.

Faivre, F., Joshi, A., Bezard, E., Barrot, M., 2019. The hidden side of Parkinson's disease: studying pain, anxiety and depression in animal models. Neurosci. Biobehav. Rev. 96, 335–352.

Fan, K.Y., Baufreton, J., Surmeier, D.J., Chan, C.S., Bevan, M.D., 2012. Proliferation of external globus pallidus-subthalamic nucleus synapses following degeneration of midbrain dopamine neurons. J. Neurosci. 32, 13718–13728.

Fénelon, G., Alves, G., 2010. Epidemiology of psychosis in Parkinson's disease. J. Neurol. Sci. 289, 12–17.

Fénelon, G., Goujon, C., Gurruchaga, J.-M., Cesaro, P., Jarraya, B., Palfi, S., Lefaucheur, J.-P., 2011. Spinal cord stimulation for chronic pain improved motor function in a patient with Parkinson's disease. Parkinsonism Relat. Disord. 18, 213–214.

Fieblinger, T., Graves, S.M., Sebel, L.E., Alcacer, C., Plotkin, J.L., Gertler, T.S., Chan, C.S., Heiman, M., Greengard, P., Cenci, M.A., Surmeier, D.J., 2014. Cell type-specific plasticity of striatal projection neurons in parkinsonism and L-DOPA-induced dyskinesia. Nat. Commun. 5, 5316.

Flora, E.D., Perera, C.L., Cameron, A.L., Maddern, G.J., 2010. Deep brain stimulation for essential tremor: a systematic review. Mov. Disord. 25, 1550–1559.

Foffani, G., Priori, A., Egidi, M., Rampini, P., Tamma, F., Caputo, E., Moxon, K.A., Cerutti, S., Barbieri, S., 2003. 300-Hz subthalamic oscillations in Parkinson's disease. Brain 126, 2153–2163.

Fogelson, N., Pogosyan, A., Kühn, A.A., Kupsch, A., van Bruggen, G., Speelman, H., Tijssen, M., Quartarone, A., Insola, A., Mazzone, P., Di Lazzaro, V., Limousin, P., Brown, P., 2005. Reciprocal interactions between oscillatory activities of different frequencies in the subthalamic region of patients with Parkinson's disease. Eur. J. Neurosci. 22, 257–266.

Fox, S.H., Visanji, N., Reyes, G., Huot, P., Gomez-ramirez, J., Johnston, T., Brotchie, J.M., 2010. Neuropsychiatric behaviors in the MPTP marmoset model of Parkinson's disease. Can. J. Neurol. Sci. 37, 86–95.

Freeman, A., Ciliax, B., Bakay, R., Daley, J., Miller, R.D., Keating, G., Levey, A., Rye, D., 2001. Nigrostriatal collaterals to thalamus degenerate in parkinsonian animal models. Ann. Neurol. 50, 321–329.

Fuentes, R., Petersson, P., Siesser, W.B., Caron, M.G., Nicolelis, M.A.L., 2009. Spinal cord stimulation restores locomotion in animal models of Parkinson's disease. Science 323, 1578–1582.

Fuentes, R., Petersson, P., Nicolelis, M.A.L., 2010. Restoration of locomotive function in Parkinson's disease by spinal cord stimulation: mechanistic approach. Eur. J. Neurosci. 32, 1100–1108.

Gaspar, P., Berger, B., Febvret, A., Vigny, A., Henry, J.P., 1989. Catecholamine innervation of the human cerebral cortex as revealed by comparative immunohistochemistry of tyrosine hydroxylase and dopamine-beta-hydroxylase. J. Comp. Neurol. 279, 249–271.

Gatto, E.M., Aldinio, V., 2019. Impulse control disorders in Parkinson's disease. a brief and comprehensive review. Front. Neurol. 10, 351.

German, D.C., Manaye, K., Smith, W.K., Woodward, D.J., Saper, C.B., 1989. Midbrain dopaminergic cell loss in Parkinson's disease: computer visualization. Ann. Neurol. 26, 507–514.

Gittis, A.H., Kreitzer, A.C., 2012. Striatal microcircuitry and movement disorders. Trends Neurosci. 35, 557–564.

Gittis, A.H., Hang, G.B., LaDow, E.S., Shoenfeld, L.R., Atallah, B.V., Finkbeiner, S., Kreitzer, A.C., 2011. Rapid target-specific remodeling of fast-spiking inhibitory circuits after loss of dopamine. Neuron 71, 858–868.

Goda, S.A., Piasecka, J., Olszewski, M., Kasicki, S., Hunt, M.J., 2013. Serotonergic hallucinogens differentially modify gamma and high frequency oscillations in the rat nucleus accumbens. Psychopharmacology (Berl.) 228, 271–282.

Groiss, S.J., Wojtecki, L., Südmeyer, M., Schnitzler, A., 2009. Deep brain stimulation in Parkinson's disease. Ther. Adv. Neurol. Disord. 2, 20–28.

Halje, P., Tamtè, M., Richter, U., Mohammed, M., Cenci, M.A., Petersson, P., 2012. Levodopa-induced dyskinesia is strongly associated with resonant cortical oscillations. J. Neurosci. 32, 16541–16551.

Halje, P., Brys, I., Mariman, J.J., da Cunha, C., Fuentes, R., Petersson, P., 2019. Oscillations in cortico-basal ganglia circuits: implications for Parkinson's disease and other neurologic and psychiatric conditions. J. Neurophysiol. 122, 203–231.

Hammond, C., Bergman, H., Brown, P., 2007. Pathological synchronization in Parkinson's disease: networks, models and treatments. Trends Neurosci. 30, 357–364.

Hassan, S., Amer, S., Alwaki, A., Elborno, A., 2013. A patient with Parkinson's disease benefits from spinal cord stimulation. J. Clin. Neurosci. 20, 1155–1156.

Hebb, A.O., Zhang, J.J., Mahoor, M.H., Tsiokos, C., Matlack, C., Chizeck, H.J., Pouratian, N., 2014. Creating the feedback loop: closed-loop neurostimulation. Neurosurg. Clin. N. Am. 25, 187–204.

Hely, M.A., Reid, W.G.J., Adena, M.A., Halliday, G.M., Morris, J.G.L., 2008. The Sydney multicenter study of Parkinson's disease: the inevitability of dementia at 20 years. Mov. Disord. 23, 837–844.

Herron, J.A., Thompson, M.C., Brown, T., Chizeck, H.J., Ojemann, J.G., Ko, A.L., 2017. Cortical brain–computer interface for closed-loop deep brain stimulation. IEEE Trans. Neural Syst. Rehabil. Eng. 25, 2180–2187.

Herz, D.M., Siebner, H.R., Hulme, O.J., Florin, E., Christensen, M.S., Timmermann, L., 2014. Levodopa reinstates connectivity from prefrontal to premotor cortex during externally paced movement in Parkinson's disease. Neuroimage 90, 15–23.

Hirschmann, J., Hartmann, C.J., Butz, M., Hoogenboom, N., Ozkurt, T.E., Elben, S., Vesper, J., Wojtecki, L., Schnitzler, A., 2013. A direct relationship between oscillatory subthalamic nucleus-cortex coupling and rest tremor in Parkinson's disease. Brain 136, 3659–3670.

Hirschmann, J., Butz, M., Hartmann, C.J., Hoogenboom, N., Özkurt, T.E., Vesper, J., Wojtecki, L., Schnitzler, A., 2016. Parkinsonian rest tremor is associated with modulations of subthalamic high-frequency oscillations. Mov. Disord. 31, 1551–1559.

Hirschmann, J., Schoffelen, J.M., Schnitzler, A., van Gerven, M.A.J., 2017. Parkinsonian rest tremor can be detected accurately based on neuronal oscillations recorded from the subthalamic nucleus. Clin. Neurophysiol. 128, 2029–2036.

Hjorth, J., Blackwell, K.T., Kotaleski, J.H., 2009. Gap junctions between striatal fast-spiking interneurons regulate spiking activity and synchronization as a function of cortical activity. J. Neurosci. 29, 5276–5286.

Hoang, K.B., Cassar, I.R., Grill, W.M., Turner, D.A., 2017. Biomarkers and stimulation algorithms for adaptive brain stimulation. Front. Neurosci. 11, 564.

Holgado, A.J.N., Terry, J.R., Bogacz, R., 2010. Conditions for the generation of beta oscillations in the subthalamic nucleus-globus pallidus network. J. Neurosci. 30, 12340–12352.

Holsheimer, J., Buitenweg, J.R., 2015. Review: bioelectrical mechanisms in spinal cord stimulation. Neuromodulation 18, 161–170.

Horn, A., Neumann, W.-J., Degen, K., Schneider, G.-H., Kühn, A.A., 2017. Toward an electrophysiological "sweet spot" for deep brain stimulation in the subthalamic nucleus. Hum. Brain Mapp. 38, 3377–3390.

Horn, A., Wenzel, G., Irmen, F., Huebl, J., Li, N., Neumann, W.-J., Krause, P., Bohner, G., Scheel, M., Kühn, A.A., 2019. Deep brain stimulation induced normalization of the human functional connectome in Parkinson's disease. Brain 142, 3129–3143.

Hubsch, C., D'Hardemare, V., Ben Maacha, M., Ziegler, M., Patte-Karsenti, N., Thiebaut, J.B., Gout, O., Brandel, J.P., 2019. Tonic spinal cord stimulation as therapeutic option in Parkinson disease with axial symptoms: effects on walking and quality of life. Parkinsonism Relat. Disord. 63, 235–237.

Hunt, M.J., Kasicki, S., 2013. A systematic review of the effects of NMDA receptor antagonists on oscillatory activity recorded in vivo. J. Psychopharmacol. 27, 972–986.

Ivica, N., Richter, U., Sjöbom, J., Brys, I., Tamtè, M., Petersson, P., 2018. Changes in neuronal activity of cortico-basal ganglia-thalamic networks induced by acute dopaminergic manipulations in rats. Eur. J. Neurosci. 47, 236–250.

Jenkins, A., 2013. Self-oscillation. Phys. Rep. 525, 167–222.

Jenkinson, N., Kühn, A.A., Brown, P., 2013. Gamma oscillations in the human basal ganglia. Exp. Neurol. 245, 72–76.

Johnson, L.A., Nebeck, S.D., Muralidharan, A., Johnson, M.D., Baker, K.B., Vitek, J.L., 2016. Closed-loop deep brain stimulation effects on parkinsonian motor symptoms in a nonhuman primate—is beta enough? Brain Stimul. 9, 892–896.

Kammermeier, S., Pittard, D., Hamada, I., Wichmann, T., 2016. Effects of high-frequency stimulation of the internal pallidal segment on neuronal activity in the thalamus in parkinsonian monkeys. J. Neurophysiol. 116, 2869–2881.

Kobayashi, R., Kenji, S., Taketomi, A., Murakami, H., Ono, K., Otake, H., 2018. New mode of burst spinal cord stimulation improved mental status as well as motor function in a patient with Parkinson's disease. Parkinsonism Relat. Disord. 57, 82–83.

Kühn, A.A., Volkmann, J., 2017. Innovations in deep brain stimulation methodology. Mov. Disord. 32, 11–19.

Kühn, A.A., Kupsch, A., Schneider, G.-H., Brown, P., 2006. Reduction in subthalamic 8-35 Hz oscillatory activity correlates with clinical improvement in Parkinson's disease. Eur. J. Neurosci. 23, 1956–1960.

Kühn, A.A., Kempf, F., Brücke, C., Gaynor Doyle, L., Martinez-Torres, I., Pogosyan, A., Trottenberg, T., Kupsch, A., Schneider, G.-H., Hariz, M.I., Vandenberghe, W., Nuttin, B., Brown, P., 2008. High-frequency stimulation of the subthalamic nucleus suppresses oscillatory beta activity in patients with Parkinson's disease in parallel with improvement in motor performance. J. Neurosci. 28, 6165–6173.

Kühn, A.A., Tsui, A., Aziz, T., Ray, N., Brücke, C., Kupsch, A., Schneider, G.-H., Brown, P., 2009. Pathological synchronisation in the subthalamic nucleus of patients with Parkinson's disease relates to both bradykinesia and rigidity. Exp. Neurol. 215, 380–387.

Kumar, R., Lozano, A.M., Sime, E., Lang, A.E., 2003. Long-term follow-up of thalamic deep brain stimulation for essential and parkinsonian tremor. Neurology 61, 1601–1604.

Kupsch, A., Benecke, R., Müller, J., Trottenberg, T., Schneider, G.-H., Poewe, W., Eisner, W., Wolters, A., Müller, J.-U., Deuschl, G., Pinsker, M.O., Skogseid, I.M., Roeste, G.K., Vollmer-Haase, J., Brentrup, A., Krause, M., Tronnier, V., Schnitzler, A., Voges, J., Nikkhah, G., Vesper, J., Naumann, M., Volkmann, J., Deep-Brain Stimulation for Dystonia Study Group, 2006. Pallidal deep-brain stimulation in primary generalized or segmental dystonia. N. Engl. J. Med. 355, 1978–1990.

Kwon, J.S., O'Donnell, B.F., Wallenstein, G.V., Greene, R.W., Hirayasu, Y., Nestor, P.G., Hasselmo, M.E., Potts, G.F., Shenton, M.E., McCarley, R.W., 1999. Gamma frequency-range abnormalities to auditory stimulation in schizophrenia. Arch. Gen. Psychiatry 56, 1001–1005.

Lalo, E., Thobois, S., Sharott, A., Polo, G., Mertens, P., Pogosyan, A., Brown, P., 2008. Patterns of bidirectional communication between cortex and basal ganglia during movement in patients with Parkinson disease. J. Neurosci. 28, 3008–3016.

Landi, A., Trezza, A., Pirillo, D., Vimercati, A., Antonini, A., Sganzerla, E.P., 2012. Spinal cord stimulation for the treatment of sensory symptoms in advanced Parkinson's disease. Neuromodulation 16, 276–279.

Lindén, H., Pettersen, K.H., Einevoll, G.T., 2010. Intrinsic dendritic filtering gives low-pass power spectra of local field potentials. J. Comput. Neurosci. 29, 423–444.

Lindenbach, D., Conti, M.M., Ostock, C.Y., George, J.A., Goldenberg, A.A., Melikhov-Sosin, M., Nuss, E.E., Bishop, C., 2016. The role of primary motor cortex (M1) glutamate and GABA signaling in l-DOPA-induced dyskinesia in parkinsonian rats. J. Neurosci. 36, 9873–9887.

Lindvall, O., Björklund, A., 1979. Dopaminergic innervation of the globus pallidus by collaterals from the nigrostriatal pathway. Brain Res. 172, 169–173.

Little, S., Pogosyan, A., Neal, S., Zavala, B., Zrinzo, L., Hariz, M., Foltynie, T., Limousin, P., Ashkan, K., FitzGerald, J., Green, A.L., Aziz, T.Z., Brown, P., 2013. Adaptive deep brain stimulation in advanced Parkinson disease. Ann. Neurol. 74, 449–457.

Little, S., Beudel, M., Zrinzo, L., Foltynie, T., Limousin, P., Hariz, M., Neal, S., Cheeran, B., Cagnan, H., Gratwicke, J., Aziz, T.Z., Pogosyan, A., Brown, P., 2016a. Bilateral adaptive deep brain stimulation is effective in Parkinson's disease. J. Neurol. Neurosurg. Psychiatry 87, 717–721.

Little, S., Tripoliti, E., Beudel, M., Pogosyan, A., Cagnan, H., Herz, D., Bestmann, S., Aziz, T., Cheeran, B., Zrinzo, L., Hariz, M., Hyam, J., Limousin, P., Foltynie, T., Brown, P., 2016b. Adaptive deep brain stimulation for Parkinson's disease demonstrates reduced speech side effects compared to conventional stimulation in the acute setting. J. Neurol. Neurosurg. Psychiatry 87, 1388–1389.

Litvak, V., Jha, A., Eusebio, A., Oostenveld, R., Foltynie, T., Limousin, P., Zrinzo, L., Hariz, M.I., Friston, K., Brown, P., 2011. Resting oscillatory cortico-subthalamic connectivity in patients with Parkinson's disease. Brain 134, 359–374.

Litvak, V., Eusebio, A., Jha, A., Oostenveld, R., Barnes, G., Foltynie, T., Limousin, P., Zrinzo, L., Hariz, M.I., Friston, K., Brown, P., 2012. Movement-related changes in local and long-range synchronization in Parkinson's disease revealed by simultaneous magnetoencephalography and intracranial recordings. J. Neurosci. 32, 10541–10553.

Liu, C., Zhu, Y., Liu, F., Wang, J., Li, H., Deng, B., Fietkiewicz, C., Loparo, K.A., 2017. Neural mass models describing possible origin of the excessive beta oscillations correlated with parkinsonian state. Neural Netw. 88, 65–73.

Lofredi, R., Neumann, W.-J., Bock, A., Horn, A., Huebl, J., Siegert, S., Schneider, G.-H., Krauss, J.K., Kühn, A.A., 2018. Dopamine-dependent scaling of subthalamic gamma bursts with movement velocity in patients with Parkinson's disease. eLife, 7, e31895.

Lofredi, R., Neumann, W.-J., Brücke, C., Huebl, J., Krauss, J.K., Schneider, G.-H., Kühn, A.A., 2019a. Pallidal beta bursts in Parkinson's disease and dystonia. Mov. Disord. 34, 420–424.

Lofredi, R., Tan, H., Neumann, W.-J., Yeh, C.-H., Schneider, G.-H., Kühn, A.A., Brown, P., 2019b. Beta bursts during continuous movements accompany the velocity decrement in Parkinson's disease patients. Neurobiol. Dis. 127, 462–471.

Lopez-Azcarate, J., Tainta, M., Rodriguez-Oroz, M.C., Valencia, M., Gonzalez, R., Guridi, J., Iriarte, J., Obeso, J.A., Artieda, J., Alegre, M., 2010. Coupling between beta and high-frequency activity in the human subthalamic nucleus may be a pathophysiological mechanism in Parkinson's disease. J. Neurosci. 30, 6667–6677.

Loucif, K.C., Wilson, C.L., Baig, R., Lacey, M.G., Stanford, I.M., 2005. Functional interconnectivity between the globus pallidus and the subthalamic nucleus in the mouse brain slice. J. Physiol. 567, 977–987.

Magill, P.J., Bolam, J.P., Bevan, M.D., 2000. Relationship of activity in the subthalamic nucleus-globus pallidus network to cortical electroencephalogram. J. Neurosci. 20, 820–833.

Malekmohammadi, M., Herron, J., Velisar, A., Blumenfeld, Z., Trager, M.H., Chizeck, H.J., Brontë-Stewart, H., 2016. Kinematic adaptive deep brain stimulation for resting tremor in Parkinson's disease. Mov. Disord. 31, 426–428.

Marreiros, A.C., Cagnan, H., Moran, R.J., Friston, K.J., Brown, P., 2013. Basal ganglia-cortical interactions in parkinsonian patients. Neuroimage 66, 301–310.

McCarthy, M.M., Moore-Kochlacs, C., Gu, X., Boyden, E.S., Han, X., Kopell, N., 2011. Striatal origin of the pathologic beta oscillations in Parkinson's disease. Proc. Natl. Acad. Sci. U. S. A. 108, 11620–11625.

Meibach, R.C., Katzman, R., 1979. Catecholaminergic innervation of the subthalamic nucleus: evidence for a rostral continuation of the A9 (substantia nigra) dopaminergic cell group. Brain Res. 173, 364–368.

Meidahl, A.C., Tinkhauser, G., Herz, D.M., Cagnan, H., Debarros, J., Brown, P., 2017. Adaptive deep brain stimulation for movement disorders: the long road to clinical therapy. Mov. Disord. 32, 810–819.

Meidahl, A.C., Moll, C.K.E., van Wijk, B.C.M., Gulberti, A., Tinkhauser, G., Westphal, M., Engel, A.K., Hamel, W., Brown, P., Sharott, A., 2019. Synchronised spiking activity underlies phase amplitude coupling in the subthalamic nucleus of Parkinson's disease patients. Neurobiol. Dis. 127, 101–113.

Milosevic, L., Kalia, S.K., Hodaie, M., Lozano, A.M., Fasano, A., Popovic, M.R., Hutchison, W.D., 2018a. Neuronal inhibition and synaptic plasticity of basal ganglia neurons in Parkinson's disease. Brain 141, 177–190.

Milosevic, L., Kalia, S.K., Hodaie, M., Lozano, A.M., Popovic, M.R., Hutchison, W.D., 2018b. Physiological mechanisms of thalamic ventral intermediate nucleus stimulation for tremor suppression. Brain 141, 2142–2155.

Molina, R., Okun, M.S., Shute, J.B., Opri, E., Rossi, P.J., Martinez-Ramirez, D., Foote, K.D., Gunduz, A., 2018. Report of a patient undergoing chronic responsive deep brain stimulation for Tourette syndrome: proof of concept. J. Neurosurg. 129, 308–314.

Molnar, G., Barolat, G., 2014. Principles of cord activation during spinal cord stimulation. Neuromodulation 17, 12–21.

Nambu, A., Tachibana, Y., Chiken, S., 2015. Cause of parkinsonian symptoms: firing rate, firing pattern or dynamic activity changes? Basal Ganglia 5, 1–6.

Neumann, W.-J., Kühn, A.A., 2017. Subthalamic beta power-Unified Parkinson's disease rating scale III correlations require akinetic symptoms. Mov. Disord. 32, 175–176.

Neumann, W.-J., Jha, A., Bock, A., Huebl, J., Horn, A., Schneider, G.-H., Sander, T.H., Litvak, V., Kühn, A.A., 2015. Cortico-pallidal oscillatory connectivity in patients with dystonia. Brain 138, 1894–1906.

Neumann, W.-J., Degen, K., Schneider, G.-H., Brücke, C., Huebl, J., Brown, P., Kühn, A.A., 2016a. Subthalamic synchronized oscillatory activity correlates with motor impairment in patients with Parkinson's disease. Mov. Disord. 31, 1748–1751.

Neumann, W.-J., Staub, F., Horn, A., Schanda, J., Mueller, J., Schneider, G.-H., Brown, P., Kühn, A.A., 2016b. Deep brain recordings using an implanted pulse generator in Parkinson's disease. Neuromodulation 19, 20–24.

Neumann, W.-J., Horn, A., Ewert, S., Huebl, J., Brücke, C., Slentz, C., Schneider, G.-H., Kühn, A.A., 2017a. A localized pallidal physiomarker in cervical dystonia. Ann. Neurol. 82, 912–924.

Neumann, W.-J., Staub-Bartelt, F., Horn, A., Schanda, J., Schneider, G.-H., Brown, P., Kühn, A.A., 2017b. Long term correlation of subthalamic beta band activity with motor impairment in patients with Parkinson's disease. Clin. Neurophysiol. 128, 2286–2291.

Neumann, W.-J., Huebl, J., Brücke, C., Lofredi, R., Horn, A., Saryyeva, A., Müller-Vahl, K., Krauss, J.K., Kühn, A.A., 2018a. Pallidal and thalamic neural oscillatory patterns in Tourette's syndrome. Ann. Neurol. 84, 505–514.

Neumann, W.-J., Schroll, H., de Almeida Marcelino, A.L., Horn, A., Ewert, S., Irmen, F., Krause, P., Schneider, G.-H., Hamker, F., Kühn, A.A., 2018b. Functional segregation of basal ganglia pathways in Parkinson's disease. Brain 141, 2655–2669.

Neumann, W.-J., Turner, R.S., Blankertz, B., Mitchell, T., Kühn, A.A., Richardson, R.M., 2019. Toward electrophysiology-based intelligent adaptive deep brain stimulation for movement disorders. Neurotherapeutics 16, 105–118.

Nini, A., Feingold, A., Slovin, H., Bergman, H., 1995. Neurons in the globus pallidus do not show correlated activity in the normal monkey, but phase-locked oscillations appear in the MPTP model of parkinsonism. J. Neurophysiol. 74, 1800–1805.

Nishioka, K., Nakajima, M., 2015. Beneficial therapeutic effects of spinal cord stimulation in advanced cases of Parkinson's disease with intractable chronic pain: a case series. Neuromodulation 18, 751–753.

Oswal, A., Beudel, M., Zrinzo, L., Limousin, P., Hariz, M., Foltynie, T., Litvak, V., Brown, P., 2016. Deep brain stimulation modulates synchrony within spatially and spectrally distinct resting state networks in Parkinson's disease. Brain 139, 1482–1496.

Pan, M., Kuo, S., Tai, C., Liou, J., Pei, J., Chang, C., Wang, Y., Liu, W., Wang, T., Lai, W., Kuo, C., 2016. Neuronal firing patterns outweigh circuitry oscillations in parkinsonian motor control. J. Clin. Invest. 126, 4516–4526.

Pavlides, A., Hogan, S.J., Bogacz, R., 2015. Computational models describing possible mechanisms for generation of excessive beta oscillations in Parkinson's disease. PLoS Comput. Biol. 11, e1004609.

Pedrosa, D.J., Quatuor, E.-L., Reck, C., Pauls, K.A.M., Huber, C.A., Visser-Vandewalle, V., Timmermann, L., 2014. Thalamomuscular coherence in essential tremor: hen or egg in the emergence of tremor? J. Neurosci. 34, 14475–14483.

Pedrosa, D.J., Brown, P., Cagnan, H., Visser-Vandewalle, V., Wirths, J., Timmermann, L., Brittain, J.-S., 2018. A functional micro-electrode mapping of ventral thalamus in essential tremor. Brain 141, 2644–2654.

Petersson, P., Halje, P., Cenci, M.A., 2019. Significance and translational value of high-frequency cortico-basal ganglia oscillations in Parkinson's disease. J. Parkinsons Dis. 9, 183–196.

Phookan, S., Sutton, A.C., Walling, I., Smith, A., O'Connor, K.A., Campbell, J.C., Calos, M., Yu, W., Pilitsis, J.G., Brotchie, J.M., Shin, D.S., 2015. Gap junction blockers attenuate beta oscillations and improve forelimb function in hemiparkinsonian rats. Exp. Neurol. 265, 160–170.

Picconi, B., Piccoli, G., Calabresi, P., 2012. Synaptic dysfunction in Parkinson's disease. In: Kreutz, M., Sala, C. (Eds.), Advances in Experimental Medicine and Biology. Springer, pp. 553–572.

Piña-Fuentes, D., Little, S., Oterdoom, M., Neal, S., Pogosyan, A., Tijssen, M.A.J., van Laar, T., Brown, P., van Dijk, J.M.C., Beudel, M., 2017. Adaptive DBS in a Parkinson's patient with chronically implanted DBS: a proof of principle. Mov. Disord. 32, 1253–1254.

Plenz, D., Kital, S.T., 1999. A basal ganglia pacemaker formed by the subthalamic nucleus and external globus pallidus. Nature 400, 677–682.

Pogosyan, A., Yoshida, F., Chen, C.C., Martinez-Torres, I., Foltynie, T., Limousin, P., Zrinzo, L., Hariz, M.I., Brown, P., 2010. Parkinsonian impairment correlates with spatially extensive subthalamic oscillatory synchronization. Neuroscience 171, 245–257.

Pontone, G.M., Williams, J.R., Anderson, K.E., Chase, G., Goldstein, S.A., Grill, S., Hirsch, E.S., Lehmann, S., Little, J.T., Margolis, R.L., Rabins, P.V., Weiss, H.D., Marsh, L., 2009. Prevalence of anxiety disorders and anxiety subtypes in patients with Parkinson's disease. Mov. Disord. 24, 1333–1338.

Prensa, L., Parent, A., 2001. The nigrostriatal pathway in the rat: a single-axon study of the relationship between dorsal and ventral tier nigral neurons and the striosome/matrix striatal compartments. J. Neurosci. 21, 7247–7260.

Priori, A., Giannicola, G., Rosa, M., Marceglia, S., Servello, D., Sassi, M., Porta, M., 2013. Deep brain electrophysiological recordings provide clues to the pathophysiology of Tourette syndrome. Neurosci. Biobehav. Rev. 37, 1063–1068.

Quinn, E.J., Blumenfeld, Z., Velisar, A., Koop, M.M., Shreve, L.A., Trager, M.H., Hill, B.C., Kilbane, C., Henderson, J.M., Brontë-Stewart, H., 2015. Beta oscillations in freely moving Parkinson's subjects are attenuated during deep brain stimulation. Mov. Disord. 30, 1750–1758.

Quiroga-Varela, A., Walters, J.R., Brazhnik, E., Marin, C., Obeso, J.A., 2013. What basal ganglia changes underlie the parkinsonian state? The significance of neuronal oscillatory activity. Neurobiol. Dis. 58, 242–248.

Raz, A., Vaadia, E., Bergman, H., 2000. Firing patterns and correlations of spontaneous discharge of pallidal neurons in the normal and the tremulous 1-methyl-4-phenyl-1,2,3,6-tetrahydropyridine vervet model of parkinsonism. J. Neurosci. 20, 8559–8571.

Raz, A., Frechter-Mazar, V., Feingold, A., Abeles, M., Vaadia, E., Bergman, H., 2001. Activity of pallidal and striatal tonically active neurons is correlated in mptp-treated monkeys but not in normal monkeys. J. Neurosci. 21, RC128.

Reijnders, J.S.A.M., Ehrt, U., Weber, W.E.J., Aarsland, D., Leentjens, A.F.G., 2008. A systematic review of prevalence studies of depression in Parkinson's disease. Mov. Disord. 23, 183–189.

Reis, C., Sharott, A., Magill, P.J., van Wijk, B.C.M., Parr, T., Zeidman, P., Friston, K.J., Cagnan, H., 2019. Thalamocortical dynamics underlying spontaneous transitions in beta power in parkinsonism. Neuroimage 193, 103–114.

Restrepo, J.G., Ott, E., Hunt, B.R., 2006. Synchronization in large directed networks of coupled phase oscillators. Chaos 16, 015107.

Richter, U., Halje, P., Petersson, P., 2013. Mechanisms underlying cortical resonant states: implications for levodopa-induced dyskinesia. Rev. Neurosci. 24, 415–429.

Rodriguez-Oroz, M.C., López-Azcárate, J., Garcia-Garcia, D., Alegre, M., Toledo, J., Valencia, M., Guridi, J., Artieda, J., Obeso, J.A., 2011. Involvement of the subthalamic nucleus in impulse control disorders associated with Parkinson's disease. Brain 134, 36–49.

Rommelfanger, K.S., Wichmann, T., 2010. Extrastriatal dopaminergic circuits of the basal ganglia. Front. Neuroanat. 4, 139.

Rosa, M., Arlotti, M., Ardolino, G., Cogiamanian, F., Marceglia, S., Di Fonzo, A., Cortese, F., Rampini, P.M., Priori, A., 2015. Adaptive deep brain stimulation in a freely moving parkinsonian patient. Mov. Disord. 30, 1003–1005.

Rosa, M., Arlotti, M., Marceglia, S., Cogiamanian, F., Ardolino, G., Di Fonzo, A., Lopiano, L., Scelzo, E., Merola, A., Locatelli, M., Rampini, P.M., Priori, A., 2017. Adaptive deep brain stimulation controls levodopa-induced side effects in parkinsonian patients. Mov. Disord. 32, 628–629.

Rosin, B., Slovik, M., Mitelman, R., Rivlin-Etzion, M., Haber, S.N., Israel, Z., Vaadia, E., Bergman, H., 2011. Closed-loop deep brain stimulation is superior in ameliorating parkinsonism. Neuron 72, 370–384.

Ryapolova-Webb, E., Afshar, P., Stanslaski, S., Denison, T., de Hemptinne, C., Bankiewicz, K., Starr, P.A., 2014. Chronic cortical and electromyographic recordings from a fully implantable device: preclinical experience in a nonhuman primate. J. Neural Eng. 11, 016009.

Samotus, O., Parrent, A., Jog, M., 2018. Spinal cord stimulation therapy for gait dysfunction in advanced Parkinson's disease patients. Mov. Disord. 33, 783–792.

Sanchez-Gonzalez, M.A., García-Cabezas, M.A., Rico, B., Cavada, C., 2005. The primate thalamus is a key target for brain dopamine. J. Neurosci. 25, 6076–6083.

Santana, M.B., Halje, P., Simplicio, H., Richter, U., Freire, M.A., Petersson, P., Fuentes, R., Nicolelis, M.A., 2014. Spinal cord stimulation alleviates motor deficits in a primate model of Parkinson disease. Neuron 84, 716–722.

Scheller, U., Lofredi, R., van Wijk, B.C.M., Saryyeva, A., Krauss, J.K., Schneider, G.-H., Kroneberg, D., Krause, P., Neumann, W.-J., Kühn, A.A., 2019. Pallidal low-frequency activity in dystonia after cessation of long-term deep brain stimulation. Mov. Disord. 34, 1734–1739.

Schuepbache, W.M.M., Rau, J., Knudsen, K., Volkmann, J., Krack, P., Timmermann, L., Hälbig, T.D., Hesekamp, H., Navarro, S.M., Meier, N., Falk, D., Mehdorn, M., Paschen, S., Maarouf, M., Barbe, M.T., Fink, G.R., Kupsch, A., Gruber, D.,

Schneider, G.-H., Seigneuret, E., Kistner, A., Chaynes, P., Ory-Magne, F., Brefel Courbon, C., Vesper, J., Schnitzler, A., Wojtecki, L., Houeto, J.-L., Bataille, B., Maltête, D., Damier, P., Raoul, S., Sixel-Doering, F., Hellwig, D., Gharabaghi, A., Krüger, R., Pinsker, M.O., Amtage, F., Régis, J.-M., Witjas, T., Thobois, S., Mertens, P., Kloss, M., Hartmann, A., Oertel, W.H., Post, B., Speelman, H., Agid, Y., Schade-Brittinger, C., Deuschl, G., EARLYSTIM Study Group, 2013. Neurostimulation for Parkinson's disease with early motor complications. N. Engl. J. Med. 368, 610–622.

Schuurman, P.R., Bosch, D.A., Bossuyt, P.M.M., Bonsel, G.J., van Someren, E.J.W., de Bie, R.M.A., Merkus, M.P., Speelman, J.D., 2000. A comparison of continuous thalamic stimulation and thalamotomy for suppression of severe tremor. N. Engl. J. Med. 342, 461–468.

Schwab, B.C., Heida, T., Zhao, Y., van Gils, S.A., van Wezel, R.J.A., 2014. Pallidal gap junctions-triggers of synchrony in Parkinson's disease? Mov. Disord. 29, 1486–1494.

Sharott, A., Grosse, P., Kuhn, A.A., Salih, F., Engel, A.K., Kupsch, A., Schneider, G.-H., Krauss, J.K., Brown, P., 2008. Is the synchronization between pallidal and muscle activity in primary dystonia due to peripheral afferance or a motor drive? Brain 131, 473–484.

Sharott, A., Vinciati, F., Nakamura, K.C., Magill, P.J., 2017. A population of indirect pathway striatal projection neurons is selectively entrained to parkinsonian beta oscillations. J. Neurosci. 37, 9977–9998.

Sherman, M.A., Lee, S., Law, R., Haegens, S., Thorn, C.A., Hämäläinen, M.S., Moore, C.I., Jones, S.R., 2016. Neural mechanisms of transient neocortical beta rhythms: converging evidence from humans, computational modeling, monkeys, and mice. Proc. Natl. Acad. Sci. U. S. A. 113, E4885–E4894.

Shimamoto, S.A., Ryapolova-Webb, E.S., Ostrem, J.L., Galifianakis, N.B., Miller, K.J., Starr, P.A., 2013. Subthalamic nucleus neurons are synchronized to primary motor cortex local field potentials in Parkinson's disease. J. Neurosci. 33, 7220–7233.

Shreve, L.A., Velisar, A., Malekmohammadi, M., Koop, M.M., Trager, M., Quinn, E.J., Hill, B.C., Blumenfeld, Z., Kilbane, C., Mantovani, A., Henderson, J.M., Brontë-Stewart, H., 2017. Subthalamic oscillations and phase amplitude coupling are greater in the more affected hemisphere in Parkinson's disease. Clin. Neurophysiol. 128, 128–137.

Silberstein, P., Kühn, A.A., Kupsch, A., Trottenberg, T., Krauss, J.K., Wöhrle, J.C., Mazzone, P., Insola, A., Di Lazzaro, V., Oliviero, A., Aziz, T., Brown, P., 2003. Patterning of globus pallidus local field potentials differs between Parkinson's disease and dystonia. Brain 126, 2597–2608.

Singer, W., 2018. Neuronal oscillations: unavoidable and useful? Eur. J. Neurosci. 48, 2389–2398.

Smith, Y., Kieval, J.Z., 2000. Anatomy of the dopamine system in the basal ganglia. Trends Neurosci. 23, S28–S33.

Smith, Y., Villalba, R., 2008. Striatal and extrastriatal dopamine in the basal ganglia: an overview of its anatomical organization in normal and parkinsonian brains. Mov. Disord. 23 (Suppl. 3), S534–S547.

Steiner, L.A., Neumann, W.-J., Staub-Bartelt, F., Herz, D.M., Tan, H., Pogosyan, A., Kuhn, A.A., Brown, P., 2017. Subthalamic beta dynamics mirror parkinsonian bradykinesia months after neurostimulator implantation. Mov. Disord. 32, 1183–1190.

Steiner, L.A., Barreda Tomás, F.J., Planert, H., Alle, H., Vida, I., Geiger, J.R.P., 2019. Connectivity and dynamics underlying synaptic control of the subthalamic nucleus. J. Neurosci. 39, 2470–2481.

Swann, N.C., de Hemptinne, C., Miocinovic, S., Qasim, S., Wang, S.S., Ziman, N., Ostrem, J.L., San Luciano, M., Galifianakis, N.B., Starr, P.A., 2016. Gamma oscillations in the hyperkinetic state detected with chronic human brain recordings in Parkinson's disease. J. Neurosci. 36, 6445–6458.

Swann, N.C., de Hemptinne, C., Thompson, M.C., Miocinovic, S., Miller, A.M., Gilron, R., Ostrem, J.L., Chizeck, H.J., Starr, P.A., 2018. Adaptive deep brain stimulation for Parkinson's disease using motor cortex sensing. J. Neural Eng. 15, 046006.

Tamtè, M., Brys, I., Richter, U., Ivica, N., Halje, P., Petersson, P., 2016. Systems-level neurophysiological state characteristics for drug evaluation in an animal model of levodopa-induced dyskinesia. J. Neurophysiol. 115, 1713–1729.

Terman, D., Rubin, J.E., Yew, A.C., Wilson, C.J., 2002. Activity patterns in a model for the subthalamopallidal network of the basal ganglia. J. Neurosci. 22, 2963–2976.

Thelin, J., Halje, P., Nielsen, J., Didriksen, M., Petersson, P., Bastlund, J.F., 2016. The translationally relevant mouse model of the 15q13.3 microdeletion syndrome reveals deficits in neuronal spike firing matching clinical neurophysiological biomarkers seen in schizophrenia. Acta Physiol (Oxf.) 220, 124–136.

Tinkhauser, G., Pogosyan, A., Little, S., Beudel, M., Herz, D.M., Tan, H., Brown, P., 2017a. The modulatory effect of adaptive deep brain stimulation on beta bursts in Parkinson's disease. Brain 140, 1053–1067.

Tinkhauser, G., Pogosyan, A., Tan, H., Herz, D.M., Kühn, A.A., Brown, P., 2017b. Beta burst dynamics in Parkinson's disease OFF and ON dopaminergic medication. Brain 140, 2968–2981.

Tinkhauser, G., Torrecillos, F., Duclos, Y., Tan, H., Pogosyan, A., Fischer, P., Carron, R., Welter, M.-L., Karachi, C., Vandenberghe, W., Nuttin, B., Witjas, T., Régis, J., Azulay, J.-P., Eusebio, A., Brown, P., 2018. Beta burst coupling across the motor circuit in Parkinson's disease. Neurobiol. Dis. 117, 217–225.

Tort, A.B.L., Scheffer-Teixeira, R., Souza, B.C., Draguhn, A., Brankačk, J., 2013. Theta-associated high-frequency oscillations (110–160 Hz) in the hippocampus and neocortex. Prog. Neurobiol. 100, 1–14.

Trager, M.H., Koop, M.M., Velisar, A., Blumenfeld, Z., Nikolau, J.S., Quinn, E.J., Martin, T., Bronte-Stewart, H., 2016. Subthalamic beta oscillations are attenuated after withdrawal of chronic high frequency neurostimulation in Parkinson's disease. Neurobiol. Dis. 96, 22–30.

Tseng, K.Y., Kasanetz, F., Kargieman, L., Riquelme, L.A., Murer, M.G., 2001. Cortical slow oscillatory activity is reflected in the membrane potential and spike trains of striatal neurons in rats with chronic nigrostriatal lesions. J. Neurosci. 21, 6430–6439.

van Wijk, B.C.M., Pogosyan, A., Hariz, M.I., Akram, H., Foltynie, T., Limousin, P., Horn, A., Ewert, S., Brown, P., Litvak, V., 2017. Localization of beta and high-frequency oscillations within the subthalamic nucleus region. Neuroimage Clin. 16, 175–183.

Velisar, A., Syrkin-Nikolau, J., Blumenfeld, Z., Trager, M.H., Afzal, M.F., Prabhakar, V., Bronte-Stewart, H., 2019. Dual threshold neural closed loop deep brain stimulation in Parkinson disease patients. Brain Stimul. 12, 868–876.

Vidailhet, M., Vercueil, L., Houeto, J.-L., Krystkowiak, P., Benabid, A.-L., Cornu, P., Lagrange, C., Tézenas du Montcel, S., Dormont, D., Grand, S., Blond, S., Detante, O., Pillon, B., Ardouin, C., Agid, Y., Destée, A., Pollak, P., 2005. Bilateral deep-brain stimulation of the globus pallidus in primary generalized dystonia. N. Engl. J. Med. 352, 459–467.

Vidailhet, M., Vercueil, L., Houeto, J.-L., Krystkowiak, P., Lagrange, C., Yelnik, J., Bardinet, E., Benabid, A.-L., Navarro, S., Dormont, D., Grand, S., Blond, S., Ardouin, C., Pillon, B., Dujardin, K., Hahn-Barma, V., Agid, Y., Destée, A., Pollak, P., 2007. Bilateral, pallidal, deep-brain stimulation in primary generalised dystonia: a prospective 3 year follow-up study. Lancet Neurol. 6, 223–229.

Villalba, R.M., Smith, Y., 2018. Loss and remodeling of striatal dendritic spines in Parkinson's disease: from homeostasis to maladaptive plasticity? J. Neural Transm. 125, 431–447.

Volkmann, J., Mueller, J., Deuschl, G., Kühn, A.A., Krauss, J.K., Poewe, W., Timmermann, L., Falk, D., Kupsch, A., Kivi, A., Schneider, G.-H., Schnitzler, A., Südmeyer, M., Voges, J., Wolters, A., Wittstock, M., Müller, J.-U., Hering, S., Eisner, W., Vesper, J., Prokop, T., Pinsker, M., Schrader, C., Kloss, M., Kiening, K., Boetzel, K., Mehrkens, J., Skogseid, I.M., Ramm-Pettersen, J., Kemmler, G., Bhatia, K.P., Vitek, J.L., Benecke, R., DBS Study Group for Dystonia, 2014. Pallidal neurostimulation in patients with medication-refractory cervical dystonia: a randomised, sham-controlled trial. Lancet Neurol. 13, 875–884.

Wang, X.-J., 2010. Neurophysiological and computational principles of cortical rhythms in cognition. Physiol. Rev. 90, 1195–1268.

Wang, J., Hirschmann, J., Elben, S., Hartmann, C.J., Vesper, J., Wojtecki, L., Schnitzler, A., 2014. High-frequency oscillations in Parkinson's disease: spatial distribution and clinical relevance. Mov. Disord. 29, 1265–1272.

Wang, Q., Chen, J., Li, M., Lv, S., Xie, Z., Li, N., Wang, N., Wang, J., Luo, F., Zhang, W., 2019. Eltoprazine prevents levodopa-induced dyskinesias by reducing causal interactions for theta oscillations in the dorsolateral striatum and substantia nigra pars reticulate. Neuropharmacology 148, 1–10.

Weaver, F.M., Follett, K., Stern, M., Hur, K., Harris, C., Marks, W.J., Rothlind, J., Sagher, O., Reda, D., Moy, C.S., Pahwa, R., Burchiel, K., Hogarth, P., Lai, E.C., Duda, J.E., Holloway, K., Samii, A., Horn, S., Bronstein, J., Stoner, G., Heemskerk, J., Huang, G.D., CSP 468 Study Group, 2009. Bilateral deep brain stimulation vs best medical therapy for patients with advanced Parkinson disease: a randomized controlled trial. JAMA 301, 63–73.

Wichmann, T., Bergman, H., DeLong, M.R., 1994. The primate subthalamic nucleus. III. Changes in motor behavior and neuronal activity in the internal pallidum induced by subthalamic inactivation in the MPTP model of parkinsonism. J. Neurophysiol. 72, 521–530.

Wichmann, T., Bergman, H., DeLong, M.R., 2018. Basal ganglia, movement disorders and deep brain stimulation: advances made through non-human primate research. J. Neural Transm. 125, 419–430.

Williams, D., Tijssen, M., Van Bruggen, G., Bosch, A., Insola, A., Di Lazzaro, V., Mazzone, P., Oliviero, A., Quartarone, A., Speelman, H., Brown, P., 2002. Dopamine-dependent changes in the functional connectivity between basal ganglia and cerebral cortex in humans. Brain 125, 1558–1569.

Wilson, C.L., Cash, D., Galley, K., Chapman, H., Lacey, M.G., Stanford, I.M., 2006. Subthalamic nucleus neurones in slices from 1-methyl-4-phenyl-1,2,3,6-tetrahydropyridine-lesioned mice show irregular, dopamine-reversible firing pattern changes, but without synchronous activity. Neuroscience 143, 565–572.

Yang, A.I., Vanegas, N., Lungu, C., Zaghloul, K.A., 2014. Beta-coupled high-frequency activity and beta-locked neuronal spiking in the subthalamic nucleus of Parkinson's disease. J. Neurosci. 34, 12816–12827.

Printed in the United States
By Bookmasters